XXIII International Conference on Manufacturing (Manufacturing 2018)

IOP Conference Series: Materials Science and Engineering Volume 448

Kecskemet, Hungary
7 – 8 June 2018

ISBN: 978-1-5108-7661-3
ISSN: 1757-8981

Printed from e-media with permission by:

Curran Associates, Inc.
57 Morehouse Lane
Red Hook, NY 12571

proceedings.com

Some format issues inherent in the e-media version may also appear in this print version.

Printed by Curran Associates, Inc. (2019)

Institute of Physics
Dirac House, Temple Back
Bristol BS1 6BE UK

Phone: 44 1 17 929 7481
Fax: 44 1 17 920 0979

techtracking@iop.org

Additional copies of this publication are available from:

Curran Associates, Inc.
57 Morehouse Lane
Red Hook, NY 12571 USA
Phone: 845-758-0400
Fax: 845-758-2633
Email: curran@proceedings.com
Web: www.proceedings.com

TABLE OF CONTENTS

IOP Conf. Series: Materials Science and Engineering **448** (2018) 011001 doi:10.1088/1757-899X/448/1/011001

XXIIIrd

MANUFACTURING - 2018 CONFERENCE

Preface

The Scientific Society for Mechanical Engineering (GTE in Hungarian) has traditionally been the largest scientific society in Hungary for several decades. It was founded 67 years ago and counted more than 15.000 Hungarian engineer members in the '60s. Although the membership has decreased significantly since, GTE's value and ambitions are constantly high.

Among other active divisions of GTE, the Manufacturing Systems Division organizes the Manufacturing conferences with national and international participation mostly in a biannual sequence, usually but exclusively in Budapest. Now, at the XXIIIrd conference on manufacturing, Kecskemét City was chosen, where the John von Neumann University GAMF Faculty campus offered to host and support this event in early June 2018.

The conference had invited participants from universities, from manufacturers, end-users and specialists from the industry sectors from countries across Europe. The aim of the Conference was to show up the newest scientific and professional achievements related to planning, designing, implementing, evaluating products and manufacturing technologies, to highlight research work performed by students and industry-experts, and offer presentation and demonstration opportunity for all. An important effect was also welcome, i.e. to meet face-to-face with experts, to share and dispute on best-practices, evaluate and compare results with other solutions. The organized laboratory and factory visits offered excellent experiences how theory and practice can both show interrelated research results.

The seamless flow of the events and activities of all this had depended heavily on the support of the John Von Neumann University, GAMF Faculty of Engineering and Computer Science, Department of Material Sciences and Vehicle Technology from the very beginning, with support received from the Program Committee members, and from members of the Organizing Committee and contributions from many others.
The outstanding professional supporters and sponsors of the Conference were:
GTE, Scientific Society for Mechanical Engineering and the
John von Neumann University -GAMF Faculty of Engineering and Computer Science;
PADMA, Pallas Athene Foundation, and
MTA SZTAKI, Institute for Computer Science and Automation.
Last but not least, the #1 contributors to the success of the conference were the lecturing participants and the poster-paper presenters. With the edited PROCEEDINGS of the Manufacturing-2018 Conference, the editors and the organizers offer the readers a valuable set of papers with excellent ideas and topics to remember and to quote for further scientific papers.

Now let me express my thanks to all of the supporters and to all of my colleagues for putting their heart and soul into their work to prepare the Proceedings. Furthermore, I wish all the participants success in their careers, fruitful further work, new professional contacts and inspiration to prepare for participating at the next upcoming XXIVth MANUFACTURING Conference of the GTE Society.

Kecskemét, 2018

Prof. Dr. János Takács
Chairman of the Conference

MTA **SZTAKI**

Hungarian Academy of Sciences
Institute for Computer Science and Control

PADMA

PALLAS ATHÉNÉ
DOMUS MENTIS
FOUNDATION

GTE®

SCIENTIFIC SOCIETY
FOR MECHANICAL
ENGINEERING

JOHN VON NEUMANN UNIVERSITY

AMF

IOP Conf. Series: Materials Science and Engineering **448** (2018) 011002 doi:10.1088/1757-899X/448/1/011002

Chair of the conference
Dr. János Takács
Budapest University of
Technology and Economics

Organizers
Scientific Society for Mechanical Engineering
Responsible persons:

- Dr. János Kodácsy
- Dr. János Takács

John von Neumann University,
GAMF Faculty of Engineering
and Computer Science
Responsible persons:

- Dr. Zoltán Weltsch
- Dr. Pál Lukács

Further members of Organizing Committee

- Dr. Géza Haidegger
- Dr. Zsolt Csaba Johanyák
- Dr. János Líska
- László Valenta
- Dr. Balázs Vehovszky

Scientific committee
Chair:
Prof. Dr. János Kundrák
University of Miskolc

- Dr. Piroska Ailer
- Dr. Angelos P. Markopoulos
- Dr. Athanasios G. Mamalis
- Dr. Balázs Mikó
- Dr. Béla Illés
- Dr. Béla Palotás
- Dr. Bernhard Karpuschewski
- Dr. Csaba Gyenge
- Dr. Gábor Kalácska
- Dr. Gábor Stépán
- Dr. György Takács
- Dr. Gyula Mátyási
- Dr. Ibolya Zsoldos
- Dr. Ildikó Mankova
- Dr. Imre Czinege
- Dr. Imre Horváth
- Dr. István Gyurika
- Dr. János Jósvai
- Dr. János Lukács
- Dr. József Váncza

IOP Conf. Series: Materials Science and Engineering **448** (2018) 011002 doi:10.1088/1757-899X/448/1/011002

- Dr. Lóránt Kovács
- Dr. Kutin Andrey Anatolevich
- Dr. Mátyás Horváth
- Dr. Nicolae Balc
- Dr. Numan M. Durakbasa
- Dr. Péter János Szabó
- Dr. Raul Turmanidze
- Dr. Tibor Szalay
- Dr. Vencel József Csibi
- Dr. Wojciech Zebala
- Dr. Zsolt János Viharos
- Dr. Zsolt Maros

Chief Patron
Dr. László Monostori
Director of Research Laboratory on Engineering & Management
Institute for Computer Science and Control, Hungarian Academy of Sciences

IOP Conf. Series: Materials Science and Engineering **448** (2018) 011003 doi:10.1088/1757-899X/448/1/011003

Peer review statement

All papers published in this volume of *IOP Conference Series: Materials Science and Engineering* have been peer reviewed through processes administered by the proceedings Editors. Reviews were conducted by expert referees to the professional and scientific standards expected of a proceedings journal published by IOP Publishing.

IOP Conf. Series: Materials Science and Engineering **448** (2018) 012001 doi:10.1088/1757-899X/448/1/012001

Joining of non-weldable AA7075 and weldable AA6082 aluminium alloy sheets by Friction Stir Welding

Zsolt Kovács[1], Ferenc Hareancz[2]

1,2 John von Neumann University, GAMF Faculty of Engineering and Computer Science, Dep. of Vehicle Technology, Kecskemét H-6000, Hungary

kovacs.zsolt@gamf.uni-neumann.hu

Abstract. In this work, an attempt has been made to join the AA6082-T651 and AA7075-T651 aluminium alloys by friction stir welding (FSW). The FSW is a solid-state joining process which primarily used on aluminium because the metal is not melted during the process so the base material can not soften. For the welding were using cylindrical pin tool, marble table to avoid the harmful heat conduction, two rotation rate of 500 and1000 rpm and different traverse speeds of 15, 20, 25 and 30 mm/min to weld butt joints. The aim of present study were to experimentally explore the dissimilar bounds. For evaluations, hardness testing were using to create hardness profiles across the joint in through thickness direction and optical microscope to classify the microstructures and crystallographic textures of base materials (BM). Based on the results obtained it can be stated that the FSW suitable for AA7075 and AA6082 welding by further optimization of the process parameters.

1. Introduction

The difficulty of making high-strength, fatigue and fracture resistant welds in aerospace aluminum alloys, such as highly alloyed 6XXX or 7XXX series, has long inhibited the usability of welding for joining aerospace and military applications such as light combat aircraft (LCA), light combat vehicle (LCV), future main battle tank (FMBT), bridge layer tank (BLT), armored ambulance, submarine torpedo, etc. [1, 2].

Aluminium can generally welded by fusion welded processes, for example the TIG or MIG welding technologies. But in case, the 7XXX series particularly the alloy AA7075 is known as non-weldable, because of the poor solidification microstructure and porosity in the fusion zone. After the invention of Friction Stir Welding (FSW) process in 1991 aluminum alloys that were regarded unweldable were found to be weldable by the FSW, furthermore the welded bonds are very good mechanical properties [3]. "Friction Stir Welding (FSW) is a simple, clean and innovative joining technology for light metals invented by The Welding Institute (TWI), England, U.K. in 1991" [4]. The FSW tool is a non-consumable rotating tool with a specially designed pin and shoulder is inserted into the abutting edges of sheets or plates to be joined and traversed along the line of joint. The tool have two primary functions: heating of the workpiece, and movement of material to produce the joint [1]. The Figure 1 represents the tool' parts and welding process.

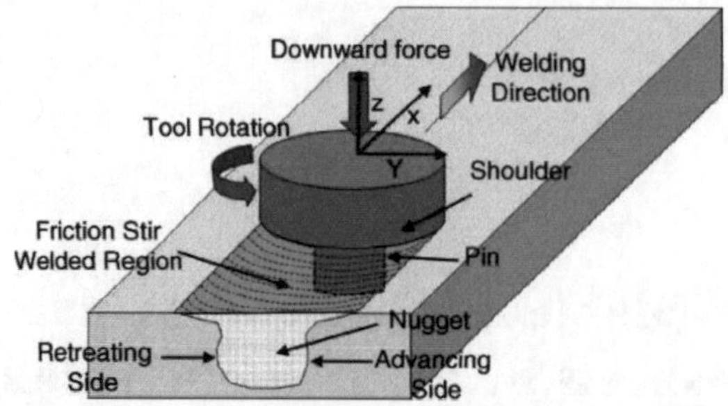

Figure 1. Schematic drawing of friction stir welding [1]

After the invention of the FSW there are several studies which investigated the relationships between process parameters and resulting joint integrity [5] or between process parameters and the welded bonds defects [6] and the effect of the tool geometry have also been studied [7]. Based on the studies can be stated that the rotational speed, traverse speed, and vertical pressure on the plates during welding are the main process parameters of FSW. Also the tool geometry has significant characteristic which affects the weld strength. S. M. Bayazida et al. investigated the effect of tool geometry (triangle, square and cylindrical). About their results all of the tool geometry occurs defects but the cylindrical can made the most acceptable bond.

The FSW process generates three distinct microstructural zones: the weld nugget (WN), the thermo-mechanically- affected zone (TMAZ) and the heat-affected zone (HAZ). The WN is the region through which the tool pin passes, and thus experiences high deformation and high heat. It generally consists of fine equiaxed grains due to full recrystallization. The TMAZ adjacent to the nugget is the region where the metal is plastically deformed and also heated, but this is not sufficient to cause recrystallization [3].

In this study, the effects of various rotational speed and traverse speed on microstructural characteristics and hardness parameters of the dissimilar welded joint made from a weldable (AA6082-T651) and non-weldable (AA7075-T651) aluminium alloys were studied.

2. Experimental procedure

Two type of 4 mm thickness rolled plates of AA6082-T651 and AA7075-T651 aluminium alloys were used in the present investigation (Table 1. shows the chemical compositions).

Table 1. Chemical composition (wt %) of aluminium alloys used in this investigation

Alloys	Si	Fe	Cu	Nm	Mg	Cr	Zn	Ti
AA6082-T651	0,4-08	0,7	0,15-0,40	0,15	0,8-1,2	0,04-0,35	0,25	0,15
AA7075-T651	0,4	0,5	1,2-2,0	0,3	2,1-2,9	0,18-0,28	5,1-6,1	0,2

Many trial experiments were conducted by varying tool rotational speed between 500 - 1000 rpm to find the right value. Finally the 1000 rpm was applied, because in case the right bond was created. To avoid the heat removal the specimens were fixed on a marble table (Figure 2.).

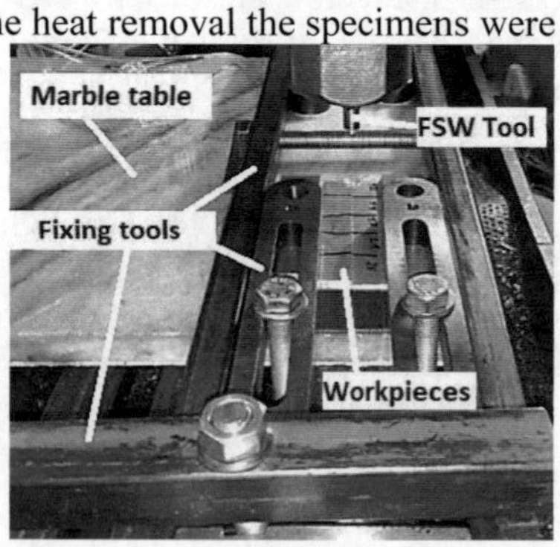

(a)

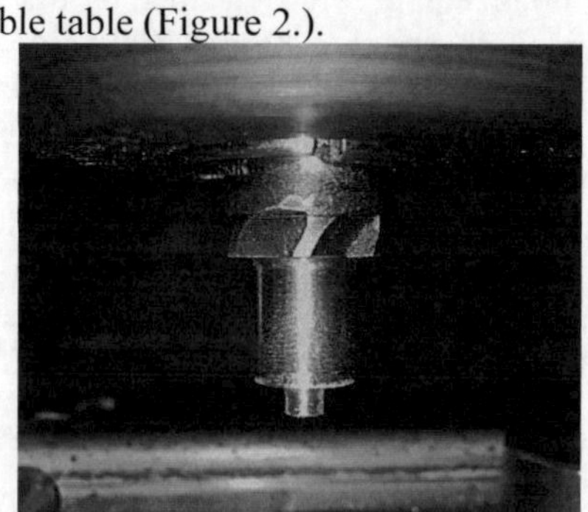

Ab)

Figure 2. (a) Setup of the experience and (b) the tool

The dimensions of base metal plates were 60 mm width by 120 mm length. The AA6082-T651 alloy was placed in the retreating side and AA7075-T651 plate was placed in the advancing side because AA7075-T651 has higher flow strength than AA6082-T651 aluminium alloy and the high frictional force and high material excavation force created in the advancing side.

The welding conditions and process parameters used to fabricate the joints are given in Table 2.

Table 2. Welding condition and process parameters

Parameters	
Tool rotational speed (rpm)	1000
Tool shoulder diameter (mm)	10
Tool pin diameter (mm)	4
Tool pin length (mm)	3,7
Tool material	High speed steel
Tool traveling speed (mm/min)	12, 14, 17, 21, 25

The microstructure of the specimens was studied by using optical microscope (ZEISS AXIO) and the hardness of the bond by Vickers microhardness tester (Wilson Wolpert 401MVD). The specimens after the polishing were etched using standard Keller's reagent (5 ml HNO_3, 3 ml HCl, 2 ml HF and 190 ml distilled water) to reveal the microstructure of the bond, the acid treatment takes 20 sec.

3. Results and discussion

3.1. Macrostructure

The macroscopic morphology of friction stir welded AA6082-T651 and AA7075-T651 dissimilar joints by five different feed rate (shown in Figure 3.).

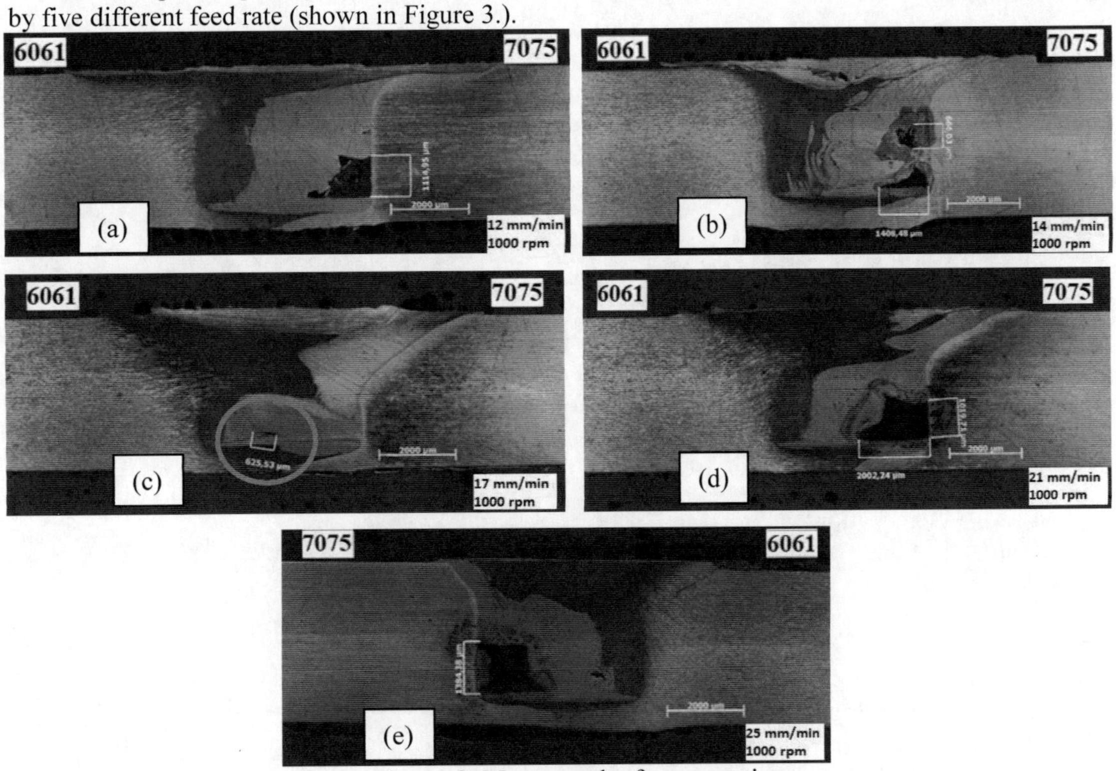

Figure 3. Macrograph of cross sections

IOP Conf. Series: Materials Science and Engineering **448** (2018) 012001 doi:10.1088/1757-899X/448/1/012001

FSW are known to have individual defects like cracks, pin hole or tunnel defect. It seems that all of the cross sections contains some tunnel defects and pin hole. The Figure 3. (c) specimen has the smallest defect "pin hole" (marked with red circle) which size of 10x625,53 μm while the other one have tunnel defects and pin hole which are higher than 600x1000 μm . The top and bottom surface of the weld region shows defect free and has equally spaced material ripples. There were large microstructural changes during welding. The interfaces of various regions like HAZ, TMAZ and WN were clearly visible in the Figure 4.

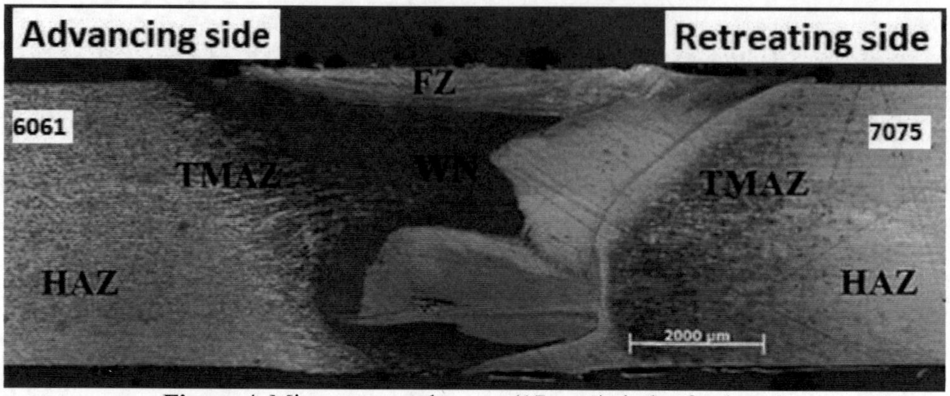

Figure 4. Microstructural zones (17 mm/min by feed rate)

In the FZ zone the grains finer and orientated perpendicular to the surface. The grains are greatly deformed in the AS-TMAZ region.

3.2. Microhardness

Because between the welded specimens the smallest mistake was produced at 17 mm/min by feed rate, so the hardness measurements were done on this one.

The micro Vickers (HV0.1) measurements were carried out across the weld at three different depths from the surfaces (1 mm, 2 mm and 3 mm) to associate the hardness variations among the welding zones. The hardness profiles are shown in Figure 5.

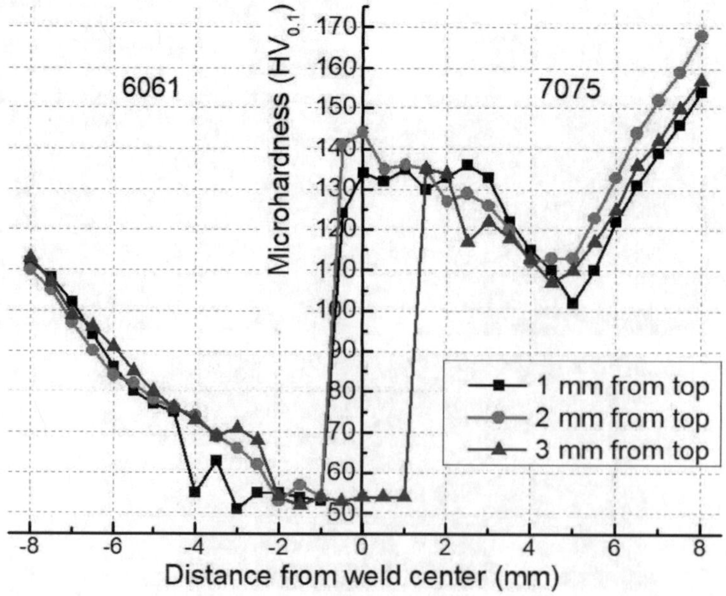

Figure 5. Hardness measurement of cross section at three different depths from top (17 mm/min by feed rate)

At 1 and 2 mm below the top of the surface hardness values of the weld material seems to be similar while 3 mm from the top, the values moved to the right in the stir zone, so the hardness of the workpiece the lowest under the pin. The hardness is further decreased in TMAZ regions. The microhardness values in the stir zone result in higher hardness values than in TMAZ regions.

4. Conclusions

The effects of feed rates on microstructure and microhardness of friction stir welded dissimilar AA7075-T651 - AA7075-T651 aluminium alloy joints were investigated and the following conclusions are derived:

- the friction stir welding creates a weld nugget zone under the tool' shoulder which has a fine microstructure
- tunnel defects generated in the retreating zone and a pin hole about 3 mm from top at 17 mm/min by feed rate
- the material hardness was decreased near end of the pin about the microhardness measurements

Acknowledgement

This research is supported by EFOP-3.6.1-16-2016-00006 "The development and enhancement of the research potential at John von Neumann University" project. The Project is supported by the Hungarian Government and co-financed by the European Social Fund.

References

[1] R.S. Mishra, Z.Y. Ma 2005 Friction stir welding and processing, *Materials Science and Engineering R* **50** 1 – 78

[2] M. Ilangovan, S. Rajendra Boopathy, V. Balasubramanian 2005 Effect of tool pin profile on microstructure and tensile properties of friction stir welded dissimilar AA 6061-AA 5086 aluminium alloy joints, *Defence Technology* **11** 174 - 184

[3] T. S. Rao, G. M. Reddy, S.R. K. Rao 2016 Studies on variations in microstructure and hardness of AA7075-T651 aluminum alloy friction stir welds, *La Metallurgia Italiana* **108** 29-35

[4] Bharat Raj Singh *2012 A Hand Book on Friction Stir Welding, LAP Lambert Academic Publishing,* UK 1-70

[5] Van Haver, W., Stassart, X., de Meester, B., Dhooge, A. 2008 Friction Stir Welding of Aluminum High Pressure Die Castings: Parameter Optimisation and Gap Brigeability, *Welding in the World* **52** 20 - 29

[6] Kayode Oyedemi, Patrick McGrath, Hannalie Lombard, Balázs Varbai 2017 A Comparative Study of Tool-Pin Profile on Process Response of Friction Stir Welding of AA6082-T6 Aluminium Alloy, *Periodica Polytechnica, Mechanical Engineering* **10** 6082 - 6088

[7] S. M. Bayazid, H. Farhangi, A. Ghahramani 2015 Effect of Pin Profile on Defects of Friction Stir Welded 7075 Aluminum Alloy, *Procedia Materials Science* **11** 12 – 16

IOP Conf. Series: Materials Science and Engineering **448** (2018) 012002 doi:10.1088/1757-899X/448/1/012002

The effects of machining strategies of magnetic assisted roller burnishing on the resulted surface structure

Zsolt Kovács[1, 4], **Zsolt János Viharos**[2, 3], **János Kodácsy** [1]

1 John von Neumann University, GAMF Faculty of Engineering and Computer Science, Dep. of Vehicle Technology, Kecskemét H-6000, Hungary

2 Institute for Computer Science and Control of the Hungarian Academy of Sciences, Kende u. 13-17, Budapest H-1111, Hungary

3 John von Neumann University, Faculty of Economics and Business, Department of Economics and Law, Kecskemét H-6000, Hungary

4 Department of Manufacturing Science and Technology, Budapest University of Technology and Economics, Budapest H-1111, Hungary

kovacs.zsolt@gamf.uni-neumann.hu

Abstract. The aim of this paper is to investigate the effect of different burnishing strategy on a C45 steel machining. Burnishing is a well-known, cold working surface improving technology, it is usually applied for cylindrical workpiece. In the reported research a novel developed magnetic assisted ball burnishing (MABB) tool was applied which was designed for MABB machining of flat and harmonic surfaces. To increase its efficiency, different types of machining strategies were applied. After the burnishing process the machined surfaces were measured by surface roughness tester to determine the most important Rsk and Rku tribological parameters. According to the results it can be stated that all different burnishing strategies have special application areas where they can be used efficiently (e.g. sliding surface, moulding tool and after- or pre-machining for ultra-precision machining).

1. Introduction

Nowadays, industry is expecting high efficiency and quality of machining surfaces. There are many types of finishing technologies e.g. superfinishing [1], polishing and burnishing [2]. In the reported research high quality surfaces are produced by ball burnishing, to enhance the workpiece characteristics like wearing [3], tribology [4] or corrosion [5]. Generally, the burnishing is an applied technology for machining cylindrical workpiece in order to improve the surface [6]. But in this published research the authors used a novel developed burnishing tool which was designed especially for flat and harmonic surfaces. Different types of machining strategies were applied to increase the tools' efficiency of the burnishing process.

Some of studies report on the advantages of various machining strategies, but most of them was used for milling. These strategies can reduce the machining time, force and surface roughness while increase the tolerance, efficiency and sliding properties [7, 8, 9] not to mention the discerning tool path [9]. Based on these aspects, similarly positive results can be expected for ball burnishing, too. Because the novel MABB tool and operation is similar to a face milling tool, almost the same machining strategies can be applied as were used in milling process. Burnishing has been successfully applied to many materials like titanium alloys [3], stainless steels [11], polyethylene [12], aluminum alloys [13, 14], magnesium alloys [5] and steels [6]. But because the MABB tool works by magnetic force, it requires a workpiece also being magnetizable to ensure the necessary burnishing force between the tool and workpiece.

R. Jerez-Mesa et al. [15], investigated the influence of different strategies of ultrasonic burnishing on the resulted surface. Their tool was a 10 mm diameter ball type in a house and it was attached to a piezoelectric module which can be excited through an external generator with a 40-kHz electrical signal. They burnished a pre-milled specimen with three different strategies (parallel, 45 degrees and perpendicular), the Fig. 1. shows their result [15].

Fig. 1. Burnished workpiece by R. Jerez-Mesa et al. [15]

Based on their results it can be stated that the improvements of surface roughness were equal to the applied feed directions for burnishing, so the parallel feed direction improves the y direction, the perpendicular improves the x direction and the 45 degrees improves both directions [15].

Because the novel magnetic assisted ball burnishing (MABB) tool is able to burnish having a width of 45 mm (thanks to the design), the authors can use all types of strategies which are available in CAM programs. The paper explores the burnished surfaces which were produced by various machining strategies.

2. The MABB technology & tool

The Magnetism Aided Machining or Magnetic Assisted Machining (MAM) technologies are relatively novel industrial machining processes. The newly developed MABB tool works with a NdFeB magnet and it generates the magnetic fluxes which pushes the balls into the machined surface [16]. As represented in Fig. 2., the fluxes are focused in the tool-workpiece contact zone and as a result the magnetic flux density (B) is high.

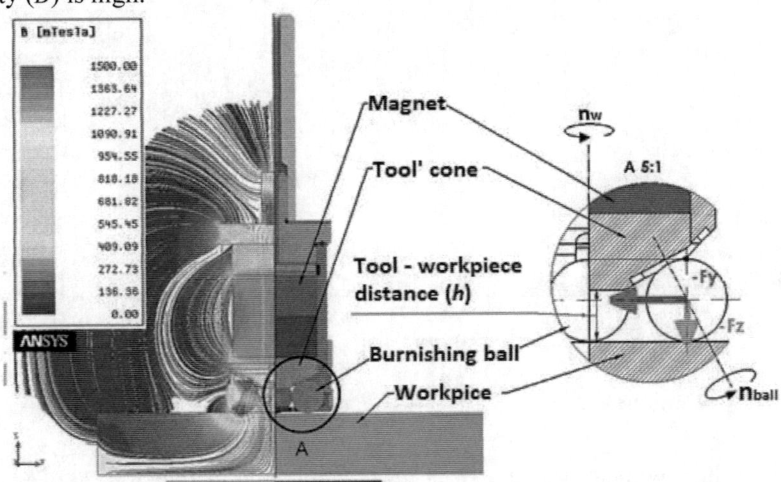

Fig. 2. FEM simulation about the magnetic fluxes around the MABB tool

The Fig. 2. also shows that the tool-workpiece distance has very important role, for ensuring the optimal rolling pressure [17]. The authors determined in their previous research the optimal working gap by calculation and experience, too [18]. This optimum *h* gap is 10 mm where the F_z force is ~350N as shown in Fig. 3.

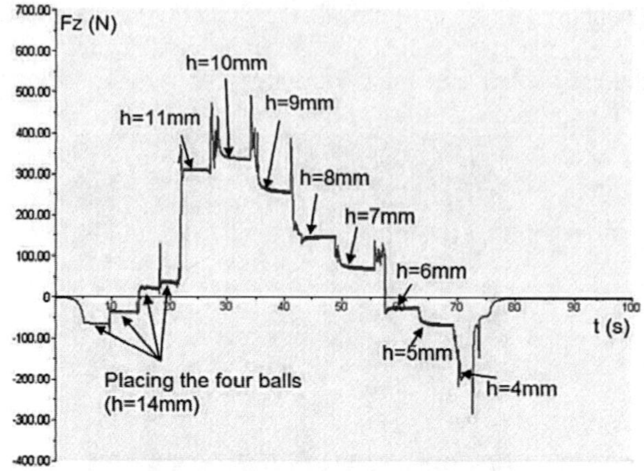

Fig. 3. Magnetic force depending on the tool distance [18]

3. Burnishing experiments

An C45 specimen was fixed on a CNC milling machine. The surface was pre-machined by a 8 teeth face milling cutter, with the cutting velocity of v_c = 120 m/min; feed speed rate of v_f = 200 mm/min, the cutting deep was: a_p = 2 mm, and APMX 160408TR-M14 MP2500 wiper geometrical inserts were used. This previous milling operation allowed homogenizing the surface of workpiece before the burnishing. Fig. 4-5. show some of the used machining strategies which were applied for burnishing. Both sides (A and B) of the workpiece were burnished with three-three strategies, respectively (dimensions of the workpiece surface were 300x160 mm).

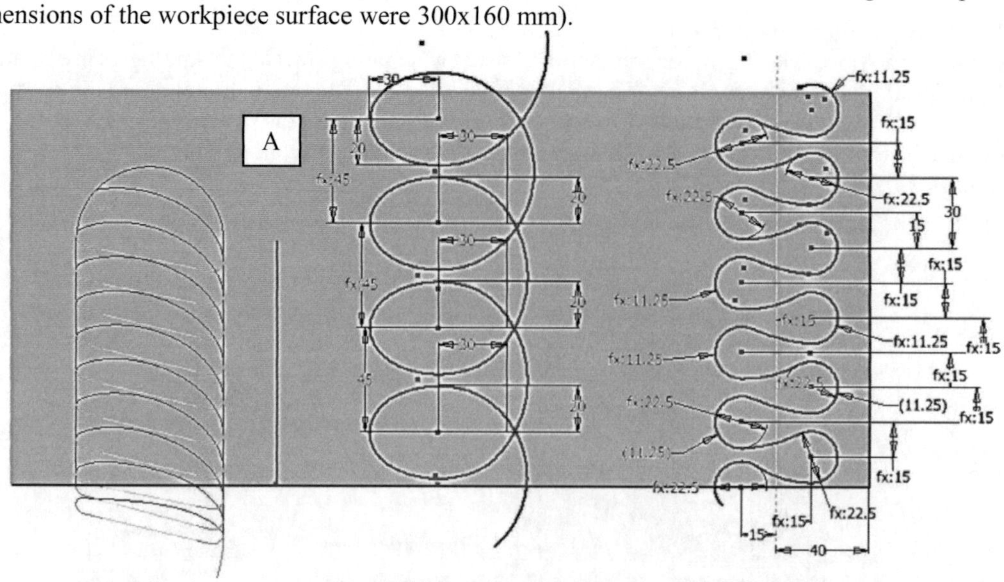

1- Adaptive 2- Cycloid 3- All-round

Fig. 4. Burnishing strategies on the A side

During a burnishing with an strategy, all of the technological parameters were kept constant to ensure homogenized and comparable surfaces (velocity: v_b = 40 m/min, feed speed: f_b = 50 m/min and tool gap: h = 10 mm).

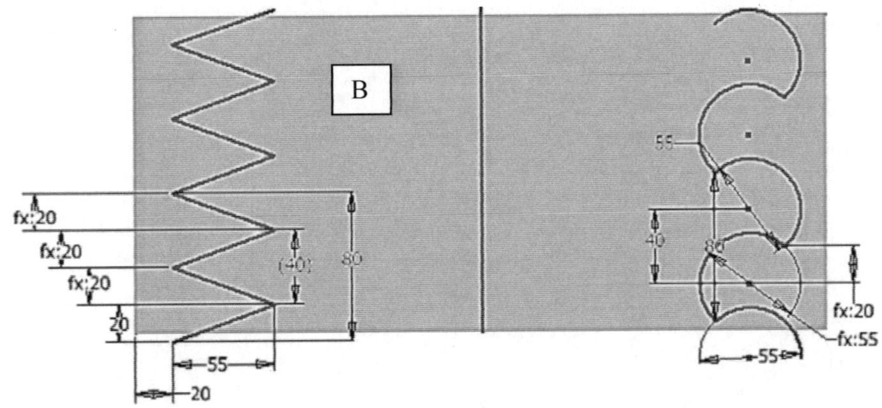

4- Zig-zag 5- Line 6- Stepping-swinging

Fig. 5. Burnishing strategies on the B side

As shown in Fig. 4. there is one adaptive machining strategy which can be found in the Inventor HSM software. The other ones (Fig. 4-5.), were prepared in the same software by the authors, because the adaption of this path is not available for flat surfaces. This software contains many strategies but all of them require some kind of set-up (e.g. slot or pocket). But the Inventor HSM software can make (by post processing) CNC programs by own created strategies, so these four strategies and also a linear path were tested. Fig. 6. shows the resulted, burnished surfaces produced by different strategies.

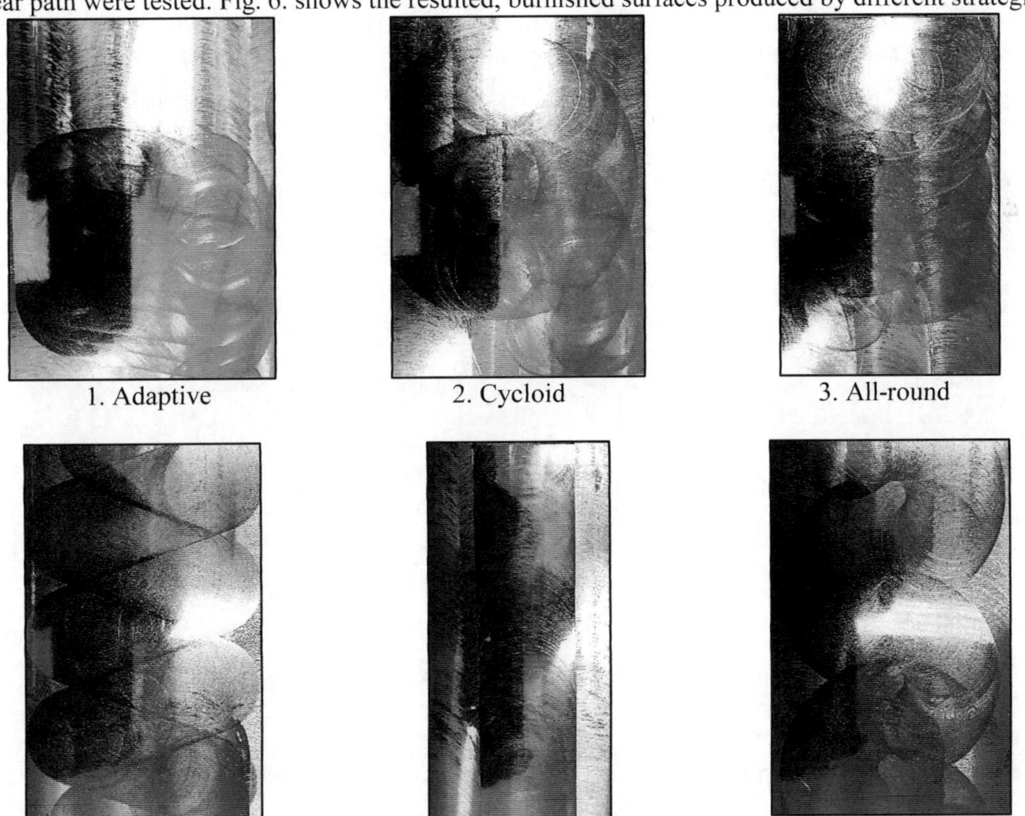

1. Adaptive 2. Cycloid 3. All-round

4. Zig-zag 5. Line 6. Stepping-swinging

Fig. 6. The surfaces of the burnished workpieces produced by six different strategies

4. Results and discussion

All burnished surfaces and the original one were measured along the x (parallel to the machining) and y (perpendicular to the machining) directions by a MITUTOYO Formtracer SV-C3000 surface measurement equipment (Fig. 7.). The pre-milled surface was marked with '0' and it was also measured in both directions.

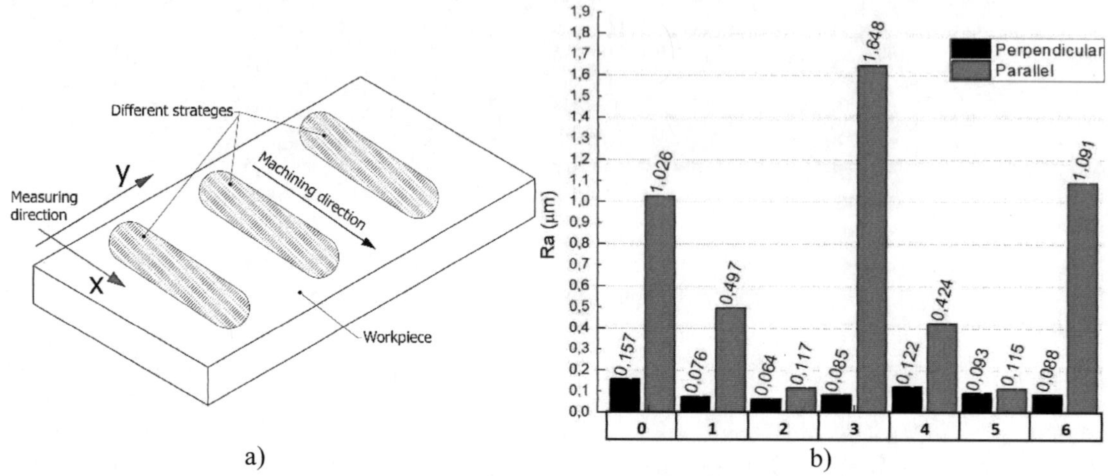

Fig. 7. The a) measuring directions and b) measured R_a roughness parameters Fig. 7. b) respresents that the higher R_a roughness values were measured in parallel directions.

To show the surfaces integrity, too, a summary was prepared about the surface roughness profile (R-profile) which is represented in Fig. 8.

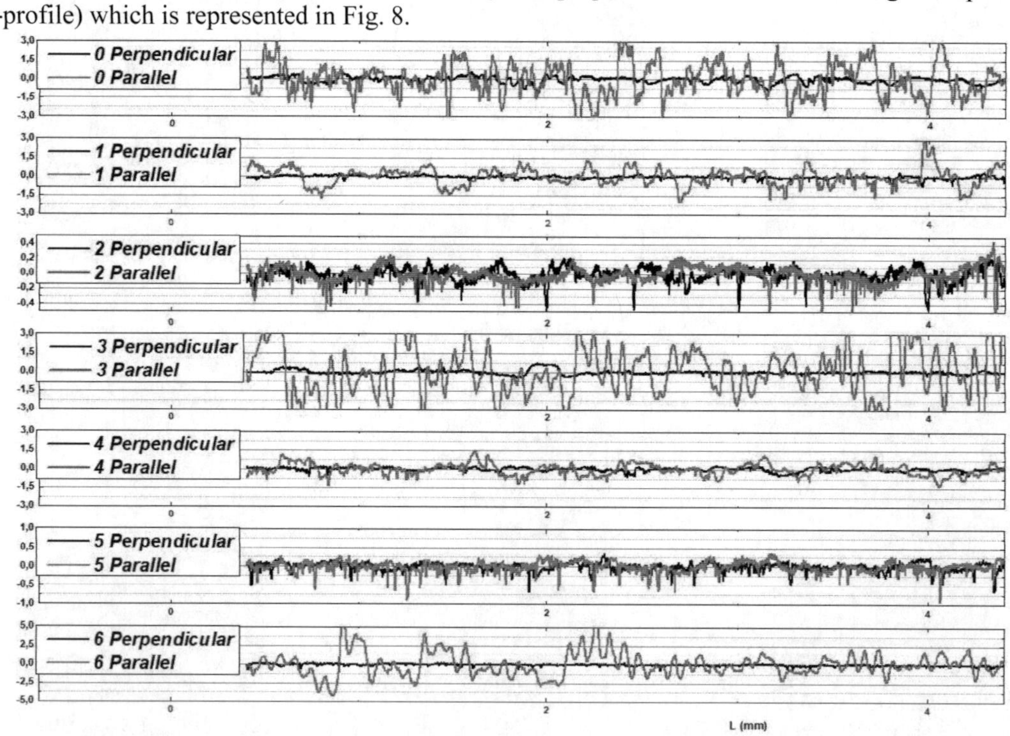

Fig. 8. Comparison of the milled (0) and burnished surface (1-6) roughness profiles (the perpendicular measuring direction is black, the parallel is red while the y-axes have different scaling

The Fig. 8. well represents that some of this R-profiles mirror advantageous features according to tribological aspects. For example the surface number 2 contains relative harmoniously arranged oil pockets. This can be said about the surface number 5, too. But based only on the R-profile it cannot be determined which surface is the most advantageous according to tribological aspects. So, the authors also measured the classical tribological indicators Rsk and Rku.

The skewness (Rsk) is a measure of the symmetry of the profile according to the mean line, giving information about the asymmetry of profiles having the same Ra values. Negative values of Rsk indicate a predominance of roughness (material), while positive ones are observed for surfaces with peaks [20].

The kurtosis (Rku) is a measure of the sharpness of the profile according to the mean line that provides information about the distribution of spikes above and below the mean line. Thus, spiky surfaces will have a high kurtosis value (Rku > 3) and bumpy surfaces a low value (Rku < 3) [20].

These indicators represented in one tribological diagram can form a so-called topological map. The Fig. 9. represents this topological map and indicates the places of different machining strategies according to tribology aspect.

The measurement of the Rsk and Rku parameters does not require any special instrument because its collection is part of the roughness measurement. So, the resulted surfaces were placed on the topological map, see Fig. 9. b).

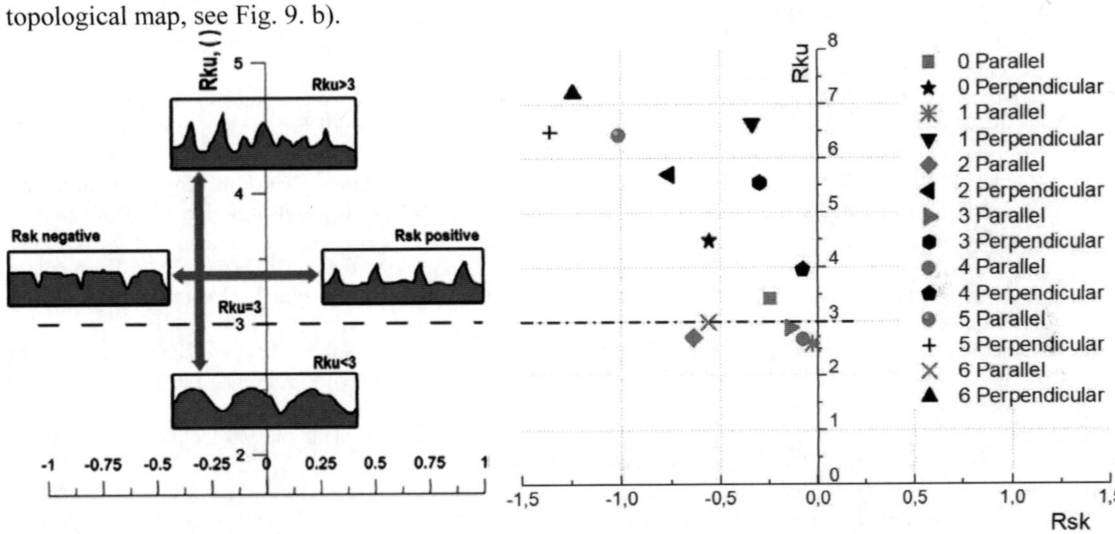

a) b)

Fig. 9. a) Graphical representation of Rsk and Rku topological map [20] and the b) topological map of the analysed burnishing according to different strategies

As the Fig. 9. b) shows all of the measured values are negative in Rsk axis and there are five value which are below the value 3 in the Rku axis. These are: 2 Parallel, 6 Parallel, 3 Parallel, 4 Parallel and 1 Parallel.

Overall, surface numbered by 2 (cycloid) has the lowest Ra value (Ra = 0,064) and it is the best in tribological aspect in parallel direction, too. This means that the MABB toll is able producing high quality surfaces with good sliding properties by cycloid strategy in the machining direction.

5. Conclusions

A flat C45 specimen has been burnished by the novel introduced Magnetic Assisted Ball Burnishing tool. Instead of the ordinary linear machining direction five different types of machining strategies has been successfully tested and analysed. Based on this detailed analysis the following statements can be concluded:

- The direction of the measurement has a significant effect on the tribological evaluation, those directions resulted the best tribological criterions which were measured parallel to the machining.
- The cycloid, all-round, zig-zag and stepping-swinging strategies are suitable to produce well working surfaces according to tribological aspects.
- Cycloid strategy can create the lowest Ra roughness values in both measuring directions.
- The cycloid strategy resulted the best surfaces overall, so it will be examined further in the optimisation of the technological parameters.

Acknowledgement

This research is supported by EFOP-3.6.1-16-2016-00006 "The development and enhancement of the research potential at John von Neumann University" project. The Project is supported by the Hungarian Government and co-financed by the European Social Fund and European Commission through the H2020 project EPIC under grant No. 739592.

References

[1] Jan Liska, Zdenko Lipa, Ivan Baranek, Katarina Siketova 2009 Superfinishing theory *Annals of DAAAM for 2009 & Proceedings of the 20th International DAAAM Symposium* 1297 – 1298

[2] Varga G. 2014 Effects of technological parameters on the surface texture of burnished surfaces *Key Engineering Materials* **581** 403-408

[3] Goutam Devaraya Revankar, Raviraj Shetty, Shrikantha Srinivas Rao,Vinayak Neelakanth Gaitonde 2017 Wear resistance enhancement of titanium alloy (Ti–6Al–4V) by ball burnishing process *Journal of Materials Research and Technology* **6** 1 13 – 32

[4] N.S.M. El-Tayeb, K.O. Low, P.V. Brevern 2007 Influence of roller burnishing contact width and burnishing orientation on surface quality and tribological behaviour of Aluminium 6061 *Journal of Materials Processing Technology* **186** 272 – 278

[5] M. Salahshoor, Y.B. Guo, C. Li 2017 Surface Integrity and Corrosion Performance of Biomedical Magnesium-Calcium Alloy Processed by Hybrid Dry Cutting-Finish Burnishing *Procedia Manufacturing* **10** 467 – 477

[6] Masato Okada, Shohei Suenobu, Kei Watanabe, Yorihiro Yamashita, Naoki Asakawa 2015 Development and burnishing characteristics of roller burnishing method with rolling and sliding effects *Mechatronics* **29** 110 – 118

[7] Michael Gerstenmeyer, Frederik Zanger, Volker Schulze 2016 Complementary Machining–Machining Strategy for Surface Modification *Procedia CIRP* **45** 247 – 250

[8] I. Szalóki, S. Csuka, S. Csesznok, S. Sipos 2012 Can trochoidal milling be ideal? The XXI. Conference of GTE on Manufacturing and related technologies Budapest, Hungary, Paper S6 08. (ISBN:978-963-9058-35-4)

[9] Adam Jacso, Tibor Szalay, Juan Carlos Jauregui, Juvenal Rodriguez Resendiz 2018 A discrete simulation-based algorithm for the technological investigation of 2.5D milling operations *Proceedings of the Institution of Mechanical Engineers part C-Journal of Mechanical Mngineering Science* 1-13

[10] M. Otkur, I. Lazoglu 2007 Trochoidal milling *International Journal of Machine Tools & Manufacture* **47** 1324 – 1332

[11] V. Chomienne, F. Valiorgue, J. Rech, C. Verdu 2016 Influence of ball burnishing on residual stress profile of a 15-5PH stainless steel *CIRP Journal of Manufacturing Science and Technology* **13** 90–96

[12] Łukasz Janczewski, Daniel Toboła, Witold Brostow, Kazimierz Czechowski, Haley E. Hagg Lobland, Marcin Kot, Krzysztof Zagórski 2016 Effects of ball burnishing on surface properties of low density polyethylene *Tribology International* **93** 36 – 42

[13] Vaibhav Pandey, K. Chattopadhyay, N.C. Santhi Srinivas, Vakil Singh 2017 Role of ultrasonic shot peening on low cycle fatigue behavior of 7075 aluminium alloy *International Journal of Fatigue* **103** 426 – 435

[14] Varga, G., Ferencsik, V 2017 Analysis of surface topography of diamond burnished aluminium alloy components, *Springer International Publishing AG 2017, Eds: K. Jármai and B. Bolló, Vehicle and Automative Engineering, Lecture notes in Mechanical Engineering* 143-154, ISSN 2195-4356, ISSN 2195-4364 (electronic), doi: 10.1007/978-3-319-51189-4_15

IOP Conf. Series: Materials Science and Engineering **448** (2018) 012002 doi:10.1088/1757-899X/448/1/012002

[15] R. Jerez-Mesa, G Gomez-Gras and J.A. Travieso-Rodriguez 2017 Surface roughness assessment after different strategy patterns of ultrasonic ball burnishing *Procidia Manufacturing* **13** 710 – 717
[16] „Apparatus and method for deburring and roller-burnishing machine parts" International patent. 2012. Patent No.:172 7648 Notifiers: KF GAMF Faculty. Inventor: Dr. János Kodácsy (78%), Dr. József Danyi (22%)
[17] Loh NH, Tam SC. 1988 Effects of ball burnishing parameters on surface finish-a literature survey and discussion *Precis Eng* **10** 215 – 20
[18] Zsolt F. Kovács, Zsolt J. Viharos, János Kodácsy 2018 Determination of the working gap and optimal machining parameters for magnetic assisted ball burnishing *Measurement* **118** 172 – 180
[19] Varga, G., Kundrák, J. 2017 Effects of technological parameters on surface characteristics in face milling *Solid State Phenomena* **261** 285-292
[20] E.S. Gadelmawlaa, M.M. Kourab, T.M.A. Maksoucf, I.M. Elewaa, H.H. Solimand 2002 Roughness parameters *J. Mater. Process. Technol.* **123** 133 – 145

Direct monitoring of hole damage in carbon fibre-reinforced polymer (CFRP) composites

N Geier[1,*], Gy Póka[1] and T Szalay[1]

[1] Budapest University of Technology and Economics, Department of Manufacturing Science and Engineering, Hungary

E-mail: geier@manuf.bme.hu

Abstract. Carbon fibre-reinforced polymer (CFRP) composite laminates have good specific material properties, these composites are therefore widely used by the high tech industries. Machining operations are often necessary in order to achieve dimensional requirements. However, CFRP is a difficult-to-cut material due to its non-homogenous and anisotropic features and the strong wear-effect of carbon fibres on the cutting tool. Increasing hole numbers causes the increase of cutting edge radius of the cutting tool, therefore hole damages (uncut fibres, delamination etc.) occur more often. The main objective of the present paper is to minimize uncut fibres in CFRP using direct monitoring of hole damage. Firstly, numerous orbital drilling experiments were conducted and the characteristics of uncut fibres were analysed. Secondly, the orbital drilling experiments were repeated with direct monitoring of uncut fibres and process control was applied in order to decrease the size of the burr area. Results show that hole damage decreased with the application of the used process control.

1. Introduction

Carbon fibre-reinforced plastic (CFRP) composite laminates are widely used by the aerospace, marine, sport and automobile industries due to its excellent special material properties [1]. Manufacturers aspire to laminate CFRP parts directly into the final shape, but machining is often necessary in order to remove waste parts left by the laminating process. Furthermore, to achieve dimensional requirements and to assembly parts, it is also necessary to drill holes in CFRP composites. However, CFRP is a difficult-to-cut material due to its non-homogenous and anisotropic features and the strong wear-effect of carbon fibres on the cutting tool [2].

Many researchers [3–5] investigated the influence of process parameters (feed rate, cutting speed, crew pitch of the feeding helix etc.) on hole quality parameters in CFRP in order to optimise the machining process. Sorrentino et al. [6] showed that feed rate has the most significant effect on uncut fibres during conventional drilling of CFRP. Orbital drilling (helical milling) produces less delamination and less uncut fibres than conventional drilling, according to Voss et al. [7].

Wang et al. [8] were researching the influence of cutting edge radius on the cutting-induced surface damages when machining unidirectional carbon fibre-reinforced plastic (UD-CFRP) composites. They found that if the cutting edge radius increases, the surface quality decreases significantly. If the cutting edge radius is high, the burr formation mechanisms is bending dominated. In the other hand, if the cutting edge radius is small (cutting edge is sharp), the burr formation mechanisms is crushing dominated. Ramirez et al. [9] carried out drilling experiments in UD-CFRP and analysed wear mechanisms of the cutting tool. They showed that carbon fibres and high cutting speed significantly influence the wear of the tool radius. Furthermore, wear can be indirectly measured by monitoring the cutting force. They and [10] showed that tool wear rapidly increases with the increase of the number of holes.

The main objective of the present study is to decrease hole damage using direct monitoring and diagnostics of characteristics of uncut fibres in UD-CFRP. Numerous orbital drilling experiments were carried out with fixed parameters and the characteristics of uncut fibres were analysed. Based on these experiments and on previous researches, a decision algorithm was developed and tested with new machining experiments.

2. Experimental setup and methods

2.1. Experimental setup

Unidirectional carbon fibre-reinforced vinyl ester matrix composite laminate (of 6 mm thickness) was machined with a HSS Ø8 TIVOLY 80308810808 end mill. The helix angle of the tool is 30° and it has only one cutting edge, as can be seen in figure 1. This end mill is suggested by the tool producer to machine aluminium and plastic materials due to its sharp cutting edge (small cutting edge radius and positive rake angle).

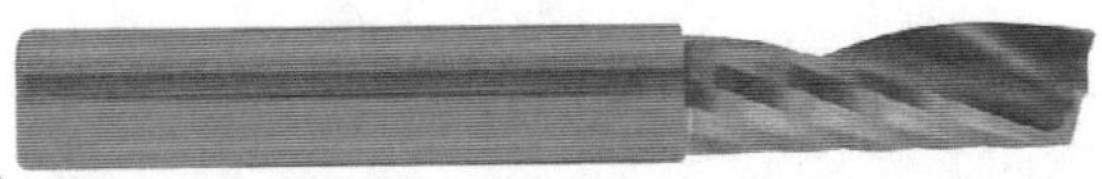

Figure 1. TIVOLY Ø8 80308810808 end mill

Orbital drilling (helical milling) technology was used to machine holes in CFRP because it produces less delamination, fibre pull-out and uncut fibres than conventional drilling processes [5]. The machining experiments were carried out on a Kondia B640 machining centre, equipped with a Nilfisk GB733 vacuum cleaner. The CFRP specimen was fixed between two support plates in order to decrease push-down delamination. Images of the drilled holes were taken by a Dino-Lite AM4013MT digital microscope (Magnification: 10-70x; Resolution: 1.3 Megapixel; Maximum frame rate: 30 fps). The experimental setup can be seen in figure 2.

Figure 2. Experimental setup: (a) digital microscope (b) support plate (c) fixture (d) end mill and (e) UD-CFRP plate

2.2. Digital image processing

Digital image processing (DIP) was used in this study in order to analyse characteristics of uncut fibres. A unique LED light-source was located under the CFRP specimen and it lighted the hole from underneath, as can be seen in figure 3(c). With this lighting solution it was possible to increase the contrast of images of the holes. The workpiece became a shadow and the background became highlighted, so the shape of the hole became sharper, as can be seen in figure 3(a). The taken images were then segmented using a colour histogram (figure 3(b)). The white pixels of the image were then summed and compared with the sum of the white pixels of an image with a nominal hole on it.

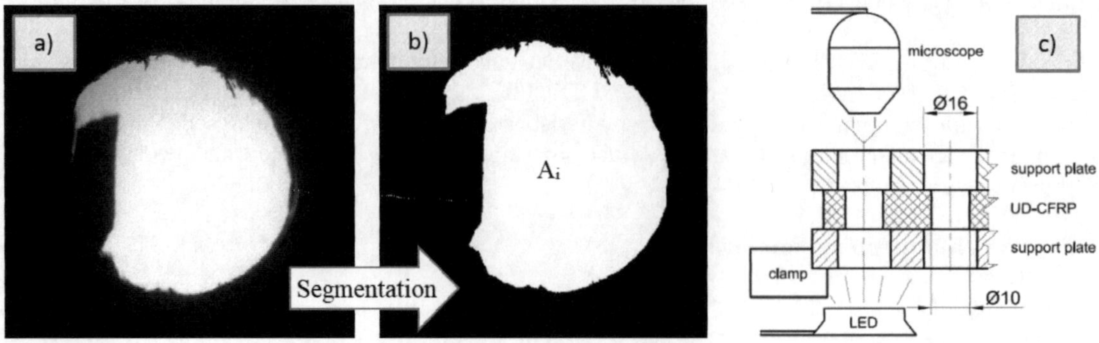

Figure 3. Digital image processing (a) image taken by the digital microscope (b) segmented image using colour histogram (c) measurement setup

In this study, the uncut fibres are characterised by the area factor (α), expressed by Eq. (1).

$$\alpha = \frac{1}{n}\sum_{i=1}^{n}\frac{A_i}{A_0} \quad (1)$$

where n is the number of holes in one evaluation, A_i is the number of white pixels on the analysed image and A_0 is the number of white pixels on an image of a damage-free hole. If the fibres are cut properly (the hole is damage-free) the ratio of A_i/A_0 is 1. The deviation of the results of DIP of CFRP holes is usually high, using the average of *n* ratios is therefore necessary.

2.3. Monitoring and diagnostics

The flowchart of the monitored orbital drilling process can be seen in figure 5. Input parameters are the following: A_0 is the number of the white pixels on an image of a damage-free hole, α_1=0.9 and α_2=0.8 are the critical values of the area factor, *h* is the screw pitch of the feeding helix, *z* is the drilling depth, and I is the variation interval of the parameters, as can be seen in figure 4. Based on previous research [5] the influence of *h* on hole-damage was the most significant, therefore this parameter is used as the first factor of the diagnostics process. With the second factor (*z*), it is possible to control deeper the cutting tool, so a wear-free helical flute can cut the fibres, which was not used before.

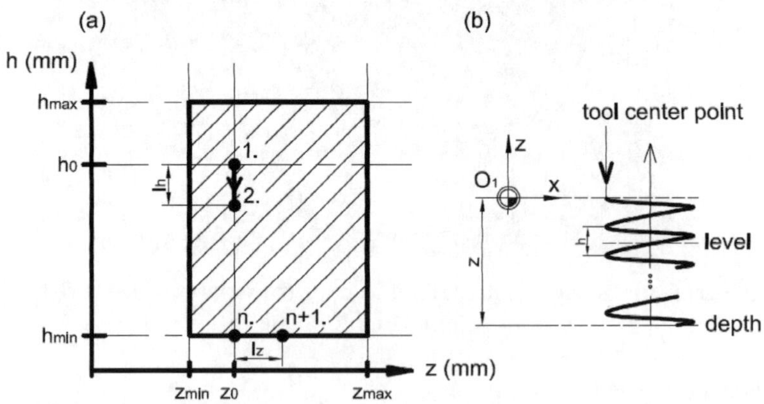

Figure 4. (a) Factor space and (b) tool patch of the monitored orbital drilling process

The explanation of theory of the method used in the tests are as follows. The first orbital drilling experiment is conducted with h_0=3 mm and z_0=14 mm parameters. In the case when good quality holes

IOP Conf. Series: Materials Science and Engineering **448** (2018) 012003 doi:10.1088/1757-899X/448/1/012003

($\alpha>\alpha_1$) were machined, h could be increased with I_h. On the other hand, if $\alpha_1>\alpha>\alpha_2$, then process parameters should not be changed. Furthermore, if $\alpha<\alpha_2$, then h has to be decreased with I_h until $h\geq h_{min}$. If $\alpha<\alpha_2$ and $h=h_{min}$, then z is increased with I_z until $z\geq z_{max}$. If $\alpha<\alpha_2$ and $h=h_{min}$, and $z=z_{max}$, then the tool is worn and it has to be changed.

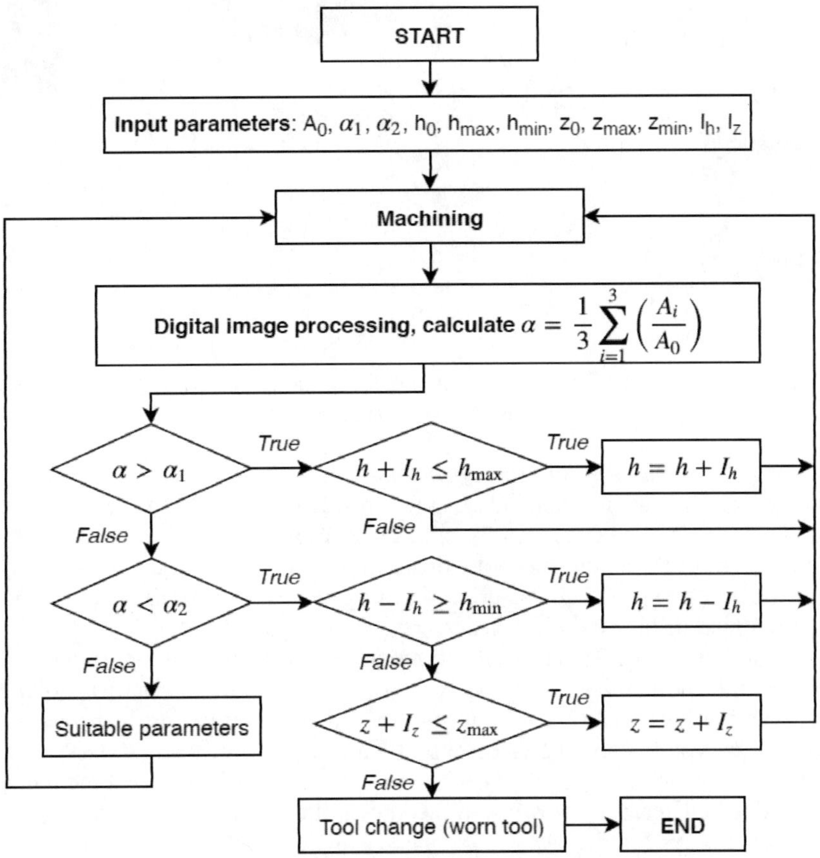

Figure 5. Flowchart of the monitored orbital drilling process

3. Results and discussion

3.1. Results of the drilling experiments when parameters were fixed

80 orbital drilling experiments were conducted with fixed process parameters: cutting speed (v_c=150 m/min), feed rate (v_f=358 mm/min), screw pitch of the feeding helix (h=3 mm) and climb milling without any lubricant liquid was applied. Process parameters were chosen based on previous studies [4, 5] and suggestions by tool manufacturers. Due to the high deviation of DIP of CFRP hole damages, area factor was calculated after each third drilling process from the mean value of the three drilled holes. Number of drilled holes decreases the area factor, as can be seen in figure 6. The possible reason of this is that the cutting edge radius is increased by the number of holes due to tool wear, according to [8]. Furthermore, the fibre fracture mode became bending dominated, the fibres are not cut properly.

Fibre cutting angle (angle between the vector of the cutting speed and the vector of the fibre orientation [8]) has a significant effect on surface quality of CFRP, according to Jia et al. [11]. It can be seen from the result that the middle point of the burr area (where uncut fibres were not cut properly) is located at fibre cutting angle of 130°. Furthermore, it is also clear from the results that the number of holes increases the burr area when orbital drilling CFRP.

IOP Conf. Series: Materials Science and Engineering **448** (2018) 012003 doi:10.1088/1757-899X/448/1/012003

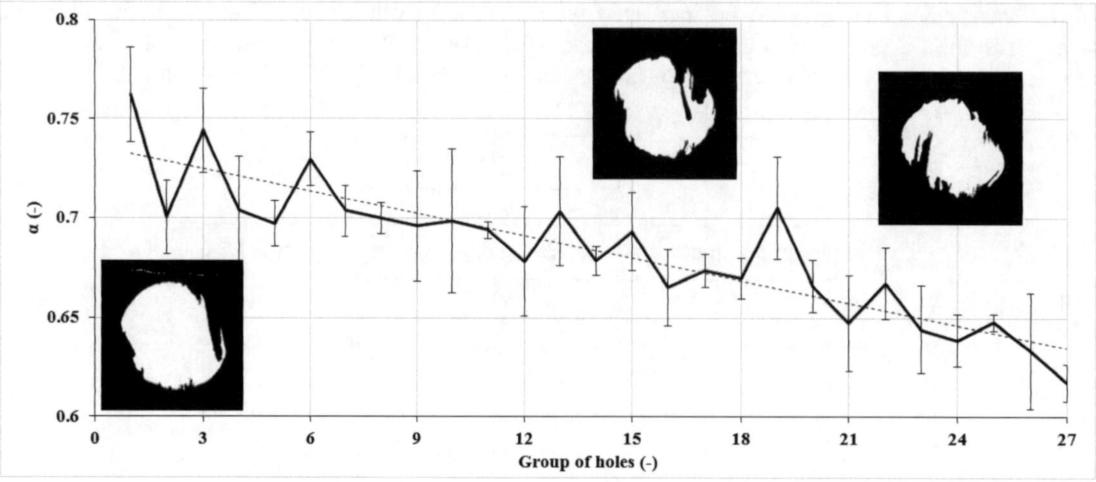

Figure 6. Group of holes vs area factor when drilling of UD-CFRP experiments were conducted with fixed parameters

3.2. Results of the monitored drilling experiments and discussion

According to the experiments described above and previous researches, it was found that the wear of the cutting edge radius is increasing rapidly with large *h* parameters. Furthermore, it was found that the helical flute length of the end mill is not exhausted totally. Based on these two findings, a decision algorithm was made in order to diagnose orbital drilling process, as can be seen in figure 5. Results of direct monitored orbital drilling experiments can be seen in figure 7.

Results show that in the early period of the monitored experiments (1-5 group of holes) the area factor decreased with the number of holes. However, after the fifth group of holes the area factor started to increase (the hole quality started to improve) due to optimal process parameters. As can be seen in the figure, six good quality holes (group of 8 and 9) were drilled (α>0.8). These better quality holes were achieved by the deeper control of the tool. However, in the case of the tenth group of holes the area factor decreased again due to worn of the helical flute length of the end mill.

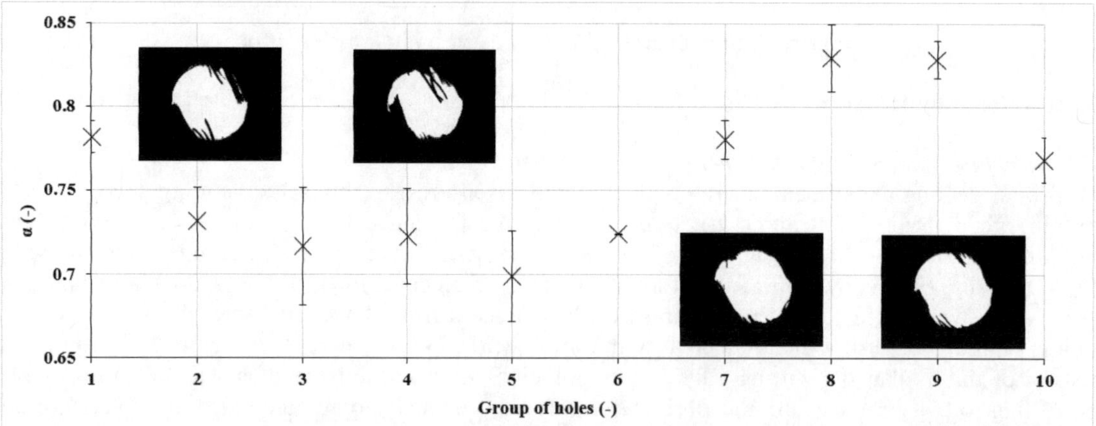

Figure 7. Group of holes vs area factor when monitored drilling of UD-CFRP experiments were conducted

The results of the original and the repeated orbital drilling experiments can be seen on figure 8. It can be seen from the diagram that the monitored drilling process could provide better quality holes than the original one. However, the machining time of the monitored experiments are longer due to longer cutting tool path. Based on the deviation of the area factor the size of the groups of holes (it was n=3 in this study) can be changed in order to optimise the monitoring process. However, in the future, the algorithm has to be optimised based on the present study in order to minimise the number of holes, where α<0.8.

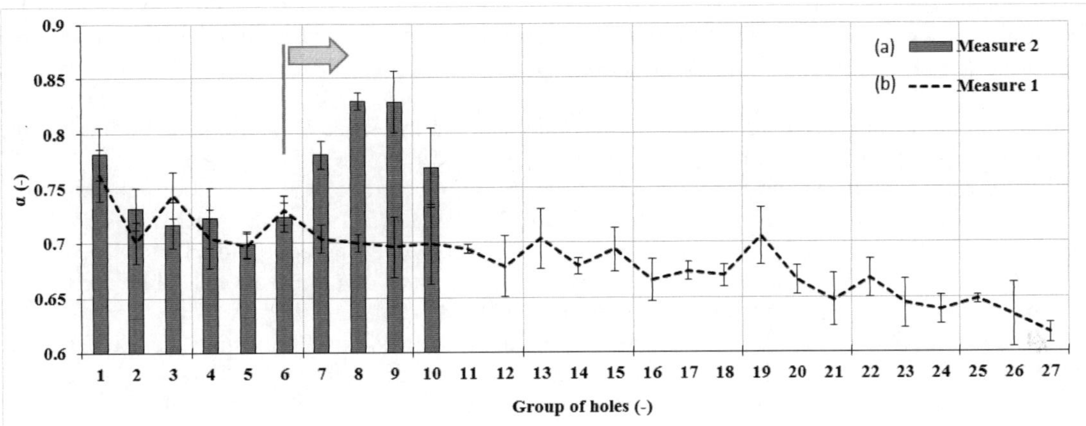

Figure 8. Group of holes vs area factor when (a) drilling of UD-CFRP experiments with fixed parameters and when (b) monitored drilling of UD-CFRP experiments were conducted

4. Conclusions

In the present study, numerous orbital drilling experiments were conducted and the characteristics of uncut fibres were analysed in UD-CFRP. Secondly, the orbital drilling experiments were repeated with direct monitoring of uncut fibres and process control was applied. According to the present study, the following conclusions can be drawn:

- A decision algorithm was developed in order to diagnose orbital drilling process.
- Number of orbital drilled holes decreases the area factor due to tool wear. Furthermore, number of holes increases the burr area too, when orbital drilling UD-CFRP.
- Monitored orbital drilling process could provide better quality holes than the conventional one. However, the machining time of the monitored experiments are longer due to longer cutting tool path.

Acknowledgement

This work was partly supported by the Higher Education Excellence Program of the Ministry of Human Capacities in the frame of Nanotechnology and Material Science research area of Budapest University of Technology and Economics (BME FIKP-NANO). This research was partly supported by the EU H2020-WIDESPREAD-01-2016-2017-TeamingPhase2-739592 project "Centre of Excellence in Production Informatics and Control" (EPIC). Furthermore, the authors acknowledge to prof. Gy. MÁTYÁSI and to the Ph.D. student B.Z. BALÁZS for their participation in the experimental work.

References

[1] Karataş M A and Gökkaya H, A review on machinability of carbon fiber reinforced polymer (CFRP) and glass fiber reinforced polymer (GFRP) composite materials, 2018 *Defence Technology*

[2] Li H, Qin X, He G, Jin Y, Sun D and Price M, Investigation of chip formation and fracture toughness in orthogonal cutting of UD-CFRP, 2016 *Int J Adv Manuf Technol*, **82** (5–8) 1079–1088.

[3] Mathivanan N R, Mahesh B S and Shetty H A, An experimental investigation on the process parameters influencing machining forces during milling of carbon and glass fiber laminates, 2016 *Measurement*, **91** 39–45.

[4] Geier N and Szalay T, Analysis of the cutting forces in machining of uni-directional carbon fiber-reinforced polymers (UD-CFRP), 2016 Poceedings of 7th International Technological Conference of CVUT. 42–46.

[5] Geier N and Szalay T, Optimisation of process parameters for the orbital and conventional drilling of uni-directional carbon fibre-reinforced polymers (UD-CFRP), 2017 *Measurement*, **110** 319–334.

[6] Sorrentino L, Turchetta S and Bellini C, A new method to reduce delaminations during drilling of FRP laminates by feed rate control, 2018 *Composite Structures*, **186**, 154–164.

[7] Voss R, Henerichs M and Kuster F, Comparison of conventional drilling and orbital drilling in machining carbon fibre reinforced plastics (CFRP), 2016 *CIRP Annals*, **65**, (1) 137–140.

[8] Wang F, Yin J, Ma J, Jia Z, YangF and Niu B, Effects of cutting edge radius and fiber cutting angle on the cutting-induced surface damage in machining of unidirectional CFRP composite laminates, 2017 *Int J Adv Manuf Technol*, **91** (9–12) 3107–3120.

[9] Ramirez C, Poulachon G, Rossi F and M'Saoubi R, Tool Wear Monitoring and Hole Surface Quality During CFRP Drilling, 2014 *Procedia CIRP*, **13**, 163–168.

[10] Poulachon G, Outeiro J, Ramirez C, André V and Abrivard G, Hole Surface Topography and Tool Wear in CFRP Drilling, 2016 *Procedia CIRP*, **45**, 35–38.

[11] Jia Z, Fu R, Niu B, Qian B, Bai Y and Wang F, Novel drill structure for damage reduction in drilling CFRP composites, 2016 *International Journal of Machine Tools and Manufacture*, **110**, 55–65.

IOP Conf. Series: Materials Science and Engineering **448** (2018) 012004 doi:10.1088/1757-899X/448/1/012004

Obtaining of light biocompatible magnesium alloys using levitation equipment under controlled argon atmosphere

V Geantă[1], I Voiculescu[1], H Kelemen[2] and G Kelemen[3]

[1] University POLITEHNICA of Bucharest, Splaiul Independenței 313, 6 District, Bucharest, Romania
[2] University of Medicine and Pharmacy of Târgu Mureș, Gheorghe Marinescu Street 38, Tîrgu Mureș, Romania
[3] AAGES SA, 16 Agricultorilor Street, Sângeorgiu de Mureș, Romania

E-mail: victorgeanta@yahoo.com, ioneliav@yahoo.co.uk, kelemen_h@yahoo.com, gy.kelemen@aages.ro

Abstract. The use of resorbable materials to make orthopedic implants is a new direction, with major benefits for both the patient and the surgeon. The purpose of a bioresorbable implant is to support the regeneration and healing of tissues, followed by completely dissolving after its purpose is achieved, being characterized by the biodegradability and biocompatibility capability. Magnesium is a natural element of the human body, the magnesium-based implant having the ability to be fully resorbed without inducing local or systemic toxic effects. Experimental alloys were obtained from 99.5% pure magnesium powders in which alloying elements (Ca) or micro-alloying (La, Ce) were added in the 0.5-5% range to identify the best bio-compatible alloy recipes. The metallurgical process was conducted in an induction melting equipment under a controlled atmosphere of argon, through levitation. The melting time for each alloy weighing 15 g was about 3 minutes, and the electrical parameters varied in the range: Voltage U_{mf} = 480 - 500 V; Current I_{mf} = 55 - 65 A; Frequency f_{mf} = 60 - 70 kHz; Power P_{mf} = 20 - 35 kW. Following the process, about 10% magnesium losses were obtained, due to the vaporization process. The samples obtained were characterized in terms of chemical composition, microstructure and micro-hardness. The micro-hardness values were ranging from 49 to 87.4 $HV_{0.2}$.

1. Introduction

At present, new generations of Mg-based metal materials appear to be particularly promising in the field of biocompatible implantable devices due to their ability to dissolve slowly in living tissues, without the need for further surgery to extract the implant. Furthermore, magnesium is a benign macro-element for living organisms, and the clinical trials show good resorbability and high biocompatibility of implants used for bone fixation [1]. The main application involving biodegradability is the execution of temporary implants which can be inserted into the human body in order to support problem areas (fractures, facial or maxillofacial reconstruction, artificial grafts, etc.) for a period of time before their slow dissolution [2-6].

The special mechanical properties (density and Young's modulus values similar to the bones' values, E = 10-30 GPa) and the controllable corrosion rate (resorbability in physiological environments) make magnesium alloys acceptable for use in cardiovascular interventions [7] and in

IOP Conf. Series: Materials Science and Engineering **448** (2018) 012004 doi:10.1088/1757-899X/448/1/012004

bone fracture repairs [8-11]. Since the properties of pure magnesium are not good enough to be used for prostheses, it is often alloyed with various other chemical elements [12-19]. The use of alloying elements such as Al, Ca, Li, Mn, Y, Zn, Zr and RE in Mg alloys can significantly improve the physical and mechanical properties of the alloy by: grain-size finishing, improving corrosion resistance, and increasing mechanical characteristics by forming intermetallic compounds [9, 17]. The metal ions released by biodegradable medical implants can affect the cells of the surrounding tissue, producing biological effects that occur over time, depending on the type and sensitivity of each organism. Typically, the estimator used for expressing toxicity is the toxic dose 50% (TD50) for associated bone cells (the MC3T3E1 and MG63 cell lines). In terms of toxicity to tissues in living organisms, the chemical elements are considered to feature mild, moderate and severe toxicity.

The elements with mild toxicity include Mg, Ca, Li, Al and Zr. However, some scientific reports have shown that certain elements featuring mild toxicity such as Al (500 nM), even at very low concentrations, induce the proliferation of associated bone cells [3].

Ca, Mn and Zn are oligo-elements essential to life and some rare earths with anti-carcinogenic properties should be the first choice for introducing in the chemical composition of magnesium alloys [2, 8, 14 and 15].

Calcium acts as a grain-size finishing agent in Mg alloys, stabilizing the grain size to concentrations up to 0.5% Ca, slightly decreasing in case of further additions. Binary Mg-Ca alloys were studied in particular by modifying the concentration from 0.5% Ca to 20% Ca [13]. The increase in the calcium concentration involves a higher Mg_2Ca secondary phase content, mainly distributed over the grain boundaries. The Mg_2Ca secondary phase is brittle and reduces the ductility of Mg-Ca alloys with increased Ca concentration. It also influences the corrosion resistance properties of Mg alloys. A high volumetric fraction of the Mg_2Ca secondary phase reduces the corrosion resistance of Mg-Ca alloys due to the formation of galvanic microcells. Therefore, it can be observed that an excessive concentration of Ca accelerates the corrosion rate of the Mg-Ca alloys, its optimal value being $\leq$ 1%wt Ca.

The breaking resistance of the Mg-Ca alloys ranges between 90 and 280 MPa and the elongation at break ranges between 3 and 20%. The rapid reduction of resistance due to degradation does not represent a decisive factor for a degradable implant when used in tissue engineering applications; however, when used as a substitute for bone engineering, it is expected to provide temporary support for a sufficiently long period [2].

Generally, magnesium alloys can be machined by casting and plastic deformation (rolling or hot extrusion) [20]. The main disadvantage of using magnesium in medical engineering is the rapid corrosion rate and the release of hydrogen gas during corrosion. These can induce local and systemic toxic phenomena by modifying the pH or the gas accumulation [2]. In order to improve the corrosion behaviour, with positive effect on the biocompatibility of magnesium and its alloys, two main methods are envisaged, i.e. the adjustment of the microstructure by alloying and the mechanical machining and surface treatments by the use of coatings [19].

Mg-based alloys are extremely difficult to obtain due to the strong oxidation phenomenon in ordinary atmosphere, continued with the burning of magnesium and its alloys. Therefore, it is necessary to produce magnesium alloys in controlled environments, using special installations designed for this purpose, i.e. in an inert atmosphere of argon [2]. Thus, in the research, a number of magnesium alloys were obtained using an installation for melting in argon environment by electro-magnetic levitation. Because of the violent reactivity of the magnesium with the oxygen at 600 °C, calcium was added to the granular mixture, after which it was compacted by cold pressing. The results demonstrated the possibility of obtaining magnesium alloys with different Ca contents, highlighting problems related to calcium vaporisation, ignition of granular mixtures or lack of chemical homogeneity of the alloys.

2. Obtaining experimental Mg-Ca alloys

The Mg-Ca biodegradable alloys were obtained by means of melting installation in levitation equipped with a medium frequency converter under inert atmosphere (Figure 1). The following materials were used:

- Mg, purity 99.5%, grains with diameter of 2-5 mm, product code HP.Mg.2N5.100000;
- Ca, purity 98.5%, grains with diameter of 2-6 mm, product code HP.Ca.100000.

In order to obtain alloy recipes, technological calculations were carried out for a batch weight of 15 g, corresponding to the volume of the casting mould, taking into account the evaporation and oxidation losses specific to each element of approximately 10% wt.

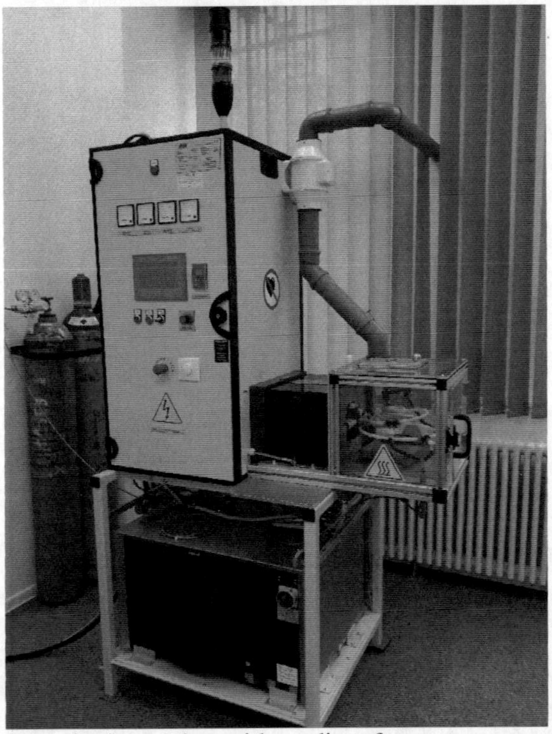

Figure 1. Levitation melter with medium frequency converter and inert atmosphere.

For the compaction of granular mixtures and reduction of oxidation effects, cold pressing was performed using a 20 ft hydraulic press, obtaining 15 mm diameter and 10 mm height buttons (Figure 2).

Figure 2. Mg-Ca alloy buttons obtained by cold pressing.

The buttons obtained by cold pressing were subjected to induction melting under an inert atmosphere of argon, purity 5.3, at a pressure of 2.4 atm. The metallurgical melting process of a batch lasted between 2 and 3 minutes, depending on the working parameter values adopted: medium frequency voltage U_{mf} = 480 - 500 V; medium frequency current I_{mf} = 55 - 65 A; working frequency f_{mf} = 60 - 70 kHz; power P_{mf} = 20 - 35 kW. The molten alloy was cast freely, in argon atmosphere, into copper moulds, protected inside with refractory paint to avoid the adhesion of the liquid metal (Figure 3). After solidification, the end zones of the mini-ingots were removed and the cylindrical surfaces were mechanically machined by turning to remove the zones of contact with the casting mould (Figure 4). Disc shape samples of 1.2 mm in thickness and 11 mm in diameter were taken from each alloy type for cell proliferation tests and metallographic analysis.

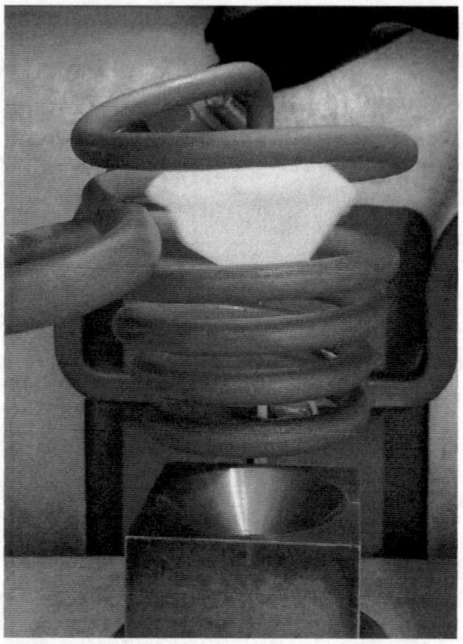

Figure 3. The Mg-Ca alloy button levitation melting inside the inductor.

Figure 4. The appearance of the alloy during solidification and after extraction from the copper mould.

IOP Conf. Series: Materials Science and Engineering **448** (2018) 012004 doi:10.1088/1757-899X/448/1/012004

3. Characterization of experimental magnesium alloys

The magnesium alloys samples were used to determine the local chemical micro-composition in different areas, to establish the level of homogenization and to evaluate the tendency of segregation of the alloying and micro-alloying elements.

For the metallographic analyses, the samples were polished using metallographic abrasive paper of progressive grain sizes: 360, 400, 600, 800, 1000, followed by polishing on textile support impregnated with fine polishing solutions with alpha alumina particles (1, 0.7 and 0.25 μm grain sizes). No chemical attack solutions were used for electronic microscopy microstructural analysis in order to prevent any influence on the chemical composition values. The samples were examined using a QUANTA INSPECT F50 electronic scanning microscope, fitted with field emission gun (FEG) - 1.2 nm resolution, with energy dispersive X-ray Spectrometer (EDX) - MnKα 133 eV resolution, and an Olympus GX51 optical microscope. The values of the chemical compositions, determined by micro-zone analysis, for the set of 21 samples made of Mg-Ca experimental alloys are shown in Table 1.

Table 1. Chemical composition of Mg-Ca alloy samples

Sample code	**Chemical composition**					
	Mg		**Ca**		**O / Si / La / Ce**	
	Atomic percentage, at%	Weight percentage, wt%	Atomic percentage, at%	Weight percentage, wt%	Atomic percentage, at%	Weight percentage, wt%
0	100	100	-	-	-	-
1	86.13	87.49	3.22	5.39	10.65	7.12
2	95.83	96.88	0.35	0.58	3.82	2.54
4	97.97	98.4	0.26	0.42	1.78	1.17
5	95.19	96.49	0.31	0.51	4.5	3
6	95.85	96.93	0.31	0.52	3.84	2.55
7	97.08	97.63	0.45	0.74	2.47	1.64
8	93.23	95.02	0.43	0.72	6.34	4.25
9	91.34	93.75	0.39	0.66	8.26	5.58
10	84.99	87.22	2.6	4.4	12.41	8.38
11	91.84	94.05	0.44	0.75	7.71	5.2
12	96.94	97.45	0.53	0.88	2.53	1.67
13	95.24	96.21	0.62	1.04	4.13	2.75
14	94.14	95.7	0.37	0.62	5.49	3.67
15	85.83	89.09	1.2	2.05	12.98	8.86
16	92.63	94.37	0.68	1.14	6.69	4.48
17	95.16	95.5	1.31	2.17	3.53	2.33
18	89.9	92.26	0.9	1.52	9.2	6.21
18 inclusion	69.22	58.06	29.25	40.45	**Si** = 1.53	**Si** = 1.48
18 central zone	49.35	45.27	0.38	0.57	**La** = 1.48 **Ce** = 3.62 **O** = 45.17	**La** = 7.70 **Ce** = 19.12 **O** = 27.27
19	94.16	94.78	1.36	2.25	4.48	2.97
20	93.9	95.1	0.83	1.39	5.27	3.51
21	98.68	97.84	1.32	2.16	-	-

The microstructural analysis revealed that the metal matrix of the Mg-Ca alloys generally has a dendritic appearance, and the easily fusible Mg_2Ca eutectic is located in the inter-dendritic zone (Figure 5).

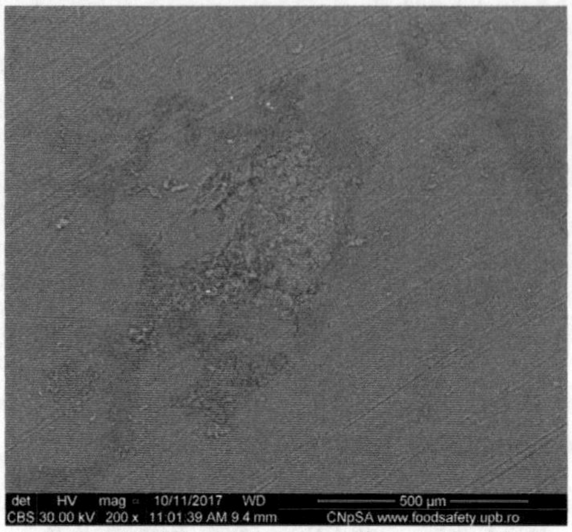

a) Central zone of sample 18

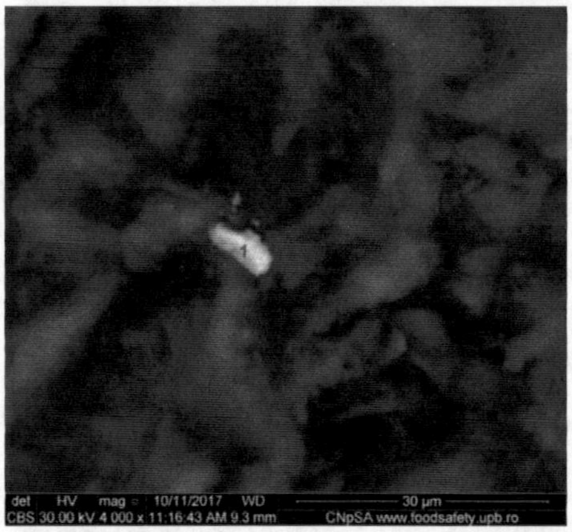

b) Inclusion in central zone of sample 18

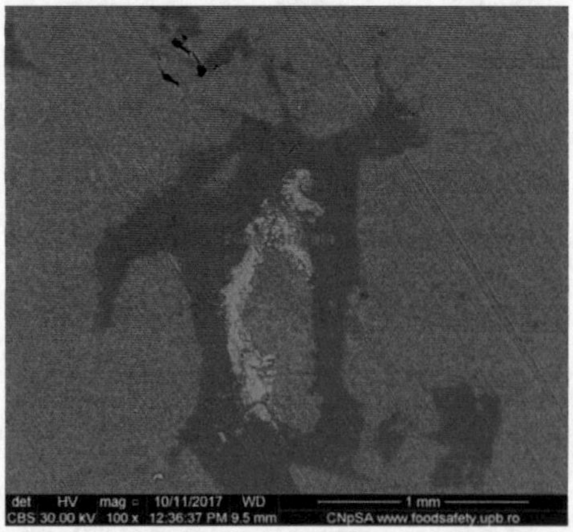

c) Segregation in central zone of sample 20

Figure 5. The micro-structural aspect of Mg-Ca alloys in the analyzed range.

The mechanical characteristics of the new experimental Mg-Ca alloys obtained in levitation were determined by microhardness measurements in different zones (centre, edge) using the Shimadzu HMV 2T microhardness Vickers tester. The $HV_{0.2}$ microhardness measurements were performed on cross-sectioned samples of the Mg-Ca experimental alloys after grinding with metallographic paper. The measurements were made in line, from the edge to the centre of the samples, maintaining a minimum distance between indentations of 300 μm. The pushing force of the penetrator was 200g, and the penetration time was 10 seconds. A set of 5 successive measurements was performed for each sample and the point values, the mean value and the coefficients of variation were recorded; the data obtained is presented in Table 2.

Table 2. Hardness values measured on Mg-Ca experimental alloys samples

Sample code	Values	Mean value	Variation coefficient	Standard error
0	47,1; 48; 51; 57; 63,9	53.4	13.15	7.02
1	51,3; 48,8; 48,8; 47,5; 48,6	49.0	1.39	2.85
2	59; 41; 46,8; 48,4; 52,8	49.6	6.74	13.59
4	58,4; 58,8; 61,1; 52,6; 51,4	56.5	4.22	7.48
5	56,2; 50,8; 69,4; 66,1; 72	62.9	9.04	14.37
6	72,4; 63,8; 66,5; 63,5; 72,4	67.7	4.43	6.54
7	73,1; 87,4; 8,6; 9,5; 94,2	87.4	8.51	9.74
8	70,3; 72; 66,7; 60,3; 64,1	66.7	4.71	7.06
9	71,6; 71,4; 65,6; 72,8; 76,9	71.7	4.05	5.65
10	76,2; 64,7; 69,1; 78,8; 61,7	70.1	7.31	10.42
11	65,4; 67,3; 60,3; 59,7; 59,1	62.4	2.98	5.06
12	85; 58,3; 76,8; 75,8; 59,5	71.1	11.69	16.44
13	60,9; 62,5; 66,8; 68,2; 66,8	65.0	3.15	4.85
14	68,9; 66,3; 66,8; 56,9; 63	64.4	4.94	7.28
15	67,6; 61,7; 62,4; 72,7; 67	66.3	4.46	6.73
16	60; 56,4; 60,3; 70,8; 59,9	61.5	5.45	8.86
17	69,3; 69,8; 70,4; 65; 61,3	67.2	3.91	5.82
18	57,4; 65,5; 68,5; 68; 61,3	64.1	4.72	7.36
19	69,6; 76; 68,4; 76,3; 71,3	72.3	3.65	5.04
20	73,8; 68,3; 67,1; 66,6; 60,8	67.3	4.64	6.89
21	65,5; 65,2; 71,1; 66,4; 71,6	68.0	3.11	4.58

Samples were collected from each set of Mg-Ca alloys for testing the cell proliferation behaviour of some live bone cells collected by surgery (Figure 6). The time interval between bone tissue collection and use on metal matrices submerged in compatible solutions was less than 2 hours. After this time interval, the results can no longer be considered conclusive as the process of irreversible cell degradation is installed. The results of the cell proliferation tests showed a tendency for the formation and proliferation of live cells on the surface of the Mg-Ca system experimental alloys only if the Ca percentage is less than 2.5% Ca.

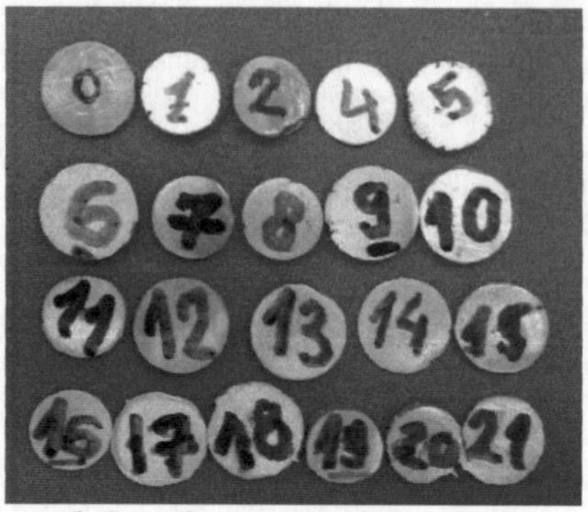

Figure 6. Set of samples prepared for the cell proliferation test.

4. Conclusions

- Mg-Ca alloys can be obtained by melting in the levitation installation under an inert environment, consisting of argon, for protection against rapid oxidation.
- The purity level of the alloys depends on the quality of the metal materials used and on the working conditions (inert gas protection for the melt, the use of perfectly clean solidification moulds, constant heating parameters).
- The degree of homogeneity of the Mg-Ca alloys depends on their Ca concentration. At concentrations up to 2wt%Ca, the homogeneity is acceptable, as no clusters of Mg_2Ca compounds are noted in the central zones of the ingot. The tendency to separate in the form of compounds (Mg_2Ca or oxides) occurs at higher Ca concentrations.
- The hardness of the Mg-Ca alloy metal matrix increases with the Ca content increase, the values ranging from 49 to 87.4 $HV_{0.2}$. For some samples, the increase in hardness was also determined by the presence of inclusions (Si, La, Ce) coming either from paint contamination on the surface of the solidification mold or from the metallic materials used.
- Low homogeneity zones occur as a result of the formation of compounds (Mg_2Ca or oxides), leading to relatively large hardness oscillations (10-20% HV), depending on the measuring zone.

Acknowledgements: This paper was supported by the Romanian Ministry of Research and Innovation, CCCI - UEFISCDI, through the project number PTE 14/2016, "Equipment to obtain by levitation in inert gas of light bio-compatible alloys with high purity - BIOLEV" and project PN-III-P1-1.2-PCCDI-2017-239 / 60PCCDI 2018 - "Obtaining and expertise of new biocompatible materials for medical applications - MedicalMetMat, within PNCDI III.

References

[1] Chen Q and Thouas G A 2015 *Metallic implant biomaterials, Materials Science and Engineering: R: Reports* **87** pp 1-57.
[2] Geanta V and Voiculescu I 2018 *Tratat de obtinere a materialelor metalice biocompatibile,* Ed. Printech, Bucuresti.
[3] Breme H J and Helsen J A 1998 *Selection of materials, In: Metals as Biomaterials*, John Wiley and Sons Ltd, Baffins Lane, Chichester, West Sussex PO19 IUD, England.
[4] Muralidharan P and Seeram R 2015 *Biodegradable materials for clinical applications: A review, Advanced Sciences and Engineering* **4(3)** pp. 221-38.

IOP Conf. Series: Materials Science and Engineering **448** (2018) 012004 doi:10.1088/1757-899X/448/1/012004

[5] Witte F 2015 *Reprint of: The history of biodegradable magnesium implants: A review. Acta Biomater.* **23** Suppl: S28-40.

[6] Walker J, Shadanbaz S, Woodfield T B, Staiger M P and Dias G J 2014 *Magnesium biomaterials for orthopedic application: a review from a biological perspective, J. Biomed. Mater. Res. B Appl. Biomater.* **102(6)** pp 1316-31.

[7] Kusniar D P, Suhaila S and Bambang A 2014 *A review of magnesium alloys for use in biodegradable cardiovascular stents, World Applied Sciences Journal 30 (Innovation Challenges in Multidisciplinary Research & Practice)* pp 375-81.

[8] Ding Y, Wen C, Hodgson P and Li Y 2014 *Effects of alloying elements on the corrosion behavior and biocompatibility of biodegradable magnesium alloys: a review, J. Mater. Chem. B* **2** pp 1912-33.

[9] Yang L 2013 *Development of Mg-RE alloys for medical applications* Doctoral Thesis, Clausthal University of Technology, Liaoning/VR China.

[10] Vojtěch D and Kubásek J 2013 *Structure, mechanical and corrosion properties of magnesium alloys for medical applications. Acta Metallurgica Slovaca - Conference* **3** pp 82-89.

[11] Agarwal S, Curtin J, Duffy B and Jaiswal S 2016 *Biodegradable magnesium alloys for orthopaedic applications: A Review on corrosion, biocompatibility and surface modifications, Mater. Sci. Eng. C Mater. Biol. Appl.* **68** pp 948-63.

[12] Song G 2005 *Corrosion and protection of magnesium alloys: An overview of research undertaken by CAST, Materials Science Forum, Trans Tech Publication, Switzerland* **488-489** pp 649-52.

[13] Li Y C, Li M H, Hu W Y, Hodgson P D and Wen C E 2010 *Biodegradable Mg-Ca and Mg-Ca-Y alloys for regenerative medicine, Materials Science Forum, Trans Tech Publications* **654-656** pp 2192-95.

[14] Yin D S, Zhang E L and Zeng S Y 2008 *Effect of Zn on mechanical property and corrosion property of extruded Mg-Zn-Mn alloy, Transactions of Nonferrous Metals Society of China* **18(4)** pp 763-68.

[15] Song Y, Han E H, Shan D, Yim C D and You B S 2012 *The effect of Zn concentration on the corrosion behavior of Mg–xZn alloys, Corrosion Science* **65** pp 322-30.

[16] Witte F, Hort N, Vogt C, Cohen S, Kainer K U, Willumeit R and Feyerabend F 2008 *Degradable biomaterials based on magnesium corrosion, Curr. Op. Sol. St. Mater. Sci.* **12** pp 63–72.

[17] Hort N, Huang Y, Fechner D, Störmer M, Blawert C, Witte F, Vogt C, Drücker H, Willumeit R, Kainer K U and Feyerabend F 2010 *Magnesium alloys as implant materials – Principles of property design for Mg-RE alloys, Acta Biomater.* **6(5)** pp 1714-25.

[18] Bettles C and Barnett M 2012 *Advanced in wrought magnesium alloys Fundamentals of processing, properties and applications* ed Woodhead Publishing Ltd. (Oxford/Cambridge/Philadelphia/New Delhi: Elsevier).

[19] Istrate B 2016 *Cercetări asupra influenței unor depuneri superficiale pe aliaje metalice biodegradabile utilizate în domeniul medical* PhD Thesis, "Gheorghe Asachi" Technical University of Iași (Iași).

[20] Friedrich H E and Mordike B L 2011 *Magnesium technology: metallurgy, design data, applications* Ed Springer (Berlin/London).

IOP Conf. Series: Materials Science and Engineering **448** (2018) 012005 doi:10.1088/1757-899X/448/1/012005

Hob mill for trilobed rotor – Graphical method in CATIA

Nicuşor Baroiu, Virgil Gabriel Teodor, Florin Susac, Nicolae Oancea

111 Domnească Street, Galaţi, România, 800201

florin.susac@ugal.ro

Abstract. The machining of ordered curl of surfaces, associated to a centrode, is completed in very good yield and precision conditions with hob mill tool. In this case, the issue is to determine the enwrapping worm of the ordered curl of profiles, as a fundament for hob mill tool's construction. Based on the "generating trajectories" method, an algorithm for profiling the generating hob mill for machining the trilobed rotor of a peristaltic pump is presented. A graphical example, developed in CATIA design environment, shows the profiling technique.

1. Introduction

Some methods can be used for profiling the enwrapping worm for an ordered curl of profiles: fundamental method of Gohman for surfaces enveloping [1, 2]; complementary method of "minimum distance" [3] or the method of "substitutive circles" [3].

Advances in graphical design environment allow elaborating graphical methods for profiling a helical surface reciprocally enwrapping with an ordered curl of surfaces, associated to a circular centrode.

The hob mill enveloping an ordered curl of profiles is a classical type of enwrapping generation with a single point of contact. The solution of this problem, determining the primary peripheral surface of hob mill tool, is identified based on the second theorem of Olivier [1], as so as, using the "intermediary surface" method, based on the Gohman theorem [1, 2].

Others complementary methods were proposed, as the "minimum distance" method [3, 8]. This method allows approaching problems for a single point contact between enwrapping surfaces.

Issue of profiling worm tools, as so as, quality of these tools, was studied by Radzevich [9], referring to impact of the major design parameters of an involute hob on deviations of the hob tooth profile, by developing an analytical method for computation of the deviation, for optimal configuration of worm tool.

Hsieh et al. [10] propose a novel hub cutter design of straight side hob cutter with multiple pressure angles, in order to simplify the manufacturing process.

Jiang [11] use a CNC hob machine in order to minimize the tooth flank's deviation, demonstrates that the hob flank can be approximated to the given tooth flank by adjusting the coefficients of the polynomials based on their sensitivity.

In this paper, a problem concerning the hob mill profiling, for manufacturing a non involute profile, based on a complementary theorem of "generating trajectories family" [6] is presented. It is also presented an analytical development and a graphical application in CATIA.

The results for an axial profile of hob mill, determined using both methods, analytical and graphical, are compared.

2. Surface to be generated and generating rack gear tool

The rotor of a peristaltic pump is a composed cylindrical surface. The frontal profile is presented in figure 1, where a presented the reference systems too.

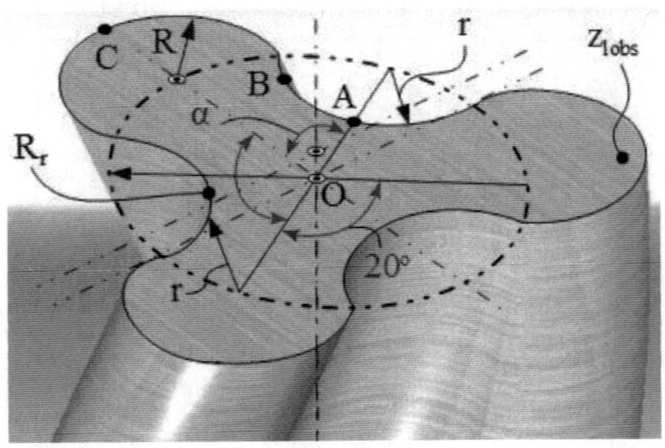

Figure 1. Frontal profile of the peristaltic pump's rotor

The rotor's profile is composed by an assembly of circle's arcs, AB, BD, DE, with radii r and R, fillet to each other.

The reference systems are defined, see Figure 2:

xyz is the world reference system, with z axis of rotor;

XYZ – mobile reference system, joined with the frontal profile of the rotor and with the associated centrode C_1;

$\xi\eta\zeta$ – mobile reference system, joined with the C_2 centrode of the future rack gear tool.

The analytical models of the frontal profiles of rotor are defined: AB, arc with radius r:

$$AB \left| \begin{array}{l} X = -R\cos\alpha - r\cos\theta; \\ Y = -R\sin\alpha + r\sin\theta; \\ Z = 0, \end{array} \right. \tag{1}$$

with θ variable parameter in limits $\theta_A = \alpha$; $\theta_B = \beta$, where:

$$\beta = \arcsin\left(\frac{R_r}{r+R}\cdot\sin\alpha\right), \tag{2}$$

and, similarly, the arc BC, with radius R:

$$BC \left| \begin{array}{l} X = -R_r + R\cos\eta; \\ Y = -R\sin\eta; \\ Z = 0, \end{array} \right. \tag{3}$$

with η as a variable parameter:

IOP Conf. Series: Materials Science and Engineering **448** (2018) 012005 doi:10.1088/1757-899X/448/1/012005

$$\eta_C = 0;\ \eta_B = \beta. \tag{4}$$

2.1. Generating movements of the rack gear – rack gear form
The relative motion between $\xi\eta\zeta$ and XYZ reference systems is writing:

$$\xi = \omega_3^T(\varphi)X - a;\ a = \begin{Vmatrix} -R_r \\ -R_r \cdot \varphi \\ 0 \end{Vmatrix}, \tag{5}$$

and the profiles' family:

$$(\Sigma_{AB})_\varphi : \begin{vmatrix} \xi = (-R\cos\alpha - r\cos\theta)\cos\varphi - (-R\sin\alpha + r\sin\theta)\sin\varphi + R_r; \\ \eta = (-R\cos\alpha - r\cos\theta)\sin\varphi + (-R_r\sin\alpha + r\sin\theta)\cos\varphi + R_r\varphi; \\ \zeta = t. \end{vmatrix} \tag{6}$$

Similarly, for arc:

$$(\Sigma_{BC})_\varphi : \begin{vmatrix} \xi = (-R_r + R\cos\eta)\cos\varphi - R\sin\eta\sin\varphi + R_r; \\ \eta = (-R_r + R\cos\eta)\sin\varphi + R\sin\eta\cos\varphi + R_r\varphi; \\ \zeta = t. \end{vmatrix} \tag{7}$$

For both (6) and (7) equations, the t parameter is defined along the cylindrical surface's generatrix – the rotor's flanks.

The enveloping of (6) and (7) profiles family, in the rolling movement of the two centrodes C_1 and C_2, with the rolling condition:

$$\lambda = R_r \cdot \varphi, \tag{8}$$

representing the generating rack-gear's profile.

2.2. The enveloping condition at contact between cylindrical surface and generating rack-gear
The establishing of enwrapping condition based on the "generating trajectories family" method [3] it is proposed. The method assumes to know the normals trajectories families. For the arc, from equations (1):

$$\vec{n}_{\Sigma_{AB}} = \begin{vmatrix} \vec{i} & \vec{j} & \vec{k} \\ r\sin\theta & r\cos\theta & 0 \\ 0 & 0 & 1 \end{vmatrix} = \cos\theta\cdot\vec{i} - \sin\theta\cdot\vec{j}, \tag{9}$$

which represents the directrix parameters for the normal to the profile from which, in the current point onto profile, the normal form results:

$$\vec{N}_{\Sigma_{AB}} = \left[(-R\cos\alpha - r\cos\theta) + \lambda\cos\theta\right]\vec{i} + \left[(-R\sin\alpha - r\sin\theta) - \lambda\sin\theta\right]\vec{j}. \tag{10}$$

Similarly, for BC profile:

$$\vec{N}_{\Sigma_{BC}} = \left[\left(-R_r + R\cos\eta\right) - \lambda\cos\eta\right]\vec{i} + \left(-R\sin\eta + \lambda\sin\eta\right)\vec{j}, \tag{11}$$

with λ variable scalar parameter.

According to the "generating trajectories" theorem, [3], the characteristic curve onto Σ_{AB} and Σ_{BC} surfaces, is the geometric locus of points belonging to these surfaces, where the normals' families, in the generating motion (5), pass through the generatrix with direction $\vec{t}$ drawn from the gearing pole:

$$P\begin{vmatrix} \xi_p = 0; \\ \eta_p = R_r\varphi. \end{vmatrix} \tag{12}$$

From (5) and (10), the normals family for the AB arc results:

$$\left(N_{\Sigma_{AB}}\right)_\varphi \begin{vmatrix} \xi = \left(-R\cos\alpha - r\cos\theta + \lambda\cos\theta\right)\cos\varphi - \left(-R\sin\alpha - r\sin\theta - \lambda\sin\theta\right)\sin\varphi + R_r; \\ \eta = \left(-R\cos\alpha - r\cos\theta - \lambda\cos\theta\right)\sin\varphi + \left(-R\sin\alpha - r\sin\theta - \lambda\sin\theta\right)\cos\varphi + R_r\varphi. \end{vmatrix} \tag{13}$$

From condition that the $\left(N_{\Sigma_{AB}}\right)_\varphi$ normals family (13) pass through the gearing pole (12) results condition:

$$\varphi = -\alpha. \tag{14}$$

In this way, the (6) and (14) equations assembly represents the enwrapping of profiles family $\left(\Sigma_{AB}\right)_\varphi$, the S_{AB} profile of rack-gear tool.

Similarly, the normals trajectories family is established for profile Σ_{BC}, see form (3):

$$\left(N_{\Sigma_{BC}}\right)_\varphi \begin{vmatrix} \xi = \left(-R_r + R\cos\eta + \lambda\cos\eta\right)\cos\varphi - \left(-R\sin\eta - \lambda\sin\eta\right)\sin\varphi + R_r; \\ \eta = \left(-R_r + R\cos\eta + \lambda\cos\eta\right)\sin\varphi + \left(-R\sin\eta - \lambda\sin\eta\right)\cos\varphi + R_r\cdot\varphi. \end{vmatrix} \tag{15}$$

From condition that the gearing pole coordinates check the equations for (15) family, see (12), the specific enwrapping condition results:

$$\varphi = 0. \tag{16}$$

In this way, for this case, the crossing profile equations for generating rack-gear, see relations (7), become:

$$C_{BC}\begin{vmatrix} \xi = R\cos\eta; \\ \eta = R\sin\eta. \end{vmatrix} \tag{17}$$

IOP Conf. Series: Materials Science and Engineering **448** (2018) 012005 doi:10.1088/1757-899X/448/1/012005

3. Peripheral primary surface of hob mill

In figure 2, the reference systems are presented, as so as the positions of the future primary peripheral surface of hob mill, $\vec{V}$ axis.

In order to determine the peripheral primary surface of hob mill, reciprocally enwrapping with the cylindrical surface of generating rack-gear, the principles of helical motion decomposition are used [7].

We can consider that the helical motion with $\vec{V}$ axis and p helical parameter, when the peripheral primary surface of hob mill is generated, is decomposed in two motions:

- translation along the generatrix of the cylindrical surface of rack-gear flank, the direction $\vec{t}$, see figure 2;

- rotation around the $\vec{A}$ axis, parallel with axis $\vec{V}$ of the helical motion and at the a distance from $\vec{V}$,

$$a = p \cdot tg\theta. \tag{18}$$

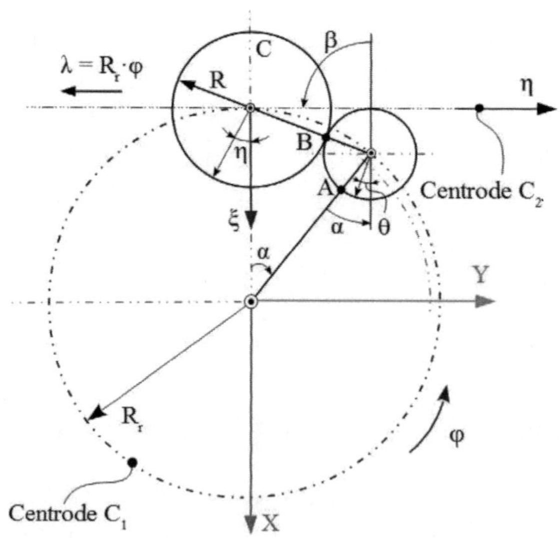

Figure 2. The positions of the future primary peripheral surface of hob mill; reference systems

In relation (18), with p was denoted the helical motion's parameter and with θ the angle between cylindrical surface's generatrix and the axis of hob mill.

The decomposition of the helical motion $(\vec{V}, p)$ may be symbolized:

$$(\vec{V}, p) \approx T(v, \vec{t}) + A(\vec{A}, \varphi). \tag{19}$$

The v and φ values are the moving parameters, with the component motions: v is scalar parameter along $\vec{t}$ direction and φ is angular parameter in rotation motion around the $\vec{A}$ axis, see figure 2.

In this way, the determining of rack-gear surface's characteristic $\left(S_{\Sigma_{AB}}\right)$ – equations (13) and (14) and $\left(S_{\Sigma_{BC}}\right)$ – equations (16) and (17) do not depend from the translation component $\left(T\left(v,\vec{t}\right)\right)$, when the $\left(S_{\Sigma_{AB}}\right)$ and $\left(S_{\Sigma_{BC}}\right)$ surfaces are self-generated.

So, the rack-gear surface's characteristic curve will depends on rotation around the $\vec{A}$ axis, only.

Regarding the a value, we can state: If the circular pitch of peristaltic pump rotor's teeth is defined as:

$$P_c = \frac{2\pi R_r}{z_{lobs}}, \tag{20}$$

the normal pitch of generating rack-gear is defined as, see figure 3:

$$P_{n_{cr}} = \frac{2\pi R_r}{z_{lobs}} \cdot \cos \beta_d, \tag{21}$$

where β_d is the angle between the cylindrical surface's generatrix and rotor's axis.

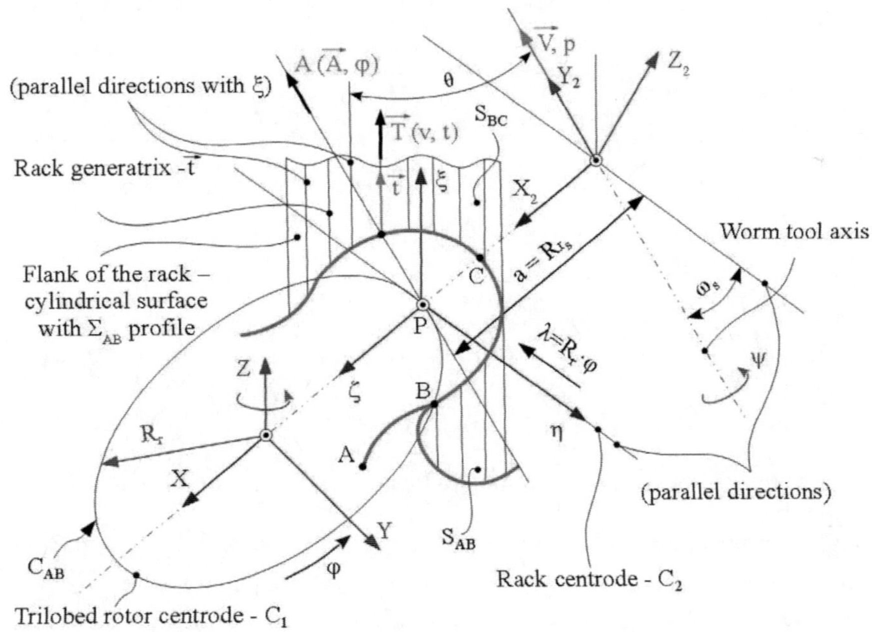

Figure 3. Hob mill axis; reference systems; helical movement decomposition, $\left(\vec{V}, p\right)$

The angle of hob mill helix is:

$$tg\omega_s = \frac{P_{ax_m}}{2\pi R_{r_s}}, \tag{22}$$

IOP Conf. Series: Materials Science and Engineering **448** (2018) 012005 doi:10.1088/1757-899X/448/1/012005

with P_{ax_m} axial pitch of hob mill and R_s rolling radius of hob mill.

In figure 3, is obviously the equality:

$$\theta = \frac{\pi}{2} - \omega_s. \tag{23}$$

From the condition that the normal pitch of rack-gear to be equal to the normal pitch of the hob mill, it results:

$$\sin \omega_s = \frac{R_{r_p}}{R_{r_s}} \cdot \frac{1}{z_{lobs}} \cdot \cos \beta_d. \tag{24}$$

The helical parameter of the hob mill is defined as:

$$p = \frac{P_{ax_m}}{2\pi} = \frac{R_{r_p}}{z_{lobs}} \cdot \frac{\cos \beta_d}{\cos \omega_s} \cdot \frac{1}{2\pi}, \tag{25}$$

from where it results:

$$a = p \cdot tg\theta = \frac{R_{r_p}}{z_{lobs}} \cdot \frac{\cos \beta_d}{\sin \omega_s} \tag{26}$$

and

$$tg\omega_s = \frac{p}{R_{r_s}}, \tag{27}$$

or

$$a = R_{r_s}. \tag{28}$$

In this manner, it is obviously that the rolling radius value depends on the characteristics of helical motion decomposition.

Determining the primary peripheral surface of the hob mill, enwrapping for generating rack-gear, previously determined for section BC (and similarly for circle's arc AB) leads to determining the characteristic curve at contact between rack-gear and helical surface, with respect to the helical motion decomposition particularities (see equation (19)).

3.1. Enwrapping condition at contact between rack gear and hob mill's helical surface

The enwrapping condition will depends only by rotation around the $\vec{A}$ axis.

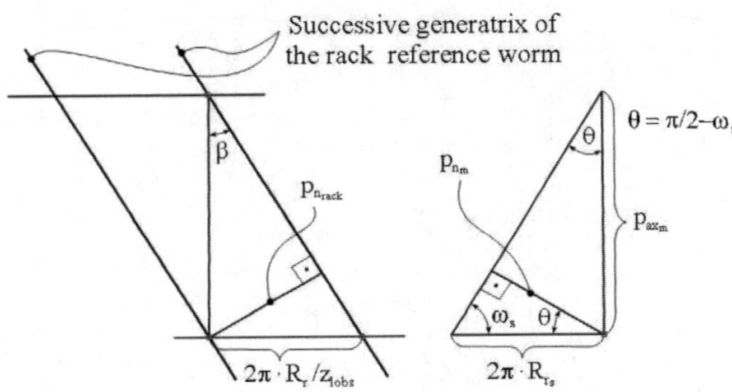

Figure 4. Normal and circular pitch of rack-gear; characteristic angles

So, for generating rack-gear's surfaces it can be used equations:

$$S\left|\begin{array}{l}\xi=\xi(\upsilon);\\ \eta=\eta(\upsilon);\\ \zeta=\zeta(t),\end{array}\right. \qquad (29)$$

with v and t independent parameters, expressed in $\xi\eta\zeta$ reference system, see figure 4, where the $\vec{A}$ and $\vec{V}$ axis are presented, as the reference systems:

- $\xi\eta\zeta$ is the reference system joined with the generating rack-gear;
- $\xi_1\eta_1\zeta_1$ – reference system joined with $\vec{A}$ axis.

The normal to the S surface (rack gear flank) has directrix parameters:

$$\vec{N}_S=\begin{vmatrix}\vec{i} & \vec{i} & \vec{k}\\ \dot{\xi}_{(v)} & \dot{\eta}_{(v)} & 0\\ 0 & 0 & 1\end{vmatrix} \qquad (30)$$

or

$$\vec{n}_s=\dot{\eta}_{(v)}\vec{i}-\dot{\xi}_{(v)}\vec{j}. \qquad (31)$$

In this way, the normals in the current points onto S, has equations:

$$\vec{N}_S=\left[\xi(v)+\lambda\dot{\eta}_{(v)}\right]\vec{i}+\left[\eta(v)-\lambda\dot{\xi}_{(v)}\right]\vec{j}, \qquad (32)$$

which, by coordinates transformation,

$$\begin{Vmatrix}\xi_1\\ \eta_1\\ \zeta_1\end{Vmatrix}=\begin{Vmatrix}1 & 0 & 0\\ 0 & \cos\omega_s & \sin\omega_s\\ 0 & -\sin\omega_s & \cos\omega_s\end{Vmatrix}\cdot\begin{Vmatrix}\xi\\ \eta\\ \zeta\end{Vmatrix}, \tag{33}$$

may lead to the following form of normals to S profile:

$$N_S \begin{vmatrix}\xi_1=\xi(v)+\lambda\dot{\eta}_{(v)};\\ \eta_1=\left[\eta(v)+\lambda\dot{\xi}_{(v)}\right]\cos\omega_s;\\ \zeta_1=-\left[\eta(v)+\lambda\dot{\xi}_{(v)}\right]\sin\omega_s.\end{vmatrix} \tag{34}$$

The trajectories family of N_S normals in the rotation around $\vec{A}$ axis is:

$$\begin{Vmatrix}\xi_1\\ \eta_1\\ \zeta_1\end{Vmatrix}=\begin{Vmatrix}\cos v & 0 & -\sin v\\ 0 & 1 & 0\\ \sin v & 0 & \cos v\end{Vmatrix}\cdot\begin{Vmatrix}\xi(v)+\lambda\dot{\eta}_{(v)}\\ \left[\eta(v)+\lambda\dot{\xi}_{(v)}\right]\cos\omega_s\\ -\left[\eta(v)+\lambda\dot{\xi}_{(v)}\right]\sin\omega_s\end{Vmatrix}. \tag{35}$$

It is imposed that the normals family intersect the $\vec{A}$ axis:

$$\vec{A}=\vec{j}. \tag{36}$$

Therefore, the equations system results

$$\begin{cases}\xi_{(v)}\cos v+\eta_{(v)}\sin\omega_s\sin v=-\lambda\left[\dot{\eta}_{(v)}\cos v-\dot{\xi}_{(v)}\sin\omega_s\sin v\right]\\ \xi_{(v)}\sin v-\eta_{(v)}\sin\omega_s\cos v=-\lambda\left[\dot{\eta}_{(v)}\sin v+\dot{\xi}_{(v)}\sin\omega_s\cos v\right],\end{cases} \tag{37}$$

and, starting from this, by eliminating the λ parameter, the following equation can be obtained:

$$\xi(v)\dot{\xi}_{(v)}+\eta(v)\dot{\eta}_{(v)}=0, \tag{38}$$

which represents the specific enwrapping condition.

The condition (38) has the geometric significance: the points belonging to the characteristic curve, represents the $\vec{A}$ axis projection onto the S surface (surface of the generating rack gear).

The $\xi_{(v)}$ and $\eta_{(v)}$ equations represents a generic form of the rack gear flank.

3.2. Primary peripheral surface of hob mill

In principle, the equations of the characteristic curve onto the generic surface of generating rack gear S (29) is a curve representing the projection of the $\vec{A}$ axis, in principle known in form:

$$C_S \begin{vmatrix} \xi_1 = \xi_1(v_A); \\ \eta_1 = \eta_1(v_A); \\ \zeta_1 = \zeta_1(v_A), \end{vmatrix} \tag{39}$$

with v_A parameter for which is accomplished the condition(38).

Giving to the curve (39) a helical motion with axis $\vec{V}$ and helical parameter p (see equation (25)), the primary peripheral surface of the hob mill is generating.

In figure 4, it is defined, the $X_2Y_2Z_2$ reference system, with Y_2 axis overlapped to the axis $\vec{V}$ and with axis parallel to the $\xi_1\eta_1\zeta_1$ reference system:

$$\begin{Vmatrix} X_2 \\ Y_2 \\ Z_2 \end{Vmatrix} = \begin{Vmatrix} \xi_1 \\ \eta_1 \\ \zeta_1 \end{Vmatrix} - \begin{Vmatrix} -R_{r_s} \\ 0 \\ 0 \end{Vmatrix}. \tag{40}$$

By transforming equation (40), the C_S curve (39) is defined in the new reference system joined with hob mill:

$$C_{S_{X_2Y_2Z_2}} \begin{vmatrix} X_2 = X_2(v_A); \\ Y_2 = Y_2(v_A); \\ Z_2 = Z_2(v_A). \end{vmatrix} \tag{41}$$

In the helical motion,

$$E_{X_2Y_2Z_2} \begin{Vmatrix} X_2 \\ Y_2 \\ Z_2 \end{Vmatrix} = \omega_2^T(\psi) \begin{Vmatrix} X_2(\upsilon_A) \\ Y_2(\upsilon_A) \\ Z_2(\upsilon_A) \end{Vmatrix} + P\psi\vec{j}, \tag{42}$$

with p helical parameter defined with equation (25).

The axial section of the helical surface $E_{X_2Y_2Z_2}$ with a plane ($X_2 = 0$) represents the axial profile of the hob mill.

4. Graphical method in CATIA

In order to be able to profile the hob mill using the graphic method, the intermediate surface method will be used.

This method consists in determining an intermediate surface which is the primary peripheral surface of a rack type tool.

Subsequently, on the intermediate surface, the characteristic curve by projecting the hob mill axis on this surface is determined.

The characteristic curve is generated with a helical movement with $\vec{V}$ axis and p helical parameter, resulting in the primary peripheral surface of the hob mill.

The determination of the rack gear tool's profile using the method of “generating trajectories family” consists in generating in CATIA the mechanism that reproduces the kinematic of the rake tool generation.

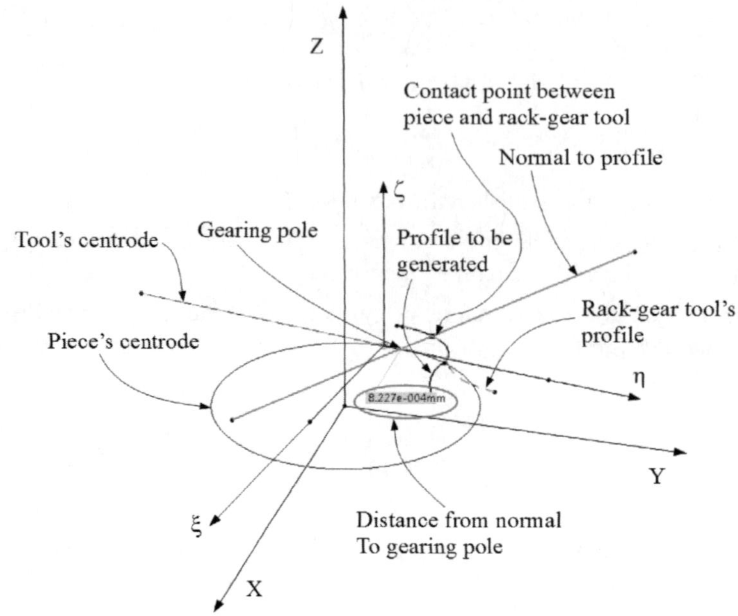

Figure 5. Profiling the rack gear tool

The mechanism is composed of three elements. The first element is the reference file containing the profile to be generated, the axis of the reference system associated with the part and the centrode of the piece, which in this case is a circle.

The second element is also a part file, but it will contain the axis of the reference system associated to the tool.

The third element has only a basic role and contains the axes of the reference systems associated to the piece and the tool, represented in the fixed reference system. In this way, the relative positioning of the two mobile systems can be made.

Onto the curve representing the profile to be generated a current point by which a straight line is drawn perpendicular to that profile is defined. In the assembly drawing, the operation of the mechanism is simulated, considering the rack reference as the fixed element, and the distance from the perpendicular to the pole of the gear is monitored. When this distance is zero, the contact condition between the track profile and the tool profile is met. The current point on the track profile is transferred to the tool reference system, obtaining a point on its profile.

Change the position of the current point and resume the mechanism simulation until sufficient points on the tool profile are obtained so that it can be obtained as a spline curve. This curve will represent the profile of the rack.

Subsequently, in the file where the toothed rack profile was obtained, the primary peripheral surface is thereof obtained by using the command "*extrusion*", which is applied to the curve in the direction of the rack generator version, see figure 3.

On the surface thus obtained, the tool axis is projected (see figure 3), resulting in the characteristic curve, which is the contact curve between the primary peripheral surface of the rack and the primary peripheral surface of the worm mill.

Applying the characteristic curve to a helical motion with $\vec{V}$ axis and p helical parameter, it produces the primary peripheral surface of the worm gear.

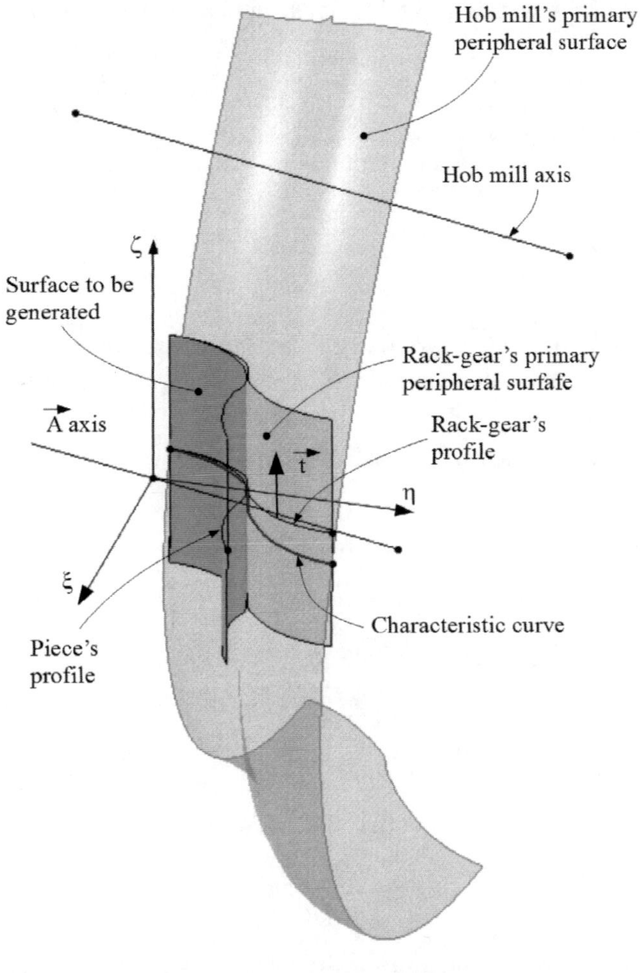

Figure 6. Hob mill's primary peripheral surface

The advantage of using CATIA software in order to determine the rack gear flank relies in the availability, inside CATIA, of all instruments that are needed for tool profiling.

In addition, the method for graphical design, through its specific characteristics and using numerical algorithms especially dedicated to graphical manipulation, helps avoiding the use of large mathematical equations which would normally lead to a large computer time.

Moreover, any error may occur in graphical design would be very visible and easy to identify, the program also offers solutions for identifying the correction possibilities.

5. Conclusions

The graphical method, developed in the CATIA graphics design environment, based on the complementary theorem of generating trajectories, leads to a simple expression of the enwrapping condition – the projection of the $\vec{A}$ axis on the component surfaces of the generating rack gear flank.

The method of determining the characteristic curve at the contact of the primary peripheral surface of the hob mill with the rack flank is rigorous and easy to achieve using a CATIA-specific command.

The characteristic point of contact of the helical surfaces of the hob mill with the helical flank of the rotor to be generated is obtained as the intersection point of the C_S characteristic curve, mainly in the shape (39), with the rectilinear generatrix of the rack.

From numerical point of view, this method is simple, intuitive and rigorous.

Appendix

It is proposed an analysis of the constitutive elements from the relation (19), regarding the decomposition of the helical motion, see Figure 4.

The circular pitch of the tooth, p_C, of the compressor rotor, in the frontal plane,

$$P_c = \frac{2\pi R_{r_p}}{z_{lobs}}, \tag{43}$$

where z_{lobs} is the number of lobs of the rotor ($z_{lobs} = 3$, for the driving rotor and $z_{lobs} = 6$ driven rotor);

R_{r_p} – radius of the associated centrode: $R_{r_p} = R_{r_{P1}}$, for driving rotor and $R_{r_p} = R_{r_{P2}}$, for driven rotor.

The rack gear normal pitch is defined,

$$p_{n_c} = p_c \cdot \cos\beta_d, \tag{44}$$

hence

$$p_{n_c} = \frac{2\pi R_{r_p}}{z_{lobs}} \cdot \cos\beta_d. \tag{45}$$

The angle of the hob mill helix

$$tg\omega_s = \frac{p_{ax_m}}{2\pi R_{r_s}}, \tag{46}$$

where: p_{ax_m} is the axial pitch of the hob mill;

R_{r_s} - rolling radius of the hob mill.

Considering the relation (18), from the decomposition of the helical motion, it results:

$$a = p \cdot tg\theta, \tag{47}$$

where: p is the helical parameter of the hob mill;

θ – the angle of cylindrical surface generatrix with the hob mill axis,

$$\theta = \frac{\pi}{2} - \omega_s. \tag{48}$$

From the equality condition between the generating rack gear normal pitch and the normal pitch of the hob mill, the following condition results:

$$\frac{2\pi R_{r_p}}{z_{lobs}} \cdot \cos\beta_d = 2\pi R_{r_s} \cdot tg\omega_s \cdot \cos\omega_s. \tag{49}$$

From the relation (49) results:

$$\sin\omega_s = \frac{R_{r_p}}{R_{r_s}} \cdot \frac{1}{z_{lobs}} \cdot \cos\beta_d. \tag{50}$$

The helical parameter of the hob mill is calculated as:

$$p = \frac{p_{ax_m}}{2\pi} = \frac{p_{n_m}}{\cos\omega_s} \cdot \frac{1}{2\pi} = \frac{R_{r_p}}{z_{lobs}} \cdot \frac{\cos\beta_d}{\cos\omega_s} \cdot \frac{1}{2\pi}, \tag{51}$$

where p_{n_m} is the normal pitch of the hob mill helix.

The a distance is calculated, from (18), as:

$$a = \frac{R_{r_p}}{z_{lobs}} \cdot \frac{\cos\beta_d}{\cos\omega_s} \cdot \frac{\sin\left(\frac{\pi}{2} - \omega_s\right)}{\cos\left(\frac{\pi}{2} - \omega_s\right)}, \tag{52}$$

resulting:

$$a = \frac{R_{r_p}}{z_{lobs}} \cdot \frac{\cos\beta_d}{\sin\omega_s}. \tag{53}$$

Considering the definition of the ω_s, the a dimension has the expression:

$$a = \frac{R_{r_p}}{z_{lobs}} \cdot \cos\beta_d \cdot \frac{R_{r_s} \cdot z_{lobs}}{R_{r_p} \cdot \cos\beta_d}, \tag{54}$$

hence,

$$a = R_{r_s}. \tag{55}$$

Obviously, the value of rolling radius R_{r_s} is equal to the dimension a resulted from the decomposition of the helical motion.

The equality of the normal pitch:

- rack gear's circular pitch:

$$p_c = \frac{2\pi R_{r_p}}{z_{lobs}}; \tag{56}$$

- rack gear's normal pitch:

$$p_{n_c} = p_c \cdot \cos\beta_d, \tag{57}$$

IOP Conf. Series: Materials Science and Engineering **448** (2018) 012005 doi:10.1088/1757-899X/448/1/012005

hence:

$$p_{n_c} = \frac{2\pi R_{r_p}}{z_{lobs}} \cdot \cos\beta_d; \tag{58}$$

- hob mill's axial pitch:

$$p_{ax_m} = 2\pi R_{r_s} tg\omega_s; \tag{59}$$

- hob mill's normal pitch:

$$p_{n_m} = p_{ax_m} \cdot \cos\omega_s, \tag{60}$$

hence:

$$p_{n_m} = 2\pi R_{r_s} tg\omega_s \cdot \cos\omega_s. \tag{61}$$

If the normal pitches are equals, then it results:

$$\frac{2\pi R_{r_p}}{z_{lobs}} \cdot \cos\beta_d = 2\pi R_{r_s} tg\omega_s \cdot \cos\omega_s. \tag{62}$$

Therefore,

$$\sin\omega_s = \frac{R_{r_p}}{R_{r_s}} \cdot \frac{1}{z_{lobs}} \cdot \cos\beta_d \tag{63}$$

and

$$a = p \cdot tg\theta, \tag{64}$$

where θ is the angle of the $\vec{t}$ generatrix with the axis of hob mill,

$$\theta = \omega + \frac{\pi}{2} - \beta_d; \tag{65}$$

p – helical parameter of the hob mill,

$$p = \frac{P_{ax_m}}{2\pi}, \tag{66}$$

Finally,

$$p = \frac{P_{n_m}}{\cos\omega_s} \cdot \frac{1}{2\pi} = \frac{2\pi R_{r_p}}{z_{lobs}} \cdot \cos\beta_d \cdot \frac{1}{\cos\omega_s} \cdot \frac{1}{2\pi}, \tag{67}$$

resulting:

$$p = \frac{R_{r_p}}{z_{lobs}} \cdot \frac{\cos \beta_d}{\cos \omega_s}. \tag{68}$$

Therefore,

$$a = p \cdot tg\theta = \frac{R_{r_p}}{z_{lobs}} \cdot \frac{\cos \beta_d}{\cos \omega_s} \cdot \frac{\sin\left(\frac{\pi}{2} - \omega_s\right)}{\cos\left(\frac{\pi}{2} - \omega_s\right)}, \tag{69}$$

and simplifying, the following is obtained:

$$a = \frac{R_{r_p}}{z_{lobs}} \cdot \frac{\cos \beta_d}{\sin \omega_s}. \tag{70}$$

If it is considered the definition of ω_s,

$$\sin \omega_s = \frac{R_{r_p}}{R_{r_s}} \cdot \frac{1}{z_{lobi}} \cdot \cos \beta_d, \tag{71}$$

so,

$$a = \frac{R_{r_p}}{z_{lobi}} \cdot \cos \beta_d \cdot \frac{R_{r_s}}{R_{r_p} \cdot \cos \beta_d} \cdot z_{lobi}, \tag{72}$$

so,

$$a = R_{r_s}. \tag{73}$$

References

[1] Litvin, F. L., *Theory of Gearing. Reference Publication*, 1212, NASA, Scientific and Technical Information Division, Washington D.C., 1984

[2] Oancea, N, *Methode numerique pour l'etude des surfaces envelopes*, Mechanism and Machine Theory, Vol. 31, no. 7, pp. 957-972, 1996

[3] Teodor, V.G., Baroiu, N., Berbinschi, S., Susac, F., Oancea, N., *A graphical solution in Catia for profiling end mill tool which generates a helical surface*, ModTech International Conference Modern Technologies in Industrial Engineering, June 14-17, Sibiu, Romania, IOP Conf. Series: Materials Science and Engineering, Vol. 227 Article Number: UNSP 012128, 2017

[4] Berbinschi, S. et al., *3D Graphical Method for Profiling Tools that Generate Helical Surfaces*, The International Journal of Advanced Manufacturing Technology, DOI: 10.1007/s00170-011-3637-3, Vol. 60, pp. 505-512, 2012

[5] Petukhov, Yu E., Domnin, P.Y., *Shaping Precision in Machining a Screw Surfaces*, Russian Engineering Research, Vol. 31 (10), pp. 1013-1015, 2011

[6] Petukhov, Yu E., Movsesyan, A. V., *Determining of Shape of the Back Surfaces of Disc Milling Cutter for Machining a Contoured Surface*, Russian Engineering Research, Vol. 22(8), pp. 519-521, 2007

[7] Kiryutin, A.S., *Mathematical Model of Profiling Cutter Tools*, Moskow State Technological University "STANKIN", Nauchno-prakticheokoi journal "Sovremenoe naychnoe issledovania I inovatzii", ISSN 2223-4888, Scientific and Practical Journal. Modern Scientific Researches and Innovations, Vol. 10, 2014

[8] Baroiu, N., Teodor, V., Oancea, N., *A new form of plane trajectories theorem. Generation with rotary cutters*, Buletinul Institutlui Politehnic Iasi, Tom LXI (LXV), Construcţii de maşini, pp. 27-36, ISBN 1011-2855, 2015

[9] Radzevich, P.S., *On the accuracy of precision involute hobs: an analytical approach*, Journal of Manufacturing Processes, Vol. 9(2), pp. 121, 2007

[10] Hsieh, J.K, Tseng, H.C., Chang, S.L., *A novel hob cutter design for the manufacture of spur-typed cutters*, Journal of Materials Processing Technology, pp. 847-855, 2009

[11] Jiang, J., Fang, Z., *High-order tooth flank correction for a helical gear on a six-axis CNC hob machine*, Mechanism and Machine Theory, 2015, pp. 227-237

IOP Conf. Series: Materials Science and Engineering **448** (2018) 012006 doi:10.1088/1757-899X/448/1/012006

Nanometric face milling simulations using MD method

A P Markopoulos[1], N E Karkalos[1] and J Kundrák[2]

[1]Section of Manufacturing Technology, School of Mechanical Engineering, National Technical University of Athens, Heroon Polytechniou 9, 15780, Athens, Greece
[2]Institute of Manufacturing Science, University of Miskolc, Miskolc-Egyetemváros H-3515, Hungary

E-mail: amark@mail.ntua.gr

Abstract. Face milling is considered as one of the most important methods in order to produce parts with high quality surfaces and considerable dimensional accuracy in a relatively short time. Investigation of the effect of process parameters on the outcome of this process is usually conducted experimentally or computationally by soft computing or numerical methods. However, when it is desired to study face milling in the nanoscale, experimental cost can be extremely high and numerical methods such as the Finite Element Method is unable to be employed, as it is a method mainly suited for continuum mechanics. Thus, a method such as Molecular Dynamics can be used to perform simulations of face milling in the nanoscale. In this work, cases with cutting tools with one and three cutters, moving at different feed values are considered and results concerning chip morphology and cutting forces are presented.

1. Introduction

Milling is essential in manufacturing industry in order to fabricate efficiently a wide variety of parts, with different levels of geometric complexity and dimensional tolerances. The rapid progress of manufacturing and the requirements of high-end industries such as the electronics industry lead to an increasing interest in micro and even nano-manufacturing processes. The capabilities of nano-milling to act as a viable alternative of established nano-fabrication techniques, such as lithography or nano-printing are considered as promising [1-3].

For the establishment of nano-milling, it is necessary to determine its characteristics in respect to the process parameters, as well as the underlying physical phenomena occurring during material removal. Afterwards, it will be able to determine the optimum process parameters, with a view to reduce machining time and prevent excessive subsurface damage to the workpieces. These investigations should be carried out experimentally, but due to the relatively high cost of performing experiments in the nanoscale or even the inability to accurately measure physical quantities during these experiments, theoretical investigations using proper computational techniques should be also performed. As the Finite Element Method (FEM) is not able to simulate explicitly the phenomena occurring in the nanoscale, the Molecular Dynamics method is adopted for nanoscale simulations.
Nano-machining simulations conducted with the aid of MD method have been reported since the early 1990s. One of the earliest works, was that of Rentsch and Inasaki [4] who studied nano-grinding with a 3D model of a diamond abrasive grain and a copper substrate. This model was created to investigate material removal mechanisms in the nanoscale. In another work [5], they studied the phenomenon of

IOP Conf. Series: Materials Science and Engineering **448** (2018) 012006 doi:10.1088/1757-899X/448/1/012006

crack initiation during nano-grinding, as well as the integrity of machined surface. In a pioneering study, Komanduri et al. [6] employed MD models with large negative rake angle tools in order to simulate nanometric grinding and were able to detect dislocations in the workpiece, determine the subsurface deformation and determine the variation of cutting forces and energy in several cases. Apart from nano-grinding, models for general machining [7] and various models of polishing [8] have been presented in the relevant literature. Nevertheless, models related to nano-milling have not been thoroughly investigated yet.

Cui et al. [9] created an MD nano-milling model in order to investigate the quality of produced grooves. In this study, a cutting tool with two cutting edges moving at a realistic way and a copper substrate were employed and the different characteristics of nano-milling in comparison to nano-cutting were determined. In another study [1], they compared the cases of end-milling and peripheral milling in nanometric level using suitable MD models. When comparing the outcome of these two processes, it was found that cutting forces were lower in end-milling and that this process was more stable; however with peripheral milling, grooves of better quality were able to be obtained. The effects of various process parameters such as depth of cut, feed rate, angular speed and shape of cutting tool on cutting force and groove quality were investigated by Cui et al. [10]. It was found that the cutting tool with four cutting edges was performing milling with less fluctuations of the cutting force and that increase of feed and depth of cut led to higher cutting forces. Furthermore, it was shown that the depth of cut was the most important parameter, which affected the quality of produced grooves.

More recently, another comprehensive work regarding nano-milling in order to study the material removal mechanisms was conducted by the same scientific group [11], in which they assessed the groove quality which was obtained. For that purpose, simulation with cutting tools with different dimensions, different feed rate and rotational speed values were performed. It was found that the most important factors affecting the groove profile are the cutting speed and groove dimension factors and that profile quality increases by decreasing the former and increasing the latter. Similar results are also summarized in the work of Cui and Zhang [12]. Besides these models, some researchers have proposed different nano-milling models as well. Olufayo et al. [13] used a box-shaped cutting tool in order to study milling on copper workpieces. Furthermore, Wu et al. [14] investigated the characteristic of nano-milling by conducting simulations with a rotating conical cutting tool. However, these models describe mainly the formation of nano-grooves by nanometric milling cutting tools, but no study on face milling with a cutting tool with multiple cutting inserts is yet presented.

In the present work, an MD model for nanometric face milling is introduced, with a view to explore the possibilities of this type of milling and study various characteristics of the process. For that reason, simulations using cutting tools with 1 and 3 cutters, at various feed rates will be performed on copper substrates. By analyzing the results regarding chip morphology and cutting forces, useful observations on the underlying phenomena of nanometric face milling are conducted.

2. Molecular Dynamics method

As it was aforementioned, due to the fact that it can more realistically and accurately simulate machining in the nanoscale, where matter is not a continuum, MD method is preferred for these simulations. Hereafter, a brief description of the method is attempted, with a view to present the basic characteristics of this method. Using MD method, the system comprising the cutting tool and the workpiece is represented by its atomistic structure. The atoms of the cutting tool and the workpiece can move freely under the influence of interatomic forces, they can be fixed in a specific position, or move with a constant speed as a rigid body. For atoms which can move freely, the forces exerted on them are summed and they move according to the second law of Newton. Interatomic forces are calculated with the aid of appropriate potential functions, such as Morse function or Embedded Atom Model (EAM) function which is suitable for metals. The atomic trajectories are calculated by integrating the Newtonian equation of motion and after the simulation is completed, atomistic forces, temperatures, stresses and subsurface damage can be determined. Although modern MD software can provide relatively lower computational cost, it is still difficult to perform simulations on systems exceeding 10^6 atoms in most cases and thus, simulations are limited to nm size substrates.

3. Methodology

In the present work, various cases of nano-milling of single-crystal copper (FCC) substrates with two different cutting tools and at three different feed rate values are studied by means of MD method. The depth of cut in all cases is the same, namely 0.6 nm, the rotational speed of the cutting tool is 50.24 rad/ps, the feed rate values are 100, 200 and 300 m/s and the total cutting length 7 nm in every case. The diameter of the cutting tool in all cases is 5 nm and the value of the rotational speed is equivalent to 5.024×10^{13} rad/s (4.797×10^{14} 1/min). Although the latter speed is quite high, it is a reasonable value regarding the dimensions of the workpiece, the feed and the final time of the nano-milling process. As the simulation lasts for several tens of ps, the use of much lower rotational speed would mean that the tool could not have performed a single revolution during the simulation; it would behave almost as a non-rotating rigid tool. Moreover, high rotational speeds have also been reported in nano-milling works in the relevant literature [10,11]. The cutting tool is composed of single-crystal diamond and consists of a number of cutting inserts like a real face milling cutting tool; half of the cases involve a cutting tool with a single cutter and the other cases involve a cutting tool with three identical cutters, each positioned at an angular distance of 120° from the next.

The axis of symmetry of the cutting tool is considered to be aligned with the axis of symmetry of the workpiece in x-axis, as can be seen in Figure 1. The cutters are considered to be rigid and their relative position does not change so that the whole tool behaves as a single body, moving towards the y-axis and rotating around its axis of symmetry at the same time. The total number of atoms in the system are almost 52,000. For the interatomic interactions between Cu and Cu atoms, a suitable EAM potential function is selected, whereas for the interactions between Cu and C, a Morse potential function is selected.

(a)

(b)

Figure 1. Initial configuration of the nano-milling model: (a) for the case with a single cutter, (b) for the case of three cutters.

As can be seen in Figure 1, the workpiece is divided into three different regions of atoms: the atoms in the inner part of the workpiece constitute the Newtonian atoms, which are allowed to move freely when a force is exerted on them by other atoms of the workpiece or the cutting tool, the atoms below the Newtonian atoms constitute the thermostat zone where velocity is rescaled in order to reduce the effect of artificial heat and finally, the boundary atoms which are positioned at the bottom and right side of the workpiece are able to prevent rigid body motion of the workpiece. In order to simulate a workpiece with larger dimensions, periodic boundary conditions are imposed on the yz plane. The cutting tool is initially positioned slightly away from the workpiece and after a short time period the cutter begins to remove material from the workpiece surface. The initial temperature of the workpiece is 293 K and time-step is set to 0.5 fs for all cases. For the simulations, LAMMPS software was employed.

4. Results and discussion

After conducting the aforementioned simulations, the characteristics of nanometric face milling are revealed. In all the examined cases, there are obvious similarities due to the common parameter values employed, such as the depth of cut or the diameter of cutting tool. The various stages of the process resemble the stages of the equivalent macroscopic process. Thus, the cutting tool is beginning to move at a short distance from the workpiece surface and after a while, the cutter comes in contact with the workpiece. As the cutting tool gets closer to the workpiece, the contact of the cutter and the workpiece surface is taking place for a longer period, until the cutter is able to perform material removal for a semi-circle. Due to the fact that the cutting tool diameter is almost equal to the width of the substrate, a large amount of material is removed and so cutting forces are relatively high. In Figures 2a and 2b, two distinct snapshots of nano-milling in the case of a single cutter are presented and in Figures 3a and 3b snapshots for the same time periods for the case of three cutters are presented. As can be seen in these figures, while the cutting tool is moving in the upper surface of the workpiece, atoms are removed and chip is piled in the area in front of the trajectory of the cutter. The differences between cases with higher feed rates or different cutting tool are observable in most cases.

(a) (b)

Figure 2. Snapshots for a case with a single cutter: (a) at 15 ps, (b) at 25 ps.

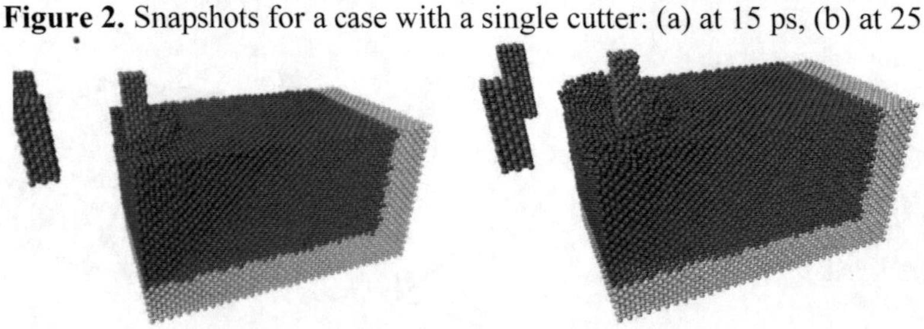

(a) (b)

Figure 3. Snapshots for a case with three cutters: (a) at 15 ps, (b) at 25 ps.

In order to assess more quantitatively the size of the pile of removed atoms in each case, measurements of the height and the length of the pile, in z- and y-directions, respectively, were conducted for the final timestep of each case, i.e. corresponding to 7 nm cutting length. For the cases with the single cutter, the main pile of removed atoms is created in front of the cutting tool. Its height varies slightly, from 21.4 to 22.25 and 24.1 Å, in the cases with feed rate of 100, 200 and 300 m/s, respectively. The length also increases with the feed rate, namely 47.96, 53.94 and 56.7 Å in the three cases. So, it can be seen that, although the same amount of material is to be removed, more atoms are added to the front pile, rather than the sides of the workpiece. However, in cases with three cutters, a different morphology of the pile of removed atoms is observed. In these cases, the height of the pile is smaller and especially in the cases with 200 and 300 m/s the pile in front of the cutting tool is not distinct from the pile of atoms in the sides of the workpiece. Between the last two cases the length is not changed but a slight increase in pile height is observed from 11.13 to 15.42 Å. Thus, it can be seen clearly that the use of a cutting tool with more cutters led to different arrangement of the chip atoms in these cases.

After the observation of process characteristics for the various cases examined in the present work, the analysis of cutting forces for each case was conducted and the components of cutting force in each direction were obtained. As it was anticipated in the case of face milling, the x- and y-component of the cutting force were fluctuating continuously as the relative position of the cutter is continuously changing, leading to variations in cutting forces. When the contact with the workpiece is longer, non-zero values of cutting force are observed for a longer time. The results regarding the time-averaged components of cutting force are presented in Table 1 and also depicted in Figure 4; in this Table, "1-100" denotes a case with 1 cutter and 100 m/s feed rate, "1-200" denotes a case with 1 cutter and 200 m/s feed rate and so on.

Table 1. Results concerning the components of cutting force in each case.

Case	Fx (nN)	Fy (nN)	Fz (nN)
1-100	37.301	38.560	51.054
1-200	44.072	51.010	60.872
1-300	49.510	59.504	68.915
3-100	35.443	40.602	82.043
3-200	39.101	48.283	84.494
3-300	42.957	55.396	89.772

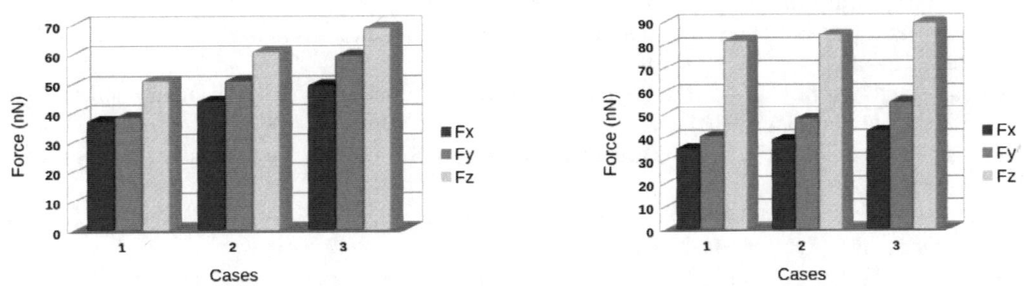

Figure 4. Variation of forces in the case of: (a) cutting tool with a single cutter and (b) with a cutting tool with three cutters.

As can be seen from Table 1, there are clear differences in the time averaged force components between cases with different feed rate, something that is in accordance with findings in the macro-scale, where the feed rate is considered as the most important factor affecting cutting forces, along with the depth of cut. In each case, Fy component is larger than the Fx component and the normal force Fz is larger than all the others. Regarding Fz force, it is observed that is larger in the case of cutting tool with three cutters, due to the more intense contact of the cutting tool with the workpiece in this case. The lower average Fx and Fy values for the cases with three cutters can be attributed to the influence of the feed per tooth to the cutting forces; as feed per tooth values are lower in the cases with larger number of cutters, it is expected that the force should be reduced. Nevertheless, this effect needs to be further investigated in a future study.

5. Conclusions

In the present work, a 3D MD model for nanometric face milling was developed, in order to investigate the characteristics of this process and conduct a preliminary evaluation of the factors

IOP Conf. Series: Materials Science and Engineering **448** (2018) 012006 doi:10.1088/1757-899X/448/1/012006

affecting its outcome. Models with a milling cutting tool with one and three cutters are presented for cases with various feed rate values. From these investigations, several conclusions are drawn.

At first, the basic characteristics of nanometric face milling were able to be observed by means of the developed model. It was confirmed that the kinematics of the model were accurate, both in the case of one and three cutters and thus, it can be employed also for more complex and thorough simulations of nano-milling. Regarding chip formation during the process, it was observed that due to the high rotational speed, a few atoms were separated from the workpiece with high speed but the rest of the removed atoms were piled in front of the cutting tool and on the sides of the workpiece. The morphology of the pile of removed atoms was shown to differ considerably between cases with one and three cutters.

Finally, the variations of the components of force were calculated for each case. It was noted that the increase of feed rate resulted in an increase of the cutting forces both for one and three cutters. Especially, for Fx and Fy, it was noted that there is an evident influence of the feed per tooth value to the results as these cutting forces were slightly larger in the case with one cutter because feed per tooth value is greater. The observed trends of cutting force variations are proposed to be further investigated in a future work with a broader range of feed rate, depth of cut and rotational speed values, in order to gain a more comprehensive view of nanometric face milling.

References

[1] Cui D D, Mylvaganam K and Zhang L C 2012 Atomic-scale grooving on copper: End-milling vs. peripheral-milling *Adv. Mater. Res.* **565** 546-551

[2] Varga G and Dudás I 2008 3D Topography for drilled surfaces *Key Engineering Materials* **381-382** 225-228

[3] Dudás I and Varga G 2005 3D Topography for environmentally friendly machined surfaces *J Phys Conf Ser* **13(1)** 24-27

[4] Rentsch R and Inasaki I 1994 Molecular Dynamics Simulation for Abrasive Processes *CIRP Ann.* **43** 327–30

[5] Rentsch R and Inasaki I 1995 Investigation of Surface Integrity by Molecular Dynamics Simulation *CIRP Ann.* **44** 295–8

[6] Komanduri R, Chandrasekaran N and Raff L M 1999 Some aspects of machining with negative-rake tools simulating grinding: A molecular dynamics simulation approach *Philos. Mag. B* **79** 955–68

[7] Alhafez I A, Gao Y and Urbassek H M 2017 Nanocutting: A Comparative Molecular-Dynamics Study of Fcc, Bcc, and Hcp Metals *Curr. Nanosci.* **13** 40–7

[8] Eder S J, Cihak-Bayr U and Pauschitz A 2015 Nanotribological simulations of multi-grit polishing and grinding *Wear* **340–341** 25–30

[9] Cui D D, Mylvaganam K and Zhang L C 2011 Nano-grooving on copper by nano-milling and nano-cutting *Adv. Mater. Res.* **325** 576–81

[10] Cui D D, Zhang L C and Mylvaganam K Nano-Milling on Copper: Grooving Quality and Critical Depth of Cut *J. Comput. Theor. Nanosci.* **11** 964–70

[11] Cui D, Zhang L, Mylvaganam K, Liu W and Xu W 2017 Nano-milling on monocrystalline copper: A molecular dynamics simulation *Mach. Sci. Technol.* **21** 67–85

[12] Cui D D and Zhang L C 2017 Nano-machining of materials: understanding the process through molecular dynamics simulation *Adv. Manuf.* **5** 20–34

[13] Olufayo O A, Otieno T, Olufayo A A and Akinlabi E T 2017 Optimization Strategy for Milling of Copper Using Molecular Dynamics Modelling *Procedia Manuf.* **7** 374–86

[14] Wu C D, Fang T H and Kuo C H 2015 Nanomilling mechanism on Cu surfaces investigated using atomistic simulation *Mol. Simul.* **41** 1159–65

IOP Conf. Series: Materials Science and Engineering **448** (2018) 012007 doi:10.1088/1757-899X/448/1/012007

Measurement and evaluation of the flatness error of a milled plain surface

Balázs Mikó

Óbuda University, Institute of Material and Manufacturing Science, 1081 Budapest Népszínház u. 8. Hungary
miko.balazs@bgk.uni-obuda.hu

Abstract – The accuracy requirement of a machine part means not only the dimensional accuracy, but the surface quality, the accuracy of shape and the position of a geometric feature. Because of the increasing industrial requirements, the meaning, measuring and manufacturing of shape and position tolerances come into the limelight. In this article the effect of the different measuring strategy on the coordinate measuring machine is investigated in case of a milled plain surface. The question is, how can be minimized the number of measured points beside the right measured value. From the viewpoint of the measuring process the aim is to create an appropriate strategy, which can ensure the real value of the errors and short measuring time.

1. Introduction

During the machine part manufacturing, there are differences between the designed and the realized properties. The reasons of these deviations are the uncertainty of the different parameters, like material, machine tool, cutting tool, process parameters, measuring devices and process, environment and human factors. The level of the deviation can be reduced by the process planning and quality control. The deviation cannot be eliminated, so the acceptable level of it can be defined as tolerance. Based on standards, dimensional tolerances [1], shape, orientation and form tolerances [2], and surface roughness [3] can be defined.

During the identification of the type and value of the tolerances, the designer must consider the nature, the use, the size, the material of the part and the manufacturing and assembly technologies [4]. The final design requirements must satisfy contradictory conditions and constrains. Dimensional tolerances don't ensure right definition of requirements, so the shape, orientation and position tolerances came into view in the last decade. This tendency increases the importance of the defining, measuring, interpretation and producing of this type of tolerances. Płowucha et al. [5] summarizes the geometric product specification (GPS) related standards and points out the importance of the GPS in the engineering work and the education. The development of implementation algorithms is an important and actual field in the GPS [6][7][8][8].

Runje et al. [10] defines the flatness by the least square method in case of a granite table. Different measurement routes were applied, and the measuring instrument was an electronic level. Wang et al. [11] describes the mathematical model of the flatness error evaluation and suggests a genetic algorithm to calculate the error based on minimum zone solution. Wen et al. [12] proposes an improved genetic algorithm (GA) for flatness error evaluation and the GUM and Monte Carlo (MCM) methods for evaluate the uncertainty.

Hadžistević et al. [13] investigates the flatness of surfaces created by turning, milling and grinding. The flatness was determined based on different number of points (10 to 2000) by least square (LS) and minimum zone (MZ) methods. Based on the LS method largest values of flatness are determined. In case of turning (3 test parts) the number of points has the largest effect, but the increasing surface roughness means increasing flatness. In case of grinding (1 test part) the number of investigated points has no effect to the flatness value. In case of same surface roughness the face milling (1 test part) ensures better flatness than turning and the number of points has smaller effect to the flatness.

Mikó and Drégelyi-Kiss [14] investigates the flatness, perpendicularity and parallelism in case of shoulder milling. Besides the cutting parameters and the machine tool, the milling direction and the position of the surface have effect to the geometric error. The quality of the milled plain surface can be described or evaluated by geometric parameters of the tool and the cutting parameters (e.g. [15]), but the question is the effect of these parameters to the geometric deviation of the surface.

The different types of form, orientation and position standards are defined by ISO 1101-2017 standard [2]. The current article focuses on the flatness. The standard defines the flatness as the distance of two parallel plains which bound the produced geometric feature (Figure 1). The position and the orientation of the two plains which create the flatness tolerance zone are independent within the dimensional tolerance zone.

During the measuring by a coordinate measuring device (coordinate measuring machine (CMM), measuring arm, 3D scanner etc.) points of the surfaces are recorded, and this point cloud substitutes the surface. The size of the point cloud and the methods of mathematical modelling aren't determined in the standards.

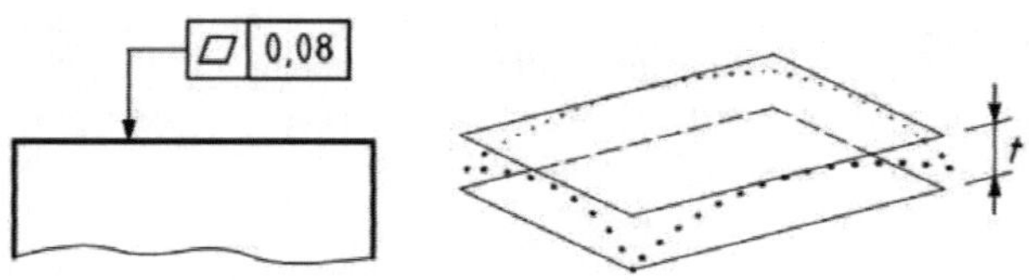

Figure 1 The standard mark of the flatness tolerace and the interpretation of the tolerance zone [2]

In this article the influence of the measuring strategy in case of coordinate measuring machine is presented in connection with the flatness of a milled plain surface. The question is how the number of measured points can be minimized, beside the right measured value. From the viewpoint of the measuring process, the aim is to create an appropriate strategy, which can ensure the real value of the errors and short measuring time.

2. The methods and devices of the test

The applied test part was made of 42CrMo4 (1.7225) pre-hardened steel; the size is 175x155 mm. The plain surface was machined by milled with a UF-231 conventional milling machine. The diameter of the milling cutter was 80 mm, the number of teeth is 7 (ISCAR HeliDo SOF 45 8/16-D080-07-27 R). The cutting parameters are: v_c = 60 m/min, f_z = 0.046 mm, n = 240 1/min, v_f = 78 mm/min, a_p = 1 mm, a_e = 40 mm. The surface roughness was measured by Mahr-Perten Concept instrument, in two perpendicular direction, at 2x(4x4) position. The measured surface roughness was Ra = 2,97 μm (σ = 0,9 μm), Rz = 14,75 μm (σ = 4,5 μm).

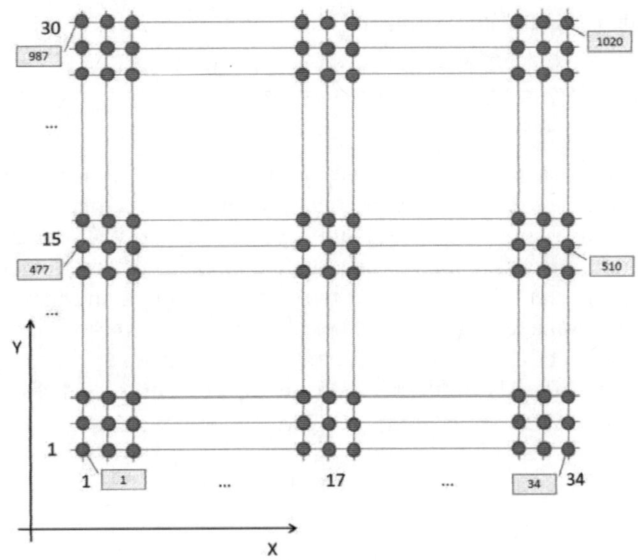

Figure 2 The pattern of the test points

The surface was measured by Mitutoyo Crysta-Apex S 776 CMM, 1020 points were recorded in 34x30 pattern with 5 mm distance (Figure 2). During the evaluation of the flatness (*FL*), the minimum zone method was used: seek the best fitted boundary plains, which have the smallest distance. The applied algorithm is showed in Figure 3.

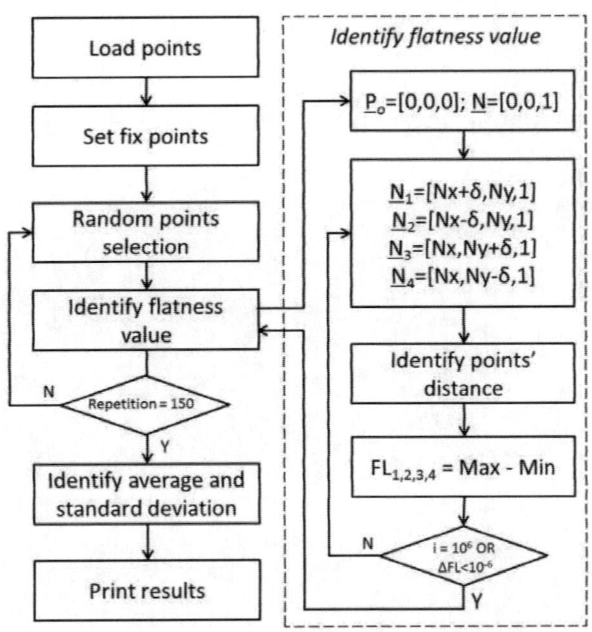

Figure 3 Evaluation algorithm

The equation of a plain is (where the $\underline{P}_0$ is a point of the plain, $\underline{N}$ is the normal vector):

$$N_x \cdot x + N_y \cdot y + N_z \cdot z - (N_x \cdot P_{0x} + N_y \cdot P_{0y} + N_z \cdot P_{0z}) = 0 \quad (1)$$

The distance between a $\underline{P}_i$ point and the plain is:

$$D_i = \frac{N_x \cdot (P_{0x} - P_{ix}) + N_y \cdot (P_{0y} - P_{iy}) + N_z \cdot (P_{0z} - P_{iz})}{\sqrt{N_x^2 + N_y^2 + N_z^2}} \qquad (2)$$

The orientation and position of the two boundary plains were determined by iteration algorithm. The initial position of the reference plain is horizontal, the normal vector is N = [0,0,1], and contains the P_0 = [0,0,0] point. During the iteration the X and Y components of the normal vector were changed with δ, and in case of the four possible new positions the distance of the elements of the point cloud were determined. The flatness is the difference between the minimum and maximum distance (Figure 4). The best value can be selected, and in the next iteration step this position was modified. The iteration stops, if the difference between two iterations (ΔFL) is less than 10^{-6} mm or the iteration step number (i) is 10^6. The δ is randomly defined in every iteration step, the value is $\delta = 10^{-6}\, r$ (where r = 0…1 random value).

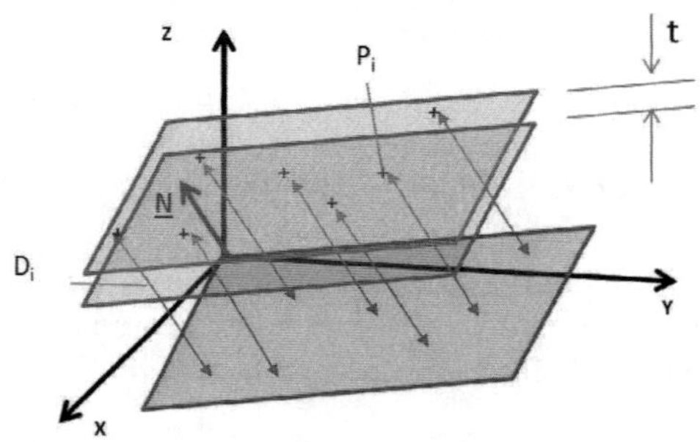

Figure 4 Interpretation of the flatness

The reference value of the flatness deviation was determined based on 1020 points by Kotem SmartProfile 4.0 evaluation software based on ISO 1101 by least square method. The reference value is FL_{1020} = 0.075 mm (Figure 5).

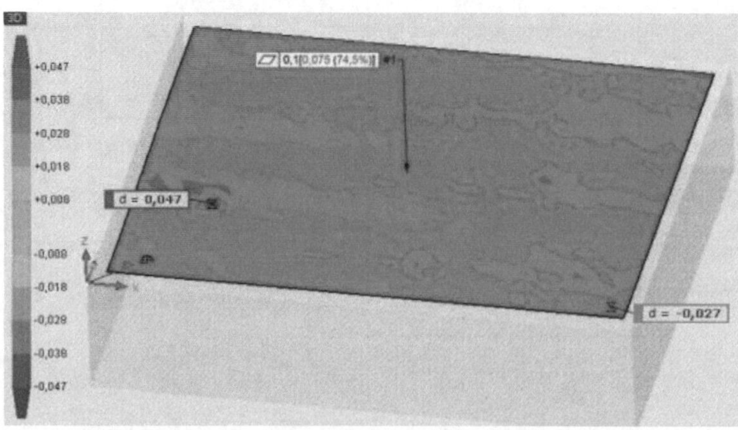

Figure 5 The reference flatness value in SmartProfile

3. The Random Points Method

In case of the Random Points Method (RPM) the points, which represent the whole plain and define the flatness are selected randomly. The random selection was repeated 150 times, and the average value of the different selection gave the result. During the research constant points were defined beside the random points, in order to increase the accuracy of the evaluation.

In case of RMP-0 there isn't a constant point, only random. In case of RPM-4, the four corners are the constant elements; at RPM-5 the centre point is added. The RPM-9, RPM-16 and RPM-25 mean 3x3, 4x4 and 5x5 fixed points (Figure 6).

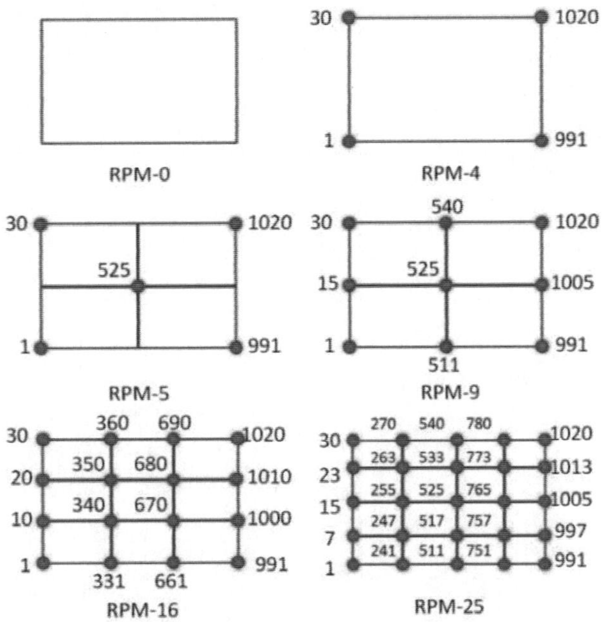

Figure 6 The location of the constant points

The number of investigated points (*NoP*) was: 6, 7, 8, 9, 10, 11, 12, 13, 14, 15, 20, 25, 30, 35, 40, 45, 50, 75, 100, 150, 200, 250, 500, and 999, and 0, 4, 5, 9, 16 or 25 points were defined in fix position and the rest were selected randomly. The random point selection causes uneven distribution of the points, the 4, 5, 9, 16 and 25 fixed points compensate this effect (Figure 7). Not every test were started from 6 points, the RPM-9 test series starts from 15 points, the RPM-16 starts from 20 and RPM-25 starts from 30 points. In the abbreviation the first number shows the number of fix points (*NoCoP*), and the second value shows the number of investigated points (*NoP*), for example the RPM-9/20 means the flatness is calculated based on 20 points and 9 points were in fix position.

IOP Conf. Series: Materials Science and Engineering **448** (2018) 012007 doi:10.1088/1757-899X/448/1/012007

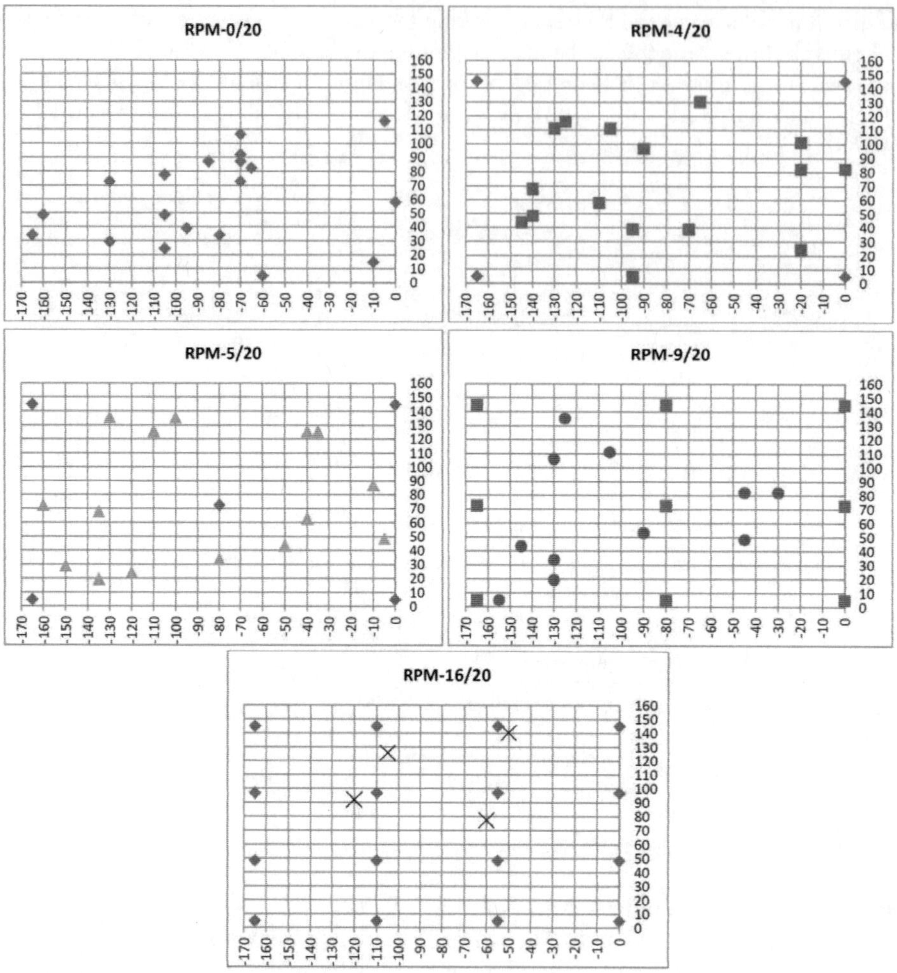

Figure 7 An example of distribution of the selected points (20 points example)

4. The results of the Random Points method

The flatness deviation is calculated by the average of 150 selection cycle in case of every number of points (*NoP*). Beside the average value, the standard deviation was determined too. In case of increasing number of points, the flatness deviation is increasing as well (Figure 8), but it never reaches the reference value (FL_{1020}). So in case of limited number of points the determined flatness is underestimated.

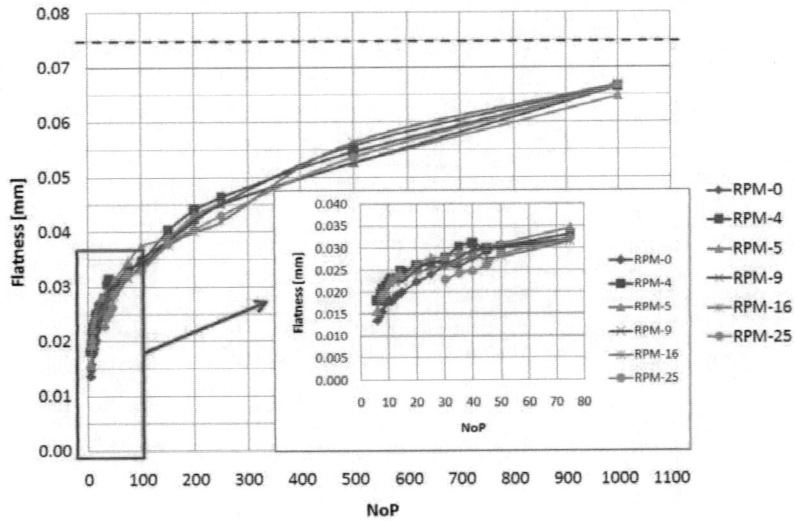

Figure 8 The value of the flatness

The standard deviation is smaller in case of less measured points, but fluctuates. Parallel with increasing number of points the standard deviation is increasing also, but at large number of points it is decreasing a little (Figure 9).

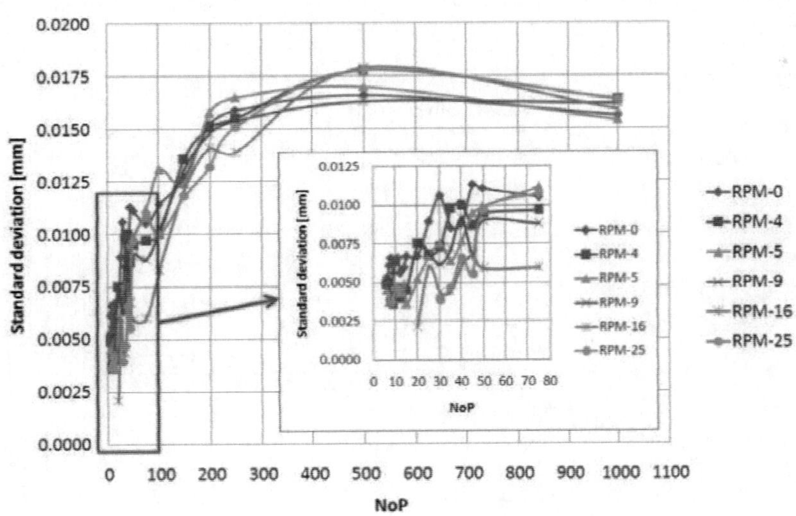

Figure 9 The values of the standard deviation of the flatness

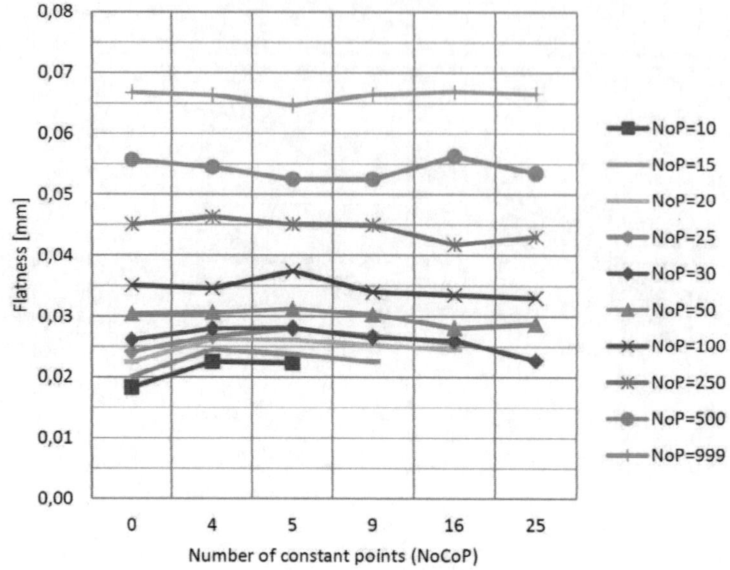

Figure 10 The values of the flatness

Based on the diagrams (Figure 10), the effect of the parameters of the sets (number of points, number of fixed points) are hard to identify, but based on the main effect plots (Figure 11) the importance of number of points (*NoP*) is clear. The number of constant points (*NoCoP*) has a smaller, but important effect.

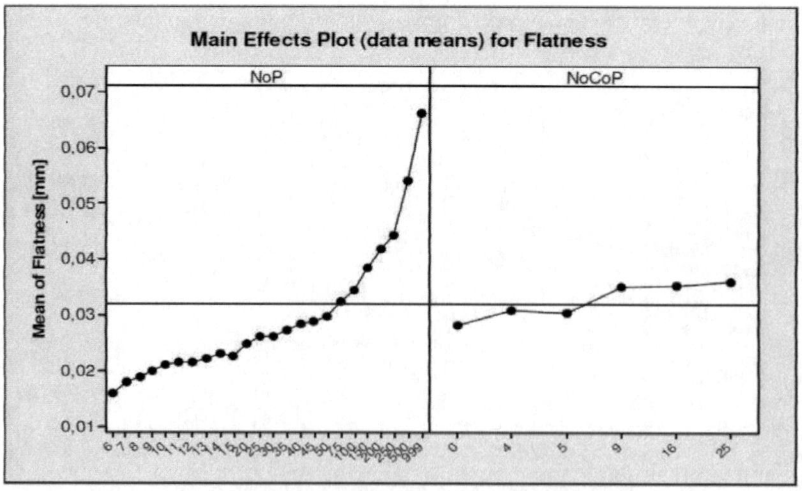

Figure 11 Main effects plot for flatness

The evaluated flatness value approximates continuously to the real (FL_{1020}) value, so the real value can be estimated by the correction of the measured value, where the FL_E is the estimated flatness, the FL_M is the measured flatness, and the C_{FL} is the correlation coefficient.

$$FL_E = C_{FL} \cdot FL_M . \tag{3}$$

The correlation coefficient depends on the number of points (*NoP*) and the number of constant points (*NoCoP*):

$$C_{FL} = f(NoP, NoCoP) \tag{4}$$

This coefficient is optimized by MS Excel Solver, by minimizing the R^2adj (#1; R^2adj = 0.862):

$$C_{FL} = 5.4311 - 0.6968 \cdot \ln(NoP) - 0.0035 \cdot NoCoP \tag{5}$$

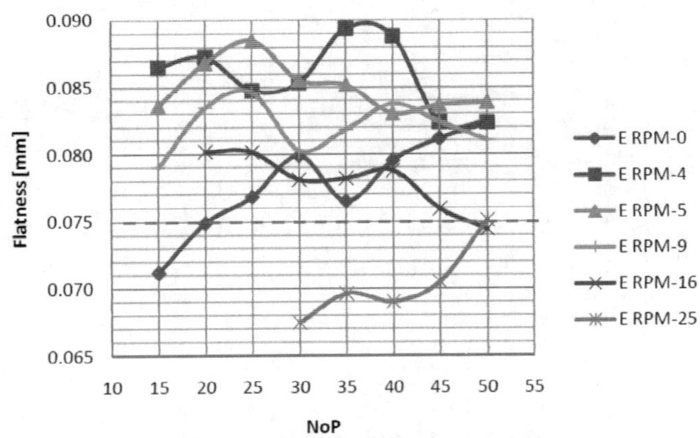

Figure 12 Estimated flatness #1

Based on this result the method underestimates the flatness at the two ends of the value range, and overestimates at the middle of the range. Less than 15 points is not enough to cover the surface, in case of more than 50 points, the measuring time is quite long. From the viewpoint of the measuring process, the 15-50 measuring points are the relevant. In this zone, there is a large difference in the estimated flatness; the error range is 0.025 mm (Figure12).

Based on this middle (15-50 points) zone the coefficient is recalculated (#2; R^2adj = 0.700):

$$C_{FL} = 5.8901 - 0.9360 \cdot \ln(NoP) - 0.0165 \cdot NoCoP \tag{6}$$

The estimated flatness values by the modified coefficient shows better results (Figure 13), the values there are in a±0.009 mm zone.

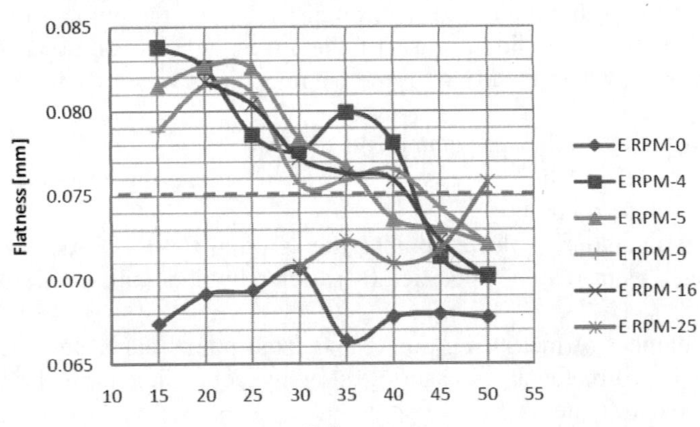

Figure 13 Estimated flatness #2

The application of RPM-0 is not realistic, because the cover of the surface is very uncertain. If the RPM-0 is excluded, the coefficient factor modifies next (#3; R^2adj = 0.902):

$$C_{FL} = 4.7991 - 0.6607 \cdot \ln(\,NoP) - 0.0238 \cdot NoCoP \qquad (7)$$

As the Figure 14 shows, the error range deceased, the zone is only ±0.005 mm

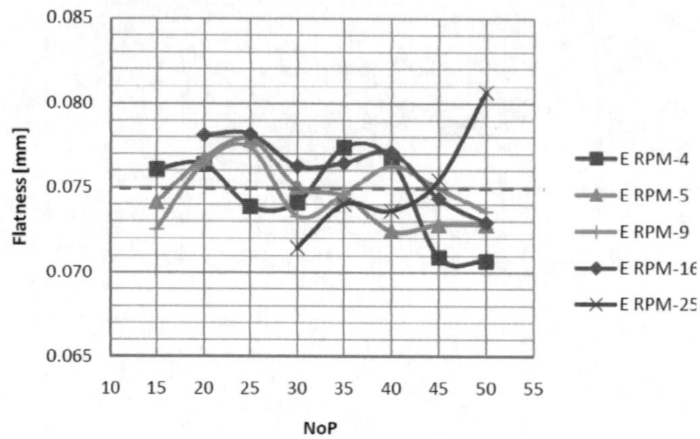

Figure 14 Estimated flatness #3

5. Results and discussion

With the definitions of standard in mind, the flatness deviation can be evaluated by different methods. In case of measuring by coordinate measuring devices, the key question is the number of measured points and the strategy of points recording. In case of the presented random points method (RPM), the evaluated flatness is increased parallel with the increasing number of points, but the real value of the flatness cannot be reached.

The constant points have important role in better cover of surface, but based on statistical analysis the influence to the flatness is small. More constant points are correct the flatness value, but in case of 4/5 points strategies and 9/16/25 points strategies there are no significant differences. Therefore 9 (3x3) constant points strategy is more productive, than 16 (4x4) or 25 (5x5) strategies, because of the faster measuring process can be ensured.

Based on the presented theory, the real flatness value can be evaluated with help of a correction coefficient, which depends on the number of measured points and the number of constant points. The presented numerical solution is valid in case of the investigated conditions, but the idea can be extended to other test parts, where the effect of the manufacturing methods, the size of the surface can be investigated.

Summarizing the results the following statements are made:

- The measurement of the flatness by coordinate measuring machine is influenced by the number of recorded points.
- Based on the randomly selected points, the value of flatness is determined by the position of the points, which can be specified by the fixed points with providing better coverage of the surface.
- Since the number of measuring points and fixed points correlates with flatness value, the result can be corrected by measuring strategy depending correction factors in order to estimate the real flatness.

Based on the results, there are several further possibilities and aims:

IOP Conf. Series: Materials Science and Engineering **448** (2018) 012007 doi:10.1088/1757-899X/448/1/012007

- The correction coefficient can be specified by further machined surfaces, this way other influence factors can be considered, like the machining technology and the surface roughness.
- The size of the surface is an important factor, so other surface sizes must be investigated.
- The evaluated points were selected randomly. A method should be developed, which ensures equable cover of the surface.

References

[1] ISO 14405-1:2010, Geometrical product specifications (GPS) — Dimensional tolerancing — Part 1: Linear sizes

[2] ISO 1101-2017 Geometrical product specifications (GPS) — Geometrical tolerancing — Tolerances of form, orientation, location and run-out

[3] ISO 4287-2002, Geometrical Product Specifications (GPS) — Surface texture: Profile method — Terms, definitions and surface texture parameters

[4] Henzold G 2006 *Geometrical Dimensioning and Tolerancing for Design, Manufacturing and Inspection* (Elsevier)

[5] Płowucha W; Jakubiec W; Humienny Z; Hausotte T; Savio E; Dragomir M; Bills P; Marxer M; Wisła M; Mathieu L 2014 Geometrical product specification and verification as toolbox to meet up-to-date technical requirements; *The 11th International Scientific Conference "Coordinate Measuring Technique" CMT2014* pp.131-139.

[6] H.S. Halkacı; Ö. Mavi; O. Yigit 2007 Evaluation of form error at semi-spherical tools by use of image processing; *Measurement* Vol.40 pp.860-867

[7] Rossi A, Antonetti M, Barloscio M, Lanzetta M 2011 Fast genetic algorithm for roundness evaluation by the minimum zone tolerance (MZT) method; *Measurement* Vol.44 pp.1243-1252

[8] Rossi A and Lanzetta M 2013 Roundness: A closed form upper bound for the centroid to minimum zone center distance by worst-case analysis; *Measurement vol.46* pp.2251-2258

[9] Mao J; Cao Y; Yang J 2009 Implementation uncertainty evaluation of cylindricity errors based on geometrical product specification (GPS); *Measurement* Vol.42 pp.742-747

[10] Runje B; Marković M; Lisjak D; Medić S; Kondić Z 2013 Integrated procedure for flatness measurements of technical surfaces; *Technical Gazette* Vol.20 No.1 pp.113-116.

[11] Wang D; Wen X; Wang F 2012 A Differential Evolutionary Algorithm for Flatness Error Evaluation; *AASRI Conference on Computational Intelligence and Bioinformatics AASRI Procedia 1* pp.238-243

[12] Wen X; Zhu X; Zhao Y; Wang D; Wang F 2012 Flatness error evaluation and verification based on new generation geometrical product specification (GPS); *Precision Engineering* Vol.36 pp.70-76

[13] Hadžistević M; Štrbac B; Spasić Jokić V; Delić M; Sekulić M; Hodolič J 2015 Factors of estimating flatness error as a surface requirement of exploitation; *Metalurgija* Vol.54 no.1 pp.239-242

[14] Mikó B and Drégelyi-Kiss Á 2015 Study on tolerance of shape and orientation in case of shoulder milling; *Development in machining technology* Vol.5. pp.136-150.

[15] Kundrák J; Felhő Cs 2018 Investigation of the topography of face milled surfaces; *Material Science Forum* Vol.919, pp.78-83

IOP Conf. Series: Materials Science and Engineering **448** (2018) 012008 doi:10.1088/1757-899X/448/1/012008

Competitive-resource information model of the machine building manufacturing system

A A Kutin[1], V A Dolgov[1], A A Kabanov[2], M I Sedykh[1,*] and S S Ivashin[1]

[1]Production Engineering Department of the Moscow State University of Technology STANKIN, Vadkovskij per. 1, Moscow, 127994, Russian Federation
[2]Department of Aerospace Engineering of the Moscow Aviation Institute (National Research University), Volokolamskoe shosse, 4, Moscow, 125993, Russian Federation

* Corresponding author. Tel.: 7-499-973-30-80; fax: +7-499-973-3080. E-mail address: sedykhmi@mail.ru

Abstract. The article discusses the method of creation a competitive-resource information model of the manufacturing system on the example of producing batch of parts, which allows expanding the field of determining the information links acting in the manufacturing process. It makes possible to determine the relationship of machining operations with elements of the manufacturing system. List of input data is selected for assessing of parts machining effectiveness for productivity parameters. A conceptual description of the competitive resource model with data groups of the information model is presented. The results of practical implementation of the methodology tested at one of Russian machine-building enterprises are presented.

1. Introduction

The modern period of machine-building enterprises development is characterized by the transition to digital production, which in turn involves the replacement of the output of a relatively constant range of products for the manufacturing of customized products [1]. Changing the paradigm leads to the need to ensure the flexibility of the manufacturing system, which are implemented as part of the production process planning. This requires a comprehensive study of technological and organizational problems.

Thus, the technological system must be developed in conjunction with control, maintenance and repair systems, forming a single design object.

The development of new approaches to the design of engineering enterprises includes the definition of a object complex design, its information model and its support processes at all stages of the manufacturing system life cycle[2].

As such a design object, the manufacturing system (MS) of the enterprise is proposed, which includes the main blocks shown in figure 1.

The efficiency of the manufacturing system at different stages of the life cycle is largely determined by the quality of information support.

The MS is regarded as a complex of organizational and technical systems intended for the manufacturing of products. Stages of the MS life cycle differ from the stages of manufactured products life cycle (aircraft, cars, etc.).

The manufacturing system of the enterprise is part of the Smart Manufacturing Ecosystem. The Smart Manufacturing Ecosystem encompasses a broad scope of systems in the manufacturing business including production, management, design, and engineering functions. The product life cycle is concerned with the information flows and controls beginning at the early product design stage and continuing through to the end-of-life of the product. The manufacturing system life cycle focuses on the design, deployment, operation and decommissioning of an entire production facility including its systems [3].

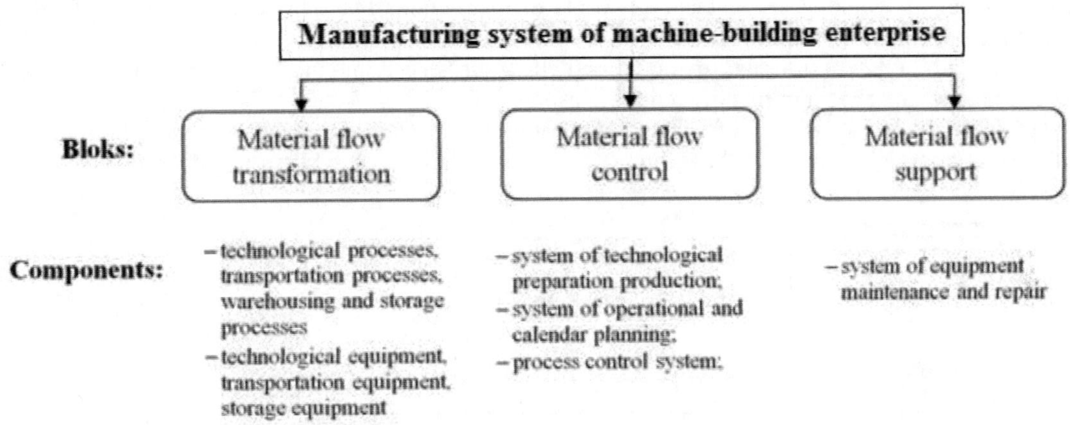

Figure 1. The main blocks and components of the manufacturing system.

The main difference is that the MS at the stage "Operation of the substation" includes the following stages: "Design of the configuration of the substation", "Utilization / Construction of the configuration of the substation" and "Operation of the configuration of the substation" (figure 2.) This is due to the need for its technical re-equipment when changing the products and / or the release program [4].

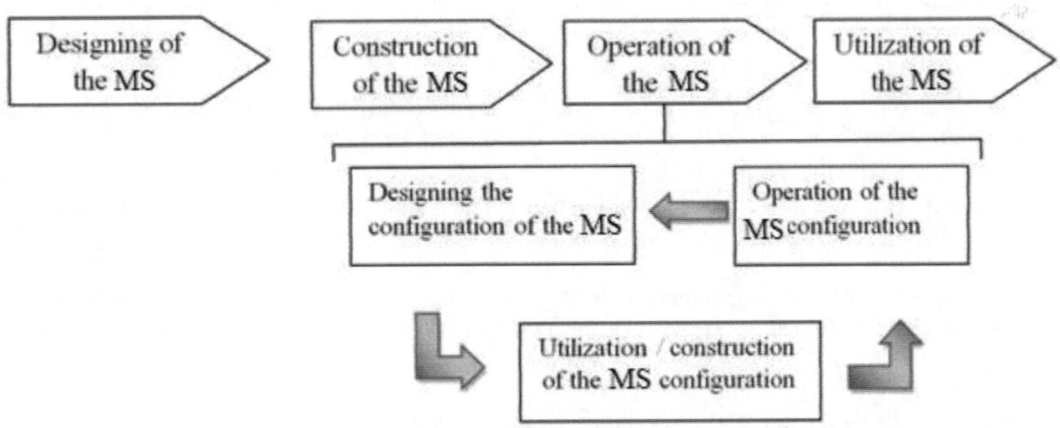

Figure 2. Content of the stage "Operation of MS" of the life cycle of the production system.

Thus, the simulation model (SM) of the manufacturing system undergoes periodic significant changes at the stage of "Operation of the substation".

The SM of the *i*-th configuration of the MS includes information blocks that are formed at different stages of the "Operation of the MS" phase.

At the stages "Design of the configuration of the substation" and "Construction / Utilization of the configuration of the substation", a project data block is formed, at the stage "Operation of the configuration of the substation" - a block of operational data.

IOP Conf. Series: Materials Science and Engineering **448** (2018) 012008 doi:10.1088/1757-899X/448/1/012008

2. Initial data for creating a competitive-resource information model

The primary sources of input data for the developed method are designing and technological documentation.

To perform the evaluation according to the developed method, only a certain part of the information that is contained in this documentation is needed. And for its choice requires considerable preparatory work.

One of the ways to reduce the preparatory work is to present the initial information in the form of a formal model of a limited set of facts, concepts or instructions that are necessary and sufficient to inform the developed method.

The availability of an information model will allow further automating the process of preparing the initial data and, thereby, significantly reducing its laboriousness.

The information model is an intermediate link in the information interaction of structural divisions that are suppliers of designing and technological documentation and users of the developed method (figure 3).

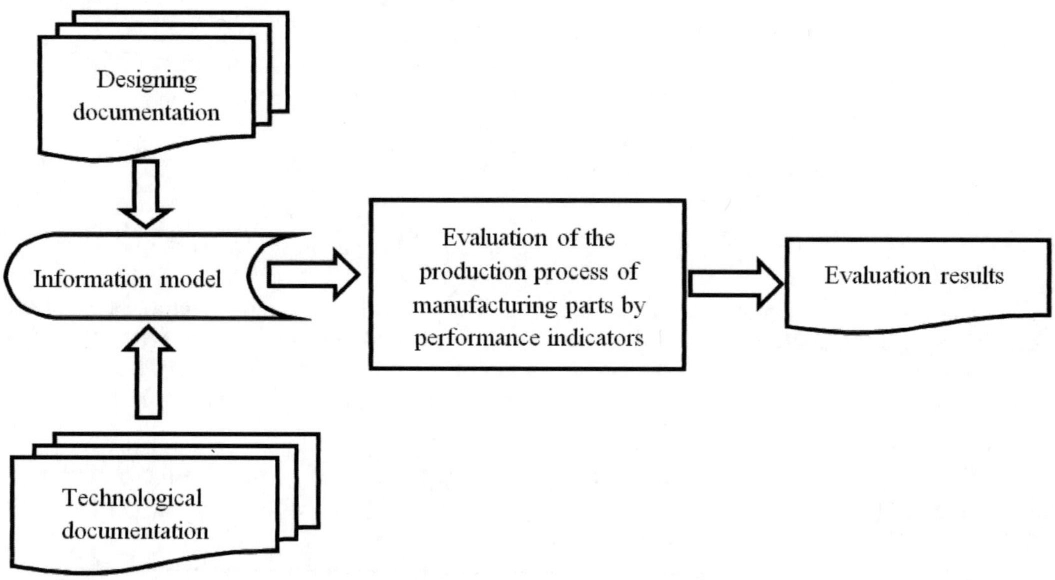

Figure 3. Diagram of the information links of the developed method with the primary source information.

In accordance with the conceptual model for assessing the production process of manufacturing part for performance indicators (figure 4), the information model should contain the following data groups [5]:

- general information about the parts produced,
- data on available resources of the technological subsystem $Tech_i^a$,
- data on the required resources of the technological subsystem $Tech(n)_i^{req}$,
- data on the available resources of the transport subsystem Tr_i^a ,
- data on the required resources of the transport subsystem $Tr(n)_i^{req}$,
- data on the duration of storage (accumulation) and the laboriousness of the storage processes of blanks $Lab_{St}(n)_i$.

Analysis of designing documentation showed that for each unit of technological and transport equipment, the number of hours per year involved in performing operations to manufacture all parts is a complex and time-consuming task. Often such information is not available [6].

Therefore, in the present paper, it is suggested that the available time for each unit of technological and transport equipment will be determined through the factor of consolidation of operations.

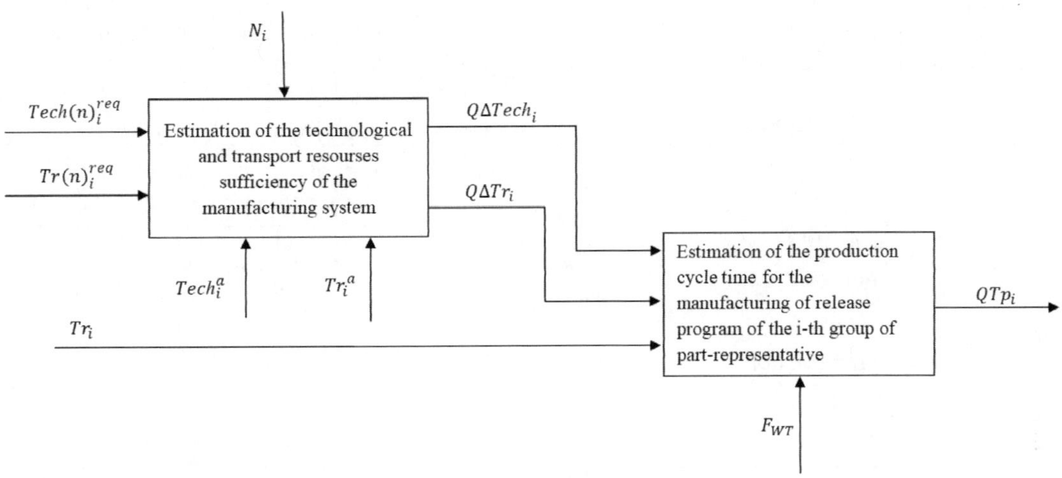

Figure 4. A conceptual model for evaluating the production process of manufacturing part of the *i*-th group in terms of performance indicators in the manufacturing system.

Then the available resources of the technological subsystem for each unit of equipment can be determined from the following relation:

$$Tech_{Mj,gi}^{a} = F_{WT\,gi} \cdot k_{fo\,gi} \tag{1}$$

where $F_{WT\,gi}$ is the year working time of the technological equipment performing the *g*-th technological operation of the *Mj*-th method of transformation of the *i*-th group part material flow;

$k_{fo\,gi}$- the coefficient of fixing the operation of the technological equipment performing the *g*-th technological operation of the *Mj* -th method of transformation of the *i*-th group part material flow.

The available resources of the transport subsystem are determined by the following relation:

$$Tr_{Mj,si}^{a} = F_{WT\,si} \cdot k_{fo\,si} \tag{2}$$

$F_{WT\,si}$ - is the year working time for the transport equipment performing the *s*-th transport operation of the *Mj*-th method of transformation of the of the *i*-th group part material flow,

$k_{fo\,si}$ - the factor of the operation fastening for transport equipment performing the *s*-th transport operation of the *Mj*-th method of transformation of the of the *i*-th group part material flow.

Thus, the formal model determines the manufacturing system of the production process of the installed batch of the part.

One of the features of this information model is the description of available resources for performing technological and transport operations of the corresponding subsystems elements, taking into account the existing competition for them in the operations of other parts producing.

In accordance with the above, the proposed information model was called a competitive-resource information model of the manufacturing system of a batch parts production process.

3. Practical implementation of the competitive-resource model in one of the shops of the Russian machine-building enterprise

Let's consider an example of the project of the manufacturing system of the shop for producing some unit.

At the first stage of the workshop project, a process model has been developed that describes the manufacturing process of the product.

Critical ways of the product manufacturing were determined. The analysis of the identified critical paths made it possible to draw a conclusion about the possibility of performing operations lying on critical paths.

To fulfill a given production program, 89 workplaces were provided in the project based on the use of the process program with a two-shift work schedule (Table.1.)

Table 1. Workplaces specification of the shop floor

Workplace name	Quantity of workplaces according to project documentation
Worktable	1
Bench	5
Dry box	1
Electrical worktable	80
Test bench	2
Total:	**89**

The calculated values of the coefficients of loading the workplaces K_L of the production system for manufacturing the product are shown in figure 5 [1].

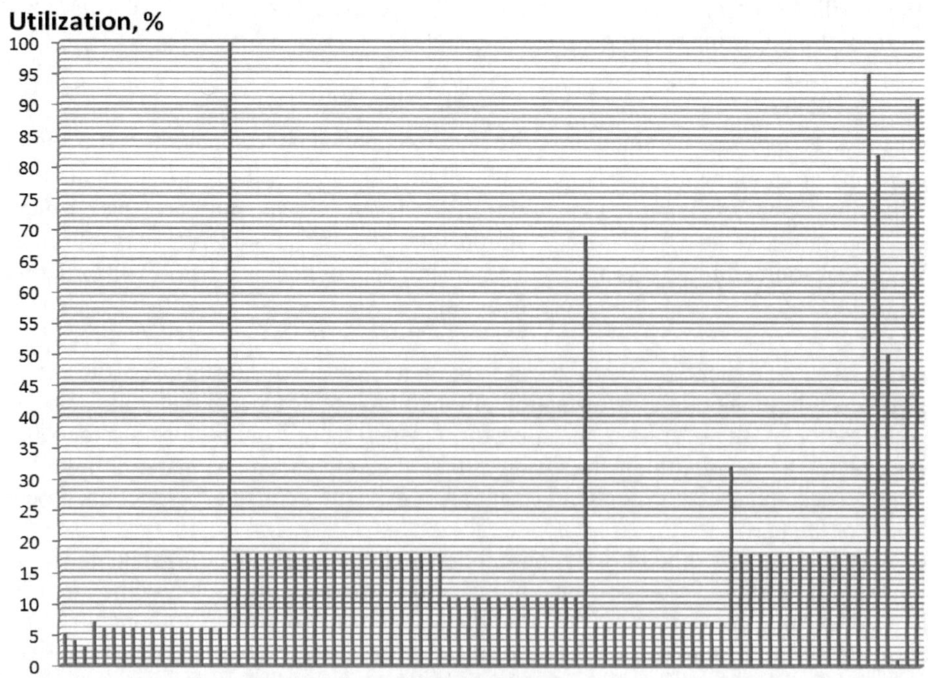

Figure 5. The diagram of workplaces loading of the manufacturing system project, obtained on the basis of the process model use.

The results of the analysis in figure 5 show that the projected manufacturing system is highly unbalanced - the lack of one jobs, with a clear surplus of others.

IOP Conf. Series: Materials Science and Engineering **448** (2018) 012008 doi:10.1088/1757-899X/448/1/012008

An analysis of the results obtained on the basis of the use of the process model made it possible to draw a conclusion about the impossibility of carrying out this program.

Next, a competitive resource model of the manufacturing system of the shop for making the product was built. A fragment of the competitive resource model is shown in figure 6, in which the dashed lines indicate the return loops of the assembled components of the product to the previous workplaces.

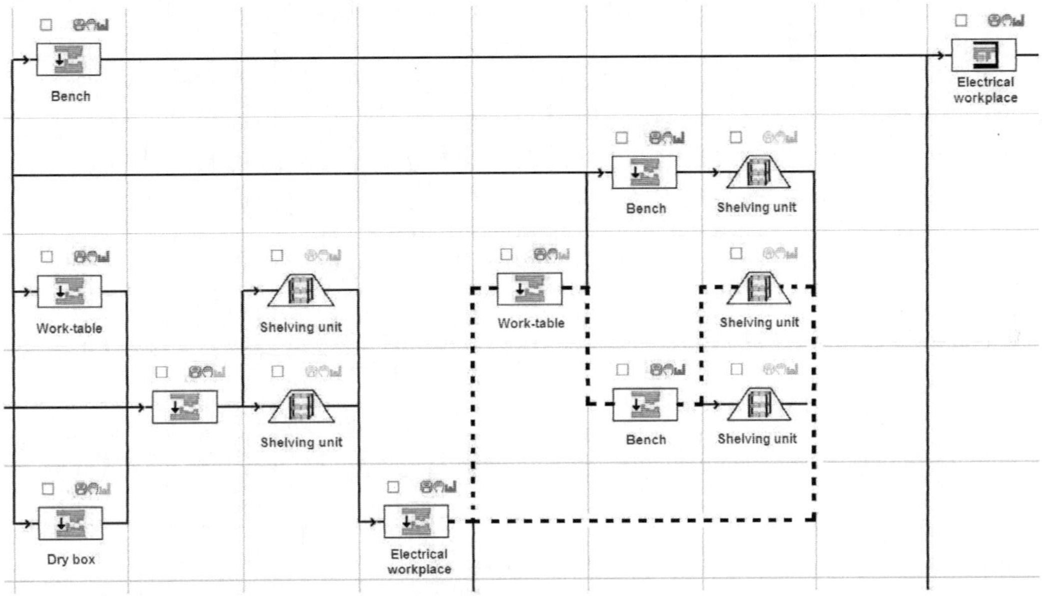

Figure 6. Fragment of the competitive-resource model of the shop manufacturing system.

Based on the results of the calculation using a competitive resource model, "bottlenecks" in the manufacturing system were identified. For their elimination, proposals were developed to adjust the specification of the workshop workplaces (Table 2).

Table 2 Adjust the specification of the workshop workplaces

Workplace name	Quantity of workplaces specified in the first stage	Quantity of workplaces specified in the second stage
Worktable	1	1
Bench	27	12
Dry box	1	1
Electrical worktable	49	36
Test bench	11	9
Automated workplace	–	3
Total:	**89**	**62**

Taking into account the developed proposals for adjusting the specification of workplaces, the design of the shop manufacturing system was adjusted and the performance indicators were calculated.

The calculated values of the coefficients of loading the workplaces of the manufacturing system for the product are shown in figure 7.

IOP Conf. Series: Materials Science and Engineering **448** (2018) 012008 doi:10.1088/1757-899X/448/1/012008

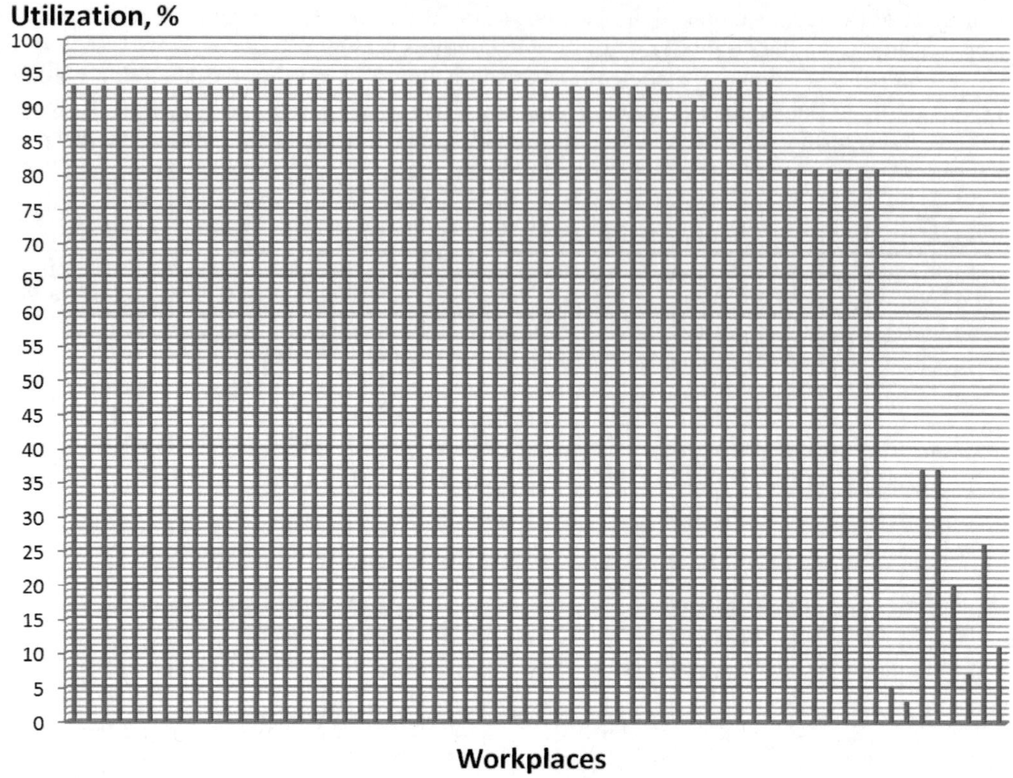

Figure 7. The diagram of loading workplaces of the corrected manufacturing system.

Calculations using a competitive resource model have shown that 62 workplaces are required to perform a given output volume with their relatively uniform loading, instead of the 89 planned.

Based on the analysis of the results obtained, a conclusion was made on the possibility of implementing the program.

4. Conclusions

To increase the efficiency of digital manufacturing systems, it is necessary to shift to the use of a competitive-resource information model of the manufacturing system, based on the use of personalized links of the different parts and elements of the manufacturing system, established by the processes of their production.

The introduced concept of a competitive-resource information model of the manufacturing system of parts production process has made it possible to expand the scope of determining the complexity of information links operating in the production process.

The developed competitive-resource information model of the manufacturing system of parts production process made it possible to determine the relationship between the operations of parts production and the elements of the manufacturing system.

Practical implementation of the model at the enterprise allowed reducing the number of workplaces at the enterprise for a given volume of output by 30%.

Acknowledgment

This research was made under financial support of Russian Ministry of education and science, project task N9.2731.2017/4.6

IOP Conf. Series: Materials Science and Engineering **448** (2018) 012008 doi:10.1088/1757-899X/448/1/012008

References

[1] Grigor'ev S N, Dolgov V A, Krasnov A V, Kabanov A A, Andreev N S 2015 *A method of technologic audit of technical re-equipment projects in aircraft production enterprises.* (*Russian Aeronautics*) vol 58 Issue 2 pp 244-250.

[2] Dolgov VA 2010 *Simulation of technological systems in custom production* (*Russian Engineering Research*) vol 30 Issue 9 pp 951-955.

[3] Yan Lu, KC Morris, Simon Frechette 2016 *Current Standards Landscape for Smart Manufacturing Systems. NISTIR 8107,* http://dx.doi.org/10.6028/NIST.IR.8107

[4] Kutin A, Dolgov V, Sedykh M, Ivashin S 2018 *Integration of different computer-aided systems in product designing and process planning on digital manufacturing* (*11th CIRP Conf. on Intelligent Computation in Manufacturing Engineering – CIRP ICME '17*) vol 67 pp 476 - 481 doi: 10.1016/j.procir.2017.12.247

[5] Kutin A A, Dolgov V A, Milkin V A 2016 *The method for the production potential evaluation of multinomenclature engineering production* (*Automation and modern technologies*) vol 6 pp 27-33

[6] Kutin A, Dolgov V, Podkidyshev A, Kabanov A 2018 *Simulation modeling of assembly processes in digital manufacturing* (*11th CIRP Conf. on Intelligent Computation in Manufacturing Engineering – CIRP ICME '17*) vol 67 pp 470 - 475 doi: 10.1016/j.procir.2017.12.246

IOP Conf. Series: Materials Science and Engineering **448** (2018) 012009 doi:10.1088/1757-899X/448/1/012009

Machining of different materials with abrasive waterjet cutting

Zs Maros

Institute of Manufacturing Science, University of Miskolc, H3515 Miskolc-Egyetemváros, Hungary

zsolt.maros@uni-miskolc.hu

Abstract. Aim of the paper is summarization of some result in the field of waterjet cutting on different materials, like steel, ceramic and polymer. Cutting experiments were accomplished to investigate the efficiency and the accuracy of the cut while machining of mentioned materials. From the result there are conclusions drawn up on the extent of technological parameters like feedrate, abrasive mass flow and the pressure.

1. Introduction

Modern waterjet cutting machine tools can adjust technological parameter of the cut almost automatically. Final user does not know why these parameters are applied. Usually these parameters are overestimated and not economical. That is why the knowledge of the choice of technological data is important nowadays as well. We need to choose the parameters with good efficiency, but according to the prescribed accuracy of the cut.

Efficiency and quality of the water jet cutting process can be characterised by different output parameters, like surface roughness of the machined surface, geometry of cutting kerf (depth, width, tapering). Depth of kerf is one of the most typical parameters for describing the efficiency of the cutting process. When making kerfing tests we do not cut through the material but produce a kerf only (Fig.1).

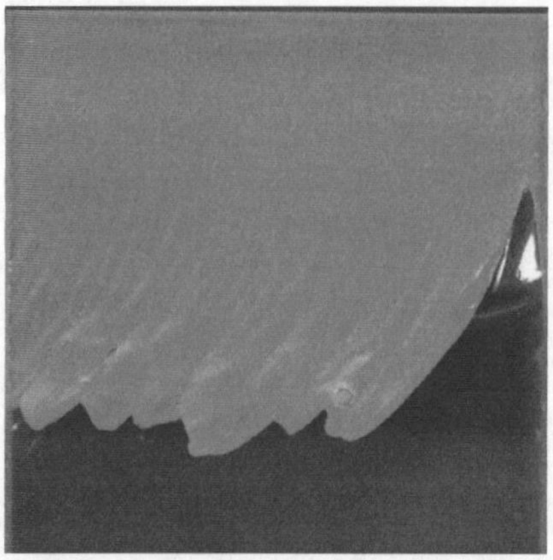

Figure 1. Kerfing test in glass material.

There are a lot of technological parameters which effect on the efficiency of waterjet cutting: diameter of the nozzle and the focusing tube, type of abrasive, stand-off-distance, pressure, traverse rate, abrasive mass flow rate etc. Recent research is focused to investigate the effect of the next process parameters: pressure, traverse speed (feed rate) and abrasive flow. Other parameters are considered constant and their value should be determined on basis of the knowledge exchange and technical possibilities of the project partners.

When cut through the material, one of the limit of application of waterjet cutting are the different accuracy problems of this machining method. Usually the form of the cutting kerf causes difficulties in the assurance of the prescribed accuracy. Form of the kerf is always very complex, but basically it can be considered like two tapered plains (Fig.2.). [1]

Kerf characteristics refer to the kerf geometrical features as kerf width, kerf taper and kerf depth. The surface topological features as represented by surface roughness and striation (or waviness) are also to as kerf characteristics. Generally, there are two types of kerfs, through cut and non-through cut. The typical geometry of a through cut kerf is shown in Fig. **Error! Reference source not found.**. It has a wider entry and its width decreases as the jet cuts into the material, so that a kerf taper is produced. In most cases the kerf is wider at the upper side (entering) then the lower side, where the jet goes out from the workpiece, so the taper can be convergent ($w_t>w_b$), divergent ($w_t<w_b$) or even divergent-convergent].

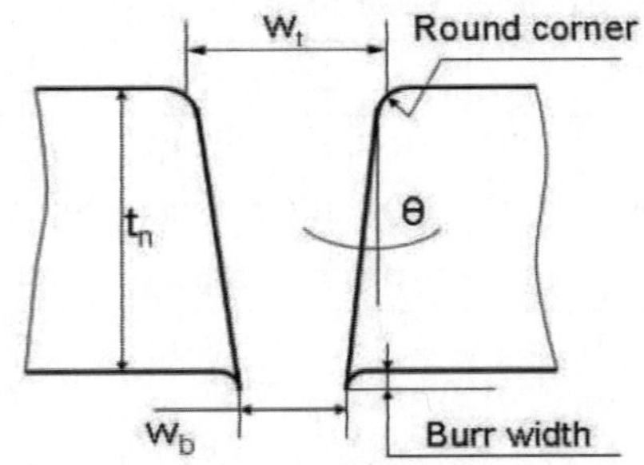

Figure 2. Taper of the kerf.

Kerf taper is normally expressed by kerf taper angle as [2]

$$\theta = \arctan(\frac{w_t - w_b}{2t_n}) \quad (1.1)$$

where w_t is the top kerf width, w_b is bottom kerf width, t_n is workpiece thickness for through cuts.

Taper of cut was investigated by more authors [3][4][5] but the conclusions are sometimes controversial.

Aim of recent research work was to get connections between the technological parameters and the taper of the kerf for finding cutting parameters with which we can get parallel cut surface.

2. Experimental conditions

We carried out two types of cutting tests. Kerfing tests (not through cut) were made for investigation of efficiency and cot through test were done for investigation of taper angle of the cut.

Main parameters and characteristics of the tests are shown in Table 1. During the experiments we changed the water pressure (p), the abrasive mass flow rate (m_a) and the feedrate (feed) speed (v_f) Other parameters can be considered constant.

Table 1. Characteristics and parameters of the cutting experiments.

Characteristic	Kerfing test	Cut through test
Type of machine tool	OMAX 120X	INNO PUMP-36HD
Workpiece material	S235JR construct. steel Polyvinylchloride PVC White marble	AlMgSi0,5 Al-alloy Polyethylene HDPE S355J2 construct. steel
Thickness of the material [mm]	is not characteristic	10
Diameter of water nozzle [mm]	0.41	0.25
Diameter of abrasive nozzle [mm]	1.07	0.8
Stand of distance [mm]	2	2
Abrasive mass flow rate [g/min]	320,550	28, 68
Water pressure [bar]	1380, 2500, 3590	2000, 2500, 3000, 3500
Feedrate speed [mm/min]	100, 120, 140, 160, 180, 200 1000, 1400, 1800, 2200, 2600 650, 750, 850, 950, 1050	50, 100, 150, 250, 300 100, 500, 1000, 1500, 2000 20, 40, 60, 80, 100

After the cutting experiments depth of kerf (k), upper width (w_t) and lower width (w_b) of the kerf were measured for determination of the depth and taper of the kerf.

Measurements of the kerf width were accomplished on a stereo microscope type ZEISS SteREO Discovery.V8 (Fig. 3).

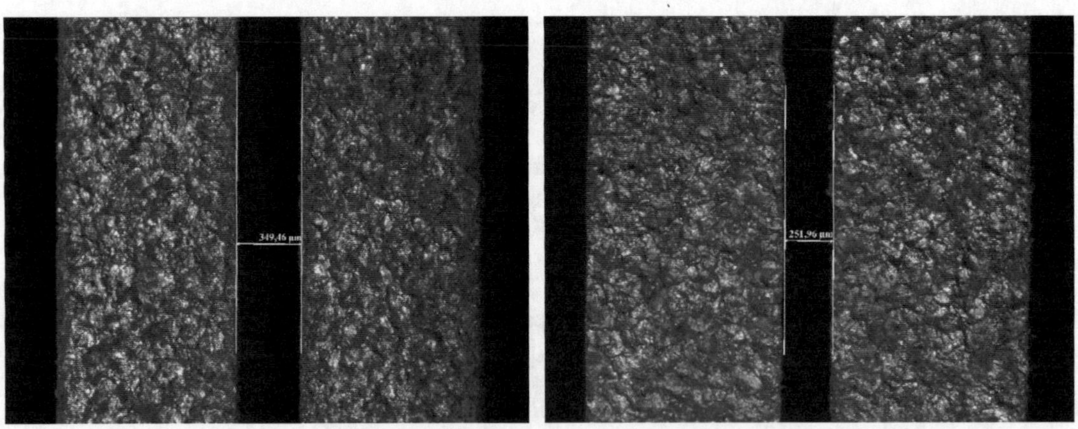

Figure 3. Measurement of the kerf with at the upper and the bottom side on microscope ZEISS SteREO Discovery.V8 (material: S355J2 steel).

3. Experiments for investigation of efficiency

Cutting through test (kerfing tests) were carried out on materials presented in Table 1. Constructional steel test part can be seen in Fig. 4. It is easy to see the different depth of the kerf machined by different technological parameters.

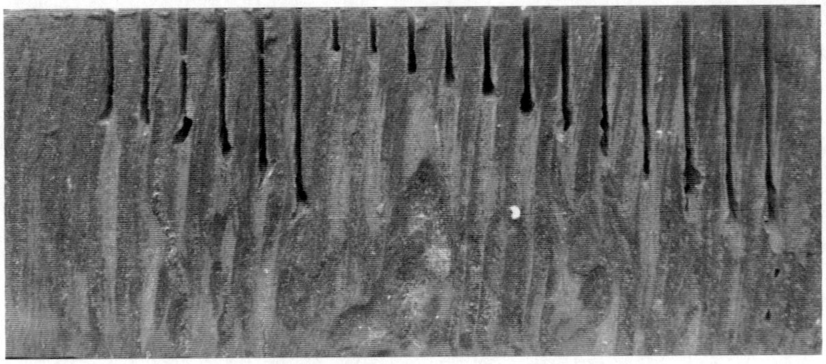

Figure 4. Kerfing test in S235JR steel.

Like an example. effect of the technological parameters on the depth of kerf (efficiency of the cut) is presented on Fig. 5 in case of PVC material.

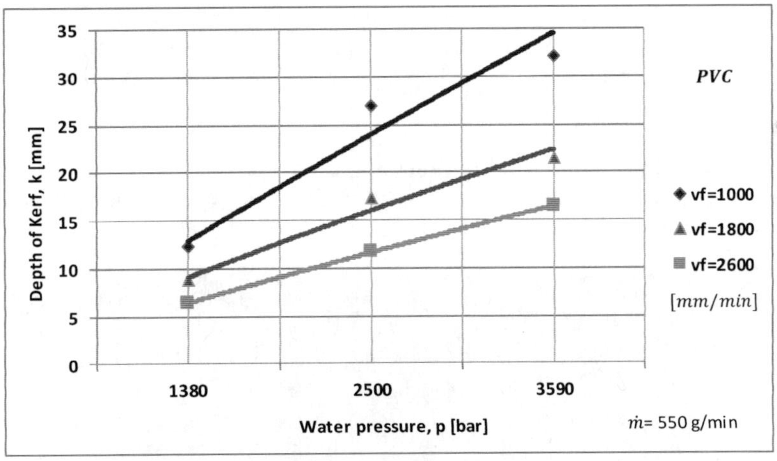

Figure 5. Influence of the pressure on the depth of kerf at different feed speed in PVC

From Fig 5 we can see that increase of the pressure increases the depth of cut and increase of the feed speed decreases it. Effect of the pressure is almost linear. At lower pressures the effect of the feed speed higher, then at high pressures. The inclination angles of the lines are higher at low pressure. This could be explained by the effect of the loading time. Increasing the feed speed the loading time is decreasing, so the higher speed of the abrasive grains – coming from the higher pressure - in the jet cannot affect totally the material removal rate. Similar trends were experienced at the marble and the steel as well.

Effect of the feed speed and the comparison of different materials can be seen on Fig. 6. From the figure we can see that the effect of the feed speed (traverse speed) is not linear, it shows exponential or power function characteristics. More interesting conclusions can be drawn by comparing the different materials

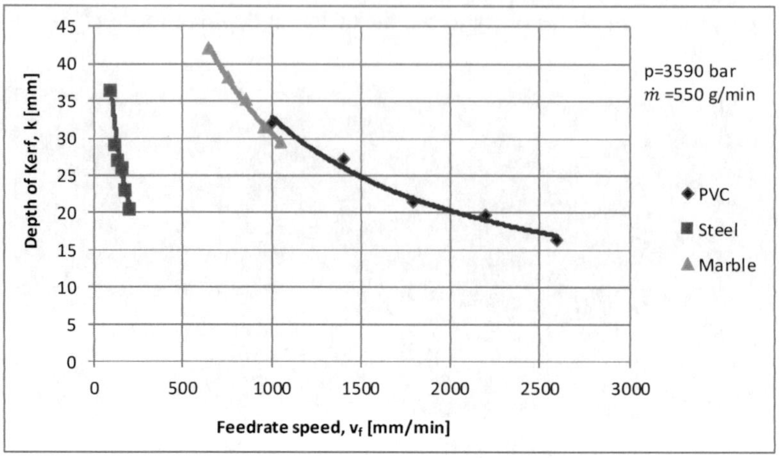

Figure 6. Influence of the feed speed on the depth of kerf at different materials

From Fig. 6 we can explain, that mist difficult to cat material is the steel and most easy is the marble. This is a little bit surprise, since aluminum alloy has the less hardness. Easier to cut a material, if the inclination angle of its curve on Fig 6 is smaller. This phenomenon can be explained with the different type of erosion process of the materials. Steel and aluminum alloy have so called ductile erosion, which is typical for ductile metals, while marble has a so called rigid erosion, which is characteristic for ceramics. If any of these two processes can be happened easy, the material is easy to cut.

4. Experiments for investigation of quality (kerf angle)

We carried cut through tests for investigation of taper angle of the kerf interpreted on Fig.2. On Fig. 7 we can see the test part from PVC.

Figure 7. Kerfing tests on PE material

Main parameters and characteristics of the tests are shown in Table 1. During the experiments we changed the water pressure (p), the abrasive mass flow rate (ma) and the feederate speed (v_f). After the tests we measured the with on the top and the bottom side of the kerf (Fig. 3.).

Effect of the feed speed and the water pressure on the width of the kerf (on the upper and the bottom side) can be seen on Fig. 8. From the figure easy to accept that the with of the upper side of the kerf decreases barely in function of the feed speed. At the bottom side of the kerf the decrease is much higher. This cause the taper angle of the cut kerf.

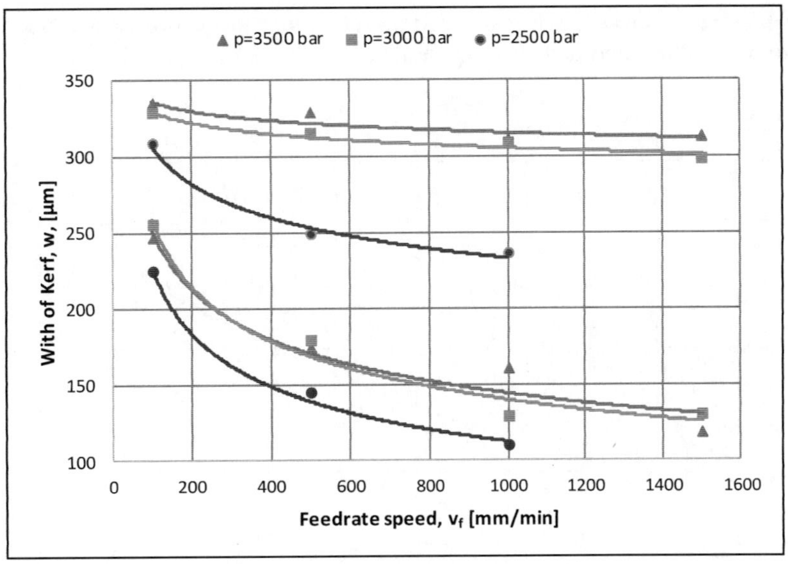

Figure 8. Effect of feed speed and the water pressure on the width of the kerf at the top and bottom side in case of PE material

Increase of the pressure basically increases the width of the kerf (Fig. 8.). It is interesting to observe that change from 3000 to 3500 bar causes minimal change in the width but change from 2500 to 3000 bar causes high change in the value of widths.

From the widths on the top and bottom side the value of the kerf angle can be calculated (1.1). Change of the calculated angels in function of feed speed and pressure can be seen on Fig. 9.

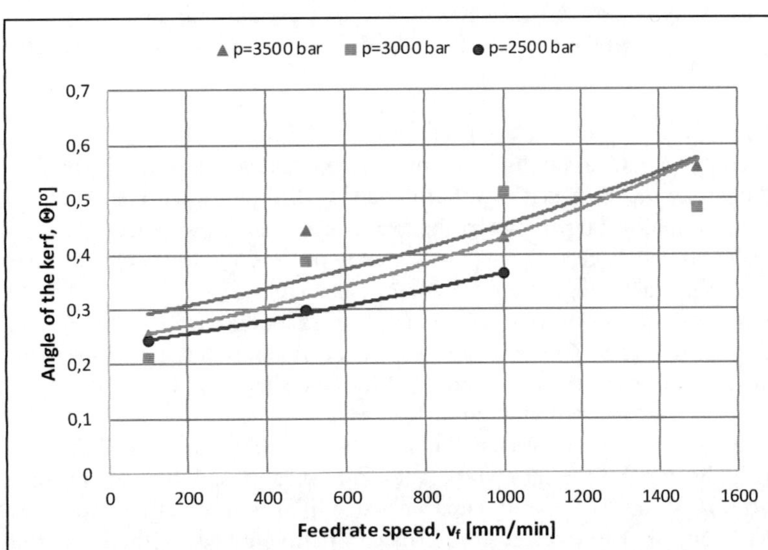

Figure 9. Effect of feed speed and the water pressure on the angle of the kerf in case of PE material

Increase of the feed speed increases the angle of the kerf, that is increases the inaccuracy of the cut. The effect of the pressure is not completely clear according to the literature [6][7][8][9]. Usually the increase of the pressure decreases the angle of the kerf while the width of the kerf increases. This statement cannot be substantiated by these experiments.

On Fig 10. we can compare the characteristics of different materials.

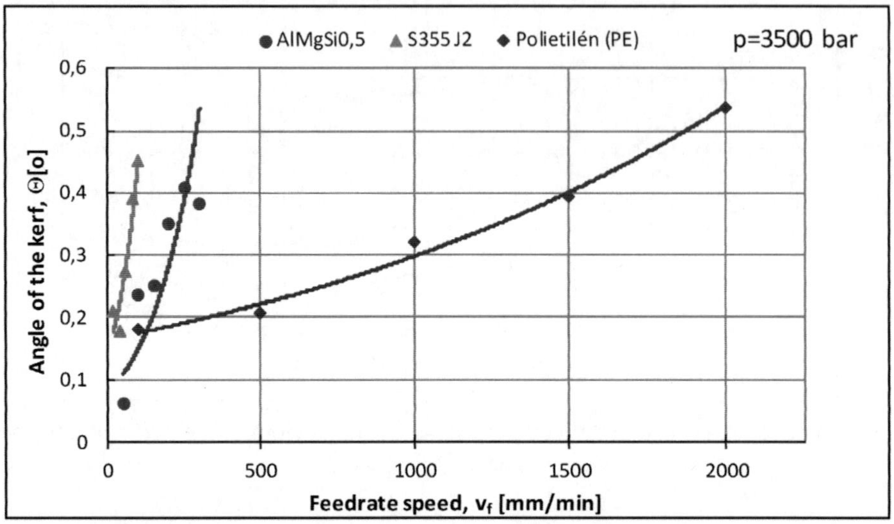

Figure 10. Effect of feed speed on the angle of the kerf in case of different material

On base of Fig 10 we can say that taper angle is the biggest problem at constructional steel from the investigated materials. In all cases the increase of the feedrate increases the angle of the kerf (inaccuracy), but slopes of the curves are very different. PE has the less difficulties. To assure parallel cut (kerf) for all materials is very difficult, it seems to be possible only at very low feedrate, which is not economical solution. As flat the curve of the material run, as easy to produce relatively accurate cut.

5. Conclusions

The following conclusions can be drawn from the experiments carried out for the investigation of cutting depth and taper angle of the kerf at waterjet cutting different materials:

- From the technological parameters the increase of the feedrate speed clearly and significantly reduces the depth of kerf that is the efficiency of the cut. However, increasing of feed speed decreases the machining time, so it is cost effective.
- Effect of the water pressure is clear. The pressure increases the depth of cut. Same effect has the abrasive flow rate. The pressure can affect efficiently if the feed speed is relatively small, because in this case there is enough time for the interaction between the jet and the workpiece. From point of view of efficiency, the pressure and the mass flow rate should be chosen as high as it is allowed by the different restrictions.
- From results of cut different materials it can be established, that steel is the most difficult to cut. Cutting of the PVC and the marble are similar in a given interval, but the PVC can be cut with very high feed speed too. This can be explained by the easy rigid erosion of the marble an easy ductile erosion of the PVC.
- On base of the cutting experiments we can point out that the accuracy of the cut surface is basically defined by the size and form of the cutting kerf. Cutting kerf is usually tapered, orientation of the taper mainly depends on the quality of machined material, at metals it is usually convergent.

- The extent of the kerf angle basically depends on the technological parameters. The feed speed (v_f) clearly increases the taper angle of the kerf. The effect of the water pressure (p) and the abrasive mass flow rate (m_a) could strongly effect on the taper mainly at lower feedrates, when there is enough loading time for the abrasive particles to remove the material from the workpiece.
- Comparing the cut of different materials, can be established that taper angle of steel is higher than at aluminum alloy or PE polyethylene. In all cases the increase of the feedrate speed increases the angle of the kerf (inaccuracy), but slopes of the curves of each material are very different. Near parallel cut can be reached in case of PE polyethylene.

From the results it can be established that the extent of taper angle and the depth of the kerf basically depends on the quantity of the energy-input and can be controlled by it. So future researches will have oriented to describe connections between the loading energy and the cut kerf geometry.

References

[1] Momber A W and Kovacevic R 1998 Principles of Abrasive Water Jet Machining *Springer-Verlag London Limited* p 394

[2] Wang J 1999 A Machinability Study of Polymer Matrix Composites Using Abrasive Waterjet Cutting Technology, *Journal of Material Processing Technology* **94** pp 30–35

[3] Öjmertz C 1997 A Study on Abrasive Waterjet Milling, *Thesis for the Gegree of PhD Chalmers University of Technology Göteborg Sweden* p 81

[4] Chung Y, Geskin, E S, Singh P 1992 Prediction of the Geometry of the Kerf Created in the Course of Abrasive Waterjet Machining of Ductile Materials *Jet Cutting Technology, Kluwer, Dordrecht* pp 525-541

[5] Arola D and Ramulu M 1996 A Study of Kerf Characteristics in Abrasive Waterjet Machining on Graphite/Epoxi Composite *Journal of Engineering Materials and Technology, Transaction of the ASNE* **118(2)** pp 256-65

[6] Gudimetla P, Wang J, Wong W 2002 Kerf formation analysis in the abrasive waterjet cutting of industrial ceramics *Journal of Materials Processing Technology* **128** pp 123-29

[7] Srinivasu D S, Axinte D A, Shipway P H, Folkes J 2009 Influence of kinematic operating parameters on kerf geometry in abrasive waterjet machining of silicon carbide ceramics *International Journal of Machine Tools and Manufacture* **49** pp 1077-88

[8] Annoni M, Monno M, Vergari A 2001 The Macrogeometrical Quality of the Kerf in the AWJ Process Parameters Selection *2nd International Conference on Water Jet Machining* Kracow Poland, pp 56-63

[9] Shanmugam D K, Wang J, Liu H 2008 Minimisation of kerf tapers in abrasive waterjet machining of alumina ceramics using a compensation technique *International Journal of Machine Tools and Manufacture* **48** pp 1527-34

IOP Conf. Series: Materials Science and Engineering **448** (2018) 012010 doi:10.1088/1757-899X/448/1/012010

Improving the efficiency of CNC machine tools with multi-pallet systems in machine-building manufacturing

A A Kutin[1,*], V A Dolgov[1], A A Kabanov[2], I V Dazuk [1] and A A Podkidyshev [1]

[1]Production Engineering Department of the Moscow State University of Technology STANKIN, Vadkovskij per. 1, Moscow, 127994, Russian Federation
[2]Department of Aerospace Engineering of the Moscow Aviation Institute (National Research University), Volokolamskoe shosse, 4, Moscow, 125993, Russian Federation

* Corresponding author. Tel.: 7-499-973-30-80; fax: +7-499-973-3080. E-mail address: aa.kutin@stankin.ru

Abstract. The paper discussed the problems of effective machining of workpieces in small batches on CNC machines with multi-pallet systems, combined into flexible manufacturing cells (FMC). In the paper was reviewed timeline of such manufacturing cells. It is shown that in order to increase the efficiency of the CNC machines with multi-pallet systems into each flexible manufacturing cell is necessary to pick up different workpieces to be machined, based on the duration of their machining cycles and installation in pallets. It allows providing machines downtime minimization. To calculate the machining cycles of workpieces in a flexible manufacturing cell, it is proposed to use the graph-analytical method, which gives an enlarged estimate, and the method of simulation of manufacturing processes, which allows obtaining a refined decision.

1. Introduction.

At present, various software and hardware for the execution of technological operations are widely introduced at domestic machine-building enterprises.

Increasing the level of technological processes automation leads to the stability of product quality, reduce the time of technological cycles of manufacturing products, and improve the efficiency of production in general [1].

Despite the great potential of using flexible manufacturing cells in technological processes, the effectiveness of their application in practice is not enough [2].

As a rule, technical problems of FMC application are mostly solved. Organizational issues are considered to a lesser extent. The effectiveness of FMC using is largely determined by the organization and maintenance factors [3].

In this paper, has been investigated an integrated approach for determining the composition of automation devices and the interaction of elements within FMC in the context of the designing of manufacturing systems.

2. Initial data for the analysis of the machines with multi-pallet systems effectiveness and statement of the problem.

As an example of a manufacturing system, a workshop of small-scale production with CNC machines equipped with multi-pallet systems and automatic pallet changers (APC) has been selected.

This solution for the automation of flexible manufacturing cells involves the completion of machines with pallet magazines and automatic feeders and pallet changers. Within the same cell can be machines equipped with APC, and without APC. Data are given for the milling operations of this workshop.

The composition of the equipment and the automation devices been used are shown in Table 1.

Table 1. Composition of milling machines

№	Type of equipment	Symbol	Automation devices
1	3-axis Milling machines designed for roughing	M3 «R»	1) CNC system. 2) Tool magazine and devices for automatic change of cutting tools 3) Devices for automatic measurement of the actual tool dimensions and correction of the control program (automatic binding devices). 4) Devices (probes) for automatic measurement of the workpiece being machined 5) The device of light signaling about a condition of the machine tool (a three-color traffic light). 6) Automatic pallet changer with a part in the working area of the machine with a pallet magazine. Storage capacity - 7 pallets, for item 2 and item 4.
2	3-axis Milling machines designed for basic finishing, equipped with APC	M3 «F» + APC	
3	5-axis Milling machines (3-axis + discrete machining on 2 additional axis)	M5 «3+2axis»	
4	5-axis Milling machines (3-axis + discrete machining on 2 additional axis) equipped with APC	M5 «3+2axis» + APC	
5	5-axis Milling machines models "S" (synchronous machining on 5-axis axes)	M5 + «S»	

Based on these types of equipment and automation devices, and on preliminary calculations of the number and types of the necessary machines, flexible manufacturing cells are formed, the structure of which is shown in Table 2. Figure 3 shows a fragment of the workshop plan with the placement of 4 units of FMC 3 (4 machines per cell with one operator) of the FMC of the milling operation.

Table 2. Structure of the flexible manufacturing cell of the milling section (basic version)

№	Types of FMC	Quantity of FMC	Composition of FMC	
			Equipment	Machine Operator
1	FMC 1	2	M3 «R» M3 «R»	CNC machine operator
2	FMC 2	4	M3 «F» + APC M3 «F» + APC	CNC machine operator
3	FMC 3	4	M3 «F» + APC M3 «F» + APC M3 «F» + APC M3 «F» + APC	CNC machine operator
4	FMC 4	4	M5 «3+2axis» M5 «3+2axis»	CNC machine operator
5	FMC 5	4	M5 «3+2axis»+ APC M5 «3+2axis»+ APC	CNC machine operator
6	FMC 6	2	M5 + «S» M5 + «S»	CNC machine operator

It is required to determine the composition and structure of the FMC taking into account the specification of the workpieces to be machined:

1) the optimal composition of the components of each FMC included in the flexible manufacturing system (FMS): the number of machines in the cell; number of machines equipped with APC;

2) distribution of the workpieces to be machined in the FMC.

3. Description of the methods used in the development of project solutions.

Frequently, it is not possible to use the traditional methods of calculation, because:

1) Algorithms of a FMS components interaction can differ significantly; therefore general used equations give an incorrect result. For example, the principle of forming a machining floor-to-floor time involves the sequential execution of of the machine adjustment for the machining of workpieces. It is mean that during the adjustment the machine does not machining of the workpiece. The time loss factor caused by organizational reasons is usually assigned based on statistical data for analogues of manufacturing systems.

2) The traditionally performed calculation of the complex coefficient (the ratio of the machine operating time to the working time of the worker) does not take into account the organizational losses of time and gives the theoretical marginal productivity of the cell.

To improve this situation it is proposed to use the following methods [4]:

1) method of graphical designing of cycle diagrams (graph-analytical method) of theFMC operation for a relatively small time interval - for analyzing the operation of a multi-station flexible manufacturing cell (Figure 1);

2) method of simulation to analyze the joint work of flexible manufacturing cells in a flexible manufacturing system for a long period of time. It is realized after the definition of strategies for forming the structure of the FMC and fixing the appropriate machining parts. It is necessary to analyze the work of the FMC for a long period of time, as well as to analyze their joint work taking into account possible deviations from the established norms of time. Based on the complexity of the manufacturing under consideration, due to the large number of elements of the manufacturing system and a wide range of logics of their interaction, the simulation model must have more flexibility, which can be realized with the help of agent or hybrid simulation techniques [5].

4. Practical implementation of the proposed integrated method for evaluating design solutions when creating flexible manufacturing systems.

Method 1

First, using the graph-analytical method, the efficiency of individual FMC is analyzed depending on the composition of cells and the type of workpieces delivered to them. Figure 1 shows a cycle diagram of parts "A" machining (conditionally medium cycle parts, processing time 7 min.) and "B" (conditionally short cycle parts, processing time 3 min.) of the selected types of parts on the milling operation in a flexible manufacturing cell consisting of two machines.

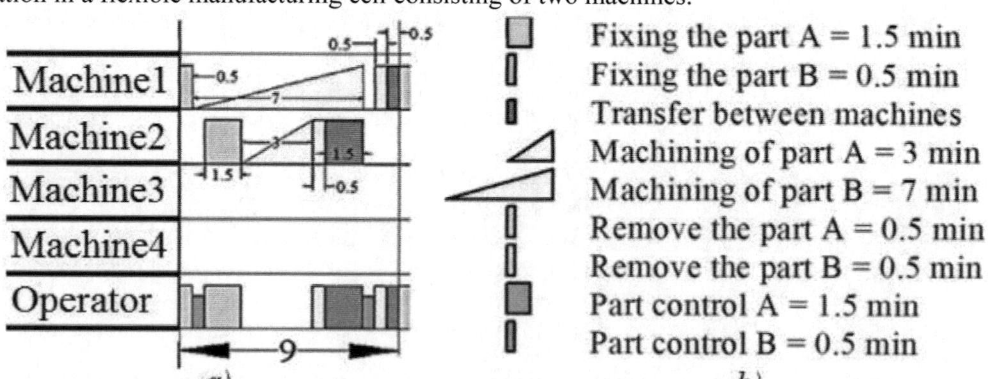

Figure 1. Graph-analytical method: a) a cycle diagram of operation of two-machine FMC while machining of parts "A" and "B"; b) the parameters of the production operations used in the cycle diagram

We perform a comparative analysis of FMC configurations. In this analysis the quantities of batches of parts are taken to be the same.

Figure 2 shows the cycle diagram for the machining of parts "A" and "B", built for a comparative analysis of the efficiency of various options for the composition of the FMC components, use of APC for each of the machines: two-machine FMC (view "a"), three-machine FMC (view "b"), four-machine FMC (view "c").

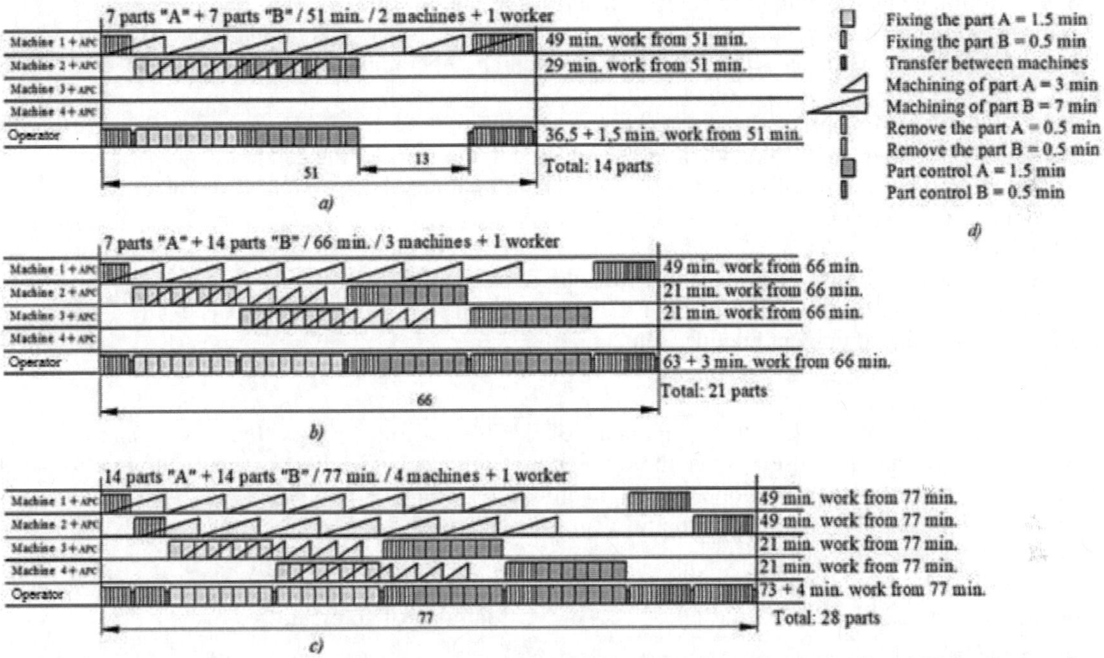

Figure 2. Cycle diagrams of the FMC operation for the machining of parts "A" and "B" using each machines with APC: a) two-machine FMC, b) three-machine FMC; c) four-machine FMC; d) the parameters used in the cycle diagrams.

From the diagrams shown in figure 2, one can see that the four machines FMC has the highest productivity - 0.36 parts per minute. For the machining of a minimum batch of 50 parts "A" and 50 parts "B" will take about 278 minutes. In this case, the shutdown time of the machine tools adjustment will be about 30 minutes. Parallel to this, loading of pallets into the shop of APC (28 minutes) and assembly of the tool magazine of the machine tool is carried out. Thus, the processing time is almost greater than the machine's shutdown time for setup.

Analysis of the cycle diagrams (see Figure 2) showed:

1) if it is not possible to provide external adjustment of tools, as well as loading control programs into the machine, setting up pallets, the tool by one worker, the time for stopping the machines will be commensurate with the machining time. In this case, the gain from increasing the productivity due to the application of the APC systems is offset by the increased time for setting up the cell, i.e. the cost of more expensive equipment is not justified;

2) the maximum theoretical performance of the FMC is obtained for the 4-machine configuration of the cell (Figure 2 "c"), the same option is most beneficial from the point of view of the minimum requirements in human resources;

3) the 4-machine configuration of the cell (Figure 2 "c") will be difficult to implement in practice, since the operator will not be able to work without intervals for a long time, besides, in this case, there are no reserves for smoothing the inevitable deviations from the established time. As a result, equipment in such a cell will be idle due to the operator's employment with the maintenance of other machines;

4) rational organizational and technological solutions for the formation of the FMC are:

- three machine configuration of the FMC provided that two machines are loaded with "A" type parts (with a longer machining cycle) and the remaining one with "B" type parts,
- two machine configuration of the FMC, in which both machines are engaged in processing short-cycle parts of type "B". The final composition of the FMC in the flexible manufacturing system (FMS) will be determined by the ratio of the output quantities of the "A" and "B" type parts.

Currently, methods of external adjustment of tools have been worked out and successfully used, the interfaces of CNC systems have become much simpler, and operational control systems have appeared [6].

This allows implementing the above-mentioned techniques for increasing the efficiency of the multi-step FMC and make recomendations for organizing the work of the cell:

1) during the machining of parts, the CNC machine operator keeps track of the machines. At this time other worker prepares the tools;

2) during the re-adjustment of machine for new part, the service personnel, workers and operators are involved, while simultaneously loading the tool, devices, programs;

3) using of group technologies in small-scale production also significantly reduces the amount and laboriousness of the equipment readjustment.

Method 2

To analyze the joint work of the components of flexible manufacturing cells, as well as to analyze the joint work of the FMC itself, a simulation model of the mechanical workshop of the multiproduct small-batch production was developed. A fragment of the simulation model's plan of FMC 3 (see table 2) is shown in Figure 3. According to the results of the method 1, for FMC 3 a 4-machine FMC layout was selected (Figure 2 "c").

It is required to determine the optimum composition of the FMC, namely: the optimal number of workers for this flexible manufacturing cell; the optimal number of automation tools is APC.

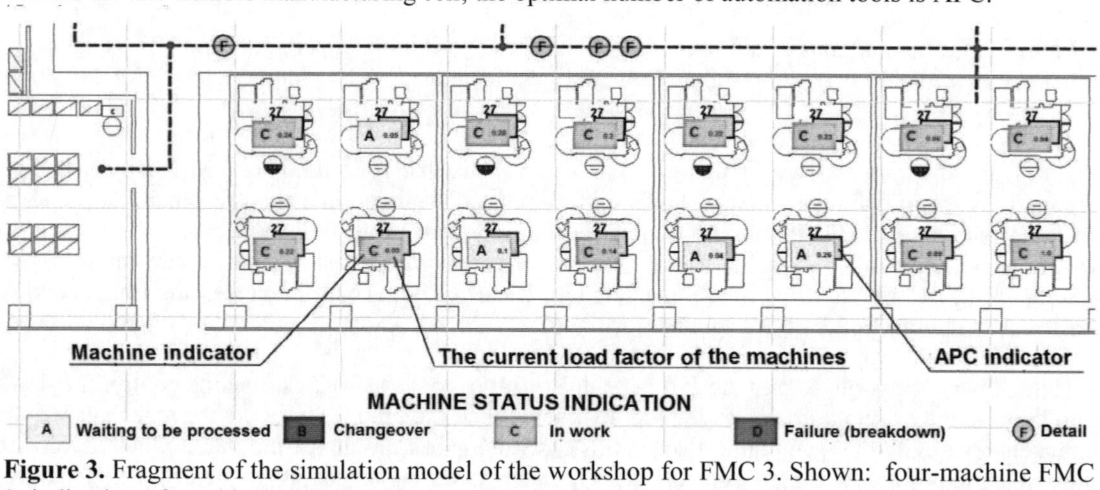

Figure 3. Fragment of the simulation model of the workshop for FMC 3. Shown: four-machine FMC 3; indication of machines status, machines utilization; indication of APCs for machines; material flow indication.

The agents of the simulation model are the following elements of the manufacturing system of the workshop: parts, technological equipment (CNC machines, thermal furnaces, workbenches, etc.), buffers (storage tanks), automatic pallet changers, operators, etc. Each of the agents has its own logic of functioning and interaction with other agents of the model. Also in the model there are: the schedules of the elements of the manufacturing system, the schedules of repair and maintenance of equipment, the probability of equipment failure, the parts machining, the schedules of the operators, multi-stop service and other.

Based on the results of the cycle diagrams of operation of the FMC of the first stage of the analysis (see Figure 2), the following logic is adopted for the operation of a multi-cell with APC.

When a batch of N parts arrives for machining, they are loaded into the APC free equipment store until the magazine of the M capacity is fully loaded (7 positions in the case under consideration). If N <M, then the whole batch of N parts or the remainder of the batch with the volume L is loaded into the store - the number of unhandled parts from the lot N, where L <M.

Further, in the automatic mode, the parts loaded into the store are machined to the full exhaustion of raw parts, after which an operator is called on the signal lamp to unload the finished parts from the APC drive and load the next batch of new parts. Thus, for the case under consideration, the operator with the APC is deliberately excluded when the machine is running, which is done to avoid mistakes made by the operator, the probability of which increases many times when mixing short-cycle parts from different batches within the same processing cycle.

The plan of the experiment provides for the development of five modeling of the script (Table 3). Of the script 4, 5 are used the results of modeling the first three of the script.

Table 3. List of modeling plans and their characteristics

№	Name of the script	Characteristic of the script	Purpose of the Script development
1	«Basic»	The initial FMS in the basic version of the FMC configurations	Verification of the model, determination of the capabilities of the initial FMS for the execution of the production program
2	«Maximum»	Elimination of multi-stop service	Determination of the maximum performance of FMS
3	«Possible solution 1»	The layout of the cells in accordance with the recommendations of the first stage of the analysis - FMC: 2 machines with APC	Determination of rational FMC layouts
4	«Possible solution 2»	Exclusion of APC with negligible effect according to scenario 3	Optimization of the FMC and FMS structure
5	«Without automation devices»	The complete exclusion of APC systems	Determination of the effect of using automation devices

Examples of output data for the analysis of the joint operation of the FMC components on the basis of experiments of the simulation model are shown in figures 4 and 5.

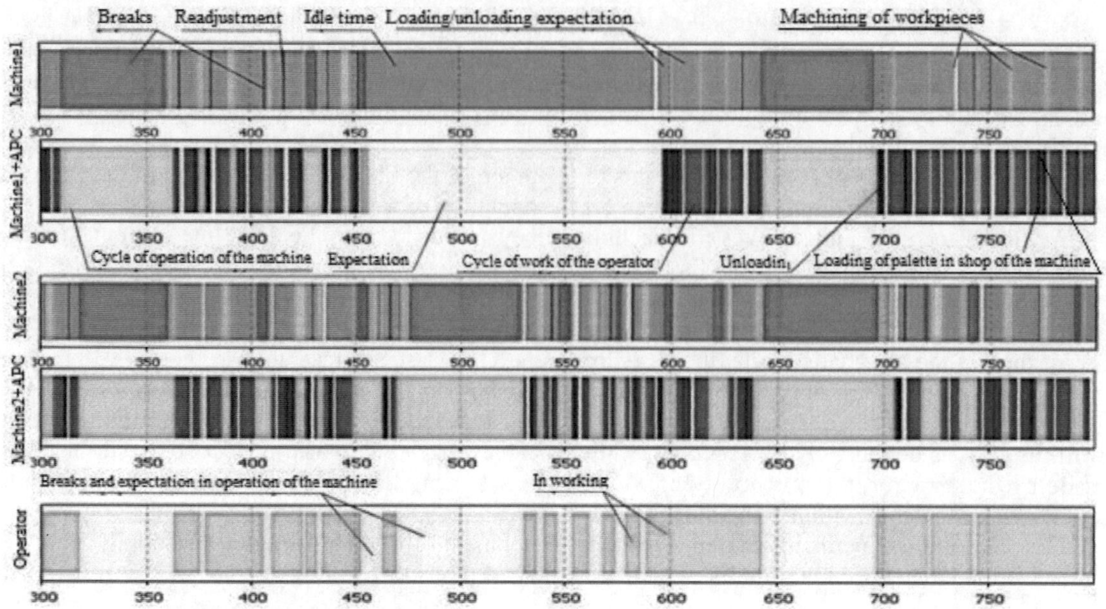

Figure 4. Cycle diagrams of the joint operation of the components of two-machines FMC during the simulation period from 300 to 800 hours. Machining of parts "A".

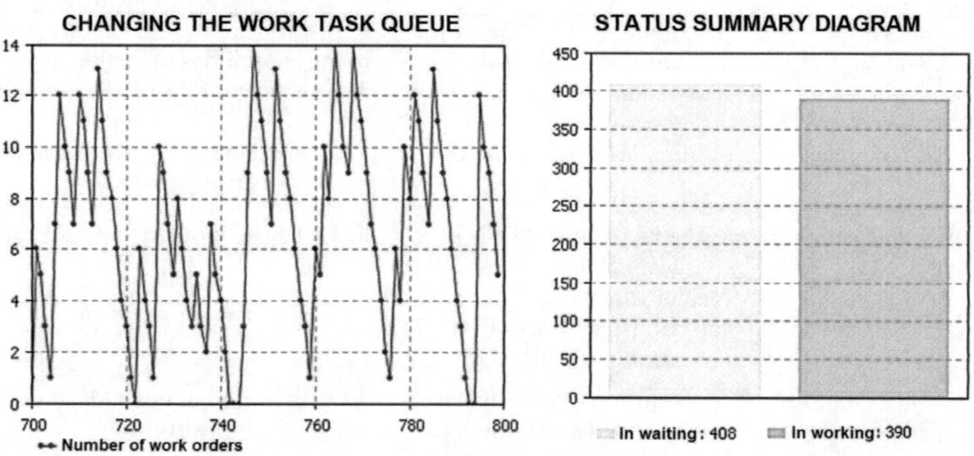

Figure 5. Real-time analytical reports for assessing the performance of a operator in two machine configurations of a FMC

5. The simulation results.

A comparative evaluation of the simulation results according to the developed scripts is shown in Table 4 (the results are shown for a FMC 3).

Table 4. Comparative characteristics of simulation results according to the developed scripts for a flexible manufacturing module of four 4-machine FMC 3

Model scripts	Script 1 «Base»			Script 2 «Max»			Script 3 «Opt 1»			Script 4 «Opt 2»			Script 5 «W/o APC»		
Components of FMC	M	W	A	M	W	A	M	W	A	M	W	A	M	W	A

Model scripts	Script 1 «Base»			Script 2 «Max»			Script 3 «Opt 1»			Script 4 «Opt 2»			Script 5 «W/o APC»		
FMC 3_1	4	1	4	4	4	0	4	2	4	4	2	4	4	2	0
FMC 3_2	4	1	4	4	4	0	4	2	4	4	2	4	4	2	0
FMC 3_3	4	1	4	4	4	0	4	2	4	4	2	4	4	2	0
FMC 3_4	4	1	4	4	4	0	4	2	4	4	2	4	4	2	0
Total	16	4	16	16	16	0	16	8	16	16	8	16	16	8	0
Machining of all parts, days	119.5			73.4			77.8			78.4			95.4		
Machine utilization, %	12.062			19.938			18.000			15.063			15.063		
Managing of material flow	No			Yes			Yes			Yes			Yes		
Changing the composition of other related FMC	No			No			No			Yes			No		

M – Machine; W – Worker; A – APC

As a result of the first script, it is established that the production program of the workshop cannot be performed due to low utilization of equipment, which was the result of the four-machine FMC.

To determine the maximum performance of the FMC and calculate the necessary machines for the production program, an experiment was conducted in the script 2, taking into account the recommendations of method 1 for fixing the selected types of parts type "A" and "B" for machines (see Table 4, line "Managing of material flow").

Taking into account the results of modeling the script 1 and 2, changes were made to the FMC structure. After that, simulation modeling of production processes of a FMC was carried out within the framework of the script 3. For the manufacturing system under script 3, one of the possible design solutions was obtained, characterized by a significant reduction in the production cycle of manufacturing the entire volume of parts: 77.8 days against 119.5 in the "script 1" and an increase in machine utilization of 18.000% . Further optimization of the manufacturing system and its analysis within the framework of the script 4, make possible to reduce the number of operators and exclude inefficiently used automation devices with a slight increase in the production cycle (78.4 days). At the same time, inefficiently used APC were excluded from the FMC 5 (see Table 2, line "Changing the composition of other related FMC") associated with the FMC 3 with the general material flow.

For a comparative evaluation of the effect of the introduction of automation devices, the 5th series of experiments was performed with the complete exclusion of APC.

6. Conclusions

1) Using the proposed approach for sharing the graph-analytical method and the simulation method significantly improve the quality of design solutions, as well as to reduce the complexity and time for the design of the manufacturing system by means of:

- simulation modelling of the joint operation of the FMC as part of a flexible manufacturing system;
- a significant limitation of possible solutions to the flexible manufacturing system, by identifying possible FMC structures and processing strategies within the framework of graph-analytical analysis of the operation of individual FMC.

2) This allowed us to solve the following practical problems in the presented flexible manufacturing system:

- clarify the FMC structure, including: machine tools, operators, APC;
- determine the optimal ratio of the number of operators and other workers;
- to develop recommendations on fixing the parts for the corresponding FMC;
- to develop recommendations on the layout of the FMC.

3) The proposed approaches can also be used in the simulation of other production systems characterized by a complex interaction of its elements.

Acknowledgment
This research was made under financial support of Russian Ministry of education and science, project task N9.2731.2017/4.6

References

[1] Grigor'ev SN, Dolgov VA, Krasnov AV, Kabanov AA, Andreev NS2015 *A method of technologic audit of technical re-equipment projects in aircraft production enterprises*. (*Russian Aeronautics*) vol 58 Issue 2 pp 244-250.
[2] DolgovVA 2010*Simulation of technological systems in custom production*(*Russian Engineering Research*)vol 30 Issue 9 pp 951-955.
[3] Monostori L. Cyber-physical production systems: Roots, expectations and R&D challenges.47th CIRP Conference on Manufacturing Systems. Procedia CIRP 17 (2014) P.9-13.
[4] Kutin A, Dolgov V,Podkidyshev A, Kabanov A 2018*Simulation modeling of assembly processes in digital manufacturing* (*11th CIRP Conf. on Intelligent Computation in Manufacturing Engineering – CIRP ICME '17*) vol 67 pp 470- 475 doi: 10.1016/j.procir.2017.12.246
[5] Borshcev A., Filippov A. FromSystem Dynamics and Discrete Event to Practical Agent Based Modeling: Reasons, Techniques, Tools. In: Proceedings of the 22nd International Conference of the System Dynamics Society, Oxford, 2004.
[6] Kutin A, Dolgov V,Sedykh M,Ivashin S 2018*Integration of differentcomputer-aided systems in product designing and process planning on digital manufacturing* (*11th CIRP Conf. on Intelligent Computation in Manufacturing Engineering – CIRP ICME '17*) vol 67 pp 476 - 481doi: 10.1016/j.procir.2017.12.247.

IOP Conf. Series: Materials Science and Engineering **448** (2018) 012011 doi:10.1088/1757-899X/448/1/012011

Application possibilities of the Big Data concept in Industry 4.0

P Dobos, P Tamás, B Illés and R Balogh

Institute of Logistics, University of Miskolc, Hungary

E-mail: peterdobos30@gmail.com

Abstract. Nowadays the flexibility and specific cost of manufacturing have a relevant role in the competitiveness of the companies. In our opinion, the most important objective of Industry 4.0 is the realization of intermittent manufacturing at mass production's productivity and specific cost. This aim can be only reached by creating more complex manufacturing systems. We can collect new types of data that enable the improvement of product and process service quality. Due to the big amount of collected data's is essential to have an application possibilities of the Big Data to be able to evaluate and transform the data's in usable information. This paper introduces the essence of Industry 4.0, as well as the new application possibilities of the Big Data for increasing the efficiency of logistics processes. Possible research directions for overcoming challenges are also presented.

1. Introduction

The satisfaction mode of the unique customer needs (quality, deadline, cost) significantly influences the competitiveness of companies. In production area, the most important objective of Industry 4.0 is the satisfaction of unique customer needs with the specific costs of mass production [1]. Complexity of manufacturing processes significantly increases with expansion of the product types, which requires elaboration of new solutions in the fields of development and quality assurance of processes. This provides several basic research topics [2,3]. This objective can be seen futuristic, but with the continuous improvement of the technology we get closer and closer to the target. The presence of cyber-physical systems and the big data conception have created new possibilities in the improvement of product and service quality [4]. We can collect and process data for the manufacturing processes that has not been gathered so far. We can determine the correlations among data, in order to make several conclusions regarding failure forecast of products, material handling equipment or the technological equipment. In addition, rationalization of production system operation can become more effective. To be able to reach the mentioned efficiency increase, we have to reveal the data types to be recorded and determine the technological consequences for data recording, as well as elaborate possible methods and procedures for evaluation and encroach of the data. This paper introduces the characteristics of the industrial revolutions development and covers the most important tools of the 4th industrial revolution. Afterwards possibilities of the Big Data practices will be presented, also their advantages to be expected.

2. Development of industrial revolutions

Basically, industrial resolutions are related to social, economic and technological changes since the appropriate economic and social environment is necessary for the invention and spread of technologies.

We can define the beginning of the first industrial revolution from the invention of modern steam engine. The biggest novelty of the second industrial revolution was the invention of electricity. Nowadays, cyber-physical systems have been formed as result of the increase in the cohesion of the information technology and automation. This new technology has induced the beginning of a new time period, so called Industry 4 [5, 6]. Figure 1 introduces several important characteristics of the industrial revolutions.

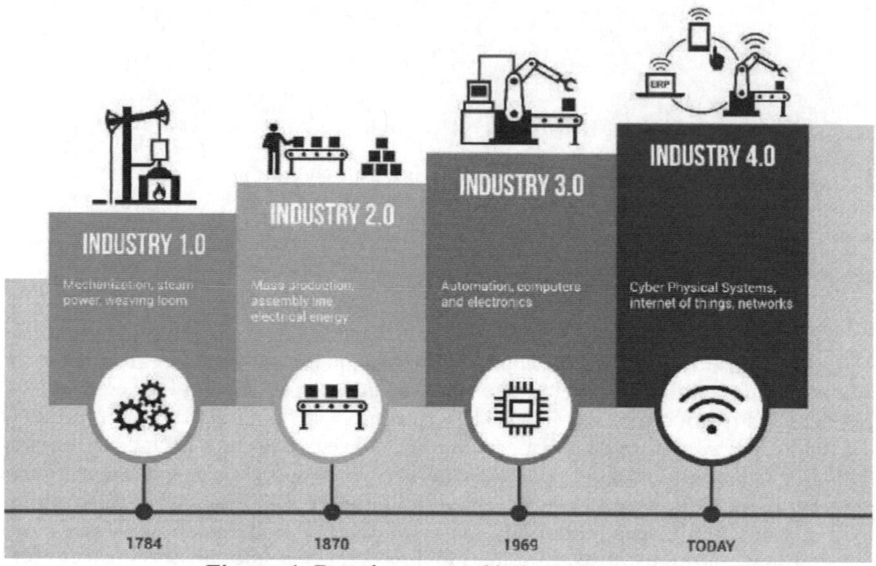

Figure 1. Development of industrial revolations [5]

3. Introduction of most important tools in Industry 4.0

Several tools have been appeared in connection to industry 4.0 Their application enables new opportunities for the production and service companies [4, 2]. The section presents an explanation of these tools. Most important tools of the 4th industrial revolution:

- Internet of Things (IoT): The term: „Internet of Things" was used first by Kevin Ashton in 1999 [7]. Several Hungarian expressions were created in the translations, but „dolgok internete" might be the most eligable to point out the concept behind. IoT enables the access of different equipment (car, fire place, safety system, parts, material handling equipment etc.) through the internet/some networks, as well as the communication between tools in certain cases. In the past decades people have recorded the majority of the data can be accessed on the internet, which has significantly limited the type and quantity of available data. To enhance the system efficiency operated by humanity, date recording and manipulation in the operating system are essential based on the collected data.
- Cyber-physical systems [8]: The development of informatics and automation, as well as their increasing cohesion, have enabled the application of cyber-physical systems (if an electronic device contains a control and network connection then we can call this system a cyber-physical system). These systems are able to collect data from their environment by using sensors and act based on the analysis of their situation. Cyber-physical systems are connected through a network, and their significant parts are also connected with each other; because of this we can apply swarm intelligence, which can result in more efficient work.
- Big Data concept [4]: The amount of data in the world approximately doubles every two years, which results in huge amounts of data in the different areas of life (astronomy, logistics, trade, stock exchange, etc.). We can create useful conclusions and new services based on the huge amount of data by analysing data correlations. An example for such a service is a software, which

IOP Conf. Series: Materials Science and Engineering **448** (2018) 012011 doi:10.1088/1757-899X/448/1/012011

forecast flight prices based on ticket purchase of a previous time-period and determines the adequate purchase time without letting us know the calculation method behind. Big data's essence is forecasting, which means operation with huge amount of data by using mathematical methods and procedures to gain probabilities. Big data will significantly change the future in many people's opinion, because we can make adequate decisions very likely based on the huge amount of data without knowing the causality. Expressions explained above are strongly connected to each other. Therefore, without IoT we cannot speak neither about cyber-physical systems, nor about big data.

4. Most important application fields of Big Data conception

By determining connections between huge amount of data, we can realize advantages in several fields. The most important are listed as following [4].

Astronomy: in 2000, with the start of Sloan Digital Sky Survey program (SDSS), more data were collected during the first few weeks by operating telescope placed in New-Mexico than the amount were aggregated during the entire astronomy history so far. Since 2016, the same amount of data is gained in every fifth day by Large Synoptic Survey Telescope placed in Chile. By elaborating these data, we are able to make more and more precise forecasting regarding changes in our planet environment.

- atmospheric pressure of wheels (influences the duct planning),
- life time of parts (influences the duct planning),
- observation of vibration frequency by determined parts (influences maintenance planning),
- observation of humidity by determined parts (influences maintenance planning).

Technological tools: In course of using technological tools, the most important target is to ensure the maximal readiness similarly as in case of material handling tools, and to execute the technological operations on adequate level. Besides that, continuous improvement is expected, which requires continuous improvement of tooling procedures [14, 16], and elaboration of alternative procedures [10, 15]. Big Data conception gives great support to speed up these developments and their physical implementation. Collecting the following data may provide a help to increase efficiency in technological tool operation:

- tyre pressure needed for operation, electricity consumption (influences the ideal operation of a tool),
- environment temperature, humidity (influences the ideal opration of a tool),
- forces to product (influences the product quality),
- forces to the tooling machine [11] (influances the maintanance planning),
- life time of technological tool parts (influances the maintanance planning),
- vibration frequency of determined parts (influances the maintenance planning),
- observation of humidity in case of determined parts (influences the maintenance planning).

Human resources: In case of human resource application, important target is to ensure the maximal readiness and the work according to standards. Supporting this, the following information might be helpful:

- environment temperature, humidity, brilliancy (influences the efficiency of work),
- speed, acceleration, pulse number of movement (influences the efficiency of work),
- data related to the specific work (standard keeping can be supervised).

Collecting so far unknown data makes necessary to introduce new data collecting tool and online data transferring tools. These tools shall be fixed on raw material or their direct carriers. Expectation for data collecting tools [3]:

- recording method does not influence raw material quality,
- damage resistance,

- recycling,
- capability for continuous data transmission,
- insurance of possibility for far distance data transmission,
- availability of in-built carries.

By applying Big Data conception, development of following important services will be possible:

- Product failure prognosis system: By collecting of technological and product data and by determining the relations among data, product failures can be forecasted with a greater probability.
- Technological tool failure prognosis system: The failures can be or casted on a greater probability by collecting different data of technological tools (e.g. temperature, vibration, pressure etc.), and by finding relations between data and failure incidents
- Material handling tool failure prognosis system: The material handling tool failures can be forecasted by collecting different data of material handling tools (e.g. pressure, level of battery charge, vibration data etc., and finding relations between data and failures.
- Application to provide prognosis for stock level in inventory: Impact of demand changes on stock level can be forecasted by collecting data regarding costumer needs and stock level and by finding relations among them.
- Application to support encroaching in logistics processes: We have the possibility to reorganize logistics processes in time based on failure prognosis of product, technological and material handling tool, which results to loss reduction.

Several production and logistics tasks need to be solved smoothly from the beginning till the end of the production process to make the costumers satisfied with the specific product or service provided [12]. Thanks to artificial intelligence, these processes can be even more faster and precise. We can define artificial intelligence as intelligence presented by a machine, program or artificial conscience. common future of Big Data, production and logistics is artificial intelligence, to be more exact the computer/machine learning, which influences automatization and the evaluation process quality. More and more companies see the potential in this technology. Thanks to self-learning algorithm, computers can analyses data on their own, and they are able to tailor-cut models based on changing phenomenon's and requirements.

5. Summary

The paper has given an overview about the development of fourth industrial revolution, its important tools and challenges. Science fields have been presented, where application of Big Data conception has especially significant role (e.g. astronomy, finance, logistics etc.). It can be stated that we can make remarkable steps in the planning and real production processes to forecast and avoid failures by collecting data which were not collected at all before or just in a limited amount and by determining relations among them with the help of mathematical and statistical methods. Efficiency of earlier applied process optimization algorithm can increase significantly by forecasting the future status more precisely, also new possibilities will arise regarding running supply chain optimization.

Acknowledgement

The described article was carried out as part of the EFOP-3.6.1-16-00011 "Younger and Renewing University – Innovative Knowledge City – institutional development of the University of Miskolc aiming at intelligent specialisation" project implemented in the framework of the Szechenyi 2020 program. The realization of this project is supported by the European Union, co-financed by the European Social Fund." This project has received funding from the European Union's Horizon 2020 research and innovation programme under grant agreement No 691

References

[1] Tamás P 2017 Application of value stream mapping at flexible manufacturing systems *Key Engineering Materials* 686 pp 168-173 doi:10.4028/www.scientific.net/KEM.686.168

[2] Gubán Á and Kása R 2016 Conceptualization of fluid flows of logistificated processes *Advanced logistic systems* 7 pp 27-24

[3] Cselényi J and Illés B 2006 Planning and controlling of material flow systems, *Textbook,* Miskolci Egyetemi Kiadó

[4] Mayer V and Kenneth S Cukier Big Data *A revolution that will transform how we live, work, and think, ISBN 978-0-544-00269-2*

[5] Husi G Ipar 4.0, *jegyzet* (http://rinti.jedlik.eu/plc/jegyzet/industry4.pdf)

[6] Industry 4.0 2014 Challenges and solutions for the digital transfomation and use of technologies *Deloitte, http://www2.deloitte.com/*

[7] Ashton A 2009 That 'Internet of Things' Thing, in the real world things matter more than ideas. *RFID Journal*

[8] Illés B and Trohák A and Bíró Z 2013 Can message filter algorithms for remote diagnostics of vehicles, *APPLIED MECHANICS AND MATERIALS* 309: pp. 213-220.

[9] Illés B and Glistau E and Machado N. I. C 2007 Logistik und Qualitätsmanagement, Miskolc, ISBN 978-963- 87738-1- 4

[10] Kundrak J and Molnar V and Deszpoth I 2018 Comparative Analysis of Machining Procedures,

[11] Machines, 6(2), 13, p. 10.

[12] Kundrák J and Szabó G and Markopoulos A P 2016 "Numerical investigation of the influence of tool rake angle on residual stresses in precision hard turning" *Key Engineering Materials,* Vol. 686, 68-73, DOI: 10.4028/www.scientific.net/KEM.686.68

[13] Womack J P and Jones D D. T. 2008 Lean thinking, Simon & Schuster Inc.,

[14] Balogh R 2018 A Big Data koncepció alkalmazása a logisztikában,

[15] Kundrák J and Deszpoth I and Molnár V 2014 Comparative study of material removal in hard machining of bore holes, *Tehnicki Vjesnik,* 21(1), Pp. 183-189.

[16] Kundrak J and Mamalis AG and Markopoulos A 2004 Finishing of hardened boreholes: Grinding or hard cutting?, *MATER MANUF PROCESS* 19 (6): 979-993

[17] Felho C and Karpuschewski B and Kundrak J 2015 Surface roughness modelling in face milling. *Procedia CIRP* , 31, 136–141

IOP Conf. Series: Materials Science and Engineering **448** (2018) 012012 doi:10.1088/1757-899X/448/1/012012

Productivity analysis of machining milled surfaces

J Kundrák[1], **V Molnár**[2*], **I Deszpoth**[1]

[1] Institute of Manufacturing Science, University of Miskolc, Hungary
[2] Institute of Management Science, University of Miskolc, Hungary

*szvmv@uni-miskolc.hu

Abstract. Milling is a widely applied procedure because its kinematic versions and applied tools allow the machining of several surfaces and surface combinations. In this study the efficiency of material removal performed by face milling was analyzed when machining aluminum alloys by diamond tools. The analyzed material is the material of an automotive industrial component that is produced in large scale. When producing components in large scale the analysis of material removal efficiency and profitability of machining is particularly important. The efficiency of material removal has a direct impact on the time parameters characterizing the production. The focus of the paper is how the productivity can be increased by the selection of technological data that reduce the machining time of the procedure or increase the values of material removal rate or surface rate. Experiments and analyses were carried out for machining prismatic components among operating conditions. It is found that in a given workpiece-gadget-machine-tool system, a significant reduction in production time and therefore cost savings can be reached by appropriately chosen cutting data while the geometric accuracy and surface quality of the component remain as specified in the part drawing.

1. Introduction

Super hard tools such as diamond tools allow efficient chip removal owing to their special characteristics. Artificial diamond tools have a polycrystalline material structure; therefore, damage (e.g. cracks) from the load and stress do not affect the whole diamond volume. Among the tool materials, diamond is among the hardest and has high wear-resistance, a low friction coefficient and high heat conduction capacity. These characteristics, mainly the heat resistance and chemical activity, determine the chip removal process. The friction coefficient of the diamond tool is extremely low and is independent of the cutting speed. The extent of chip deformation is low, there is no secondary chip deformation and the friction within the chip is negligibly low. Owing to the lower heat production which is the consequence of these properties, and to the high heat conduction capacity of the tool, the cutting temperature is low. Therefore cutting by diamond tool is also called ‘cold cutting’. Tool wear is mechanical wear and independent of the cutting speed because of the hardness and the low temperature, depending only on the machined path (area of the machined surface). This means that at different cutting speeds the same surface area can be machined. In machining soft and non-ferrous metals and their alloys, in contrary to the changes experienced in cutting by carbide tools, increasing the cutting speed produces essentially no differences in chip deformation, the friction coefficient, cutting forces, the machined length and the surface roughness, all of which remain almost constant. Owing to this the speed can be increased to the extent that is allowed by the rigidity of the system of

IOP Conf. Series: Materials Science and Engineering **448** (2018) 012012 doi:10.1088/1757-899X/448/1/012012

machine-tool / gadget / tool. In our experiments analyses were carried out for cutting of aluminum alloy by diamond coated insert. The excellent properties of diamond tool can be utilized in cutting non-ferrous metals if they are considered in the technology design. The effects of alteration of cutting data are analyzed from an economical point of view in this paper. While there are several calculation methods for economic efficiency. In our research the efficiency of material removal was analyzed.

To calculate the theoretical value of material removal rate (MRR, Q_w) the most typical method is to consider cutting data and the dimension of the parameter is given in mm^3/min. Mukherjee et al. carried out experiments for turning mild steel in order to determine the optimum value of MRR on the basis of cutting data [1]. Zhong et al. carried out turning experiments and calculated the theoretical MRR values for carbide steel. In their paper MRR served as a decision parameter for minimizing the energy consumption of machining [2]. Mia et al. optimized the MRR on the basis of cutting data in their hard turning experiments with the Taguchi method [3]. Satyanatayana et al. applied a genetic algorithm in their analyses when turning super alloys [4].

Table 1. Application possibilities of MRR.

Procedure	MRR calculation	Dimension of MRR	Application of MRR, workpiece material, source
Turning	Theoretical formula	cm^3/min	Optimization of cutting data on the basis of MRR (Ni-alloy) [9], [10]. Tool life analysis on the basis of MRR (Heat resistant super alloy [11]
Turning	Mass and density of removed material	mm^3/min	Effect of cutting data on MRR (AL-alloy) [12]
Turning	Specific change in workpiece volume (machining time considered)	mm^3/s	Simulation and experiment for determining MRR (Inconel) [13]
Turning	Removed mass	g/min	Multi-attribute optimization of cutting data (EN353 alloy) [14]
Milling	Theoretical formula	mm/min	Cutting data selection [15]
Milling	Theoretical formula, modified by cycle time	mm^3/min	Effect of cutting data on MRR (En19) [5]
Milling	Change in mass and density of workpiece	mm^3/min	Determination of optimum MRR, relational analysis coupled with fuzzy logic (Al-alloy) [16]
Micro milling	Theoretical formula	mm^3/min	Effect of cutting data on MRR (Al-alloy) [17]
Plasma cutting	Mass difference before and after machining	g/s	Optimization of cutting data on the basis of MRR [18]
Honing	Removed volume, total area of honing stones, honing time	cm^3/min	Effect of cutting data on MRR (steel) [19]
Polishing	Luo-Dornfeld model for abrasive machining	-	Construction of a theoretical model [20]

Table 1 summarizes further results beyond turning for milling and other procedures. Among these, the research of Parashar and Purotrit is particularly relevant; they considered the cycle time of machining in calculation of the material removal rate [5]. In efficiency analyses of machining the surface quality [6] and accuracy [7] are important limiting factors of cutting because these requirements have to be fulfilled by the surfaces among every circumstance. The efficiency of machining was described by the practical parameter of material removal rate. In the practical parameter not only the cutting data but also the time data characterizing the machining are considered [8].

2. Methodology

One method for efficiency analysis of machining procedures is the comparison on the basis of efficiency parameters (material removal rate, Q_w [mm³/s]; surface rate, A_w [mm²/s]). These parameters provide information about the intensity of cutting technology. The parameter Q_w provides information about the specific material volume removed and A_w about the specific surface removed with a certain procedure. The theoretical values of these parameters can be calculated on the basis of the cutting data characterizing the procedure. For face milling Q_w and A_w can be calculated as:

$$Q_w = a_p \cdot a_e \cdot v_f \text{ (mm}^3\text{/s)}, \quad (1)$$

$$A_w = a_e \cdot v_f \text{ (mm}^2\text{/min)}, \quad (2)$$

where a_p is depth of cut [mm]; a_e is width of cut [mm]; v_f is feed rate [mm/min].

From the formulas of Q_w and A_w it can be seen that the calculation of the two parameters differs only in the consideration of depth-of-cut (DoC). In our case DoC is considered basically as fixed, therefore it is enough to analyze only the factors influencing surface rate. In machining the plain surfaces of a given prismatic workpiece the length of the tool path remain the same even if the cutting data change. Therefore, in machining these types of workpieces the material removal rate (thus the efficiency of machining) is basically determined by the v_f feed rate. Calculation of feed rate in face milling is:

$$v_f = f_z \cdot z_s \cdot n_s \text{ (mm/min)}, \quad (3)$$

where f_z is feed per edge [mm/edge]; z_s is number of edges [-]; n_s is revolution per minute (rpm) of the tool [1/min].

From equation (3) it is clear that an increase in the material removal rate when machining the analyzed prismatic workpiece can be obtained by increasing three parameters:

1. Increasing the f_z feed per edge;
2. Increasing the rpm (thus the cutting speed) of the tool;
3. Increasing the number of teeth (cutting edges).

In this research the first two parameters were chosen for increasing the efficiency. The first parameter (f_z) is determined on the basis of literature results and experiments. The increase of rpm of the tool (n_s) should be selected on the basis of the cutting characteristics of the diamond tool when cutting aluminum alloys, considering of the limits of the machine tool in use. The increase of teeth number (z_s) can be ensured by the increase of tool diameter when a given tool construction is applied. The number of teeth is considered as fixed in the experiment described in this paper. The theoretical material removal rate provides information about the efficiency of the physical material removal, namely when the tool is cutting (other necessary movements of tool such as approach of the workpiece or positioning to another surface is not considered). Therefore it does not provide information about the whole machining process. That is why it is necessary to introduce a practical parameter that allows the measurement of the operation by both the different time parameters connected to the machining (e.g. machining time, base time, piece time, operation time, etc.) and the parameters characterizing the efficiency of material removal. In our research the value of the practical material removal rate is analyzed by considering the machining time and then the base time in the formula.
The introduced parameter can be calculated as [21]):

$$Q_w \sim L \cdot v_f / T_x \text{ [mm}^3\text{/min]}, \quad (4)$$

where T_x is the time characterizing the machining (machining time, base time).

3. Experiment design

The goal of the experiments was to determine the efficiency of the face milling operation. The machining time and the base time of the operation element were measured. The objective of this was to determine the values of the practical material removal rate taking these time values into account. The experiments were carried out in the plant where the analyzed component is produced in large-scale; therefore, the applied cutting data correspond to the technological circumstances of the manufacturing. In the experiments the effects of the change in the feed per tooth and the rpm of the tool were analyzed. Based on these we chose the feed rate, which is an additional factor (besides depth-of-cut and width-of-cut) that directly determines the machining time.

The milling experiments of the prismatic component (gearbox housing) were carried out on a DMG MORI DMC 60 H linear type horizontal machining center. The milling tool is the Lach Diamant MB-X3-063-08-Z12/P type face milling head (soldered diamond insert). The diameter of the milling head is D_s=63 mm. Number of teeth is z_s=12; tool cutting edge angle is κ_r=90°. Workpiece material is AlSi9Cu3(Fe). The cutting data applied in the experiments are summarized in table 2. In the table the fixed and the variable data are also included.

Table 2. Cutting data.

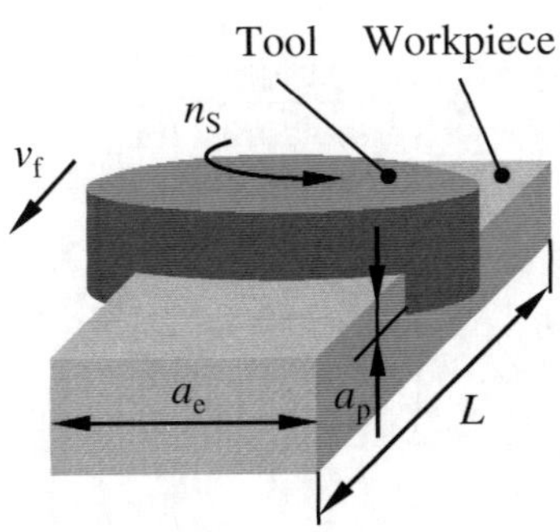

a_p [mm][a]			1.5		
a_e [mm]a			35		
n_s [1/min]	12000	12500	13000	13500	14000
v_c [m/min]	2375	2474	2573	2672	2771
f_z [mm/edge]		feed rate, v_f [mm/min]			
0.12	17280	18000	18720	19440	20160
0.13	18720	19500	20280	21060	21840
0.14	20160	21000	21840	22680	23520
0.15	21600	22500	23400	24300	25200
0.16	23040	24000	24960	25920	26880

[a] Average values

4. Discussion

The machining time and base time of machining were calculated. The machining time describes the time during which the tool is cutting, while the base time includes the supplementary times. In the experiments these supplementary times were the time of tool change, workpiece approach and overrun. The machining time varied between 3.54 and 5.50 s at different feed per teeth and feed rates (figure 1). The decreases in machining time compared to the highest value are summarized in table 3.

The base times are summarized in figure 2. The difference between the highest (19.38 s) and the lowest (17.43 s) values is greater than that of the machining times. The reason for that is the presence of the constant supplementary time. In table 4 the rates are summarized compared to the highest value.

We believe that the practical parameter of the material removal rate is more useful than the theoretical version because the other times connected to the machining are also considered in it when analyzing the efficiency. The theoretical values are summarized in table 5 and figure 3.

The machining operation element (face milling) is characterized in the most appropriate way by the base time (table 6 and figure 4).

The practical material removal rate values calculated by the base time are significantly lower than the values of the theoretical parameter when the feed per tooth (f_z) and the feed rate (v_f) combinations are the same. The reason for this is that the base times are 3.53-4.92-times higher than the machining times. The values of the theoretical parameter vary between 15120 and 23520 mm^3/s (table 5). The values of the practical parameter calculated with the base time vary between 4288 and 4768 mm^3/s (table 7). In the case of the theoretical parameter the highest value is 1.56 times higher than the lowest one, while it is 1.11 for the practical parameter. The rates of material removal rate are summarized in figures 3 and 4. While the lowest value of the theoretical parameter is 64% of the highest one, the difference in the practical parameter is smaller (90%).

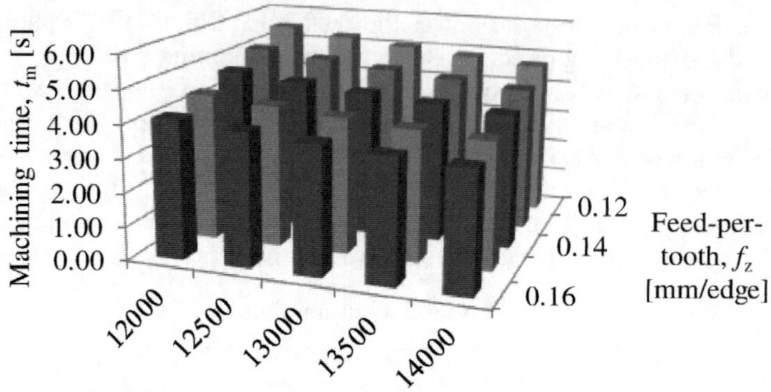

	12000	12500	13000	13500	14000
■0.16	4.12	3.96	3.81	3.66	3.54
■0.15	4.40	4.22	4.06	3.91	3.77
■0.14	4.71	4.52	4.35	4.19	4.04
■0.13	5.07	4.87	4.68	4.51	4.35
■0.12	5.50	5.28	5.07	4.89	4.71

rpm of main spindle, n_s

Figure 1. Machining times at different rpm and feed-per-tooth values.

Table 3. Decrease in machining times; basis: n_s=12000 1/min and f_z=0.12 mm/edge.

% t_m		rpm of main spindle, n_s [1/min]				
		12000	12500	13000	13500	14000
Feed-per-tooth, f_z [mm/edge]	0.12	0%	-4%	-8%	-11%	-14%
	0.13	-8%	-11%	-15%	-18%	-21%
	0.14	-14%	-18%	-21%	-24%	-27%
	0.15	-20%	-23%	-26%	-29%	-31%
	0.16	-25%	-28%	-31%	-33%	-36%

The efficiency of machining can also be expressed by the times connected to the machining / production (tables 3 and 4). When analyzing these values it is important to consider the time as the basis of the analysis that expresses mostly the main goal of the analysis. In the face milling operation of machining prismatic components that contain complex surfaces, this specific time is the base time that includes the supplementary times of machining beyond machining time. We have found that this time is significantly higher than the machining time.

To extend the efficiency analysis the material removal rate proved also to be appropriate, but its theoretical value, which is used widely in literature, does not characterize the machining in a very accurate manner. The parameter was modified by including the base time and thus describes the milling operation more exactly. This new parameter is named practical material removal rate. By applying it, we can gain a clearer picture about the efficiency of material removal of the operation. This practical parameter was analyzed by different feed per tooth and rpm of main spindle values. We demonstrated how the theoretical material removal rate and the base time-based practical parameter change when increasing the above mentioned cutting data (figures 3 and 4).

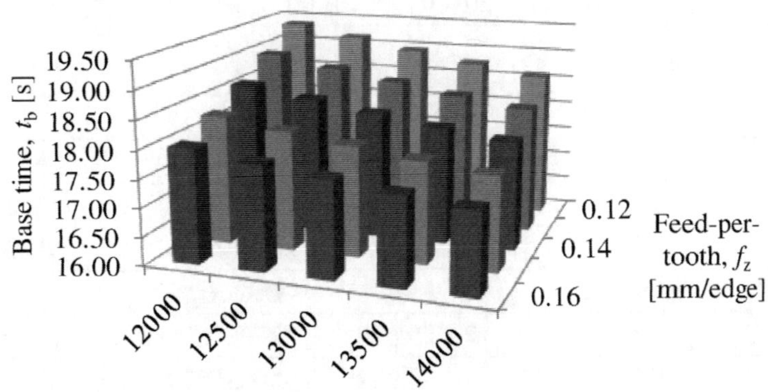

	12000	12500	13000	13500	14000
■0.16	18.01	17.84	17.69	17.55	17.43
■0.15	18.28	18.11	17.95	17.80	17.66
■0.14	18.60	18.41	18.24	18.07	17.93
■0.13	18.96	18.76	18.57	18.40	18.24
■0.12	19.38	19.16	18.96	18.77	18.60

rpm of main spindle, n_s

Figure 2. Base times at different rpm and feed-per-tooth values.

Table 4. Decrease in base times; basis: n_s=12000 1/min and f_z=0.12 mm/edge.

% t_b		rpm of main spindle, n_s [1/min]				
		12000	12500	13000	13500	14000
Feed-per-tooth, f_z [mm/edge]	0.12	0%	-1%	-2%	-3%	-4%
	0.13	-2%	-3%	-4%	-5%	-6%
	0.14	-4%	-5%	-6%	-7%	-8%
	0.15	-6%	-7%	-7%	-8%	-9%
	0.16	-7%	-8%	-9%	-9%	-10%

Table 5. Values of theoretical material removal rates.

Theoretical material removal rate, Q_w [mm^3/s]		rpm of main spindle, n_s [1/min]				
		12000	12500	13000	13500	14000
Feed-per-tooth, f_z [mm/edge]	0.12	15120	15750	16380	17010	17640
	0.13	16380	17063	17745	18428	19110
	0.14	17640	18375	19110	19845	20580
	0.15	18900	19688	20475	21263	22050
	0.16	20160	21000	21840	22680	23520

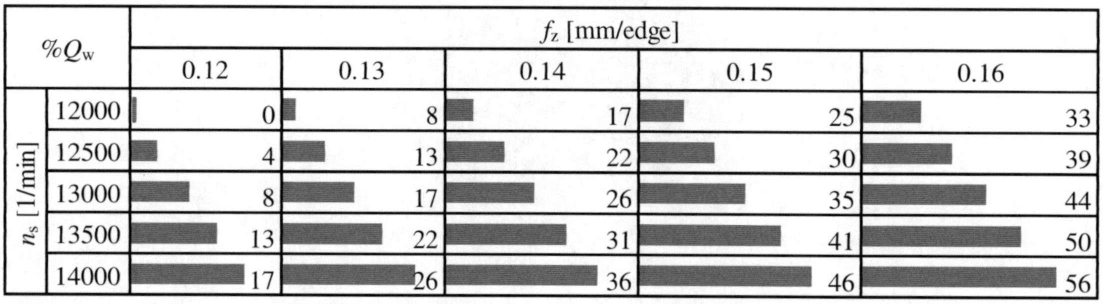

Figure 3. Increase in theoretical material removal rates at different rpm and feed-per-tooth values; basis: n_s=12000 mm/min and f_z=0.12 mm/edge.

Table 6. Values of practical material removal rates.

Practical material removal rate, $Q_{wp,b}$ [mm^3/s]]		rpm of main spindle, n_s [1/min]				
		12000	12500	13000	13500	14000
Feed-per-tooth, f_z [mm/edge]	0.12	4288	4337	4383	4427	4469
	0.13	4383	4431	4475	4517	4557
	0.14	4469	4514	4557	4598	4636
	0.15	4545	4589	4631	4670	4707
	0.16	4615	4657	4697	4735	4768

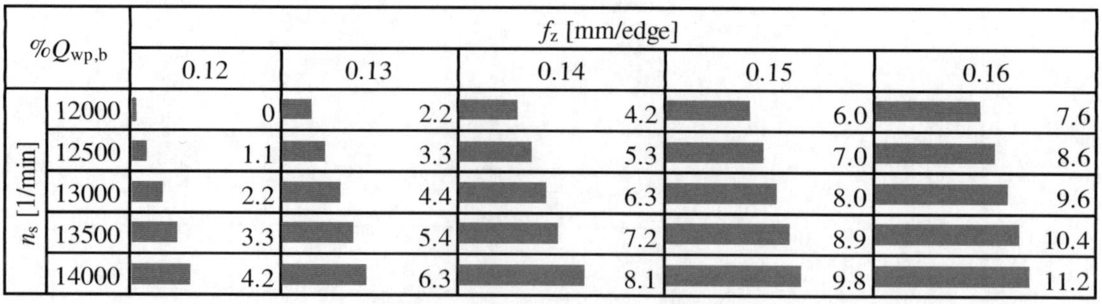

Figure 4. Increase in practical material removal rates at different rpm and feed-per-tooth values; basis: n_s=12000 mm/min and f_z=0.12 mm/edge.

5. Summary

The efficiency analysis of face milling operation element of plain surfaces of prismatic components that contains many surfaces was carried out by the calculation of times necessary for the machining. The cutting data for machining aluminum material with diamond tool were determined by the material removal characteristics of diamond cutting. On the basis of this the possible ways (strategies) of economically efficient cutting were designated and specified in the experimental setup. The possible extent of base time reduction was analyzed when increasing the feed and the feed rate. In our analyses the practical material removal rate was calculated on the basis of the base time of machining, which characterized the milling operation appropriately. This strategy proved to be efficient because it was found that in the practical value of material removal rate a 3.3-4.2% and 6.6-7.6% increase in efficiency can be reached, respectively. Considering the parallel change in the two cutting data, an 11% efficiency increase is reached. In further research aimed at increasing efficiency we plan to analyze the effects of increasing the number of tool teeth on the efficiency of material removal.

Acknowledgments
The authors greatly appreciate the support of the National Research, Development and Innovation Office – NKFIH (No. of Agreement: K 116876).

References

[1] Mukherjee S, Kamal A and Kumar K 2014 Optimization of material removal rate during turning of SAE 1020 material in CNC lathe using Taguchi technique *Procedia Engineering* **97** pp 29–35
[2] Zhong Q, Tang R and Peng T 2017 Decision rules for energy consumption minimization during material removal process in turning *J. Clean. Prod.* **140** pp 1819–27
[3] Mia M, Dey P R, Hossainc M S, Arafat Md T, Asaduzzaman Md, Ullah Md S and Zobaer S M T 2018 Taguchi S/N based optimization of machining parameters for surface roughness, tool wear and material removal rate in hard turning under MQL cutting condition *Measurement* **122** pp 380–91
[4] Satyanatayana B, Nitin P R and Reddy M D 2018 Optimization of controllable turning parameters for high speed dry machining of a super alloy *Materials Today: Proceedings* **5** pp 4878–86
[5] Parashar V and Purohit R 2017 Investigation of the effects of the machining parameters on material removal rate using taguchi method in end milling of steel grade EN19 *Materials Today: Proceedings* **4** pp 336–41
[6] Varga G and Ferencsik V 2017 Analysis of surface topography of diamond burnished aluminium alloy components *Lecture Notes in Mechanical Engineering* pp 143–54
[7] Mamalis A G, Kundrak J and Gyani K 2002 On the dry machining of steel surfaces using superhard tools *Int. J. Adv. Manuf. Technol.* **19** pp 157–62
[8] Kundrak J, Toth T and Gyani K 2004 How to make a choice of machining methods on the basis of economy: comparison between hard turning and grinding *Conference Proceddings of The Eleventh International Conference on Machine Design and Production* Turkey pp 31–45
[9] Hernandez A E B, Beno T, Repo J and Wretland a 2016 Integrated optimization model for cutting data selection based on maximal MRR and tool utilization in continuous machining operations *CIRP J. Manuf. Sci. Technol.* **13** pp 46–50
[10] Palacios J A, Olvera D, Urbikain G, Elias-Zuniga A, Martonez-Romero O, Lopez de Lacalle L N, Rodriguez C and Martinez-Alfaro H 2018 Combination of simulated annealing and pseudo spectral methods for the optimum removal rate in turning operations of nickel-based alloys *Adv. Eng. Softw.* **115** pp 391–7
[11] Hernandez A E B , Beno T, Repo J and Wretland, A 2015 Analysis of tool utilization from material removal rate perspective *Proc. CIRP* **29**, pp 109–13
[12] Moganapriya C, Rajasekar R, Ponappa K, Venkatesh R and Jerome S 2018 Influence of coating

material and cutting parameters on surface roughness and material removal rate in turning process using taguchi method *Materials Today: Proceedings* **5** pp 8532–8

[13] Yadav R K, Abhishek K and Mahapatra S S 2015 A simulation approach for estimating flank wear and materialremoval rate in turning of Inconel 718 *Simul. Model. Pract. Th.* **52** pp 1–14

[14] Kumar R, Bilga P S and Singh S 2017 Multi objective optimization using different methods of assigning weights to energy consumption responses, surface roughness and material removal rate during rough turning operation *J. Clean. Prod.* **164** pp 45–57

[15] Budak E and Tekeli A 2005 Maximizing chatter free material removal rate in milling through optimal selection of axial and radial depth of cut pairs *CIRP Annals* **54** pp 353–6

[16] Tamiloli N, Venkatesan J and Ramnath B V 2016 A grey-fuzzy modeling for evaluating surface roughness and material removal rate of coated end milling insert *Measurement* **84** pp 68–82

[17] Shinge A R and Dabade U A 2018 The effect of process parameters on material removal rate and dimensional variation of channel width in micro-milling of aluminium alloy 6063 T6 *Procedia Manufacturing* **20** pp 168–73

[18] Das M K, Kumar K, Barman T Kr and Sahoo P 2014 Optimization of process parameters in plasma arc cutting of en 31 steel based on mrr and multiple roughness characteristics using grey relational analysis *Procedia Materials Science* **5** pp 1550–9

[19] Buj-Corral I, Vivancos-Calvet J and Coba-Salcedo M 2014 Modelling of surface finish and material removal rate in rough honing *Precis. Eng.* **38** pp 100–8

[20] Wang P, Gao R X and Yan R 2017 A deep learning-based approach to material removal rate prediction in polishing *CIRP Ann. Manuf. Technol.* **66** pp 429–32

[21] Kundrak J, Molnar V and Deszpoth I 2018 Comparative analysis of machining procedures *Machines* **6** p 10

Impact of Industry 4.0 on production logistics

G Nagy, B Illés and Á Bányai

Institute of Logistics, University of Miskolc, Miskolc 3515, Hungary
altagota@uni-miskolc.hu

Abstract. The Fourth Industrial Revolution enables technological processes and digitalization to ensure the transparency of corporate processes; integrates the corporate value chain and the supply network. The aim of the study was to present the technologies of the fourth industrial revolution and to examine the importance of vertical and horizontal integration. The paper reviews the role of production logistics in the corporate logistics system. In a growing market competition, companies are only able to retain their market position by exploiting the integration opportunities between production management and logistics. Based on this, we analyze the positive impacts achieved through logistically integrated production management, providing the possible design of the information system required for implementation and the management tasks. In the context of a specific production logistic example, we present the benefits of real-time scheduling.

1. Introduction

At some stages of industrial revolutions, the evolution of communication coincided with the development of energy utilization and mobility, leading to a higher level of quality of life and changing business models [1]. Today, we are on the verge of the fourth industrial revolution, based on digitization and data. The development of technology creates a permanent network of people, machines and companies, and through continuous sharing of value-creating processes, a competitive, fully customized product is available. The source of competitive advantage, therefore, will not only be the production of coordinated or totally new funds (e.g. additive production), but the supply of products with digital services.

2. Technology background and features of Industry 4.0

Industry 4.0 utilizes the opportunities offered by digitization to maximize the transparency of processes and integrate the corporate value chain and the supply chain, bringing new value to customer value creation. Industry 4.0 technologies can be divided into three major areas: The security problems of information technology (IT) are greatly aggravated by the reuse of previously completed projects.

- Cloud Computing,
- cyber-physical systems,
- smart factory.

Cloud-based services encompass the entire computing space, including computing capacity, storage, platforms, and software. Cloud based systems enable on-demand and decentralized availability of IT resources, data, services and digital business models over the Internet.

Under the Cyber-Physical System (CPS), we mean the integration of IT, software technology, and mechanical and electronics elements where the elements communicate with each other. One of the main features of a cyber-physical system is the very high complexity. Cyber-physical systems are created by networking embedded systems through wired and increasingly wireless communication networks.

The Industry 4.0 paradigm will be widely spread in many fields of engineering and engineering related systems and processes, including manufacturing [2], assembly [3], logistics [4,5], services [6] but the most intensive development can be recognised in automotive domain [7,8] with focus on the IT and telecommunication [9].

The intelligent factory sees a production environment in which the manufacturing and control equipment largely coordinates and organizes themselves - without human intervention. Its technological basis is formed by cyber-physical systems, the elements of which are communicated with each other through "the Internet of Things" (IoT). In the smart factory, the product (e.g. the workpiece) and the production equipment communicate with each other: the product itself provides production information in the readable, readable form of the production equipment. This data is based on the product on the production line, and the production phases are based on these data.

Figure 1 illustrates the characteristics of Industry 4.0, which demonstrate the enormous capacity that exists in industry and traditional production [10]:

- vertical networking of smart production systems,
- horizontal integration via a new generation of global value chain networks,
- through-engineering across the entire value chain,
- acceleration through exponential technologies.

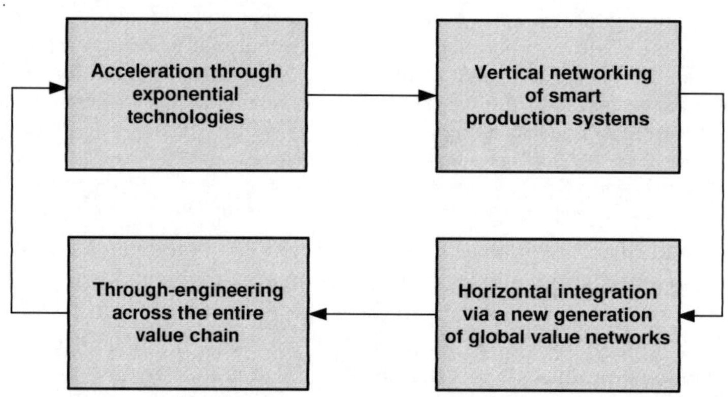

Figure 1. The four characteristics of industry 4.0 [10].

The vertical network of intelligent manufacturing systems uses cyber-physical production systems (CPPS), which integrates production units to make production more customer-oriented and unique. The design of the system requires complete information integration. Integrating data generated during production requires the further development of smart sensor technologies. It is necessary to develop methods to monitor autonomous production systems and track product life. CPPSs should focus not only on the production process, but on the management of the entire supply system. All this requires that all elements of a complex network must record, archive and make available the data generated during their operation. This data should cover all components of the manufacturing process, i.e. production, product, operators, quality change, downtime and, last but not least, resource management [11]. The latter is particularly important because the primary aspect of development is the efficient management of resources, that is, the efficient, full management of material, energy and human resources.

The second key feature of Industry 4.0 is horizontal integration through new generation global value chain networks. These new value creation networks are real-time optimized networks that allow integrated transparency and high levels of flexibility in order to respond faster to problems and failures and thus facilitate global optimization. Like the networked production systems, CPPS provides the network, from procurement logistics to warehousing, production, marketing and distribution logistics to post-tracking services. All data on any part or product is recorded and available at any time, ensuring constant traceability. This creates transparency and flexibility through the entire chain of processes - e.g. from procurement to production through to sales or from supplier to company through to the buyer. The adaptation of the buyer-specific approach is possible not only in production, but also in development, ordering, production planning, product composition and distribution, allowing to dynamically manage factors such as quality, time, risk, price and environmental sustainability, in real-time, and at all stages of the value chain.

The third key feature of Industry 4.0 is not only the engineering activity across the entire value chain, but also the engineering support and management of product and customer life cycles. This engineering approach is present in the design and development of new products and services, as well as in manufacturing. This feature of engineering is characterized by the fact that data and information are available at all stages of the product life cycle, enabling more flexible data to be formulated by modeling prototypes and product states.

The fourth major feature of Industry 4.0 is the impact of exponential technologies such as an accelerator or catalyst that allows personalized solutions, flexibility and cost reduction in manufacturing processes. Industry 4.0 already requires automation solutions to be highly cognitive and highly autonomous. The potential in artificial intelligence, advanced robotics, and sensor technology is to increase self-reliance and accelerate individualization and flexibility. Artificial Intelligence not only helps to design more driver-driven vehicle paths in factories and warehouses, but also to save time and cost in Supply Chain Management (SCM), increase reliability in production, or analyze large data volumes to find new design and design solutions or to enhance cooperation between man and machine [12].

3D Printing (Additive Production) is an exponential technology that accelerates Industry 4.0 to make it more flexible. 3D printing enables new production solutions (such as functionality, greater complexity at no extra cost), or new supply chain solutions (e.g. reduced inventory, faster delivery times), or a combination of both, leading to completely new business models (e.g. eliminating supply chain members, customer integration).

As far as Industry 4.0 is concerned, the most important requirement is that the elemental manufacturing cells must be "smart" (Figure 2).

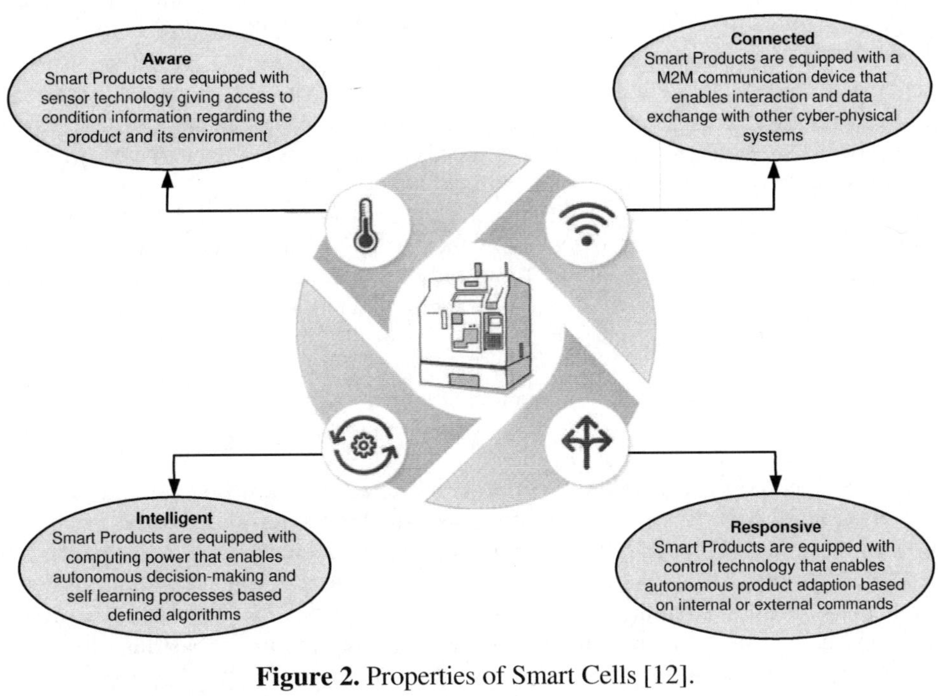

Figure 2. Properties of Smart Cells [12].

In the smart cell manufacturer, by using the appropriate sensors, the devices must have information on the qualitative and quantitative properties of the raw materials, the environmental conditions and the expected properties of the product [11]. This implies that the manufacturing unit must be integrated, that is, the cyber-physical system must be connected in the most flexible way via the appropriate communication protocol, in order to ensure the continuous exchange of information for the corresponding data. The smart unit must be intelligent, that is, the unit autonomously organizes its own work by relying on its previous experiences of decisions, that is, the unit learns. Thanks to this learning capability, the unit determines the necessary tasks and order of the production of the product, and provides for the requisite resources [12], based on the signals embedded in products and raw materials.

Through Industry 4.0, the value created for the buyer is transformed. The product should now be understood in an expanded sense, which is equipped with digital elements, e.g. a sensor or communication device (RFID), which sends a signal of its current state. However, development is in the direction of complementing the smart product with data analysis services and providing a complete solution to the customer [13].

The technology elements that are incorporated into the product make it possible for them to offer increasingly complex services to the customer. Four levels of product-based services can be distinguished: monitoring, control, optimization and autonomy, depending on the extent to which they allow autonomy to send the product or machine signal and intervene in the production process [14].

The first level of the clever product is that the product is capable of delivering data on its own, monitoring its status, working process, and external environmental impacts. In this case, in case of deviation from normal, it sends a signal to the operator who can decide on the intervention. This also includes the buyer's ability to track the product's delivery and delivery. Data is excellently utilized in product design, market segmentation, after-sale services.

The second level is for clever products, when monitoring results in an intervention (control). Based on the observed data, the system detects the deviation and either the algorithm or command installed in the device or the cloud is executed.

The third stage is optimization. At this point, not only current monitoring data and built-in control devices are available, but additional software can be used to incorporate historical data into the analysis, which greatly improves production volume, efficiency, and utilization.

The fourth grade is an autonomous product. Autonomous products use algorithms to analyze and communicate with and control themselves and their environment. The human operator now only supervises the operation. Smarter products can learn and be able to memorize their environment, diagnose their own service needs, and adapt to the needs of their users. They can work in places where it's difficult or dangerous for humans.

In addition to the products, the processes also change. The uniqueness of the product grows, while the flexibility of production processes increases, with the customer and other business partners increasingly intensifying, so the company will not only offer a high-quality product but will also be connected to a high level of service. Innovations in the product and in the production-process make it possible to optimize production, increase efficiency, and reduce downtime, errors and scraps dramatically. All of these circumstances lead us to the clever factory where products communicate with their environment and influence the production system's settings [13].

3. Logistics integrated production management

Production logistics is the most definitive system of corporate logistics; the materials and production tools required for production processes, the harmony of the production process and its sub processes, and the flow of material and related information flow processes are necessary to ensure the production of products in the production to consumers / users. In the process of value creation, its function can only be efficiently fulfilled if connections to purchasing, distribution and waste logistics logistical connections are based on key strategic issues and function optimally, as far as possible. The connections of the production logistic system are illustrated in Figure 3.

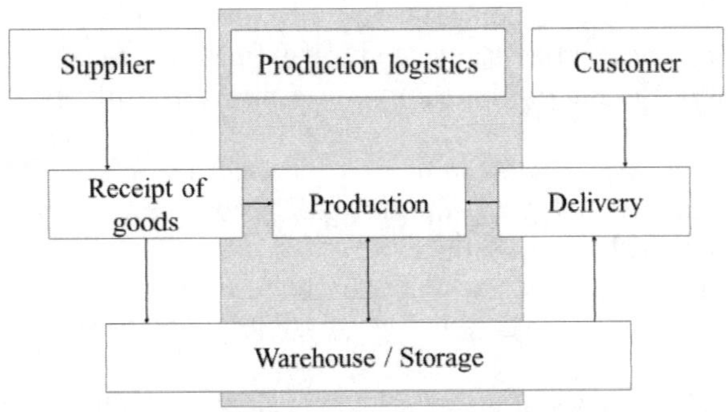

Figure 3. Production Logistics Connections [15].

In a growing market competition, companies are only able to retain their market position by exploiting the integration opportunities between production management and logistics. The efficient operation of logistics-integrated production management implies an integrated information system (Figure 4).

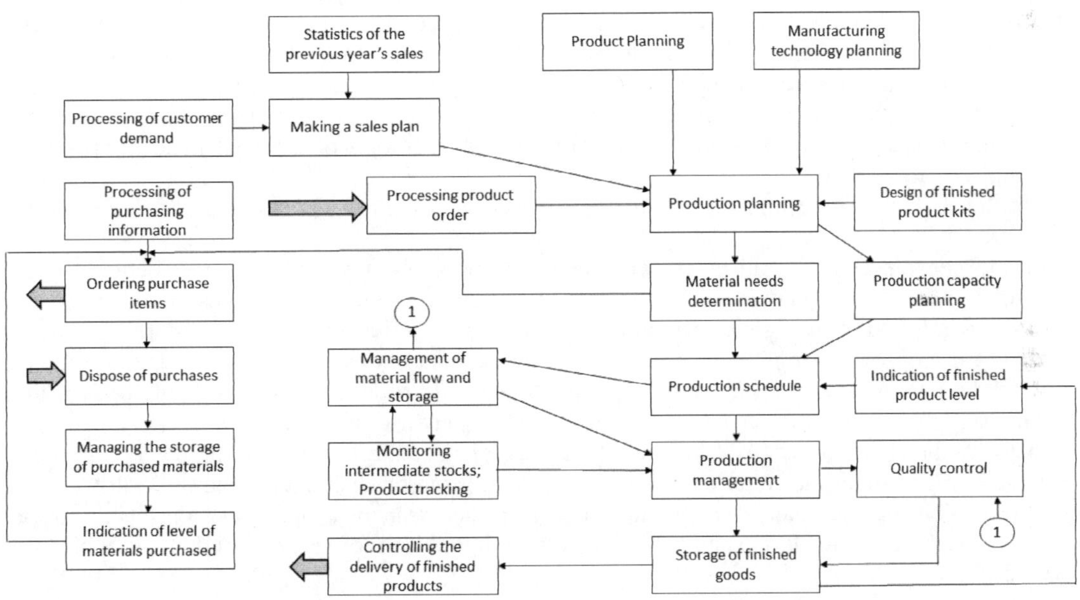

Figure 4. Integrated information management system for production management and logistics [15].

Logistics integrated production management takes the inventory level of finished products into account, it manages data from product design (product family tree, component inventory, component drawing) and production technology and process design (order of operation, order of technology data, technology equipment for operation), determines material demand and production capacity needs. Based on the demand for material and the level of readiness of the materials procured - processing of information obtained from the procurement market - is done by sending orders and asking for quotes. Following the evaluation of the quotations and the finalization of the orders, the dispensing of the purchases, the control of the storage of the purchased materials (storage location, assignment order, etc.) and the stock level of the purchased materials will be carried out. A further step in production planning is production scheduling, taking inventory levels and logistics capacities into account. That is, integrated production scheduling also reveals that, besides production resources, logistics resources are also taken into account and optimized.

IOP Conf. Series: Materials Science and Engineering **448** (2018) 012013 doi:10.1088/1757-899X/448/1/012013

The scheduling can be based on the computerized management (dispensing) of material flow and storage within the manufacturing process, the control of material flow techniques and production equipment, the monitoring of the interim inventory and follow-up of the products, and the control of the storage of finished products, the finished product level indicator. Production management is closely linked to quality assurance, which can be used to control stock frames, flow management and production control. Finally, the delivery of finished goods takes place in accordance with the order confirmation date and the finished warehousing [15].

Logistic-oriented manufacturing processes aim to create a production organization structure that will solve the logistical principles as fully as possible and thereby increase the company's competitiveness significantly.

The management of the production logistics system is hierarchically structured, that is, strategic, tactical and operational. On the strategic level, processing the order of production and service tasks, taking into account the overall strategy of logistic management, logistics-integrated production-service scheduling, delivery scheduling, stock inventory registration, procurement preparation and administrative work required for these. Logistics control data on the utilization of the device and equipment and capacities are entered to this level. After the processing of such incoming data, the dynamic information base is constantly updated at the tactical level.

The dynamic information base includes a selection of available materials handling equipment, tools and aids, as well as a collection of managed management strategies. At the tactical level, the order of service comes from the strategic level on one hand and on the other hand as a feedback on the operational level as a collection of data. An order management strategy is selected for the order, preparation for dispatching and control is carried out, that is, the delivery of the data to the control body or their manager. At operational level happens the operation of material handling machines, equipment control, acknowledgment of performed tasks and delivery of information to tactical and strategic level. Finally, the logistics system is checked at a tactical level, from which feedback is necessary.

For the cost-efficient operation of companies, the proper management of production logistics processes is of great importance as it is able to influence the company's business performance and customer satisfaction. In logistics and supply chain processes there may be losses in a number of places. The use of sensors in processes creates transparency and flexibility when data is transmitted to the cloud through the Internet and shared with the partners, in order to ensure systemic optimization. By doing so, value-creating processes can be fully tracked in real-time [16].

One of the basic requirements of Industry 4.0 is that it can connect to an electronic network through the Internet. Throughout the entire production chain, data from product data, customer order, capacity data, production logistics data, and quality control are accessed in a network - the cyber-physical system (CPS). Smart innovation is aimed at integrating digital technology into products, machines and processes.

4. Solution for Production Logistics by Industry 4.0 Principles

The Industry 4.0 paradigm should have a great impact on the performance of production logistics. One reason for this is that, in order to meet dynamically changing customer needs, we have to meet manufacturing processes that are flexible and dynamic in responding to customer needs while reducing the finished product inventory. One of the new approaches to manufacturing and installation processes in the automotive industry and the mechatronics assembly industry is the milkrun concept, which essentially means that milkruns are not only supplying material, that is, physical processes, but also providing flow of information.

Numerous well-proven methods are available in the literature for designing the milkrun but have not yet considered how it is possible to schedule new service tasks in real-time while optimizing scheduling of static servicing tasks. During the real-time scheduling of the problem is that for each milkrun scheduled maintenance tasks are assigned to time windows and the milkrun's capacity limitations not be exceeded. In this spirit, the design of dynamic candidate maintenance tasks can be defined as an NP-hard problem, where assignment and scheduling solutions integrated into the goal.

IOP Conf. Series: Materials Science and Engineering **448** (2018) 012013 doi:10.1088/1757-899X/448/1/012013

The starting point of the milkrun is usually warehouse or supermarket, so the optimum target function can be formulated as a complex cost function where each cost component is given by the following activities:

- the cost of pre-scheduled deliveries without assigning dynamic demands;
- the costs of transporting and loading the milkrun to the supermarket and the first and last service places;
- loading and transport costs of dynamic demands.

Limitations to the optimization task affect two areas:

- Limitations on the capacity constraints of millruns, and
- specifications for time windows for scheduled services.

A number of heuristic algorithms can provide acceptable results for solving this problem, and as a result of this optimization, a reliable, flexible, high availability system can be run that takes into account efforts to reduce the cost of material handling tasks.

As a result of scheduling, it can be stated in general that, due to time constraints and milkrun capacity constraints, dynamically occurring tasks may not be scheduled for the nearest service point but must be pushed in some direction. However, this shift ensures that both time window and capacity limitations can be respected.

Figure 5 shows an example of the practical implementation of the theoretical problem presented above.

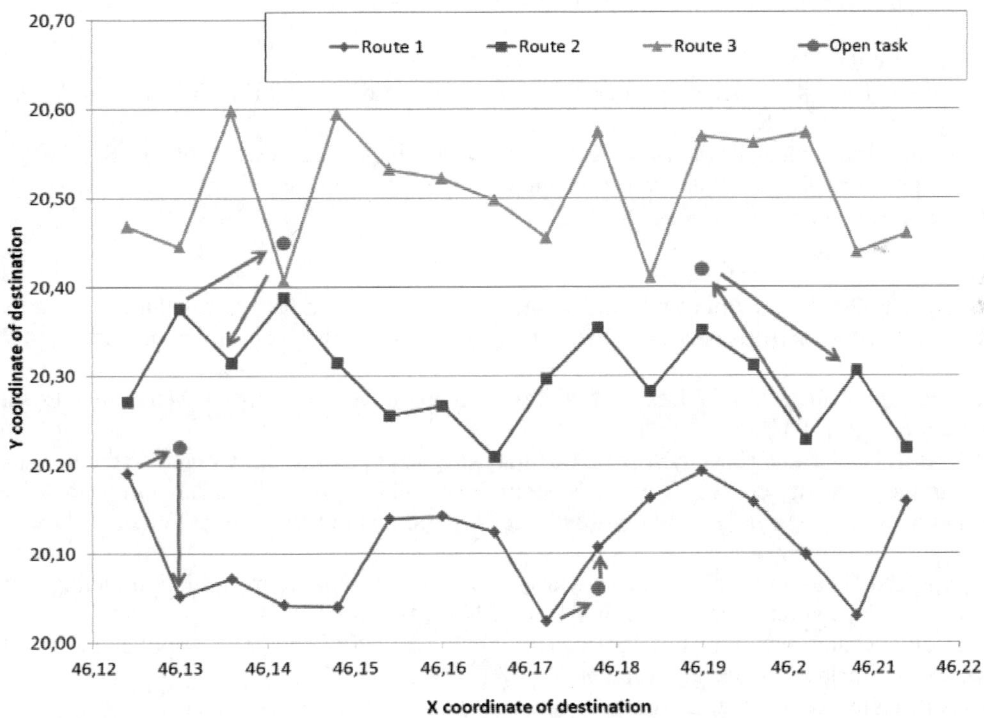

Figure 5. Scheduling Dynamic Service Tasks for Milkrun Based Serving.

Three milkrun flights are available and two dynamic requests have to be scheduled. The scheduling of Dynamic Demand #1 could be solved by not having to shift the service from the nearest service point. For Dynamic Action #2, you had to push the drop-out point for the service task that was added to maintain the capacity for milkrun and the service time windows limit. This means that the release of the Dynamic Requirement #2 was only possible at a point of delivery, with all the other delivery points behind, there are enough reserves in the time of delivery, sufficient to perform the material handling operations required to deliver a dynamic demand.

5. Summary

The factory of the future is flexible and transparent. This means that the time of workflows, preparations and switching times can be accurately calculated, so you can determine where there are free capacities. This way, when you receive a customer order, you can accurately calculate when the order is being made and what capacities, machines, tools will be assigned, and what purchases will have to take place until the start of production.

In this paper, we reviewed the technological background and main features of the fourth industrial revolution. We analyzed the competitive advantages achieved through the vertical network of intelligent manufacturing systems and horizontal integration, demonstrated the properties of the smart cell, and determined the levels of services built into the smart product.

The paper deals with the relationship system of the company's production logistic system, with logistics-integrated production management tasks and with the competitive advantages it can achieve. In solving a specific production logistic problem, we have demonstrated the competitive advantages of real-time scheduling.

The above-mentioned research has many possible directions for further development, as Industry 4.0 is characterized by multidisciplinary that allows further research not only in technical but also in economic and social fields. In relation to production logistics, our future objectives include further examination of the service processes of complex manufacturing systems, in which we intend to set up in-house supply chains that ensure the efficient management of large complex production processes through integrated management of cooperative tools and sub-processes.

Acknowledgments

The described study was carried out as part of the EFOP-3.6.1-16-00011 "Younger and Renewing University – Innovative Knowledge City – institutional development of the University of Miskolc aiming at intelligent specialisation" project implemented in the framework of the Szechenyi 2020 program. The realization of this project is supported by the European Union, co-financed by the European Social Fund.

References

[1] Holodny E. 2017 A key player in China and the EU's „third industrial revolution" describes the economy of tomorrow Business Insider http://www.businessinsider.com/jeremy-rifkin-interview-2017-6

[2] Kundrak J, Molnar V and Deszpoth I 2018 Comparative Analysis of Machining Procedures *Machines* **6**(2) 1-10

[3] Oláh B, Bányai T and Cselényi J 2002 Optimal assignment of assembly plants to the final product requirements of the end users in a cooperative assembly system Proc Int Symp on Intelligent Manufacturing & Automation: Focus on Reconstruction and Development (Vienna) (DAAAM) p 730

[4] Kota L and Jármai K 2015 Mathematical modeling of multiple tour multiple traveling salesman problem using evolutionary programming *Appl Math Mod* **39**(12) 3410-33

[5] Hartványi T and Hencz C I 2016 Economic efficiency analysis of the application of electric transport vehicles – case study *Adv Log Sys* **10**(2) 19-26

[6] Kása R, Gubán M and Gubán Á 2016 Logistical processes of service systems, with special regard to their amelioration – a model framework in: Challenges in Process Management: Decision points, network systems and strategies in practice (ed Gyenge B) (Gyöngyös: Károly Róbert Kutató-Oktató Közhasznú Nonprofit Kft) 31-51

[7] Kulcsárné Forrai M and Kulcsár G 2017 Modeling and Solving an Extended Parallel Resource Scheduling Problem in the Automotive Industry *Acta Poly Hung* **14**(4) 27-46

[8] Losonci D, Kása R, Demeter K, Heidrich B and Jenei I 2017 The impact of shop floor culture and subculture on lean production practices *Int J Oper and Prod Man* **37**(2) 205-25

[9] Konyha J, Bányai T 2017 Sensor networks for smart manufacturing processes *Solid State Phenom* **261** 456-62

[10] Deloitte 2014 Industry 4.0 study (Basel: Deloitte)

[11] Nagy J 2017 Industry 4.0: definition, elements and effect on corporate value chain http://unipub.lib.uni-corvinus.hu/3115/
[12] Bechtold J, Kern A, Lauenstein C and Bernhofer L 2014 Industry 4.0 - The Capgemini Consulting View 31 (Capgemnini Consulting)
[13] Abonyi J and Miszlivetz F 2016 Intersections of networks: Social challenges of I4.0 (Szombathely: Savaria University Press)
[14] Porter M E, Heppelmann J E 2014 How smart, connected products are transforming competition *Harvard Bus Rev* **92**(11) 64-88
[15] Illés B, Glistau E and Machado N I C 2007 *Logistik und Qualitätsmanagement* (Budapest: Budai Nyomda) p 195
[16] Nagy G Bányai Á, Illés B and Glistau E 2018 Analysis of supply chain efficiency in blending technologies *Lect Notes in Mech Eng* **2018** 280-91

IOP Conf. Series: Materials Science and Engineering **448** (2018) 012014 doi:10.1088/1757-899X/448/1/012014

Logistic aspects of Industry 4.0

R Skapinyecz, B Illés and Á Bányai

Institute of Logistics, University of Miskolc, Miskolc 3515, Hungary
altagota@uni-miskolc.hu

Abstract. Through the fourth industrial revolution, today such technological innovations and methods have become available which enable the development of complex logistics systems where the entire supply chain can be operated in an automated way. The aim of the study was to investigate how to increase the efficiency of logistics processes through the exploitation of the opportunities offered by the fourth industrial revolution. The paper presents the essence of Industry 4.0, the technological conditions of the fourth industrial revolution, its opportunities and challenges, and examines its impact on both intra-corporate and non-corporate logistics processes. We analyze the operational processes of logistics networks and the efficiency gains achieved through Industry 4.0 applications. Reliability and quality assessment of logistics networks are a complex problem. In this paper, we will present an innovative solution which is based on Industry 4.0 infocommunication solutions and the application of risk management and quality assurance tools, one that enables the optimal selection of logistics service providers in the network from a reliability point of view.

1. Introduction
The forthcoming fourth industrial revolution or otherwise Industry 4.0 means the accelerating growth of the efficiency of production systems. This includes online cyber-physical systems, the Internet of Things (IoT) and cloud-based solutions.

The fourth industrial revolution creates what we call the "smart factory". In modular smart factories, cyber-physical systems control physical processes that create an apparent copy of the physical world and make decentralized decisions. Objects, cyber-physical systems communicate over the Internet and interact with each other and with people in real time.

2. The fourth industrial revolution
An industrial revolution can be talked about when the efficiency of production systems is significantly increased due to a new technology solution. The design and operation of systems are becoming increasingly knowledge and resource-intensive, so a breakthrough can only be achieved by increasing the complexity of the systems [1].

The first industrial revolution mobilized the mechanization of production, using the power of water and steam, that is, the revolutionary developments were made possible by the emergence of mechanical production systems. The second industrial revolution, with the help of electricity, introduced mass production, followed by the digital revolution. During the third industrial revolution, electronics and information technology (IT) began to automate processes. Automation offered an alternative to the human work needed to operate, accelerating and decoupling the core manufacturing processes. It was this time when the first PLCs were released in the industry.

The fourth industrial revolution can be achieved because such technological innovations and methods have become available that enable the development of even more complex systems, where the entire supply chain can be operated in an automated way. This is based on cyber-physical systems that connect real objects with the processing of information. Cyber-physical systems can be

IOP Conf. Series: Materials Science and Engineering **448** (2018) 012014 doi:10.1088/1757-899X/448/1/012014

represented in many fields of engineering, like services [2], traditional [3,4] and networked manufacturing [5].

The prerequisite for these solutions is to create smart factories that have independent, shared intelligence and are in close contact with each other. By using detailed and interconnected models of elements, the complex system of these clever components can be optimized for comprehensive goals. To make this compilation work requires not only the "smart" elements to be networked, but also makes it necessary to coordinate these in a "cybernetic space". This is feasible with model-based optimization and development with the use of simulation tools.

In the future, the production process in the manufacturing network can be automatically adjusted to the urgency of incoming orders. According to the adaptive production, orders can be processed continuously and transferred to procurement and logistics systems. The operating network accordingly ensures the optimization of material and energy flow along the entire value chain. Intelligent machines continuously share information about current inventory levels, issues, errors, and the changes in demand. In process management, the goal is to increase efficiency, optimize lead times and increase capacity utilization [6]. Table 1 illustrates the fundamental differences between today's factory and the factory of the future.

Table 1. Comparison of today's factory and the factory of the future [6].

		Today's factory		**Factory of the future**	
	data source	**main characteristics**	**main technologies**	**main characteristics**	**main technologies**
compo-nent	**sensor**	precision	smart sensors and failure detection	knowledge of own operations, predictive ability	monitoring of all features, life expectancy forecasts
machine	**controller**	manufacturability and performance	state-based system monitoring and diagnostics	knowledge of own operation, predictive ability, comparability ability	real-time preventive status indicators
manufac-turing system	**networked**	performance and total asset efficiency	Lean operations: work and waste reduction	self-configuration, self-maintenance, self-organizing ability	risk exemption, performance

The factory of the future will produce products in an intelligent and fully integrated way, flexibly and efficiently. Separate and autonomous applications will form a unified system in an integrated network. Decentralization takes place in decision processes and enables real-time autonomous decisions at machine level and flexible decision-making about the production processes based on timely data. This requires the rethinking of the traditional business approaches and solutions. Better monitoring of products and production processes can also increase the relationship with suppliers, and controlling and financial planning can also provide greater precision, thus making more informed decision-making. Changes will be enforced by growing customer expectations, and companies will need to provide quick and tailor-made responses to these changes [7].

Cyber-Physical-Productive Systems (CPPS) do not only incorporate production tools into intelligent networks, but integrate the entire supply chain. Clever production tools share information about their utilization and status (for example, maintenance) and decide on their own issues independently. Shared information is achieved by using simulation and optimization tools, and optimally, in a completely independent manner. In order to increase efficiency, utilize capacities, reduce resources, improve quality and reduce lead times, the goal is to coordinate processes [8].

Cyber-Physical Systems are able to collect data from their environment using sensors and act after analyzing their situation [9]. Cyber-physical systems are networked, a significant part of them are interconnected, so it is possible to use swarm intelligence (the application of a common strategy in operation), which results in even more efficient operation (Figure 1).

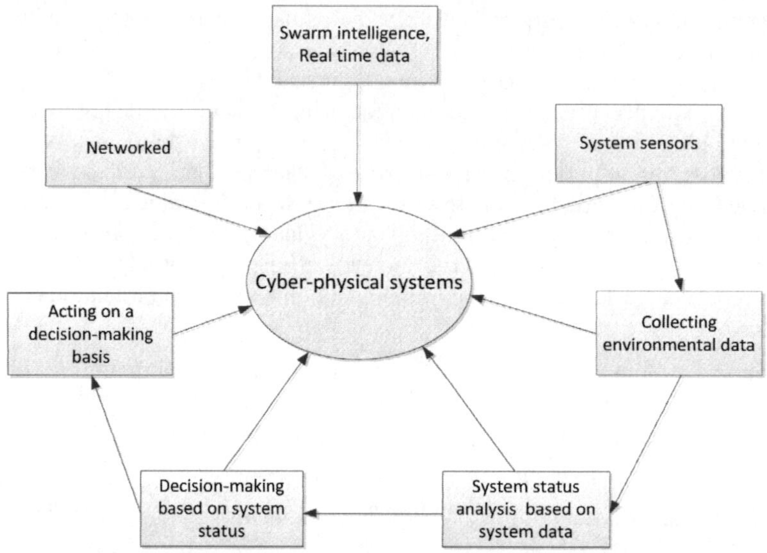

Figure 1. Cyber-Physical Systems [9].

The fourth industrial revolution has many advantages for economic operators, but at the same time, companies have to face many problems during its implementation.
Challenges in Industry 4.0:

- The security problems of information technology (IT) are greatly aggravated by the reuse of previously completed projects.
- The need for reliability and stability for critical machine-to-machine communication (M2M) including very short and stable waiting times.
- The integrity of production processes must be maintained.
- Avoid any unforeseen obstacle to information technology that can cause expensive production downtime.
- Industrial know-how must be protected.
- There is a lack of appropriate expertise that would speed up the progress towards the fourth industrial revolution.
- Redundancy in information technology (IT).
- It is difficult for affected parties to change.
- Automated and IT-driven processes result in the loss of jobs, especially in the lower-tier social strata.

It is an important hindering factor for the spread of industry 4.0 achievements that standards, norms and certificates are not yet available to ensure the interconnection of different systems. The core technologies of the digital infrastructure are still slowly spreading and are not necessarily available in any supplier or customer organization, so cooperation may be limited. Data security is a critical point for development.

Access control must be ensured, as well as the security of the network, the devices, the sensors, etc., together with the proper encryption of information. The production of smart products requires a new kind of technology, skills and processes throughout the entire value chain[7]. The company needs to realistically see what capabilities it can develop itself and when it needs to involve an external partner.

3. Impact of the fourth industrial revolution on logistics

The expansion of the important role of logistics in the economy is significantly supported by the rapid and dynamic development of informatics and cybernetics. The emergence and application of cyber-physical systems in the logistics processes leads to further significant changes in the operation of the economy. Of course, it should not be forgotten that logistics is a human-machine-based system based on material handling machines, machine systems, IT systems and networks, the techniques and technology of these and the appropriate standard of human vocational training.

Current trends in logistics are influenced by changes in the economy and society. The most important parameters affecting the economy are illustrated in Table 2.

Table 2. The most important parameters affecting the economy.

Changes in the market and in production	Globalization and localization
	Distributed intelligence based network systems
	Different solvency market needs
	The development of informatics and information technology
	Product structure change
	Decrease in production depth
	Decrease in Product Lifecycle
	Decrease in product ranges
	Concentration on core activity
	Make or buy philosophy
	Outsourcing activities
	JIT-based deliveries
	New interpretation of quality

Regarding the development of informatics and information technology, we need to highlight digitalization, which has fundamentally changed the market expectations and opportunities for logistics. In the field of logistics, the diversity of products being handled, the new techniques and technologies introduced in production and service, the operation of the various networks and the need for globalization make it necessary to increase the complexity of logistics systems while ensuring the transparency of such systems. This can be solved only with automated and optimized complex structures, processes, and subprocesses.

When designing these types of systems, the various cyber-physical systems play a decisive role. The operation of a complex logistic process is influenced by a variety of parameters. Real-time values of the parameters can be collected by sensors. By gathering as much information as possible about the operation of the system, digitalizing these and applying real-time evaluations the real state of the system/constituent can be determined. Using the right choice of optimization method, the best way of intervention between the real state of the system and the possible decision set can be implemented in the logistics system. The system or system element can be networked. Through digitalization, a large amount of real data can flow between systems and system elements, which provide the optimum operation of a complex system when the proper operating strategy based on the given criteria is applied [10].

In the next period, digitalization completely transforms the economy and life of the whole human race. In the area of production and services, more and more operational parameters of logistics activities become accessible, while in the frame of "the Internet of Things", the ensuring of the management of data sets results in the strengthening of the customer controlled activities in logistics services. In the field of logistics, digitalization takes place in the direction of standardized relationships instead of individual relationships and individual business processes, then towards the direction of the development phase of bidirectional direct connections and the provision of highly flexible, large-scale information access to external partners through integrated connections.

When examining the impact of Industry 4.0 on logistics, we have to take care of both intra-corporate and non-corporate logistics processes. The planning, design and operation of intra-corporate logistics processes are significantly influenced by the achievements of the fourth industrial revolution, allowing companies to track customer needs and procurement orders in real time. Based on these, they can plan their own capacities and resources to prepare for receiving and storing the components about to be delivered, and for the steps and demand of serving the production. In the case of distribution logistics, the basis for planning the finished product stocks, the warehouse locations and the deliveries is again provided by the production plan. Thus, logistics capacity planning can be realized at the same time as production planning, relying on the real-time data from the latter. Thanks to the automation, robotization, and development of logistics processes based on Industry 4.0 technology, the performance of the personnel performing logistics operations can be increased or, where appropriate, replaced.

Freight transport is a critical point for logistics processes outside the company. Using the achievements of the fourth industrial revolution, significant efficiency gains in freight transport can be achieved. Sensors built into products ensure product tracking, fix shipping conditions, facilitate identification of goods at loading and unloading, significantly reducing their time. Logistics processes, including freight transport, are fundamentally altered by the appearance and spread of driverless vehicles. These devices have many benefits: they will make transport safer, fuel consumption and pollution will be reduced, while there will be no need for mandatory rest periods for carriage which increase downtime, thus reducing lead times. Many other factors in road transport will also change with the emergence of autonomous vehicles: transport distances will be shortened due to the relocation of production; reduced shipping rates and consequently increased frequency; city logistics is more emphasized; Drivers can avoid traffic jams by using RFID technology and intelligent applications; applications will suggest the best route based on user community information rather than on GPS data; smart solutions improving road safety will be further developed.

3.1. Increasing the efficiency in logistics networks

The digitalization and transparency of production processes create the possibility for the procurement to rely on the prepared forecast and production plan, and based on these obtain the necessary raw materials and sub-assemblies. With the help of plans and production information that are constantly up-to-date based on real-time data, it can be accurately calculated that how much raw materials will be needed and where, which can increase the performance of the company's procurement logistics system. Further efficiency gains can be achieved through the integration and coordination of activities within and between companies, which is the essence of the supply chain approach (Figure 2).

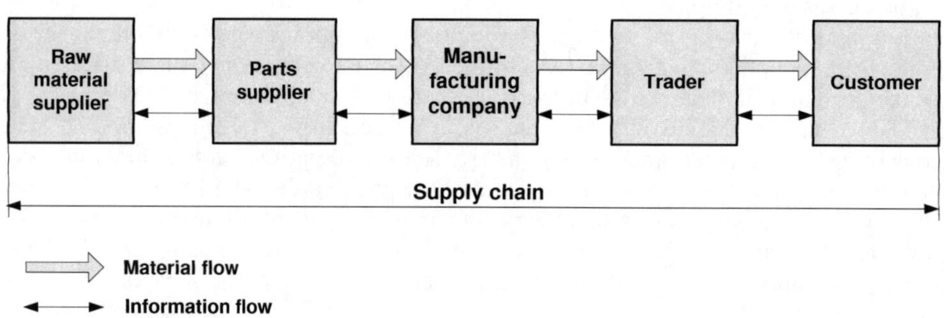

Figure 2. Material and information flow in the supply chain.

Nowadays, not isolated companies, but rather more or less co-operative supply chains compete with each other in order to meet the customer needs ever more efficiently. The need to organize supply chains is justified by the increase in the economic role of supply [11].

One of the most cost-effective solutions to enhancing competitiveness and reducing inventories is provided by networking and network-like operation. Today, due to the complexity of the supply chain, it is more appropriate to use the term of a supply network, since chain members must build a complex network of activities to produce a single product. Accordingly, the supply chain is a process chain of the network along which the product can be traced from the supplier to the final consumer. Thus, the

IOP Conf. Series: Materials Science and Engineering **448** (2018) 012014 doi:10.1088/1757-899X/448/1/012014

processes go beyond the boundaries of the company, the goal is to enhance the overall chain performance.

Tracking a cross-company relationship system is a very difficult task. Appropriate and quick response to any mistakes on the market is a great advantage as costs can be better planned. The collection and processing of good quality data in appropriate quantities requires a high level of IT and automation. Industry 4.0 provides an effective solution to this problem, as in addition to manufacturing systems, it aims to increase the efficiency of supply networks as well [12].

It assumes high level of flexibility and transparency in real time, in other words, the operation of real-time optimized new value creation networks. The smart factory adapts to the new circumstances of its environment at all times and optimizes the production processes by itself. The sharing of information on the production systems is assumed by this concept, which requires the development of a new business model for cooperation between organizations. This is achieved through integration with suppliers and customers in the value chain. The new business models should focus primarily on developing new collaboration models with business partners and handling customer specific expectations, which will require new forms of information sharing with customers and suppliers. It is important to mention that continuous communication is required throughout the product's lifecycle, i.e. the management of customer requirements, research and development, procurement, production, and market data all have to be covered. Smart products have information on their own manufacturing processes and collect and transmit data from the production and use phases of their life cycle. The development of smart logistics systems, in addition to information supply over the supply chain, means the development of independent and flexible external and internal logistics solutions.

The quality of the relationship between members of the supply network is basically determined by the use of integrated information systems and advanced forecasting methods. The effectiveness of the link depends on the accuracy of the sales information provided by the buyer and the supplier's willingness to keep the inventory:

$$M_h = f(I_m;\ K_k) \tag{1}$$

where

M_h - the effectiveness of the link between members of the network,

I_m - the quality (accuracy) of the sales information provided by the buyer,

K_k - the inventory keeping ability of the supplier.

Based on the above, we can state that the level of cooperation between the members of the network can be increased by providing the supplier with more accurate sales information from the buyer, as the supplier's willingness to maintain the inventory increases in a straightforward manner.

4. Increasing efficiency through the optimization of logistics networks

Ensuring the proper operation of modern factories today is almost unimaginable without the existence of such logistic systems which operate in a harmonized manner with the production processes. This is even more true for smart factories resulting from the fourth industrial revolution, where real-time product tracking and JIT (Just In Time) principles are virtually binding on both the delivery and distribution sides. In order to ensure the above conditions, logistics service providers involved in the supply chain must, of course, have to meet a number of often difficult to quantify quality requirements. As a result, the performance assessment of modern, multi-player logistics networks is a distinct complex problem. In the followings, an example for an advanced methodology will be presented, one that makes possible the optimal selection of logistics service providers in the network from a reliability point of view, building on the infocommunication solutions of Industry 4.0 and on the tools of risk management and quality assurance.

One of the key pillars of the methodology that is going to be presented is the application of the process capability concept. The concept of process capability is derived from statistical process control (SPC), which first appeared in the 1920s and 30s in the field of manufacturing and has now become one of the essential tools of quality assurance at a large number of production companies. It is important to emphasize that the process capability approach is increasingly being used in the service

IOP Conf. Series: Materials Science and Engineering **448** (2018) 012014 doi:10.1088/1757-899X/448/1/012014

sector as well, especially as part of Six Sigma, so its use is also gaining momentum in the field logistics [13]. This is supported by the increasingly wider range of logistics applications of Six Sigma, in particular within supply chains [14-19].

The concept of process capability is based on the recognition that fluctuations in characteristic parameters of industrial processes can usually be described by means of normal distribution (where this is not achieved, the approach can generally still be applied after appropriate normalization). This also means that from a sufficiently large sample of the relevant parameter in the process, the standard deviation (σ) and the expected value (μ) for the parameter can be deduced. The essence of the concept is that the statistical indicators thus obtained are compared to the technically permissible limits for the relevant parameter (LSL - lower specification limit, USL - upper specification limit), thereby giving an accurate picture of the future performance of the process. From a practical point of view, the so-called critical process capability (c_{pk}) is particularly important, which is defined by the following relationships [13]:

$$c_{po} = \frac{USL - \mu}{3\sigma} \tag{2}$$

$$c_{pu} = \frac{\mu - LSL}{3\sigma} \tag{3}$$

$$c_{pk} = min\{c_{po}; c_{pu}\} \tag{4}$$

The central idea of the procedure being presented is that with the knowledge of the right historical data, a cumulative critical process capability index can be calculated for any arbitrary combination of logistics service providers and logistics processes. The goal is to select the provider-process combination in such a way that this value is maximized. This is mathematically expressed by the following goal function:

$$kc_{pk} = \sum_{j=1}^{m} \frac{q_j \cdot p_j}{P} \cdot \sum_{i=1}^{n} \sum_{k=1}^{r} c_{pk}(i,j,k) \cdot pr_k \cdot x_{ijk}^{stk} \rightarrow Max \tag{5}$$

The meanings of the variables which are present in the goal function are the following:

- q_j: the quantity of the goods that has to be handled in the j-th process,
- p_j: the average value of a single unit of the goods which has to be handled in the j-th process,
- P: the total value of the goods which are going to be handled in the examined processes during the relevant time interval: $\mathrm{P} = \sum_{j=1}^{m} p_j$
- $c_{pk}(i,j,k)$: the calculated value of c_{pk} for the j-th process at the i-th provider in relation to the k-th risk parameter,
- pr_k: the weight of the k-th risk parameter (this can be determined with the use of the risk model which is going to be presented later),
- x_{ijk}^{stk}: the binary decision variable that is used for the selection of the relations included in the solution (from this, it can be seen that the problem in this form esentially can be solved with the use of linear programming).

From the foregoing, it can be seen that, for the application of the outlined concept, a great deal of historical data is needed, which are primarily necessary for the determination of statistical parameters. Furthermore, it is necessary to collect these data both from the service providers and from the production companies (which fulfill the role of clients) in the network, thus assuming the application of a network model based around information sharing. The scope of the information to be collected, processed and produced in connection with the procedure and the relationships between the data objects are presented in the following data model (Figure 3):

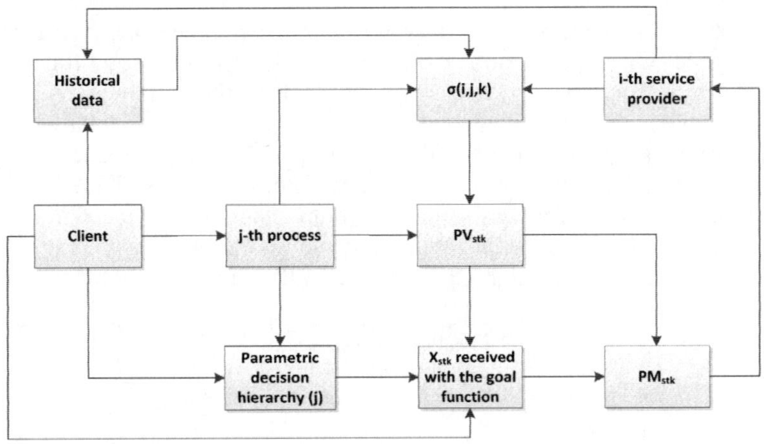

Figure 3. High-level data model defining the information sharing model.

The data model includes the following types of data objects and their relationships:

- Client type data objects represent the production companies that use the system. The information to be obtained from them is mainly necessary to determine two other data object types, the logistic processes and the parametric decision hierarchy (the latter will be discussed later).
- The j-th process data objects are assigned to the clients. The main data elements of the data type are naturally the tolerance intervals which the client sets for the process in relation to each risk parameter.
- The Parametric Decision Hierarchy data object can be considered as a hierarchical data model itself, which forms the risk model within the procedure. Essentially, it is a decision model created by using the AHP (Analytic Hierarchy Process) method. The parameterization of the model can be carried out by the particular client, thus enabling the procedure to be widely applicable (later the actual decision model itself is also going to be presented).
- The i-th service provider type data object corresponds to the i-th service provider that joins the system. Among the information related to the service provider, the most important are those $\sigma(i,j,k)$ standard deviation values which characterize the given service provider from the aspect of maintaining the k-th parameter of the j-th process. These are also necessary for calculating the respective $c_{pk}(i,j,k)$ values.
- σ(i,j,k) is both connected to the i-th service provider and the j-th process data objects, and hence itself has been represented as a separate object. Determining the value of $\sigma(i,j,k)$ can be done in two ways, from which the simpler solution is possible if the provider has already provided the process. If this criterion is not met, or simply because of the nature of the parameter k, a greater amount of data is needed, than the value of $\sigma(i,j,k)$ can be also calculated on the basis of the existing σ values of the other processes in the same category at the given service provider.
- The Historical data object contains all the data (past values of the relevant logistics parameters), based on which the value of $\sigma(i,j,k)$ can be determined for each service/ process/risk factor relation.
- The PV_{stk} data object represents a three-dimensional matrix that contains the relevant c_{pk} (i,j,k) values for each provider/process/risk factor relation (that is, the critical process capability that can be achieved in a particular relation).
- The X_{stk} received with the goal function data object contains the provider/process/risk factor relations that have been selected using the goal function (the data object here is also a three-dimensional matrix, but its elements in this case are binary values).

- Finally, the PM_{stk} data object contains the $c_{pk}(i,j,k)$ values for the selected provider/process/risk factor relations (therefore it is also a three-dimensional matrix).

As mentioned several times, the weighting of each risk factor in the target function is determined using a separate risk model that was created using the AHP method. AHP itself is one of the most widely used decision making methods that is widely used in a number of areas, including logistics, with particular regard to supplier, site and route selection [20-22]. In the current procedure, the purpose of setting up the AHP model was therefore to determine the weighted ranking of the risk factors for the particular client. The basis for the comparison is the question of how much the individual risk factors contribute to the realization of the additional risk costs on the side of the client, from the aspects of each decision criteria? The structure of the decision hierarchy, i.e. the risk model, is shown in Figure 4:

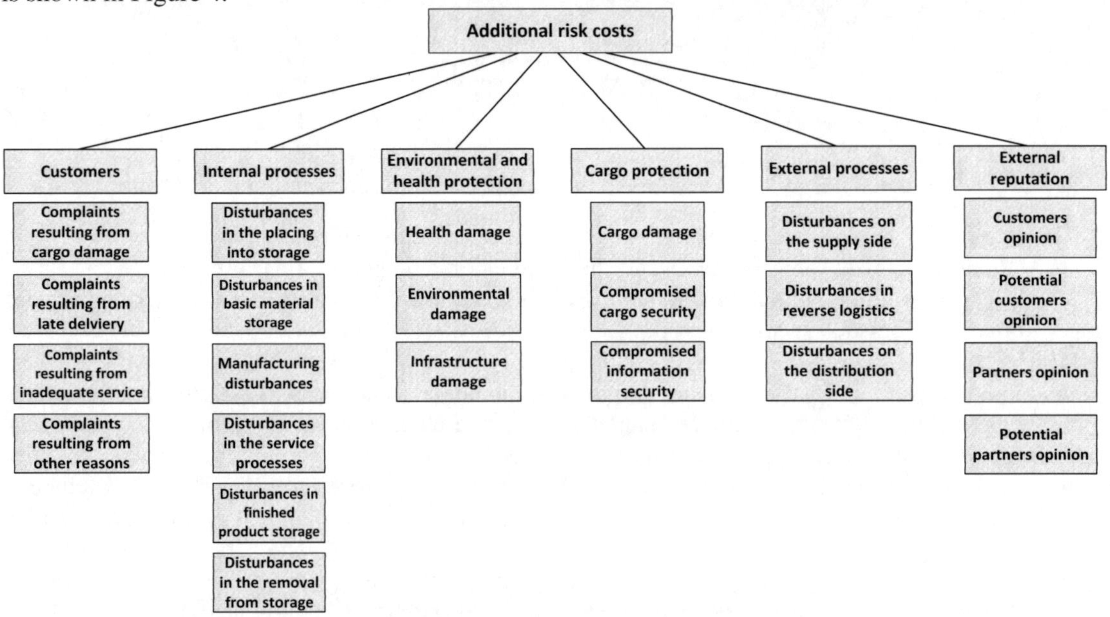

Figure 4. The risk model implemented as a decision hierarchy.

As can be seen from the foregoing, the process outlined above thus largely builds on information sharing between the parties involved in the logistics network. Of course, the use of Industry 4.0 infocommunication solutions is necessary in order to implement the required level of information exchange. One of the most promising foundations in this respect could be the so-called RAMI 4.0 - Reference Architectural Model Industrie 4.0 - concept, which has been recently developed by participant organizations of the Platform Industrie 4.0 in Germany. RAMI 4.0 basically identifies six main horizontal layers, which are from top to bottom: the "Business" layer, the "Functional" layer, the "Information" layer, the "Communication" layer, the "Integration" layer, and the "Asset" layer. For the outlined process, the most important issue is, of course, the implementation of the Information and Communication layers. As can be seen in Figure 5, one of the most obvious solutions for this is the use of the OPC UA - Open Platform Communications Unified Architecture - standard. It is very important that OPC UA is an existing standard that allows a wide range of industrial equipment to be interconnected (the specification is found in IEC 62541). Another advantage is that it is also a widely applicable standard, thanks to the applicability of companion specifications and extended information models. It is also very important that OPC UA is compatible with both the IPv4 and IPv6 standards. For these reasons, it could be useful in many ways to apply it to logistics systems that work in accordance with the Industry 4.0 concept. In case of the described methodology, it may offer a particularly advantageous solution for the sharing of information between the actors in the logistics network. The communication patterns of OPC UA are illustrated in the following figure [23].

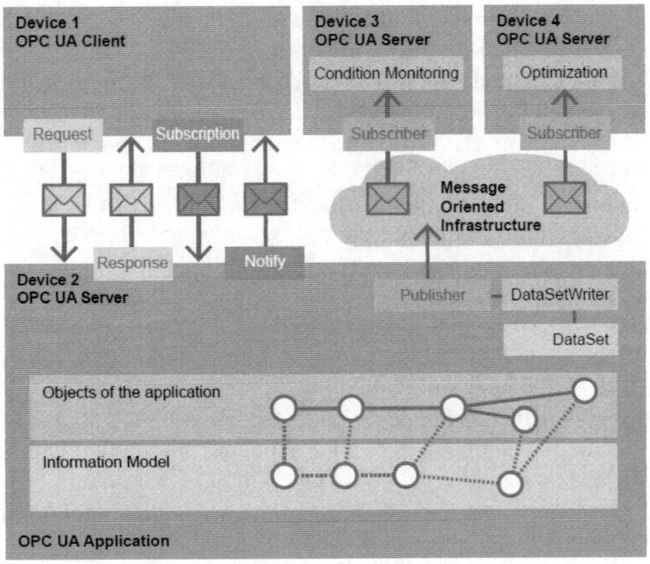

Figure 5. OPC UA Communication Patterns [23].

All in all, it can be stated that RAMI 4.0 and OPC UA could already jointly offer solutions to the application of the presented procedure in an Industry 4.0 environment. At the same time, it is also important to note that the development of the process was justified by the fact that the high degree of digitalization and extensive information sharing, which go hand in hand with the Industry 4.0 concept, will require the use of such risk management tools in logistics networks that can facilitate effective decision-making based on an increasingly large number of data.

5. Summary

Today, due to increasing complexity in supply chains and uncertain supply and demand, ensuring flexibility in future supply chain planning will be one of the most important tasks. One of the most important prerequisites for supply chain flexibility is the development of network-based collaborations and the use of Industry 4.0 developments.

The study analyzed the operational processes of logistics networks and examined how the efficiency of logistics processes could be increased by exploiting the opportunities offered by the fourth industrial revolution. Based on Industry 4.0's infocommunication solutions, we introduced a methodology that allows the optimization of logistics networks using the tools of risk management. During the development of the presented research work, we have set the goal of developing a risk management process that is suitable for optimally assigning the service providers and the logistics processes to each other within a given network from a risk-based perspective.

Acknowledgments

This project has received funding from the European Union's Horizon 2020 research and innovation programme under grant agreement No 691942. This research was partially carried out in the framework of the Center of Excellence of Mechatronics and Logistics at the University of Miskolc.

6. References

[1] Heng S 2014 Industry 4.0: Upgrading of Germany's Industrial Capabilities on the Horizon https://ssrn.com/abstract=2656608

[2] Kása R, Gubán Á, Gubán M, Hua N S and Molnár L 2014 The concept of perception driven service process reengineering by entropy reduction *Pannon Man Rev* **3**(1) 11-54

[3] Kundrák J, Deszpoth I and Molnár V 2014 Comparative study of material removal in hard machining of bore holes *Tehnicki Vjesnik* **21**(1) 183-89

IOP Conf. Series: Materials Science and Engineering **448** (2018) 012014 doi:10.1088/1757-899X/448/1/012014

[4] Kulcsár G and Kulcsárné Forrai M 2013 Detailed Production Scheduling based on Multi-objective Search and Simulation *Prod Sys and Inf Eng* **6** 41-56
[5] Bányai T 2002 Clusters vs virtual enterprises *Mechanics and Mechanical Eng* **6**(2) 17-24
[6] Lee J, Bagheri B and Kao H A 2014 Recent Advances and Trends of Cyber-Physical Systems and Big Data Analytics in Industrial Informatics *Proc. Int. Conf. on Industrial Informatics* (Porto Alegre) (IEEE)
[7] Nagy J 2017 Industry 4.0: definition, elements and effect on corporate value chain http://unipub.lib.uni-corvinus.hu/3115/
[8] Abonyi J and Miszlivetz F 2016 Intersections of networks: Social challenges of I4.0 (Szombathely: Savaria University Press)
[9] Tamás P and Illés B 2016 Smart logistics in manufacturing *Gépgyártás* **55**(2) 51-6
[10] Illés B, Varga A K and Czap L 2018 Logistics and digitization *Lect Notes in Mech Eng* **2018** 220-25
[11] Szegedi Z 2012 Supply chain management (Budapest: Kossuth Kiadó)
[12] Nagy G, Bányai Á, Illés B and Glistau E 2018 Analysis of supply chain efficiency in blending technologies *Lect Notes in Mech Eng* **2018** 280-91
[13] Illés B, Glistau E and Machado N I C 2007 *Logistik und Qualitätsmanagement* (Budapest: Budai Nyomda) p 195
[14] Garg D, Narahari Y and Viswanadham N 2004 Design of six sigma supply chains *IEEE Trans on Autom Sci and Eng* 38-57
[15] Narahari Y, Viswanadham N and Bhattacharya R 2000 Design of synchronized supply chains: A six sigma tolerancing approach *Proc Int Conf on Robotics and Automation* (IEEE) 1151-56
[16] Christopher M and Rutherford C 2004 Creating supply chain resilience through agile six sigma *Critical eye* **24** 28
[17] Knowles G, Whicker L, Femat J H and Canales F D C 2005 A conceptual model for the application of Six Sigma methodologies to supply chain improvement *Int J of Log: Res and Appl* **8**(1) 51-65
[18] Dasgupta T 2003 Using the six-sigma metric to measure and improve the performance of a supply chain. *Total Quality Management and Business Excellence* **14**(3) 355-366
[19] Lee M-C and Chang T 2010 Developing a lean design for Six Sigma through supply chain methodology *Int J of Prod and Qual Manag* **6**(4) 407-34
[20] Vaidya O S and Kumar S 2006 Analytic hierarchy process: An overview of applications *Eur J of Oper Res* **169**(1) 1-29
[21] Ho W 2008 Integrated analytic hierarchy process and its applications–A literature review *Eur J of Oper Res* **186**(1) 211-28
[22] Sipahi S and Timor M 2010 The analytic hierarchy process and analytic network process: an overview of applications Manag Dec **48**(5) 775-808
[23] VDMA 2018 Industrie 4.0 Communication Guideline Based on OPC UA https://industrie40.vdma.org/documents/4214230/20743172/Leitfaden_OPC_UA_Englisch_1506415735965.pdf/a2181ec7-a325-44c0-99d2-7332480de281

IOP Conf. Series: Materials Science and Engineering **448** (2018) 012015 doi:10.1088/1757-899X/448/1/012015

The Influence of Process Parameters of Injection on Nano-mechanical Properties of Polypropylene

Martin Ovsik, Lenka Hylova and Tomas Fiala

Tomas Bata University in Zlin, TGM 5555, 760 01 Zlin, Czech Republic

ovsik@utb.cz

Abstract. Injection of polymers is one of the most widely used technology in the production of plastic parts. During injection moulding it is important to analyse the individual factors of the injection process so that the optimal injection parameters can be set from these established process conditions. When injecting, it is important to achieve the optimal production process to reach optimal production costs, increasing the quality of the product and eliminating possible defects in the products. The presented work deals with the investigation of the influences of the process parameters of the injection process (holding pressure and mould temperature) on resulting nano-mechanical properties. To measure nano-mechanical properties by the DSI method, a polypropylene sample was chosen. Measurements were performed on the Nano-Combi-Tester. From the results of this study it is clear that optimal nano-mechanical properties are required to optimize the process parameters of the whole injection moulding process.

1. Introduction

Polypropylene is a common thermoplastic polymer, largely used in industrial applications for the variety of its properties, which makes it versatile. PP components are semi-rigid, translucent, fatigue and heat resistant, tough and chemically durable. The main manufacturing processes for polypropylene parts are extrusion and injection moulding (IM). Through these technologies a numerous kind of products are available: buckets, bowls, crates, toys, medical components, washing machine drums, battery cases, bottle caps, etc. In the large-scale industrial production world, plastic parts made by injection moulding re getting a key role, because of their affordability and lightness, especially when there are no requirements for mechanical properties. Although the process optimization often focuses on reducing the cycle time (in accordance with dimensional and aesthetic specs), more and more frequently it's fundamental that the manufactured parts have mechanical performances. These an be affected by process parameters, which can induce residual stresses to the component due to deformation at high shear rates, temperature and pressure. Right combination of these parameters can optimize the properties of the component. In general, research activity on injection moulded polymers' characterization by process parameters has been developed by several groups and different approaches have been used. Typically, IM input parameters are: injection speed, melt temperature, mould temperature, packing time, packing pressure, cooling time. [1]

Injection Moulding is one of the most advanced polymer processing technologies. It enables the production of finished products, which do not require any further operations. The tools used for their production - the injection moulds - are very complicated assemblies that are made using several technologies and materials. [2]

The fluidity of all polymers during injection moulding cycle is affected by many parameters (mould design, melt temperature, mould temperature, injection rate, pressures, etc.) and by the flow properties of polymers. A plastic nucleus is formed by due to character of laminar flow, which enables the compression of the melt in the mould and consecutive creeping. A constant flow rate given by the axial movement of the screw is the choice for most of the flows. During filling of the mould cavity the plastic material does not slide along the mould surface but it is rolled over. This type of laminar flow is usually described as a “fountain flow” (Figure 1). [2]

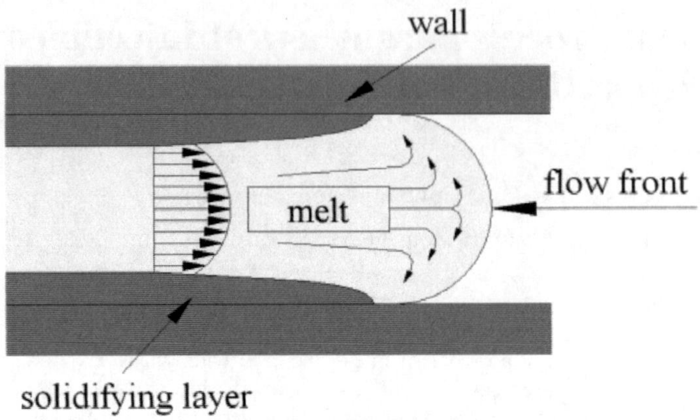

Figure 1. Fountain flow.

The article deals with the influence of processing parameters of the injection moulding technology (holding pressure and mould temperature) on nano-mechanical properties (hardness, modulus) of the surface layer of tested PP.

2. Experimental

Measurement of all properties mentioned above was performed 10 times to ensure statistical correctness.

2.1. Material

Polypropylene PP IM 5580 TST was used for this experiment.

2.2. Injection molding process

The samples were made using injection moulding technology on an Arburg Allrounder 470H injection moulding machine (Loßburg, Germany). The normalized specimens, with dimensions of (80 x 10 x 4) mm, were used (Figure 2). The process parameters were set according to the manufacturer’s recommendations; see Table 1 and Table 2.

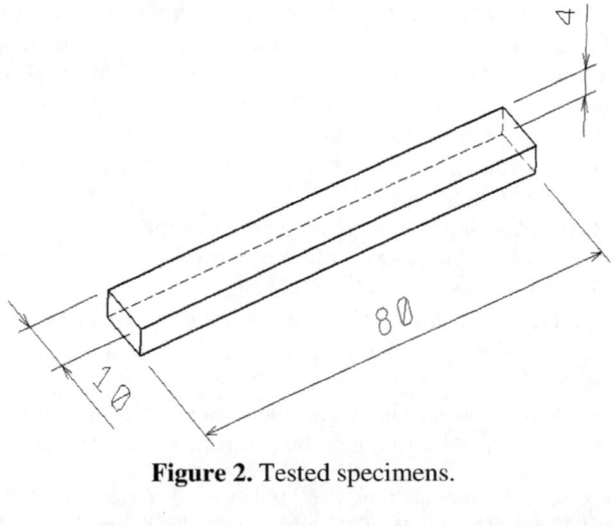

Figure 2. Tested specimens.

Table 1. Process parameters – changing of holding pressure.

Parameter	Unit	Value
Injectionpressure	bar	900
Holding pressure	bar	300, 400, 500, 600, 700, 800
Holding time	s	15
Coolingtime	s	35
Fill speed	mm/s	40
Temperature T1	°C	200
Temperature T2	°C	210
Temperature T3	°C	220
Temperature T4	°C	230
Mouldtemperature	°C	80

Table 2. Process parameters – changing of mould temperature.

Parameter	Unit	Value
Injectionpressure	bar	900
Holding pressure	bar	700
Holding time	s	15
Coolingtime	s	35
Fill speed	mm/s	40
Temperature T1	°C	200
Temperature T2	°C	210
Temperature T3	°C	220
Temperature T4	°C	230
Mouldtemperature	°C	30 , 40, 50, 60, 70, 80, 90

2.3. Nano-indentation test

Nano-indentation properties were measured by means of Nano Combi Tester NHT[3] (Figure 3) made by Anton Paar (Graz, Austria), according to the CSN EN ISO 14577 standard. On tested material at least ten indents were made and the results were statistically processed. Standard simple loading-unloading mode was used. From the obtained "load vs. depth of penetration" curves values of nano-indentation properties were calculated according to Oliver and Pharr. The measured parameters were the values of the maximum load 50mN and the loading rate (and unloading rate) was 100mN/min. The holding time was 90 s at the indentation.

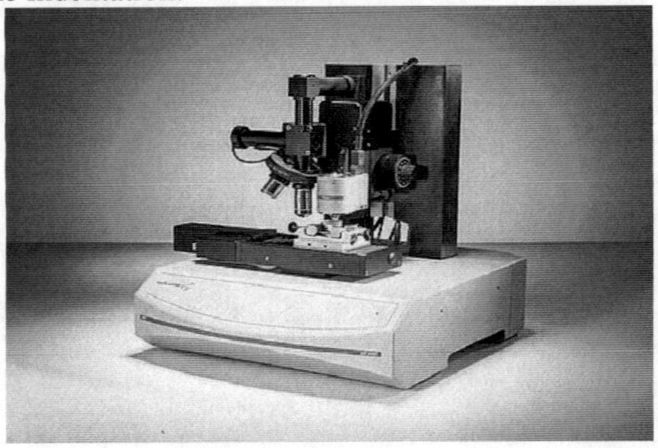

Figure 3. Nano-indentation tester.

The indentation hardness (H_{IT}) was calculated as maximum load (F_{max}) to the projected area of the hardness impression (A_p) and the indentation modulus (E_{IT}) was calculated from the Plane Strain modulus (E*) using an estimated sample of Poisson's ratio (ν) according to [3-10]:

$$H_{IT} = \frac{F_{max}}{A_p} \quad (1)$$

$$E_{IT} = E^* \cdot (1 - \nu_s^2) \quad (2)$$

3. Results and discussion

Holding pressure is one of the most important processing parameters of the polymer injection moulding technology, which significantly influences the quality of the injected product. After the phase of cavity filling the phase of holding pressure follows, which has the task to compensate the lowering of the volume (shrinkage) of the product during the polymer cooling in the mould. It is only possible to perform this process until it is impossible to fill the injection moulding cavity with the polymer melt. A bad choice of the holding pressure value can influence the mechanical properties of the resulting product.

Looking at the holding pressure results in the dependence on the nano-indentation hardness (Figure 4) it is clear, that holding pressure significantly influences the properties of the product surface layer. At the lowest value of holding pressure (300 bar), the nano-hardness was 60 MPa. The higher value of holding pressure, the higher values of nano-hardness up to 85 MPa. Higher holding pressure than 85 MPa causes the drop of nano-hardness. As it can be seen from the figure, as the optimal value of holding pressure is in range from 600 bar to 700 bar where the highest values of nano-hardness of the surface layer were measured.

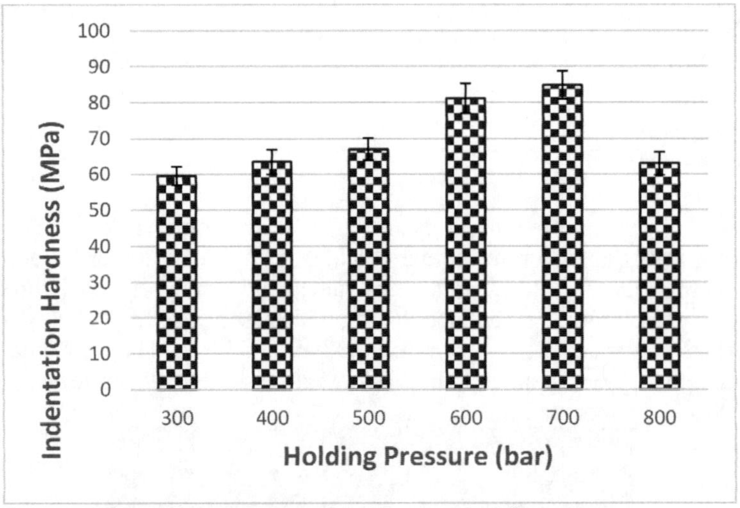

Figure 4. Indentation hardness vs. holding pressure.

Indentation modulus values that were measure yielded similar results to indentation hardness, Indentation modulus characterizes the stiffness of the surface layer as it is shown in figure 5. During instrumental test of nano-hardness the lowest value of modulus aas measured at holding pressure 300 bar (1.12 GPa), while the highest value was at holding pressure 700 bar (1.63 GPa). The difference between the lowest and the highest value of modulus is 45 %. The higher value of holding pressure causes lower indentation modulus.

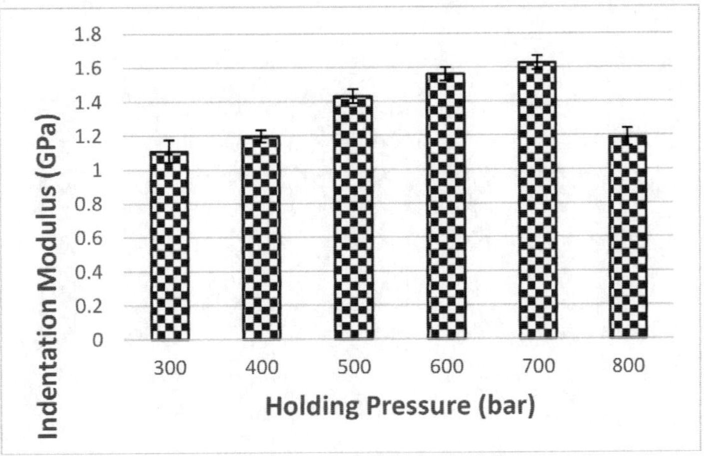

Figure 5. Indentation modulus vs. holding pressure.

Appropriate choice of tempering agent, its construction and dimensioning can achieve the required quality of the injected products, dimensionally accurate parts, quality surface, required physical and mechanical properties, minimal deformation, shortening of the injection cycle as a result of shorter cooling times and last but not least optimal production costs. The temperature and its time course influence the behaviour of the melt, (during and after process) including the resulting properties (mechanical, shrinkage, surface quality, degree of crystallinity, mass and density). Temperature over time also affects magnitude of internal stress and last but not least (both for amorphous and semicrystalline plastics) the total cycle time of the injection cycle, where the heat dissipation (cooling phase) occupies its predominant part.

As well as the pressure, the temperature of the mould has a significant effect on nano-mechanical properties of the surface layer of tested PP. As can be seen in figure 6, the low mould temperature causes indentation hardness reduction. Conversely, the higher mould temperature results in an increase in indentation hardness. The lowest indentation hardness value was measured at a mould temperature of 30 °C (60 MPa), on the other hand the highest at 80 °C (102MPa). As an optimal mould temperature range for PP injection, 60 °C to 80 °C appears to have the highest measured hardness value. The difference between mould temperature at 30 °C and 80 °C is 70%.

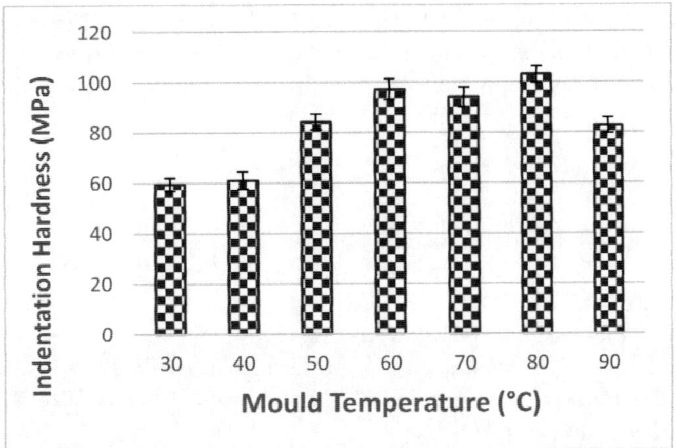

Figure 6. Indentation hardness vs. melt temperature.

From the results of the indentation modulus of elasticity, it is obvious that the optimal value of the mould temperature for tested PP is 80 °C (1.76 GPa). At a lower temperature, the modulus decreased significantly.

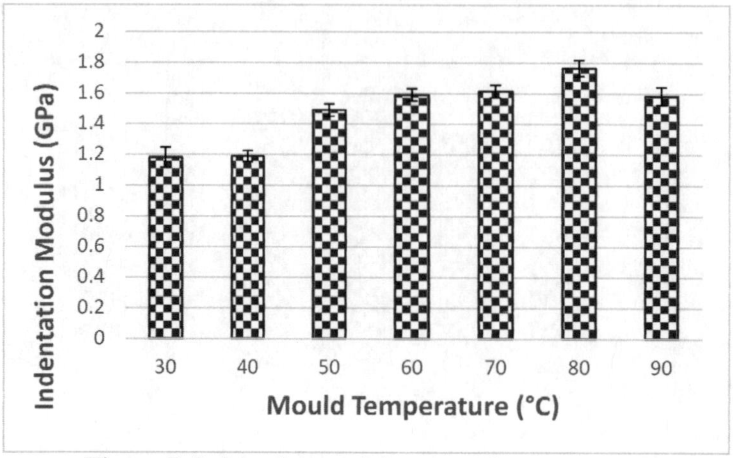

Figure 7. Indentation modulus vs. melt temperature.

4. Conclusions

To measure nano-mechanical properties by the DSI method, polypropylene sample was selected. Measurements were performed on CSM's Nano-Combi-Tester. When changing the pressure and temperature of the mould to resulting mechanical properties of the product was measured.

When changing the pressure, the nano-mechanical properties changed. The low holding pressure has manifested in indentation hardness and modulus values by decreasing the values compared to the higher holding pressure. This means that the ideal holding pressure is in range between 600 and 700 bar. Below 700 bar, its mechanical properties are the highest. At low pressure, the mould may be insufficiently filled, resulting in hardness reduction. The high value impact was manifested by extruding the melt out of the mould cavity.

Changes in mould temperature change the properties determined by the DSI method. At low mould temperatures, the hardness and modulus values are very low compared to the increasing mould temperature. The ideal mould temperature is at 80 ° C, here it achieves maximum of its mechanical properties.

Acknowledgments

This work was supported by the European Regional Development Fund under the project CEBIA-Tech Instrumentation No. CZ.1.05/2.1.00/19.0376 and by the Ministry of Education, Youth and Sports of the Czech Republic within the National Sustainability Program project no. LO1303 (MSMT-7778/2014). Moreover, it was supported by the Internal Grant Agency of TBU in Zlin: no. IGA/FT/2018/012.

References

[1] Farottia E and Natalinib M 2018 Procedia Structural Integrity **8** 256
[2] Stanek M, Manas D, Manas M, Skrobak A, Janostik V and Senkerik V 2016 *MATEC Web of Conferences* **76** 2020
[3] Pharr G M. 1998*Materials Science and Engineering*
[5] Oliver W C and Pharr, G M 2004*Journal of Materials Research* **19**1564
[6] Behalek L and Dobránsky J 2013*Applied Mechanics and Materials* **308** 127
[7] Dobransky J 2015*Metalurgija* **54** 477
[8] Manas D, Ovsik M, Mizera A, Manas M, Hylova L, Bednarik M, Stanek M 2018 *Polymers* **10** 5
[9] Manas D, Manas M, Mizera A, Navratil J, Ovsik M, Tomanova K, Sehnalek S 2018 *Polymers* **10** 6
[10] Fischer C, Seefried A, Merle B, Göken M, Drummer D 2017*Polymer Engineering and Science* **57**

IOP Conf. Series: Materials Science and Engineering **448** (2018) 012016 doi:10.1088/1757-899X/448/1/012016

Investigation of the influence of different burnishing parameters on shape correctness and residual stresses

G Varga[1], V Ferencsik[2]

[1,2] Institute of Manufacturing Science, University of Miskolc, Miskolc – Egyetemváros, Hungary

[1]gyula.varga@uni-miskolc.hu
[2]ferencsik.viktoria@uni-miskolc.hu

Abstract. Cold plastic procedures such as surface burnishing play an important role in life-enhancing machining methods. Application of this process the surface roughness improves, shape correctness also changes and the increases lifetime due to the compressive residual stress remains in subsurface are of the workpiece. This article deals with the change of these attributes and the correlation of these with the burnishing parameters: burnishing speed, feed rate, burnishing force and number of passes on diamond burnished low alloyed aluminium shafts. The experiments were executed by the Taguchi type full factorial experimental design method, the measurements of the cylindricity of the specimens were done with a circularity and position error machine while the measurement of residual stress was performed by X-ray diffraction method. The aim of this study is to define the appropriate range of technological parameters which result high improvements.

1. Introduction

In designing machining, it is increasingly common to choose methods for solving the manufacturing process by which the lifetime of the parts can be increased compared to the previous ones [1], while providing additional benefits as well [2]. Conventional machining processes such as milling and turning lead to a surface with inherent irregularities, while finishing processes such as grinding, lapping, polishing and honing commonly being employed to improve the surface finish. Unlike the above-mentioned traditional methods, which depend on chip removal, the burnishing process produces a smooth and work-hardened surface by plastic deformation of irregularities [1].

Burnishing differs from other cold-working surface treatment processes such as shot peening and sand blasting etc. produces a good surface finish, shape correctness and induces residual compressive stresses at the surface layers [4], thus quality of these layers has influence on fatigue life and work safety. The procedure of burnishing is performed for final finishing manufacturing of inner and outer cylindrical surfaces, the main areas of its applications are automotive and aeronautics industries.

The Taguchi type full factorial experimental design was used in this research [5-6], which is valid in between the minimum and maximum values of the input parameters. In the present experiments input parameters were: number of tool passes (i), feed (f), burnishing force (F) and burnishing speed (v_b), while the output parameters are: cylindricity (CYLt), tangential (σ_t) and axial (σ_a) residual stress.

2. Burnishing of outer cylindrical surfaces

Cold plastic processes, like burnishing, have always played an important role in industry and the first patent about this chipless procedure was published in the 1910s [7]. It has undergone tremendous progress over more than a century, adapting the ever-growing industrial and user needs.

Burnishing of outer cylindrical surfaces can be executed on conventional universal lathes or up-to-date CNC lathes.

In diamond burnishing the improving of shape correctness of the surface is caused by the interaction in between the tool - which material much harder than the workpiece - and the workpiece while performing rotational and linear movement with sliding friction between them [8]. The ball applies a normal force high enough to produce a plastic deformation on the surface which displaces the material from ridges to depressions on superficial micro-irregularities [9] (Fig.1.).

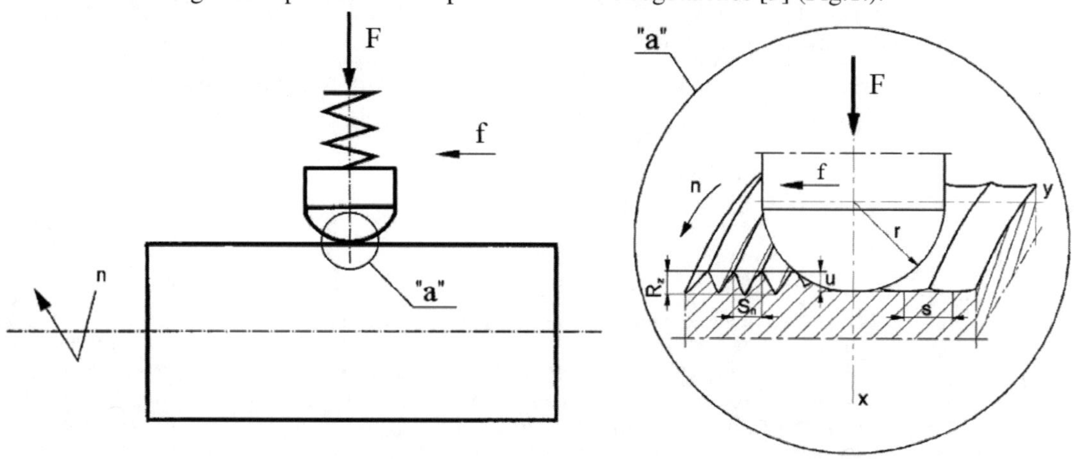

Figure 1. Schematic illustration of burnishing process [10]
F: burnishing force, n: workpiece rotational speed, f: feed, u: depth of tool penetration, Sn: spacing of irregularities of the treated surface, r: radius of the tool, Rz: previous surface roughness

This plastic deformation is realised in 0.01-0.2 mm thick layer in the subsurface of the workpiece because of the static contact shaping element (burnishing tool) and the outer surface of the workpiece [11-12].

3. Experimental conditions

3.1. Features of the investigated workpiece

For this experiment we have chosen lightly alloyed aluminium material. The reason of the decision of the examination of aluminium was supported by the fact that automobile, aeronautics and astronautics industries more and more claims for the application of non-ferrous materials due to its low density and good mechanical properties [11].

Table 1. Chemical composition of the aluminium alloy

Elements	Al	Si	Fe	Cu	Bi	Pb
Averaged wt% (weight percent)	92.11	0.19	0.84	5.65	0.46	0.74

Fig. 2 shows the geometric design of the workpiece which is justified by the chosen experimental design method, it will be detailed in a later chapter and we considered the risk of bending during machining by examination of the so-called l/d relationship that is the proportion of the length and diameter of the piece. Based on these, the capture of the specimen was clamped into a three jaw chuck and supported by a tailstock.

IOP Conf. Series: Materials Science and Engineering **448** (2018) 012016 doi:10.1088/1757-899X/448/1/012016

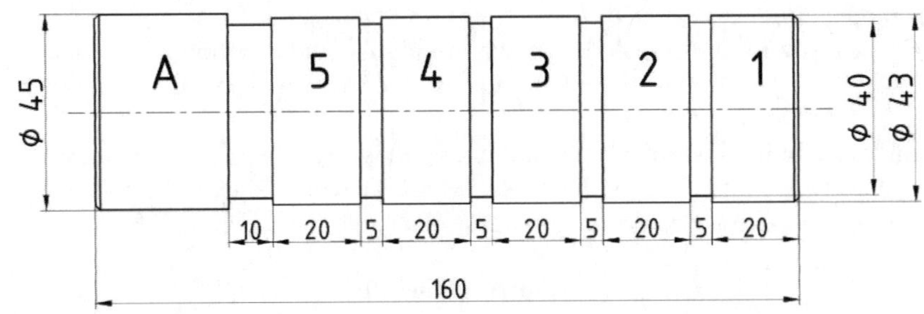

Figure 2. Geometrical dimensions of the workpiece

3.2. The execution of the burnishing process

During the experiments, a CNC lathe with flatbed by firm OPTIMUM type OPTIturn S600 was used which is in the workshop of Institute of Manufacturing Science at University of Miskolc. The tool tip was PCD (polycrystalline diamond) with 3.5 mm radius and the kinematic viscosity of the applied oil was 70 mm2/s. Fig. 3 shows the process in progress.

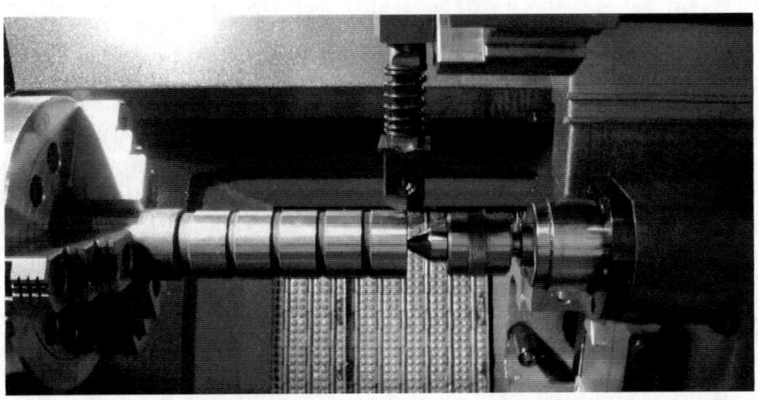

Figure 3. The burnishing process

The following burnishing parameters were examined: number of tool passes, feed rate, burnishing force and speed, all of it have a lower and a higher value according to the matrix of the Taguchi type Full Factorial Experimental Design can be seen on Table 2, which contains the burnishing parameters in natural dimensions and their transformed values.

Table 2. The applied burnishing parameters

Sign of specimen	Parameters of burnishing			Transformed parameters		
	i [ø]	f [mm/rev]	F[N]	x_1	x_2	x_3
1	1	0.001	10	-1	-1	-1
2	3	0.001	10	+1	-1	-1
3	1	0.005	10	-1	+1	-1
4	3	0.005	10	+1	+1	-1
5	1	0.001	20	-1	-1	+1
6	3	0.001	20	+1	-1	+1
7	1	0.005	20	-1	+1	+1
8	3	0.005	20	+1	+1	+1

3.3. Measuring of the cylindricity

Measurement of the cylindricity was done with a circularity and position error measuring machine type Talyrond 365 [14]. In this investigation an inductive sensor was applied for measuring before and after burnishing in 2 mm distance.

In all 16 cylindricity indices was analysed and one of them was chosen and examined which mostly determines operating properties. This is the so called CYLt which shows the total distance between peaks (CYLp) and valleys (CYLv) according to a reference cylinder (Fig. 4) [15].

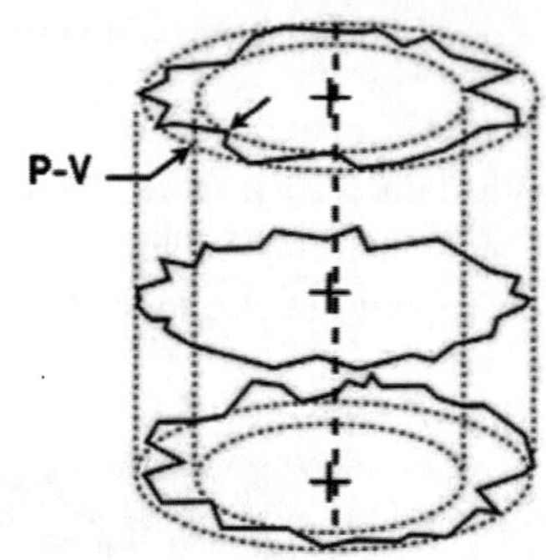

Figure 4. Cylindricity parameters: CYLp, CYLv and CYLt

3.4. Measuring of the residual stress

Residual stress is a process-induced stress, frozen in a moulded part, that exists in a body in the absence of external loading or thermal gradients. In a structural material or component, residual stresses exist in the object without the application of services or other external loads. They affect a part similarly to externally applied stresses.

In materials due to their macroscopic residual stress the atomic cores in the lattice points are put of their balance. In the point of view of crystallography, it means that the parameters of the lattice do changes. Distance between lattice planes are changed because of the residual stress, these distances can be measured, so the values of the stress can be calculated from them [16].

So, during measuring of residual stress, the wave-length of the X-ray is known, and the dislocation of the Bragg-angle is measured. The dislocation has been caused by the changing of lattice planar distances. The formula (1) was applied what is called Bragg-equation [17]:

$$n\lambda=2d_{hkl}\sin\Theta, \quad (1)$$

where:

n:	integer determined by order given
λ:	wave-length of X-ray
dhkl:	spacing between the planes in the atomic lattice
Θ:	angle between the incident ray and the scattering planes

Residual stress distributions in cold formed sections are dependent primarily on the plastic deformation in the manufacturing process. Thus, it is necessary to know the manufacturing process before measuring the residual stress. These methods can be classified into three types including destructive, non-destructive and semi-destructive one [18].

In this investigation the measuring process was executed with an X-ray diffraction measuring machine type Stresstech Xstress 3000 G3R, which realizes non-destruction test and complex shapes can be measured providing rotation of the measuring head is not restricted. Fig. 5 illustrates the basic principles of the described method.

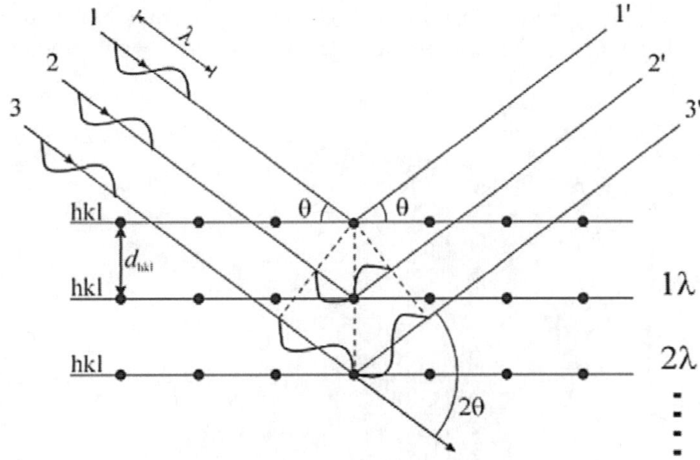

Figure 5. The X-ray principle

Measuring was done on 3 points at 60° in tangential and axial direction too because it was supposed that the direction of the processing has influence on the changing of the stress condition.

4. Results and evaluations

For evaluation of measured data improvement ratios were introduced, which are shown in formula (2) and (3):

$$\rho_{\sigma/CYLt} = \frac{|x_b|}{|x_t|} \quad (2)$$

$$\rho\% = (\rho - 1) \cdot 100\%, \quad (3)$$

where: ρ_σ Improvement ratios of residual stresses (σ) and cylindricity (CYLt). This is a dimensionless ratio, which textures the changes occurring because of manufacturing,

x_t Residual stress and cylindricity remain after turning,

x_b Residual stress and cylindricity remain after burnishing,

ρ% The percentage value of the improvement ratio.

The higher the value of ρ_σ, the greater the improvement due to burnishing, while in the point of view of cylindricity (ρ_{CYLt}), the smallest values of the ratios mean positive effect.

The measured data and the calculated improvement ratios are summarized in Table 3.

Table 3. Measured values and calculated improvement ratios

Sign of spec.	σ_t [MPa]		$\rho_{\sigma t}$ [%]	σ_a [MPa]		$\rho_{\sigma a}$ [%]	CYLt [μm]		ρ_{CYLt} [%]
	turned	burnished		turned	burnished		turned	burnished	
1	-9.5	-187.0	1868	74.5	-217.0	391.3	14.32	14.69	2.583
2	-9.5	-142.3	1397	74.5	-202.7	372.1	13.04	173.99	1234
3	-9.5	-194.1	1943	74.5	-300.3	503.1	15.83	13.01	-17.81
4	-9.5	-183.5	1831	74.5	-307.9	513.3	13.2	24.11	82.65
5	-9.5	-152.8	1508	74.5	-212.9	385.8	18.88	64.6	242.2
6	-18.5	-115.1	522.2	67.0	-167.1	349.4	16.41	201.54	1128
7	-18.5	-198.7	974.1	67.0	-308.9	561	12.36	23.7	91.75
8	-18.5	-132.5	616.2	67.0	-198.9	396.9	14.96	76.62	412.2

IOP Conf. Series: Materials Science and Engineering **448** (2018) 012016 doi:10.1088/1757-899X/448/1/012016

Application of Factorial Experiment Design method empirical formulas (4-6) were created from the calculated values. Calculations and axonometric figures (Fig. 6-8) were prepared using „Mathcad 16.0" software.

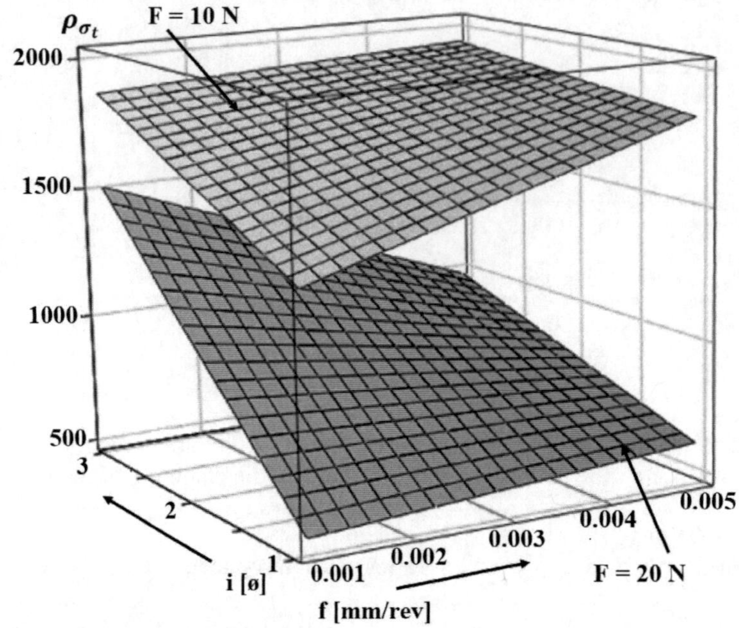

Figure 6. Changing of improvement ratio of tangential residual stress

$$\rho_{\sigma_t} = 2.0463 \cdot 10^3 + 10.637 \cdot i + 1.597 \cdot 10^5 \cdot f + 8.324 \cdot F + 1.126 \cdot 10^4 \cdot i \cdot f - 29.101 \cdot i \cdot F - -1.858 \cdot 10^4 \cdot f \cdot F + 3.361 \cdot i \cdot f \cdot F \quad (4)$$

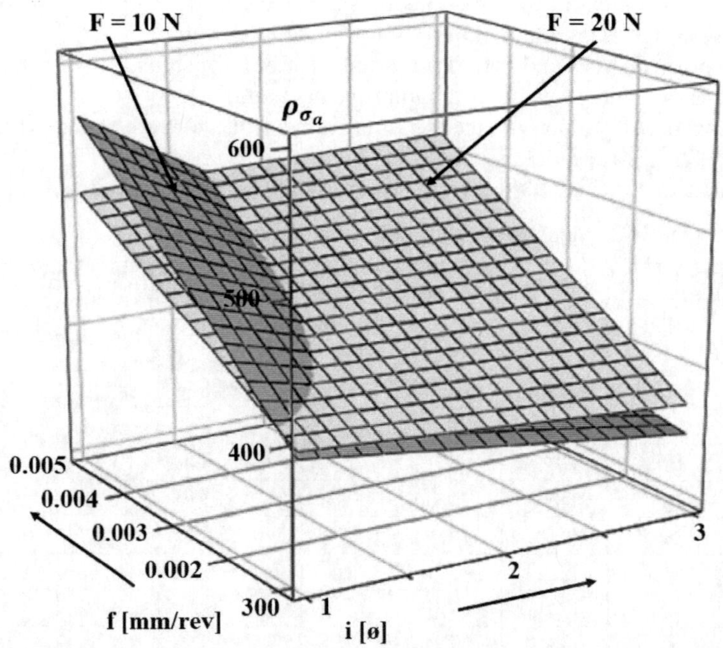

Figure 7. Changing of improvement ratio of axial residual stress

$$\rho_{\sigma_a} = 409.0125 - 24.312 \cdot i - 1.121 \cdot 10^4 \cdot f - 3.239 \cdot F + 2.331 \cdot 10^4 \cdot i \cdot f + 1.104 \cdot i \cdot F + \\ +3.549 \cdot 10^3 \cdot f \cdot F - 1.964 \cdot 10^3 \cdot i \cdot f \cdot F \quad (5)$$

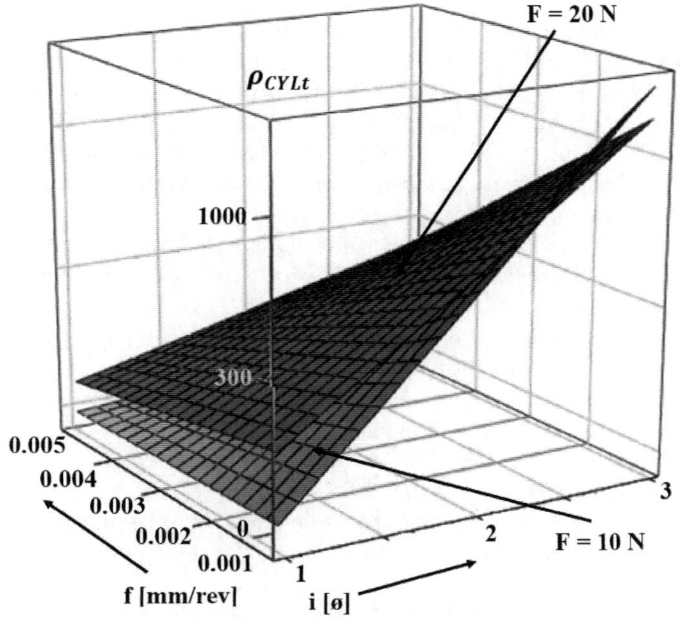

Figure 8. Changing of improvement ratio of CYLt cylindricity parameter

$$\rho_{\sigma_{CYLt}} = -1.265 \cdot 10^3 + 1.001 \cdot 10^3 \cdot i + 2.395 \cdot 10^5 \cdot f + 51.564 \cdot F - 2.121 \cdot 10^5 \cdot i \cdot f - 24.351 \cdot i \cdot F - \\ -1.032 \cdot 10^4 \cdot f \cdot F + 7.07 \cdot 10^3 \cdot i \cdot f \cdot F \quad (6)$$

5. Summary

Following statements can be asserted on the base of executed and evaluated experiments:

- In tangential direction, perpendicular to the direction of manufacturing, the value of the improvement ratios of residual stresses were the highest, $\rho_{\sigma t}$ = 1943 %, when the burnishing parameters were: f = 0.005 mm/rev; and F = 10 N and single number of passes (i = 1). Using these burnishing parameters, in axial direction the maximum value of the improvement ratio of residual stresses is $\rho_{\sigma a}$ = 503 %.
- While the residual stresses (both in tangential and axial direction) were and greatly improved due to the burnishing process, in cylindricity in only one workpiece surface point of view was experienced a beneficial change signed "3", in which the most favourable tangential residual stress value was measured too.
- Examining the evaluated results can be stated that applying the smaller burnishing force with single number of passes is more positive, while feed rate has not considerable effect to the improvement ratios in these burnishing parameter ranges.

Acknowledgments

"The described study was carried out as part of the EFOP-3.6.1-16-00011 "Younger and Renewing University – Innovative Knowledge City – institutional development of the University of Miskolc aiming at intelligent specialisation" project implemented in the framework of the Szechenyi 2020 program. The realization of this project is supported by the European Union, co-financed by the European Social Fund."

"Project no. NKFI-125117 has been implemented with the support provided from the National Research, Development and Innovation Fund of Hungary, financed under the K_17 funding scheme."

References

[1] AG Mamalis, J Kundrak, K Gyani 2002*On the dry machining of steel surfaces using superhard tools*, vol.19 (3) Int. Journal of Advanced Manufacturing Technology pp 157-162

[2] J Kundrak, AG Mamalis, A Markopoulos 2004 *Finishing of hardened boreholes: Grinding or hard cutting?* vol 19 (6) Material Manufacturing Processes, pp 979-993

[3] A M Hassan, H F Al-Jalil, A A Ebied 1997 *Burnishing force and number of ball passes for the optimum surface finish of brass components (Irbid)* vol 83 Journal of Materials Processing Technology (Amsterdam: Elsevier) pp 176-179

[4] L Wang, X Yu 1999 *Effect of various parameters on the surface roughness of an aluminium alloy burnished with a spherical surfaced polycrystalline diamond tool* vol 39 International Journal of Machining Tools Manufacturing pp 459-469

[5] G Taguchi 1987 *System of experiment design (New York)* UNIPUB Kraus International Publications

[6] L Fridrik 1987 *Chosen chapters from the topics of experimental design of production engineering (Budapest)* Műszaki Könyvkiadó (in Hungarian)

[7] I A Scibner 1916 *Burnishing-Tool* United States Patent Office 1,171,146, Serial No. 811,261 https://google.com/patents/US1171146 (29.05.2018)

[8] A Akkurt 2011 *Comparison of roller burnishing and other methods of finishing treatment of the surface of openings in parts from tool steel D3 for cold forming (Russia)* Vol 53 Metal Science and Heat Treatment pp 145-150

[9] A Saldana-Robles, H Plascencia-Mora, E Aguilera-Gómez, A Saldana-Robles, A Marquez-Herrera, J A Diosdado-De la Pena 2018 *Influence of ball-burnishing on roughness, hardness and corrosion resistance of AISI 1045 steel (Mexico)* Vol 339 Surface & Coatings Technologies (Elsevier) pp 191-198

[10] M Korzynski 2007 *Modelling and experimental validation of the force-surface roughness relation for smoothing burnishing with a spherical tool (Poland)* Vol 47 International Journal of Machine Tools & Manufacture (Elsevier) pp 1956-1964

[11] L Luca, S Neuga-Ventzel, I Marinescu 2005 *Effects of working parameters on surface finish in ball-burnishing of hardened steels (Russia)* vol 29 Precision Engineering (Elsevier) pp 253-256

[12] T A El-Tawel, M H El-Axir 2009 *Analysis and optimization of the ball burnishing process through the Taguchi technique,* Vol 41The Int. Journal of Adv. Manufacturing Technology pp 301-310

[13] H Luo, J Liu, L Wang, Q Zhong 2005 *The effect of burnishing parameters on burnishing force and surface micro-hardness (London)* International Journal of Advanced Manufacturing Technologies Vol 28 pp 707–713

[14] G Varga 2014 *Possibility to increase the life time of surfaces on parts by the use of diamond burnishing process (Miskolc)* Vol 686 Key Engineering Materials pp 100-107

[15] N N 2013 *Measurement of circularity*, Study aid, Taylor Hobson Precision p 34 (in Hungarian)

[16] V Mertinger, J Sólyom 2017 *Examination of residual stress by X-Ray diffraction (Miskolc)* Study aid pp 3-39

[17] M E Fitzpatrick, A T Fry, P Holdway, F A Kandil, J Shackleton, L Suominen 2005 *Determination of residual stresses by X-ray diffraction (Middlesex)* No. 52 Measurement good practice guide p 68

[18] B Zheng, G Shu, Q Jiang 2017 *Experimental study on residual stresses in cold rolled austenitic stainless steel hollow sections (Jiangsu)* Journal of Constructional Steel Research pp 1-11

IOP Conf. Series: Materials Science and Engineering **448** (2018) 012017 doi:10.1088/1757-899X/448/1/012017

The effect of conventional metal spinning parameters on the spun-part wall thickness variation

M Frnčík[1], J Šugárová[1], P Šugár[1] and B Ludrovcová[1]

Slovak University of Technology, Faculty of Materials Science and Technology in Trnava, Institute of Production Technologies, Jána Bottu 25, 917 24 Trnava, Slovakia

martin.frncik@stuba.sk, jana.sugarova@stuba.sk

Abstract: Conventional metal spinning technology is based on the sequential shaping of a circular sheet over a mandrel through the action of a roller that produces localised pressure to produce a hollow axisymmetric product. The wall thickness of the formed part in this process is, in general, considered to be nearly constant, but in fact, a non-uniform distribution of wall thickness is observed. The wall thickness variation is a major defect found in produced parts that can cause part failure and therefore the spinning process must be properly optimized. This paper deals with the analysis of the chosen conventional metal spinning parameters (tool path profile, mandrel speed and the tool feed rate) on the wall thickness variation of cylindrical shaped spun-parts made of DC01 low carbon steel, while a three-level full factorial design of experiment method (DoE) and ANOVA (Analysis of Variance) were used. The experimental results revealed that tool path profile and the tool feed affects the wall thickness variation in a significant way. The effect of the mandrel speed was not clearly observed.

Key words: conventional metal spinning, tool path profile, mandrel speed, tool feed, wall thickness, ANOVA

1. Introduction

Wall thickness variation is a major defect found in parts made from metal sheets that influences the intensity of part defects and possibly causes the cup failure [1]. The wall thickness of the formed part in conventional spinning is, in general, considered to be nearly constant [2]. But in fact, the wall thickness of conventionally spun cup is not uniform [3, 4]. Therefore, the spinning process must be carefully optimized in order to produce components with a minimal thickness variation [5].

The cup edge is much thicker than the original blank thickness; what is more, cup wall has two nearby necks. The first neck is at the cup bottom and the second one is at the cup wall. Cup wall fractures occur due to oversized blanks at the location of the second neck [6]. Conventional spinning process produces thinner cup wall with smaller roller feeds [3, 4], smaller roller nose radii [3, 4] and smaller radial clearances [3]. The inner profile of the final spun cup is larger than the mandrel profile due to springback. This difference gets wider with smaller roller nose radii, higher roller feeds, and higher conical roller angles. The final spun cup by conventional single roller doesn't closely fit to the forming mandrel. The cup is rather shaped with bulged form having a gap between the cup inner diameter and the mandrel. This gap is not uniform and increases with cup height with its minimum at the cup bottom [4].

In the work of Wang and Long [7] three key process parameters were selected: spindle speed, feed rate and type of material as the experimental input factors and the wall thickness, depth, and inside diameter variations of the spun cups was measured as the output factors. Authors concluded that the type of material has the most significant effects on the variations of thickness and depth of the spun parts, followed by the spindle speed and the feed ratio. In addition, thinner wall thickness is also obtained when a higher spindle speed is applied. It is shown that in this experiment a higher level of feed rate leads to thinner spun parts.

These experimental results are partially supported by the experimental results of Essa and Hartley [8], it was demonstrated that the mandrel rotational speed did not show any effect on the formability of spun part (average cup wall thickness, thickness variation and springback of the cylindrical cups). In order to obtain a more uniform thickness distribution, a low feed rate and a negative relative clearance should be used. High feed rates increase the contact area between the roller and workpiece that tends to decrease the applied stresses [4]. Therefore, the workpiece deformation decreases and thus, a high thickness variation is found.

Šugár et al. [5] analyzed the influence of mandrel speed feed ratio and tpp on the spun parts wall thickness distribution via a statistical analysis. Authors concluded that mandrel speed has minimal effect on the thinning of the component wall. These findings can be expressed as follow: mandrel speed has negligible influence on axial and radial force components, therefore radial and thickness strain is minimal.

On the other hand, Wong et al. [9] suggested that a low feed ratio would result in excessively thinning of the cup wall. This conclusion was supported by the work of Zhan et al. [10]. Šugár et al. [5] and Šugár et al. [11] explains this feature as follows: a lower feed ratio means that the tool acts on the workpiece over more revolutions, which in turn leads to higher shearing effects, an intensive flow of material to the open end of the part and an appreciable thinning of the wall. Material hearing between the roller and workpiece due to frictional effects may be the main reason of the sheet thinning. Wang and Long [7] adds that high feed ratios can help to maintain the original wall thickness unchanged but cracking failures may occur if a larger feed ratio is applied.

Design of the toolpath strongly influences the quality of the final part, and is especially important in conventional multi-pass spinning. Various toolpaths such as linear, concave, convex was studied. Careful choice and design of the toolpath leads to a high quality, wrinkle and crack-free product [12]. The tendency to buckle and cause wrinkles as well as cracking can be avoided by introducing the correct roller path [9].

Up to now, most investigations of roller path and passes in conventional spinning have focused on single pass or multi-pass with no more than 3 passes; mainly linear paths [13]. During multi-pass conventional spinning process, the number of spinning passes, the spacing between two consecutive passes and the shape of the part are determined by the roller trace [14].

Kang et al. [15] studied the deformation mechanism of a single-pass conventional spinning, where three types of roller trace have been applied: straight line, concave curve and convex curve. The thinning rate of the wall thickness should be well taken into account in the design of the roller path of the first pass, because the deformation in the first path plays a decisive part in the final wall thickness distribution of the workpiece. In the first pass, the deformation mode of the blank is quite different from that of other passes due to the great change of blank shape from plate to shell.

Wang and Long [16] carried out a FE analysis of the effects of roller paths (linear, concave, convex, and combined curve) on the variation of wall thickness and tool forces. The results show that the concave path produces highest tool forces among these four roller path profiles. Using the concave roller path tends to cause higher reductions of wall thickness of the spun part and using the convex roller path helps to maintain the original wall thickness unchanged. A greater curvature of the concave path would result in more thinning in wall thickness of the spun part.

Polyblank and Allwood [17] proposed a parameterized toolpath design based on a quadratic Bezier curve. They experimentally investigated how tool force, part geometry and various failure modes evolve with the varying of key features of the tool path. They also proposed a set of rules that are useful for automatic toolpath generation, for instance: the maximum deformation in forwards pass (hence the maximum tool force) is limited by the acceptable thinning, and the appearance of wrinkles or foldback.

This paper deals with the experimental analysis of the cups wall thickness variation during the incremental material deformation process applying the conventional spinning technology, while the spun cups are produced of DC01 low carbon steel blanks. The influence of tool path profile, feed rate and mandrel speed on the spun cups wall thickness distribution was identified using the statistical analysis method.

2. Experimental setup

The input circular sheet metal blanks necessary for the production of the experimental samples were prepared by the AWJ cutting technology; the outer diameter D_0 of the blank is calculated to 142 mm, initial sheet thickness t_0 is 1 mm. The blanks, made of a low carbon steel DC01 (1.0330; chemical composition and the calculated basic mechanical properties, according to ISO 6892-1:2016 standard for uniaxial tensile tests, are specified in Table 1), were used for the production of the hollow cylindrical shaped spun cups; the inner diameter of the spun cup is 91±0.5 mm, the sidewall height is 33±1 mm and the sidewall thickness of 1±0.1 mm (see Figure 1).

Table 1. Chemical composition of DC01 steel [18] and its basic mechanical properties

Element	**C max**	**Mn max**	**P max**	**S max**	**Si max**	**Al max**
wt (%)	0.12	0.6	0.045	0.045	0.03	0.02

Yield strength (MPa)	**Tensile strength** (MPa)	**Elongation** (%)	**Degree of planar anisotropy** Δr (-)
187 ± 11.2	321 ± 8.5	44 ± 1.6	+ 0.34

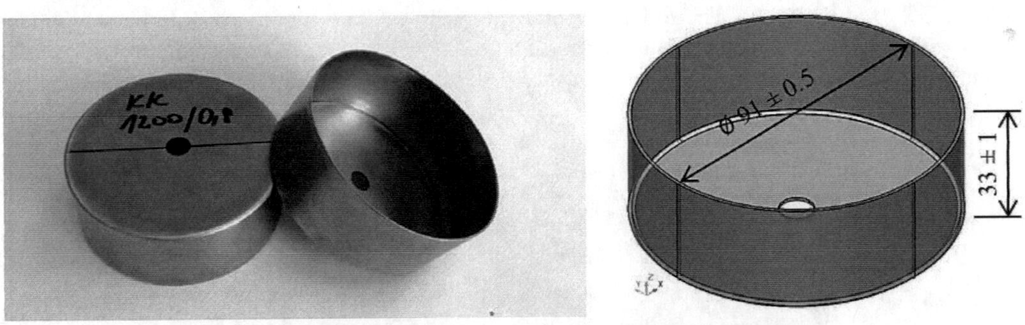

Figure 1. Cylindrical-shaped spun parts

Spun cups were produced under various combinations of CNC metal spinning process parameters according to the Table 2, using a 3-level full factorial Design of Experiment, where a tool path profile (*tpp*), mandrel speed (n) and the tool feed rate (f) were chosen as the input process parameters. Mandrel rotational speed varies from 400 to 1200 min^{-1}, tool feed rate varies from 0.4 to 1.2 mm in a combination with tree most applied tool path profiles: linear, convex and concave. Labeling of the tool path profiles is given with the respect to the initial metal blanks. The designed tool path profiles are depicted in Figure 2. For this purpose a DENN ZENN 80 CNC conventional metal spinning machine equipped with a Sinumeric Siemens 840D control unit was used during the whole experimental research. A multi-pass spinning was applied for incremental material deformation, using 16 forward (movement of the tool is towards the blank edge) tool passes in total, consisting of 15 forming passes and one calibration pass. 27 spun cups were produced in total according to the above defined DoE (one cup for each process parameters combination) at a room temperature. Production of spun cups was carried out with the application of liquid oil lubricant.

Table 2. Studied input metal spinning parameters

Parameter	Level 1	Level 2	Level 3
Tool-path profile ***tpp***	Convex	Linear	Concave
Tool feed rate ***f*** (mm)	0.4	0.8	1.2
Mandrel speed ***n*** (min^{-1})	400	800	1200

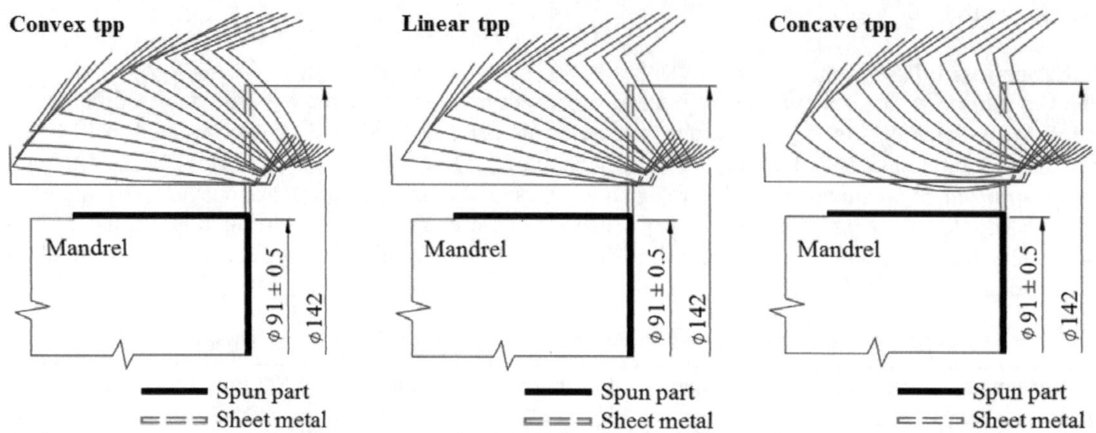

Figure 2. Studied tool path profiles: convex, linear and concave

The dimensional parameters of the applied forming tool – roller are chosen as follows: diameter d_r of 100 mm, roller nose radius r_n of 8 mm and the angle between the axis of the roller and the axis of the blank is 35° (see Figure 3). The roller is produced of 90MnCrV8 tool steel (1.2842) which is heat-threated to 60±1 HRC (quenching and tempering processes); since this type of material is widely applied in production of blank metal forming tools.

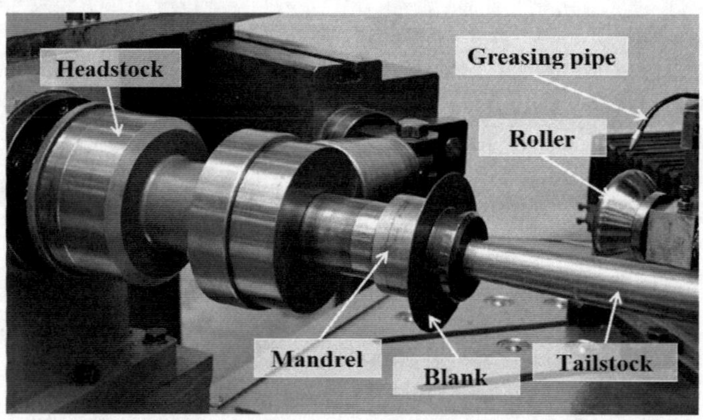

Figure 3. Setup of the conventional metal spinning process

After the production of spun parts, according to the DoE method, it was possible to measure the sidewall thickness in three different directions of the component: 0°, 45° and 90° directions relative to the blank material rolling direction (as depicts Figure 4), because the effect of the blank material planar anisotropy is expected. What is more, thickness was measured in three zones on the spun cup sidewall: at a distance of 3mm (transition radius), 8 mm (wall) and 30 mm (open end of the cup) from the cup bottom.

The measured data revealed that the effect of the material planar anisotropy on the thickness variation is only minimal and therefore can be negligible; this means that the mean values of the sidewall thickness measured in three directions were taken into account for the final evaluation.

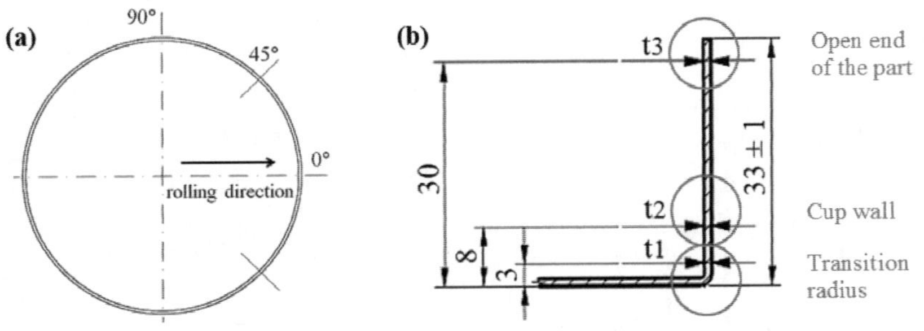

Figure 4. Spun cup sidewall thickness measuring plan: a) measured directions, b) measured zones

3. Results and discussion

After the implementation of the experimental measuring process it was possible to analyse the effect of the input process parameters on the thickness variation of the blank sheet during the deformation process. The measured thickness data are summarized in the Table 3 for the individual combinations of the studied parameters. The percentage thickness variation data, which is calculated relative to the initial thickness of the blank sheet (*t0* equals to 1 mm), are specified in Table 4. It can be clearly seen that the measured thickness values are varying in the wide interval from 0.903 mm to 1.138 mm. The experimental results also show that the most massive thickness reduction is observed on the transition radius zone (*t1*) and on the cup wall *t2* (for the concave tpp). On the open end of the spun part (*t3* zone), a thickening effect dominates over a thinning effect. Only a few thinning cases were observed in this zone.

Table 3. Measured sidewall thickness for different tpp and various process parameters

Mandrel speed (min^{-1})	Tool feed rate (mm)	Convex tpp			Linear tpp			Concave tpp		
		t1 (mm)	*t2* (mm)	*t3* (mm)	*t1* (mm)	*t2* (mm)	*t3* (mm)	*t1* (mm)	*t2* (mm)	*t3* (mm)
400	**0.4**	0.949	0.964	1.076	0.956	0.948	1.018	0.931	0.903	0.938
	0.8	0.970	0.991	1.106	0.973	0.975	1.103	0.941	0.917	1.004
	1.2	0.980	1.011	1.124	0.953	0.987	1.085	0.953	0.929	1.037
800	**0.4**	0.958	0.975	1.062	0.956	0.950	1.045	0.932	0.909	0.945
	0.8	0.984	1.009	1.111	0.959	0.977	1.094	0.947	0.931	0.982
	1.2	0.975	1.015	1.138	0.976	0.988	1.122	0.962	0.946	1.028
1200	**0.4**	0.944	0.964	1.063	0.949	0.949	1.074	0.943	0.918	0.905
	0.8	0.962	0.990	1.085	0.963	0.977	1.089	0.972	0.952	1.035
	1.2	0.979	1.006	1.120	0.976	0.994	1.109	0.955	0.940	1.060

Thickness standard deviation values did not exceed the 0.010 mm value in any combination of input parameters

The graphs in Figure 5 show the evolution of sidewall thickness in consideration of the tool feed and the applied tool path profile at a mandrel rotation speed of 800 min^{-1}. It is clear that the parameters of tool path profile as well as the size of its feed rate significantly affect the variation in the thickness of the spun-cups sidewall. These graphs also indicate that the higher the roller tool, the more significant sidewall thickness variation (thinning means the negative thickness variation or thickening-positive variation of thickness) occurs.

Table 4. Thickness variation relative to the initial sheet thickness

Mandrel speed (min^{-1})	Tool feed rate (mm)	Convex tpp			Linear tpp			Concave tpp		
		tv1 (%)	*tv2* (%)	*tv3* (%)	*tv1* (%)	*tv2* (%)	*tv3* (%)	*tv1* (%)	*tv2* (%)	*tv3* (%)
400	**0.4**	- 5.1	- 3.6	+ 7.6	- 4.4	- 5.2	+ 1.8	- 6.9	- 9.7	- 6.2
	0.8	- 3.0	- 0.9	+ 10.6	- 2.7	- 2.5	+ 10.3	- 5.9	- 8.3	+ 0.4
	1.2	- 2.0	+ 1.1	+ 12.4	- 4.7	- 1.3	+ 8.5	- 4.7	- 7.1	+ 3.7
800	**0.4**	- 4.2	- 2.5	+ 6.2	- 4.4	- 5.0	+ 4.5	- 6.8	- 9.1	- 5.5
	0.8	- 1.6	+ 0.9	+ 11.1	- 4.1	- 2.3	+ 9.4	- 5.3	- 6.9	- 1.8
	1.2	- 2.5	+ 1.5	+ 13.8	- 2.4	- 1.2	+ 12.2	- 3.8	- 5.4	+ 2.8
1200	**0.4**	- 5.6	- 3.6	+ 6.3	- 5.1	- 5.1	+ 7.4	- 5.7	- 8.2	- 9.5
	0.8	- 3.8	- 1.0	+ 8.5	- 3.7	- 2.3	+ 8.9	- 2.8	- 4.8	+ 3.5
	1.2	- 2.1	+ 0.6	+ 12.0	- 2.4	- 0.6	+ 10.9	- 4.5	- 6.0	+ 6.0

+ positive thickness variation (thickening),
- negative thickness variation (thinning).

The thickness reduction was observed for all combinations of input parameters in the transition radius measured zone (*t1*). The maximal reduction in thickness was recorded using the concave tpp and the mandrel rotational speed of 400 min^{-1}, with the thickness reduction ranged from -6.9% (tool feed of 0.4 mm) to -4.7% (tool feed of 1.2 mm). Vice-versa, the minimal reduction in thickness was noticed using a convex tpp at the mandrel rotational speed of 800 min^{-1} when the thickness reduction value ranges from -4.2% (tool feed of 0.4 mm) to -2.5% (tool feed of 1.2 mm).

For the spun part wall (*t2* zone), both the positive and the negative variation in the wall thickness of the workpieces is already observed. Similarly to *t1* zone, the maximal reduction in thickness was recorded using a concave tpp and a mandrel speed of 400 min^{-1} when the reduction value ranged from -9.7% (tool feed of 0.4 mm) to -7.1 % (tool feed of 1.2 mm). On the other hand, the minimal reduction and even increase in wall thickness were observed using a convex tpp at a mandrel rotational speed of 800 min^{-1} when the reduction value ranged from -2.5% (feed of 0.4 mm) to + 1.5% (1.2 mm feed).

In the measured zone at the open end of the spun parts (*t3* zone), an increase in wall thickness was observed using almost all combinations of input parameters. The maximal increase was recorded applying the convex tpp at 800 min^{-1} rotational speed, where the thickness increased from +6.2% (tool feed of 0.4 mm) to +13.8% (tool feed of 1.2 mm). This indicates that the material flows towards the rim of blank through the forward passes in the spinning process, which is also observed in [7, 19, 20]. On the contrary, the minimal thickness changes were recorded using the concave tpp at 800 min^{-1} rotational speed, while this variation ranges from -5.5% (tool feed of 0.4 mm) to +2.8% (feed 1.2 mm).

Therefore it can be concluded, that the concave profile contributes to the maximal reduction in thickness in the *t1* and *t2* measured zones; however at *t3* zone, if sidewall thickening occurs, the thickness increasing is only minimal. Hence, this type of tool path profile appears to be the most improper from the originated thickness variation point of view. The convex tpp contributes to the minimal thickness reduction in the *t1* zone, but with the increasing measured distance there is a thickening of the component sidewall observed; while the thickness acquires the maximal values at the open end of the component using this tpp type. Using the linear tpp, the thickness reduction in the *t1* and *t2* zones is approximately constant; however there is a slight increase in the sidewall thickness in the *t3* zone.

The influence of the of the tool feed size on the sidewall thickness in the individual measured zones is as follows: with the increasing feed values, the thickness reduction decreases, but the thickness of the sidewall increases towards to the open end of the spun parts. The reason is that the higher forming velocity causes a higher material strain rate, this result in a larger material work-hardening during the deformation process. Mechanical properties of the deformed material are increasing, however, the formability of the material is worse and the thickness reduction is therefore reduced [8-10, 21].

Using a high feed rate leads to an increase in the compressive tangential stress and accordingly, compressive deformation and thickness strain increase. The final average thickness will therefore deviate away from the initial thickness [8].

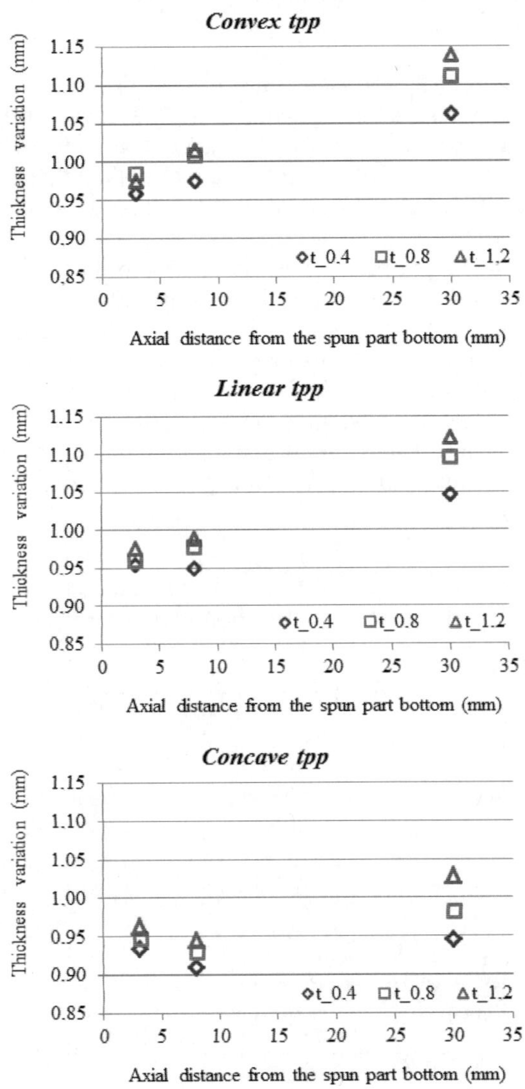

Figure 5. Measured thickness variation for different tool feed rates (mandrel speed of 800 min^{-1})

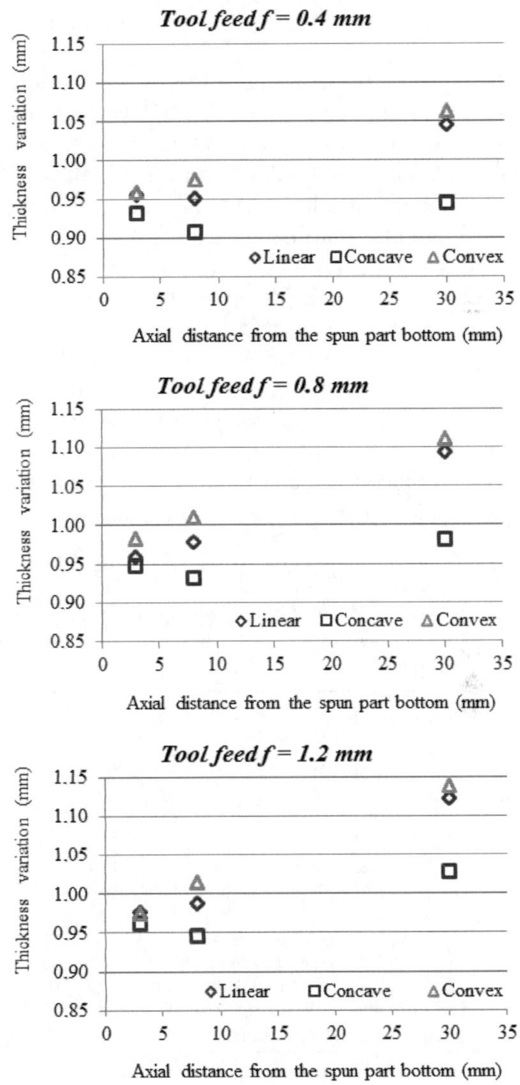

Figure 6. Measured thickness variation for different tpp (mandrel speed of 800 min^{-1})

Figure 6 shows the effect of the tool path profile parameter on the thickness values evolution for the mandrel rotational speed of 800 min^{-1} and a given tool feed. For instance, for the tool feed f = 0.8 mm it can be clearly seen, that the thickness reduction is equal to 1.6% (application of convex tpp), 4.1% (application of linear tpp) and 5.3% by use of concave tpp in the $t1$ measured zone. In the $t2$ measured zone the thickness increasing of about 0.9% for the convex tpp was observed and thickness reduction of about 2.3% and 6.9% for the linear and concave tpp was spotted. The thickening effect is characteristic for the open end of the parts on $t3$ measured zone. The thickness increase of about

11.1% and 9.4% is typical for application of convex and linear tpp respectively. Thickness reduction of 1.8% was recorded for the concave tpp in this zone. It is possible to notice at the same time that the most massive thickness variation is achieved using the concave tool path profile and vice versa, application of convex tpp helps to minimise the thickness variation. According to [16] using concave roller paths in spinning tend to cause higher reduction of the wall thickness of the spun part and using the convex roller path helps to maintain the original wall thickness unchanged. A greater curvature of the concave path would result in more thinning in wall thickness of the spun part.

In order to quantify the effects of the studied factors and 2-factor interactions on the thickness variation of the spun cups, the analysis of variance (ANOVA) using Minitab v.17 software was performed. The Fisher's ratio (*F*-ratio), which is the ratio between the variance due to the effect of a factor and the variance due to an error term, was used to measure the significance of the factor at the desired significance level. If the *F-value* is greater than the tabulated F_{crit} value, the process parameter is considered as significant.

The ANOVA results (Table 5) revealed that the studied input parameters: tool path profile *tpp* and tool feed *f* can be considered as statistically significant in the zones of transition radius (*t1*- thinning) and extremely significant in the zone of the part open end (*t3*- thickening). The effect of the mandrel speed on thickness variation is only minimal in these zones, which is supported by other experimental researches [5, 8, 22]. This can be explained as follows: mandrel speed has the negligible effect on the axial and the radial force components, therefore radial and thickness strain is minimal [23].

In the part wall zone (*t2*- thinning or thickening occurs) it was found that besides the tool path profile and tool feed rate, the mandrel speed and 2-factor interaction of *tpp*n* can be considered also as significant factors on the thickness variation.

Table 5. ANOVA for studied sidewall thickness in three different zones

	Thickness variation *t1*			**Thickness variation *t2***			**Thickness variation *t3***		
Source	DF	*F*-ratio	*p*-value	DF	*F*-ratio	*p*-value	DF	*F*-ratio	*p*-value
tpp	2	**7.18**	***0.016**	2	**289.12**	***0.000**	2	**52,03**	***0.000**
n	2	0.53	0.610	2	**5.44**	***0.032**	2	0.15	0.861
f	2	**9.90**	***0.007**	2	**98.10**	***0.000**	2	**25.07**	***0.000**
*tpp*n*	4	0.90	0.508	4	**4.22**	***0.040**	4	0.57	0.694
*tpp*f*	4	0.31	0.861	4	1.41	0.313	4	1.26	0.361
*n*f*	4	0.17	0.950	4	0.41	0.797	4	0.24	0.909
Error	8			8			8		
Total	26			26			26		

where: *tpp* - tool path profile, *n* - mandrel speed, *f* - tool feed rate, *DF* - degrees of freedom.
Tabulated F_{crit} at 95 % confidence level: $F_{crit}(0.05; 2; 8) = 4.46$; $F_{crit}(0.05; 4; 8) = 3.84$
R-sq(adj) = 86.63 % for ***t1***; *R-sq(adj)* = 96.82 % for ***t2*** and *R-sq(adj)* = 84.77 % for ***t3***; replicates: 1
* statistically significant parameter

4. Conclusion

The following conclusions can be drawn according to the experimental results:

- Thickness evolution in the individual measured zones is significantly affected by the combination of input process parameters, while the massive thinning and thickening was observed.
- Tool path profile *tpp* and tool feed rate *f* can be considered as statistically significant parameters among the all studied process parameters and their 2-factor interactions. The effect of the mandrel speed is only minimal.
- From the minimal thickness reduction point of view it is recommended to apply the convex tool path profile, however this tpp leads to the most massive thickness increase on the open end zone. Contrariwise, the concave tool path profile causes the biggest reduction on first two measured zones,

what also predicted the potential zones of the generated cracking of the material. However, the thickening feature on the open end is minimal.

- It was found that higher tool feed values contribute to the lesser thickness reduction, but on the other hand to a thickness increase on the open end zone of the formed parts. This can be explained as follows: if lower tool feed is applied, the material deformation takes a longer time, which results in a considerable thinning of the spun part wall; hence the wall heights values are higher.
- Authors recommend the application of higher mandrel rotational speeds during the deformation process. There are two reasons for this: one is an increased magnitude of spinning force due to the high deformation rate; the other is that the deformation energy required per revolution is likely to decrease because the feed rate is inversely proportional to the spindle speed (mm/rev) [24].

Acknowledgement

This research paper was written thanks to the material and the financial support of FormTool Manunet project No. 2014/11283 and VEGA project No. 1/0669/15. The authors would also like to thank for financial contribution from the STU Grand Scheme for Support of Young Researchers (No. 1322, Apple2 project). The authors also thank to MSc. J. Petrovič, PhD. of Eiben Co. Ltd. for his technical support.

References

[1] Brabie G, Chirita B and Albut A 2015 *Prec. Eng.* **42** 311–20

[2] Quigley E and Monaghan J 2000 *J. Mat. Proc. Tech.* **103(1)** 114-19

[3] Hayama M and Murota T 1963 *Bull. Fac. Eng.* **12** 53–88

[4] El-Khabeery M M, Fattouh M, El-Sheikh M N and Hamed O A 1991 *Int. J. Mach. Tools Manuf.* **31(2)** 203–19

[5] Šugár P, Šugárová J and Petrovič J 2016 *Stroj. Vest. – J.Mech. Eng.* **62(3)** 171-78

[6] Ahmed K I, Gadala M S and El-Sebaie M G 2015 *Int. J. Mach. Tools Manuf.* **97** 72-85

[7] Wang L and Long H 2013 *Mat. Des.* **46(4)** 645-53

[8] Essa K and Hartley P 2010 *Proc. Inst. Mech. Eng. B: J. Eng. Manuf.* **224(11)** 1691-1705

[9] Wong C C, Dean A and Lin J 2003 *Int. J. Mach. Tool. Man.* **43(14)** 1419-35

[10] Zhan M, Yang H, Zhang J H, Xu Y L and Ma F 2007 *J. Mat. Proc. Tech.* **187-188** 486-91

[11] Šugár P, Šugárová J, Morovič L and Zemko P 2012 *App. Mech. Mat.* **217-219** 2423-26

[12] Music O, Allwood J M and Kawai K 2010 *J. Mat. Proc. Tech.* **210(1)** 3-23

[13] Wang L 2012 *Analysis of Material Deformation and Wrinkling Failure in Conventional Metal Spinning Process* (Durham: Durham University) p 163

[14] Liu J H, Yang H and Li Y Q 2002 *J. Mat. Proc. Tech.* **129(1-3)** 326-29

[15] Kang D-Ch, Gao X-Ch, Meng X-F and Wang Z-H 1999 *J. Mat. Proc. Tech.* **91(1-3)** 226-30

[16] Wang L and Long H 2011 *J. Mat. Proc. Tech.* **211(12)** 2140-51

[17] Polyblank J A and Allwood J M 2015 *CIRP Ann.- Man. Tech.* **64(1)** 301-04

[18] Arcelor Mittal 2010 Cold Rolled Steels for Drawing and Forming [online] Available at: <https://flatsteel.arcelormittalsa.com/fspcatalogue/DataSheets/UnCoated/Web_datasheet_b3.1.pdf>

[19] Arai H and Kanazawa T 2018 *J. Mat. Proc. Tech.* **260** 66-76

[20] Sangkharat T and Dechjarern S 2017 *Proc. Eng.* **207** 1713-18

[21] Watson M, Long H and Lu B 2015 *Int. J. Adv. Man. Tech.* **78(5-8)** 981-95

[22] Long H and Hamilton S 2008 *The 9th International Conference on Technology of Plasticity* South Korea **9** 735-40

[23] Wang L, Long H, Ashley D, Roberts M and White P 2011 *J. Eng. Man.* **225(11)** 1991-2006

[24] Wang Q, Wang T and Wang Z R 1989 *Proceedings of the Fourth International Conference of Rotary Forming* Beijing **4** 103-08

IOP Conf. Series: Materials Science and Engineering **448** (2018) 012018 doi:10.1088/1757-899X/448/1/012018

Process-based planning of material handling in manufacturing systems

Péter Telek

PhD, University of Miskolc, H3515 Miskolc-Egyetemváros

E-mail: alttelek@uni-miskolc.hu

Abstract. Manufacturing processes are more and more complicated which requires complex and flexible material handling solutions. The design of material handling for the advanced manufacturing processes can be realized in different ways, but none of them offers generally usable solutions. An integrated design process gives the best device for the planning, but its very complex characterisation makes the application too hard. To make an easier useable design method, which is closer to the operation process of the manufacturing system, we introduce a new planning concept for the material handling. This new concept uses the integrated, task-based approach however; it is mainly focusing to the handling processes. The objective of this concept is not the selection of the optimal solution; it targets to build effective, followable handling processes. In this paper, we describe the new concept and present a scenario to understand its operation.

1. Introduction

Suited to the higher requirement of the customers, the production processes are more and more complicated, so all the solutions (control, material handling, maintenance, etc.) which are applied related to them also have to be more complex. Among others, new, advanced handling processes required to build highly productive, flexible manufacturing systems.

Of course, these changings influence not only the applied machines, but the planning process of the handling solutions.

The design of material handling for manufacturing processes can be realized in different ways, but none of them offers generally usable solutions. An integrated design process gives the best device for the planning, but its very complex characterization makes the application too hard. The objective of our research to make an easier useable design method, which is closer to the operation process of the manufacturing system and let the users to follow the method in every detail. This task requires a new concept in the aspect of the planning process, which will be presented in this paper.

2. Planning of material handling in manufacturing systems - literature review

There are many proved and frequently applied methods for planning of material handling devices, in generally with computer aid [1]. In case of certain machines (forklifts, cranes, etc.), the producers offer standard, exact variations to select, so the planning process is simplified to a selection procedure and a simple application [2]. This is also valid for complex handling systems (monorails, conveyors, etc.), where the required solutions can be built from exact, predefined modules [3]. So, we can say, that the planning or selection of a single material handling machine or a simple handling system is easily realized by the known methods and devices (dimensioning processes, catalogues, software, etc.).

The problem, that most of the production processes need complex material handling solution, which applies more than one machine types, where the planning process is much more complicated. In this case, the application of a simple planning method is not suitable, because the integration and harmonization of the different machines also must be solved during the planning process. The solution of this situation requires the application of the system concept and system planning [4].

2.1. Material handling systems

Material handling system means a handling solution where more than one handling machines or module elements are applied for different handling processes. These systems, in generally, integrate the specifications of the individual material handling machines, but in certain cases, new specifications can be appeared during the integrated operation [5].

The handling systems can be significantly different depends on the system elements and their relations, but the main difference is the physical scale of the system [6], which can be

- internal handling systems (intra-logistics),
- external handling systems (inter-logistics) and
- workplace handling systems.

Because of the differences among the system types, the planning and operation processes are also different, at this paper we are focusing on the internal handling systems, the others are out of our scope.

In internal handling systems, handling processes operate within a restricted (usually indoor) area and the distances among the objects (manufacturing machines, workplaces, etc.) are short. Important characterization of the system is the exact input and output point, where it can connect to other systems. In these systems, transport routes and rules are flexible and suited to the requirements of the production process [7].

2.2. Planning of handling systems

During the planning of material handling systems, we are looking for suitable handling devices for complex handling processes and harmonize their operation. In the literature two basic solutions exist, which are the task-based and system-based approaches.

In system-based approach, we are analyzing the whole system and looking for a similar, existing one to apply its handling solutions to our system [8]. Similarity of the system can be found in objects, handling tasks, technology or handling devices. During the planning, we are searching in a special database for a similar system-structure and trying to adapt its handling solutions. To realize this planning concept, we need a special database, which are built on certain system types [9]. Typical application of this approach is the duplication of production processes, which is used by multinational companies to multiply their production [10].

In task-based approach, we realize the planning with the using of different individual planning tasks (facility planning, route planning, etc.). Main advantage of this approach is the using of the material-flow parameters, which enables exact, mathematically described calculations. This approach is much better published and used in the practice, but the application in complex systems is not so easy.

Main problems of this approach are the large number of the planning tasks, their complexity and the iterative solution process (Figure 1), which result three different application cases in the practice: single-task planning, multiple-task planning and integrated planning process.

During single-task planning we can solve only one planning task (e. g. [11]) in any kind of system. The complexity of the solution method depends on the scale of the handling system. There are three advanced methods to realize single-task planning in complex systems: Knowledge-Based Engineering [12], Computer Aided Design [13] and Simulation Software [14].

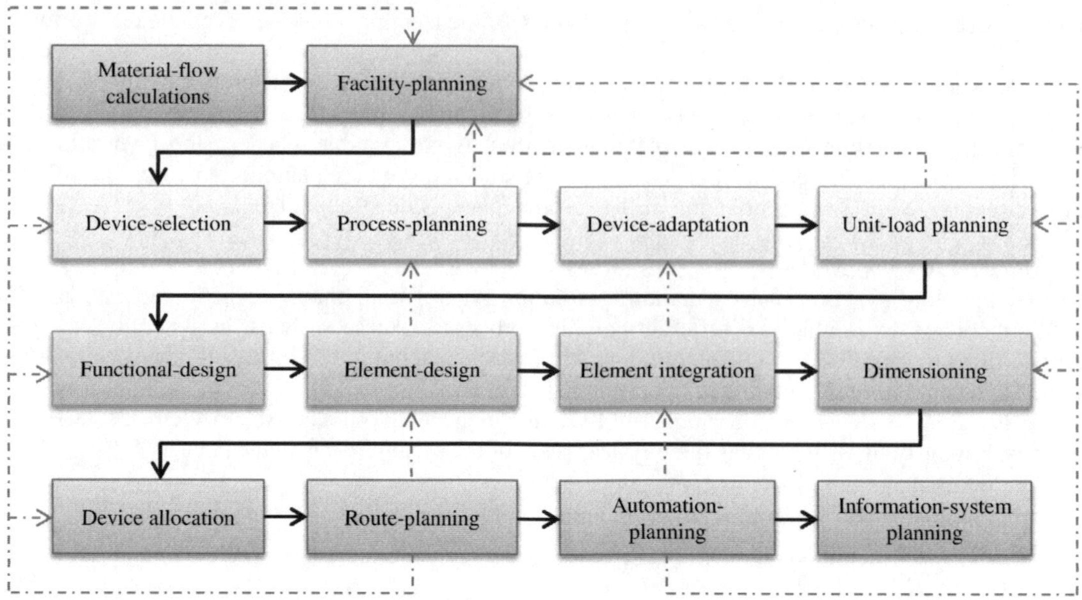

Figure 1. Structure and steps of the integrated planning

Multiple-task planning means the solving of a group of planning tasks, where usually the methods for the single-task are used, but this concept is basically more complex for any system [15].

Integrated planning tries to solve all the planning tasks in one process, but because of the large number of the tasks and the iterative procedure it is usable only for simple handling systems [16].

Large, complex production systems require advanced, computerized procedures to plan the handling system, which means in generally, an automated optimization. However, as it was described above, because of the complexity of the production system and the planning processes, there is no chance to make a generally usable method for this purpose. There are different initiatives to solve these problems using different software [17], but non-of them gives general solution, and because of the complexity level, users cannot look into their operation method and understand its steps.

An integrated design process gives the best device for the planning, but its very complex characterization makes the application too hard. To make an easier useable design method, which is closer to the operation process of the manufacturing system, we introduce a new planning concept for the material handling.

3. Process-based planning: a new concept

To avoid the disadvantages of the above-mentioned planning processes, we are looking for another solution, which is easily understandable and user friend, and can be applied in every handling situation. This new concept uses the integrated, task-based approach however; it is mainly focusing to the handling processes. The objective of this concept is not the selection of the optimal solution; it targets to build effective, followable handling processes.

During this concept, we use the steps of the task-based planning approach but develop a simple procedure for all of them and link them into a chain as it can see in Figure 2.

The individual steps of the planning concept can be linked into different blocks, which determines certain parameters of the material handling system. The main blocks are: determination of the relations, device preselection, building of handling processes, dimensioning and scheduling.

As the methods and realizations of the dimensioning and scheduling of the handling machines are the same as in any other planning concept, in this paper we are dealing only with the first three blocks, the others are out of our research scope.

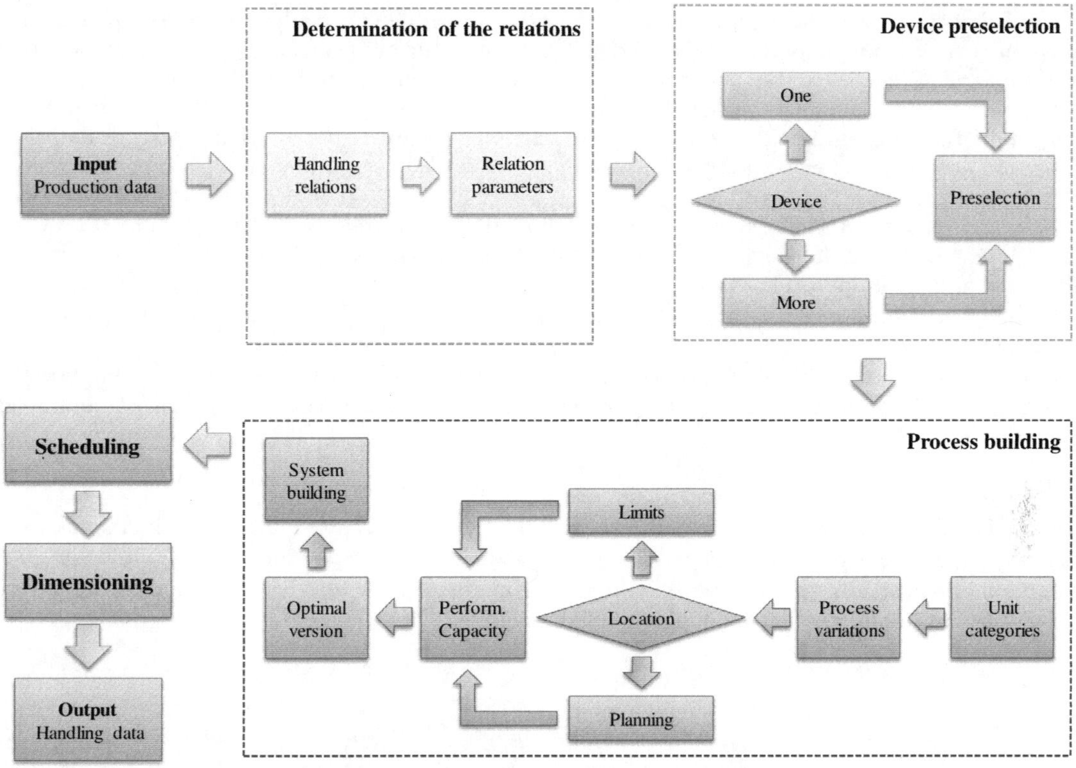

Figure 2. Algorithm of the process-based planning

3.1. Determination of the handling relations

Materials handling relation means a special connection between two objects (production or service objects) which involves any kind of handling activity. Handling relations can be defined by the two linked objects and at least one handling parameter (quantities, distances, costs, etc.) existing among them [6].

In the aspect of handling requirements, handling relations can be always described between one source and one destination object. There are some differences within the relations depend on the role of the handling process, because at the realization of the handling tasks we can also determine relations between two sources or two destination objects (e.g. collection systems) [18].

Materials handling relations can be defined and demonstrated by different methods, e. g. handling process charts, material handling graphs, material handling matrices, etc. In generally material handling matrix is used to involve the data of the handling relations, but to demonstrate the structure we also can use a graph. Main advantage of the materials handling matrices, that certain parameters of the handling process can be easily calculated by matrix-operations (e.g. the analysis of the rows of the matrix gives information about the source objects, etc.) [19]. The best way for the analytical determination of the handling relations is the application of the method defined in [20], but other solutions can also be used.

3.2. Device preselection

Main objective of the device selection is to choose the best material handling solution for all the individual handling relations. During the selection process the optimal solution is searched along a given objective function with the comparison of parameters of the materials handling devices and the given relation. Materials handling parameters can be exclusion-type, limitation-type or numerical parameters [21].

IOP Conf. Series: Materials Science and Engineering **448** (2018) 012018 doi:10.1088/1757-899X/448/1/012018

Exclusion-type parameters can exclude the application of certain equipment types (for example: roller conveyor cannot be used for bulk solids). They can be unambiguous exclusions (function, goods type, etc.) and definable exclusions (operation characteristic, handling method, track-line, etc.).

Limitation-type parameters do not exclude equipment types, but they can narrow their practical application field (for example: forklifts cannot be used for small boxes). They can be numerical limitations (goods parameters, task parameters, etc.) and not numerical limitations (object parameters, track parameters, etc.).

Numerical parameters are the bases of the analytic design process, their values can be different at the different materials handling machines. Equipment can also be qualified based on numerical parameters using special formulas (see [16]).

In this phase of the process-based planning, we cannot find the best version, but we can narrow the applicable types.

In certain cases, depend on the handled goods, different devices can be applied for the different units linked to the given relation. In this case, parallel independent preselection or uniform unit loads have to be applied.

3.3. *Building of handling processes*

Material handling process means a given set of handling tasks which are linked to each other suited to a certain logic. The logic of the allocation can be production order, handling specifications, location limits, etc. In generally the material handling system of a manufacturing procedure is built form different linked or independent handling processes (Figure 3).

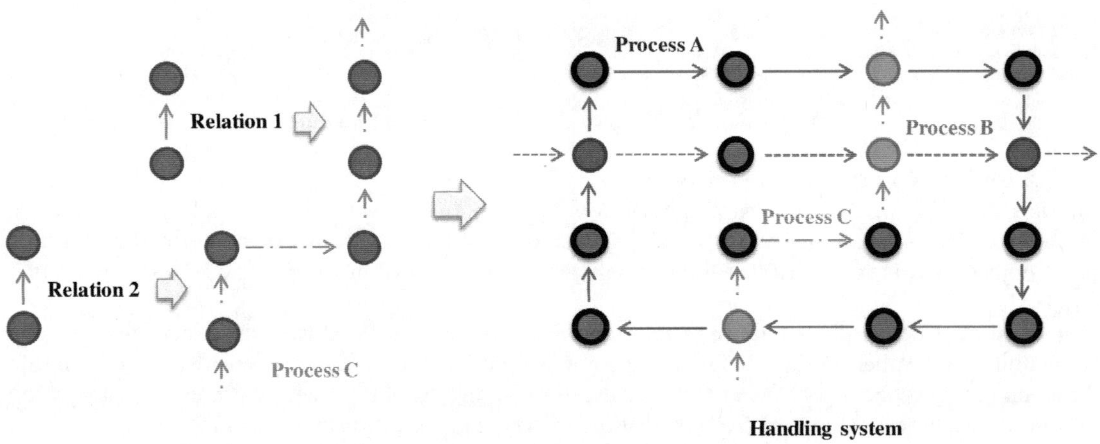

Figure 3. Building of a material handling system

During the building of handling processes we are looking for those relations which can be linked into one process. The selection procedure is based on different parameters of the handling relations:

- continuity of the handling process,
- parameters of the handled units,
- location of the objects,
- time-requirement of the relation,
- applicable machines, etc.

Based on the parameters of the material handling relations, the process-building can be realized in the following steps:

1. Building of handling unit categories
2. Building of process variations based on the device types

IOP Conf. Series: Materials Science and Engineering **448** (2018) 012018 doi:10.1088/1757-899X/448/1/012018

3. Application of location limits/location planning
4. Calculation of the required performances/capacities
5. Selection of the best solutions/optimization
6. System building from the processes

3.3.1. Building of handling unit categories. The first step of the process building is a grouping based on the parameters of the handled units. The main objective of this activity is to discover the similarities of the handled unit types which can help to realize a uniform handling concept.

The most characteristic differences exist among the three main handled material/element categories (Figure 4), which are

- the raw materials and elements for the manufacturing objects,
- the interoperation semi-finished units produced by the manufacturing objects,
- the finished products.

Units related to these categories are handled in independent processes usually individual handling machines, which can be significantly different (e. g. forklifts for the elements and roller conveyors for the semi-finished units).

Within the categories above, further subgroups can be created depends on the handling parameters, e. g. size, handling solution, etc. These sub-categories can help to select the suitable devices for a given handling process. Another result of this grouping can be the uniformization of the unit-loads, which can reduce the inhomogeneity of the goods and the handling devices.

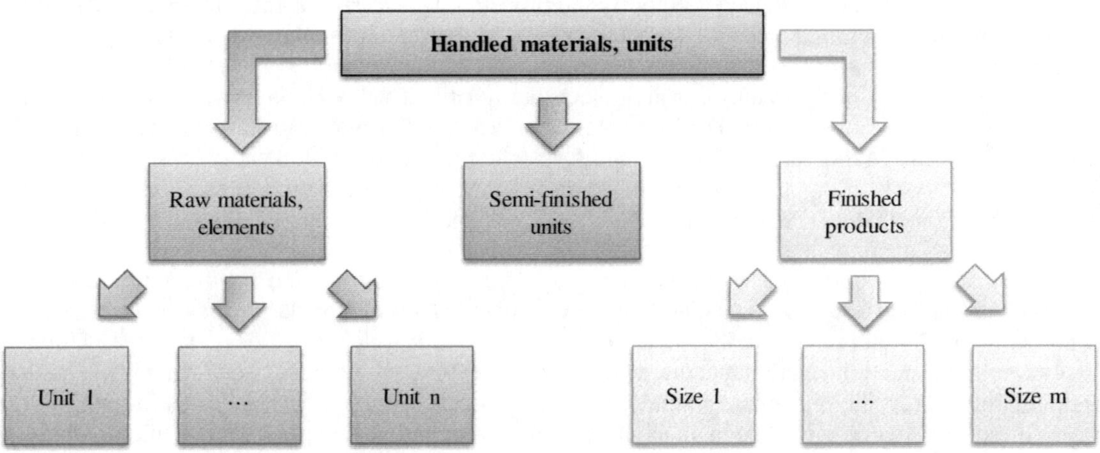

Figure 4. Example for the handling unit categories

3.3.2. Building of process variations. Within the above determined unit categories, we can build process variations using the material handling solutions preselected to the individual handling relations. This activity largely depends on the number of the handling relations and the complexity of the manufacturing system. In case of a simple system, it can be easy to realize, and we can get effective solutions, but complex systems require complicated optimisation methods.

In the aspect of the continuity, handling processes can be continuous or discontinuous. Discontinuous processes contain individual handling tasks where waiting phases can be appeared between the working phases. In this case machines move unit loads and workplace stores are required. During a continuous handling process, there are no waiting between the phases, and the moving of the units (usually pieces or bulk solids) is constant, periodic or stochastic in time.

In the aspect of the process building, handling machines have to be fitted in time and handling specifications.

Handling processes can apply three different machine allocations:

- linking one device to one process,
- linking one device to more than one processes,
- linking more than one device to one process.

The best solution is when one machine works in one handling process, so during the process-building we try to find such solution. However, in many of the cases, material handling machines have too few or much capacity to fulfill this expectation, so we need to take other relation parameters into consideration during the process building.

Important element of the process building is the capacity analyses of the machines, which can reduce the number of the applicable equipment types, because of the capacity limits and efficiency. Usually the main goal of the process building is the cost requirement of the processes, but it depends on different changing factors (e. g. local environment, economic situations, fuel prices).

3.3.3. Application of location limits. The location of the manufacturing objects influences the allocation of the process elements. During the planning of the handling system we can meet two different cases:

- locations of the objects are known,
- locations of the objects must be determined.

If we know the location data during the planning process, then we have to take them as a limit into consideration. It means that we have to modify the process allocations, if some relations are out of the effective area of a handling machine. In this case, we have to repeat the process allocation in all affected processes.

If the location data of the manufacturing objects are not fix at the beginning, we have to determine them during the planning process. The best solution if we select locations for the objects suited to the handling processes. In this case the handling performance requirement will be low. Other possibility is to apply a facility planning optimization procedure [8], which gives an optimal result for its objective function, but cannot take the handling processes into consideration.

3.3.4. Calculation of the required performances. In the knowledge of the handling processes, the required capacity of the applied machines can be calculated. The calculation procedure is different for continuous and discontinuous handling machines.

At continuous material handling equipment (e. g. conveyors), where usually one machine is applied for one process, capacity means a minimal transport performance along the track-line, in many of the cases these are the velocity and a machine specific transport parameter (cross section area, car distance, etc.) of the handling equipment.

At discontinuous machines (forklift, crane, etc.), the handling capacity of the equipment is a constant maximal value, so the objective of the calculation is to determine the required number of the machines for all the handling processes. In generally, the processes cannot use all the total capacity of all machines allocated to them, so some free capacity remained, which can be shared among the other processes if required.

3.3.5. Selection of the best solutions. As a result of the process-building, we have got several process variations, which fulfil the expected requirements, but only one of them can be applied in the manufacturing system. Next step of the planning process is to compare the variations and select the optimal one. For the comparison, numerical parameters must be applied, which can be

- material handling performance,
- handling costs,
- transport distances, etc.

Using the required parameter as an objective function, the calculation can be realized, and the results can be compared to choose the optimal version. If there are no significant differences among the variations in the numerical values, other parameters can be considered to select the best processes.

3.3.6. System building from the processes. Handling systems consist of different handling processes, which can be different relation in each other. To operate the system, we need to determine the relations among the individual processes. Processes can be linked differently, so at the locations, where they meet (objects, locations, tasks, etc.) the parameters of the different processes must be harmonized.

In the meeting points of the handling processes, depends on the parameters of the applied machines, some special technical solution must be used to avoid joining problems (transfer, storing, etc.).

In another case, when one handling machine is used in two or more handling processes, the relation can affect the operation of the individual processes.

4. Scenario for the process-based planning

To demonstrate the process-based planning, in this chapter we present an example to determine the details of the handling system for a production hall. Of course, there is only a simple manufacturing process with static production requirement and easy understandable relations to explain the operation of this new planning concept.

4.1. Manufacturing objects and production data

In our simple example we analyze a production process with 8 manufacturing machines, 3 final product types and unit handling. Figure 5 shows the graph-structure [22] of the manufacturing objects and the scaled production processes for the different products.

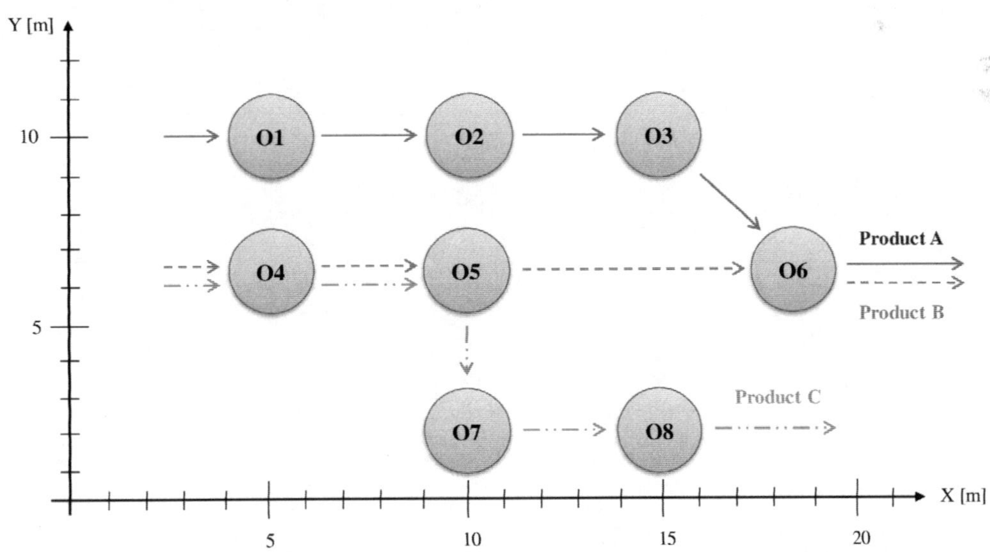

Figure 5. Manufacturing objects and their relations

The production sequence for all the final products and the required element types can be followed in Figure 6, Table 1 involves the data for the objects and the products.

IOP Conf. Series: Materials Science and Engineering **448** (2018) 012018 doi:10.1088/1757-899X/448/1/012018

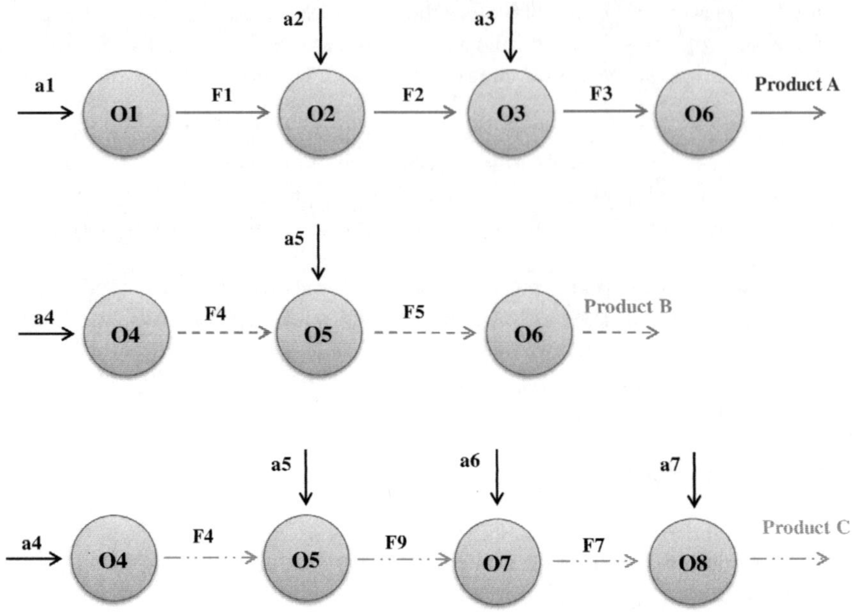

Figure 6. Production sequence for the three product types

Table 1. Production parameters

Products		**Produced quantities** [pcs/h]	**Element needs at the objects** [pcs/unit]							
			O1	O2	O3	O4	O5	O6	O7	O8
1.	**Product a)**	**30**	x	x	x			x		
	elements		a1	a2	a3			-		
	quantities		1	2	1					
2.	**Product b)**	**50**				x	x	x		
	elements					a4	a5	-		
	quantities					1	1			
3.	**Product c)**	**20**				x	x		x	x
	elements					a4	a5		a6	a7
	quantities					1	1		3	1

4.2. Material handling relations

Based on the production data, using the method defined in [20], we can calculate the parameters of the material handling relations among the system objects. At our scenario we used the MHRelCalcV1.0 software (described in [23]) to calculate the relation matrix of the analyzed system (Table 2). To follow the material flow of the different element/unit types, a material handling graph also presented in Figure 7.

Table 2. Relation matrix
(ES – elemet/material store, PS – final product store)

Objects	ES	O1	O2	O3	O4	O5	O6	O7	O8	PS
ES		30	60	30	70	70		60	20	
O1			30							
O2				30						
O3							30			
O4						70				
O5							50	20		
O6										80
O7									20	
O8										20
PS										

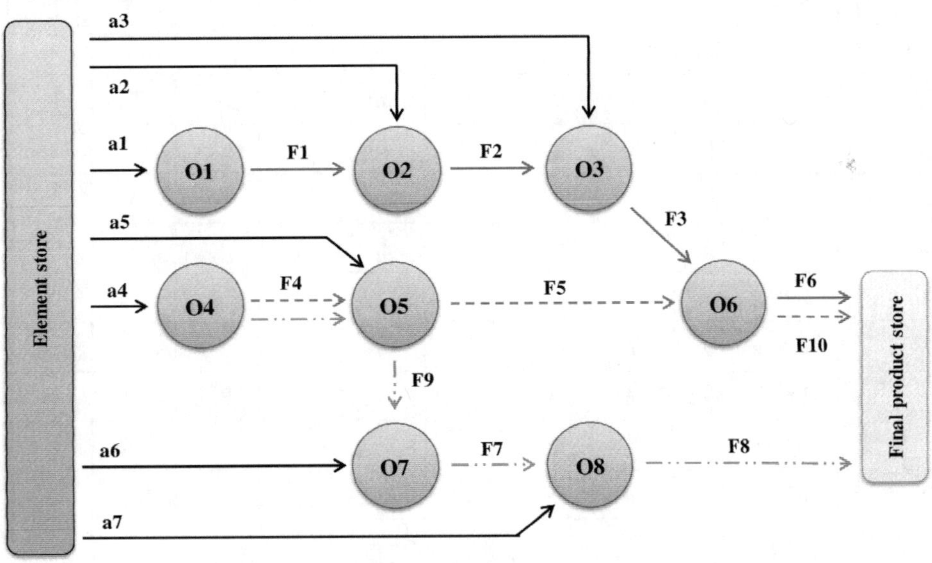

Figure 7. System objects and their relations

4.3. Device preselection

To determine the applicable solutions for the handling process, at first, we have to define the parameters of the individual relations, which described in Table 3.

Required operations of the given relation are depends on the technical structure of the related objects, at internal handling systems they can be

- transportation (if the distance is longer than the range of a loader machine),
- loading (if any or both objects requires loading activities),
- storing (if any or both objects cannot accept the goods directly).

Table 3. Parameters of the relations
(Y – yes, N – no, US – upper-side, DS – down-side, FS – front-side, AS – all-side, S – supporting, H – hanging, P – pressing, Box – box limit, Pal – palette limit, Any – no described)

Relations	Goods		Operations			Handling			Route		
	Type	Quantity [pcs]	Transport	Load	Store	Grabbing	Reaching	Size limit	Distance [m]	Fall	Complexity
ES→O1	a1	30	Y	Y	Y	Any	FS	Bx	20	N	1
ES→O2	a2	60	Y	Y	Y	Any	FS	Bx	20	N	1
ES→O3	a3	30	Y	Y	Y	Any	FS	Bx	25	N	1
ES→O4	a4	70	Y	Y	Y	Any	FS	Bx	15	N	1
ES→O5	a5	70	Y	Y	Y	Any	FS	Bx	15	N	1
ES→O7	a6	60	Y	Y	Y	Any	FS	Bx	20	N	1
ES→O8	a7	20	Y	Y	Y	Any	FS	Bx	20	N	1
O1→O2	F1	30	Y	N	N	Any	FS	Bx	5	N	1
O2→O3	F2	30	Y	N	N	Any	FS	-	5	N	1
O3→O6	F3	30	Y	N	N	Any	FS	-	5	N	1
O4→O5	F4	70	Y	N	N	Any	FS	Bx	5	N	1
O5→O6	F5	50	Y	N	N	Any	FS	-	8	N	1
O5→O7	F9	20	Y	N	N	Any	FS	-	5	N	1
O7→O8	F7	20	Y	N	N	Any	FS	-	5	N	1
O6→PS	F6	30	Y	Y	Y	S	FS	Pal	15	N	1
	F10	50	Y	Y	Y	S	FS	Pal	15	N	1
O8→PS	F8	20	Y	Y	Y	S	FS	Pal	15	N	1

The determinative handling characterizations of the objects within a given relation can be

- grabbing (supporting, hanging, pressing, etc.),
- reaching (usually upside, downside or frontside),
- size-limits (can be minimal, maximal, palette-size, box-size, etc.).

The route parameters of the relations can be physical (distance and fall) or complexity characterizations. Complexity of a relation depends on the number, directions (horizontal, vertical, slopes, arcs, etc.), angles and connections of the line-sections of the routes. It can be described by a number (e. g. 1-9), which are defined previously based on the system parameters.

In our scenario, two size-limits exist:

- at all the objects workplace stores are required, where the elements can be placed only in boxes with a given size,

- final products are transported on standard palettes into the product store.

If we described the parameters of the relations, next step is to compare them to the parameters of the handling machines (Table 4).

Table 4. Important selection parameters of material handling machines for unit handling
(Y – yes, N – no, US – upper-side, DS – down-side, FS – front-side, AS – all-side, S – supporting, H – hanging, P – pressing, Box – box limit, Pal – palette limit)

Handling machines	Direction		Functions			Handling			Line		
	Horizontal	Vertical	Transport	Loading	Storing	Grabbing	Reaching	Size limit	Free	1 track	More track
Forklifts	Y	Y	Y	Y	N	S	FS	Pal	Y	N	N
Platform trucks	Y	N	Y	N	Y	-	-	-	Y	N	N
Travelling cranes	Y	Y	Y	Y	N	H	US	-	N	N	Y
Hoists	Y	Y	Y	Y	N	H	US	-	N	Y	N
Jib cranes	Y	Y	N	Y	N	H	US	-	Y	N	N
Monorails	Y	Y	Y	N	Y	-	US	-	N	Y	N
Robots	Y	Y	N	Y	N	P	AS	-	Y	N	N
Storage-retrieval machines	Y	Y	N	Y	N	S	FS	Bx	N	N	Y
Elevators	N	Y	Y	N	N	-	-	-	N	Y	N
Tow conveyors	Y	Y	Y	N	N	-	US	-	N	Y	N
Trolley conveyors	Y	Y	Y	N	N	-	US	-	N	Y	N
Roller conveyors	Y	Y	Y	N	Y	-	-	-	N	Y	N
Belt conveyors	Y	Y	Y	N	N	-	-	-	N	Y	N

Using the data in Table 3 and 4 we can determine the applicable equipment types for the individual relations. Table 5 contains the best solutions for our test system.

4.4. Process building

As it described in Chapter 3.3, the first step of the process building is the analysis of the units handled in the system. In our scenario we have three main unit types (elements – a1-a7, processed units – F1-F5, F7, F9 and finished products – F6, F8, F10). Elements and products have uniform handling parameters, but we can determine differences within the half-finished units:

- F1 and F4 are handled in given sized boxes,
- F2, F3, F5, F7 and F9 can be handled in boxes or in palettes.

The simplest case is the handling of the final products, where only one machine type is suitable (forklift), so at this phase, we can define one handling process which contain both relations (O6→PS and O8→PS).

IOP Conf. Series: Materials Science and Engineering **448** (2018) 012018 doi:10.1088/1757-899X/448/1/012018

Table 5. Applicable machines for the handling relations

Relations	Applicable machines												
	Forklifts	Platform trucks + ML	Travelling cranes	Hoists	Jib cranes	Monorails	Robots	Storage-retrieval machines	Elevators	Tow conveyors	Trolley conveyors	Roller conveyors + ML	Belt conveyors + ML
ES→O1		x										x	x
ES→O2		x										x	x
ES→O3		x										x	x
ES→O4		x										x	x
ES→O5		x										x	x
ES→O7		x										x	x
ES→O8		x										x	x
O1→O2		x										x	x
O2→O3	x	x										x	x
O3→O6	x	x										x	x
O4→O5		x										x	x
O5→O6	x	x										x	x
O5→O7	x	x										x	x
O7→O8	x	x										x	x
O6→PS	x												
O8→PS	x												

4.4.1. Handling processes of the elements. As we can observe in Table 5, we can use three types of equipment for the transportation of the elements to the manufacturing objects. All three (platform truck, belt conveyor and roller conveyor) is suitable for the handling with manual loading of the boxes. To make process variations, we have to analyse the locations of the related objects (Figure 8.) and the transportation parameters (Table 6).

It can be seen in Figure 8 that the location of the supplied objects is dispersed on a larger area and they cannot be linked by a simple transport line, so the application of fixed-line equipment is not the best solution, because of the complex transport routes.

It is easy to see, if we apply a palette car with 0,8m x 1,2m (placed only 6 boxes on it in one route) platform, the total amount of the boxes (see Table 6) can be easily transported within one hour using even the simplest direct routes (one transport and one empty return per relation) for the supply. As the total length of the routes is over 40 m, the application of any of the continuous handling equipment (belt and roller conveyors) could be more expensive than a single palette car. Beside it, the complexity and manual handling service of them can also be problematic.

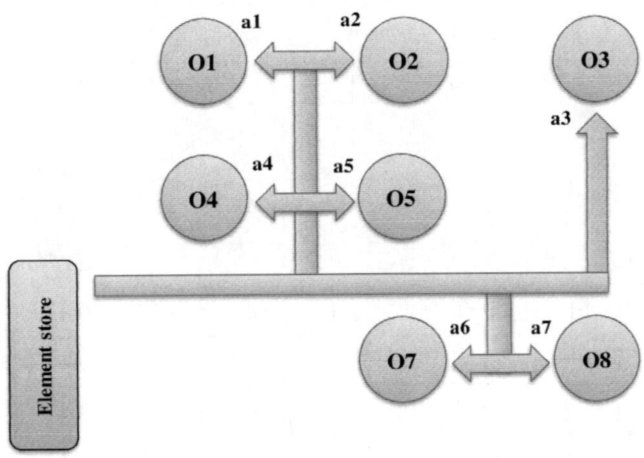

Figure 8. Objects and a road-variation in the element supply process

Table 6. Transportation parameters in the element supply process

Relation	Element		Unit						Distance
	Type	Quantity	Type	Content	Size A	Size B	Size C	Quantity	
		[pcs]		[pcs]	[m]	[m]	[m]	[pcs]	[m]
ES→O1	a1	30	Bx	10	0,4	0,4	0,4	3	20
ES→O2	a2	60	Bx	20	0,4	0,4	0,4	3	20
ES→O3	a3	30	Bx	10	0,4	0,4	0,4	3	25
ES→O4	a4	70	Bx	10	0,4	0,4	0,4	7	15
ES→O5	a5	70	Bx	7	0,4	0,4	0,4	10	15
ES→O7	a6	60	Bx	20	0,4	0,4	0,4	3	20
ES→O8	a7	20	Bx	5	0,4	0,4	0,4	4	20
Total								**33**	

4.4.2. Handling processes of half-finished units. As it can be seen in Table 5, theoretically there are four device alternatives for the interoperation handling. Of course, we can apply different combinations of them, but at first, we are analysing the one-device handling possibilities.

If we use only one machines for the handling, we have three main solution possibilities:

- handling boxes in all the relations,
- handling boxes (in two relations) and palettes (in five relations),
- handling boxes (in two relations) and pieces (in five relations).

As our main goal is the minimalization of the number of the handling machines, the starting case is the uniform handling in boxes. Table 7 shows the transportation parameters of the handling of the boxes among the system objects.

Table 7. Transportation parameters of the interoperation handling

Relation	Unit		Unit						
	Type	Quantity	Type	Content	Size A	Size B	Size C	Quantity	Distance
		[pcs]		[pcs]	[m]	[m]	[m]	[pcs]	[m]
O1→O2	F1	30	Bx	10	0,4	0,4	0,4	3	5
O2→O3	F2	30	Bx	10	0,4	0,4	0,4	3	5
O3→O6	F3	30	Bx	6	0,4	0,4	0,4	5	7
O4→O5	F4	70	Bx	10	0,4	0,4	0,4	7	5
O5→O6	F5	50	Bx	10	0,4	0,4	0,4	5	8
O5→O7	F9	20	Bx	5	0,4	0,4	0,4	4	14
O7→O8	F7	20	Bx	5	0,4	0,4	0,4	4	5
Total								**31**	

As in the element supply, we can apply a palette car with 0,8m x 1,2m platform, the total amount of the boxes (see Table 6) can be transported within one hour using even the simplest direct routes. The total length of the routes (see Figure 9) is fewer than at the element supply, but the application of a continuous handling equipment could be more expensive than a single palette car.

As one palette car is enough to handle all the relations, it is not worth to analyze the other versions, and the best solution will be finalized during the system building phase.

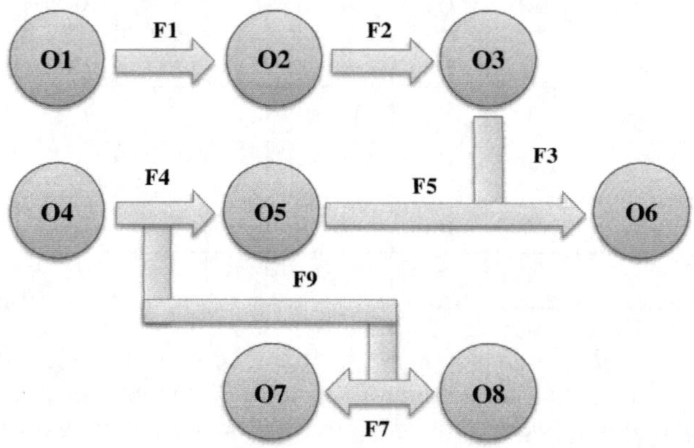

Figure 9. Objects and a road-variation for the interoperation handling

4.5. System building

During the system building, usually the final handling processes are detailed and harmonized, but in our scenario the finalization will be actualized.

The result of the process building for the analyzed system is three different process using two different machines (forklift and palette car). As two of the processes use the same machine type and their capacity use are near 15%, so it is possible to realize them with the same machine.

Of course, the planning will only be completed, if the dimensioning of the machines and the scheduling of the routes are made, but these procedures are out of the scope of our paper.

The applied process of our simple system can be seen in Figure 10.

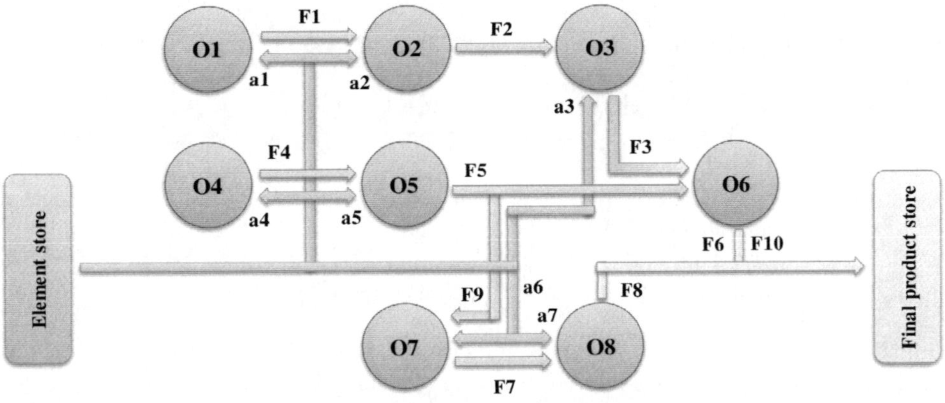

Figure 10. Handling system for the example production procedure

5. Conclusions

As the production processes are more and more complicated, which requires complex and flexible material handling solutions, the used design methods of material handling cannot offer generally usable solutions. Many computer methods try to follow the increased requirements, but they are dealing with only some part of the design tasks and in generally their procedures are not enough clear and understandable.

The objective of our research to develop a new planning concept for the materials handling, which is clear and followable for the users and gives effective results. Of course, this method cannot give global optimal solutions, but tries to find the best variation among many analyzed versions. The new concept uses the integrated, task-based approach however; it is mainly focusing to the handling processes.

In this paper, we described the new concept and presented a simple scenario to understand its operation.

The critical point of this concept is the process building, so to use this method in the practice, we have to develop an exact procedure which can be applied for the design of most of the handling systems. After it the next step will be to use computer methods to realize the process steps.

Acknowledgements

This project has received funding from the European Union's Horizon 2020 research and innovation programme under grant agreement No 691942. This research was partially carried out in the framework of the Center of Excellence of Mechatronics and Logistics at the University of Miskolc.

References

[1] Wang J. Y., Zhao J. S., Chu F. L. and Feng Z. J. 2010 Innovative design of the lifting mechanisms for forklift trucks. *Mechanism and Machine Theory* **45** 12, 1892-96

[2] Olearczyk J., Al-Hussein M. and Bouferguène A. 2014 Evolution of the crane selection and on-site utilization process for modular construction multilifts. *Automation in Construction* **43**, 59-72
[3] Fonseca D. J., Uppal G. and Greene T. J. 2004 A knowledge-based system for conveyor equipment selection. *Expert Systems with Applications* **26** 4, 615-23
[4] Apple J. M. 1972 *Materials Handling Systems Design*. Ronald Press: New York
[5] Cselényi J. and Illés, B. (eds.) 2004 *Logistic systems (in Hungarian).* University Press: Miskolc
[6] Cselényi J. and Illés, B. (eds.) 2006 *Design and control of material flow systems I. (in Hungarian).* University Press: Miskolc
[7] Telek P. 2012 Variations of material handling systems (in Hungarian). *GÉP* **63** 4, 23-26
[8] Bányai T. 2012 Structured modelling of integrated material flow systems (in Hungarian). *GÉP* **63** 4, 83-86
[9] Telek P. 2011 Characteristic solutions of material flow systems. *Advanced Logistic Systems - Theory and Practice* **5,** 57-62
[10] Antoniolli I., Guariente P., Pereira T., Pinto Ferreira L. and Silva F. J. G. 2017 Standardization and optimization of an automotive components production line. *Procedia Manufacturing* **13**, 1120-27
[11] Jiang S. and Nee A. Y. C. 2013 A novel facility layout planning and optimization methodology. *CIRP Annals - Manufacturing Technology* **62**, 483-86
[12] Landschuetzer C., Jodin D. and Wolfschluckner A. 2011 Knowledge Based Engineering – an approach via automated design of storage/retrieval systems. *Proceedings in Manufacturing Systems* **6** 1, 3-10
[13] Sugimoto Y., Seki H., Samo T. and Nakamitsu N. 2016 4D CAD-based evaluation system for crane deployment plans in construction of nuclear power plants. *Automation in Construction* **71** 1, 87-98
[14] Baizid K., Ćuković S., Iqbal J., Yousnadj A., Chellali R., Meddahi A., Devedžić G. and Ghionea I. 2016 IRoSim: Industrial Robotics Simulation Design Planning and Optimization platform based on CAD and knowledgeware technologies. *Robotics and Computer-Integrated Manufacturing* **42**, 121-34
[15] Ortner-Pichler A. and Landschützer C. 2017. Improving geometry manipulation capabilities of Knowledge-based Engineering applications by the versatile integration of 3D-CAD systems. In *CD proc. of MultiScience - XXXI. microCAD Int. Multidiscipl. Sci. Conf.* (C1: Log. section, 3.), University of Miskolc
[16] Telek P. 2013 Equipment preselection for integrated design of materials handling systems. *Advanced Logistic Systems - Theory and Practice* **7** 2, 57-66
[17] Welgama P. S. and Gibson P. R. 1995 A Hybrid Knowledge Based/Optimisation System for automated Selection of Materials Handling System. *Computers Ind. Engng.* **28** 2, 205-17
[18] Cselényi J. and Telek P. 1999 Mathematical modelling of collecting logistic system of used electronic products, In: *Bányai T. and Cselényi J. (eds.) Modelling and optimization of logistic systems: Theory and practice.* University of Miskolc, 1-12.
[19] Obádovics J. Gy. and Szarka, Z. 2009 *Mathematics at higher levels (in Hungarian).* Scholar Publishing: Budapest
[20] Telek P. 2016 Material flow relations in the design process of materials handling. *Advanced Logistic Systems - Theory and Practice* **10** 1, 53-64
[21] Telek P. 2014 Application of device-preselection for discontinuous unit handling. *Advanced Logistic Systems - Theory and Practice* **8** 1, 93-102
[22] Kulcsár B. 1998 *Industrial logistics (in Hungarian).* LSI Education Center: Budapest
[23] Telek P. 2017 Computer Method for the Determination of Materials Handling Relations In: *CD proc. of MultiScience - XXXI. microCAD Int. Multidiscipl. Sci. Conf.* (C1: Log. section, 5.), University of Miskolc

IOP Conf. Series: Materials Science and Engineering **448** (2018) 012019 doi:10.1088/1757-899X/448/1/012019

Micromachining of cold-worked tool steel by nanosecond laser

P Šugár[1], B Ludrovcová[2], J Šugárová[3] and M Frnčík[4]

[1] Professor, Faculty of Materials Science and Technology, Slovak University of Technology, Slovakia
[2] PhD student, Faculty of Materials Science and Technology, Slovak University of Technology, Slovakia
[3] Associate Professor, Faculty of Materials Science and Technology, Slovak University of Technology, Slovakia
[4] PhD student, Faculty of Materials Science and Technology, Slovak University of Technology, Slovakia

E-mail: peter.sugar@stuba.sk

Abstract. The paper brings the results of experimental study of the laser beam milling applying the industrial grade fiber nanosecond laser operating at wavelength of 1064 nm and 62SiMnCr4 cold-work tool steel as working material. The laser pulse intensity and both lateral and transverse pulse overlaps influence on the material removal rate (MRR) and surface roughness are analysed and discussed. The experimental results reveal that for high ablation efficiency in combination with good surface finish the higher values of laser pulse intensities and lower to moderate lateral pulse overlap distances should be applied.

1. Introduction

In recent years, we have seen progress and development trends in micromachining technologies that are characterized by a wide range of applications, with the ability to machine almost all kinds of materials with high precision, good surface quality and relatively low cost. One of the most common micromachining technologies is the Laser Beam Machining (LBM). It has, among other things, a wide range of applications in the field of machining of materials used in tooling industry. The successful application of laser beam machining operations (laser beam drilling, laser beam milling and laser beam texturing) of these materials requires the perfect knowledge of material ablation mechanisms and the generation of new surfaces. Therefore this topic is of the interest of several academic and research centers in the world. Some of them put emphasis on the research of laser machining of different types of advanced tool materials, including cemented carbides, ceramics and polycrystalline diamond as well. A lot of earlier works were carried out also on tool steel used in the production of dies and molds [1 – 10].

Texidor et al. [1], using statistical analysis, evaluated the effects of parameters (scanning speed, pulse intensity and pulses frequency) on surface roughness and dimensional accuracy of machined micro-channels. In this case a pulsed, Nd:YAG laser was used for machining hardened tool steel AISI H13.

The study [9] focuses on Cr12MoV alloy steel, with high carbon content, machined by a picosecond, pulsed Nd:YVO4 laser. The authors reported the effect of laser beam energy density on ablation and morphology of the machined surface for wavelengths 1064 nm, 532 nm and 355 nm. A logarithmic increase of the ablation intensity up to 41 nm per pulse was recorded using the wavelengths of 1064 nm and 355 nm. For the wavelength of 532 nm, the material ablation rate with increased beam energy density increased up to a level of 54 nm per pulse at laser fluency of $25 \times 10\ J \cdot cm^{-2}$ and then dropped to almost zero as the energy density increase above $25 \times 10\ J \cdot cm^{-2}$. With increased energy density, the roughness of the machined surface deteriorated exponentially.

The changes of the laser drilled surface microstructure of the M2 tool steel surface depending on the different wavelengths and pulse intensity, were analyzed in [10]. Based on the SEM, the authors found that the greatest material removal rate (MRR) and ablation occurred at higher pulse intensities. At a wavelength 1064 nm, using a shielding gas, typical structures of the hardened material were generated on machined surface due to the plasma effect. The direction of flow of the molten material, was identical with the direction of flow of the shielding gas above the surface of the workpiece.

Fleischer et al. [11] examined a similar field for tool steel X38CrMoV5-1 and reported that pulse energy and pulse overlaps have the greatest influence on the machined surface quality.

The laser machining parameters influences on roughness and depth of micro-channels were investigated in [12] at 30 W power on tool steel AISI H13. The results showed, that scanning speed has high impact on surface roughness and the depth of the machined micro-channels. Frequency does not affect both the surface roughness and the depth of micro-channels.

The influence of pulse overlaps on AISI H13 steel surface roughness was analyzed in [13]. By applying higher levels of pulse overlap the surface roughness parameter Sa was improved by 86.7 % from the initial roughness 1.35 μm to polished 0.18 μm.

A literature review shows, that exploring and improving the quality of laser-machined tool surfaces is a field, where it is still necessary to examine input process parameters, and optimize them for different materials, in order to achieve the best possible results. Therefore, the aim of this work is to investigate laser beam machining of the 62SiMnCr4 tool steel, designed for cold-worked molds and dies production. As the tool, a fiber nanosecond laser beam, with a wavelength of 1064 nm is used. The goal is to analyze the influence of pulse intensity and pulse overlaps on surface morphology, basic roughness parameters of machined surface and MRR.

2. Materials and methods

A laser machining centre Lasertec 80 Shape equipped with the low-power nanosecond ytterbium-doped fiber laser (wavelength of 1064 nm, maximal average output power of 100 W, pulse duration of 120 ns) was used for laser machining of cold-work tool steel 62SiMnCr4 without using a shielding gas. The working material is a low-alloy tool steel suitable for cold-worked forming tool components, with very good toughness, relatively high hardness, good resistance to dynamic alternating stress and impact stress and very good wear resistance. The chemical composition and selected physical properties of this material are specified in Table 1 and Table 2, respectively.

Table 1. Chemical composition of 62SiMnCr4 steel (wt %)

C	**Si**	**Mn**	**P**	**S**	**Cr**	**Ni**
0.55 – 0.65	1.5 – 1.9	0.6 – 0.9	≤ 0.03	≤ 0.035	0.7 – 1.0	≤ 0.35

Table 2. Selected physical properties of 62SiMnCr4 steel

Specific heat capacity	**Thermal conductivity**	**Linear expansion coefficient**	**Density**
$460\ J \cdot kg^{-1} \cdot K^{-1}$	$31.0\ W \cdot m^{-1} \cdot K^{-1}$	$12{,}7 \times 10^{-6}\ K^{-1}$	$7.8\ g \cdot cm^{-3}$

27 cavities (with square shape and 5 mm long sides) (Fig. 1) were machined by cross hatching milling strategy, according to the full factorial experimental plan (3^3 trials). The surfaces to be machined have been prepared by grinding and polishing so that the roughness was Ra 0.02. The layer-by-layer method

IOP Conf. Series: Materials Science and Engineering **448** (2018) 012019 doi:10.1088/1757-899X/448/1/012019

of material removal during laser beam machining was used with total number of 10 layers.

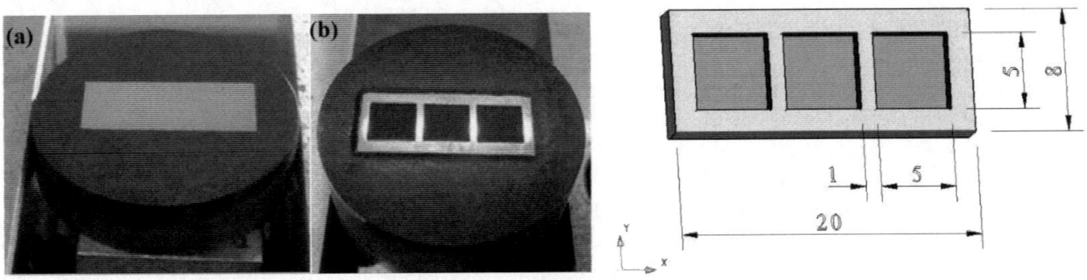

Figure 1. Experimental sample
a – before machining, b – after machining

The influence of laser pulse intensity and lateral and transverse pulse overlap on the surface roughness of machined surface and MRR has been studied. The experimental factors and their levels are listed in Table 3. For each experimental design the pulse frequency was set at 80 kHz. Laser beam with spot size of 50 μm was focused on the top of surface to be machined.

Table 3. Factors of the experiment and their levels

Experimental factor	Level		
	1	2	3
Laser pulse intensity (**PI**) [$\times 10^8$ W · cm^{-2}]	2.12	3.18	4.25
Lateral overlap distance (**DL**) [μm]	12.5	20	27.5
Transverse overlap distance (**DT**) [μm]	12.5	20	27.5

The principle of lines and machined areas formation applying the lateral and transverse pulse overlap is illustrated in Figure 2.

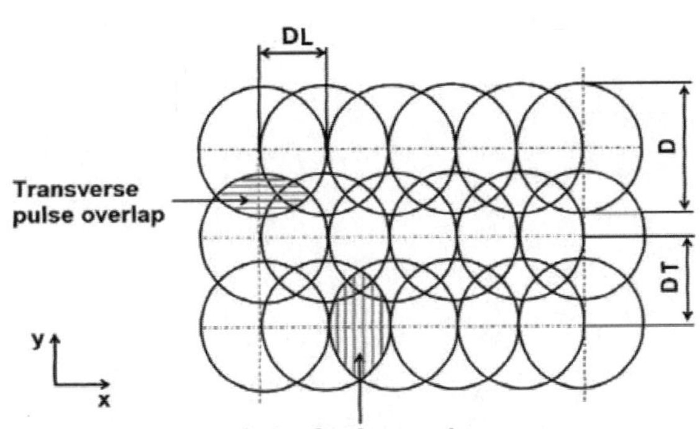

Figure 2. Schematic of lateral and transverse pulse overlap
(D – spot diameter, DL – lateral overlap distance, DT – transverse overlap distance)

Roughness parameters of machined surfaces were measured using a contact-gauge profilometer Zeiss Surfcom 5000. The profile parameters Ra and the Abbotte-Firestone curve parameters Rpk, Rvk were elaborated and reported in the perpendicular direction to the direction of laser beam movement in the last machined layer. The measurements were taken according to the standards [14], [15], [16].

The MRR for each experimental trial has been calculated based on the dimensions of the machined cavities (the depth of cavities was measured by Z-measurement probe installed on the laser machine tool) and the total time of machining, recorded by the machine tool control system.

In order to quantify the effects of the studied factors and interactive influences among them on the surface profile parameters and MRR, the analysis of variance (ANOVA) using Minitab v. 17 software was performed. The significance of factors has been evaluated at the significance level of 0.5.

3. Experimental results and discussion

The statistical significance and the main effect of the process parameters were detected by ANOVA tables and main effect plots. Because ANOVA assumes that the analysed data are normally and independently distributed with the same variance for each treatment or factor level, therefore these hypotheses were checked by normality test and test for equal variances.

3.1 Material removal rate

The results of ANOVA and main effect plots of the MRR are depicted in Fig. 3. It shows that pulse energy (PI) of laser beam has strong, directly proportional effect on the material removal rate. The effect of the lateral overlap distance, that is controlled by pulse frequency and scanning speed, is also statistically significant, but with lower effect on the response variable, compared with the pulse intensity. The opposite effect is produced regarding the transverse pulse overlap distance, where no significant impact on the volume of removed material was documented. Also no significant influence of process parameter interactions on the MRR was observed.

Source	Sum of squares	DF	Mean square	F - value
PI	20.247	2	10.124	*** 28.57**
DL	4.167	2	2.084	*** 5.88**
DT	0.18	2	0.09	**0.25**
PI*DL	0,359	4	0.0897	0.25
PI*DT	0.652	4	0.163	0.46
DL*DT	0.982	4	0.246	0.69
Error	2.835	8	0.354	
Total	29.2422	26		

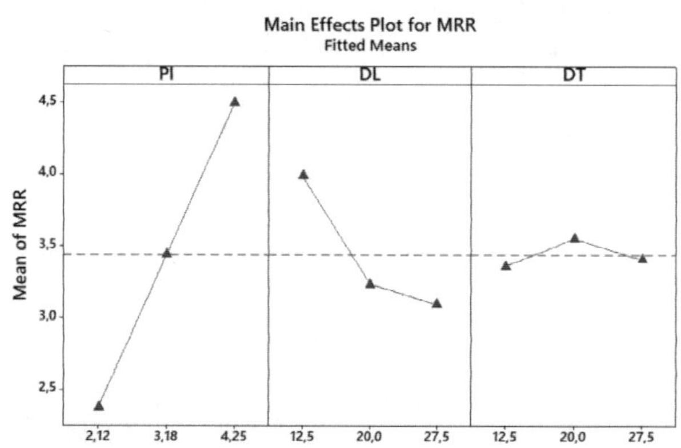

Tabulated F-test values at 95 % confidence level: F (0.05; 2.8) = 4.46; F (0.05; 4.8) = 3.84; R^2_{adj} = 68.69 %
* significant parameter

Figure 3. ANOVA table and main effect plots for MRR

3.2 Surface roughness

The results of the evaluation of the response variables Ra, Rpk and Rvk as a function of the process parameters are summarized in Figs. 4 – 6. The arithmetical mean deviation of the assessed profile Ra was documented due to its simplicity of its geometrical meaning and because it is the most common acceptation by researchers and production engineers. The surface roughness parameters Rpk and Rvk were assessed because these parameters brings information about the tribological properties of the surface - Rpk characterizes protruding peaks that might be eliminated during function and Rvk characterizes the valleys that will retain lubricant or worn-out materials.

As can be seen, the increasing of pulse energy reveals increasing of all evaluated surface parameters. On the other hand the roughness decreases when both lateral and transversal overlap distance increase, i. e. the pulse overlap decreases. The significant influence of the process parameters interactions has not been observed for all evaluated response variables.

IOP Conf. Series: Materials Science and Engineering **448** (2018) 012019 doi:10.1088/1757-899X/448/1/012019

Source	Sum of squares	DF	Mean square	F - value
PI	9.442	2	4.721	* **38.86**
DL	12.457	2	6.228	* **51.27**
DT	6.799	2	3.400	* **27.99**
PI*DL	1,779	4	0.448	3.66
PI*DT	0.653	4	0.163	1.34
DL*DT	0.356	4	0.089	0.73
Error	0.972	8	0.121	
Total	32.458	26		

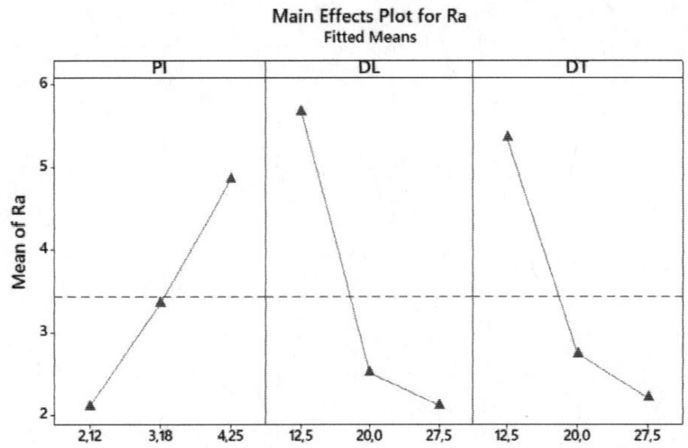

Tabulated F-test values at 95 % confidence level: F (0.05; 2.8) = 4.46; F (0.05; 4.8) = 3.84; R^2_{adj} = 90.27 %
* significant parameter

Figure 4. ANOVA table and main effect plots for Ra

Source	Sum of squares	DF	Mean square	F - value
PI	3.215	2	1.608	**3.34**
DL	6.423	2	3.212	* **6.67**
DT	3.790	2	1.895	**3.93**
PI*DL	2,404	4	0.601	1.25
PI*DT	0.982	4	0.245	0.51
DL*DT	2.783	4	0.696	1.44
Error	3.854	8	0.482	
Total	23.450	26		

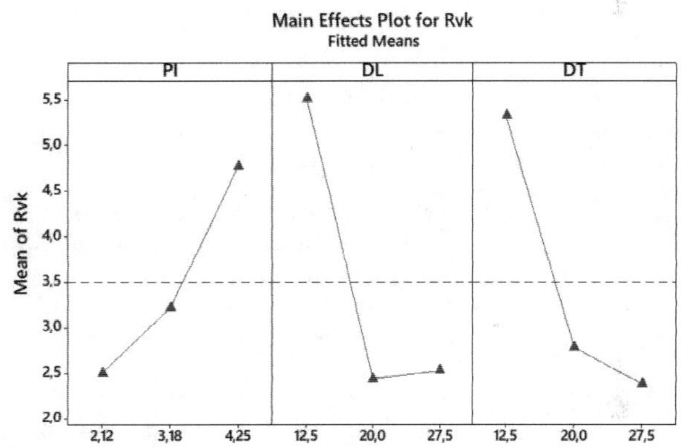

Tabulated F-test values at 95 % confidence level: F(0.05; 2.8) = 4.46; F (0.05; 4.8) = 3.84; R^2_{adj} = 46.59 %
* significant parameter

Figure 5. ANOVA table and main effect plots for Rvk

Source	Sum of squares	DF	Mean square	F - value
PI	7.441	2	3.705	* **46.09**
DL	11.091	2	5.545	* **68.99**
DT	3.700	2	1.850	* **23.02**
PI*DL	0,120	4	0.030	0.37
PI*DT	0.566	4	0.142	1.76
DL*DT	0.378	4	0.095	1.18
Error	0.643	8	0.080	
Total	23.909	26		

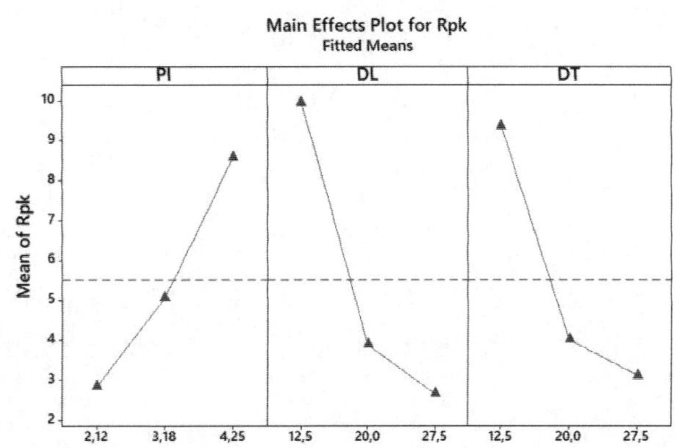

Tabulated F-test values at 95 % confidence level: F (0.05; **2.8**) = 4.46; F (0.05; **4.8**) = 3.84; R^2_{adj} = 91.26 %
* significant parameter

Figure 6. ANOVA table and main effect plots for Rpk

IOP Conf. Series: Materials Science and Engineering **448** (2018) 012019 doi:10.1088/1757-899X/448/1/012019

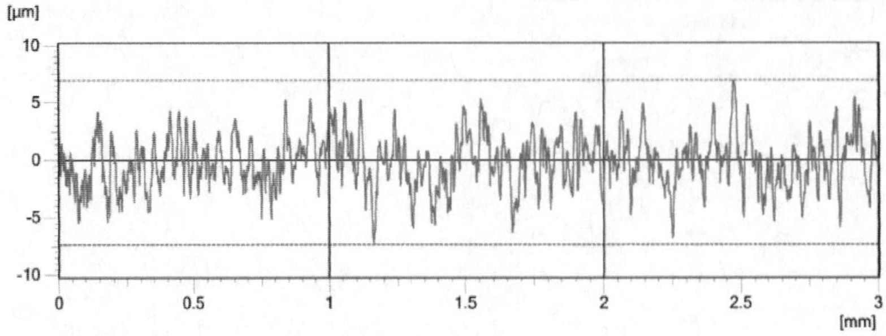

a

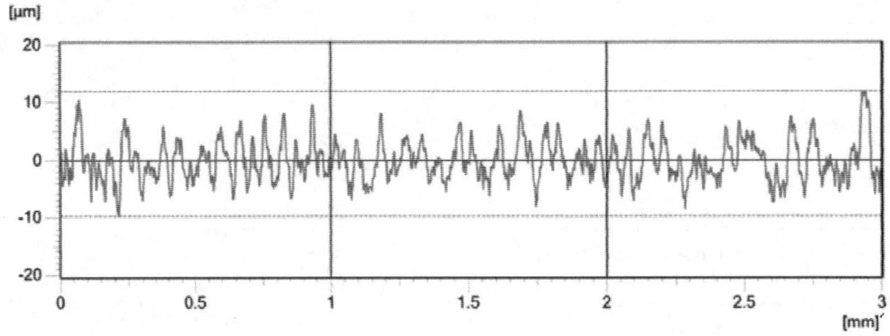

b

Figure 7. Machined surface microprofiles (DL = DT = 20 µm)
a – low level of energy (PI = 2,12 × 10^8 W · cm^{-2}): Ra 1.80 µm, Rpk 2.30 µm, Rvk 2.17 µm
b – high level of energy (PI = 4,25 × 10^8 W · cm^{-2}): Ra 2.83 µm, Rpk 4.15 µm, Rvk 2.23 µm

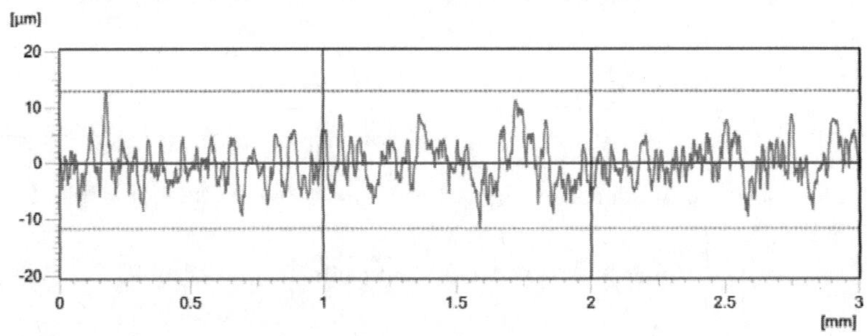

a

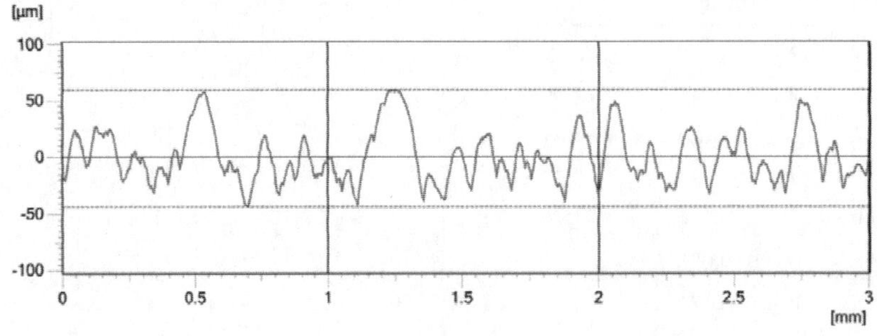

b

Figure 8. Machined surface microprofiles (PI = 4,25 × 10^8 W · cm^{-2}, DT = 12,5 µm),
a – low level of lateral overlap (DL = 27.5 µm): Ra 2.99 µm, Rpk 4.39 µm, Rvk 2.81 µm
b – high level of lateral overlap (DL = 12.5 µm): Ra 18.28 µm, Rpk 34.85 µm, Rvk 10.51 µm

IOP Conf. Series: Materials Science and Engineering **448** (2018) 012019 doi:10.1088/1757-899X/448/1/012019

The strong effect of pulse intensity and lateral overlap distance on the surface finish is visible in the Figs. 7 – 8, where the significantly different surface microprofiles, affecting the surface tribology properties are seen.

The combinations of the process parameters for achieving the maximal MRR and good surface finish are obvious from the 4D bubble plot (Fig. 9). The high MRR and low Ra area is associated with higher level of laser pulse intensities and low to moderate lateral overlap distances.

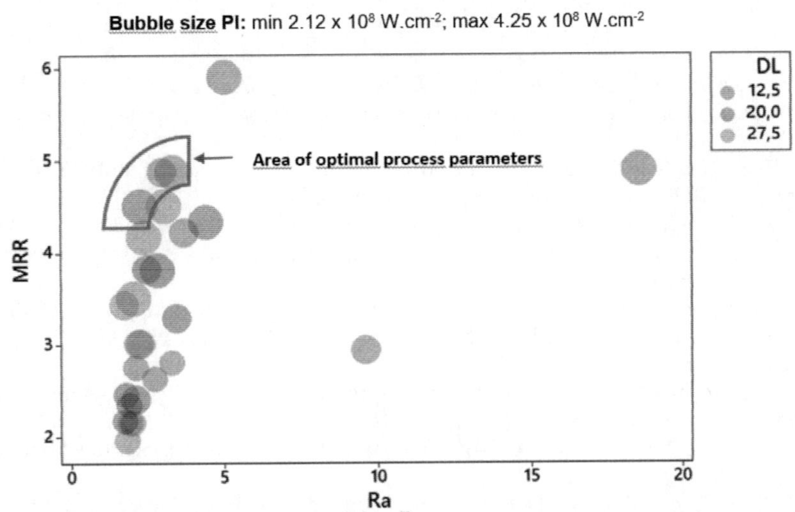

Figure. 9 A 4D bubble plot showing the design points obtained with variables of MRR, surface roughness Ra, lateral overlap distance DL (represented by color) and laser pulse intensity PI (represented by buble size)

Conclusions

The effect of nanosecond laser pulse intensity, lateral and transverse pulse overlap on ablation efficiency and machined surface quality of 62SiMnCr4 cold work steel have been studied. On the basis of results obtained from this investigation and statistical analysis, the following conclusions can be drawn:

(1) Significant effect of the laser pulse intensity on the MRR and surface morphology and surface roughness has been observed. It was confirmed that high laser pulse intensities lead to the worst surface finish. The head accumulated in the irradiated region in the case of higher values of pulse intensities causes a rapid increase in temperature which in turn increases the ablation efficiency. However, the phenomena such as stronger plasma irradiation and explosions of molten material during machining by high pulse intensities resulted in worst surface finish.

(2) The lower values of lateral pulse overlap, given by scanning speed and pulse frequency, lead to decreasing the material ablation efficiency and also surface roughness. Strong effect of both type of pulse overlaps on surface roughness parameters Ra and Rpk has been recorded while the roughness parameter Rvk has been slightly affected only by lateral pulse overlap. Similar trend has been observed in [17] during laser machining of the similar workpiece material by picosecond laser using frequencies lower than 1 MHz. However, according to this study, when the higher pulse frequencies (above 1 MHz) in combination with ultra-short pulses are applied, the increasing of lateral pulse overlap leads to better surface roughness.

(3) The only minimal effect of transverse pulse overlap on MRR has been reported.

(4) For high ablation efficiency and good surface finish the higher values of laser pulse intensities in combination with lower lateral pulse overlap should be applied.

Acknowledgments

This research was supported by an international research project under the MANUNET 2014 call, entitled: Innovative methods of sheet metal forming tools surfaces improvement - R&D (2014/11283); VEGA project: Laser surface texturing technology research for an optimizing of tribology conditions in the sheet metal forming processes (1/0669/15). The authors express their sincere thanks for financial contributions.

References

[1] Texidor D, Ferrerr I, Ciurana J and Özel T 2012 Optimization of process parameters for pulsed laser milling of micro-channels on AISI H13 tool steel *Robotics and Computer-Integrated Manufacturing* **29** 209-218

[2] Popp U and Engel U 2005 Microtexturing of cold-forging tools – influence on tool life *Special Issue Paper Proc. IMechE* **220** Part B*: J. Engineering Manufacture* 27-33

[3] Steyn J, Naidoo K and Land K 2007 Improvement of the surface finish obtained by laser ablation with a Nd: YAG laser on pre-ablated tool steel. *International Conference on Competitive Manufacturing*, Stellenbosch, XVI, 516 pp.

[4] Ciruana J, Arias G and Ozel T. 2009 Neural Network Modeling and Particle Swarm Optimization (PSO) of Process Parameters in Pulsed Laser Micromachining of Hardened AISI H13 Steel. *Materials and Manufacturing Processes* ***24*** 358-368.

[5] Šugár P, Šugárová J and Frnčík M. 2016 Laser surface texturing of tool steel: textured surfaces quality evaluation. *Open Engineering* ***6*** 90-97

[6] Demir A G, Maressa P and Previtali B 2013 Fibre laser texturing for surface functionalization *Physics Procedia* **41** 759-768

[7] Sokolová J 2018 The analysis of laser machined surfaces [Diploma Thesis] - Slovak University of Technology in Bratislava. Faculty of Material Science and Technology in Trnava, 82 pp.

[8] Lauer B, Jäggi B and Neuenschwander B 2014 Influence of the pulse duration onto the material removal rate and machining quality for different types of steel *Physics Procedia* **56** 963-972

[9] Wu B, Liu P, Duan J, Deng L, Zeng X and Wang X 2016 Study on picosecond pulse laser ablation of Cr12MoV cold work mold steel *Materials and Design* **110** 549-557

[10] Jackson M J and O´Neill W O 2003 Laser micro-drilling of tool steel using Nd:YAG lasers *Journal of Materials Processing Technology* **142** 517-525

[11] Fleischer J, Halvadjiysky G and Haupt S 2008 Process parameter analysis in ablating micro-mold manufacturing *Microsyst Technol* **14** 1367-72

[12] Kasman S and Saklakoglu I E 2012 Determination of process parameters in the laser micromilling application using Taguchi method: A case study for AISI H13 tool steel *Int J Adv Manuf Technol* **58** 201-209

[13] Hafiz A M K, Bordatchev E V and Tutunea-Fatan R O 2012 Influence of overlap between the laser beam tracks on surface quality in laser polishing of AISI H13 tool steel *Journal of Manufacturing Processes* **14** 425-434

[14] ISO 4287:1997 Geometrical product specifications (GPS) – Surface texture: Profile method – Terms, definitions and surface texture parameters

[15] ISO 4288:1996 Geometrical product specifications (GPS) – Surface texture: Profile method – Rules and procedures for the assesment of surface texture

[16] ISO 3274:1996 Geometrical product specifications (GPS) – Surface texture: Profile method – Nominal characteristics of contact (stylus) instruments

[17] Wu B, Deng L, Liu P, Zhang F, Duan J and Zeng X. 2017 Effect of picosecond laser repetition rate on ablation of Cr12MoV cold work mold steel Applied surface science **409** 403-412

Measurement uncertainty of surface roughness measurement

G Farkas[1], **Á Drégelyi-Kiss**[2]

Department of Manufacturing Technology, Bánki Donát Faculty of Mechanical and Safety Engineering, Óbuda University, H-1081 Budapest Népszínház u. 8.

[1] farkas.gabriella@bgk.uni-obuda.hu
[2] dregelyi.agota@bgk.uni-obuda.hu

Abstract. There are several measurement methods to microgeometric examination of the surfaces machined by different cutting processes. At Department of Manufacturing Technology, Institute of Materials and Manufacturing Sciences, Óbuda University it has a traditional contact (stylus) instrument with related software which is widely used in the industrial practice. All measurements such as the surface roughness are characterized by uncertainties which are essential for accurate measurement of the result since the surface quality is characterized by the obtained roughness measurements. The measurement and the characterization of surface roughness and the method of determining measurement uncertainty are standardized. This article describes the determination and analysis of measurement uncertainty for 2D roughness parameters on surfaces prepared with different cutting methods.

1. Introduction

Characterization of surface irregularities is described by a number of measurement methods which examine the surface from different points of view and provide different information. There are three measuring techniques distinguished divided into two types according to the result such as quantitive and visualization. There are the contact (stylus) measurements, optical measurements and the other technical solutions. Typical measurement methods are to compare with standards, the contact stylus and the optical methods. Laser surface test equipment is used especially for cases where the touch probe would cause surface deformation. [1] The tactile measuring technique plays a central role in technical practice where the surface roughness can be detected with a properly configured touch probe. [2]

Other division for methods of examining surface roughness can be the subjective and the objective method. In the subjective comparison the test surface is compared to the standard one with naked eye, touch, or microscope. This method has only a subordinate role because of its subjectivity. During the comparison it was necessary to ensure that the roughness standard has produced by the same machining method as the surface of the test. Nowadays primarily the profiling testers are used for the objective method which are capable of supplying the characteristics used to determine the surface quality and is also capable of computer data processing. It can be distinguished from the point of view of operation: the fine mechanics, optical, pneumatic roughness measuring instruments.

1.1. The contact stylus measurement

IOP Conf. Series: Materials Science and Engineering **448** (2018) 012020 doi:10.1088/1757-899X/448/1/012020

The contact measuring technology plays a central role in technical practice. In this process, the properly configured probe slides over the surface and the signal generated by its motion transformed into an electric signal. For profilometric surface roughness gauges (in case of profiling), the diamond tip with a very small (about 1 to 10 μm) round radius is moving at a uniform speed and thus smoothes the unevenness of the surface. The mechanical design of measuring probes has several types. Manufacturers offer a plurality of probes (whole series) for a roughness measurement, which can be used to measure the widest range of surfaces. The selection of the most appropriate touch probe for the specific task requires a great deal of experience and theoretical knowledge of the person performing the measurement. The probes are mostly conical or spherical with a few micrometers of rounded end. Preferably selected enlarged image graphs reflect the nature of the surface and, with sufficient experience and practice, can be the source of much important information. Advanced surface roughness tools allow you to draw roughness and ripple profiles as well as unfiltered, undistorted profile views. The so-called diagrams are almost indispensable for providing a comprehensive microgeometric evaluation. When evaluating, however, it is important to note that due to the significant difference between vertical and horizontal magnifications, the representation is distorted.

Like all measurement techniques, this also has its own sources of error that are needed to properly evaluate the results of the instrument, such as rounding the tip of the probe head, the applied load force, the touch probe slipping [3].

The rounding radius of the probe tip is given (0.5, 2, 5, 10 μm). This small value also means that the touch probe is unable to detect smaller openings, gaps and faults, so overwhelm them. Thus, the resulting surface does not exactly match the original. This error is negligible if the unevenness of the surface texture is significantly greater than the peak punctuation of the touch probe, but it may be remarkable to use a high-peak circumferential touch probe for a very fine surface. This can smooth out the surface to such an extent that our results are practically insurmountable [4].

For probes that are rigidly attached to the drive structure, the fault may occur - especially for soft surfaces, under the touch probe. Small load on a small surface can also result in high surface pressure. Touch probes that are free to move only by self-weight cannot have the same effect [5].

A possible slipping of the touch probe from the test surface may occur as a malfunction. This means that the non-perfectly rigid probe slides out of the large protrusions, steep edges, and is not perfectly aligned but is different from that. Thus, the highest measured peaks will be smaller than in reality [6].

These errors are mechanical in nature and only refer to the uncertainties of scanning.

1.2. Microgeometric parameters

The task of roughness metrics is to characterize the surface. The requirement is that if the two measurements are the same, the "quality" of the two surfaces is the same. Separately, the roughness measurements are analysed, none of which is appropriate to determine the surface and its operating conditions. It follows that two, or three, separate metrics describing different characteristics are necessary to allow the two surfaces to be considered as identical.

The microgeometric characterization of working surfaces is practically measured in two-dimensional measurement techniques. The concept definitions and characteristics associated with plane (2D) microgeometric parameters have been standardized at international level. [7] The rules for drawing specifications for machine parts are contained in ISO 1302 [8]. Although there are national standards, these are identical to the ISO 4287 [9] and ISO 4288 [10] standards with insignificant technical differences. Based on these standards, the characteristics can be divided into following main categories:

- amplitude parameters (Ra, Rt, Rz, Rq, Rsk, Rku Rq, Rp, Rv, Rc)
- spacing parameters (RSm)
- hybrid parameters (RΔq)
- curves and related parameters (Rmr(c), Abott Firestone curve, Rδc, Rmr, profile height amplitude curve).

The determination of 2D parameters is the most widely used in technical practice, and mainly the elevation parameters are used to determine the surface quality. [11]

IOP Conf. Series: Materials Science and Engineering **448** (2018) 012020 doi:10.1088/1757-899X/448/1/012020

Roughness average (Ra) according to the ISO 4287 [9] "*is the arithmetic average of the absolute values of the roughness profile ordinates*". It contains low information since the value of Ra is not particularly sensitive to the peaks and recesses of the profile. It does not allow the definition of surface characteristics and provides the least amount of grounding for surface performance. For finely machined surfaces, Ra is unusable as it only shows the average of surface roughness as if the surface roughness of peaks and regular variations of recesses would shape that length. In contrast, this is often used in industrial practice to characterize technical surfaces. [9]

Maximum height of profile (Rz) according to the ISO 4287 [9] "*is sum of height of the largest peak height and the largest profile valley depth within a sampling length*". The value of the maximum height of the Rz profile is used to evaluate the homogeneity of the surface but still does not allow the detection of surface characteristics and working conditions.

Mean width of the profile elements (RSm) is spacing parameter. It means according to the ISO 4287 [9] "*value of the profile element widths Xs within a sampling length*". This parameter is more closely related to the functional nature of the surface. It shows the specify of the surface from the aspect of adhesion and surface treatment.

The kurtosis of the assessed profile (Rku) is amplitude parameter. It is according to the ISO 4287 [9] "*quotient of the mean quartic value of the ordinate values and the fourth power of Rq respectively within a sampling length*". This parameter expresses good the pointing of the height distribution. $Rku = 3$ means normal distribution, if $Rku > 3$ means the height distribution is sharp and if $Rku < 3$ the height distribution isn't sharp.

1.3. Measurement uncertainty

The measurement uncertainty is the quantitative value of the quality of the measurement result, allowing comparability of the measurement result with other results, references, technical characteristics or standards. Moreover, each measurement is charged with error, the measurement result differs from the true value of the measured characteristic. The goal of each measurement is to determine the value of the quantity to be measured. The measurement result is the estimate of the quantity to be measured. This will only be complete if the uncertainty of the measurement is also estimated. Measurement error and measurement uncertainty are not the same definitions. According to the [13] the measurement error is "*measured quantity value minus a reference quantity value*" and the measurement uncertainty is a "*non-negative parameter characterizing the dispersion of the quantity values being attributed to a measurand, based on the information used*". Measurement uncertainty contains in general many components. There are two type of the calculation in the GUM [14]:

- Type A evaluation of measurement uncertainty from the statistical distribution of the quantity values from series of measurements and can be characterized by standard deviations.
- Type B evaluation of measurement uncertainty can also be characterized by standard deviations, evaluated from probability density functions based on experience or other information.

The steps of the determination of measurement uncertainty in generally shows the figure 1. The purpose of the calculation is to determine the expanded measurement uncertainty, *U*. The value of the *k* coverage factor is usually 2 which refers to the approximately 95% probability range.

IOP Conf. Series: Materials Science and Engineering **448** (2018) 012020 doi:10.1088/1757-899X/448/1/012020

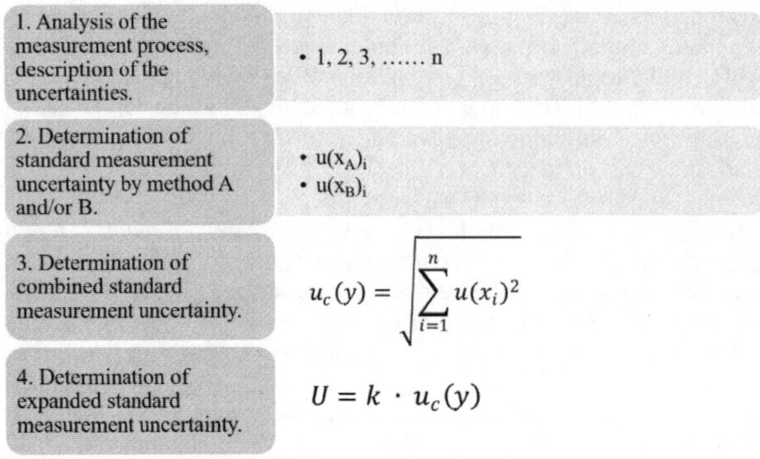

Figure 1. Steps of the determination of measurement uncertainty.

The VDI/VDE 2602 [15] describes the method for calculating measurement uncertainty of contact stylus instruments. The evaluations are consists of two parts:

- Measurement uncertainty of the vertical parameters.
- Measurement uncertainty of the horizontal parameters.

In this study several surface roughness parameters were analysed based on their measurement uncertainties. Turned and milled surfaces with the same nominal Ra values were compared with each other based on the measured Ra, Rz, RSm and Rku parameters and their associated expanded measurement uncertainty values (*U*). The *U* of the surfaces turned and milled having different surface quality is determined and compared in pair. The aim of this study it to describe the properties of the uncertainty in case of surface roughness measurements for several manufacturing processes having different cutting parameters.

2. Materials and methods

Comparative standards were chosen for the studies obtained by cutting processes (table 1). Measurements were carried out on milling and turning surfaces, where measurements can be repeated several times. Surfaces had a specific Ra value.

Table 1. Ra values of the comparative standards (turning and milling surface).

Sign of surface	Ra (μm)
N5	0.4
N6	0.8
N7	1.6
N8	3.2
N9	6.3

2D-la roughness examinations of surfaces machined with Mahr Perthometer-Concept feeler type instrument have been carried out in the Topography Laboratory of the Department. MFB-250 type feeler was used, which parameter is 90°/2 μm. The setup used at the evaluation of surfaces according to ISO 4288:1996 standard [10]:

- evaluation length, lm = 4 mm but in case of N9 lm = 12.5 mm
- the prescribed filter, lc = 0.8 mm but in case of N9 lc = 2.5 mm

The measurements were repeated twelve times for each surface under the same conditions and settings. In this study the examined microgeometric parameters were: Ra, Rz, RSm, Rku.

3. Measurement results and evaluation of data

The results of the examinations are presented for each microgeometric characteristic. The mean and the standard deviation of the measured Ra, Rz, RSm and Rku values for turned and milled surfaces can be found in table 2.

Table 2. Mean and standard deviation values of the 12 measurements for each sample (the Ra Diff. data show the difference values from the nominated value of each sample).

Cutting type	Part	Ra [μm]			Rz [μm]		Rsm [μm]		Rku [-]	
		Mean	StDev	Diff.	Mean	StDev	Mean	StDev	Mean	StDev
Turned	N5	0.3242	0.00097	-0.0758	1.6633	0.0273	45.1238	0.4670	2.1632	0.0201
Milled	N5	0.3468	0.00083	-0.0532	2.1804	0.0154	123.4524	11.9323	2.8171	0.0161
Turned	N6	0.8625	0.00174	0.0625	4.4334	0.0718	66.5180	0.9619	2.1091	0.0062
Milled	N6	0.7658	0.00117	-0.0342	4.8271	0.0197	196.0958	10.4236	3.0574	0.0102
Turned	N7	1.6240	0.00097	0.0240	6.1956	0.0342	126.7379	0.0201	1.6994	0.0010
Milled	N7	1.6576	0.00148	0.0576	8.0157	0.0155	177.4028	0.0245	2.3729	0.0015
Turned	N8	2.9751	0.00106	-0.2250	13.8532	0.0397	264.2753	0.1070	2.6757	0.0024
Milled	N8	3.4848	0.00157	0.2848	15.3823	0.0349	537.7273	0.0000	1.8322	0.0007
Turned	N9	7.1840	0.00132	0.8840	29.0472	0.0241	307.5000	0.0000	2.1042	0.0007
Milled	N9	5.9220	0.01551	-0.3781	32.2546	0.4966	674.1176	0.0000	2.5343	0.0157

In case of Ra it is seen that the standard deviation is quite low for N5-N8 surfaces for both turned and milled process output. The standard deviation is quite large in case N9 comparing with the previous ones. This is because the Ra values continuously increase during the repeated measurements in case of N9 sample, ranging between 5.9033-5.9301 μm (figure 2).

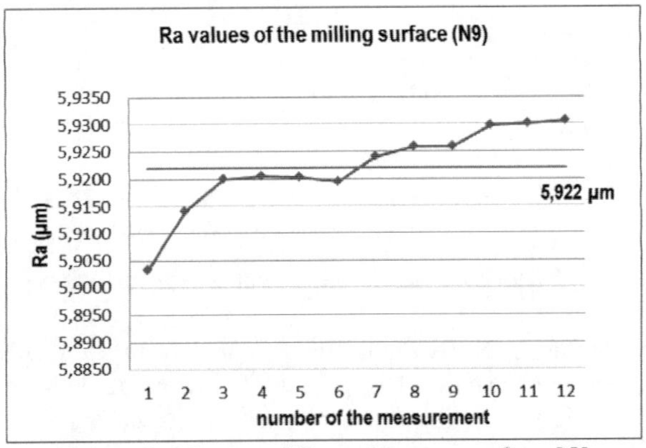

Figure 2. Measured Ra values on surface N9.

The mean values of the measured surface roughness in case of Ra can be examined comparing with the reference or nominal values of the samples, standards (see table 1). There are differences related to the manufacturing type i.e. the surface was produced by turning or milling machine. The largest differences are in case of N8 and N9 samples which have the largest roughness as well (figure 3).

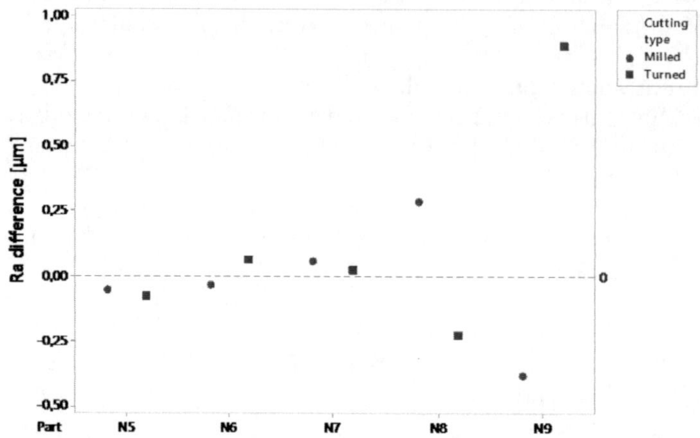

Figure 3. Measured Ra values minus the nominal Ra values on the examined surfaces.

In case of Rz it is seen that the standard deviation is quite low for N5-N8 surfaces for both turned and milled process output. The standard deviation is quite large in case N9 comparing with the previous ones. The StDev value on the N9 surface increased the reason being that the Rz values continuously increased during the repeated measurements, ranging between 31.6486-32.4661 µm (figure 4).

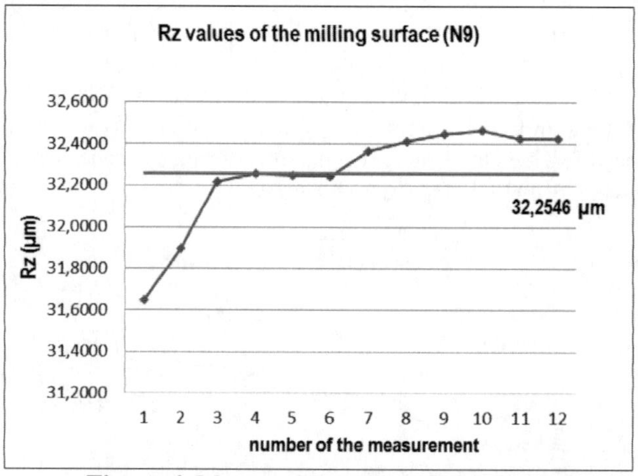

Figure 4. Measured Rz values on surface N9.

In the case of turning surfaces (N5-N9) the RSm ranged from 45 to 307 µm and the deviation of repeated measurements was rather small. The values of Rku parameter were between 1.7-2.1 on the turning surfaces. The average deviation of repeat measurements was small (0.003).

3.1. Measurement uncertainty calculations

The measurement uncertainty calculation is based on the GUM [14]. In case of this study the Type A uncertainty was taken into account. It is supposed that the Type B uncertainty components such as calibration uncertainty, temperature effect, effects on the environment, measurement set and measurement process have small impact on the data of combined uncertainty. Therefore the standard uncertainty of the measurement is calculated by the statistical analysis of repeated measurements.

The standard uncertainty is as follows:

$$u_c = u_A = \frac{StDev}{\sqrt{n}} \tag{1}$$

where *StDev* is the standard deviation of the repeated measurement, n is the number of repetition. In the investigated cases n=12, therefore it has not to take into account the Student t-factor.

The expanded measurement uncertainty (U) is as follows:

$$U = 2 \cdot u_A \tag{2}$$

Figure 5 shows the calculated extended measurement uncertainty of Ra on the examined surfaces. For turning surfaces, the U value was broadly constant (U = 0.0004 μm) as the standard deviation of repeated measurements were low. In the case of milling surfaces, the U value showed a roughly constant value on the N5-N8 surfaces and the standard deviation of repeated measurements was small.
It can be seen in figure 6 the calculated extended measurement uncertainty of Rz on the examined surfaces. For turning surfaces, the U value was broadly constant (U = 0.0045 μm) as the standard deviation of repeated measurements were low. It can be observed that on the milling surfaces the U value showed a roughly constant value on the N5-N8 surfaces and the standard deviation of repeated measurements was also small.

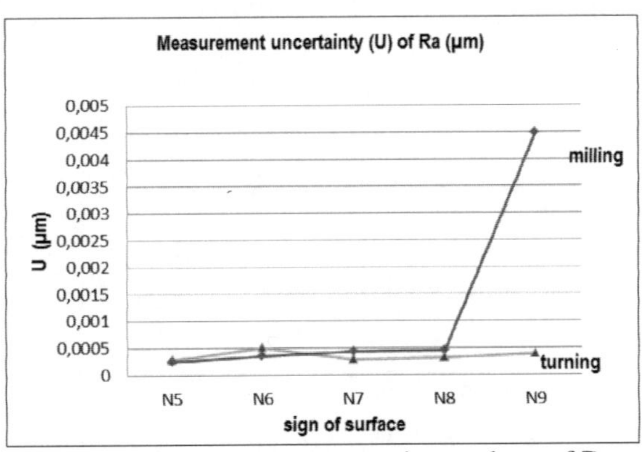

Figure 5. Measurement uncertainty values of Ra.

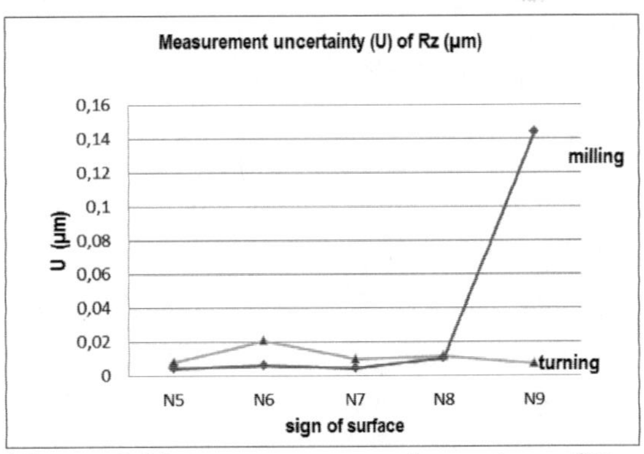

Figure 6. Measurement uncertainty values of Rz.

The calculated extended measurement uncertainty of RSm didn't change on the examined turning surfaces. For the milling surfaces the U value was broadly constant (U = 0.015 μm) only the N7-N9 as the standard deviation of repeated measurements were small. The U value on the N5-N6 surfaces increased (3.0-3.5 μm) the reason being that the RSm values definitely changed during the repeated measurements (figure 7).

It can be seen in the diagram (figure 8) that the extended measurement uncertainty is almost constant (0.0007-0.005). In the case of milling surfaces (N5-N9) the Rku ranged from 1.8 to 3 and the deviation

IOP Conf. Series: Materials Science and Engineering **448** (2018) 012020 doi:10.1088/1757-899X/448/1/012020

of repeated measurements was rather small (0.0003-0.008) accordingly the value of extended measurement uncertainty has only slightly changed (0.0007-0.005).

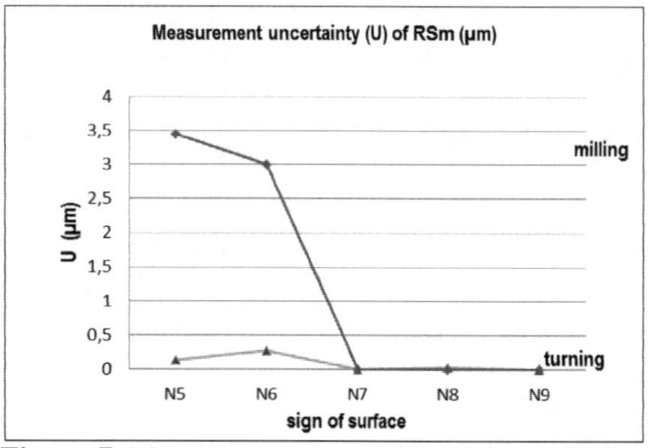

Figure 7. Measurement uncertainty values of RSm.

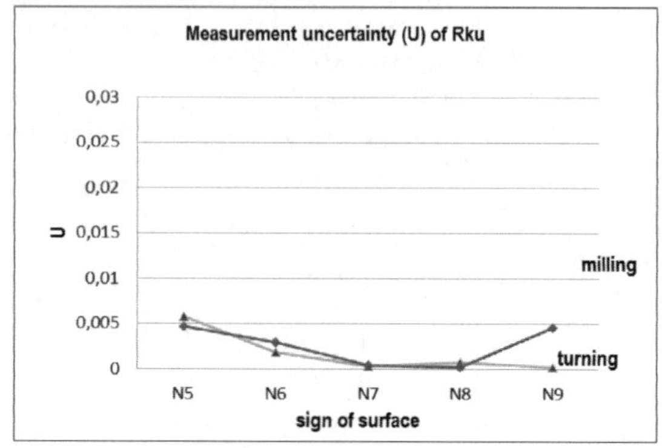

Figure 8. Measurement uncertainty values of Rku.

3.2. *Statistical comparison of roughness values and uncertainties*

There are significant differences between the samples (N5...N9) milled and turned. The results of the two sample t-probe show this deviations for all the investigated Ra, Rz, RSm and Rku values. It is seen in figure 9 (for Ra) and in figure 12 (for Rku) that the direction of the deviation is not the same for the each sample. On the N6 and N9 surfaces, due to the difference in the cutting marks, the result of the two roughness parameters differs from that seen on the other surfaces. Since weights repeated the same measurement length twelve times the measurement results didn't qualify the surface. The reason for this discrepancy is rather to find that the selected line was a surface error that gave this result. In case of Rz (figure 10) and RSm (figure 11) the direction is the same, therefore the value of the roughness parameter is larger significantly in case of milled surfaces.

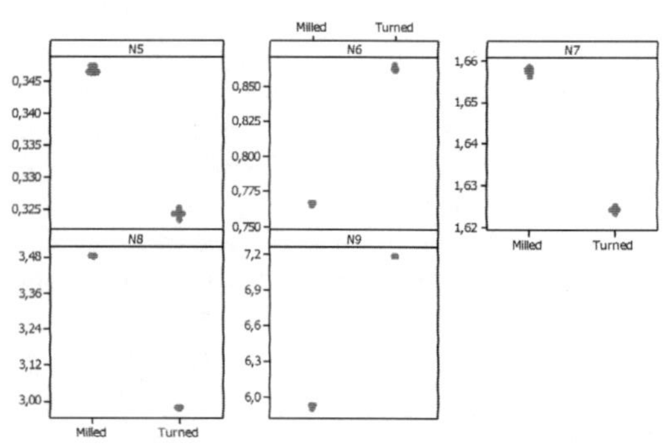

Figure 9. Measured Ra values [µm].

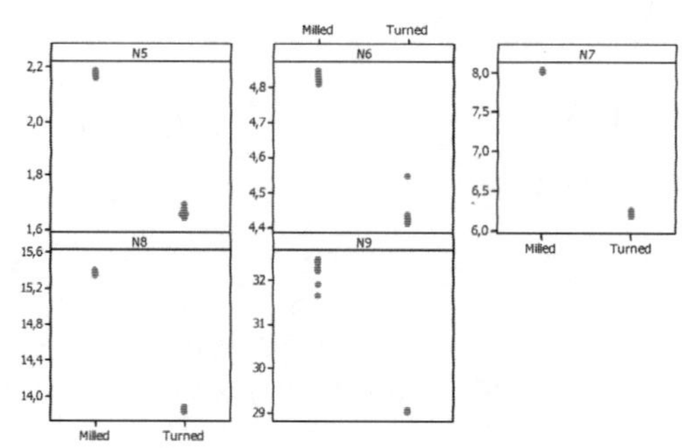

Figure 10. Measured Rz values [µm].

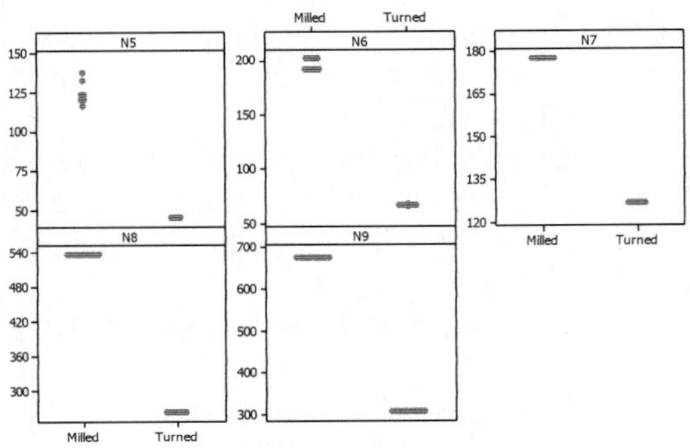

Figure 11. Measured RSm values [μm].

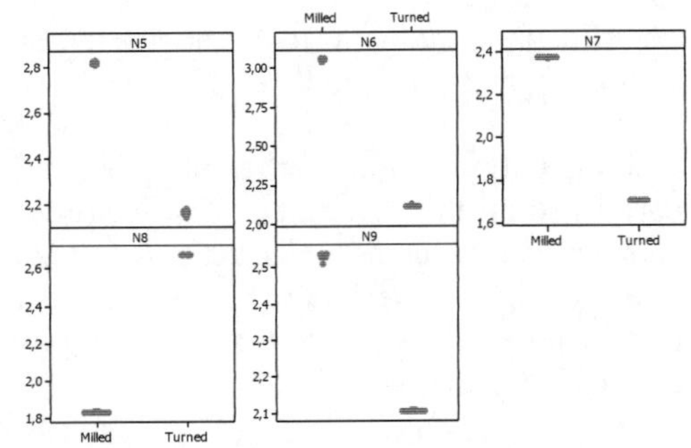

Figure 12. Measured Rku values [μm].

Nonparametric paired two sample tests (Wilcoxon signed rank test) were calculated (table 3) for the investigation of the effect of the cutting process in case of the expanded measurement uncertainty values (*U*). The Wilcoxon test can show whether there are differences between U values regarding the different manufacturing processes. This test is a nonparametric test therefore could be used for the comparison of these data. The results show that there is no significant difference at 95% level between the cutting processes (figure 13).

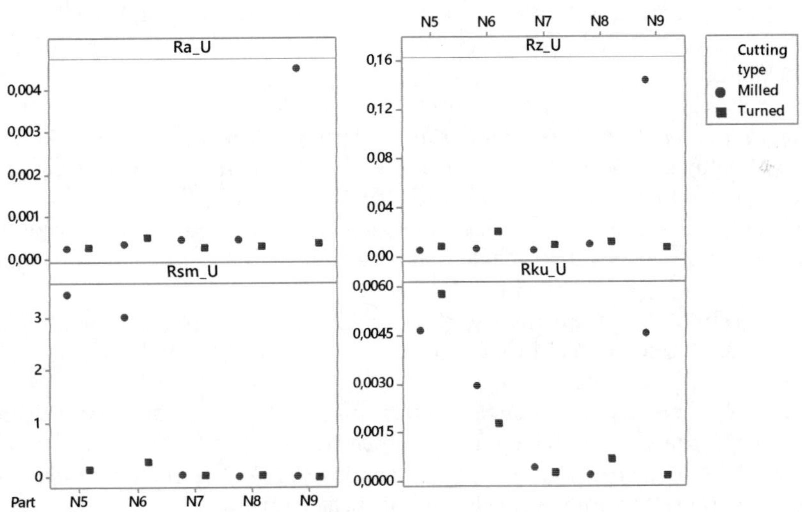

Figure 13. Calculated *U* values for Ra, Rz, RSm and Rku parameters.

Table 3. The results of the Wilcoxon test for the expanded measurement uncertainty values in case of Ra, Rz, RSm, Rku between milled and turned surfaces.

Test of median = 0,000000 versus median ≠ 0,000000

	N	N for Test	Wilcoxon Statistic	P	Estimated Median
Diff Ra_U	5	5	5,0	0,590	-0,0001465
Diff Rz_U	5	5	10,0	0,590	0,003391
Diff Rsm_U	5	4	2,0	0,361	-1,366
Diff Rku_U	5	5	5,0	0,590	-0,0003416

4. Conclusion

The aim of the study was to investigate measurement uncertainty in surface roughness, which also shows tangible results for practice. The available probing roughness tester, which is commonly used in industrial practice, is typically used in research tasks. Therefore, it was important to examine the exact reproducibility of the metrics supplied by the weighing machine. Previously, such tests were not performed on this equipment. Two cutting surfaces (turning, milling) were examined on the selected comparative standards. From the processing of the measurement results, we can conclude that the values of extended measurement uncertainty have characterized the individual roughness parameters in a very narrow range. This means that during repeated measurements the measured roughness characteristics are precisely provided.

Acknowledgement

The research was supported by the ÚNKP-17-IV-III New National Excellence Program of the Ministry of Human Capacities.

References

[1] Kim, H. Y., Shen, Y. F., Ahn, J. H.: Development of surface roughness measurement system using reflected laser beam, *Journal of Materials Processing Technology* 130-131, 2002, pp. 662-667

[2] Jensen, M. F., Christensen, L. H., Prichystal, J., Kjeldsteen, P., Benzon, M. E.: Evaluating the 3D-Topography of Engineering Surfaces, *Stucture magazine* 39 2001, pp. 3-7

[3] Vrabel, M., Maňková, I., & Beňo, J.: Surface roughness modeling and prediction by ANN when drilling Udimet 720. *In Key Engineering Materials* (2014, Vol. 581, pp. 366-371). Trans Tech Publications.

[4] Czifra, Á.; Barányi, I: Tribological Aspects for Reliable Operation of Engineering Surfaces. In: *Critical Infrastructure Protection Research*. Springer, Cham, 2016. p. 35-45.

[5] Beňo, J., Maňková, I., Vrabel, M., & Kottfer, D. : Roughness measurement methodology for selection of tool inserts. *Measurement*, 2003, 46(1), p. 582-592.

[6] Drégelyi-Kiss, Á.; Czifra, Á.: Measurement uncertainty and gauge capability of surface roughness measurements in the automotive industry: a case study. *Surface Topography: Metrology and Properties*, 2014, 2.3: 034001.

[7] ASME B46.1-2002: Surface texture (surface roughness, waviness, and lay) 2002.

[8] ISO 1302:2002 - Geometrical Product Specifications (GPS) — Indication of surface texture in technical product documentation

[9] ISO 4287:1997 Geometrical product specifications (GPS) – Surface texture: Profile method – Terms, definition and surface texture parameters.

[10] ISO 4288:1996 Geometrical product specifications (GPS) – Surface texture: Profile method – Rules and procedures for the assessment of surface texture.

[11] Kayahan, E., Oktem, H., Hacizade, F., Nasibov, H., Gundogdu, O.: Measurement of surface roughness of metal using binary speckle image analysis, Tribology International 2010, pp. 307-311

[12] Csaba Felho, János Kundrak: Comparison of theoretical and real surface roughness in face milling whit octagonal and circular inserts. *Key Engineering Materials* (2014, Vol. 581, pp. 360-365). Trans Tech Publications.

[13] JCGM 200:2012 International vocabulary of metrology – Basic and general concepts and associated terms (VIM)

[14] JCGM 100:2008 Evaluation of measurement data — Guide to the expression of uncertainty in measurement More references

[15] VDI/VDE 2602:2014 Part 4 Surface measurement. Roughness measurement using contact (stylus) instruments. Calibration, uncertainty of measurement.

IOP Conf. Series: Materials Science and Engineering **448** (2018) 012021 doi:10.1088/1757-899X/448/1/012021

Polymer-based nanocomposites with nanoclay

Dorottya Nagy[1] *, Eszter Kókai[1]

[1]Department of Materials Technology, GAMF Faculty of Engineering and Computer Science, John von Neumann University, Kecskemét, Hungary

*nagy.dorottya@gamf.uni-neumann.hu

Abstract. In the last few years a huge number of papers were published in the topic of nanocomposites. Mixing some kind of nanoscale additive into a polymer matrix provides an extra advantage to create polymer-based nanocomposites with better properties. The clay particles, like montmorillonite, can improve thermal, mechanical, rheological and barrier properties and conductivity, although the structure of the platelets greatly influence their behaviour. Moreover, nucleating and synergetic effect can occur, which also accomplish the properties. This review discusses the properties changes of the polymer nanocomposites reinforced with montmorillonite (MMT).

1. Introduction

Polymer nanocomposites can be widely used in many applications in the vehicle industry. The nanoadditive have at least one dimension smaller than 100 nm. The nanoscale dimensions make the macroscopic properties dimension-depending ones. The properties can be changed easily with only few percent of nanoparticles [1-3]. Some parts of the mechanism of the nanomaterial effect have been revealed, but not at all.

The influential factors come from both the matrix and additive properties, their proportion in the mixture, the processing technology and parameters, e.g. processing temperature, applied pressure [4-7]. The cohesion between the matrix and the additive materials is essential. For this reason surface treatment is required of the nanoadditives [8-10]. Carbon nanotube, graphene and several sort of clay particles (ceramics) are used as nanoscale additives. Montmorillonite is mostly used and also kaolinite, bentonite, hectorite, sepiolite, laponite etc. are used as nanoclay [11-12].

This work also summarises the main property changes in the nanocomposites and marks out a new way for our research.

2. Structure of MMT

MMT is a very soft, fine-grained phyllosilicate, member of the smectite group. It has sandwich shape from silica and alumina sheets. In Figure 1. its structure is shown in scanning and transmission electron microscopic (SEM, TEM) pictures.

IOP Conf. Series: Materials Science and Engineering **448** (2018) 012021 doi:10.1088/1757-899X/448/1/012021

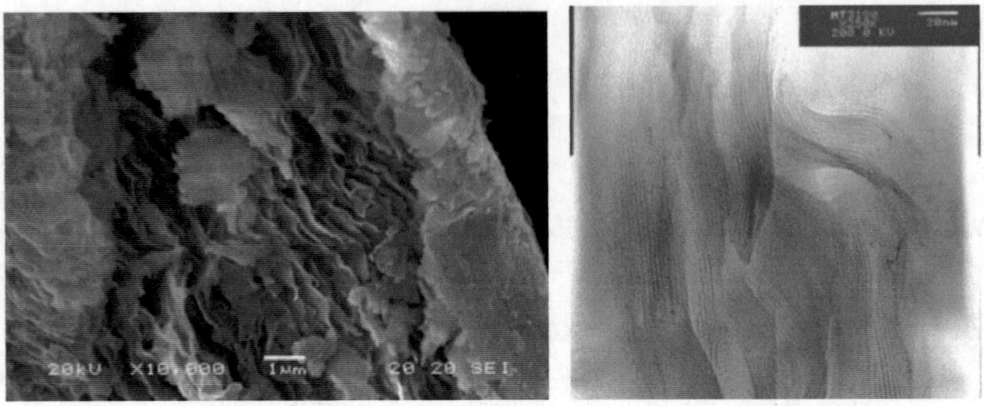

a) b)

Figure 1. Pictures of MMT: a) SEM, b) TEM [13-14]

It can form by precipitation from water solution as microscopic crystals. Chemically, it is hydrated sodium calcium aluminium magnesium silicate hydroxide $(Na,Ca)_{0.33}(Al,Mg)_2(Si_4O_{10})(OH)_2 \cdot nH_2O$. It can contain other cations, like potassium or iron [15].

The structure and dispersion of the clay will affect the properties of the nanocomposite. According to the state of dispersion, the clay can be in immiscible, intercalated or exfoliated form (Figure 2.) [16].

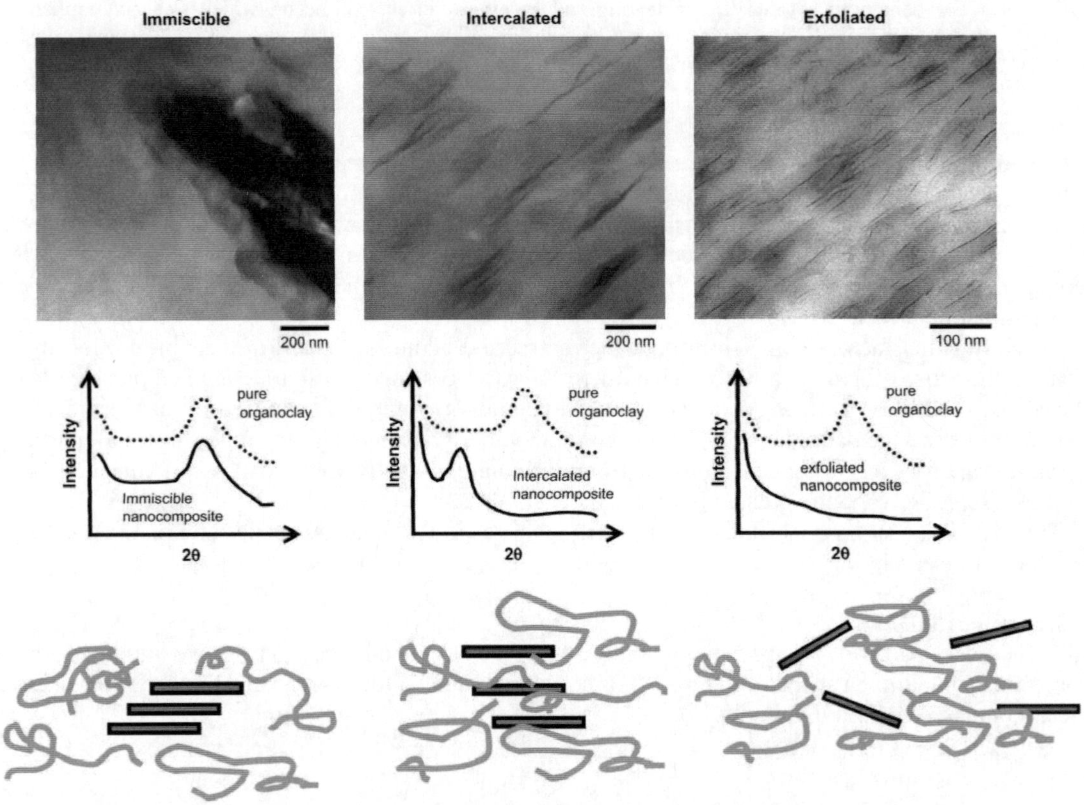

Figure 2. The structure and dispersion of the clay in polymers [16]

The immiscible particles have the worst properties, because the matrix material cannot get into the clay platelets. Intercalated form means a well-ordered structure with interaction between the layers and the matrix material. Exfoliated montmorillonite has delaminated form, the additive can disperse in the matrix well. The state of the clay can be organophilized or as it is (native). The clay is mined, and it can be modified by organic substances in order to achieve better interaction between the materials [3, 6].

3. Matrix polymers

Matrix material can be thermoset (epoxy or polyester resin) or thermoplastic [17-18]. In the field of thermoplastics, the most used polymers are polystyrene (PS), poly(lactic acid) (PLA), polyethylene terephthalate (PET), polypropylene (PP), polyamide (PA), polymethyl methacrylate (PMMA), poly e-caprolactone (PCL), polyvinylidene fluoride (PVDF), polyethylenglykol (PEG), polyethylene (PE), acrylonitrile butadiene styrene (ABS), polyvinyl acetate (PVA), polyethylene oxide (PEO), polyimide (PI), etc. [19-24]. Matrix polymers influence the processing technology and determine the application field (e.g. the operating temperature).

Preparing polymer-based nanocomposites can be in different methods. Precipitation means that a suspension or solution of modifier is added to the polymer solution. The solution should not dissolve the polymer. In sol-gel reaction, the nanocomposites are made by crosslinking from polymer/oligomer solution. With in-situ polymerisation, layered silicate nanocomposites can be prepared. In this reaction the monomers transform into polymer chains among the layers. Melt intercalation means that polymer powder or granulates and silicate are mixed together while the polymer is melting. An often-used method is also the exfoliation, in which nanoadditives are mixed to the polymer melt. The shear forces separate the silicate layers, so the cohesion between the matrix polymer and additive are assured [16].

4. Properties of the nanocomposites

Montmorillonite-based nanocomposites are relatively new materials, so the behavioural changes of the properties should be investigated.

4.1. Thermal properties

Generally, MMT increases the heat stability, while the incorporation of nanoclays reduces the mobility of the polymer chains [25]. Nanoclays also hinder the out-diffusion of the volatile substances [2]. As it is shown in Figure 3., the same percent of weight loss occurs at higher temperature adding MMT to polystyrene (PS). The number after PS means the MMT content, and FPS5 was made by another preparing method (dry mixing) with 5 wt% of MMT. The other three sample were prepared by solution blending method.

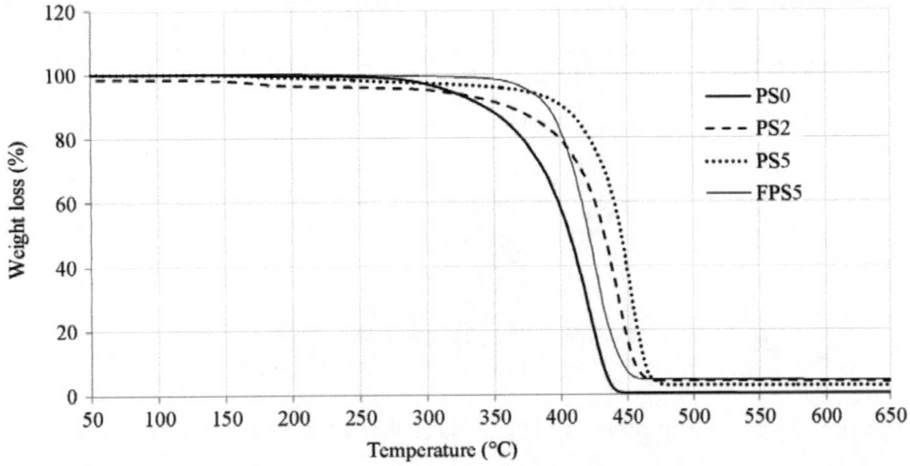

Figure 3. TGA curves of MMT filled PS [2]

In several cases intercalated MMT accelerates biodegradation of compostable materials (PLA), because the clay accumulates heat, so it functions as heat source [26]. The characteristic temperature values can also change, e.g. glass transition temperature (Tg) or melting temperature (Tm). The segmental motion occurs only at higher temperatures at an increased Tg, so the matrix material will be stiffer in this region.

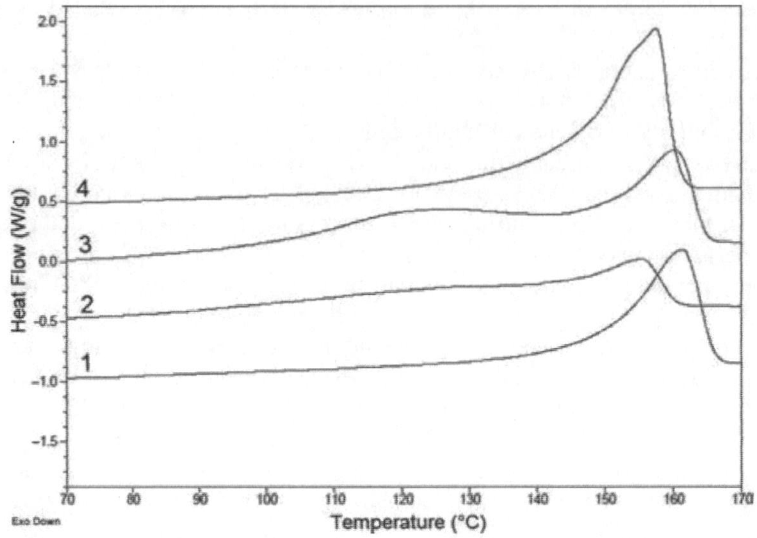

Figure 4. DSC curves of polypropylene (PP) with different clays [26]

In Figure 4. the melting ranges can be seen by Differential Scanning Calorimetry (DSC) method. **1** is for PP without clay, the other three curves for PP with clay: **2** for 16.9 wt% of Cloisite 30B, **3** for 1.1wt% of Cloisite 15A, **4** for 9.9 wt% of Claytone HY. Adding clay to the matrix polymer, the melting range is broadening and also the crystalline structure changes [26].

4.2. Barrier properties

The increased barrier property means reduction in gas (oxygen or carbon dioxide) or water vapor permeability [27]. The oxygen permeability plays an important role in oxidative degradation.

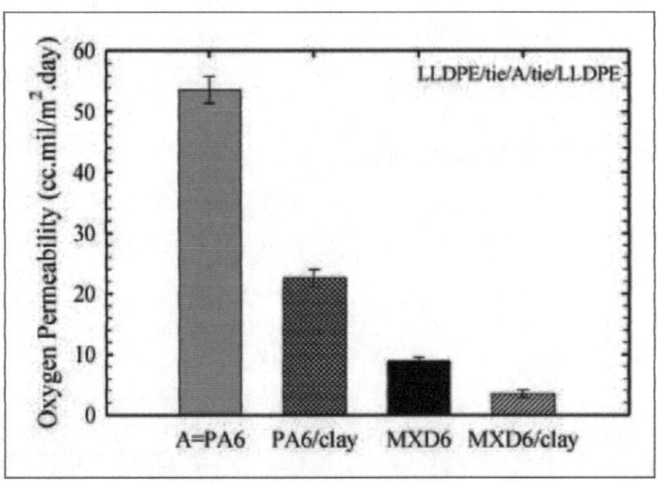

Figure 5. Oxygen permeability of PA/LLDPE blends with and without clay [28]

In Figure 5 and 6, it can be seen that with addition of clay, the oxygen permeation and the water vapor permeation of polyamide 6 (PA6)/linear low density polyethylene (LLDPE) and poly(m-xylene adipamide) (MXD6)/LLDPE blends reduces at least by 50% and by 27%, respectively [28].

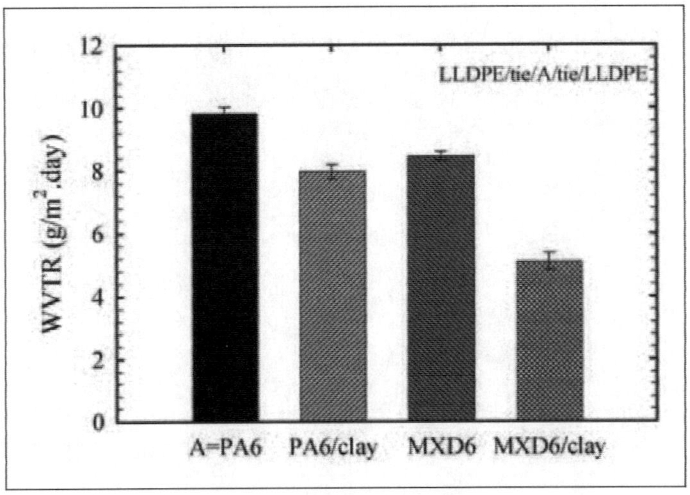

Figure 6. Water vapour transmission rate of PA/LLDPE blends with and without clay [28]

PVA membranes with chitosan and MMT content show improved methanol barrier properties [29]. Adding MMT is also an efficient method to prevent water uptake, which tenderize plastics [30]. Clay can block the flow of water the same way, as it does with heat.

4.3. Nucleation effect

Montmorillonite can function as nucleation agent. In general, crystallinity in polymers will be more dispersed and occurs at higher temperature [26, 31, 32]. In some cases, crystallinity slightly increases [32-33]. It is revealed that nanofiller content decrease the characteristic time of crystallisation of the polyamide (PA11) matrix in isothermal conditions, and that the nucleation efficiency of organophilized montmorillonite (OMMT) is higher than of sepiolite (Figure 7.) [34].

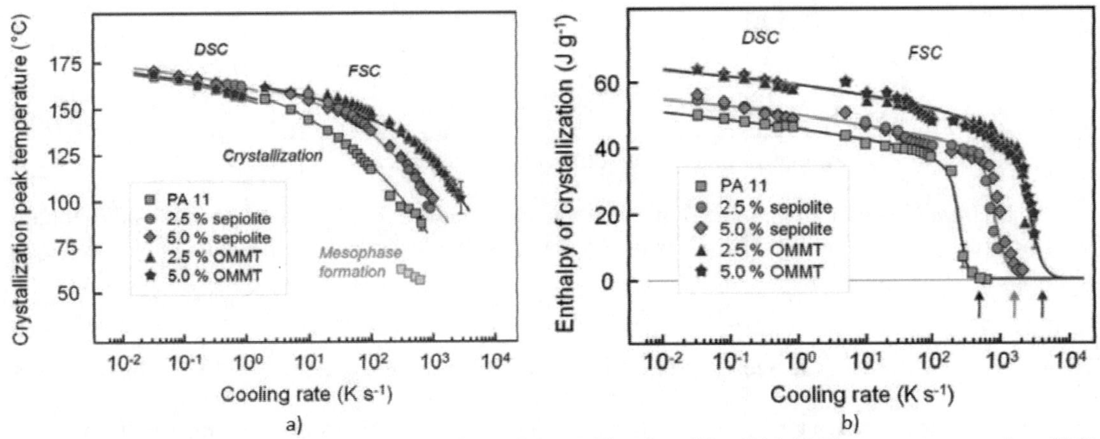

Figure 7. Peak temperature (a) and enthalpy of crystallisation (b) of PA11/clay nanocomposites [34]

During non-isothermal measurements, nanoclays are also effective nucleating agents, so their presence reduces the degree of super cooling required for crystallisation [35].

4.4. Mechanical properties

Mechanical properties of polymer/MMT nanocomposites are the most investigated ones. Tensile properties significantly depend on the dispersion characteristics of montmorillonite particles in polymer matrix [36]. The MMT improves the macro- and microstructure of the composite. The tensile properties will be better, both tensile strength and elongation at break increases (Figure 8.) [37].

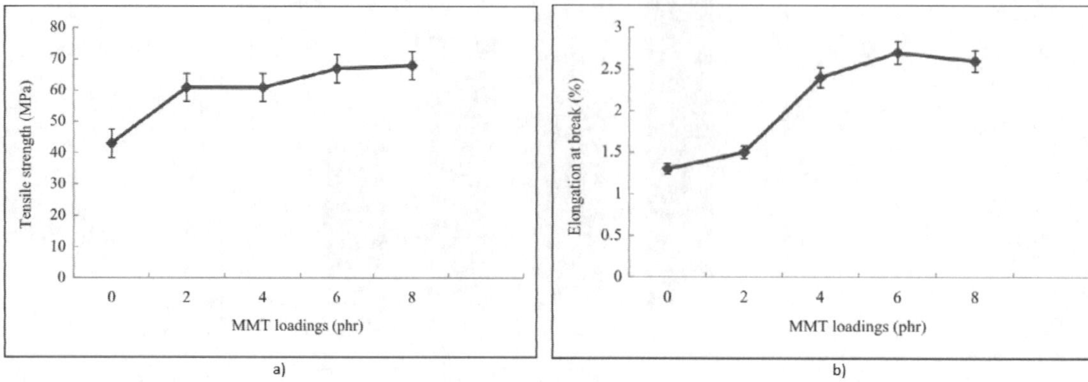

Figure 8. Tensile strength (a) and elongation at break (b) of MMT loaded PLA [37]

Adding MMT into PA6, the creep recovery, storage modulus and also loss factor decreases, explained by the reduced molecular motion [38]. Properties, like Young's modulus is seem to have a maximum value, which occurs at 3-5 wt% MMT content (Figure 9.) [37, 39].

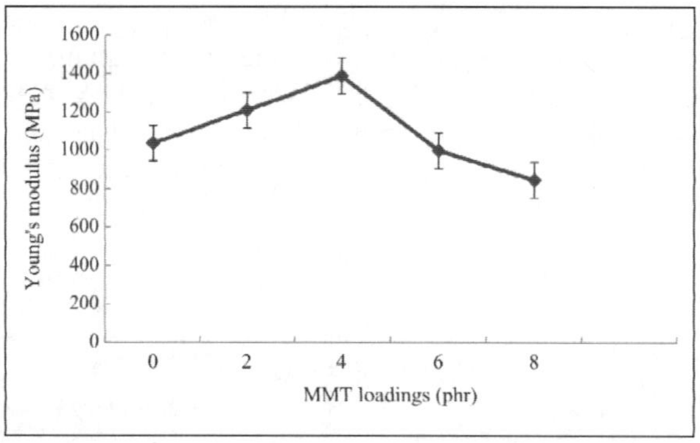

Figure 9. Young's modulus of MMT loaded PLA [37]

The structure is likely to transform above a certain percentage of clay, from intercalated or exfoliated into aggregated parts. However, MMT is a soft clay, the hardness and the fracture toughness of the composites changes [40-42].

4.5. Rheological properties

The melt flow index (MFI) decreases with the increased filler content (Figure 10.) [31]. The nanofillers were bentonite (Nanoclay 1) and montmorillonite (Nanoclay 2). It means that the viscosity increases in this region. In low frequency region, higher complex viscosity of the nanocomposites is observed compared to the neat polymer, while there is strong interaction between the matrix and the clay [21]. These interaction hinder the motion of polymer chains. The nanocomposites show dominant shear thinning behavior. Therefore at higher frequencies, the viscosity curve sharply decreases.

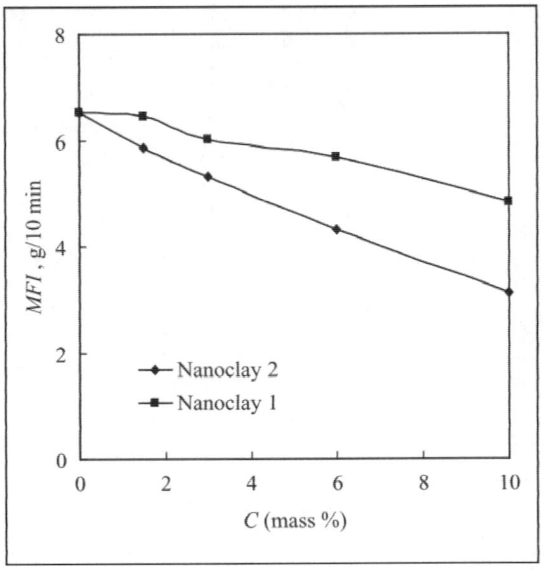

Figure 10. MFI values of poly(vinylidene fluoride-co-hexafluoropropylene with clay content [31]

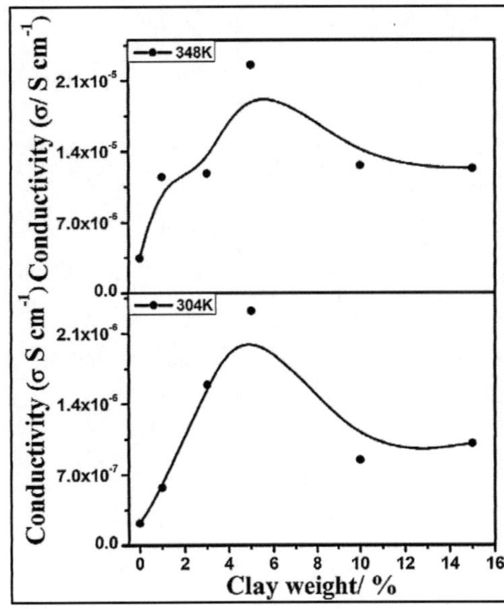

Figure 11. Dc-conductivity of PEG-PU/PMMA-$LiClO_4$/MMT composite with clay content [43]

4.6. Conductivity

Ionic conductivity increases up to 5 wt% MMT content, then it decreases (Figure 11.) [43]. In solid polymer-salt-clay electrolytes, ionic transport is associated with the segmental motions [44]. The intercalated MMT structure is more favorable for high conductivity, while exfoliated MMT can be used for low permittivity ion conducting electrolytes [5, 45]. Also proton conductivity increases adding montmorillonite [30].

4.7. Synergistic with other additives

Montmorillonite is often used with other additives, e.g. flame retardant, carbon nanotube, graphene, silver to reach the synergetic effect. Addition of secondary filler can provide additional functionalities to the matrix [24]. For instance, MMT improves the compatibility between the polymer and alumina trihydrate (flame retarder), so the flammability is ameliorated [46]. Moreover the antibacterial effect of silver (Ag) containing film can be improved [47]. The same way, electrical, barrier or mechanical properties can be modified by synergetic effect [11, 48].

In Figure 12. the voltage amplitude is shown in function of time. Sepiolite and montmorillonite was added to low density polyethylene (LDPE). The overall clay content was 5 wt%. The lifetime of the nanocomposites increases with clay content. Moreover with the use of more types of clay (with the same overall clay content) higher values are shown [11].

IOP Conf. Series: Materials Science and Engineering **448** (2018) 012021 doi:10.1088/1757-899X/448/1/012021

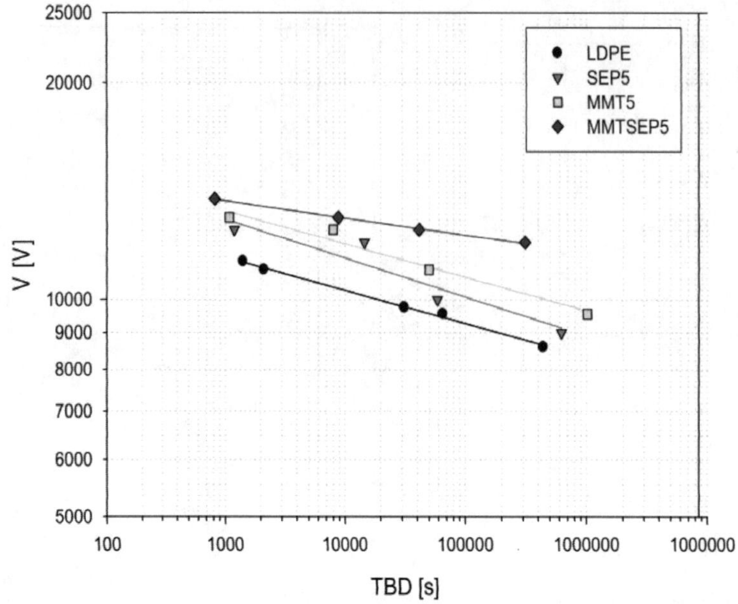

Figure 12. The lifetime diagrams of LDPE/clay composites [11]

5. Conclusion

Clay particles, like MMT, can improve thermal, mechanical, rheological, barrier properties and conductivity, although the structure of the platelets greatly influences their behaviour. Moreover, nucleating and synergetic effect can occur, which also improve the properties. The evaluated papers provide a solid foundation for the investigations. In conclusion, 3-5 wt% additive is enough for the modification, above this percentage the structure of MMT can transform. The most used polymers are polyamide (PA) and polypropylene (PP).

References

[1] Naderi-Samani H et al. 2017 The effect of organoclay on the morphology and mechanical properties of PAI/clay nanocomposites coatings prepared by the ultrasonication assisted process *Ulrasonics Sonochemistry* Vol. 38. 306-316.

[2] Bilgic C and Karakehya N 2016 Inverse gas chromatographic characterization of polystyrene and organo-montmorillonite/ polystyrene nanocomposites *Journal of Adhesion Science and Technology* Vol. 30. 1945-1956.

[3] Carrasco F et al. 2015 Kinetics of the thermal degradation of poly(lactic acid) obtained by reactive extrusion: Influence of the addition of montmorillonite nanoparticles *Polymer Testing* 48. 69-81.

[4] Chatterjee U et al. 2017 High energy ball milling for the processing of organo-montmorillonite in bulk *Applied Clay Science* Vol. 140. 10-16.

[5] Dhatarwal P et al. 2017 Effect of intercalated and exfoliated montmorillonite clay on the structural, dielectric and electrical properties of plasticized nanocomposite solid polymer electrolytes *Composites Communications* Vol. 5. 1-7.

[6] Lei F et al. 2014 Exfoliation of organic montmorillonite in iPP free of compatibilizer through the multistage stretching extrusion *Polymer Bulletin* Vol. 71. 3261-3273.

[7] Shishavan S M et al. 2014 Investigation of the effect of nanoclay and processing parameters on the tensile strength and hardness of injection molded Acrylonitrile Butadiene Styrene-organoclay nanocomposites *Materials and Deign* Vol. 58. 527-534.

[8] Li X et al. 2014 Influence of organic Intercalants on the morphology and dielectric properties of XLPE/ montmorillonite nanocomposite dielectrics *IEEE Transaction on Dielectrics and Electrical Insulation* Vol. 21, No. 4. 1705-1717.

[9] Liao X et al. 2015 Nanocomposite membranes based on quaternized polysulfone and functionalized montmorillonite for anion-exchange membranes *Journal of Power Sources* Vol. 286. 258-263.

[10] Weltsch Z and Lovas A 2013 The relation between the surface tension and the bulk properties of diluted silver based melts *Material Science Forum* Vol. 729. pp. 19-24.

[11] Guastavino F et al. 2014 Insulation properties of LDPE nanocomposites obtained by the dispersion of different nanoparticles *IEEE Transaction on Dielectrics and Electrical Insulation* Vol. 21. No. 2. 444-451.

[12] Mauroy H et al. 2015 Anisotropic clay-polystyrene nanocomposites: Synthesis, characterization and mechanical properties *Applied Clay Science* Vol. 108. 19-27.

[13] Belessi V, Lambropoulou D, Konstantinou I, Katsoulidis A, Pomonis P, Petridis D, Albanis T 2007 Structure and photocatalytic performance of TiO2/clay nanocomposites for the degradation of dimethachlor, *Applied Catalysis B: Environmental* 73 292-299

[14] http://nanotech.sc.mahidol.ac.th/center/research.html

[15] https://www.mindat.org/min-2821.html

[16] Paul D R et al. 2008 Polymer nanotechnology: Nanocomposites *Polymer* Vol. 49. 3187-3204.

[17] Guevara-Morales A et al. 2013 Mechanical and dielectric properties of epoxy-clay nanocomposites *Journal of Materials Science* Vol. 49. 1574-1584.

[18] Laatar F et al. 2016 Preparation of nanocomposite by adding of Tunisian nanoclay in unsaturated polyester matrix: mechanical and thermal study *Journal of Polymer Research* Vol. 23:239

[19] Achilias D S et al. 2015 Effect of organoclays type on solid-state polymerization (SSP) of Poly(ethylene terephthalate): Experimental and modeling *European Polymer Journal* 63 156-167.

[20] Beuguel Q et al. 2016 Influence of clay mineral structure and polyamide polarity on the structural and morphological properties of clay polypropylene/polyamide nanocomposites *Applied Clay Science* Vol. 135. 253-259.

[21] Fereydoon M et al. 2013 Rheological, crystal structure, barrier, and mechanical properties of PA6 and MXD6 nanocomposite films *Polymer Engineering and Science* Vol. 54. 2617-2631.

[22] Maisanaba S et al. 2017 Development, characterization and cytotoxicity of novel silane-modified clay minerals and nanocomposites intended for food packaging *Applied Clay Science* Vol. 138. 40-47.

[23] Mansoori Y and Masooleh T M 2014 Polyimide/organo-montmorillonite nanocomposites: A comparative study of the organoclays modified with aromatic diamines *Polymer Composites* Vol. 36. 613-622.

[24] Nasiri A et al. 2016 Effect of nanoclay on the transfer properties of immanent additives in food packages *Journal of Materials Science* Vol. 51. 9732-9748.

[25] Arunachalam S et al. 2015 An investigation of mechanical and thermal properties of polypropylene clay nanocomposites containing different nanoclays *Macromolecular Materials and Engineering* 300. 966-976.

[26] Almeida L A et al. 2015 Synthesis of polypropylene/organoclay nanocomposites via in situ polymerization with improved thermal and dynamic-mechanical properties *Journal of nanoscience and Nanotechnology* Vol. 15. 2514-2522.

[27] Annamalai P K and Martin D J 2014 Can clay nanoparticles accelerate environmental biodegradation of polyolefins? *Materials Science and Technology* Vol. 30. No. 5. 593-602.

[28] Fereydoon M et al. 2014 Properties of co-extruded nanoclay-filled aliphatic nylon (PA6)/linear low-density polyethylene and aromatic nylon (MXD6)/ linear low-density polyethylene multilayer films *Journal of Plastic Film and Sheeting* Vol. 31. 1-33.

[29] Gaur S S et al. 2014 Prospects of poly (vinyl alcohol)/chitosan/poly (styrene sulfonic acid) and montmorillonite Cloisite® 30B clay composite membrane for direct methanol fuel cells *Journal of Renewable and Sustainable Energy* Vol. 6.

[30] Hu Z et al. 2013 Montmorillonite-reinforced sulfonated poly(phthalazinone ether sulfone ketone) nanocomposite proton exchange membranes for direct methanol fuel cells *Journal of Applied Polymer Science* Vol. 131.

[31] Atanassov A et al. 2012 Properties of clay nanocomposites based on poly(vinylidene fluorid-co-hexafluoropropylene) *Journal of Thermoplastic Composite Materials* 1-16.
[32] Benhacine F et al. 2015 Development of long-term antimicrobial poly (ε-caprolactone)/silver exchanged montmorillonite nanocomposite films with silver ion release property for active packaging use *Polymer Bulletin* Vol. 73. 1207-1227.
[33] Arslan M et al. 2015 Poly(epsilon caprolactone)/ clay nanocomposites via host-guest chemistry *European Polymer Journal* Vol. 71. 259-267.
[34] Jariyavidyanont K et al. 2016 Crystallization kinetics of polyamide 11 in the presence of sepiolite and montmorillonite nanofillers *Colloid and Polymer Science* Vol. 294. 1143-1151.
[35] Sahoo R K et al. 2015 Effect of nanoclay on the nucleation, crystallization and melting behavior of polypropylene: A study on non-isothermal crystallization kinetics *Journal of Thermoplastic Composite Materials* Vol. 29.
[36] Ejder-Korucu M et al. 2016 Poly(ethylene oxide)/clay nanocomposites: Thermal and mechanical properties *Applied Surface Science* Vol. 378. 1-7.
[37] Jalalvandi E et al. 2013 Effects of montmorillonite (MMT) on morphological, tensile, physical barrier properties and biodegradability of polylactic acid/starch/MMT nanocomposites *Journal of Thermoplastic Composite Materials* Vol. 28.
[38] Magniez K et al. 2014 Structure-property relationships in Nylon 6 nanocomposites based on octaphenyl-, dodecaphenyl-POSS, montmorillonite, and their combinations *Polymer Composites* Vol. 36. 153-160.
[39] Malkappa K et al. 2016 Functionalized polybutadiene diol based hydrophobic, water dispersible polyurethane nanocomposites: Role of organo-clay structure *Polymer* Vol. 99. 404-416.
[40] Francis V and Jain P K 2016 Experimental investigations on fused deposition modelling of polymer-layered silicate nanocomposite *Virtual and Physical Prototyping* Vol. 11. 109-121.
[41] Gao G et al. 2015 Self-healable, tough, and ultrastretchable nanocomposite hydrogels based on reversible polyacrylamide/montmorillonite adsorption *Applied Materials and Interfaces* Vol. 7. 5029-5037.
[42] Cumkur E A et al. 2015 Poly(lactic acid)-layered silicate nanocomposites: The effect of modifier and compatibilizer on the morphology and mechanical properties *Journal of Applied Polymer Science* 132. 42553.
[43] Chilaka N and Ghosh S 2014 Dielectic studies of poly (ethylene glycol) -polyurethane/poly (methylmethacrylate)/montmorillonite composite *Electrochimica Acta* Vol. 134. 232-241.
[44] Dam T et al. 2014 Observation of ionic transport and ion-coordinated segmental motions in composite (polymer-salt-clay) solid polymer electrolyte *Ionics* Vol. 21. 401-410.
[45] Sengwa R J and Choudhary S 2014 Dielectric properties and fluctuating relaxation processes of poly(methyl methacrylate) based polymeric nanocomposite electrolytes *Journal of Physics and Chemistry of Solids* Vol. 75. 765-774.
[46] Bee S-T et al. 2014 Dispersion and roles of montmorillonite on structure, flammability, thermal and mechanical behaviours of electron beam irradiated flame retarded nanocomposite *Composites: Part B* 61. 41-48.
[47] Savas L A and Hancer M 2015 Montmorillonite reinforced polymer nanocomposite antibacterial film *Applied Clay Science* Vol. 108. 40-44.
[48] Kelnar I et al. 2016 Influence of clay-nanofiller geometry on the structure and properties of poly(lactic acid)/thermoplastic polyurethane nanocomposites *RSC Advances* 6. 30755-30762.

Acknowledgments
This research is supported by GINOP-2.2.1-15-2017-00077 "New developed materials: the use of nanocomposites in compressed air breaking and auxiliary systems" project. The Project is supported by the Hungarian Government and co-financed by the European Social Fund.

IOP Conf. Series: Materials Science and Engineering **448** (2018) 012022 doi:10.1088/1757-899X/448/1/012022

Formability Investigations of Advanced High Strength Steels

Y Mehdi[1], **M Tisza**[1]

[1]Institute of Materials Sciences & Technology, University of Miskolc, Miskolc, Hungary
e-mail : om08@live.fr

Abstract. Car manufacturing has been significantly influenced by sheet metal forming developments, thus the requirements and developments in car manufacturing have decisive role in the development of sheet metal forming, too. The automotive industry is facing challenges of reducing body weight in consideration of environmental problems and higher collision safety. These requirements are being addressed by the application of various dual-phase steels, ultra and extra advanced high strength steels (UHSS, XAHSS). Forming Limit Diagrams (FLD) are the most appropriate tools to characterize the formability of sheet metals. Theoretical and experimental investigations of forming limit diagrams are in the forefront of todays' research activities. In this paper, an up-to-date research methodology elaborated and applied at the Department of Mechanical Technology at the University of Miskolc will be shown.

1. Introduction

Sheet metal forming is one of the most important manufacturing processes. This is particularly valid for the automotive industry, where sheet metal forming has an even more important key position. The automotive industry is the leading sector in many countries and the main driving force behind the sheet metal forming developments as well [1].

As automakers are challenged to improve safety and fuel economy, they search for new materials to meet higher standards. Advanced high-strength steels (AHSS) help engineers meet requirements for safety, efficiency, lower emissions, manufacturability, durability, and quality at a low cost [2].

AHSS are a newer generation of steel grades that provide extremely high-strength and other advantageous properties, while maintaining the necessary formability required for manufacturing. They have been on the road for many years, but with additional research and development, automakers are using these newer grades in more applications [2].

2. Advanced High Strength Steels

The AHSS may be distinguished based upon the strength properties that roughly can be defined: 1500 MPa < Rp0.2 (tensile strength). As opposed to the conventional high strength steels, in which ductility decreases by the increase of strength but they have better formability even at higher strength properties, modern AHSS steels combine high strength and formability/ductility [3].

General classification of these steels is as follows:

- Highs strength steels with a high energy absorption potential (DP and TRIP steels with UTS < 1000 MPa), for dynamic loading occurring during car crashes or collisions.
- Extremely high strength steels, typically martensitic steels, with a very high UTS (>1200 MPa), providing high stiffness, anti-intrusion, load-transferring barriers for the protection of automotive passengers.
- The rationales for increased use of the AHSS in the automotive industry are as follows:
  - The reduction of the car weight resulting from the use of high strength thinner gauge sheet steel, reducing the fuel consumption.
  - Increased passenger safety by an improved crash worthiness.
  - The strong competition from the light-weight materials, such as Al and Mg alloys and plastics [3].

IOP Conf. Series: Materials Science and Engineering **448** (2018) 012022 doi:10.1088/1757-899X/448/1/012022

AHSS derive their properties from multi-phase complex microstructure. Since these steels are relatively new, their classification differs from conventional high strength steels and was developed by Ultra-Light Steel Automotive Body – Advanced Vehicle Concept (ULSAB-AVC) Consortium [1]. The accepted practice involves specification of both yield strength (YS) and ultimate tensile strength (UTS) in the following way: XX aaa/bbb, where XX is type of steel, aaa is minimum YS in MPA and bbb is minimum UTS in MPA.

The Advanced High Strength Steels family includes Dual Phase (DP), Complex-Phase (CP), Ferritic-Bainitic (FB), Martensitic (MS), Transformation-Induced Plasticity (TRIP), Hot-Formed (HF), and Twinning-Induced Plasticity (TWIP). These 1st and 2nd Generation AHSS grades are uniquely qualified to meet the functional performance demands of certain parts. For example, DP and TRIP steels are excellent in the crash zones of the car for their high energy absorption. For structural elements of the passenger compartment, extremely high-strength steels, such as Martensitic and boron-alloyed Press Hardening Steels (PHS) result in improved safety performance. Recently there has been increased funding and research for the development of the "3rd Generation" of AHSS. These are steels with special alloying and thermo-mechanical processing to achieve improved strength-ductility combinations compared to present grades, with potential for more efficient joining capabilities, at lower costs [4].

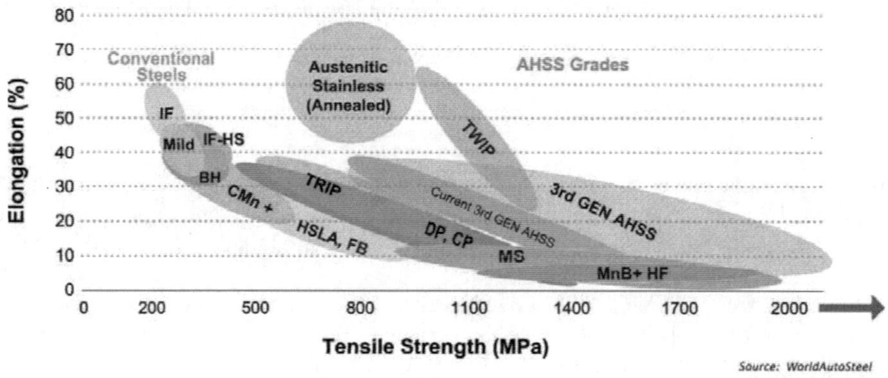

Figure 1. Steel Strength Ductility Diagram, illustrating the range of properties available from today's AHSS grades [4].

The broad range of properties is best illustrated by the famous Steel Strength Ductility Diagram (often cited as the banana diagram, captured in Figure 1. Where we have the elongation percentage in function of tensile strength, and in which we could recognize the difference between the low strength (which are known as conventional steels e.g: Mild steels and interstitial free), and advanced high strength steels family (transformation induced plasticity, Dual phase) [4].

Generally, steels with yield strength exceeding 750 MPa are referred as HSS, and in case these steels have a tensile strength exceeding 1500 MPa are called UHSS.

3. Formability of sheet metals

In general formability is the ability of sheet metal to undergo shape changes (plastic deformation) without failure by necking or tearing.

The most appropriate tool to characterize the formability of sheet metals is the Forming Limit Diagram (FLD).

Applying the Forming Limit Diagram, we can reliably estimate the forming behaviour of metallic materials under various stress- and strain state as shown in Figure 2.

IOP Conf. Series: Materials Science and Engineering **448** (2018) 012022 doi:10.1088/1757-899X/448/1/012022

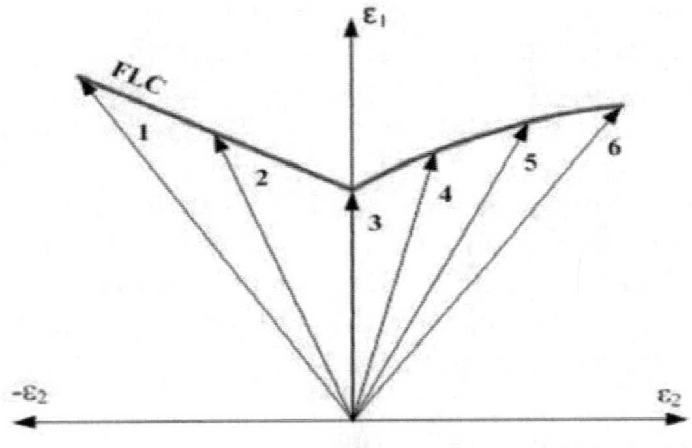

Figure 2. Forming Limit Diagram for various strain paths (1-pure shear, 2-uniaxial tension; 3-plain strain; 4, 5 6-different biaxial tension) [5].

4. Basic understandings concerning the forming limit diagrams

Forming limit diagrams (FLD's) offer a convenient and useful tool in sheet products manufacturing analysis. They show the critical combinations of major strain and minor strain in the sheet surface at the onset of necking failure. Formability in the context of multiple phase operations strongly depends of the deformation history and therefore demands an investigation of every particular case. This makes the experimental determination of FLD's unappreciative expensive and causes the necessity to develop an accurate and efficient theoretical method for formability prediction [6].

The basic concept of the FLD was first introduced by Keeler and Backofen, who developed the right-hand side of the FLD. Goodwin extended this diagram to the left-hand side. Figure 3 illustrates the FLD which is divided into two regions separated by a curve, called the forming limit curve (FLC). The region below the FLC corresponds to safe strain states whereas that above the FLC represents failure strain states. Even though the strain states of the sheet metal forming processes are complex, FLDs are often constructed using tensile and biaxial stretch tests.

The diagram attempts to provide a graphical description of material failure tests, such as a punched dome test. In order to determine whether a given region has failed, a mechanical test is performed. The mechanical test is performed by placing a circular mark on the workpiece prior to deformation, and then measuring the post-deformation ellipse that is generated from the action on this circle. By repeating the mechanical test to generate a range of stress states, the formability limit diagram can be generated as a line at which failure is onset [6].

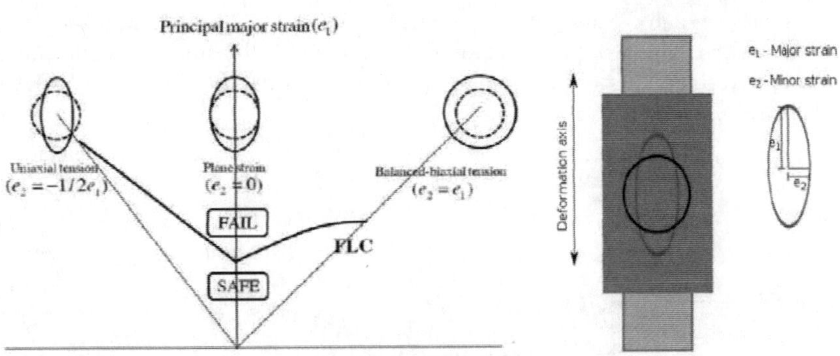

Figure 3. Forming limit diagram and the definition of the deformation axes in forming limit diagram measurement [6].

5. Experimental determination of Forming Limit Diagrams

The sheet metal forming particularly in the vehicle industry, the computer aided technology and tool design, requires more and more precisely characterized formability of the sheet materials to be processed.

Therefore, an integrated Sheet Metal Formability Testing System (SMFTS) was installed at the Department of Mechanical Engineering at the University of Miskolc with the financial support of several national and international projects. It consists of an electro-hydraulic, computer-controlled testing machine with an optical strain measurement system as shown in Figure 4. According to its nominal loadability (Fmax = 600 kN), this system is suitable to perform sheet metal testing up to 3 mm thickness.

The applied punch diameter is a standard d = 100 mm, the velocity range is v = 0÷5 mm/s. The computer control can provide the harmonized operation of the formability and optical strain measurement system [7].

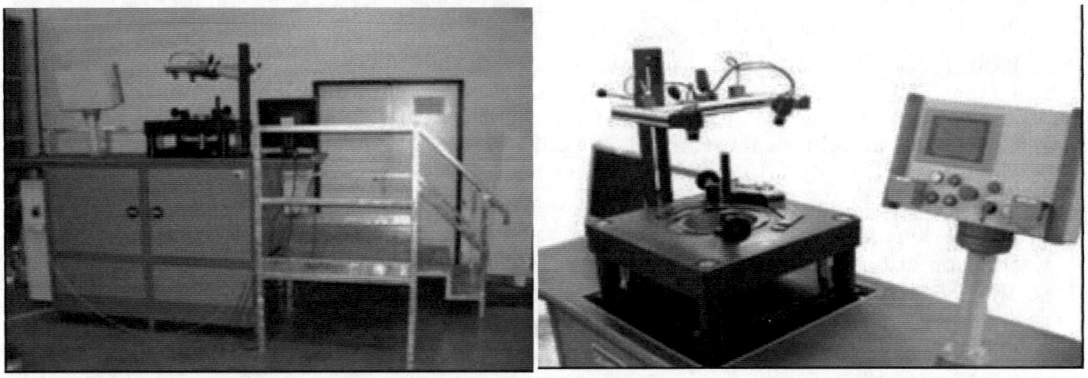

Figure 4. The Universal sheet metal formability testing system equipped with optical strain measurement system.

For the measurement of the strain distribution over the whole part, a printed grid is applied on the surface of the part before the forming operation. In our experiments, regular, square grid is applied as the carrier of the measuring information in the AutoGrid® strain measurement system.

The applied Vialux-AutoGrid optical strain measurement system is capable for inprocess and post-process evaluation of grid deformation. In the determination of forming limit diagrams, usually the post-process evaluation is applied; however, it is often combined with the in-process capabilities of the system for more precise detection of the onset of necking or the rupture itself.

5.1. Experimental specimen

The specimens proposed by Nakazima are generally used for the determination of forming limit diagrams. In our experiments, we applied a modified version of Nakazima specimens of for DC04, DC05 and DD14. They are steels for cold forming, having good formability. These modified specimens are shown in Figure 5.

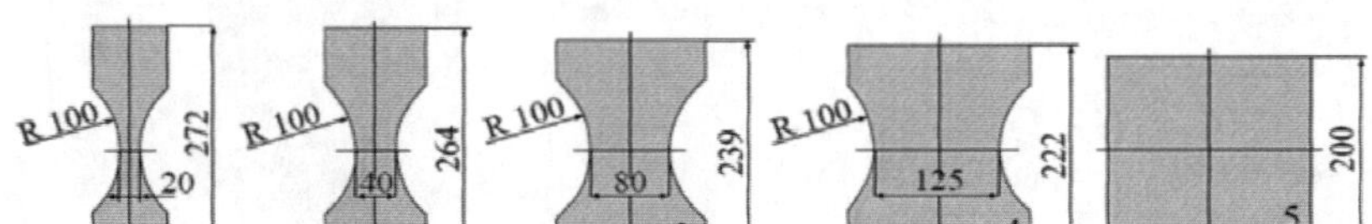

Figure 5. Modified Nakazima specimen

A printed regular square grid is applied on the surface of the specimen as the carrier of the measuring information.

5.2. The experimental procedure

As it was already mentioned, before the examinations a square grid was made on the surface of the specimens applying an offset printing technique. The deformation of the square grid was measured by the Vialux optical strain measurement system. The evaluation of the deformed grid is based upon the recognition of identical grid-crosses from the different views obtained by the four CCD cameras mounted on the testing machine.

After printing the grid, the sheet specimen is firmly clamped between the die and the blank holder, while the hemispherical punch forces the specimen to deform in the space.

Performing the evaluation for all the applied specimens having different width, various and different strain path history (ε1, ε2) can be obtained depending on the width of the specimen [7].

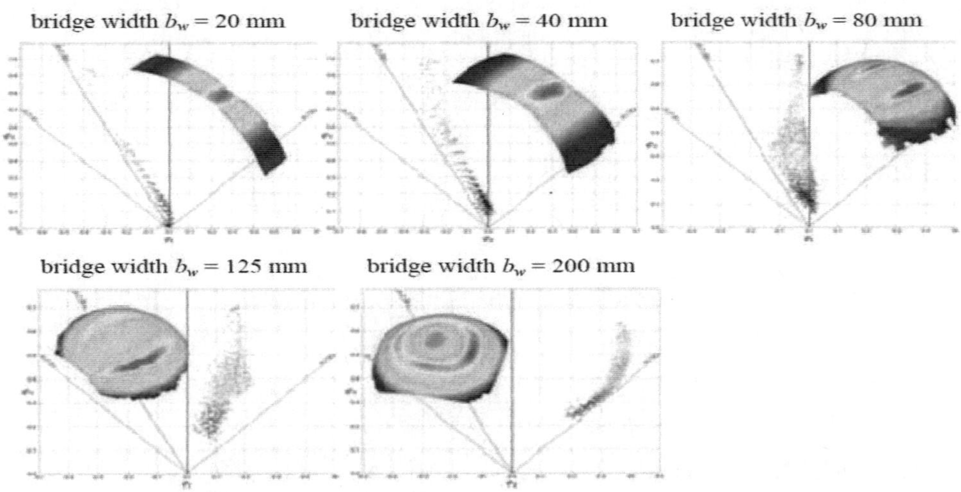

Figure 6. Results of strain measurements for various strain paths.[7]

When all the measurements and the evaluations are completed, they are simultaneously loaded into ε1, ε2 coordinate system. To get the Forming Limit Curve (FLC) we have to connect the fracture points on each deformation path (Figure 7.) [7].

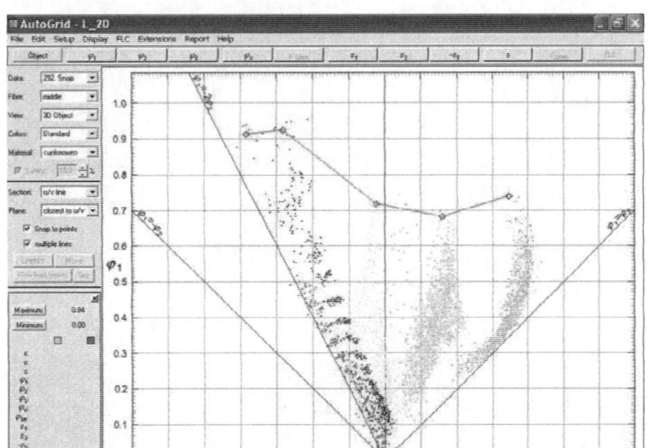

Figure 7. Graphical generation of FLC from the measured strains.[7]

6. Conclusion

A substantial progress has been achieved during last years in the development of AHSS for applications in the auto body construction. Forming AHSS involves several challenges, mainly due to its higher strength, lower formability and inconsistency of material properties [8].

The future of AHSS for automotive applications is bright. Many groups are researching these new steels to better understand their properties and to continue tailoring unique sets of characteristics. Others are focused on improving the technologies necessary for manufacturing parts made of AHSS. The steel and automotive industries have forged numerous partnerships to develop the materials and technologies necessary to put the next generation of safer and more environmentally friendly vehicles on the road [2].

Forming limit diagrams can be regarded as the most appropriate tools in the evaluation of sheet metal formability. Thus, theoretical and experimental investigations of forming limit diagrams as a special field of the formability of sheet metallic materials are in the forefront of todays' research activities.

7. References

[1] M. Tisza, P. Z. Kovács, Z. Lukács, A. Kiss, and G. Gál, "Formability of High Strength Sheet Metals with Special Regard to the Effect of the Influential Factors on the Forming Limit Diagrams," *Mater. Sci. Forum*, vol. 812, pp. 271–275, 2015.

[2] C. M. Tamarelli, "AHSS 101: the evolving use of advanced high-strength steel for automotive applications," *Steel Mark. Dev. Inst.*, p. 42, 2011.

[3] R. Kuziak, R. Kawalla, and S. Waengler, "Advanced high strength steels for automotive industry," *Arch. Civ. Mech. Eng.*, vol. 8, no. 2, pp. 103–117, 2008.

[4] S. Keeler, M. Kimchi, and P. J. Mooney, "Advanced High-Strength Steels Apllication Guidelines Version 6.0," no. September, p. 314, 2017.

[5] Prof. Dr. Miklós Tisza, "Formability investigations of Advanced High Strength Steels," *Res. plan*, no. University of Miskolc, 2017.

[6] S. G. G. Singhji, "Study The Methods To Get Forming Limit Diagram From Experimental Data."

[7] M. Tisza and Z. P. Kovács, "New Methods for Predicting the Formability of Sheet Metals," *Prod. Process. Syst.*, vol. 5, no. 1, pp. 45–54, 2012.

[8] E. Billur and P. D. T. Altan, "Challenges in Forming Advanced High Strength Steels," *New Dev. Sheet Met. Form. Proc.*, pp. 285–304, 2010.

Collaborating robots in Industry 4.0 conception

R Benotsmane [1], L Dudás [1] and Gy Kovács [2]

[1]Institute of Information Science, University of Miskolc, Miskolc, Hungary
[2]Institute of Logistics, University of Miskolc, Miskolc, Hungary
E-mail: benrabab1@gmail.com

Abstract. Nowadays the industrial model is developing and improves the capabilities of its platform through the emergence of new technologies that transform the industry to an interconnected global system in which machines and products communicate together continuously. This global evolution known as Industry 4.0 characterized by a merger between the Internet and factories, it is a concept that aims to make factories smarter and to create a digital platform that as a place of production, using modern technologies. In this article, we will discuss two parts: the first part describes this future paradigm, its advantages by identifying the challenges of this implementation and its basic elements, while the second part discusses the new technique which includes the collaborating robots via artificial intelligence which is an essential point in Industry 4.0.

1. Introduction

The industrial evolution has known three revolutions till these days: the first was the mechanization characterized by the appearance of the steam engine, the second was industrialization through production lines on the base of the electricity, and the third one was industrial revolution determined by computer integrated manufacturing [1]. These three revolutions brought the ability to change the conception of existing production. Today this industrial model is developing and improves the capabilities of its platform during the appearance of new technologies such as digitization that transforms the industry to an interconnected global system in which machines, systems and products communicate together [2]. This global evolution known as Industry 4.0 or in other words the industry of the future. In the age of Industry 4.0 the sector of manufacturing enters its fourth revolution, characterized by a merger between the Internet and factories. It is a concept that aims to make factories smarter and to create a digital platform that as a place of production brings together the main tools that are already existing as: Sensors, Big data, Internet of things, Cloud computing, Artificial intelligence, Collaboration between robots, etc. [3].

In this article, we will analyze two parts: the first part describes this future generation, its architecture and its basic elements, while the second part discusses one of the new techniques "autonomous system" which includes the technology of collaboration between robots via artificial intelligence which is an essential point in Industry 4.0 [4].

2. Industry 4.0 conception

Industry 4.0 refers to the Fourth Industrial Revolution which is the origin of a strategic project of the German government, that was highlighted for the first time at the Hanover Fair in 2011 [5], to support the digital revolution of the global industry. It is a concept that focuses on a new way of production, which is used to set up a smart factory, where everything works in interaction. The products and the machines, linked in a network itself connected to the all supply chain [2], based on the information technologies, communicating sensors, software tools for simulation, information processing, controlling, using the highest performance devices in order to transforming industry into an interconnected global system, by applying many innovations and creating a new dynamic market [4]. Basically, Industry 4.0 includes the partial transfer of autonomy and autonomous decisions to cyber-physical systems and machines, applying information systems, it is a digital transformation that is changing up the manufacturing business by bringing radical changes not only to systems and processes, but also to business models and the workforce and management styles [7] as shown in figure 1 which represents the stages of industry development throughout history.

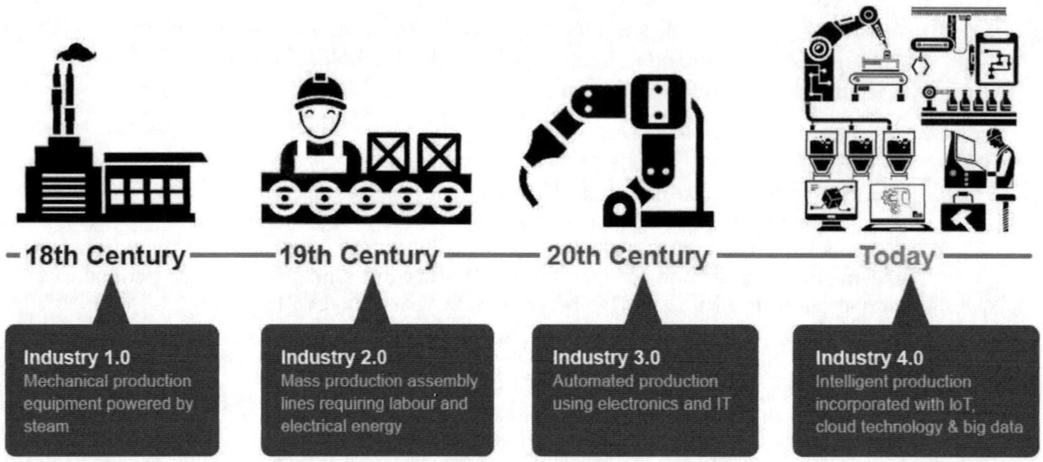

Figure 1. Stages of industry development [6].

2.1. Industry 4.0 concept in manufacturing
In the growing market globalization, where the customer demands are changing continuously the manufacturing companies have to improve efficiency, focus on cost reduction and profitability in order to increase their competitiveness. Manufacturing systems have to become more complex and flexible to respond to rapidly changing customer demands and economic environment.

In the production the resources (materials, machines, humans, space and other facilities) are limited. The current practice in the manufacturing is not sustainable and poses many challenges to the production. Therefore, it is very important at the manufacturing companies to produce cost effective and high quality products which can be achieved by maximized utilization of resources and minimized production cost.

The manufacturing is going through a paradigm shift, which will change the production conception. The current centrally controlled manufacturing processes will be replaced by decentralized control. The essence of Industry 4.0 concept in the manufacturing processes is the introduction of network-linked intelligent systems, which is based on the self-regulating ability of machines, equipment, components, workpieces and products that communicate with each other.

The essential goal of Industry 4.0 is to create smart factories and make the related industries such as logistics more efficient, more flexible, and more customer-oriented.

The following goals have to be achieved in the Industry 4.0 concept [7]:

- optimization of production processes which increase productivity,
- maximal utilization of resources,
- efficiency improvement by automation of processes,
- increase of flexibility to respond to rapidly changing customer demands,
- use of environmental-friendly materials and technologies,
- create new opportunities and business models.

Achieving the main goal can result in many benefits in manufacturing processes that are the following:

- higher productivity,
- more customized and unique finished products, higher customer satisfaction,
- higher utilization of resources,
- ensuring and enhancing the quality,
- reducing lead times,
- saving costs and reducing waste,
- realizing sustainable production, etc.

2.2. Pillars of Industry 4.0 concept

The future state of industry 4.0 represents a globally interconnected world defined by the total digitization of economic and production flows. It requires horizontal integration at each stage of the production process, interacting with the machines that communicate all together [8].

The Boston Consulting Group has identified the essential base of Industry 4.0 by nine technological pillars as shown in figure 2:

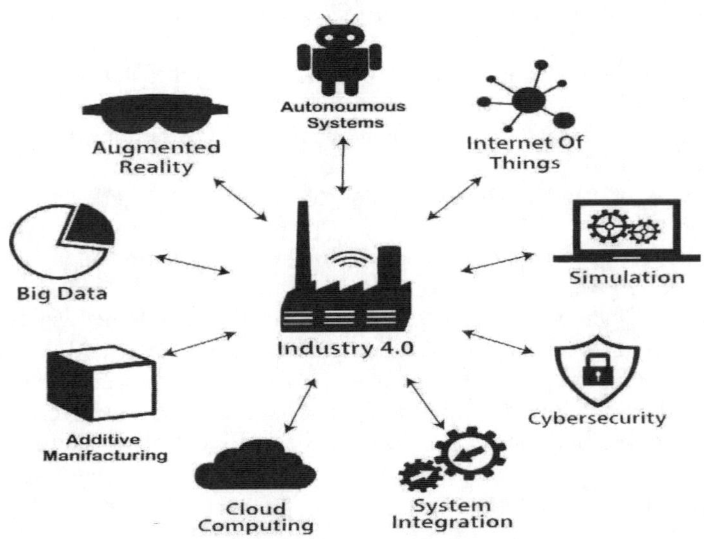

Figure 2. Pillars of Industry 4.0.

2.2.1. Autonomous systems. An automated system, including sensors, actuators and controller, capable for performing tasks continuously and designed to be applied in industrial field. Autonomous systems are providing an ever-wider range of services and becoming flexible and cooperative. Autonomous systems can be:

Multi agent system is defined as a system composed of several agents that interact together. Usually an agent is a computer system located in a dynamic environment such in robots or in intelligent machines and capable of executing independent or intelligent actions, using tools and techniques of artificial intelligence.

Intelligent industrial robot is a useful combination of a manipulator arm, sensors and intelligent controllers, which replaces a human worker and can complete tasks and resolve the problems. Eventually, it will be able to learn from humans at first. The use of these machines in industrial automation can improve productivity, product quality, creating smart industry.

2.2.2. Industrial Internet of Things (IoT). It means the ability of machines to communicate, control and collaborate with other systems, this system collects and share huge amounts of data inside the network that ensures the exchange of information between the system elements. The collected data is sent to a central cloud service where it is accumulating with other data and then shared with end users in a useful way.

2.2.3. 3D Simulation tool. 3D simulation of processes is a way for providing real-time data to observe the physical world in a virtual surrounding that will include machines, products and humans. It is a way to make scenarios for system operation and based on it to optimize the production and maximize the resources (human, equipment, etc.), thereby increasing productivity and reducing wastes and improving quality.

2.2.4. Cybersecurity. Seen that IoT has an essential effect on Industry 4.0 while it shares a huge data on network, this information should be protected, for this reason we find the concept of Cyber security comprising technologies, processes and controls that are designed to protect systems, networks and data from cyber-attacks.

2.2.5. Horizontal and vertical systems integration. To succeed in creating a smart factory that adapts with the concept of Industry 4.0, it is necessary to implement specific technologies mainly in the digital area which provides the integration of different vertical and horizontal system elements [7] defined as follows:

Vertical integration of system elements aims to optimize the reconfiguration of production processes by connecting all Cyber-Physical Systems (sensors, actuators, etc.) with the various production management tools (planning, inventory, etc.) which allows easily reconfiguring the production process according to customer demand.

Horizontal integration of system elements aims to optimize the product in supply chain, by connecting all the elements of the production chain (suppliers, customers, service providers, etc.), beyond the scope of the company, it also concerns the organizational problem.

2.2.6. Cloud computing. Cloud computing is just a metaphor for the Internet which means storing and accessing data and programs over the Internet instead of our computer's hard drive. The cloud offers industry flexibility and unlimited computing power. To receive, store and analyze Big Data, this information is secured during transfer and storage, it is using high-level encryption in terms of collaboration, the cloud remains the ideal tool to manage Big Data in Industry 4.0. It allows all machines, wherever they are, to synchronize their actions and work on shared documents and applications simultaneously in real time.

2.2.7. Additive manufacturing. Additive manufacturing is a terminology widely used in the industrial world, based on the manufacturing of a work piece used in industry known as 3D printing. It makes it possible to go beyond the limits of traditional manufacturing (injection molding, machining, cold forming, assembling, etc.) and offers the possibility of manufacturing of parts of more complex component, unachievable by other techniques, thus widening the potential for innovation. In the concept of Industry 4.0, 3D printing technology will be chosen for its very high quality in producing small batches of customized products.

2.2.8. Big data and analysis. According to the IoT principle that we have already defined, all machines and systems connected to the production plant must be capable of collecting, exchanging and backing up massive volumes of information in a completely autonomous way. Without the need of human resources, the analysis of this amount of information will optimize the operations of systems and saving time and capacities of resources, which is needed to real-time decision-making.

2.2.9. Augmented reality. Augmented reality provides visualization by transforming the real environment to a virtual model. Virtual reality solutions analyze images or detect motion and correlate everything with information from various sensors, due to this technology the operator can control the entire process and detects the anomaly in case of emergency.

2.3. Challenges of Industry 4.0 concept

Since appearance of digital technology, companies real changes that have significant effects on their competitiveness [8], their investments and the development of employers skills. The challenges of this revolution are the following:

- all Cyber physical systems must be digitalized,
- satisfying the consumer who demands more complex and unique products in small quantities,
- the traditional centrally controlled and monitored processes must be replaced by decentralized controlling,
- factories must be self-regulating that should optimize their own operation,
- ensure security and confidentiality in the development of Internet of Things,
- realizing a true connection between operation fields inside the company,
- increase productivity depending on changing customer demands,
- being able to find solution faster in case of production problems.

3. Collaborating robots via artificial intelligence

The emergence of new technologies has an important effect on improving the concept of Industry 4.0 to create a smart factory, in the next part of this article we will focus on three of the mentioned technologies: autonomous systems, simulation and system integration.

Typically, multi agent systems integrate autonomous systems which are endowed by artificial intelligence, this concept aims to apply intelligent machines collaborating together to build a flexible environment [4] as shown in figure 3 which represents collaborating manipulator arms in industry.

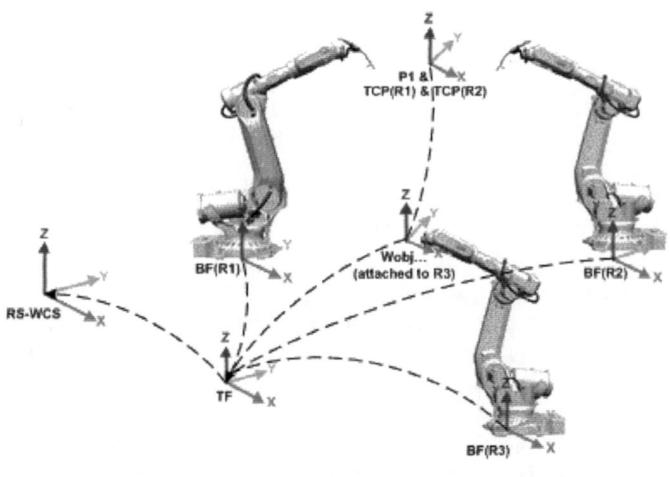

Figure 3. Collaborating robots in Industry 4.0.

3.1. Multi-agent systems

It is a paradigm that contains multiple agents that interact with each other in a common environment, some of which may be machines, computer programs, etc.

In the industry, these agents represent robots, sensors, controllers which can apply a common language that structures the rules of cohabitation and collective work. To design a multi-agent system, we must know the model of each agent that will come into action and define their environment and their interactions and their essential objectives to achieve [9]. A classic opposition has been drawn between reactive and cognitive agents: the reactive agents are those that have just reflexes while the cognitive agents are those that can form plans for their behaviors. Communication between these agents plays a vital role in this area which can show two forms [10]:

IOP Conf. Series: Materials Science and Engineering **448** (2018) 012023 doi:10.1088/1757-899X/448/1/012023

3.1.1. Implicit communication. Presents a method whose agent receives information about other agents installed in the same system through the appropriate environment by integrating different types of sensors, this type of communication can be divided into passive implicit which means the agents interact via their environment, or the second one active implicit based on sensing.

3.1.2. Explicit communication. Called the direct communication presented by the exchange of information between robots, which can be done in the form of unicast or intentional broadcast messages. This often requires a dedicated communication module edge. Existing coordination methods are mainly based on the use of explicit communication.

The agents alternate the active and passive roles and exchange sets of messages according to specific protocols defined by the following points [11, 12]:

Coordination strategy is the proper organizing in space and in time for all the tasks needed to solve a specific problem. Coordination protocols help agents to manage their commitments by defining the conditions and the actions to be taken, in the case where the circumstances in which they were developed.

Cooperation strategy is a phenomenon that occurs when a task can not be performed by a single agent but requires a set of them. This concept consists in breaking down the tasks into subtasks and then spread them among the different agents, there are several possible decompositions, the decomposition process must therefore consider available resources and skills of the agents.

Negotiation strategy occurs when agents interact to make common decisions while pursuing different goals, in this concept we can find:

- **Negotiation language** that presents the way to analyze the primitives of communication, their semantic, and their use in protocols,
- **Process of negotiation** proposes general models about the behavior of agents in negotiation.

3.2. Artificial Intelligence

In the past several years, AI techniques have become more and more robust and complex. AI is a branch of science and engineering which deals with building intelligent machines and helping them to find solutions to complex problems in a more human like fashion, this generally involves borrowing characteristics from human intelligence [11, 13], and applying them as algorithms to computing the behaviors of each machine. This concept has two classic limitations:

The anchoring of A.I.L.: Artificial Intelligence Located, the relation of each agent to his environment, to realize by adaptation, learning.

Distribution of problems D.A.I.: Distributed Artificial Intelligence realized by cooperation between agents [9].

Artificial Intelligence has conveyed to the multi-agent systems the concept of the use of logical formalisms in the manner of Knowledge-Based Systems. Following this approach, an agent has a knowledge base representation of his current state and an inference mechanism modeling his reasoning. Among these possibilities we intend to deal with three tasks.

3.2.1. Application example of AI in the tasks of cooperation between several manipulator robots. The essential goal of our research is creating a virtual simulation model of an assembly line that includes a few robot manipulators equipped with sensors, cameras and intelligent controllers. These robots as members of a multi-agent system can help each other and cooperate to finalize the tasks defined in this line, if some malfunction or another problem occurs in the production line, the robots can reconfigure themselves and reorganize the steps of the same task. One sketch of a similar system is shown in figure 4, our process is presented in two scenarios:

I. First scenario: Presents the normal process or the static mode including tasks, processes that take place as were planned. In this scenario the planning methods of AI are used for creating the optimal scheduling for the tasks taking into the parallel work of the robots. These AI methods mainly mean search algorithms and first order logic.

IOP Conf. Series: Materials Science and Engineering **448** (2018) 012023 doi:10.1088/1757-899X/448/1/012023

II. Second scenario: Presents the disturbed process including some problems, random happenings, errors, etc. when intelligent reordering, rescheduling of the steps needed. In this scenario the AI is intended for finding quasi-optimal strategies for continuing the work after the random happening. In this case the mentioned AI methods are completed with probability algorithms.

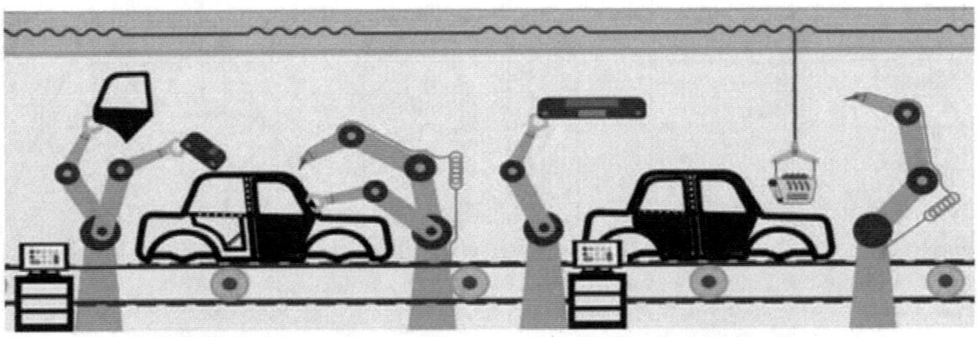

Figure 4. Cooperating manipulator arms in an assembly line

As we can create larger systems virtually only, using simulation, smaller tasks were created to demonstrate the robot cooperation in reality. These will be introduced in the followings. In the real tasks we intend to use our two Mitsubishi robots shown in figure 5.

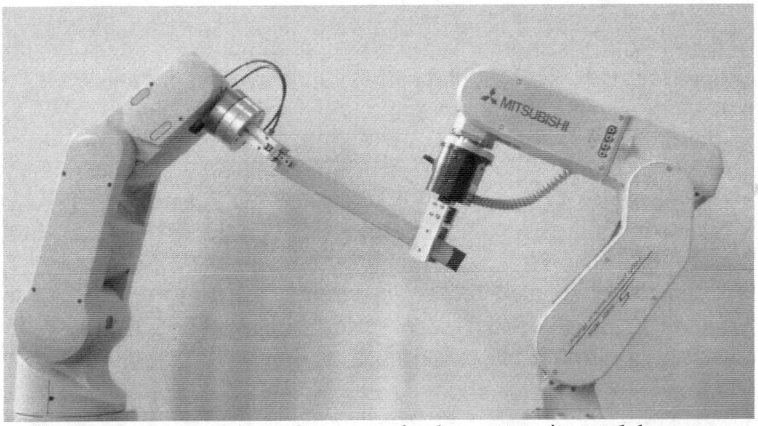

Figure 5. Cooperating of two manipulator arms in our laboratory

3.2.2. First application example for demonstrating robot cooperation. The task is the following: providing some building elements ("cards"), the two robots have to build up a house of cards. This task needs the cooperation trivially if there is no some sort of supporting element, because to build an A shape from the two cards needs two hands, in our task it needs two robots. The drawing of one "card" building element is given in figure 6(a). The structure of the house of cards to be built is shown in figure 6(c). If one of the robots fails then the other has to use the supporting element that is depicted in figure 6(b). Using this for supporting a card and removing this supporting element later needs extra time opposite to cooperating robots, so the demonstration can prove the advance of robot cooperation.

3.2.3. Second application example for demonstrating robot cooperation. In this task we would like to apply robot cooperation and AI methods. The task is to solve a 3D packaging job in the case of two co-operating robots, that is to put the bricks in a 3D work space (on one another) into a larger rectangular space. Concrete example could be to fill a large, over-open container box with different sized boxes having top-holders, for example, suction pads. To make easier the task for the first the boxes have the same height, so layers of boxes can be formed in the container.

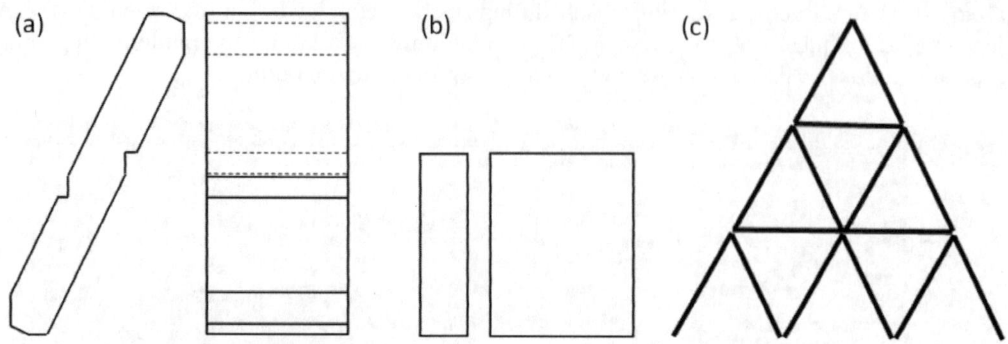

Figure 6. a., A card element in two views b., The supporting element in two views c., The house of cards structure has to be built.

Target function: the spatial location of the common center of gravity of the put in boxes should be at a minimum height. Unfortunately, this value can be determined after creating a ready filling. Due to it the problem is NP-hard, combinatorially exploding task.

Starting conditions:

Any box is less then the container. The boxes can only be rotated 90 degrees around the vertical axis (due to the holders on the top of the boxes).

Limits:

1. Only the top down function works, no horizontal pushing.
2. The boxes in the second, third, etc. layers have to be well-underpinned, so without side forces they have to sit stabile on one or on more boxes.

To divide the hard task, filled layers are created in the first subtask, then the search for the best solution among the overplacement variations of the layers is executed. In the first subtask when creating layers, boxes are put into the extreme "corner" position, to leave the largest rectangular shape free for the next boxes.

This NP-hard task can easily result in practically infinity long computer run. To avoid this first the algorithm has to calculate the estimated calculation time. If it is too large then looking for quasi-optimal solution is allowed, and the selection of AI search methods resulting in local optimum is accepted.

AI methods that can be used:

- Heuristics based on symbolic logic (to make easier the task and create heuristic target functions)
- Dijkstra's search algorithm (to find optimal solution)
- Iterative Deepening search algorithm (to find optimal solution)
- Depth first search algorithm (to find quasi-optimal solution)
- Best first search algorithm (to find quasi-optimal solution)
- Tabu-search. (to find optimal solution).

Second phase:

If the optimal or quasi-optimal computer solution is ready, the generation of the robot programs follows where the cooperation of the two robots is preferred. Opportunities for cooperation:

A., The two robots are looking for boxes and placing them alternately, so long as one of them is in the container the other is looking for new box. If one robot breaks down, the other will continue the task substituting the failed robot.

B., If there is not interference between the robot hands, the two robots can place in boxes parallel.

4. Conclusion

The main purpose of this paper is presented in two parts, firstly it explored the concept of Industry 4.0 by discussing its essential goal and its advantages, then highlighted the base that brings together the nine pillars of this industrial revolution, which can be defined as a cyberspace that controls everything from demand to product design. It is characterized by intelligent automation and integration of digital technologies: 3D printer, cloud computing, augmented reality, Internet of Things, etc. Secondly it discussed the autonomous systems that represent a useful factor for successful digital transformation of the manufacturing and create a smart factory, known as multi agent systems. In the industrial field these agents can be manipulators arms, sensors, controllers endowed by artificial intelligence. This approach is one of the technologies supported by Industry 4.0, it regroups different fields of research such as collaboration between these agents based on the communication among them and their environment and the principle of collective work. The third part of the paper introduced three demonstrative example tasks that can be used for collaborating robots. The first is intended for 3D simulation only, while the two others will be realized using our two Mitsubishi robots.

Acknowledgments

This research was partially carried out in the framework of the Center of Excellence of Mechatronics and Logistics at the University of Miskolc. The described study additionally was carried out as part of the EFOP-3.6.1-16-00011 "Younger and Renewing University – Innovative Knowledge City – institutional development of the University of Miskolc aiming at intelligent specialization" project implemented in the framework of the Széchenyi 2020 program. The realization of this project is supported by the European Union, co-financed by the European Social Fund. The financial support is acknowledged.

References

[1] Ahuett-Garza H and Kurfess T 2018 A brief discussion on the trends of habilitating technologies for Industry 4.0 and Smart manufacturing *Manuf. Lett.* **15** 60–3.

[2] Tjahjono B, Esplugues C, Ares E and Pelaez G 2017 What does Industry 4.0 mean to supply chain? *Procedia Manuf.* **13** 1175–82

[3] Zhong R Y, Klotz X ,Xu E and Newman S T 2017 Intelligent Manufacturing in the Context of Industry 4.0: A Review *Engineering-London* **3** 616–30

[4] Robla-Gomez S, Becerra V M, Llata J R, Gonzalez-Sarabia E, Torre-Ferrero C and Perez-Oria J 2017 Working Together: A Review on Safe Human-Robot Collaboration in Industrial Environments," *IEEE Access* **5** 26754–773

[5] Sung T K 2017 Industry 4.0: A Korea perspective *Technol. Forecast. Soc. Change* **132** 40–5, 2017.

[6] Sambamoorthy M 2002 Industry 4.0 *[Online]* Available at: http://www.cans2u.com/industry4.php [Accessed 15 May 2018]

[7] Gubán M and Kovács G 2017 Industry 4.0 Conception *Acta Technica Corviniensis Bulletin of Engineering* **1**(2017) 19.

[8] Vaidya S, Ambad P and Bhosle S 2018 Industry 4.0 - A Glimpse *Procedia Manuf.* **20** 233–8

[9] Stone P, Ave P, and Park F 1995 Multiagent Systems : A Survey from a Machine Learning Perspective *Robotics* **8** 345–83

[10] Yan Z, Jouandeau N and Cherif A A 2013 A survey and analysis of multi-robot coordination *Int. J. Adv. Robot. Syst.* **10** 399

[11] Zoghby N E, Loscri V, Natalizio E and Cherfaoui V 2014 Robot Cooperation and Swarm Intelligence *World Scientific Publishing Company* (2014)168–201

[12] Nicolle A 2000 L'organisation dans les systèmes multi-agents *Journées Colline* (Ch. 1)

[13] Lima P U and Custodio M. M 2004 Artificial Intelligence and Systems Theory: Applied to Cooperative Robots *Int. J. Adv. Robot. Syst.* **1** 141–8

IOP Conf. Series: Materials Science and Engineering **448** (2018) 012024 doi:10.1088/1757-899X/448/1/012024

FEM analysis of saw blade

O Bílek[1] and J Baďurová[2]

[1] Department of Production Engineering, Faculty of Technology, TBU in Zlín, Vavrečkova 275, 76001 Zlín, Czech Republic
[2] Department of Physics and Material Engineering, Faculty of Technology, TBU in Zlín, Vavrečkova 275, 76001 Zlín, Czech Republic

E-mail: bilek@utb.cz

Abstract. Article deals with identification of the critical points of the saw blade designed for wood cutting which teeth are made from stellite material. Cutting parameters and shape of blade were under investigation. FEM analysed the influence of cutting speed, size of cutting forces, clamping and geometry of saw blade. Factors that significantly affect the strength of the saw blade have been identified and can contribute to identification of critical conditions leading to the destruction of the saw blade. The results can also be applied when sawing wood with a band saw.

1. Introduction

Machining processes generally and particularly wood machining processes are complicated by many factors influencing each other. Unlike artificial materials wood is inhomogeneous biological material with significantly changeable mechanical qualities. Therefore; observing interaction between material and machining tool in conditions of different technological parameters is inevitable to understand the whole process and leads to higher quality of final product, improves effectiveness of production and machining process. In this article the saw blade made from tool steel 75Cr1 with the stellite tool tip, two dilatation slots and two ejecting slots is analysed.

Cutting tool plays the main role in cutting process [1-2]. It affects process productivity, cutting stability, and machining accuracy. It must also meet the safety requirements [3-6]. Thin cutting saw blades are preferred for its smaller cut and smaller waste. On the other hand, thin saw tools are less tolerant to changing cutting conditions and are more difficult to maintain. [7-10]

The preload of the blade is the result of effect of an external force through a hammer or rolling roller on the surface of the saw blade. In the experimental part it is determined as a rolled groove. It is a volumetric deformation that results in a local shape change resulting in a flat groove and the introduction of residual pressure stresses.

If the saw blade is used without preload, friction forces that occur during machining cause the edge of the blade to expand. The circuit thus becomes loose and machining is not straight. This phenomenon is also referred to as circling the wheel. On the other hand, the saw blade with a pronounced preload gets a plate shape, but it balances at certain speeds.

Preload compensates adverse thermal state of stress which occur as a result of friction of teeth during cutting process. Higher preload allows the blade to machine with higher speed. Preload effects on the smooth running and hence the quality of the cut. The preload allows reducing the thickness of

IOP Conf. Series: Materials Science and Engineering **448** (2018) 012024 doi:10.1088/1757-899X/448/1/012024

the saw blade, while the other cutting conditions (circumferential speed and feed rate) remain unchanged.

Cutting tool with radial grooves can compensate thermal expansion of the blade edge resulting in machining, thus maintaining their flatness. Radial grooves also reduce the noise level by interrupting the peripheral part of the disc. However; the conditions generally are: the length of the grooves is at least 0.1 × ØD and the number of grooves (for SC blades) 3-4. Radial grooves are also recommended for saw blades with smaller tooth pitch and in case of thick blade.

The knowledge of lateral stiffness contributes to the choice of the appropriate thickness of the blade. According to GOST 980-80, lateral stiffness should be greater than 40 N / mm in sweep and greater than 25 Nmm^{-1} in cross-cutting. If the values are lower, it is necessary to use the guide pins. [11-14]

Von Mises' stress σ is judged in this article. They are defined as the equivalent tensile stress (reduced stress), used in resistant materials such as steel to predict multiaxial load using the results of simple tensile tests. In this paper this parameter is used as a benchmark for the strength of the saw blade.

2. Methodology

The finite element calculation method (FEM) was chosen for this research project. A static analysis is considered when a 3D saw blade model is used. The shape and geometry of the disc are shown in figure 1. The geometry of the tool and its description presents table 1.

Figure 1. The basic shape of saw blade

Table 1. Description of geometry of saw blade

Geometric feature	Value
Diameter of the saw blade	350 mm
Number of teeth	24
Diameter of the clamping hole	80 mm
Thickness of the wheel body	2.6 mm
Width of the tooth	4 mm
Tooth height	4 mm
Radial compensating slots	2
Clearing slots	2
Compensating rolling	0.66R
Clearance angle	14°
Rake angle	18°
Wedge angle	58°
Tooth shape	FZ

IOP Conf. Series: Materials Science and Engineering **448** (2018) 012024 doi:10.1088/1757-899X/448/1/012024

The material of the saw blade was tool steel 75Cr1 with the mechanical properties presented in table 2. The material is considered homogeneous even in the area of teeth, although they are made of stellite. This material has greater strength and hardness than the saw blade material. This is a simplification for FEM analysis and is considered in the discussion of the results, but it does not affect the results of the experiment.

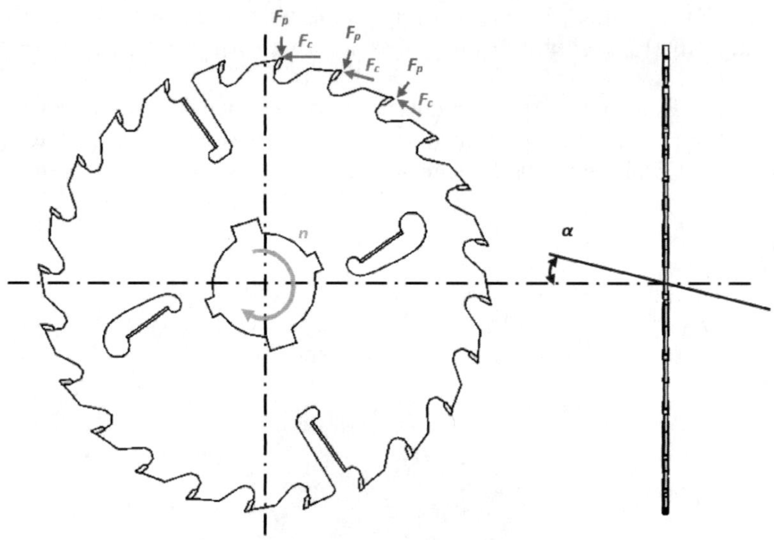

Figure 2. Scheme of loading of the model by the cutting forces F_c, F_p, revolutions per minute n and axis deviation α

Table 2. Mechanical properties of tool steel 75Cr1 (1.2003)

Property	Value
Density	7850 kg/m^3
Poisson's number	0.28
Modulus of elasticity	200 GPa
Tensile strength σ_p	650-880 MPa
Yield strength σ_k	350-550 MPa
Fatigue strength	275 MPa

Table 3. Experiment input conditions

Property	Value
Clamping	Key
Revolutions n	3500 min^{-1}
Tangential force F_c	110 N
Passive force F_p	55 N
Axis deviation α	1°
Thickness of the saw blade body	2.6 mm

3. Results

The experiments serve to clarify the behavior of the saw blade, depending on the different technological conditions. The following chapter presents variants of static analyzes of the basic shape of blade compared to blade with a rolled-out groove. The influence of clamping on the flange and the strap is monitored. Moreover; dependence of cutting forces and gears on the overall distribution of the Von Mises tension in the blade together with the feed size is expressed in the graph. In addition to the strength of the blade and its critical areas, stiffness for lateral force is determined in individual variants.

3.1. The strength of the saw blade in basic shape in case of clamping on a key

Technological conditions are displayed in table 3 and input constrains in the figure2. Circular saw blade model was clamped on a key, where the revolution speed n was 3500 min^{-1}, the size of cutting forces $F_c = 110$ N and $F_p = 55$ N, while axis deviation was $\alpha = 1°$

Result of FEM analyses showed that maximal stress according the scale on the right reach 450 MPa close to the place of clamping key way (see figure 3). Generally, the saw blade keeps low stress except weak spots between the clearing slots and the key way hole.

Figure 3. Detail of the area of the key way in the saw blade

Furthermore, the wheel displacement size was evaluated - figure 4. The maximum values of 5.44 mm are achieved in the marginal areas of the blade by deflecting the saw teeth and the entire blade body. Due to the character of the clamp, this maximum is asymmetrically oriented to the opposite side of the clamping strap.

In addition, aggregate characteristics have been determined. They influence the strength of the saw blade and thus the Von Mises stress. Similar to previous results, the higher the value, the more likely the critical limits may be exceeded. Either a permanent deformation may occur at the yield strength, or breakage of the disk can occur within the strength limits of the material. The results are presented in the following chapters.

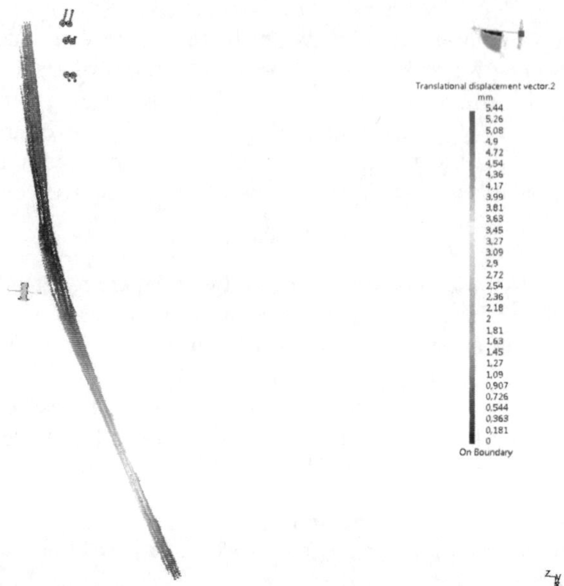

Figure 4. Deformation and circular saw blade displacement

3.2. Determination of the influence of revolution on the strength of the saw blade

Experiment was done in same input conditions. The effect of rotational speed on the strength of the saw blade as the Von Mises stress is examined. According to figure 5, it can be assumed that increasing gears evoke quadratic increase of Von Mises stress in the saw blade. Above lower yield point σ_k a permanent deformation of the saw blade may occur. Considering the equation; a permanent deformation may occur at a speed above 2900 min^{-1}.

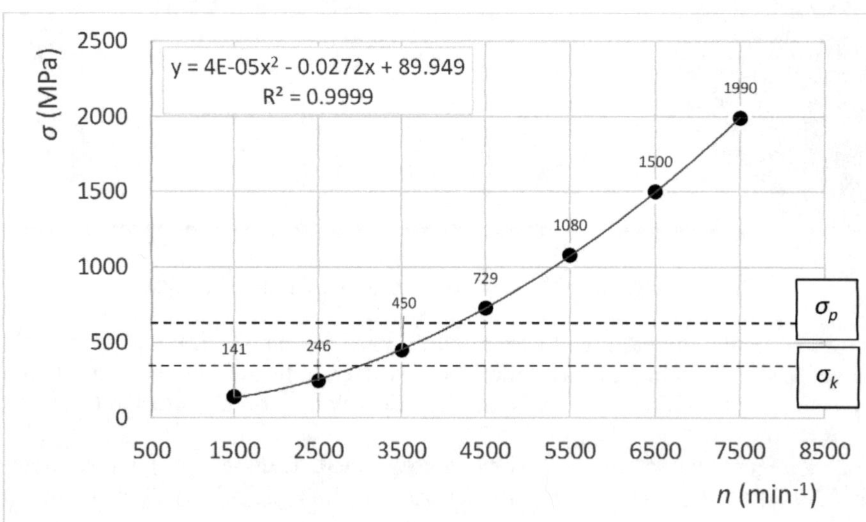

Figure 5. The effect of rotational speed on the strength of the saw blade ($\alpha = 1°$, F_c = const.)

The following figure 6 shows the size of displacement - blade deformation depending on the speed. Its trend is comparable to the Von Mises stress. According to figure 5 the largest size of the deformation is at the opposite ends from the clamp strap hole of the saw blade. Its size in units of tens of millimetres is determined by the method of clamping the blade and deflecting it in the axial direction. Compared to reality, it is possible to evaluate the trend of deformation, but in practice the clamp of blade is not only a strap but also a clamping on flange which, in the axial direction significantly prevents deformations and thus undesirable cutting of the material.

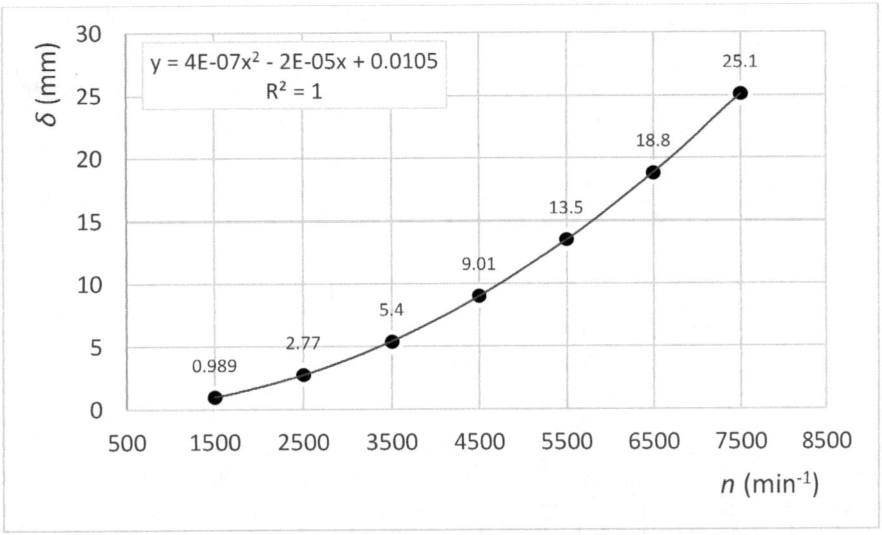

Figure 6. Effect of rotational speed on size of displacement of the saw blade ($\alpha = 1°$, F_c = const.)

3.3. Determination of the effect of the cutting forces to the saw blade strength

Results of rotating saw blade (n = cons. = 3500 min^{-1}) and its dependence on size of cutting forces are important. They simulate effect of cutting process and may serve to predict result of cutting process in dependence of machined material (hard wood etc.) Linear dependence was determined between Von Mises stress and tangential force F_c and has mathematical interpretation:

$$\sigma = 0.7326 \times F_C + 340.12 \quad (1)$$

as demonstrated in the figure 7.

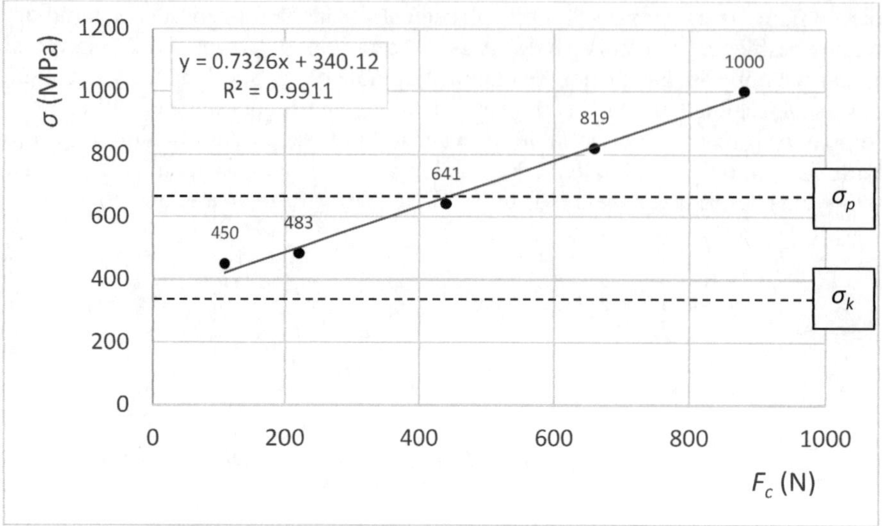

Figure 7. Effect of tangential force F_c on Von Mises stress (n = 3500 min^{-1}, α = 1°, $F_c : F_p = 2 : 1$)

4. Conclusion

At a certain moment, with the simultaneous effects of both cutting and rotating forces, a significant tool deflection may occur, which may cause the disc to break. This extreme deflection was observed in the FEM analysis of the rotating saw blade. In the case of the circular saw blade, the tool continues to permanently deform above 2900 min^{-1}. Clamping is the significant factor influencing stability and strength of the saw blade. In the case of clamping on the key is then the critical area of weakening in the key way of the saw blade.

References

[1] Cristovao L 2013 *Machining properties of wood, tool wear, cutting force and tensioning of blades* (Luleå : Luleå tekniska universitet) p 144
[2] Csanády E and Magoss E 2013 *Mechanics of Wood Machining* (Berlin: Springer) p 202
[3] Echeverri R 2003 *Modeling of static and dynamic components of cutting force in sawing*. B.m. (Colombia : Universidad de Norte) p 191
[4] Fritz K 2011 *Manufacturing Processes 1: Cutting* (Wien : Springer) p 506
[5] Kováč J and Mikleš M 2010 *J. For. Sci.* **56**, 271–7
[6] Markopoulos A P 2013 *Finite Element Method in Machining Processes* (London : SpringerBriefs in Applied Sciences and Technology) p 92
[7] Stakhiev Y M 2003 *Holz als Roh- und Werkst* **61**, 13–22
[8] Wojtowicki J, Beslin O and Nicolas J 1994 *Can. Acoust.* **1**, 105–6
[9] Mikó B, Szegh I and Kutrovácz L 1999 *Period. Polytech. Mech. Eng.* **43** 97–106
[10] Otko T and Zębala W 2016 *Precision Machining VIII* **686** 33–8
[11] Kandráč L, Maňková I, Vrabeľ M and Beňo J 2014 *Appl. Mech. Mater.* **474**, 192–9
[12] Kandráč L, Maňková I, Vrabeľ M, Beňo J, Stahovec J, Durakbasa M N, Bas G 2014 *Appl. Mech. Mater.* **693**, 358–63
[13] Matras A. 2017 *Proc. SPIE 10445, Photonics Applications in Astronomy, Communications, Industry, and High Energy Physics Experiments 2017* (Wilga : SPIE) 1044515
[14] Oravcová J, Košťál P, Delgado Sobrino D R and Holubek R 2013 *Appl. Mech. Mater.* **309** 20–6

IOP Conf. Series: Materials Science and Engineering **448** (2018) 012025 doi:10.1088/1757-899X/448/1/012025

Numerical modelling of forming high strength aluminium

V Gál[1] and Zs Lukács[2]

3515 Miskolci Egyetemváros - Anyagszerkezettani és Anyagtechnológiai Intézet

metgv@uni-miskolc.hu

Abstract: In recent decades the developments of vehicle materials and technologies have been influenced by the environmental regulations. Reducing the mass of vehicles is important to reduce the air pollution emissions. In order not to break the safety standards with this reduction, the strength of the used materials should be increased. Nowadays the high strength steels have not enough formability, so the base materials of the vehicle need to be changed. The density of aluminium is significantly lower than that of steels, which is positive for weight reduction, but the strength of the traditional aluminium alloys is low, so it needs to be enlarged. However, due their different alloying elements and the rigid disperse precipitates phase they create, the ductility of high strength aluminium is not sufficient in most cases. The lower formability is a serious technical challenge for automotive engineers. The solution is in the newest dedicated FEM codes. With the help of these software the engineers can more easily plan the technical background of manufacturing a part, or making formability test in virtual environment without using specimen. Because of this, the objective of this paper is the numerical modelling of forming high strength aluminium.

1. Introduction

Nowadays there are many new materials in the automotive industry. Many of these new materials are made from high strength metals, because of the environmental regulations. The new aluminium alloys have a growing potential in industrial applications. Due to their alloying they have high strength and also due to the rigid disperse precipitates they create, they have lower formability. The traditional forming technologies are not suitable for these cases, therefore new technologies have to be developed. With the Hot Forming and Quenching (HFQ®) process patented by the Imperial College London (ICL) and the Impression Technologies (ITL) aluminium parts with complex geometry can be produced. Accordingly the manufacturing process is given, but along with this, the software used to simulate the production must also keep up with these developments. They must have the new materials and must have the possibility to model the new technologies: a further problem that these high strength aluminium alloys can be successfully formed at elevated temperature. For this reason the concept of this paper is to give insight into the background of a simulation of a forming process at high temperatures. We will mainly focus on the difficulties of hot forming simulation, but it is reasonable to make an overview of the cold forming simulations too.

2. Input parameters of a cold forming simulation

The relationship of the stress and strain is the most important part of describing the mechanical behaviour of a material. For cold forming in sheet metal forming elastic-non-linear hardening material rules are widely applied. These hardening rules describe the material that undergo elastic deformation until yield, beyond yield stress; elastic and plastic deformations occur simultaneously. The stress value needed for further plastic deformation changes, as the material deforms, i.e. it hardens.

2.1. Plastic behaviour in cold forming simulations

The yield criteria define the condition for the elastic behaviour limit, after which the material continues deforming plastically, showing a hardening behaviour. Over the years, the modelling of the sheet metal yield criteria has evolved to describe more accurately the anisotropic behaviour, based on the yield criteria proposed for isotropic materials by Tresca, Huber and Von Mises [1]. One of the first yield criteria for anisotropic materials was made by Hill, usually referred as Hill 48, which takes the anisotropy coefficients into account in several directions (0°, 45°, and 90° to the rolling direction). The deficiency of Hill classic model was already discussed in several papers, and during the last decades, new theories were published better describing the anisotropic behaviour of sheet materials: here we just mention the Barlat'89 model developed for aluminum and the BBC2005 model elaborated for steels by Banabic [2].

As the material hardens with the increasing plastic deformation, the yield surface - which was described by the yield criterion- also will change. The four types of the changing the yield surface are shown on the Figure 1.

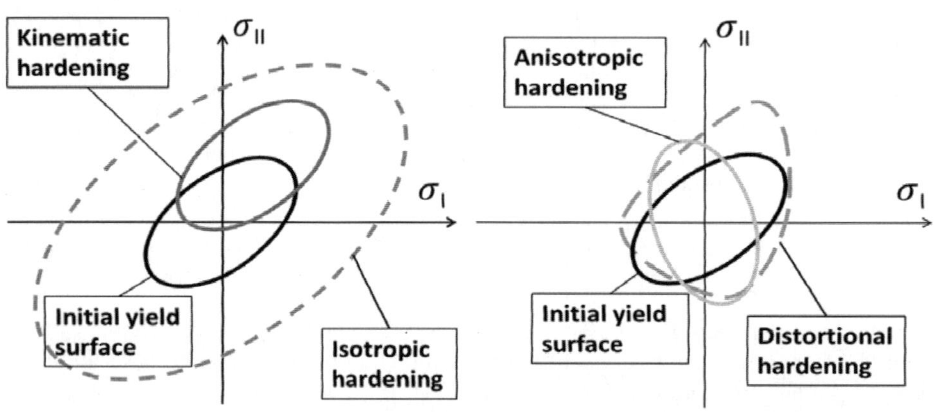

Figure 1. Evolution of yield locus according hardening types. [1]

Isotropic hardening, which refers to the proportional expansion of the initial yield surface; (ii) kinematic hardening, if the deforming material shows a yield surface that does not change its form and size, but translates in the stress space under deformation; (iii) rotational hardening, which causes the yield locus to rotate; (iv) distortional hardening, which causes the yield locus to distort [1].

To make a forming simulation usually the yield surface and the flow curves are needed as input parameters. In a dedicated FEM code, the isotropic hardening model is the default setting, but it is possible to use another model.

2.2. The limit of forming in cold forming simulations

With the Yield surface and the flow curve the behaviour of the material can be described but can't determine the occurrence of its failure. In sheet metal forming the limits of formability is usually defined by the Forming Limit Diagrams (FLD). This diagram indicates the combination of the major and minor strains that can be applied to a metal sheet without failure. Prediction the failure of the sheet based on the FLD, is widespread in the industry and the finite element simulations, mainly thanks to its simplicity. Despite its prevalence, it has several drawbacks: The FLDs depend on many things, for example the sample thickness, the strain history, etc. For non-linear strain paths (Figure 2.), it changes significantly. These drawbacks (especially the dependence from the applied strain path) have pushed the development of alternative approaches for predicting the sheet failure. There are new diagrams that was designed to ignore the effect of the different strain path. The Forming Limit Stress Diagram was one of the first developments which was later improved by Stoughton [3]. The FLSD can be derived from the FLD with finite element methods, but in case of complicated strain paths the independence from strain path is still questionable.

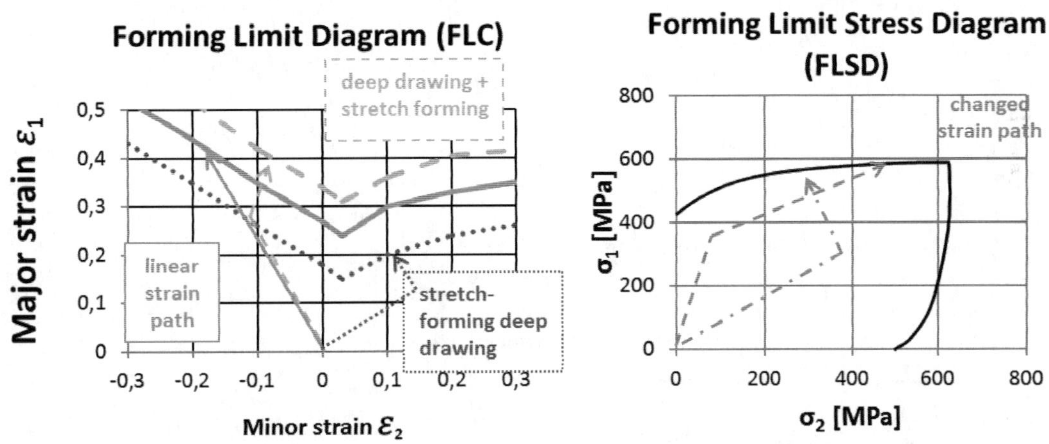

Figure 2. From a path-dependent FLD to a non-path dependent FLSD [1]

Because of the drawbacks of the earlier mentioned models, the researcher put large effort to create another way to predict the failure, which is based on analytical formulation. These Fracture Mechanism (FM) models can use fracture criteria or can be void-growth-based. FM models like McClintock or Rice and Tracey usually has a damage indicator variable. The cracking occurs when this variable reaches a defined value.

While the dedicated fem codes based on the above described parameters are very useful tool for the industry for cold forming simulations, the simulations of forming at elevated temperature has not yet developed completely. Therefore, in the following chapter I will mainly focus on it.

3. The problems in hot forming simulations

In hot forming, the material properties and relationships should be known as a function of temperature, as well. For the description of material behaviour for hot conditions besides the temperature dependence of basic physical properties we have to know the material relationships between the stress and strain components (usually given by the hardening rule)[2].

Higher strain rates at the same temperature lead to greater stress, however the higher the temperature is the lower the reached stress level is. Along with this, the yield surface and the flow curves changes as the function of temperature. Because the temperature and the strain rate are usually not constant during the forming process, describing the material behaviour becomes more complex. Most of the dedicated FEM packages are not able to handle the complexity of the above described process, even though it is well suited for simulation of cold forming process.

IOP Conf. Series: Materials Science and Engineering **448** (2018) 012025 doi:10.1088/1757-899X/448/1/012025

3.1. Plastic behaviour in hot forming simulations

As it was mentioned earlier, modelling of hot forming processes is more complicated from the viewpoint of flow curves, since in this case, the visco-plastic material model should be used. Therefore, the dependence of flow curves on the temperature and strain rate should be taken into account which requires the flow curves determination at different temperatures and strain rates. Describing the yield surfaces are also difficult tasks, because the temperature dependence of anisotropy coefficients has to be taken into consideration, as well [2].

Based on the different temperature and strain rates for a correct simulation more flow curves and Yield surfaces are required. It can be easily performed in FEM codes with a flow curve matrix that have temperature and forming velocity axes (Figure 3).

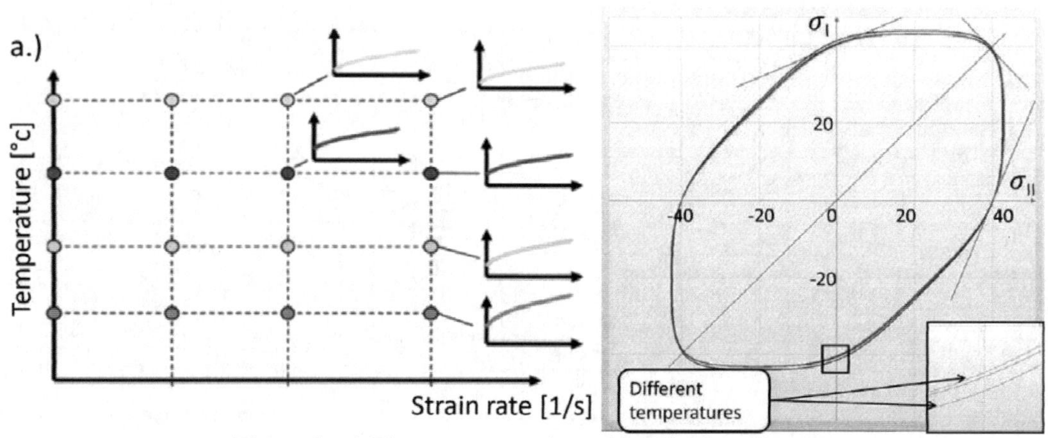

Figure 3. a.) Flow curve matrix and b.)Yield surfaces

The two input parameters that I show on the Figure 3. are not independent. The Yield surface - based on the Yield criteria - can be derived from the flow curve. Due to this reason the flow-curve-matrix is enough to describe the behaviour of the material while it deforms.

3.2. Problems with predicting the failure in hot forming simulations

The limits of formability can be regarded as one of the most important input parameters in any forming simulation. In the viewpoint of failure, the Forming Limit Diagrams are not suitable in hot forming simulations. On the one hand because of its drawbacks, on the other hand, because their strain rate and temperature dependence. In hot forming simulation, where the temperature of sheet and also the strain rate can change from point to point, which means that different Forming Limit Curves should be applied theoretically from point to point too, which is obviously impossible and also the FLC determined for a fixed temperature and strain rate may be used with acceptable accuracy in a limited range of parameters [2]. It is possible to add the strain rate or the temperature to the FLD as a third axis but not at the same time. Maybe this model might be inserted to a dedicated FEM code, but usually the temperature and strain rate are not constant during production.

Due to this reason researches want to create another way to predict the failure of the sheet. The FLSD is based on the forming limit diagram and its independence from strain path is also questionable cannot be applied either, accordingly this model also not appropriate. The Fracture Mechanism model works with a constant damage variable, however this variable changes significantly while the material deforms at elevated temperature.

3.3. New damage model

Most of the works on damage mechanics uses state variables to represent the effects of damage on the structural stiffness, remaining life and failure of the material, but damage evolution does not progress spontaneously after initiation, thus requiring a damage evolution model. Lin, Liu and Dean (2005) introduced continuum damage mechanics theories to model damage evolution in a wide range of metal forming processes, particularly for hot metal forming. One of the newest models based on the aggregation of the damage, which means that under plastic deformation discreet voids or cracks may nucleate and grow within the material and eventually these defects would coalesce to form macro cracks, leading to material failure. The effective stress forms the basis of the Continuum Damage Model (CDM): [4]

$$\sigma_{eff} = \frac{\sigma}{1-D} \quad (1)$$

Where σ is nominal stress, and D is material damage variable. This model connects the strain with an inner damage variable, which is zero before the deformation, and one in the moment of failure. It can be seen, that this damage variable changes as the deformation progresses. As the deformation correlate the temperature, and the changing of the stress level, the CDM model is currently the best option to modelling a hot forming simulation like the HFQ^{TM}. Along with the positives the CDM also have some drawbacks. It can't be yet applied in dedicated FEM codes, and need many complicated measurements to define.

4. Conclusions

In this paper we have shortly summarized the present situation of the numerical simulations and modelling. We have made an overview over the parameters of the cold forming simulations. Based on this, easy to ascertain about the excellent applicability of the dedicated FEM codes in supporting the cold forming production. The parameters that are needed by the simulations should be available and easy to define. Despite nowadays we can usefully apply these parameters to the numerical prediction of the limit of forming it needed to be upgraded in the future.

Based on the example of cold forming simulation we have made an overview about the current state of hot forming simulations. The most disadvantages of this kind of simulations come from the difficulty to defining the limit of forming. We have shown the new models that can meet the expectation, but they applicability in dedicated FEM codes is questionable.

Acknowledgments
The authors express their grateful thanks to the European Commission for their support on the H2020 project "Low Cost Materials Processing Technologies for Mass Production of Lightweight Vehicles (LoCoMaTech)", Grant No: H2020-NMBP-GV-2016 (723517).

References

[1] S. Bruschi, T. Altan, D. Banabic, P.F. Bariani, A. Brosius, J. Cao, A. Ghiotti, M. Khraisheh, M. Merklein, A.E. Tekkaya, Testing and modelling of material behaviour and formability in sheet metal forming 2014 CIRP.

[2] M Tisza, Zs Lukács, P Kovács and D Budai: Some recent developments in sheet metal forming for production of lightweight automotive parts 2017 J. Phys.: Conf. Ser. 896 012087

[3] Stoughton TB (2000) A General Forming Limit Criterion for Sheet Metal Forming. International Journal of Mechanical Sciences 42:1–27.More references

[4] Fundamentals of Materials Modelling for Metals Processing Technologies. Jianguo Lin, Imperial College London, 2015.

IOP Conf. Series: Materials Science and Engineering **448** (2018) 012026 doi:10.1088/1757-899X/448/1/012026

Last mile logistics: an integrated view

János Juhász[1] and Tamás Bányai[2]

[1] PhD student, University of Miskolc
[2] Associate professor, University of Miskolc

E-mail: juhaszj@uni-miskolc.hu

Abstract. Production and service companies try to increase their efficiencies through the application of different state-of-the-art solutions offered by the Industry 4.0 paradigm. The networking of manufacturing companies and logistics providers led to an increased complexity of supply chain solutions, especially in first mile and last mile logistics. This stream is remarkable in two different areas of the economy. The first one is represented by the cooperative supply of manufacturing companies, especially in the field of automotive industry, while the second field is the package delivery service. Within the frame of this paper, authors are focusing on the last mile solutions and give an overview of the mainstream of research. There are a lot of articles related to last mile logistics. The methodology of this structured literature review includes four main steps including the search for articles in databases and other sources, reduce the number of articles by reading them and identify the main topics, define a methodology to analyze the chosen articles and describe the main scientific results and identify the scientific gaps and bottlenecks.

1. Introduction

The last mile's typology is addressed as an innovative concept in logistics. The main question of this research is to specify the main characteristics of the last mile logistics. Companies work out these solutions to increase efficiency and decrease costs. Manufacturing and service participants try to increase their efficiency through the application of different state-of-the-art solutions offered by the Industry 4.0 paradigm.

The networking of manufacturing companies and logistics service providers led to an increased complexity of supply chain problems' solutions, especially in first mile and last mile logistics. The traditional supply chain can be optimized by innovative and modern methods and strategies. This stream is remarkable in two different areas of economy. The first is represented by the cooperative supply of manufacturing companies, especially in the field of automotive industry, while the second field is the package delivery service including effective managerial practices and technology to enhance the competitiveness. Competitiveness can be improved through better management of logistics networks and can lead to the development of new models for different sectors. The last mile challenge defines different aspects of the solutions to figure out logistics problems such as build the concepts, manage the requirements of a changing customers' demands from supply chain, change industrial policies for efficient production and movement of goods and services (Neeraja et al., 2014).

This paper is organized as follows: Section 2 presents a literature review, which summarizes the research results related to last mile logistics and gives an overview about the mainstream of research.

IOP Conf. Series: Materials Science and Engineering **448** (2018) 012026 doi:10.1088/1757-899X/448/1/012026

Section 3 shows the impacts of last mile logistics concepts. Conclusions and future research directions are discussed in Section 4.

2. Literature review

There are a lot of papers related to determining the optimal decisions in manufacturing industrial processes. To build a link between literature and the research work, we are focusing on the previous research results to answer the following questions: Who is doing what? Who has done what? Who first did it or published it? What are the research gaps? Our used methodology of structured literature review includes four main steps (Lage & Filho, 2010):

- search for articles in Scopus,
- reduce the number of articles by reading the abstract and identify the main topic,
- define a methodology to analyze the chosen articles,
- describe the main scientific results and identify the scientific gaps and bottlenecks.

Firstly, the relevant terms were defined. It is a crucial phase of the review because there are excellent review articles in the field of manufacturing optimization and we didn't want to produce an almost similar review, but we applied the presented methodology. We used the following keyword to search in the Scopus database: "Last mile". Initially, 2147 articles were identified.

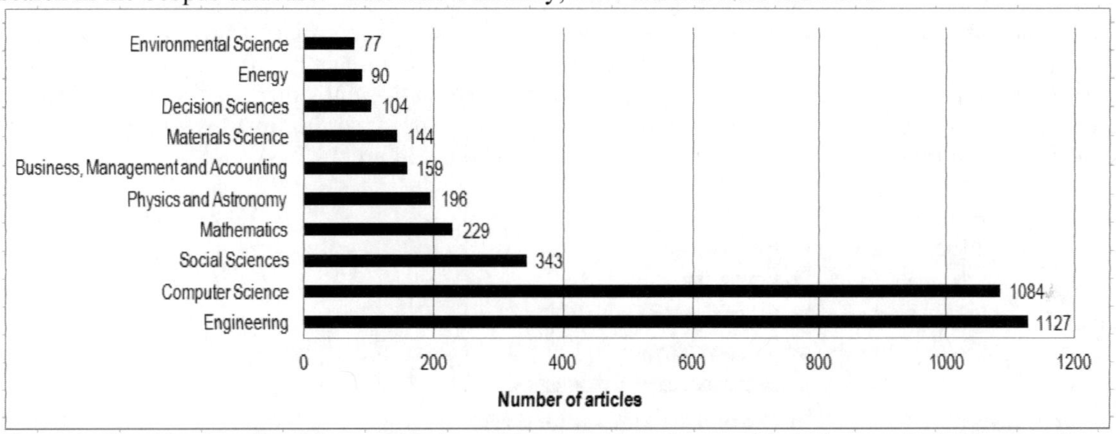

Figure 1. Classification of articles considering subject areas based on search in Scopus database using "Last mile" keyword (without refined results).

As Figure 2 demonstrates, the support of decision making in manufacturing processes has been researched in the last decades. The first publication in this field was published in 1999 for an integrative freight market simulation (Clark, E., 1999) and it was focusing on framework of pricing the last mile services. The number of published papers has been increased in the last 10 years; these show the importance of this research field.

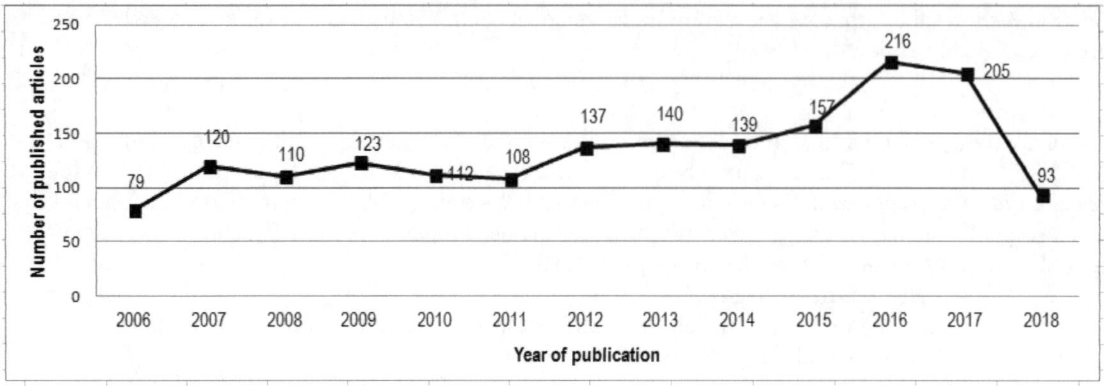

Figure 2. Classification of articles by year of publication.

We refined results by limited the following: Year: 2018; Document type: Article; Source type: Journals; and Language: English. After this, we got 39 articles. Our search was conducted in May 2018; therefore, new articles may have been published since then.

In the following step, the articles were reduced after reading the articles. We excluded articles, which topic did not find our interest and can't be addressed the research of first mile and last mile supply solutions. After this reduction, we got 33 articles. We added one other paper selected through separated search (Lance L. Putong & Marlene M. De Leon, 2018), so the final list for classification and evaluation from the point of view scientific results includes 34 articles.

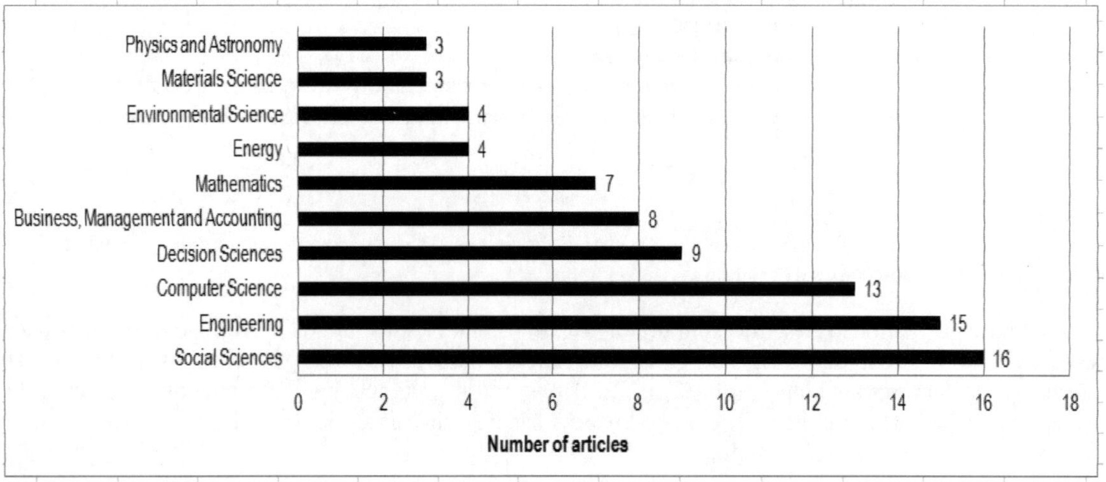

Figure 3. Classification of articles considering subject areas based on search in Scopus database using "Last mile" keyword.

The distributions of the most frequently used keywords are depicted in Figure 3. As the keywords show, the last mile solutions are complex systems; and their optimization can be solved through heuristic methods and AI algorithms. The objective function of design and operation of last mile solutions is the cost function.

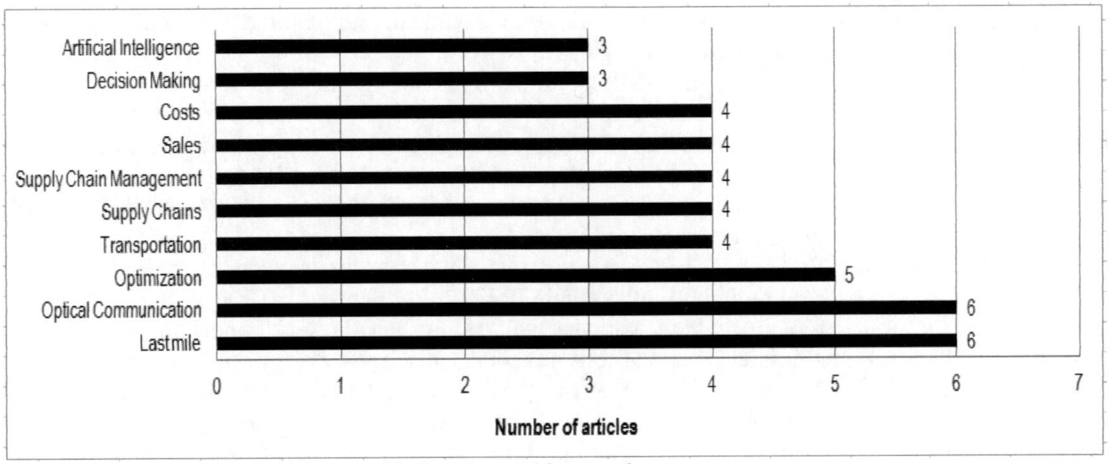

Figure 4. Classification of articles considering the used keywords.

Logistic is very important for companies because it is concentrated in several service support. The aim of logistic is serving the production area and the customer's demands. Logistics services must be performed in right time, from the right place, right products, in right condition, to the right user and with reasonable cost.

As we can find in the literature, logistic has undergone many changes based on science in the last few years. New aspects allow implementing modern and innovative strategies of supply chain. There are several impacts of supply chain such as resources, which indicate restructuring the whole logistics supply chain. Also, the Just-in-sequence (JIS) supply is mentioned as a reinterpretation of Just-in-Time (JIT) principles. The competition determines to make the manufacturing- and their processes more and more efficient through the quality and customer needs.

The industrial participants are focused to reduce operational costs, and stay competitive in the market. There are several advantages such as controlling costs, reducing risk of supply chain (Marujo et al., 2018) and minimizing energy utilization besides on travelling distance (Rabta et al., 2018).

Industry 4.0 touches the entire supply chain management, intelligent manufacturing, methods, and Logistic 4.0. The application of new technological approaches and elements become possible even with more complex systems. Smart solutions align processes to reduce resources, utilize capacities, and increase efficiencies. The logistic challenges and tendencies are representing changes in the supply chain. Supply Chain Management can improve the distribution processes by applying strategies such as Crowdsourced Logistics, CSL (Castillo et al., 2018). The supply chain management (SCM) development is a key challenge for logistics area, especially in production area, health care and in related logistics services. It means these systems need to serve society' demands and focus on network asset such as transportation. The network configuration can support the whole supply chain as synchromodal supply chain (Farahani et al., 2018). Also, the logistic service providers were analyzed in terms of parcel tasks, routing and transportation, and last-time delivery (Yi & You, 2018).

Other results show to advantage and application possibilities to analyze the customer's intention to use self-collection as last-mile delivery method (Yuen et al., 2018). Most of the studies considered to delivery tasks, and its additional issue as solving last-mile logistics problems. Delivery service providers should concentrate on their marketing force and customize their services for consumer groups, who have specific individual characteristics, such as optimism and innovation (Chen et al., 2018).

A number of books and articles have been published which addressed the available methods including analytic and heuristic development. These solutions can solve logistics issues for example profit maximization modelling, genetic algorithm, optimization, decision making, and simulation.

IOP Conf. Series: Materials Science and Engineering **448** (2018) 012026 doi:10.1088/1757-899X/448/1/012026

Algorithms are supporting the logistic enterprises to establish and resolve the fulfilment of e-commerce orders in distribution centres (Leung et al., 2018).

The Travelling Salesman Problem (TSP) solution obtained by the Concorde solver and the Randomized Variable Neighbourhood Descent (RVND) heuristic is implemented the solution of the problem (des Freitas & Penna, 2018). The computational experiments show the use of drones for last mile delivery can reduce the total delivery time. The TSP-LS algorithm (TSP-LS delivers the solution of lower quality, but very quickly.) and Greedy Randomized Adaptive Search Procedure (GRASP) algorithms can maximize the quality of service (Ha et al., 2018).

The Vehicle Routing Problem (VRP) arises in real-life applications such as in logistics. The efficiency and timeliness become important. Vehicle Routing Problem with Time Window is an NP-complete optimization problem, which can be solved by ant colony algorithm (Niroomand &Nsakanda, 2018). Other model is also able to solve this problem by modelling alternative distribution set-ups, which contains mathematical (Kin et al., 2018). There are available models provide a set of recommendations on how can minimize the routing costs as well as penalty costs for unsatisfied demands (Putong & Leon, 2018), and propose modelling approach to a new stochastic last mile relief network design problem (Elçi, 2018). The new city logistics problem involve to design of the routes for a vehicle fleet located at the depots to transport the customer demands to a subset of the satellites and the routing of a vehicle fleet from the satellites to serve all of the customers solving with hybrid multi-population genetic algorithm (Zhao et al, 2018).

Decision making with agent-based simulation and dynamic routing procedures for sustainable provision in the field of food processing (Fikar, 2018). Simulation provides an ideal environment for modelled last mile logistics solutions. The optimization of large scale transportation problems can be optimized with simulation techniques, like hybrid framework simulation which makes it possible to support traffic movements (Simoni & Claudel, 2018). Travel distances and the inefficient movement can be reduced by applying optimization (Kin et al., 2018, Fikar, 2018). Multi-criteria optimization in humanitarian logistics supports the design and operation of last mile distribution in stochastic environment, so this model calculates an actual vehicle schedule while forcing vehicles to form convoys in humanitarian operations research (Ferrer et al., 2018). Crowdshipping platform was discussed to solve maximizing the platf orm's profit and fulfil requests (Behrend & Meisel, 2018). This solution serves to execute delivery demands and also support the item-sharing.

Communication network is related to design the last mile solutions. The Routing Protocol for Low Power and Lossy Network (RPL) are standardized solution in the smart grid, because this implementation reduces the complexity and controls interactions to distribute the network states and other related information in the network (Alishahi et al., 2018). The network can be controlled via federation of FiWi (Fiber-wireless) and Edge IoT software (Bellavista et al., 2018). Most of studies are focused on network performance such as two-way multiuser mixed radio frequency/free space optical (RF/FSO) relay networks (Al-Eryani et al., 2018), and wireless technology (Horibe, 2018) as millimetre wave communication. 5G mobile network project was reviewed and analyzed in the perspectives of radio engineering and implementation (Leinonen et al., 2018), which is one of the last mile network solution. There are several methods considering the function of distribution network and internal logistics network. Stochastic model was defined to solve network issues (Noyan & Kahvecioğlu, 2018). The internal logistics networks used to serve customers across channels; it means the exploratory study contains the retailers from different contexts (Wollenburg et al., 2018). Worldwide Interoperability for Microwave Access (WiMAX) networks contain several Quality of Services (QoS) structure for guaranteed multimedia services, so the efficient scheduling algorithm named Earliest Deadline First (EDF), Deficit Round Robin (DRR); Heuristic Earliest Deadline First (HEDF), and Modified Earliest Expiry First (MEEF) can reduce delay and packet loss ratio with number of Subscriber Station (SS) in increased (Sudha & Rengarajan, 2018).

One of the main streams of literature in the field of transportation addresses the development of last mile logistics and city logistics solutions. These are focusing on cost reduction in urban freight distribution (Morganti & Browne, 2018); design of parcel locker network maximizing the profit in city transportation systems (Deutsch & Golany, 2018). The innovative contributions are classified into five main categories: innovative vehicles, proximity stations or points (Chen & Wang, 2018), collaborative and cooperative urban logistics, optimization of transport management and routing, innovations in public policies (Fillone & Mateo-Babiano, 2018) and infrastructures (Ranieri et al., 2018).

3. Impacts of last mile logistics concepts

As you can find in the literature, logistics has undergone many scientific changes in the last decades. Modern and innovative strategies are released, where solutions enable the implementation of smart supply chain management (SCM). New aspects of last mile solution can redesign the traditional supply chain, and can be serve and support the whole logistics services and processes.

There are several related issues in the logistics, namely factors and costs. It can be categorized as follows: internal, external, performance, system structure, technology integration, and other expenses as shown in Table 1.

The last mile of the supply chain is focused on the most expensive and problematic area of the supply chain. The main stream of the problems is the delivery process. Delivery processes have several issues, the main issues are the followings: One of them is a home delivery, which indicates extra costs, pollutions and transportation. The second is a DTD (door-to-door) delivery, where empty routes can be found. Also, the security is important, because several parcels are required signature to fulfil the delivery success.

Table 1. Factors and costs in logistic

Internal	**External**	**Performance**	**System structure**	**Technology integration**	**Other expenses**
Customer satisfac-tion	Technology	**Productivity development**	Better relationship	**Information systems**	Transportation cost
Quality	Globalization	Quality	Supply chain	Intelligent systems	Holding cost
Networks	Working force	**Good service**	Communications	**Smart technology**	Inventory cost
SCM	Environmental	**Modify response needs**	Organization care	**Smart grid**	Order processing cost

The traditional supply chain represents the material flows, where goods are manufactured. These are shipped to distribution centres. There are two options; the first is (after buying) the direct shipping to customers, the second is through traditional supermarkets.

With this in mind the last mile can involve delivery processes, Boyler, Fohrlich & Hult (2005) used these typologies in several academic papers to describe sub-flows as Table 2 shows.

Table 2. Typology of Last Mile

Delivery	**Order fulfilment**	
	Store-based	**Distribution centre**
Indirect	semi-extended supply chain	decoupled supply chain
Direct	fully extended supply chain	centralized extended supply chain

As Table 2 shows you can find four sub-types: semi-extended supply chain, fully extended supply chain, decoupled supply chain, and centralized extended supply chain.

The semi-extended supply chain is used in collection, where the delivery is indirect, and the order fulfilment is store-based.

The fully extended supply chain can also order picking process, where the order fulfilment is storebased. It is direct home delivery type. The decoupled supply chain is applied to the distribution

IOP Conf. Series: Materials Science and Engineering **448** (2018) 012026 doi:10.1088/1757-899X/448/1/012026

centres, where the delivery is indirect. Finally products are delivered through a centralized extended supply chain to customer by using distribution centres. Last mile solutions offer several benefits in the logistics, that relationship to the above mentioned typologies are shown in Table 3:

- Time savings: Supply Chain Management interprets and manages manufacturing and service processes. Routing time can be reduced in the last mile phase.
- Valuable data control systems: Last mile can use system data, and uses the necessary information which is required by each process.
- Security: Signature must accept to the successful delivery. Security becomes more complex, and smart with logistics and human resources.
- Energy and environmental effects: It means that the right delivery method can reduce emissions, and become economically competitive.
- Customer service level: The characterized service level is based on time windows.
- Mobility: It means the shipping time and delays are reduced by increasing mobility.
- Productivity: Last mile logistic solutions can allow productivity developments.
- Customer satisfaction: It measures the customer experiences in services/products, and assessing new markets for services and products that serve the customer's additional needs.
- Manufacturing: Logistic issues can be solved where the product is ready. The manufacturing and service companies are associated with business through the operation of last mile solutions.

These components provide important information and skills to the users, while improve the whole delivery processes, efficiency and cost reduction. However machining procedures are very important for the optimized manufacturing processes (Kundrak et al, 2014 & 2018), but in the case of networking manufacturing companies the logistics get more and more importance, especially from the supply chain solutions point of view. The networking of manufacturing companies and logistics providers led to an increased complexity of supply chain solutions, especially in first mile and last mile logistics.

Table 3. Component of last mile typology and their benefits(* less important, *** necessary)

Impacts	Semi-extended supply chain	Fully extended supply chain	Decoupled supply chain	Centralized extended supply chain
Security	**	**	**	***
Inventory	***	***	***	***
Networks	**	***	**	**
SCM	***	***	***	***
Energy	**	***	**	***
Manufacturing	*	*	*	*
Delivery speed	**	**	***	***
Customer satisfaction	**	**	**	***
City logistics	**	**	***	***
Mobility	***	***	***	***

4. Conclusions and further research directions

The solutions of last mile logistics make it possible to make decisions in supply chain networks aiming financial and environmental effects, and also reduce unnecessary movements of delivery. As the relevant state-of-art showed, significant financial savings are available for participants using the concepts of last mile. This result indicates the scientific potential of this research field including the problems of last mile logistic systems. The articles are addressed to the Manufacturing and service participants to identify the logistic aspects from design and operation point of view of supply chain. Therefore, the design aspects of supply chain processes still need more attention and research. In further studies the design of first mile and last logistics need more attention to support the modelling and optimization with heuristic based algorithms.

Acknowledgements

The described article was carried out as part of the EFOP-3.6.1-16-2016-00011 "Younger and Renewing University – Innovative Knowledge City – institutional development of the University of Miskolc aiming at intelligent specialisation" project implemented in the framework of the Szechenyi 2020 program. The realization of this project is supported by the European Union, co-financed by the European Social Fund.

References

[1] Marujo L G, Goes G V, D'Agosto M A, Ferreira A F, Winkenbach M, Bandeira R A M 2018 Assessing the sustainability of mobile depots: The case of urban freight distribution in Rio de Janeire *Transp. Res. Part D Transp. Environ.* **62** pp 256-267

[2] Rabta B, Wankmüller C, Reiner G 2018 A drone fleet model for last-mile distribution in disaster relief operations *Int. J. Disaster Risk Reduct.* **28** pp 107-112

[3] Castillo V E , Bell J E, Rose W J, Rodrigues A M 2018 Crowdsourcing Last Mile Delivery: Strategic Implications and Future Research Directions *J. Bus. Logist.* **39** (1) pp 7-25

[4] Farahani N Z, Noble J S, Klein C M, Enayati M 2018 A decision support tool for energy efficient synchromodal supply chains *J. Clean. Prod.***186** pp 682-702

[5] Yi J, You H 2018 Delivery of Chinese mitten crab during peak season: Logistics dilemma for micro-e-businesses *Int. J. Logist. Syst. Manage.* **29** (3) pp 399-408

[6] Yuen K F, Wang X, Ng L T W, Wong Y D 2018 An investigation of customers' intention to use self-collection services for last-mile delivery *Transp. Policy* **66** pp 1-8

[7] Chen Y, Yu J, Yang S, Wei J 2018 Consumer's intention to use self-service parcel delivery service in online retailing: An empirical study *Internet Res.* **28** (2) pp 500-519

[8] Leung K H, Choy K L, Siu P K Y, Ho G T S, Lam H Y, Lee C K M 2018 A B2C e-commerce intelligent system for re-engineering the e-order fulfilment process *Expert. Sys. Appl.* **91** pp 386-401

[9] de Freitas, J C, Penna P H V 2018 A Randomized Variable Neighborhood Descent Heuristic to Solve the Flying Sidekick Traveling Salesman Problem *Electr. Notes in Discr. Math.* **66** pp 95-102

[10] Ha Q M, Deville Y, Pham Q D, Hà M H 2018 On the min-cost Traveling Salesman Problem with Drone *Transp. Res. Part C Emerg. Technol.* **86** pp 597-621

[11] Niroomand I, Nsakanda A L 2018 Improving collection flows in a public postal network with contractor's obligation considerations *Int. J. Prod. Econ.* **198** pp 79-92

[12] Kin B, Spoor J, Verlinde S, Macharis C, Van Woensel T 2018 Modelling alternative distribution set-ups for fragmented last mile transport: Towards more efficient a sustainable urban freight transport *Case Stud. Transp. Policy* **6** (1) pp 125-132

[13] Lance L Putong, Marlene M De Leon 2018 A Modified Balcik Last Mile Distribution Model for Relief Operations Using Open Road Networks *Proc. Eng.* volume **212** (American Elsevier) pp 133-140

[14] Elçi Ö, Noyan N, Bülbül K 2018 Chance-constrained stochastic programming under variable

reliability levels with an application to humanitarian relief network design *Comp. Oper. Res.* **96** pp 91-107

[15] Zhou L, Baldacci R, Vigo D, Wang X 2018 A Multi-Depot Two-Echelon Vehicle Routing Problem with Delivery Options Arising in the Last Mile Distribution *Eur. J. Oper. Res.* **265** (2) pp 765-778

[16] Fikar C, Hirsch P, Gronalt M 2018 A decision support system to investigate dynamic last-mile distribution facilitating cargo-bikes *Int. J. Logist. Res. Applic.* **21** (3) pp 300-317

[17] Simoni M D, Claudel C G 2018 A simulation framework for modeling urban freight operations impacts on traffic networks *Simul. Model. Pract. Theory* **86** pp 36-54

[18] Kin B, Ambra T, Verlinde S, Macharis C 2018 Tackling fragmented last mile deliveries to nanostores by utilizing spare transportation capacity-A simulation study *Sustainability* (Switzerland) **10** (3) art. no. 653

[19] Fikar C 2018 A decision support system to investigate food losses in e-grocery deliveries *Comput. Ind. Eng.* **117** pp 282-290

[20] Ferrer J M, Martín-Campo F J, Ortuño M T, Pedraza-Martínez A J, Tirado G, Vitoriano B 2018 Multi-criteria optimization for last mile distribution of disaster relief aid: Test cases and applications *Eur. J. Oper. Res.* **269** (2) pp 501-515

[21] Behrend M, Meisel F 2018 The integration of item-sharing and crowdshipping: Can collaborative consumption be pushed by delivering through the crowd? *Transp. Res. Part B Methodol.* **111** pp 227-243

[22] Alishahi M, Yaghmaee Moghaddam M H, Pourreza H R 2018 Multi-class routing protocol using virtualization and SDN-enabled architecture for smart grid Peer-to-Peer *Netw. Appl.* **11** (3) pp 380-396

[23] Bellavista P, Giannelli C, Lagkas T, Sarigiannidis P 2018 Quality management of surveillance multimedia streams via federated SDN controllers in Fiwi-iot integrated deployment environments *IEEE Access* **6** pp 21324-341

[24] Al-Eryani Y F, Salhab A M, Zummo S A, Alouini M S 2018 Two-way multiuser mixed RF/FSO relaying: Performance analysis and power allocation *J.of Opt. Comm. and Netw.* **10** (4) pp 396-408

[25] Horibe M 2018 Measurement Uncertainty in Terahertz VNAs: Using Terahertz Vector Network Analyzers for Stable, Accurate Measurement and to Evaluate Uncertainty *IEEE Microwave Mag.* **19** (2) art. no. 8283916 pp 24-34

[26] Leinonen M E, Destino G, Kursu O, Sonkki M, Pärssinen A 2018 28 GHz Wireless Backhaul Transceiver Characterization and Radio Link Budget *ETRI J.* **40** (1) pp 89-100

[27] Noyan N, Kahvecioğlu G 2018 Stochastic last mile relief network design with resource reallocation *OR Spectrum* **40** (1) pp 187-231

[28] Wollenburg J, Hübner A, Kuhn H, Trautrims A 2018 From bricks-and-mortar to bricks-and-clicks: Logistics networks in omni-channel grocery retailing *Int. J. Phys. Distrib. Logist. Manage.* **48** (4) pp 415-438

[29] Sudha P, Rengarajan A 2018 *A comparative analysis of scheduling algorithms for WiMAX networks J. Adv. Res. Dyn. Control. Syst.* **10** pp 19-25

[30] Morganti E, Browne M 2018 Technical and operational obstacles to the adoption of electric vans in France and the UK: An operator perspective *Transp. Policy* **63** pp 90-97

[31] Deutsch Y, Golany B 2018 A parcel locker network as a solution to the logistics last mile problem *Int. J. Prod. Res.* **56** (1-2) pp 251-261

[32] Chen Y, Wang H 2018 Pricing for a Last-Mile Transportation System *Transp. Res. Part B Methodol.* **107** pp 57-69

[33] Fillone A M, Mateo-Babiano I 2018 Do i walk or ride the rickshaw? Examining the factors affecting first- and last-mile trip options in the historic district of Manila (philippines) *J. Transp. Land Use* **11** (1) pp 237-254

[34] Ranieri L, Digiesi S, Silvestri B, Roccotelli M 2018 A review of last mile logistics innovations

in an externalities cost reduction vision *Sustainability* (Switzerland) **10** (3) art. no. 782
[35] Kundrak J, Deszpoth I, Molnár V 2014 Comparative study of material removal in hard machining of bore holes *Tehnicki Vjesnik* **21**(1) 183-89
[36] Kundrak J, Molnar V, Deszpoth I 2018 *Comparative Analysis of Machining Procedures Machines* **6**(2) 10

Conformal cooling with heat-conducting inserts by direct metal laser sintering

I Hatos, B Kocsis, H Hargitai

Department of Materials Science and Technology, Széchenyi István University, Győr, Egyetem square 1, Hungary

hatos@sze.hu

Abstract. With the development of layer manufacturing technologies injection mold inserts with conformal cooling channels can be manufactured. If the cooling channels can be placed along the geometry, the heat removal is uniform and effective. In tight mold regions, formation of cooling channel is not possible or not efficient. The combination of conformal cooling and heat conductive insert can be an ideal solution for the effective cooling.

1. Introduction

Injection molding is one of the most important polymer processing technologies. The cooling time of the part is a significant phase of the injection molding cycle, which can amount to more than half the whole cycle. One of the best ways to achieve a reduction in cooling time is to use mold inserts with conformal cooling. The conformal cooling channels follows the geometry of the product, which cannot be manufactured with conventional methods, such as drilling or milling. As a result, it can extract more heat and the temperature distribution is more uniform. With additive manufacturing technologies, parts with complex internal structures can be built, which can be a major advantages for injection molds. The cooling channels can have complex form, which cannot be manufactured with conventional methods, such as drilling or milling. Injection mold inserts with conformal cooling can be manufactured by additive manufacturing [1].

Selective Laser Sintering (SLS) is an additive manufacturing process that allows the building of complex parts through the solidification of multiple layers of powders. A high power laser provides the thermal energy required for the powder sintering. The metal based SLS process is called Direct Metal Laser Sintering (DMLS) [2].

In case of injection molds from steal alloys, the tool with conformal cooling channels can extract more heat than with conventional cooling, but in many cases a tool from good thermal conductive copper based alloy can extract more heat with conventional cooling compared to steel tool with conformal cooling [3].

In some cases, formation of conformal cooling channel is not possible for geometric reason, mainly in tight mold regions. These hot spots have high impact on the part quality and cycle time. Tool steels have low heat conductive capacity compared to poor aluminium, copper or silver [4]. The combination of conformal cooling channels and fitted heat conductive inserts can serve an ideal cooling solution for tools with tight mold regions [5].

IOP Conf. Series: Materials Science and Engineering **448** (2018) 012027 doi:10.1088/1757-899X/448/1/012027

In our research work an injection moulding tool insert with conformal cooling channels was designed by modifying that of having conventional cooling system (finger). As reference the model of the original tool was used (Figure 1.) and to compare the efficiency of cooling, thermal simulations were carried out. Fitting of the new tool insert to the existing cooling circuit and the linear design of the cooling channels were important aspects of the design. The original tool insert was made of 1.2343 steel.

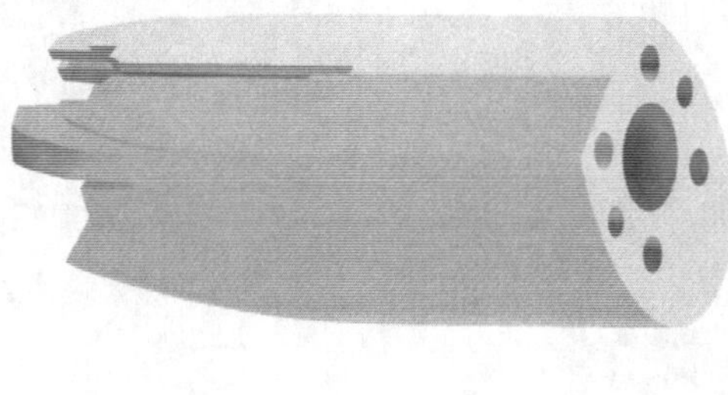

Figure 1. 3D model of the original insert.

2. Design and simulation

2.1. Conventional tool insert (original)
For the injection molding simulations Autodesk Moldflow (2014) software was used. The parameters used for the simulation derived from the production. The results of the simulation in the case of original tool were checked by thermal imaging. The thermographic image of heat distribution confirmed the accuracy of the simulation model. Based on the results of both simulation and thermographic test in terms of cooling the protruding surfaces were identified as the critical areas of the original tool (Figure 2.).

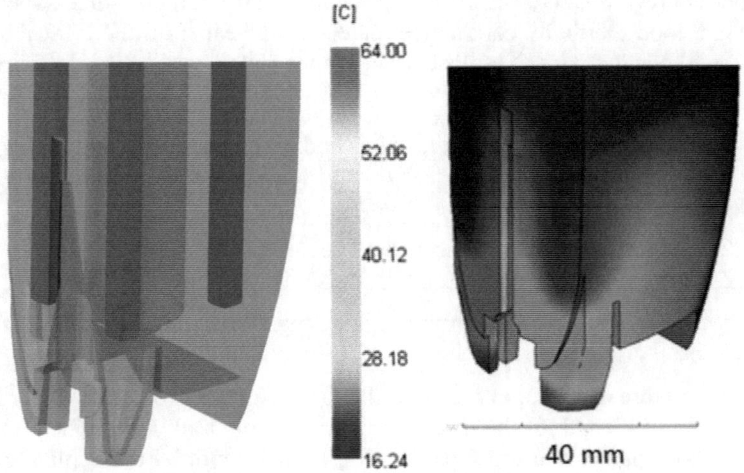

Figure 2. The 3D model and the temperature distribution of the original mold.

2.2. Tool insert with conformal cooling
Using the original model, a version with conformal cooling was designed, fitted to the previous system with Ø4 mm holes. As the minimum allowable distance between the cooling circuit and tool wall, 3 mm was determined. Direct cooling of the protrusions was discarded, even using small (Ø1.5 mm) cooling hole can serve only a partial solution, but this application can cause the clogging of the thin

channel. The own designed mold with the conformal cooling channels and the result of the thermal simulation can be seen on (Figure 3.).

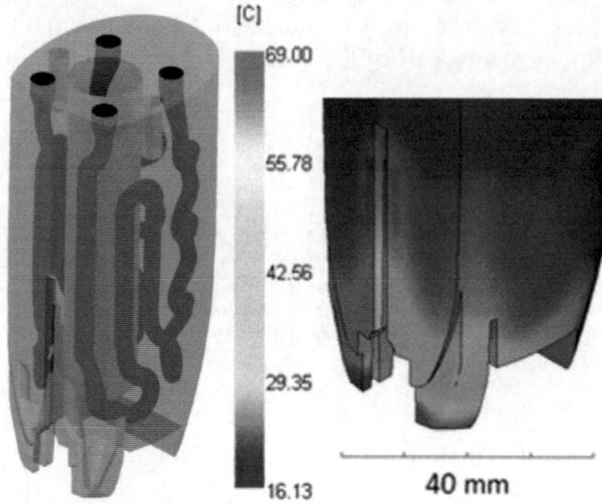

Figure 3. The mold with conformal cooling and the temperature distribution.

Comparing the simulation results of the original and conformal cooling systems it can be concluded, that by following the same initial conditions better cooling efficiency cannot be achieved by conformal cooling.

2.3. Hybrid tooling

Since we could not achieve positive result with the conformal cooling itself, a special solution was designed. By using a good thermally conductive material the heat from the critical areas was lead to the cold parts. Thermal conductivity of tool steels and other metals with high coefficient can be seen in Table 1.

Table 1. Thermal conductivity of mold steels and high conducting materials [6][7][8].

Material	Thermal conductivity at 20 °C (W/mK)
MaragingSteel MS1 (age hardened)	20
1.2343	25.3
Copper	398
Silver	418

Based on the temperature distribution image of the tool insert the formation of conformal cooling is not necessary along the full length of the insert and therefore the tool was divided into a conventional produced part and a built part. From those parts, which cannot cool by the cooling channels, the heat was removed by thermocouples. For this purpose two pieces of Ø2 mm pin and a 2 mm thick sheet were used (Figure 4.). In terms of thermal conductivity, pure copper or silver inserts are the most efficient. The application of pure silver was rejected, because of its softness and thus the low strength the tight fitting of inserts is difficult to implement.

a)

b)

Figure 4. Copper inserts (a) and the computed tomography image of the built tool with the inserts (b).

Thermal distribution of the tool containing copper inserts is shown in Figure 5. It is important to note that in the simulation there was no air gap between the inserts and the mold. To achieve the simulation results in practise a production process solution is needed to apply in that the copper inserts are continuously connected to the wall of the tool and contact with the coolant.

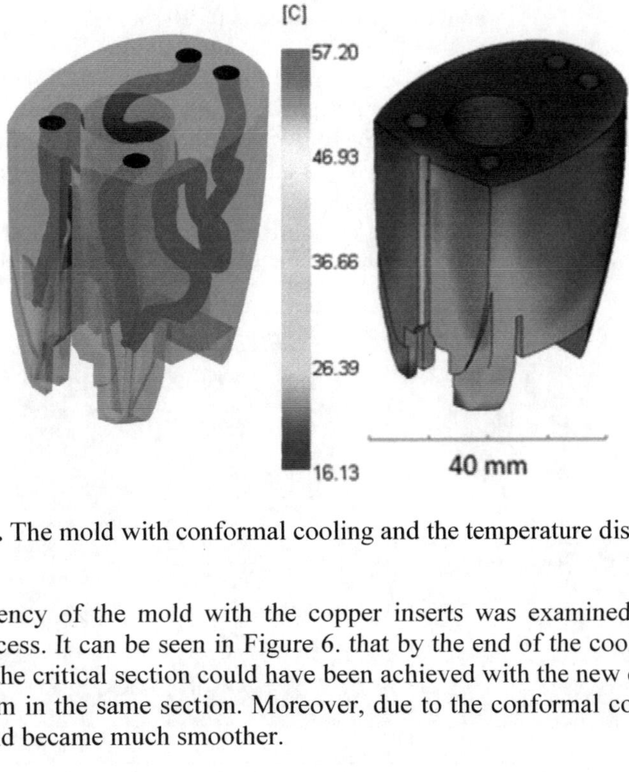

Figure 5. The mold with conformal cooling and the temperature distribution.

The cooling efficiency of the mold with the copper inserts was examined during the complete injection molding process. It can be seen in Figure 6. that by the end of the cooling time (19 s) ~9 °C lower temperature of the critical section could have been achieved with the new design than that of the original cooling system in the same section. Moreover, due to the conformal cooling the temperature distribution of the mold became much smoother.

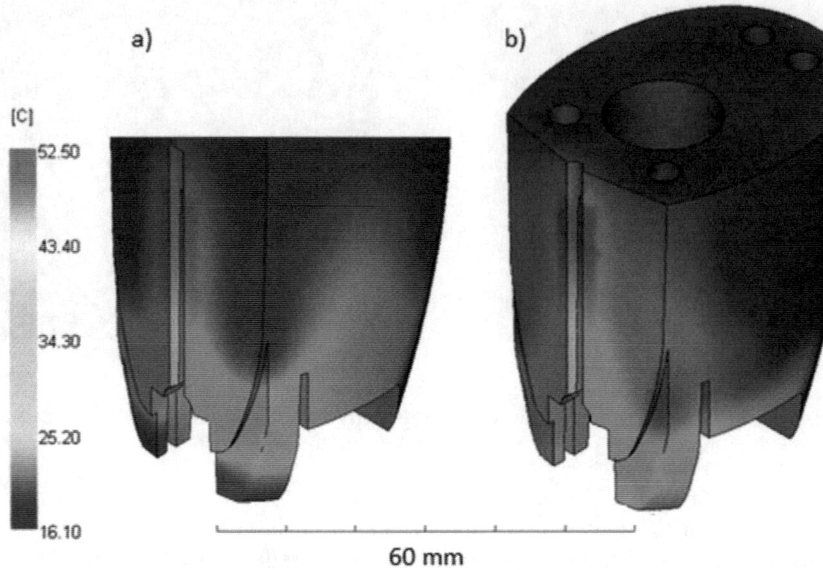

Figure 6. The temperature distribution of the original (conventionally produced) (a) and the DMLS produced (b) tool inserts (t=19 s).

3. Manufacturing of hybrid tool

The production of the tool insert with the new design was carried out by Direct Metal Laser Sintering, which is a multi-step process. As a first step, the tool is printed with the insert height and then stopped the process (Figure 7a). Then metal powder is removed from the hole and the insert is placed into that (Figure 7b). It is important that the top of the insert must be in the same plane with the last layer of the built part (Figure 7c) and then continue to build (Figure 7 d). To ensure the best filling with the insert, after reversing the tool it can be melted in the hole with a brazing process (Figure 7 e).

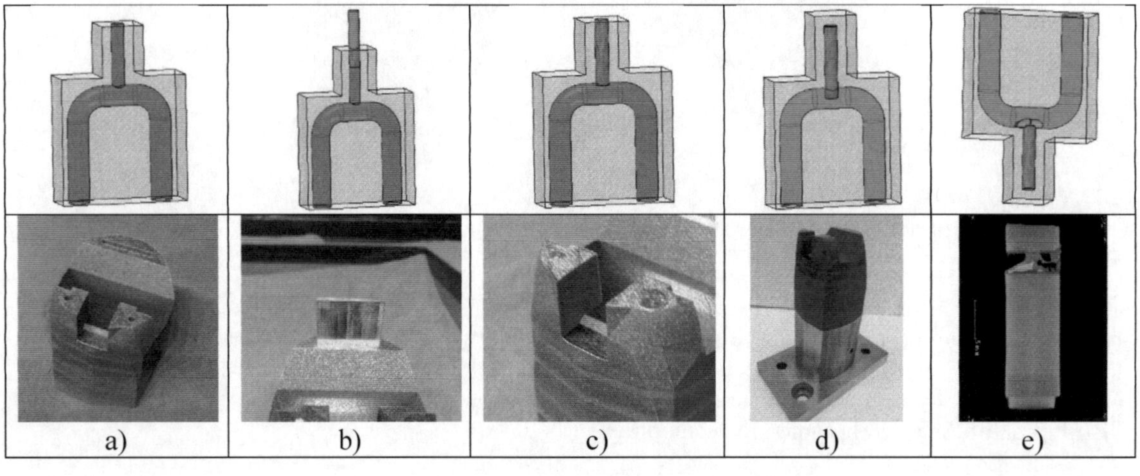

Figure 7. Manufacturing process of hybrid tool (a-d) and a computed tomography image in one section (e).

The connection of the tool and the copper insert were examined by computed tomography and cross sectional optical microscopic images. After the powder deposition the upper part of the copper insert is scanned by laser beam, so interesting to observe the interface of the two materials. In our experiments EOS M270 DMLS equipment was used with parameters applied for MS1 type maraging steel sintering. No porosity was detected in the region of steel-copper connection (interface area) by CT investigation (Figure 8.) and the same results were found by optical microscopic observation in the cross section (Figure 9.). Despite the tight fit air gap remained between the copper insert and the wall of the tool without heat treatment (Figure 8.).

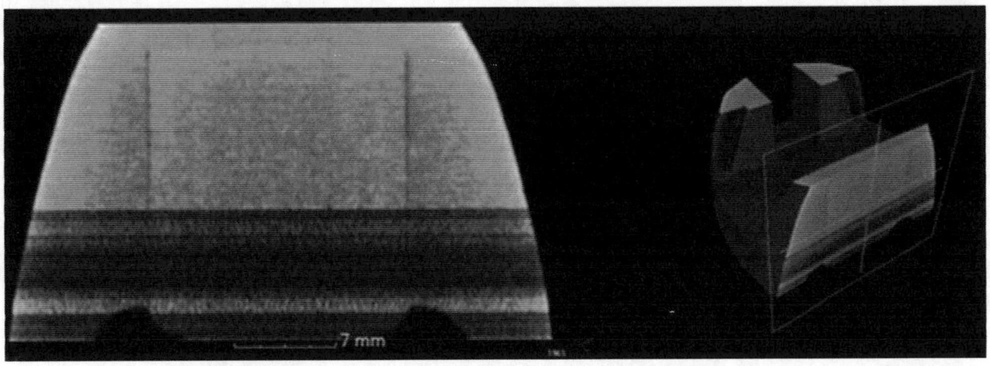

Figure 8. CT image about steel-copper connection.

By heat treatment above the melting point of the copper of the inverted parts the coherent surface air gap decreased. Based on the cross sectional optical micrographs it can be stated that there is a good wetting between the copper and the MS1 sintering steel. The molten copper is filling in the pieces of the surface roughness and micro-cavities formed in a continuous connection between the two metals (Figure 9.)

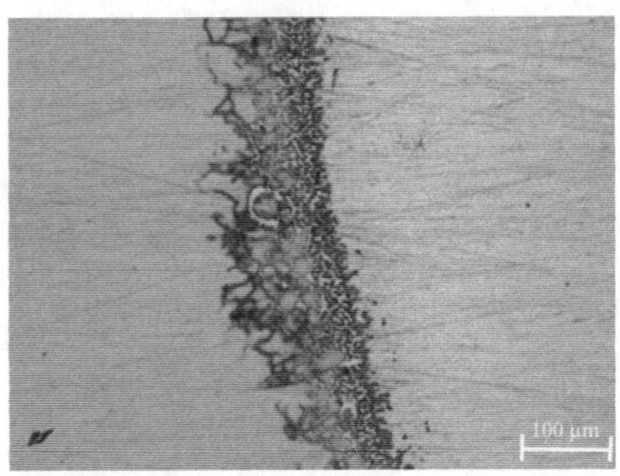

Figure 9. Contact on the side walls.

4. Conclusions

Direct metal laser sintering has growing interest in tool manufacturing due to its several advantages over conventional technologies. One of the main application filed of this technology is polymer injection molding. Providing even cooling is a challenge for certain products. Design and manufacturing of tool inserts with conformal cooling channels seems to be an effective solution for this problem.

In our paper a tool insert having conventional cooling system (finger) was redesigned with conformal cooling channels and thermal simulations were carried out. Based on the test results in terms of cooling the protruding surfaces were identified as the critical areas of the original tool.

Because of using conformal cooling system could not be achieved better cooling efficiency, a special solution was designed. Copper inserts were used to lead the heat from the critical areas. A hybrid tool insert were designed composed of a conventionally produced part and a built part with conformal cooling channels and copper inserts. The cooling efficiency of the mold was better, by the end of the cooling time lower temperature of the critical section and much smoother temperature distribution could have been achieved.

In the other part of our research work hybrid tool with the copper inserts were produced by DMLS and the connection of the tool and the insert were examined by computed tomography and optical microscopy. The cross sectional micrographs showed good wetting between the copper and the MS1 sintering steel.

Based on the experiments the new design of cooling system with combination of conformal cooling and heat-conducting pins and by combination of conventional manufacturing and additive technology seems to be an effective and recoverable investment.

Acknowledgments
The authors would like to thank to EFOP-3.6.1-16-2016-00017 'Internationalisation, initiatives to establish a new source of researchers and graduates, and development of knowledge and technological transfer as instruments of intelligent specialisations at Széchenyi István University' for the support of the research.

References

[1] Zink B and Kovács J G 2017 The effect of limescale on heat transfer in injection molding Heat and Mass Transfer **86** 101-107
[2] Lee H, Lim C H J, Low M J, Tham N, Murukesham V M, Kim Y-J 2017 Lasers in additive manufacturing: A review International Journal of Precision Engineering and Manufacturing-green Technology **4/3** 307-322
[3] Zink B, Szabó F, Hatos I, Suplicz A, Kovács N K, Hargitai H, Tábi T, Kovács J G 2017 Enhanced injection molding simulation of advanced injection molds Polymers **9(2)** 77
[4] Westhoff R 2006 Thermal balance: conformal cooling on the advance Kunststoffe International **8** 24-26
[5] Merklein M, Junker D, Schaub A, Neubauer F 2016 Hybrid additive manufacturing technologies – an analysis regarding potentials and applications Physics Procedia **83** 549-559
[6] Material data sheet: EOS MaragingSteel MS1, http://www.eos.info/ (28 May 2018)
[7] Dörrenberg: 1.2343, http://www.doerrenberg.de/ (28 May 2016)
[8] http://www.morganbrazealloys.com/downloads-3/mechanical-physical-properties (28 May 2016)

Examination of steel surfaces treated by different lasers

J Hlinka[1] and Z Weltsch[1,2]

[1] Budapest University of Technology and Economics, Department of Automotive Technologies, Budapest, Hungary
[2] John von Neumann University, Department of Materials, Kecskemét, Hungary

E-mail: jozsef.hlinka@gjt.bme.hu

Abstract. The aim of this paper is to discuss the effect of laser surface treatment with different lasers on the wettability between DC01 steel and SAC305 lead-free solder paste. The field of laser technology is in a constant state of further development. The possible applications of lasers, such as in industrial manufacturing processes and tools for material processing through welding, cutting or engraving, are plentiful. A specific laser treatment can modify the surface energy of steel and affect the wetting properties. In this work, the surfaces of steel plates were treated by Nd:YAG and CO_2 laser with different laser powers. After the laser surface treatment wetting experiments were performed between the steel plates and lead-free solder paste. The effect of laser treatment on steel surface microstructure was analyzed by optical microscopy. The wettability of solder pastes were measured with different opening time. The experimental results showed that in case of Nd:YAG laser, the wetting contact angles do not change up to 700 W laser power. Moreover, increasing CO_2 laser power results increase of the wetting angles.

1. Introduction

Soldering is the most popular joint technique in microelectronic and optoelectronic industries which technologies are widely used in the vehicle industry [1, 2]. Soldering is the process of joining two metals by the use of a solder alloy, and it is one of the oldest known joining techniques. In the soldering process, the molten solder wets the metal and reacts with a small amount of the base metal by forming intermetallic compounds. Therefore, the wetting property between solders and substrates is crucial to the reliability of the soldering joints [3-5].

During the past years, a number of researches have clarified the effects of the laser irradiation on different material surfaces, such as metals, semiconductors, polymers and multilayer materials. Previous works found that the laser surface treatments of copper lead to wettability changes [6]. Lawrence et al. [7] found that the wettability of carbon steel, alumina/silica-based oxide could be improved by high power diode laser irradiation. The surface roughness and properties of the carbon steels treated with high power diode laser irradiation changed the surface roughness became higher and partly oxidized. The changes of the wettability was attributed to three aspects: surface oxygen content; surface roughness; surface microstructure. They also found that the phase change and the increase in crystal size could lead to the changes of surface energy.

This work experimentally determines the wetting property changes between SAC solder and laser treated DC01 steel and microstructure changes in DC01 steel following the Nd:YAG or the CO_2 laser treatment.

IOP Conf. Series: Materials Science and Engineering **448** (2018) 012028 doi:10.1088/1757-899X/448/1/012028

2. Materials and methods

The investigations were carried out on DC01 steel plates. Table 1. shows the chemical composition of the used steel. The components were measured by GDA Alpha spectroscope. This type of steel is widely used for cold forming and other applications. The plates were sectioned into 15 x 15 mm squares. Before the laser treatments the steel plates were grinded and polished to remove the oxide layer and reach similar surface properties in every sample and washed with 96 % ethanol (C_2H_6O). The used solder paste was M705-GRN360-KV. It is a SAC305 solder paste with 96.5 w% Sn 3.0 w% Ag 0.5 w% Sn content.

Table 1. Chemical composition of the used DC01 steel

Fe [%]	C [%]	Mn [%]	Si [%]	P [%]	S [%]	Cr [%]	Ni [%]	Cu [%]	Al [%]	Ti [%]	Nb 2 [%]	Co [%]	Zr [%]
99.5217	0.0543	0.2630	0.0013	0.0105	0.0250	0.0257	0.0063	0.0247	0.0257	0.0003	0.0143	0.0027	0.024

Two types of lasers were used in this study a CO_2 and a Nd:YAG laser. The used CO_2 laser is an Oerlikon precision laser (LE2000T000AA) with emission of 10600 nm wavelength laser beam (Figure 1.). The laser used in continuous wave mode, its power density distribution is TEM_{00}. The movement of the X−Y table are controlled by CNC-program. CNC control unit accuracy is 0.005 mm. During the treatment the laser spot size was 0.4 mm, the beam offset was 0.3 mm and the speed of the table was 2000 mm/min. Argon gas was used with the flow-rate of 10 l/min. The treatments were made with 50 W, 100 W, 150 W and 200 W laser power.

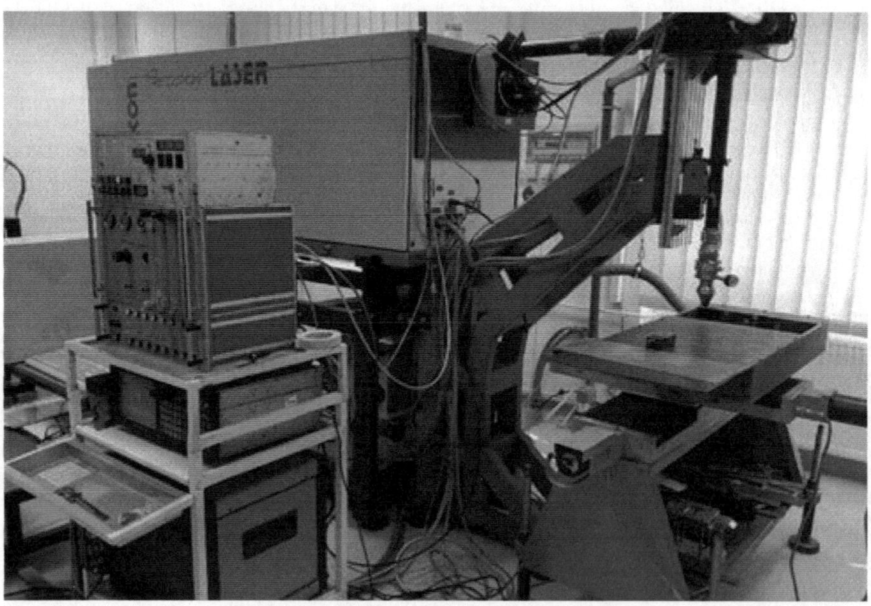

Figure 1. Oerlikon precision CO_2 laser.

The other laser used in the study was Rofin DY 027, Nd:YAG laser with emission of 1064 nm wavelength laser beam (Figure 2.). The laser beam (TEM_{00}) was focused directly onto a 0.4 mm diameter spot on the surface of steel plate. During the laser surface treating process, the laser beam was traversed across the steel samples by means of a laser scanner at speed of 500 mm/s. The laser scanning speed must be controlled properly to ensure that a 40 x 40 mm square (4 samples) area can be treated by laser. Different laser powers from 100 W to 1000 W, were used in this process. The laser was used in continuous wave. Argon gas was used as shielding gas to protect the samples from oxidation during laser irradiation with the flow-rate of 12.5 l/min.

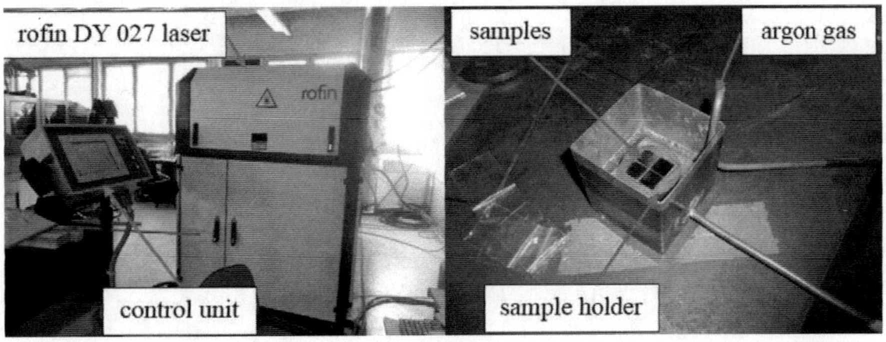

Figure 2. Rofin DY 027, Nd:YAG pulsed laser and working space with the samples.

The wetting experiments were performed in a wetting angle measuring system (Figure 3.) which can measure the wettability using sessile drop method [8], taking the methodological issues summarized by Sobczak [9] into account.

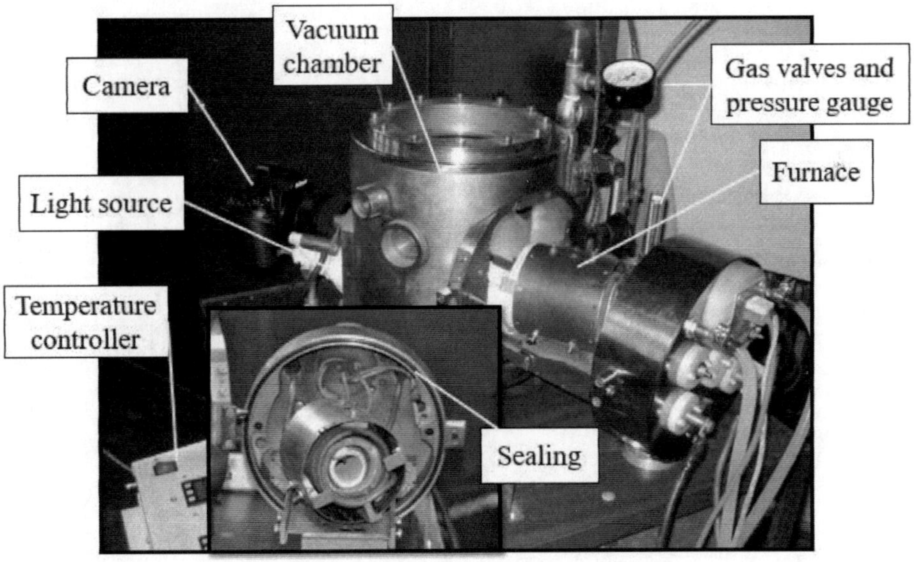

Figure 3. Wetting angle measuring equipment.

The measurements can be performed in air or inert atmosphere. The heat required for the measurements provided by a resistance-heated tube furnace which can ensure the measuring temperature up to 1473 K. In this paper the measuring temperature was not high, the melting temperature of investigated solder are below 523 K.

At the beginning of the wetting angle measurements the samples were positioned into the furnace at ambient conditions. The measurements were performed in air atmosphere. Thereafter the temperature was raised up to the measuring temperature. The contact angle values of the drop were measured directly from its images. The total uncertainty of the measured values was ±0.75°.

IOP Conf. Series: Materials Science and Engineering **448** (2018) 012028 doi:10.1088/1757-899X/448/1/012028

In order to analyze the influence of the laser treatment on microstructure, the specimens, before and after laser treatment, were sectioned with a cutting machine using a diamond-rimmed cutting blade. The sectioned specimens were mounted with methyl-methacrylate casting resin then, subsequently, grinded with SiC grinding paper and polished with diamond suspension down to 1 μm. The steel surface were examined on the cross sectioned samples by optical microscope.

3. Results and discussions

The solder paste is a mixture, which contains solder alloy powder and flux. The storage time and storage conditions have effect on the usability of the solder. The flux contains some volatile components and if these components evaporate the paste becomes worse and for example not suitable for stencil printing.

3.1. Effect of solder paste age

To get some information about the effect of holding time on the wettability between the SAC305 solder [10] and DC01 steel three different pastes were examined. The three paste conditions are fresh solder paste, old solder paste (older than the double shelf life time) and old solder paste diluted with another flux (to balance the drying of the paste). The wettability measurements were made on polished and untreated DC01 substrates. Figure 4. shows the results of 9 different measurements with each solder paste.

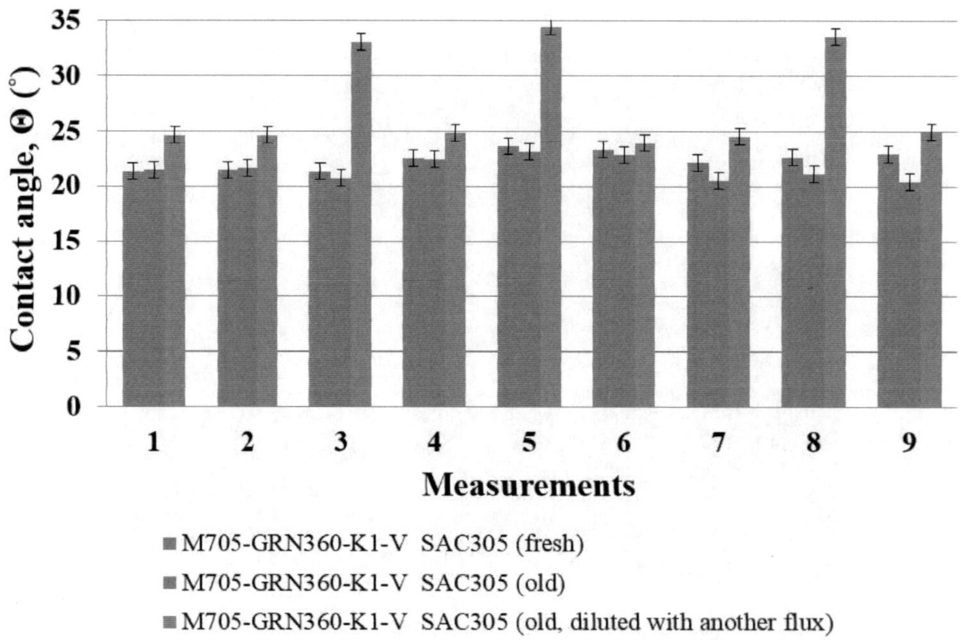

Figure 4. Measured contact angles of 3 different SAC305 solder paste and DC01 steel.

The results show that the fresh solder paste and the old solder paste do not show differences between the wettability. The contact angles are higher, when the old solder paste used which diluted with another flux. Based on the results we can say that in our measurement system the age of the measured solder paste are not affect the wettability, if it is fresher than the double shelf life of the paste.

3.2. Effect of laser treatment with CO_2 laser

The surface treatment experiments have been performed with CO_2 laser. The parameters of the treatments were described above. The treatments were made with 50 W, 100 W, 150 W and 200 W laser power. When the laser power was 200 W the edges of the samples were discolored, burned. It could be the effect of that the grinding and polishing was not the best near to the edges of the samples and the laser absorption was bigger in this points. But this has no effect on the whole treated sample because the wetting measurements were performed in the center of the square shaped steel samples. All of the samples have Ra 0.05 μm surface roughness after the laser treatment. Figure 7. presents a visual inspection on the wettability characteristics of Sn Ag Cu solder on the DC01 steel substrate after CO_2 laser surface irradiation.

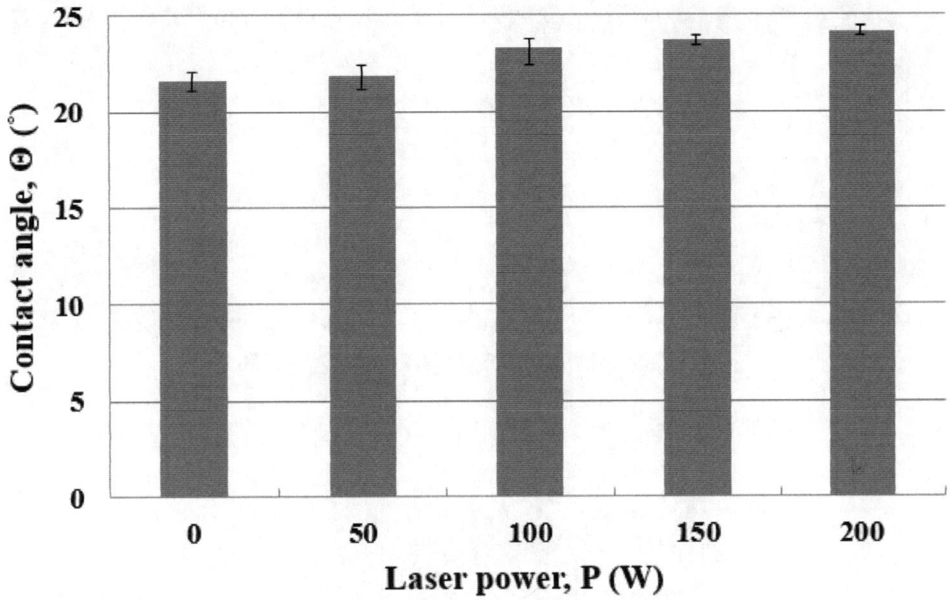

Figure 5. Measured contact angles of CO_2 laser treated DC01 steel and SAC305 solder paste.

The graph shows that the contact angles slightly increase when the laser power increase. Compared to the measurements when the samples were treated with Nd:YAG laser the difference shows a more gradual changing wetting angles. The surface layer microstructures of the optical metallographic samples of the CO_2 laser treated steel plates are shown in Figure 8. As may be seen by comparing Figure 8(a, b, c, d, e) the laser treatment did not cause noticeable change on the steel surface layer. The microstructures are look like a cold rolled steel structure.

IOP Conf. Series: Materials Science and Engineering **448** (2018) 012028 doi:10.1088/1757-899X/448/1/012028

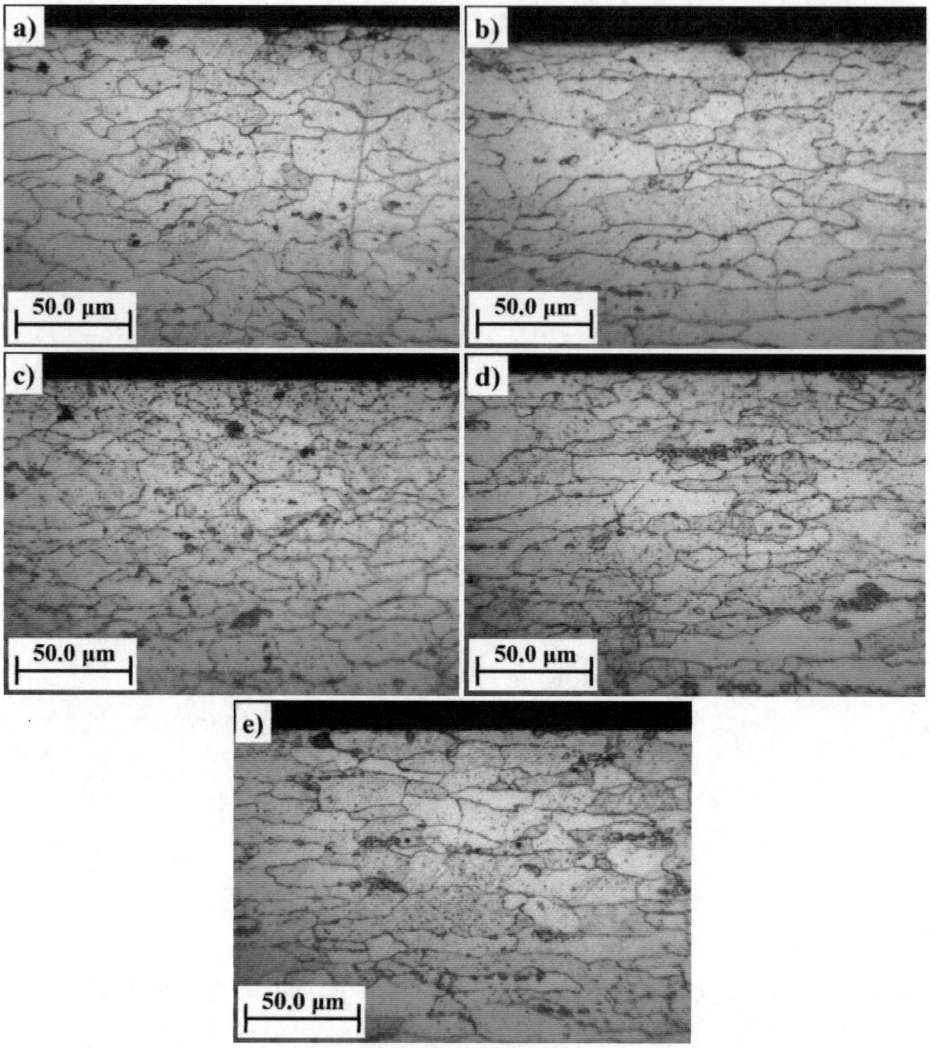

Figure 6. Optical microphotographs of the surface layer microstructures of DC01 steel: a) without laser treatment, after laser irradiation with b) 50 W, c) 100 W, d) 150 W, e) 200 W laser power (CO_2).

3.3. Effect of laser treatment with Nd:YAG laser

In this study there is a target that the laser treatment does not leave an optically visible trace on the surface. Taking this objective into account, the Nd:YAG laser treatment with the parameters described above has the upper limit at 1000 W. In this power level, a little discoloration begin. Figure 5. presents a visual inspection on the wettability characteristics of Sn-Ag-Cu solder on the DC01 steel substrate after Nd:YAG laser surface irradiation. In case of the wettability measurements there were 3 measurements and on Figure 5. the average of the 3 measurements and the differences are shown for each laser power level.

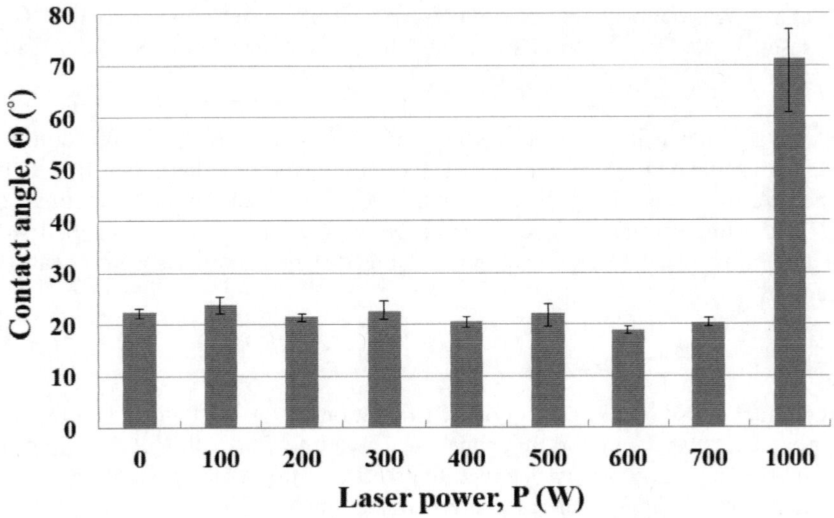

Figure 7. Measured contact angles of Nd:YAG laser treated DC01 steel and SAC305 solder paste [11].

The graph shows that the contact angles do not change definite when the steel plates treated with up to 700 W laser power. If the laser power was 1000 W the contact angles increased. On the surface of the treated samples there is no optically determinable change up to 700 W. Surface roughness measurements verify this statement. There is no difference between the surface roughness of the treated surfaces up to 700 W laser power.

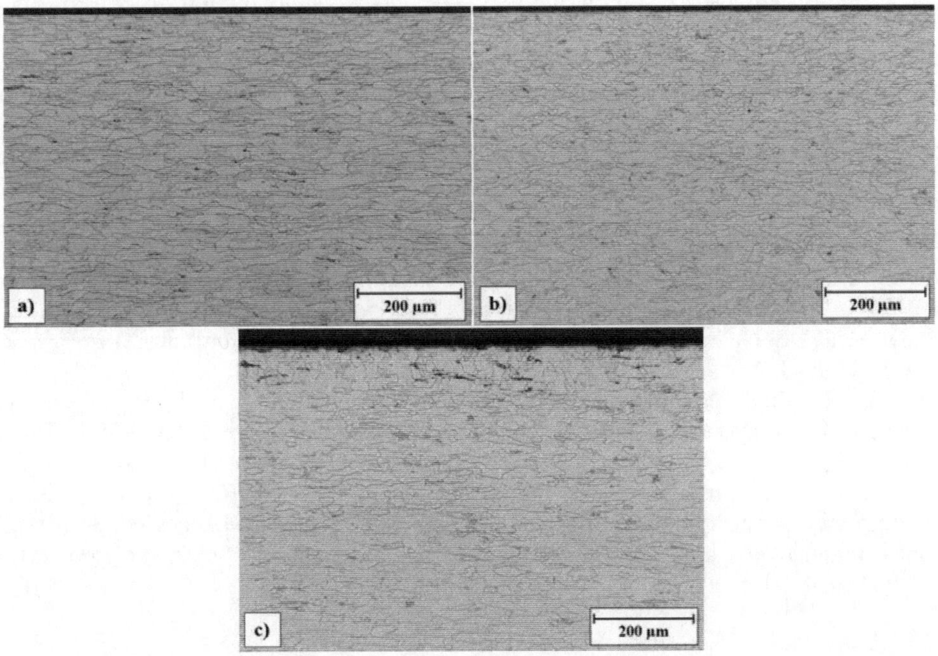

Figure 8. Optical microphotographs of the surface layer microstructures of DC01 steel: a) without laser treatment, b) after laser irradiation with 600 W laser power, c) after laser irradiation with 1000 W laser power (Nd:YAG) [11].

The measured average roughness of the samples up to 700 W laser power was Ra 0.05 μm. In case of the 1000 W laser power treatment the surface of the polished steel was optically visibly changed. The measured surface roughness of this sample was Ra 0.11 μm.

The surface layer microstructures of the optical metallographic samples of the Nd:YAG laser treated steel plates are shown in Figure 6. As may be seen by comparing Figure 6(a, b, c) the laser treatment cause noticeable change on the steel surface layer when the laser power was 1000 W. The observations indicate that when the laser power was 600 W, the heat absorbed by the steel could not lead to change of the microstructure. The microstructures are look like a cold rolled steel structure. The microstructure near to the surface layer was changed, when the laser power was 1000 W. The modified thickness was ~90 μm.

4. Conclusion

In summary, the effect of Nd:YAG and CO_2 laser irradiation on the wettability and the microstructure of DC01 steel plate was investigated in this paper. It was found that with Nd:YAG laser using a laser power of 0 W to 700 W the wettability between the steel and the solder paste do not change, namely there were no significant changes in the measured contact angles. When the laser power was 1000 W the contact angle became higher. The microstructure was influenced by the laser treatment when the laser power was 1000 W. When the samples were treated with CO_2 laser, the wetting angles of the molten solder slightly increase when the laser power increase. The microstructures do not show difference even for the used highest CO_2 laser power level.

References

[1] Tu K N 2007 *Solder Joint Technology: Materials, Properties, and Reliability* (New York: Springer)
[2] Judd M and Brindley K 1999 *Soldering in Electronics Assembly* (Oxford: Newnes)
[3] Yost F G, Hosking F M and Frear D R 1993 *The Mechanics of Solder Alloy Wetting and Spreading* (New York: Van Nostrand Reinhold)
[4] Baumli P and Kaptay G 2008 Wettability of carbon surfaces by pure molten alkali chlorides and their penetration into a porous graphite substrate, *Mater Sci Eng A.* **495** 192-196
[5] Baumli P 2015 Alacsony olvadáspontú ólommentes forraszanyagok, *ANYAGOK VILÁGA*, **1** 24-34
[6] Zhao Z, Wang C, Li M, Wang L and Kong L 2006 The effects of pulsed Nd:YAG laser irradiation on surface energy of copper, *Appl. Sur. Sci.* **252** 4257–63
[7] Lawrence J. and Li L 2001 *Laser Modification of the Wettability Characteristics of Engineering Materials*, (London and Bury St Edmunds: Professional Engineering Publishing Limited)
[8] Abtew M and Selvaduray G 2000 Lead-free Solders in Microelectronics, *Mater. Sci. Eng. R Rep.* **27** 95–141
[9] Sobczak N, Singh M, Asthana R 2005 High-temperature wettability measurements in metal/ceramic systems – some methodological issues, *Curr. Opin. Solid State Mater. Sci.* **9** 241–253
[10] *Technical Bulletin M705-GRN360-K* [Online]. Available: http://www.senju.com/images/pdf/2011-M705-GRN360-K.pdf. [Accessed: 31-Mar-2016]
[11] Hlinka J and Weltsch Z 2017 Investigation of Laser Treated Steel Surface *GRADUS* **4/2** 346-350

Acknowledgments

Supported by the ÚNKP-17-3 New National Excellence Program of the Ministry of Human Capacities.

IOP Conf. Series: Materials Science and Engineering **448** (2018) 012029 doi:10.1088/1757-899X/448/1/012029

DEVELOPING THE SUPPORT FRAMEWORK SYSTEM OF SPECIAL PURPOSE MACHINES DESIGNING

M Boleraczki[1], C Sik-Lanyi[2], Z Zsák[3] and T Papp[4]

[2]University of Pannonia, Veszprem, Department of Electrical Engineering and Information Systems
[1,3]Transmoduls Kft., Veszprem, Research and development department
[4] Transmoduls Kft., Veszprem,Application department

Corresponding author: boleraczki.miklos@transmoduls.hu

Abstract. From the customer side the requirements of the special purpose machines are getting more and more diversified and complex year by year. Could it be possible to make a design support frame-software in this field? An expert system will be shown in this paper, which is based on previously built machines. The machine parts and subassemblies are stored in a database. The possible contacts between these subassemblies are presented in a knowledge base. The purpose of this expert system is to offer a valuable help for the design engineer when a new machine plan is needed. The program output is the bill of material which contains the required subassembly, and even the cumulated price.

1. Introduction

There are many special fields in mechanical design engineering but to design and build a special purpose machine is always a challenge because the customer requirements change during the design and building or even after the delivery. In the beginning the customer has a problem - and sometimes a theoretical solution - to solve. Creativity and individual concepts are the first items of the long list of necessary skills to find fast, cheap and reliable machine designing for them. Becoming an expert mechanical engineer is an expensive and long process. In this paper you can find an idea about how to help mechanical engineer with expert systems and it shows the direction of the future of machine design.

1.1 Concept of the level of automation

If we want to step forward from an actual automation level to the next one, we need to analyze the working process of the actual manufacturing machines. To make this upgrade it is necessary to make a quantity change. There are many concepts in the literature and we embedded ours in them. Our focus is to find feasible solutions for this "quantity jump"in the field of special purpose machines. The required result by the customer is a flexible workstation on supervising level. [1,2] You can see our vision in figure 1.

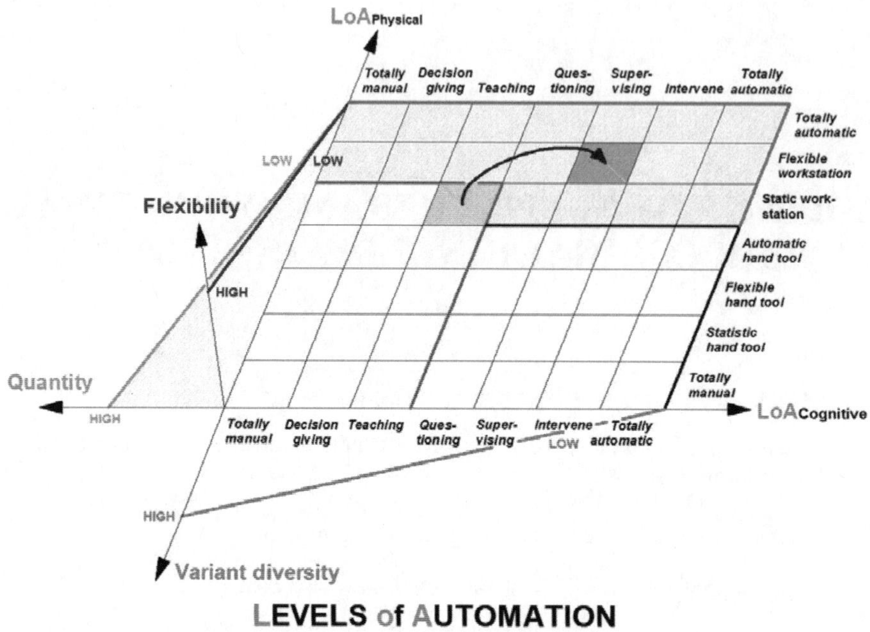

Figure 1. Level of Automation

1.2 Choose the focus of the development

For the first step we must start with a single and well definable group of machine. We have chosen welding machines. There are many products which contain a welding process, so welding machines belong to a lot of product lines. Electronic components like resistors, capacitors, contacts are required to be fixed to the leadframe or each other. Welding is an optimal solution for fixing electrical components with the required quality of the contact. Resistance welding requires less number of equipment than soldering, because no additional material is needed. Depending on the part orientation in the product it is possible to weld only from one side with two electrodes, but sometimes electrodes are needed to contact from the top and bottom as well. Defining the parts of the machines is also important to start the work. This kind of machines contains general elements, such as machine frame, machine table, welding head, fixture for the part, movement for welding head and fixture, safety cover, electrical cabinet, PLC and HMI (human machine interface). Apart from these, there are options also available for the machines like fixture ID, part fixation in the fixture, electrode ID, indexing table, PC control, operator light, automatic reject handling, automatic part ejection, quality control with camera and so on. It could be easy to make a configuration mechanism for this kind of design, but the problem is in our field (special purpose machines) that it is impossible to design every single aspect of the customer requirements in detail. We need a more complex software than a configurator. The software must find the closest "adoptable" subassembly which is available. This will be a valuable help for the machine designer and the design time will be reduced. Our expert system helps the design engineer to use the existed models and modify or change the necessary parts only.

2. Expert systems

The expert system has been in use for a long time. Analyzing data (not big data this time) and using the rules keep this concept simple and useful.

The main idea behind this kind of systems is the following: making a separation between the data and the inference engine. There are the initial data and rules with logical contacts and iteration when these rules can „fire". After this step there are updated data and different sets of rules, and a new „fire" process is coming. When there are no more data or rules, the iteration is over. [3] The remaining data is the solution. There are framework systems for expert system and we call them shells. A typical shell is a text based user interface. However, we have not implemented our system in a shell. We made a graphical interface which can be seen in figure 2.

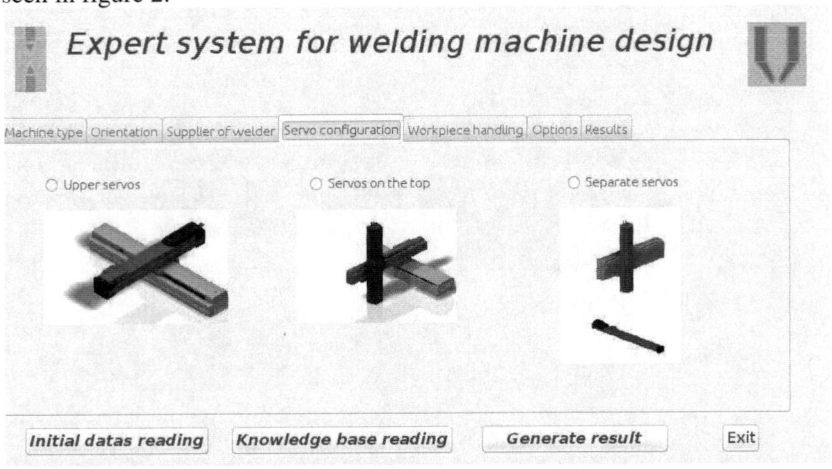

Figure 2. Screen from our developed expert system UI

Expert systems are used in a wide range of different fields. One of them called KBS (Knowledge Base System) is developed to give assistance to engineers who design agricultural machines. It begins with the design rules, standards, formulas and 3D models. Developers divided the knowledge base into two different parts: physical elements and technical experience. To store the data during the implementation, an SQL database system was used. The complete hybrid system uses an object oriented method. [4]

2.1 Modeling structures and data system

How can we make machines from a list of requirements? Actually, this is a long and costly process. If every single machine is individual - and for special purpose it is a must – the design time could be growing at an unexpected rate. Manufacturing the parts is a challenge as well. The engineer's contribution is needed even at the assembling. How can we reduce this well trained engineer's time? One concept is to use algorithms to make the variations of the structures.[5, 6] Another one is that the design process starts from the existing machines. [7] We will follow a hybrid method of these two.

IOP Conf. Series: Materials Science and Engineering **448** (2018) 012029 doi:10.1088/1757-899X/448/1/012029

To begin with the whole machine, it contains many components. These less complex structures include commercial and manufactured parts. We call them subassemblies in our terminology. A data structure for these subassemblies is required to see which part belongs to which assembly. Every single part must be an individual identifier, not just the drawing number, but the database ID, too. The object oriented method is used to store these data. Every object has methods to read the data from the knowledge base and store the unique properties. From the bottom to the top, we start with the individual parts and on the highest level with the main assembly. There are machines which contain a group of subassemblies, while other machines use totally different group of subassemblies. How could these be used to make an expert system to support the engineer? We have found a new concept to keep the flexibility but use the elements which are already designed. Figure 3 shows an overview of our framework system.

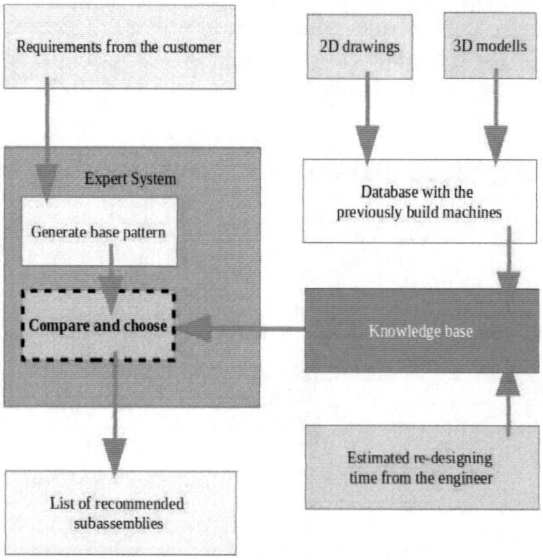

Figure 3. Concept of the framework system

2.2 Fuzzy system, knowledge base

To unlock the limitations we used the following concept. Theoretically, every subassembly could be installed in every imaginable machine. The question is how long it takes to modify the design to put the re-designed subassembly into the new machine ordered by the customer. This design period is typified by a number and defined by the design engineer of the subassembly. Actually, in our concept this number follows fuzzy logic. The fuzzy concept is the following: how much a randomly chosen subassembly belongs to the set of the new machine subassemblies. If the component design does not need to be modified, then the value of the installation is 10. If it takes only a couple of hours, this value is 8 or 9. If the complete design needs to be changed, it is 1. If it is not possible to use that certain subassembly in the current machine, the value is zero. Arranging this data into groups was necessary to organize the subassemblies. In the figure 4 it can be seen a sample of this knowledge base.

Knowledge base for Welding machine designing													
				Structure				Orientation		Welding head supplier			
	Drawing number	Price	Group	XY table	Double XY table	Indexing table	Conveyor system	Welding head top, upper and lower electrodes	Welding head on the top, with pincer electrodes	Soudax	Miyachi	Trumpf	
Machine base for XY	WM1000-101-01	-	**1**	10	5	2	1	10	10	10	10	0	
Machine base for double XY	WM1000-101-02	-	**1**	5	10	2	1	10	10	10	10	0	
Machine base for indexing table	WM1000-101-03	-	**1**	2	2	10	1	10	10	10	10	3	
Machine base for conveyor	WM1000-101-04	-	**1**	1	1	1	10	0	6	8	8	4	
Mounting for top welding head	WM1000-102-01	-	**2**	10	10	7	7	10	8	10	8	0	
Mounting for bottom electrodes	WM1000-102-02	-	**2**	10	10	3	0	10	0	10	8	0	
XY fixture mounting	WM1000-103-01	-	**3**	10	8	2	4	10	10	10	8	8	
Fixture with pneumatic releasing for indexing table machines	WM1000-103-02	-	**3**	0	0	10	0	2	10	2	3	10	

Figure 4. Detail of the knowledge base

To configure out the new machine there are some screens in the program. You must follow the screens and figure out which function is needed by the customer. After this configuration this input data is transformed into a pattern, to follow the fundamental of expert systems. This pattern contains the main components as one can seen in figure 5. It is just visualization behind the concept, but it could be very useful to understand how the system works. Yellow cells represent the required component, green cells are irrelevant in this search scope. For example "0" means that the machine structure is XY table, "1" means that the machine structure is double XY table and so on.

Pattern for the new machine, based on the requirement																											
0	1	2	3	4	5	6	7	8	9	10	11	12	13	14	15	16	17	18	19	20	21	22	23	24	25	26	27

Figure 5. Base pattern for the new machine

In figure 6 a simplified visualization of one subassembly pattern can be seen. Realizing the rules is implemented by comparing this base pattern with the patterns of the components cell by cell.

IOP Conf. Series: Materials Science and Engineering **448** (2018) 012029 doi:10.1088/1757-899X/448/1/012029

Mounting for bottom electrodes																											
0	1	2	3	4	5	6	7	8	9	10	11	12	13	14	15	16	17	18	19	20	21	22	23	24	25	26	27

Figure 6. Pattern for a component

With this realization we have made the expert systems concept understandable in our program. The yellow cells represent that this subassembly is compatible with the machines which consist the 0,1,2,4,6 and 9 properties.

2.3 Choosing methods

To choose a certain object from the knowledge base is another issue. There are attributes e.g. difficulty or price. The program can only select from the subassemblies in the same group. The concept is the following: in a cycle every subassembly compared to the base pattern. Those which contain equalities (meaning 10 to 10) are chosen and put into a temporary array, named elected array. This is similar to a rule that can "fire". If the customer wants a machine with double XY table, those subassemblies will be chosen which are compatible with this criteria and has the number 10 in cell 1. If there aren't any components with 10 (100% compatibility), the program will search for 9 and so on. After the cycle process only one step is needed over the whole knowledge base. Search through this elected array and check if there are more than one subassemblies are chosen from every group. If so, we need to choose the cheapest and delete the others. The result is: one array which contains one subassembly (the perfect one) to every required group. If the price had been added to the knowledge base, the program could have calculated the sum of the cumulated price. In the concept there is a data structure behind the knowledge base and based on that data the bill of material which contains the parts needed to manufacture for the new machine is available. Presently, this BOM is only a list, no geometric modeling is available behind these results. When the new machine has been designed, all the new subassemblies must be uploaded to the knowledge base. After each machine design, the knowledge base becomes larger and thus a growing knowledge base is available.

3. Summary

To summarize the work, we can say that the system was working with test data, however; this program is still a development platform which verifies that the theory could be working on much more difficult machines as well.

3.1 Future plans

Based on real world problems the next goal is to extend this system for many machine groups. Keeping the already designed parts and assemblies is important, but not vital. There are other cutting edge technologies in engineering, e.g. generative design and 3D printing. The next big step is, based on the necessary assembly modification, to design the new required parts directly by the program. Design engineering is not a job that could be replaced by an artificial intelligence in the near future, but engineers who use an expert system could become much more effective.

References

[1] Å. Fasth, et al., "Designing proactive assembly systems (ProAct) - Criteria and interaction between automation, information, and competence " Asian International Journal of Science and Technology in production and manufacturing engineering (AIJSTPME), vol. 2 (4), pp. 1-13, 2010.

[2] Expert systems in special machinery: Increasing the productivity of processes in commissioning Sebastian Poeschl, JannikLieb, Frank Wirth, Thomas Bauernhansl, Procedia CIRP,Volume 63, 2017, Pages 545-550, Available online 11 July 2017.

[3] STUART RUSSEL, PETER NORVIG: Mesterséges Intelligencia modern megközelítésben. Panem Könyvkiadó, 2005

[4] YU CHEN, YUEFENG DU, ZHENGHE SONG, ZHONGXIANG ZHU: A Knowledge Base System for Rapid Design of Harvesting Machinery Chassis, 54Beijing, 100083, China International Conference on Intelligent Systems Research and Mechatronics Engineering (ISRME 2015) Key Laboratory of Optimized Design for Modern Agricultural Equipment, College of Engineering, China Agricultural University, Beijing

[5] NÉMETH ISTVÁN: CAD eszköz 3-tengelyes szerszámgépek koncepcionális tervezéséhez: szintézis, analízis és optimalizálás, PhD értekezés, KatholiekeUniversiteit Leuven, 2003

[6] TAKÁCS Á.: Számítógéppel segített koncepcionális tervezési módszer, PhD értekezés, MiskolciEgyetem, Gépészmérnöki és InformatikaiKar, Miskolc, 2009.

[7] GOTTHARD VIKTOR: Moduláris felépítésű gyártósorok tervezésének elmélete és gyakorlata, PhD értekezés, Budapesti Műszaki és Gazdaságtudományi Egyetem, Budapest, 2008

IOP Conf. Series: Materials Science and Engineering **448** (2018) 012030 doi:10.1088/1757-899X/448/1/012030

Law and Economic Analysis of the Legal Environment Related to the Autonomous Vehicles – Is There a Legal Paradigm Shift?

K Pétervári1,3 **and K Pázmándi** 2,4

1 Lecturer of Business Law – Neumann János University, Faculty of Economics, & ELTE University, Institute of Business Economics, 6000 Kecskemét, Izsáki út 10.
2 Lecturer of Business Law – Neumann János University, Faculty of Economics, & ELTE University, Institute of Business Economics, 6000 Kecskemét, Izsáki út 10.
3 E-mail: petervari.kinga@gmail.com
4 E-mail: pazmandi.kinga@gtk.uni-neumann.hu

Abstract. Autonomous vehicles trigger lots of legal problems. Whether this challenge, however, adds up to a paradigm shift in legal thinking is to be analysed further. As robotics develop globally, the legal institutions need to keep pace with it, because business risk, such as liability, has to be reasonably calculable for investors. Due to the expansive applicability of robotics, the relevant legal questions are manifold. The law faces applicatory as well as regulatory challenges. What is the standard for an autonomous vehicle? What should be the safety standards in driverless cars? How many bugs are to be acceptable in a software in case of a car crash? Who should be liable for an autonomous vehicle when it follows the rule and when it does not? How to tame the regulation by design to be more like a rule of law? This paper focuses on pragmatic problems and solutions and new regulatory models, like regulation by communication. It also introduces the concept of stealthy technologies, the meaning of which is that no matter what the rule is innovative technologies are going to be overwhelmingly widespread.

1. Introduction

Since there is no consensual opinion on what the paradigm as such means in social sciences, including legal sciences, this paper does not serve with a direct answer to the question whether or not there is a paradigm shift in legal thinking. Yet, one thing is certain. There are many changes in our contemporary, specific world which could generate the feeling of radical difference. But do they touch upon basic legal principles too? The real debatable question is then whether profound changes or whether *many* profound changes, would cause paradigm shift. The exact answers will undoubtedly be delivered by the numerous decisions of the legislation and the law enforcement in the near future.

Having said that, however, in this paper it is going to be shown that the law, as it is today, operates with analogies and fictions. And as long as there are analogies and precedents or fictions available, one is not crossing the border of normal sciences, the territory of one paradigm. Obviously, there are, and has never been homologous cases, legislators and/or judges always apply the interpretations of the previous cases.

The paper here, therefore, relies on the use of analogies too. It analyses the various legal tools applicable to the relevant cases, such as the legal personality, the vicarious liability and the culpability as opposed to strict – objective – liability. Again, one thing is certain. All of these legal techniques were familiar in the course of the legal history, none of these are new, but perhaps their applicability of scale is different. Since law is a social construction, the conceptions of the various legal tools may change. But even the most radical recommendation of awarding a computer a legal personality, or constitutional fundamental rights, can found its roots in the past, like legal personality to organs created by the law, such as the companies, or the standing for the trees, which provides now legal protection to the environment.

On the other hand, one might object, that the new suggested *models* of regulation could cause a paradigm shift in legal thinking. These are the *regulation by design* or the *regulation by communications*. Regulation designers, like system designers, software engineers, and so on, need to find out the language what computers speak, and be able to speak the same language, also with the law-makers and courts and the law enforcement people.

Yet, this paper would introduce a new conception: the *stealthy automatics* or *stealthy technology*. With this term this paper argues, that whatever the regulation may be the autonomous cars are going to ameliorate and widespread.

2. The regulatory challenge

2.1. Analogies in regulating: EU v. USA

Unlike the other rivalling companies back in 2016, Uber, this innovative but fairly eccentric company, refused to apply for a permit in San Francisco when launched a test driving of its autonomous vehicles (Wired (2016/12). Although the *new* rules of California would have so required, and the authorities so enforced indeed, Uber deliberately chose to remain under the scope of the old regulations. Truly, the cars on the streets are driverless, but there is always an engineer behind the wheels, ready to intervene and so autonomous vehicles qualify as vehicles driven in a "conventional mode", Uber argued.

Unlike California, however, Pittsburgh, Pennsylvania, where Uber had its basic testing sites, did *not* have new rules on autonomous vehicles operating in the city. Since then Uber has launched there its taxi services with a driverless car fleet. Lyft followed suit quickly (Wired 2016/12).

The rivalry among manufacturing companies of autonomous vehicles is recently so tense, that one must but wonder, whether the rules matter at all.

These cases do not mean that there are no regulations whatsoever, though.

Interestingly enough, one of the first statements of the US federal authorities asserted the continuity of the existing structure of the legal institutions, namely that "the division of regulatory responsibility for motor vehicle operation between Federal and State authorities is clear… should *remain* largely unchanged for highly autonomous vehicles" (NHTSA Policy Guidance, 2016).

The Europeans in the EU have somewhat analogous experiences but with a philosophically different attitude. Whereas it is a federal competence in the USA to decide who can access the public roads, this decision belongs rather to the competence of the Member States in the EU, so far, at least. And the Member States do use their power to regulate, and they regulate fairly disparately, of course. The Hungarian version of modification, for example, deals merely with the testing of the highly automated vehicles (11/2017. (IV. 12.) NFM order), whereas the German one applies to the traffic itself, even if under strict scrutiny (Achtes Gesetz zur Anderung des Strassenverkehrsgesetzes, vom 16. Juni, 2017). In any event, due to its determinant effect on the internal market, the EU will have to act at Union level.

The EU has, however, a rather ambiguous approach to the matter. It has *shared competence* in transportation policy (Article 4.2.(g) TFEU), but transportation in its stricter sense. It means that as far as it is indispensable for the common trade policy or competition within the internal market, this area is covered by the competence of the EU. So the EU legislators are abide by the otherwise classical notion

IOP Conf. Series: Materials Science and Engineering **448** (2018) 012030 doi:10.1088/1757-899X/448/1/012030

of subsidiarity principle and proportionality principle. Under the principle of subsidiarity, "in areas which do not fall within its *exclusive competence*, the Union shall act only if and insofar as the objectives of the proposed action cannot be sufficiently achieved by the Member States,... but can rather, by reason of the scale or effects of the proposed action, be better achieved at Union level" (Article 5.3 TFEU). Further, the proportionality principle compels that the EU "shall not exceed what is necessary to achieve the objectives of the Treaties" (Article 5.4 TFEU).

On the other hand road safety does not only extend to the commercial attributes of transportation, but also to the very basic idea of the European integration. Therefore one might argue that *free movement* requirements trigger broader EU competences. Besides, the EU has again another kind of competence to act, namely "to carry out actions to support, coordinate or supplement the actions of the Member States in certain areas such as (a) the protection and improvement of human health; (b) industry or (d) tourism" (Article 6 TFEU), which all involve the very issue of the autonomous vehicles too.

These complicate structures of the EU, actually boil down to various, often soft, legal tools regarding road safety and autonomous vehicles, like declarations (Valletta Declaration on Road Safety, 2017), resolutions (Motion of the European Parliament (2015/2103(INL), guidelines but also directives (framework directive 2007/46/EC on type-approval of motor vehicles) and regulations (Regulation (EU) 2017/1154 on type-approval of motor vehicles). This structure of the legal competence in Europe is, therefore, conspicuously more complex, than that of the USA. Especially because the relevant authorities of the Member States still have further room for manoeuvre to regulate, since it is not yet preempted by the EU.

Yet, it is not preempted in the USA either. The U.S. Code, Title 49, Chapter 301-305 on motor vehicle safety, standards and compliances, which basically corresponds to the EU regulations on type-approval of motor vehicles, is not yet adjusted to the automated vehicles, therefore many states issued special regulations. Having wished to stop that regulatory wave of the States, the Obama Administration in 2016 allowed the federal supervisory regulating agency (NHTSA) to preliminarily exempt some automated vehicles from the compulsory compliance with certain requirements, until the Department of Transportation, DOT, elabourated an acceptable version of new regulations. This exemption seems to be approved in both actually, competingly, debated Senate and House Bills and even in a growing scale (Senate Bill: S.1885, House Bill HR 3388). Thus in 2020 there would be 100.000 such driverless cars in the public roads in the USA. The justification seems to be that technology becomes ever better and the driverless cars are much safer than the conventionally driven cars are.

In addition to the tiring issue of division of competences related to the topic, there are, however, other differences between the US and EU regulations: firstly the divergent structure of the EU and the USA, secondly the divergent way of approach to the regulation as such and thirdly, the divergent basic legal institutions applicable in this field.

Certainly, due to its lacking of federalism, the cross-border incidents in the EU can trigger international law further complicating the legal problems, like in the case of a cross-border accident (Kiilunen 2018, Pétervári 2017, Nagy 2010). Mostly for this reason, the EU also employs a more administrative approach towards regulation, since it lacks the real power of enforcement, so the ticking of the check box in the compliance procedures is rather convenient. But this might not serve the interest of the competing industry. The recent regulatory approach of the USA has, on the other hand, been very much innovative and competition-friendly above all. The exemption policy *puts the risk* of bad regulation, or no regulation whatsoever, on the federal regulatory agency. In the meantime not even the States may legislate on the safety requirements of a driverless car, albeit especially big cities are tempted.

Needless to point out, that the mainstream European approach is conceptually different. Already at the beginning of the 2010's various working groups in the EU suggested a "top to the bottom" approach to the legal problems of robotics (Leroux – Labruto, 2012 10). Certainly, there are rather convincing arguments supporting this attitude, like avoiding wastes by not investing in competing but failing research projects and thereby gaining time. Nevertheless there is no guarantee, that an action plan based on administrative prescriptions, no matter how deliberated it is, will serve a safer and quicker result. All such an approach can assure is, that whatever the result is, it will be expensive. This undoubtedly provides less smooth pathway for innovative manufacturers, *since the burden of risk* will be held by the industry and not the administration. Surely, however this burden of risk will ultimately be paid by the consumers, and the risk in this kind of world of innovation is generally allocated to the tax-payers. This may not be unfair but less efficient.

Not surprisingly, though, in certain fields, the stakeholders may dictate in both worlds. The driverless cars run first of all in cities all over the world and both the US and the EU concentrate the scope of regulations on automated vehicles to the lighter type of cars. Thus, for example, lorries cannot be exempted in the USA. This may seem to be counterfactual, though, since automated lorries are expected to be less dangerous rolling on the public roads (like OTTO of Uber) than the automated cars in the cities. Nevertheless due to the existing fear of employees, and the governments, the stakeholders' concerns are very strong and seemingly deserved legislative attention in this and other primarily robotics threatened industries.

The lack of the necessary skills to understand the applied technology makes the law-makers and the courts vulnerable to this controversial information asymmetry and might also have a disadvantageous impact on applying *less reasoning* and *more technical slang*. The unintended acceleration problems in car crashes in the Toyota case of 2009 are good examples to that. The jury may not be competent enough to decide on real causation, some experts claim. Whether the undeniably many bugs in a software can be proven to directly lead to fatal crashes, that is the question. In traffic accidents the driver's negligence needs to be shown to find him liable under common law. The driver can defend himself by claiming, that it was not his fault but that of the car. Then the so called proximate cause has to be evidenced, namely, that the accident is the direct causation of the existence of all those bugs in the cars' software. And then the real question is, if an expert witness explains to the jury that the standard of the numbers of the bugs allowed in "one software": is zero, or maximum 5-6, and in the "relevant software" there are 1000 of them, then if the jury can count, the software producer will definitely be found liable, no matter whether that is verified to calculate the bugs so (Cummings 2017, Parrish 2016). Although in this case Toyota finally lost generating billions of dollar damages, these sort of product liability situations do not necessarily end so. And what is more, especially in the case of driverless cars there is no "one software" or it is almost impossible to track down the "relevant software", let alone the possibility of the car to develop "own softwares" with the deep learning method.

The new challenges provoke new models and there are various possible answers for how to create an architecture of the regulation by design and how to make designers be responsible for it. Some argue for translation from the real space to the cyberspace (Lessig 1999), a commitment to moral understandings and values in technical regulations too (Urquhart – Rodden 2016), application of the rule of law considerations in the cyberspace too (Murray 2013) or language gaming, design by communications (Koops – Hildebrandt – Jaquet-Chiffelle 2010). These are all important issues so that to ascertain on whom to allocate the risk, the burden of proof, the impossibility of proving or the impossibility of foreseeability, to name but a few.

2.2. The regulatory challenge – are there robot-subjects or do the robots have legal personhoods/personalities

In fact, the regulatory challenge, related to the automated vehicles, is that it encompasses obviously many more legislative areas than just that of the transportation. Safety on roads covers a lot more than just a physical well-being or simply a being on the roads without accident.

Driverless cars create new environment. It presumes swiftly communicating softwares of others and complex infrastructures. But, most importantly, it allows on the spot decision-making without human interference. Apart from the usual technological development problems, therefore, this all requires, from a legal aspect, *special cooporations of old* market actors as well as *innovative applications of old* knowledge. Thus beside the general manufacturing questions the legal institutions seem to be tested too.

If a driverless car is to make an on-the-spot, or even a predetermined but externally generated, decision, for example, it needs to rely on special networks of databases, involving the questions of ownership rights, copyright and software issues, beside personal data protection and liability considerations. A driverless car, to be able to navigate on the public roads, needs data owned by a competitor, a driverless car's data one-two-three cars of it ahead, or needs to have access to others' special databases without the necessary legal relationship between the parties, resulting in, perhaps, the lack or in the end of the privity of a contract. Yet, vehicle to vehicle (V2V) and vehicle to infrastructure (V2I) communication may be rendered on a general duty of care in the traffic. But it triggers further concerns of protecting one's data, one's private data in a commission trip of the driverless car.

Obviously, the conception of property constantly changes. No one really thinks nowadays that without possession one could not have ownership, although possession used to be the bases of ownership: what I could possess, what I could grasp and protect should be regarded as mine. Hence come the tangible goods. The intangible things or assets, on the other hand, posit real headaches for lawyers even today. Intangible things, nonetheless, are vital to the business. Long before, came the *copyholders* of the land, then the *shareholders* in the company, which could be owned and thus became negotiable separately from others. Then even the "market position" as such was recognized as a unique subject matter of transactions thereby creating ownership and value. Afterwards the database was awarded legal protection, as a special asset, *sui generis* subject matter of the copyright and the civil law. And now the hottest debated issue is the ownership of the data in the clouds (Reed 2010).

These piecemeal changes in law basically follow the demands of the business and provide guarantees to assets there, where it is so desired by the market practices. Capitalism has always been built on information capital, thus on the efficient allocation of property of information, like business secrets, copyright, patent and know-hows (Cooter and Ulen 1989).

A self-driving car, however, generates further basic regulatory questions of robotics and law today. From a delicate legal point of view, there is an intriguing suggestion in the technology and law literature to consider (Calo 2015, Santosuosso – Bottalico 2016). Accordingly, depending on the nature of its purpose, task or function, an "intelligent" machine may be regarded either as a robot-*object* or a robot-*subject*. In a somewhat rude interpretation and summary, a robot-object, despite of its ability of information-transmission, qualifies as a thing in the eye of the law, if it is not autonomous. Autonomous in this respect would mean that the robot has independence, cognitive skills and special cognitive architecture too. Cognitive architecture would further include three special elements: teleonomy (special directed purpose), adaptiveness (capacity to adjust to the environment when challenged) and teleology (capacity to interpret the environment and adjust goals and means accordingly). Therefore, a targeting system, the argument goes, is still a robot-object, because of lack of independence: the ultimate decision is made externally, an operator pushes the "fire" button. Also, a landmine may well be independent, but lacks cognitive skills, since it explodes automatically when so programmed. A self-driving car, on the other hand, may rather be regarded as a robot-subject: (i) it is independent, it requires no external influence, (ii) makes independent decisions (iii) it has directed purpose, it understands its purpose and it is capable to adapt its actions to the environment, necessary for its goals to achieve. The final conclusion of this narrative is that the robot-subjects would be recognized differently in the legal context. Instead of calling it a thing, it would be a "non-human" agent, acting on behalf of the principal, let it be the owner or not. And in this latter case, the relevant applicable laws are different. In lieu of property and intellectual property or employment rights, products liability issues or consumer and data protection, the applicable laws or legal institutions here would rather be the agency problems, legal capacity issues, criminal law, torts, the non-contractual liability, e-personhood, intent, etc.

Consequently, driverless cars may be called also as tele-operated autonomous vehicles with cognitive skills. These cognitive skills, with special respect to the latest results in deep learning, allow better adaptivity and decision-making of own-account. Deep learning however, provokes further questions for lawyers: how to know whose software was bugged or whether it was the driverless car itself who generated the bug itself. So are these autonomous vehicles equipped also with cognitive architect, as the theory requires it from a robot-subject?

Definitely not, yet. Asserting that a robot can speak or answer the questions – even all questions – does not mean that the robot understands what it says. There are plenty of examples to that: Bina48 in the 1990s, Watson Jr. from IBM in the turn of the century/2000, or Sophia in 2018.

Driverless cars become robot-subjects, again, no matter what the rule is, when risk allocation rules for innovation so desire. The robot-objects are not at all in the visier of the law because they first need to qualify as a thing. A thing is defined by the legal literature. A general definition is that it has to be something that a man can keep under his control or power; even the wind, or the sun. As soon as the robot-object becomes more evolved, one could argue, and bursts out of this control, it becomes a robot-subject any way, even if the rules do not recognize it. Liability issues however are not at all the subject matter of pure moral and legal concerns, there can be plenty of examples to that, like the legal entity and liability or even culpability of a company.

The recent findings of neuroscience is that the free will, or the operation of a human brain maybe more pre-determined than it is generally accepted by the tort law or criminal law since at least the enlightenment (Scwaab 2014). So legal punishment might lose its ground and so it would put delicate questions to the law and robotics (Eszteri 2015).

Truly, these problems are not only legal questions but deeply rooted moral issues. The answers, therefore, more often than not, reflect one's own world view and one's tacit attitude towards legal norms as such, from a strictly legal point of view. If one accepts the view that the law is separate from any ethical concerns, than the law should be the matter of rational choice and economic interest beside the political deliberations. On the other hand, if it is not acceptable and one could not imagine a world without the moral foundations of the legal norms, then the answers to these challenges may lead to a very different path.

This paper here, focuses on only a few of the problems, namely the privacy issues, the software issues and the strict liability issues.

3. The applicatory challenges by analogies

IOP Conf. Series: Materials Science and Engineering **448** (2018) 012030 doi:10.1088/1757-899X/448/1/012030

3.1. The applicatory challenge by analogies – Privacy analogies

The construction of privacy changes. Robotics in the driverless cars are desperately data thirsty. They need various enormous databases to be able to operate. Thus driverless cars running on the streets often induce legal problems such as invasion of privacy, illegal data collection, illicit categorization and usage of them, to name but a few. Some of these include personal data, some infringe competition rules, copyrights or property rights of database-owners.

So what data and how should be protected?

The EU is highly sensitive to the personal data protection which is conspicuously demonstrated in the new General Data Protection Regulation (GDPR 2016/679). The idea of privacy by design (PbD) advocated in this piece of legal act is not at all brand new, it has been developed by the European Commission for some time, from the Privacy Establishing Technology project. However, the regulation itself was basically, though not entirely, triggered by the, allegedly, reckless use of the European citizens' personal data in the USA or US companies (e.g. Schrems v. Data Protection Commissioner C-362/14). The regulation has effect also outside of the EU in the sense that any company wishing to make any business in the EU but settled in a non-EET country should also comply (USA, China).

No doubt that both the European Union and the United States recognize privacy as a fundamental notion. The U.S. Constitution enshrines protection against state intrusions and the Charter of Fundamental Rights of the European Union (the Charter, art. 6-8) as well as the European Convention for the Protection of Human Rights and Fundamental Freedoms (the ECHR, art. 8) each mandate that law and public authorities are not to interfere with private life. Notwithstanding, lately, the state/police surveillance demands ever more invasion of these rights and a want for more data, due the recent indispensably data-addictivity of the new technologies and their reliance on mass businesses, and so allegedly mass security.

As one of the most influential judge, Richard Posner, in the USA puts it, "privacy is mainly about trying to improve your social and business opportunities by concealing the sorts of bad activities that would cause other people not to want to deal with you" (Posner 2014).

And this is perfectly demonstrated by the ECJ decision in the case of Gomez v. Google (C-131/12). The decision is wrong for its being as overbroad and over-inclusive as it is. Vesting a right in Mr. Gomez to claim a deletion of his data from Google generates serious problems with free speech and rule of law. In this case Google Spain had digitized a local newspaper in which, from that time on, if someone looked for an attorney in the region on-line in an advertisement, could bump into Advocate Gomez's house being at auction for overdue tax-depths. Upon Mr. Gomez claims the ECJ finally read the EU Treaty in a way, that Mr. Gomez was awarded the right to claim from Google to delete its link to his personal data. Hence there is a living *right be forgotten* in the EU.

Nonetheless, the right to be let alone advocated by Brandeis and Warren in 1890 was a conceptually diverging idea from the European right to be forgotten (Brandeis – Warren 1890). The *right to be let alone* meant something else, though the leading case back then might resemble to the problem of Mr. Gomez. In this legal suit Ms. Roberson, an attractive girl had had a picture made of her at a professional photo shop and in a short while she realized that her so taken picture covered the paper package of the flour in the relevant grocery shops all over New York (Roberson v. Rochester Folding-Box Co., 71 N.Y.S. 876, 876 (N.Y. App. Div. 1901). Although the majority was in favour of the photographer, there was a powerful dissent in the decision of the New York Supreme Court, even citing the famous Brandeis-Warren article on privacy. Yet, this case emerged with the rising new technology of the smaller, portable

IOP Conf. Series: Materials Science and Engineering **448** (2018) 012030 doi:10.1088/1757-899X/448/1/012030

photo camera allowing photographers to take pictures from anyone anytime anywhere, looking at the techniques of that time with a hindsight, and watching the then still man-size photo-cameras, this fear of invisible invasion of privacy by photographers must have only been a threat. However, the impression of the invasive being of photographers must have been felt real, because the new ever smaller and better versions of cameras came into the market so rapidly, that the threat must have felt very imminent even then.

Notwithstanding the internet, the crucial similarities in these two cases, more than a 100 years in between them, are that the actor who publishes or who decides to publish certain personal data is not the "owner" of the personal data. And this actor, thanks to the technical facilities, may act independently. It should be noted though, that in the early case of Ms. Roberson there were no copyright issues in the USA, whereas nowadays this aspect is not negligible at all. Ownership rights in Cyberspace, or the ownership in information is quite an important topic, since several of the basic elements of the term ownership are missing, namely the classical "thingness" of the object of the ownership and the exclusivity, or the exclusive use, of the owned "thing" (Reed 2010). On the other hand these cases show the huge difference in scale, in huge public access, which may matter the most.

Truly, the term privacy has problems and there is a discrepancy between the EU and the USA (Reidenberg 2014). Europe supports privacy even at the expense of free speech, and commercial speech too, which is the language of the business communication, the freedom of advertising and marketing. The USA does just the opposite. And indeed, legal privacy has many ramifications (Gormley 1992) and the construction of those may not always match between the American and the European one. Firstly, there should be the right to have secrets from the State even (laws against unwarranted searches), secondly, the right to have own opinion and to let it be public (free speech), thirdly, the right to be let alone (tort laws, damages by delicts), and fourthly, the right to have fundamental decisions (rule of law principles, fair hearing, fair trial). In addition, the new kind of business-models enabled by the new technologies provide further privacy issues, like the cross-border jurisdictional legitimacy problems. Whereas Brandeis and Warren dealt with basically the right to be let alone, the legal problems generated by the new technologies nowadays embrace all privacy elements.

Now, when a driverless car wishes to travel through several cities or countries in Europe, it needs to have access to a lot of databases. It uses for example, official public databases when gathers information about the street facilities. Then it uses private databases for meteorological data or pictures of the artificial world in the human environment, or outside. It may have access to personal data too when reaches out to the clouds and collects information about the exact location of a person, depending on the driverless car's mission. And again, it uses information which is owned by someone else(s). In these instances the smart things full of personal data being used and stored in one's home might become a useful, but illicit?, resource to the driverless cars.

The prescribed 'privacy by design' in the GDPR may not necessarily be safe unless privacy is legally clarified, precisely described and expressed to the designers and the design is specified in due course and in due terms. First of all however, the GDPR is a general regulation and even if it is directly applicable, certain technicalities require further elaborations. Besides, these specificities needs to be provided by the relevant authorities of the relevant Member States according to the applicable EU directives, such as the e-Privacy directive. Encoding privacy requirements into software is obviously, not at all self-evident in the EU.

Koorps and Leenes criticize therefore that the entering into force of the GDPR and thereby the privacy by design, would not solve even the problems of the personal data owners (Koorps-Leenes 2014).

IOP Conf. Series: Materials Science and Engineering **448** (2018) 012030 doi:10.1088/1757-899X/448/1/012030

According to their opinion this new regulation would not change the existing mode of operation and would not stop the pattern in which the industry tries "to achieve rule compliance by techno-regulation". The claim is, that what may be intelligible for a human, may not be as such for the machine, the driverless car.

The Turing test fails here. Even if this test can help to decide whether the other speaking party is a computer or a human, it cannot have any prediction whether this other party thinks or understands his doing as well. Searle has his relevant, even rethought, criticism with the Chinese room, in which a sole non-Chinese speaking human could answer Chinese questions with a dictionary and a grammar book without understanding the questions let alone the answers (Searle 1980). Should that now mean that the autonomous vehicle cannot be liable, and that the data collected by the driverless car on the street can be freely downloadable because it has no rights either?

A question, what data could be collected and under what purposes on the public roads, could therefore be very important. It could include to decide whether special human mental diseases may contribute to traffic accidents in the streets. Is it allowed that manufacturers collect data therefore on sensitive health issues for safety reasons or health preservation on the public roads? What is more, the transfer of such data may further create sensitive personal data infringement. In order to facilitate the designer to "translate" these kind of purposes and requirements into codes there are other possibilities. Instead of regulation by design Koorps and Leenes offer the *regulation through communication* (Koorps-Leenes 2014 9). IoTs and autonomous vehicles should use data for profiling as well, since they in fact support one's routine activities, which, again, on the other hand, is a sensitive issue in the GDPR. This to be done needs prior and clear understanding and consent and in case of circumstantial changing, the approval needs a renewal too. To translate these changes in time more thorough interpretation and understanding is suggested.

This structure of data protection does not sort out the question, though, citing Posner above, whether it would be such a problem, if certain personal data were less protected and more public, as it used to be prior to the capitalism.

Despite of its newer regulatory model of the GDPR, the very existence of the GDPR is an evidence of how the top bottom law making works in law and robotics. Advocates of the GDPR claim that this regulation is needed, because the bad practices should no longer be tolerated and thus the social conventions must be changed by the law, ultimately by legal sanctions. However, the GDPR could well create a rather administrative, check-the-box answers with a send-the-declarations or disclaimers-always-everywhere attitude.

On the other hand, designers come to the market with products, with carefully prepared regulatory design, fit in the legal framework of the laboratory times. With the shift in time, however and when they really hit the market, the law changes and so should the regulation by design again, which again, is not so trivial. Koorps and Leenes demonstrate it by referring to the simple and surely predictable problem of aging: a minor needs the approval of the parents in order to have consent to data collection, but how to monitor its changes?

Surely, laws on robotics require from lawyers the understanding of technological details. And this is the challenge the law faces and which may be described by the stealthy technologies which again would trigger a necessary interdisciplinary pragmatism (Calo 2015, Urquhart – Rodden 2017, Somkutas – Kőhidi 2017).).

3.2. The applicatory challenge by analogies – legal entity

Although it is a (statistical) fact that autonomous vehicles create safer public roads, still no one needs an extraordinarily vivid imagination to fancy how much damages an autonomous vehicle may cause, if something goes wrong, without any human interference. Certainly, as long as the machine follows the rule, the encrypted orders, it may not be held legally liable. But let's suppose that there is a fleet of driverless cars operated by a network of these cars. Is that sort of business enough for holding this network of autonomous vehicle responsible, even liable? If this business makes money, should the driverless cars also be able to make money?

Common sense, that to qualify as a legal subject, one does not need to be a natural person or a human being. And this is true the other way round, not all human beings are/were legal subjects. Good examples for the first are companies with or without legal personality, the state or its agency, like the authorities, ministries. And examples for the latter are slaves. Currently, "everyone has the right to recognition everywhere as a person before the law" (according to the UN Charter, Art. 6). Also as *ius cogens*, it is found in the Hungarian Civil Code as well (HuCC Articles 2:1, 2:4 on the legal capacity of a human person and Art 3:1 on the legal capacity, and the scope, of a legal person). Though "everyone" needs not necessarily be a human being, the preamble of the charter might be construed that way. A thing may be owned, a person acts and has self-consciousness about his doing it, about his doing it rightly or wrongly. Nonetheless not only (human) persons can cause changes to the environment, or damages for that matter. Frequent examples are animals, but plants or volcanos too, in fact.

It is a commonplace too that law is a social construction and as such, granting non-humans legal personhood (legal personality or legal entity) may be a matter of rational choice, depending upon the world view of the regulators though. For a historical analogy, companies acquired the title of legal entity geographically sporadically but almost always driven by necessity motivated by business considerations (Hansman – Kraakman 2000). Without a reliable, transparent registry, the creditors were finally awarded with the separation of the assets. In order to create a trustworthy business environment the assets needed to be separated, so that to enable the creditors to have access to the company assets without competing with other private creditors of the company owners. So the legal entity of a company was a good move, even so because soon the limited liability of the company indeed insulated the assets of the company not only from other creditors too but also from the owners and the private creditors of the owners. Credit became cheaper.

Legal personality therefore is not necessarily a moral issue. Legal personality or legal entity allows one then to be recognized by the law, so that any act of this entity be legally challengeable, have legal consequences. On the other hand this legal entity may have also access to the courts in case of its grievances. Under classical European continental law legal capacity to have rights and duties is to be distinguished from the legal capacity to act which may be denied from non-human legal entities constructed merely by the law (Smith 1928).

And so it is in the case of the companies. So is the legal capacity of a company limited, it cannot act on its own behalf. Needs an agent. Notwithstanding the limited liability of certain companies, the liability of the companies are recently extended to certain criminal activities. It means that the acts of certain employees, usually that of the management, can be attributed to the corporate entity, if this activity benefitted the company and the employee committed it as the manager of the company.

In any event this sort of development in legal thinking has ever been based on the economic interest, on property allocations. And so, even if it seems now to be a shift in the legal thinking, it is not. Such application of a fiction is not at all new and conceptually the ultimate sanction is an analogy. The termination of the existence of the legal entity may be parallel to the execution of the human person. The reason of attributing this (criminal) deed to the company is therefore a moral and a business concern rather than a legal one. Why should the company be able to get away with a tax fraud, for example, when the company had the benefit of the crime and also the collateral, perhaps, for the remedies.

The idea is that the borderlines in the fields of liability have become blurred, and there are no longer clear distinctions among the territories of the criminal and the civil law or the administrative law liability. But this is not that big a change under common law, even if it creates debates under the German civil law tradition (Bérces, 2015, Ligeti, 2002, Eszteri 2015).

The *objective liability, the liability without negligence* or intent is not at all a new legal institute. The idea, that it is not only the human that can cause harm is again not a new notion. These legal tools of interpretation, however, were designed to be applicable in environmental protection cases or anti-corruption cases and not for machines having impacts on humans' life.

And in this sense, there is a distinction. In the case of the AI, IoT or autonomous vehicle it is ultimately *not the human* whose action is being under scrutiny and then judged. Therefore, none of these cases raised as seriously the question of legal personhood as it is being discussed in relation with the IoTs nowadays.

The technological phases of the artificial intelligence (AI) obscure the borderlines of the development and so the level of capacity too. The legal capacity of these machines may be hard to fix. According to a widespread notion of conditions an AI needs to be conscious about itself in order to gain legal personhood. No doubt, it all leads us to the interpretation of the other minds problem and the predictions of others acting and doing. But neuroscience does use predictions very often even at courts, at criminal trials. And so does the judge at the court, the prosecutor and the police too (Garland ed 2004 116). Clearly, the widespread use of such technique may not qualify the method as good but certainly as something practical which cannot be replaced with something better, so far.

The breaking point in questioning of the legal capacity of an AI was provided by Lawrence Solum's highly intriguing and illuminating article in 1992 (Solum 1992). He wrote that, with due regard to the complexity of the issue, the answer needs to be a *pragmatic* one. So he put the question, whether an AI may be a trustee. Solum distinguished three different levels of an AI. Firstly, the AI absolutely routinely follows the rules/orders and serves basically as a huge library for the owner who actually makes the transaction with the AI's assistance. Secondly, the AI practically sorts out the business on its own and enters the contracts but needs the signature of the owner. And thirdly, the AI does everything alone. Solum's conclusion was that although the time was not ripe, there might come a turning point where the AI will claim for own perhaps constitutional rights: what would stop an AI to open up a Law Directory on line and commission a lawyer to represent him at the court. Nick Bostrom would certainly object, alleging that it is due to the construction, the architect of the machine, which disables a machine to "think" as a human (Bostrom 2003). Solum further argues that multiculturalism prevents us from denying this possibilities for non-humans and he predicts the necessary change in legal concept of the legal personhood.

Twenty years later, Koops, Hildebrandt and Jaquet-Chiffelle relying heavily on this article and on its findings concluded still that at the time being, a proper construction of the *existing laws would suffice* (Koops – Hildebrandt – Jaquet-Chiffelle 2010). After having screened and summarized the relevant new European (legal) literature, they argue that the laws are flexible enough and allows the interpretation of the *laws on agency* – however divergently in the USA and in the EU – to be applicable to the machines being built. On the other hand, they emphasize, that sooner or later, all of the researchers suggest modifications to the law and perhaps to the legal concept as well (e.g. Matthias 2008).

According to their model, which recommends also a three-step approach, in the short-term, at the starting point, the interpretation and extension of the existing law suffices (Koops – Hildebrandt – Jaquet-Chiffelle 2010 554.). At this level, in case of a legal transaction, the acts of the computer may be attributed to the owner/user as his general intent. In case of invalidity problems, if the computer buys more shares than should have the court would have no difficulty to decide on the intent of the owner as the intent of the human and not that of the "electronic agent", since agency rules may not yet be applicable. As a middle term, they argue for a limited personhood with strict liability. At this level agency rules could already apply, so the agents need to be registered as such in a public registry which could allow the contracting parties to find the identity of the agent's principal (555). Besides, they recommend the cautious introduction of a limited type of personhood for the e-Agent with a strict liability, so that the owner/user of the electronic agent may be found and be liable in its stead without questioning the intent. A further suggestion is that these electronic agents may receive a modest provision for their activities which may be a bases for an insurance fund in case damages occurred and they themselves may be found liable (in civil law only), mostly because the owner/user may not be located or they deny that they gave reason to the transaction be in their names. And finally or thirdly, when technology evolves further, the long term solution is a full personhood with "post-human" rights. Post-human rights would refer then also to that that this is not the "classical human rights" anymore. This should be the time though, when self-consciousness emerge. As a criticism to that smart future, it is argued too, that this sort of consciousness is rather facilitated by interconnectedness, thus not individual or "real". On the other hand, cognitive sciences pile up evidences to make us believe that "human identity itself emerges from distributed brain processes" (559.).

To translate it all to the autonomous vehicles, it is for sure, that so far we are at the threshold of level two and three. To attribute intent or *mens rea* to the driverless cars and let them prosecute or be prosecuted in criminal law, should have a long way to come so far. That would surely require a certain non-predictability or an intentional deviation from the rule, with purpose. That is not very likely yet. It is unlikely that a driverless car on mission to the airport to pick up a friend of the principal would go to the theatre instead to pick up the most celebrated actor in the region; and thereby causing a car crash in front of the entrance in the overcrowded street. However, strict liability in car accidents may be attributed to the driverless cars, since it does not demand the reasonable acting or the questioning of the reasonable care. Here there is no investigation into the proximate cause and thus the classical conditions like intent or fault, negligence or recklessness are not considered.

On the other hand due to the damages a driverless car may cause, the risk must be assessed and allocated efficiently in advance. The autonomous vehicles' liability therefore may be paralleled with the strict or objective liability of the companies (product liability) in private law and be based on a simple risk allocation mechanism and not on intent. It is not the punishment or the retribution which should count, but the desire for restitution, the restoration of the assets status quo. The restoration of the wealth distribution to the pre-accident status quo is a private law principle as opposed to the criminal law ones. So the question is to which model one reaches out. If this could cause a paradigm shift, then it seems, that the insurance lobby might object that, since with such a shift all could incalculably change.

3.3. The applicatory challenge by analogies – Software warrants and strict liability

IOP Conf. Series: Materials Science and Engineering **448** (2018) 012030 doi:10.1088/1757-899X/448/1/012030

In the case of an autonomous vehicle therefore liability issues may become somewhat complicated. Based on the three-step model of awarding legal entity to an AI an autonomous vehicle should linger somewhere between the second and the third phase, at best. That would mean that the driverless car could have a limited legal entity should be registered in a public registry and should be vested also with strict liability from the owner/user part. Driverless cars therefore should act as agents and their actions should create rights and obligations directly to the principal. Further capacities will definitely be awarded if liability on its own account will be more innovation-friendly.

In case of a car crash, the principal or the owner does not drive. This is not a big difference in most Member States in the EU, where strict liability laws apply in these cases, and, more often than not, the owner/operator of the car is liable, who might happen to be the driver but not necessarily. It can be so, since strict liability has no causal condition as in a negligent or intentional tort. Simply by operating a highly hazardous equipment triggers the stricter, non-subjective, liability. In other words in these accidents the driver/owner/operator/principal can hardly be exempted from liability if damage occurred.

On the other hand where negligent tort rules apply, like mostly in the common law, the driver himself is liable only if negligent and therefore the proximate cause of the accident has to be cleared and proven by the plaintiff in order to be able to collect. The defendant in these cases can be released if he shows that the car was defected. In the event however, that the car is autonomous, the driver is the car itself, so the manufacturer would be highly motivated to grant certain legal personhood to the driverless cars. Then the autonomous vehicle, as an electronic agent will be held liable. Should it be strict liability applied directly or should it remain a negligent tort?

In any event, the liability issues need to be determined on a case by case basis, regarding also the fact that with the driverless cars the software problems may override the hardware ones. And software failures are often harder to allocate but even harder to find the causal link with the necessary evidentiary standards. Like in the case of unintended acceleration software problems in certain Toyota types in the 2009, in which the victims of accidents managed to prove that the software of the breaking system had bugs. The case remained controversial until years later the management of the Toyota Company confessed lying. The criticism pointed out that the causal link between the admitted bugs in the software and the direct cause of the accident was not proven at the court and the jury was wrong. (Cummings 2017). "It is well known that nearly all non-trivial software has bugs. Furthermore, because there are virtually an infinite number of different ways of solving a non-trivial problem using software, one can often find many opportunities for criticizing software quality, sometimes using criteria that are highly subjective. As a result, the plaintiffs in a software trial can be expected to attack the defendant's software by looking for bugs and criticizing the software's quality. If they can find bugs, and show that more likely than not those bugs caused the accident, then they can establish causation. They can also use purported measures of software quality, which may be subjective, to argue that the defendant failed to fulfill their duty to provide software of sufficient quality" (Cummings 2017 1.).

Surely, the running of the driverless cars on the streets will trigger these sort of liabilities through the warranty problems of others' softwares, like the use of the free softwares and open source softwares. In these cases the risk and thus the liability questions need to be ascertained on a case by case basis and in case of doubts, the damages should be borne by those in whose interest the driverless cars were operated. The present rules on car crashes could well apply in the non-common-law countries, as said, where in such cases the strict liability is the rule.

Safety standards are a starting point in these cases, since consumer protection and product liability derives from the assumption that the product needs to be safe. But how safe? Obviously a driverless car is regarded as to be safer than a car without or nearly autonomous driving. So under common law jurisdiction the question is, if notwithstanding, there is an accident, then the manufacturer is liable if the driver is not at fault. Product liability consists of three elements: defect in the product, defect in the design and lack of warning or lack of instructions. To translate these into the software case, bugs should certainly qualify as defects in the software. In those autonomous vehicles though, in which the deep learning is a standard, a bug may not necessarily be programmed from the start but might have been learnt by the machine. In all these cases the state of the art defence may help out the manufacturer from liability. Since most of the car companies rely on many suppliers, the driverless cars contain various softwares from different manufacturers, embedded even. This makes even harder the proving of contributory negligence or the percentage of interaction in the operation of the hazardous equipment, the car, in the accident.

Additionally, the autonomous cars need data. These data, for cybersecurity reasons, are provided via separate, very quick Wi-Fi like facilities, like the DSRC (Dedicated Short Range Communication) in the USA, and, probably, 5G in Europe, especially in Germany. The data flow needs to be free from misunderstanding, misinformation and hacking. In case of misunderstanding the question is whether there is an interoperability problem, or the rival systems fail. In any event, the node needs to be found and decided where the faulty data came from. In-built cautious messages, designs, such as use 'break, if…', could release liability. There probably is a duty of care on the public roads among the actors, therefore, if there is a misinformation, then the sender of the information may be liable for negligence or fraudulent misrepresentation if he knew or should have known that the information was false. Besides, the receiver of the information needs to show reasonable care to rely on the information received and does not react if other information contradict, like an obvious change in the traffic order. Cybersecurity protocols require special verifications of messages and other protection but hacking remains an evergreen issue, the softwares have to be able to manage it. In the end special investigation is needed to find the very legal entity, the manufacturer, supplier, seller, repairer, owner or operator to be liable. Bugs existence needs to be accepted and the software so designed so that to react reasonably cautiously.

Despite of the Toyota case, software liabilities are generally privileged in the legal systems. In case of contractual breach, the damages are limited even above the standard of the foreseeability doctrine and in torts cases the rule of economic loss applies. These all serve the same purpose that purely economic damages cannot be recovered and that the costs of innovation be widespread. Softwares in Cyberspace are extremely intertwined and since robotics are often facilitated by/on the internet, the security of the softwares is a crucial point in driverless cars. It is because robotics, and especially the driverless cars, unlike the softwares on the internet, are designed to be active in the real space, thereby much easier and more possibly causing more physical harms.

The application of strict liability of either the driver, which can certainly be the driverless car itself according to the level two-three above, or the producer or the owner, is justified first of all because of the acceptance that there is no bug-free software and because the very existence of the bug in the software may be discovered only by use. Thus the real question boils down to the procedure or to the protocol to be followed by design in situations of emergency. It is important since strict liability might provide for exemption in cases where the accident is due to an unavoidable interference of an action from the outside of the scope of the operation of the hazardous equipment, the car.

4. Conclusions - The smart future

IOP Conf. Series: Materials Science and Engineering **448** (2018) 012030 doi:10.1088/1757-899X/448/1/012030

”Well, Senator, my position is not that there should be no regulation” said Mark Zuckerberg, the CEO of Facebook at the Senate hearing, in a dialogue with Senator Lindsey Graham (R-South Carolina) (CNET Joint Commerce and Judiciary Committee SH-216, April 2018).

Entering a search request into the search engine of the Google Scholar, one receives more than 2.660.000 findings for the technical term ‘legal paradigm’, even if there are only about 1% out of these, which really match the request. And this is so despite the fact, that the notion of paradigm, is not at all evidentially applicable in the social sciences or humanities, thus in the legal context. Thomas Kuhn used the expression of paradigm, without the adjective, primarily to describe and to understand the truthfulness of the various and sometimes contradictory scientific assessments of natural sciences. His point of view was definitely historical and argued that certain statements may well qualify as scientific in one paradigm and not at all as such in another paradigm. Paradigm, summarized rudimentarily, would mean a complex world in which scientific development is achieved with putting the tiny little verified puzzles of the whole picture together. A paradigm shift would be brought about, put it again very unsophisticatedly, when certain puzzles could not fit in and so they challenge the truthfulness of the entire picture.

But these sort of paradigms were meant to measure/commeasure sciences pure, which were computable or verifiable in its strict sense. Social sciences and humanities, on the other hand, are not based on calculations merely, even though economics, sociology, demography, statistics would so claim. In any event law, or legal studies, is evidence-based, uses interpretation, argumentation and reasoning, and it relies much much less on calculations.

But paradigm shifts, for their being a social construction too, sociologically may also be biased. Accordingly, the paradigm shift may be delayed or accelerated depending on the social embeddedness of the very advocates of the different puzzles fit or unfit into the relevant dogmatic picture of the existing paradigm.

Paradigm, nevertheless, may translate into a context: a context of rules and language, to simplify again. And if so, it may apply to the legal environment as well, since, in a way, law relies on reading and interpreting texts by using the rules of the language.

This paper discussed the problem of paradigm shifts in legal thinking through the technique of analogies. It concluded that as long as the same interpreting tools, such as fictions, analogy, distinctions, etc., are applicable in the various incidents and the court cases, there is no paradigm shift in the normal science. The answer is upto the courts.

References

[1] Aarnio A 1984 Paradigms in Legal Dogmatics *Theory of Legal Science* Springer Dordrecht Peczenik A, Lindahl L and Roermund B eds **176**

[2] Aarnio A 2011 *Essays on the Doctrinal Study of Law* Springer p 237

[3] Bérces V 2010 A magyar büntetési rendszer reformjával kapcsolatos jogalkotási kérdések *IAS* pp 83-99

[4] Bostrom N 2003 Are we living in a computer simulation? *The Philosophical Quarterly* **53** 211 pp 243-255

[5] Calo R 2015 Robotics and the Lessons of Cyberlaw *Cal. L. Rev* **103** p 513

[6] Cooter R and Ulen T 1988 *Law and Economics* HarperCollins pp 645

[7] Cox I and Wilfong G eds 1990 *Autonomous Robot Vehicles* Springer New York pp 477

[8] Cummings D 2017 Was the Jury Wrong about Toyota's Software? How Questionable Testimony on Embedded Software Tipped the Scales *IEEE Consumer Electronics Magazine* pp 103-107 [DOI: 10.1109/MCE.2017.2684939]

[9] Daithí M 2016 My Favourite Things: The Promise of Regulation by Design, JOTWELL

[10] Deibert R et al 2010 *Access Controlled - The Shaping of Power, Rights, and Rule in Cyberspace* The MIT Press

[11] de Vries U 2013 Kuhn and Legal Research. A Reflexive Paradigmatic View on Legal Research*: *Recht en Methode in onderzoek en onderwijs* **3** 1 pp 7-25

[12] Eszteri D 2015 A mesterséges intelligencia fejlesztésének és üzemeltetésének egyes felelősségi kérdései *IJ* pp 47-57

[13] Fuglinszky Á 2015 *Kártérítési jog* HVG-Orac

[14] Garland B 2004 ed *Neuroscience and the Law*. Brain, mind, and the scales of justice. Dana Press New York | Washington, DC

[15] Gelén K ed 2017 *Jog, innováció, versenyképesség.* Wolters Kluwer Budapest p 172

[16] Gillespie N 2014 Richard Posner: Privacy is "Mainly" About Concealing Guilty Behavior. *Reason. Free Minds and Free Markets*

[17] Glancy D 1979 The Invention of the Right to Privacy *Arizona Law Rev* **21**

[18] Gookins E 2012 *Achieving a Safe and Reliable Product. A Guide to Liability Prevention.* ASQ Quality Press Milwaukee, Wisconsin p 145

[19] Gormley K 1992 One Hundred Years of Privacy *Wisconsin Law Review.* WILR 1335

[20] Graziano T 2016 Cross-border traffic accidents in the EU – the potential impact of driverless cars ipol studies

[21] Hansmann H and Kraakman R 2000 The Essential Role of Organizational Law. *Yale Law Journal* **110** No 3 390-440

[22] Hildebrandt M and Rouvroy A eds 2011 *Law, human agency, and autonomic computing: the philosophy of law meets the philosophy of technology* Routledge

[23] Hildebrandt M 2015 Ambient Law, Smart Technologies and the Rule of Law 27 octobre 2015 Cycle de conférences octobre-décembre 2015 Le futur du droit : big data, algorithmes et robotisation

[24] Homicskó Á ed 2018 Egyes modern technológiák etikai, jogi és szabályozási kihívásai. *Acta Caroliensia Conventorum Scientiarum Iuridico-Politicarum* **XXII**. Budapest p 223

[25] Johnson D and Post D 1996 Law and Borders – The Rise of Law in Cyberspace. *Stan LR* **48** 1367

[26] Jürgen R 2013 *Autonomous Vehicles for Safer Driving*. SAE International p 142

[27] Kiilunen V 2018 Autonomous Vehicles, Competence and Liability in the EU –Answering the Call of the European Parliament DOI: 10.13140/RG.2.2.10863.74403

[28] Klein T ed 2018 Tanulmányok a technológia- és cyberjog néhány aktuális kérdéséről. *Médiatudományi Könyvtár*. **30**. p 166

[29] Koops B – Hildebrandt M and Jaquet-Chiffelle D 2010 Bridging the Accountability Gap: Rights for New Entities in the Information Society? *Minn. J.L. Sci. & Tech* **497**

[30] Koops B and Leenes R 2014 Privacy regulation cannot be hardcoded. A critical comment on the 'privacy by design' provision in data-protection law *International Review of Law Computers & Technology* **28** pp 159-171 preprint version

[31] Leroux C and Labruto R 2012 Suggestion For A Green Paper On Legal Issues In Robotics – *ELS issues in robotics*. Researchgate

[32] Lessig L 1999 Commentaries. The Law of the Horse: What Cyberlaw might Teach Us *Harv Law Rev* **113** pp 501-547
[33] Ligeti K 2002 Az Európai Unió büntetőpolitikája *ÁJT* pp 73-97
[34] Lunders C and Tarpley P 2017 Autonomous Vehicles, USA. in: *Autonomous Vehicles, The Legal landscape of DSRC in the US, UK and Germany. NortonRoseFulbright* p 7-11
[35] Matthias A 2008 *Automaten als Träger von Rechten. Plädoyer für eine Gesetzesänderung* Logos Berlin Verlag
[36] Murray A 2013 Looking back at the Law of the Horse: Why Cyberlaw and the Rule of Law are Important *Journal of Law, Technology & Society* Scripted 310
[37] Murray A 2007 *The Regulation of Cyberspace: Control in the Online Environment* Routledge
[38] Nagy Cs 2010 The Rome II Regulation and Traffic Accidents: Uniform Conflict Rules with Some Room for Forum Shopping – How So? *Journal of Private International Law* **6** 1 pp 93-108
[39] Parrish M 2016 The 2009 Toyota Accelerator Scandal That Wasn't What It Seemed
[40] Patterson D and Moratti S eds 2016 *Legal Insanity and the Brain Science, Law and European Courts Law and Neuroscience: revising the Legal Standard of Insanity* Hart Publishing p 336
[41] Pétervári K 2017 Autonóm járművek és jogi felelősség - a "Tesla" baleset; avagy ki vezet(het)i az autót *Gazdaság és Jog* **25**. pp 31-34
[42] Post R 1991 Rereading Warren and Brandeis: Privacy, Property and Appropriation. *Faculty Scholarship Series.Paper* 206
[43] Reed C 2010 Information 'Ownership' in the Cloud *Queen Mary School of Law Legal Studies Research Paper* 45
[44] Reidenberg J 2014 The Data Surveillance State in Europe and the United States, *Wake Forest L. Rev.* **49** 583-608
[45] Rudgard S 2011 Origins and Historical Context of Data Protection Law *European Privacy* IAPP publication
[46] Russen Spears V 2008 The Case That Started It All: Roberson v. The Rochester Folding Box Company *Privacy & Data Security Law Journal* pp 1043-1050.
[47] Sanitt A 2017 Autonomous Vehicles – The Legal Landscape of DSRC in the UK. in: *Autonomous Vehicles, The Legal landscape of DSRC in the US, UK and Germany. NortonRoseFulbright* p 35-40
[48] Santosuosso A, Goodenough O and Tomasi M eds 2015 *The Challenge of Innovation in Law. The Impact of Technology and Science on Legal Studies and Practise* Pavia University Press
[49] Santosuosso A and Bottalico B 2016 Robotics and the Law: A European Perspective *Autonomous Systems: Why Intelligence matters. Conference Lecture,* München, Automatik Industrie 4.0
[50] Sárközy T 2002 Büntetőjogi intézkedések a jogi személyekkel szemben? *MJ* pp 449-456
[51] Searle J 1980 Minds, Brains and Programs. *Behavioral and Brain Sciences*, 3: 417–57
[52] Smith B 1928 Legal Personality **37** *Yale L.J.* 3. pp 283-299
[53] Solove D 2004 *The Digital Pperson. Technology and Privacy in the lnformation Age* NYU Press p 296
[54] Solove D, Rotenberg M and Schwartz P ed 2006 *Privacy, Information, and Technology*, Aspen Publishers p 322
[55] Solum L 1992 Legal Personhood for Artificial Intelligences, *N.C. L. Rev.* **70** pp 1231
[56] Somkutas P and Kőhidi Á 2017 Az önvezető autó szoftvere magas szintű szellemi alkotás vagy kifinomult károkozó. *In Medias Res* **2** p 232–269
[57] Stallman R 2002 *Free Software, Free Society*. Selected Essays, GNU Press. Free Software Foundation Boston, MA USA pp 230

[58] Sundquist M 2012 Oonline Privacy Protection: Protecting Privacy, The Social Contract, and the Rule of Law in the Virtual World *Regent Univ Law Rev* **25** p 153-184
[59] Swaab D 2014 *We Are Our Brains. From the Womb to Alzheimer's*. Penguin Books Ltd p 432
[60] Tóth A ed 2016 *Technológia jog - Új globális technológiák jogi kihívásai.* Budapest p 294
[61] Urquhart L and Rodden T 2017 Towards User Centric Regulation
[62] Urquhart L and Rodden T 2016 Working Title: A Legal Turn in Human Computer Interaction? Towards 'Regulation by Design' for the Internet of Things *Working_Paper_March*
[63] Van Egmond Van P, Nijkamp P and Vindigni G 2003 A comparative analysis of the performance of urban public transport systems in Europe. *Int'l Social Science Journal* **55** Issue 176, p 235-247
[64] Wadhwa I and Salkever A 2017 *The Driver in the Driverless Car How our Technology Choice will Create the Future* Berrett-Koehler Publishers
[65] Warren S and Brandeis L 1890 The Right to Privacy. *Harvard Law Rev*, **4**, No. 5 pp 193-220
[66] WIRED (2016/12).
[67] Zittrain J 2006 The Generative Internet, *Harv L Rev* **119** p 1974
[68] Ződi Zs 2017 Az információs társadalom legújabb kihívásai a jog számára - Horizontális platformok. *Gazdaság és Jog* **25**. pp 27-30

Regulations

The USA

i. NHTSA: Policy (Guidance) of best practices for the safe design, development and testing of automated vehicles prior to commercial sale or operation on public roads, 2017.
ii. USA House Bill HR 3388 on "Safely Ensuring Lives Future Development and Research in Vehicle Evolution or SELF-DRIVE." 2017
iii. USA Senate Bill: S.1885 — 115th Congress (2017-2018) [Report No. 115–187]: "American Vision for Safer Transportation through Advancement of Revolutionary Technologies Act" AV START Act, 2017
iv. Autonomous Vehicles / Self-Driving Vehicles Enacted Legislation. (05.18.2018) NCSL, National Conference of State Legislatures, USA
v. California regulation 2014
vi. California regulation 2018
vii. US Department of Transportation: Fact Sheet: AV Policy Section I: Vehicle Performance Guidance for Automated Vehicles 2016 12687g2-091916-v1
viii. Federal Guidance on Autonomous Vehicles, Washington State, October 18, 2016

The EU – The European Parliament

ix. Motion for a European Parliament Resolution with recommendations to the Commission on Civil Law Rules on Robotics (2015/2103(INL))
x. Resolution of 1 June 2017 on digitalising European industry 2016/2271(INI)
xi. Resolution of 18 May 2017 on road transport in the European Union; 2017/2545(RSP)
xii. Automated vehicles in the EU, EPRS briefing, January 2016
xiii. EU strategy on cooperative intelligent transport systems, EPRS briefing, September 2017
xiv. 5G network technology - Putting Europe at the leading edge, EPRS briefing, January 2016

The European Council

xv. Discussing the Commission Communication - Digitising European Industry Reaping the full benefits of a Digital Single Market COM(2016) 180 final, 20 April 2016, 8100/16

IOP Conf. Series: Materials Science and Engineering **448** (2018) 012030 doi:10.1088/1757-899X/448/1/012030

xvi. Valetta declaration by Ministers of Transport at the High-level Conference on Road Safety(Malta 28–29 March 2017) - Way forward /Valletta Declaration on Road Safety

The European Commission

xvii. Europe on the move: An agenda for a socially fair transition towards clean, competitive and connected mobility for all, COM(2017) 283 final, 31 May 2017

xviii. European Commission/European Urban Mobility: Policy Context. 2017

xix. Final Report Summary - ROBOLAW (Regulating Emerging Robotic Technologies in Europe: Robotics facing Law and Ethics), CORDIS, 2015

Member States of the EU

xx. Guidance, (UK) The key principles of vehicle cyber security for connected and automated vehicles. Published 6 August 2017

xxi. UK: Automated and Electric Vehicles Bill 2017-19, dealing notably with the issue of insurance for automated vehicles, House of Commons Library, 2017

xxii. Directive on security of network and information systems 2016 –, https://assets.publishing.service.gov.uk/government/uploads/system/uploads/attachment_data/file/567242/national_cyber_security_strategy_2016.pdf UK version

xxiii. DE: Achtes Gesetz zur Anderung des Straßenverkehrsgesetzes vom 16. Juni 2017. Bundesgesetzblatt Jahrgang 2017 Teil I Nr. 38, ausgegaben zu Bonn am 20. Juni 2017

OECD – International Transport Forum

xxiv. Automation of the Driving Task: Some possible consequences and governance challenges 20 July 2017

xxv. Human Factors, User Requirements, and User Acceptance of Ride-Sharing in Automated Vehicles 20 July, 2017.

IOP Conf. Series: Materials Science and Engineering **448** (2018) 012031 doi:10.1088/1757-899X/448/1/012031

Application and development trends in high strength steel and aluminium

Zs Lukacs, M Tisza

University of Miskolc, Institute of Material Science and Technology

lzsolt@kugli.met.uni-miskolc.hu

Abstract. Due to increasing competition globally among car manufacturing companies cost optimization is one of the major concerns in the automotive industries right from inception stage to the final product delivery along with maintaining its highest quality. Nowadays, rising demand for concept car, automobiles with latest technology, aesthetics of vehicles with safety concern, high strength and light weight results technological advancement in sheet metal forming and innovative material development like different grades of high strength steels and lightweight aluminium alloys. In 2015 with the initiative of the Imperial College London a research consortium was established under the umbrella *Low Cost Materials Processing Technologies for Mass Production of Lightweight Vehicles.* The primary aim of this project is to provide affordable low cost weight reduction in mass production of vehicles considering the entire life-cycle. The technological process solution for HFQ™ of Al-alloys is presented which greatly helped in reducing weight and increasing the strength.

1. Introduction

The globally raising awareness toward CO_2 reduction and resource efficiency concerning to climate change have significantly increased the importance of this topic over the last few years. Lightweight materials development and design have always been an important topic in product design across several industries. Developments of different high strength steel and aluminium help in the improvement of powertrain efficiency like small turbo-charged engines, alternative powertrain/ hybridization, efficient transmission. The global automotive industry is more challenging than people think. In addition, the automotive industry has benefited as developed economies, have enacted many monetary policies to stimulate growth, pushing interest rates lower. As interest rates decreasing, consumers were more preferring to purchase a new vehicle due to the more affordable financing options available and due to rising incomes of the working class. In the last few years, due to the increased demands of vehicles in the market there were very intense developments concerning the applied materials in car body manufacturing especially different grade of high strength steel and aluminium. AHSS is one of the latest and fastest growing steels in the automotive sector. It leads not only to mass reduction but also reduces greenhouse gas emissions of vehicles. Stringent fuel efficiency regulations have facilitated R&D to develop lightweight & cost effective materials for the automotive sector. Affordability and excellent properties of newly developed high strength steel grades are likely to have positive effect its utilization in the automotive sector over the upcoming years. Aluminum has been witnessing major and crucial growth in the demand from automotive industry due to its weight reduction potential, improved strength and high recyclability as compared to steel. Aluminum can absorb twice the crash energy as compared to mild steel and provides weight reduction of up to 50%, which is expected to propel its demand over the upcoming years [1]. According to the European Union progress report 2017 (United4climate), European Union members states agreed on commitment to reduce their domestic emissions by at least 40 % between 1990 and 2030. But according to the recent projected data based on existing measures, the 20 % target for 2020 will be met. In 2030, emissions are expected to be 30 % which is lower than in 1990 if there will be no additional policies are implemented. Therefore, EU is currently negotiating new legislation to ensure that it will meet its objective of reducing emissions by at least 40 % between 1990 and 2030 [4]. Following the adoption of the Paris Agreement at COP 21, efforts and resources will be further concentrated on fighting climate change and decarbonizing the economy. A new EU's focus area, as known as 'Building a low-carbon, climate-resilient future', will specifically support the implementation of the Paris Agreement for 2018-2020 [4].

In the European Union, current climate targets require a 60% reduction in transport emissions by 2050. Pursuing these objectives could be facilitated by enhancing knowledge on the evolution of existing car fleets (hereafter defined as fleet dynamics), examining the impact of changes in material composition, weight and drivetrains of new cars, the energy system and consumer behavior. The greatest emissions savings seem to be achieved only by significant weight reduction and development of more efficient engines. So, advanced high strength steel is the fastest growing steel category in the automotive sector not only because it leads to significant weight reduction but also reduces the green house emission from the vehicles. Powertrain comprises of various forged metal parts like transmission shafts & gears, clutch hubs, drive shaft, and universal joints. In addition, the engine also have forged piston in order to increase its durability. Suspension and body frame also need the use of high strength materials to ensure vehicle safety [1]. Concerning the continuously increasing requirements on the improvement of crash resistance of vehicles, high strength materials like high strength steel and high strength Al-alloys are more and more widely used. Among these high strength materials Manganese-boron steels are the most often used in hot-press forming. Hot-press formed parts offer very high resistance against deformation during crash thank to their very high strength (R_m = 1500 to 2000 MPa). These high strength steels besides the excellent crash-resistance and safety properties are also important from mass reduction point of view. that is, reducing the fuel consumption and harmful CO_2 emissions.

2. Application of high strength steel of car manufacturing

There is an intensive research ongoing on the application of various grades of high strength steels. Steel dominated the global market in term of revenue share. Affordability and excellent properties of new steel grades including advanced/ultra-high strength steel are likely to positively affect its utilization in the automotive sector over the forthcoming years [1]. For several decades, the application of the conventional cold-rolled steels have dominated in the car body manufacturing. However, in the last 20-30 years due to the increasing demands and growth in the automotive sector, there were very intensive developments in the steel sector with more emphasis given on lightweight materials development with better strength, higher safety. These improvements kept in view the regulation of EU norm of CO_2 emission according the Paris agreement at COP 21. Some examples for the so-called advanced High Strength Steels (AHSS steels), X- AHSS, U-AHSS. As it can be seen in figure 1, different phases of development of steel called as first, second and third generation steel respectively because of its advancement in terms of increasing strength parameters, the ductility parameters and thus the formability is significantly decreasing. Therefore, the high strength steels have the limited extension of formability that is another important issue.

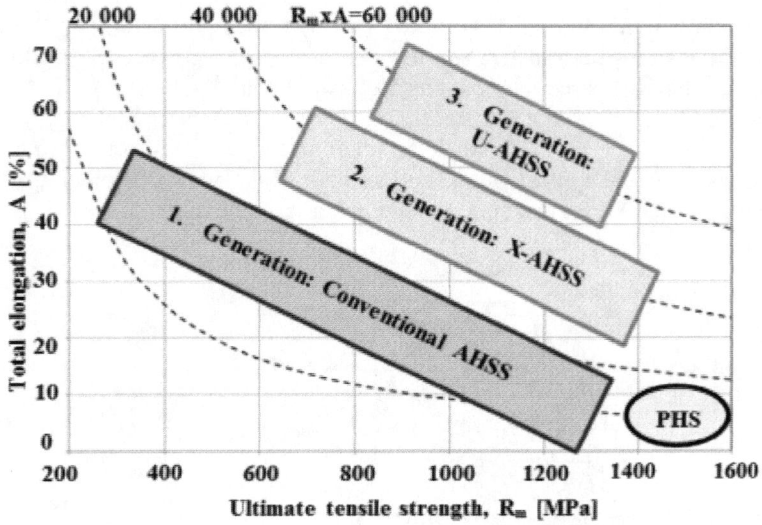

Figure 1. Total elongation vs. Ultimate Tensile Strength for various generations of high strength steels.

We can also see in figure 1 that the total elongation (A_{80}) is a function of ultimate tensile strength (R_m). Among the various new, high strength steel grades like DP steels, TRIP steel, TWIP steel, there were various other grades available and further works going on to develop innovative materials that bring revolution in the automotive sector in terms of cost, quality and sustainability. According to figure 1, we can conclude that strength increases from 1st generation to 3rd generation but elongation gets lower and lowered. Also, there is one important material developed called Press Hardening Steels (PHS) and it is widely used to produce high strength structural body elements (e.g. A and B-pillars, etc.) applying hot forming processes. From this material grade, different kinds of boron alloyed manganese steels should be mentioned that are also commonly used already in car body manufacturing in hot forming condition. That is particularly the 22MnB5 alloy which is regarded as the basic type of PHS steels. This PHS steel is important from our main point since we focus on a new hot forming process (Hot forming & Quenching-HFQ^{TM}) developed for light metals.

Even though the application of hot press forming in the automotive industry just started in the recent days, the process itself dates back to 1977 when Erland Lundström, a Swedish scientist patented a process to produce sheet steel products such as a reinforcement element in a larger structure [1].

The next important milestone in the application of hot-press forming in the automotive industry was in 1984. The Swedish SAAB company introduced this technology first time in serial production for manufacturing the side impact beam in SAAB 9000 series [2]. Since then, the use of hot-press forming was widely used in producing parts offering an extra high level of strength. Manganese-boron alloyed steel (22MnB5) as one of the most important typical grade was used as blank material.

2.1. Hot Forming of Press Hardening Steels (PHS)

Hot forming is a unique metal forming technique. Hot press forming is also known as press hardening and it is a relatively new forming process. It was developed particularly for the application of high strength steels in car body manufacturing in the automotive industry. This technology was invented in Sweden [5], then further developed in Germany [6], [7] and France [8]. Many industrial plants are working in the world among them in Hungary; the Kirchhoff Automotive is the main supplier of components produced by Press Hardening of boron alloyed manganese steels for Hungarian OEM car makers [9].

This process may be shortly summarised as follows: the material is heated up to a sufficiently high temperature up to the austenitic zone and holding temperature is selected in the homogenous γ-zone above the A_3 temperature of the given steel until the carbon fully dissolve producing homogeneous austenite microstructure. Then, hot forming occurs and the part is cooled down rapidly in the forming tool with the critical cooling rate to prevent the carbon diffusion, thus resulting in martensitic microstructure (as shown in figure 2, N=1500X) leading to significant increase of the strength parameters. Since the martensitic transformation – i.e. the hardening process – occurs in the forming tool, this is the reason that this process is often termed as press hardening.

The important process cycle of hot press forming (HPF) of 22MnB5 steel can be seen in figure 2. Boron-alloyed manganese steels are the most suitable group for hot press forming. This group includes the 22MnB5 steel as the most typical one for this application, but some other types as 8MnCrB3, 20MnB5, 27MnCrB5 and 37MnB4 can also be used in hot press forming.

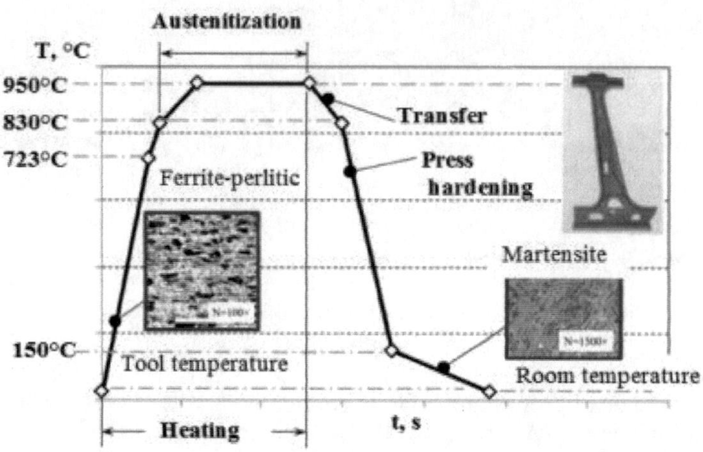

Figure 2. Temperature vs. time for hot press forming (HPF) of 22MnB5 steel.

3. High strength Aluminium alloys in the automotive industry

Significant progress has been witnessed in the demand of Aluminium from automotive industry due to its lightweight and high recyclability as compared to steel. Aluminium can absorb twice the crash energy as compared to mild steel and provides weight reduction of up to 50.0% which is expected to propel its demand over the coming years [1]. The 1st generation high strength steel are very closer properties to the high strength Al alloys, it is clearly seen from figure 3. Aluminium is also commonly used as base material for car body elements. It is very important from mass reduction point of view. However, there is an important concern that it has lower formability than steels which leads to serious formability problems in cold forming operations. It is well known that the formability is increasing significantly with the increase of the temperature. This is the main reason that significant efforts and steps are made for both in the laboratories and in industrial circumstances to apply high strength aluminium alloys in hot forming conditions.

During the recent years, several new grades of high strength aluminium alloys were developed [10]. Among them, the so called 5xxx, 6xxx, and more recently the 7xxx series are mainly used in the automotive industry. Also, with the increase of strength the formability of these alloys is also reduced. So, the formability as for most of the metals and metallic alloys can be significantly increased at elevated temperatures. Constellium is one of the major suppliers of different grades of high strength aluminium in the world and in the European automotive sector as well, providing different grade of aluminium alloys with different trade names like Sufalex®, Stronglex®, Ultralex®. These grades of aluminium alloys with different trades name have very close mechanical properties to the first generation AHSS as seen from the figure 3.

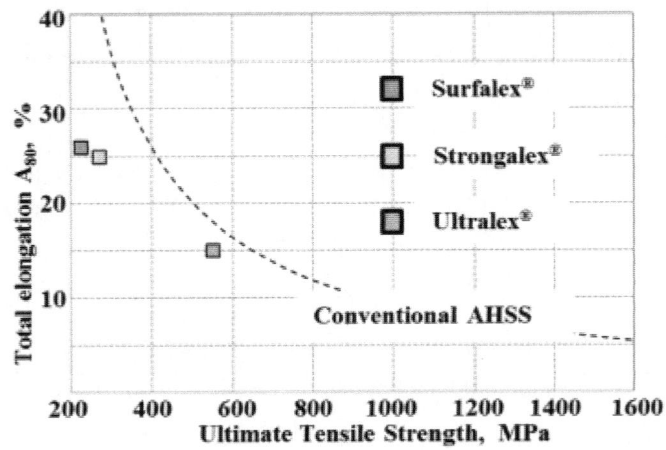

Figure 3. Compare of first generation AHSS vs. new generation high-strength Al-alloys.

The Ultralex® offers inter-granular, filiform structure and stress corrosion resistance best-in-class among the 7xxx family, provides approx. 15% weight savings vs. UHSS and formable using press hot forming & warm forming, and joinable using adapted mechanical fasteners.

Table 1. Compare of 22MnB5 vs. Ultralex® Al-alloy T6 condition.

Side impact beam product	YS, MPa	UTS, MPa	A_{80}, %	Gauge of ISO crash performance	%-light-weighting vs. 22MnB5
22MnB5	1100	1500	6	1.2	-
Ultralex®	500	550	15	2.9	-17%

In order to compare the UHSS (e.g. 22MnB5) and Ultralex® mechanical properties for side impact beam product as presented in table 1, we can conclude that Ultralex® Al-alloy in T6 condition has more than twice the total elongation and crash strength as compared to 22MnB5. The Ultralex® Al-alloy is 17% lighter than UHSS (e.g. 22MnB5) because of its lower density. Hot forming techniques helps in dissolving the precipitates.

3.1. Theoretical background of forming of high strength Al-alloys

The hot forming and quenching of high strength aluminium alloys was developed partly utilising the principle of Press Hardening of Steels but due to the very different material science background of aluminium alloys, it requires the usage of several other concepts. The basic principle behind this process is based on the unique behaviour of ageing aluminium alloys during heat treatment which is well-known from material science [11]. This principle is shown schematically in figure 4 with the usual Temperature- time (T vs t) diagram [12]. The first part of this process is a solution heat treatment (SHT) where all the precipitations are dissolved above 500 °C (usually between 525-580 °C) providing a homogenised microstructure of α-solid solution. Depending on the thickness of the part, it may take 30-60 min.

After this homogenization, a fast cooling (usually water quenching) has been required to avoid large precipitates. The result is a homogenous but super-saturated α-solid solution with very low strength and good formability parameters. The high strength is provided in the second phase by a precipitation hardening (often termed as artificial ageing) usually done about 180-190 °C, which may take several hours [13]. However, at the end of the hardening, the formed aluminium part has low strength properties due to the super-saturated solid solution state. Therefore, it is followed by a post forming precipitation hardening (artificial ageing) heat treatment to achieve the required high strength. But, it is usually a long procedure that takes several hours.

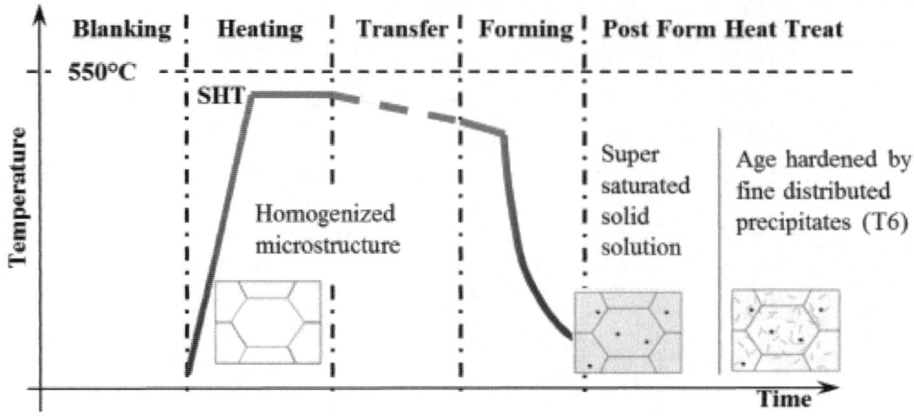

Figure 4. Basic principle of Hot Forming and Quenching™ of aluminium alloys.

3.2. Hot Forming and Quenching Technology HFQ©

The Hot Forming and Quenching (HFQ®) process was patented by the Imperial College London (ICL) and the Impression Technologies (ITL). However, this innovative new process may only be competitive if it can offer production time cycle comparable to PHS of high strength steels. The general principle of this process can be seen in figure 4. This new technique helps the manufacturers to form deep drawn complex shapes from high and ultra-high strength aluminium using a fast process with cycle time suitable for a range of volumes low to high. The technological process solution for HFQ™ of Al-alloys is shown in figure 5. In this process, blank is heated in solutionising oven, then transferred to high speed hydraulic press to make form panel and in the same high speed hydraulic press perform quenching in tool. Finally, it brought into the ageing oven that is the last stage 4. It is an important part of this process since after the solution heat treatment and fast cooling aluminium alloys have low strength properties, therefore a precipitation hardening (often termed as artificial ageing) is necessary to provide the high strength properties.

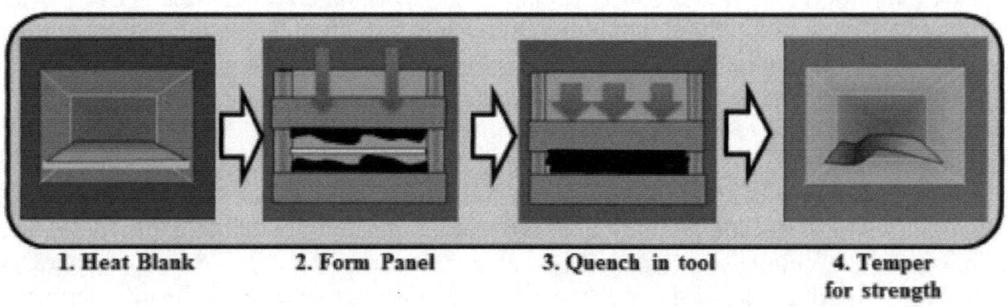

Figure 5. The technological process solution for HFQ™ of Al-alloys.

Advantages:
The major advantages of this technology are the following:

- It is easily adaptable for aluminium parts without simplifying or changing the geometry while replacing the steels.
- It is well-capable in using high-strength grades for down gauging (reduce weight, reduce material usage).
- This technology completely eliminates the need for springback-compensation in tool and part design.
- It also eliminates the use of reinforcement panels which are required to add strength, otherwise.
- This technology helps in substituting the high strength Al extrusions with high-strength HFQ® pressings.

4. Conclusions

Besides the increasing application of various advanced high strength steel materials, there are many new technological development and an intensive growth in the application of high strength aluminium alloys in the automotive sector. In this paper, we briefly summarized the application and development trends in high strength steels and aluminium alloys for producing lightweight automotive parts. The major concern besides weight reduction is to increase the strength, crashworthiness, the vehicle fuel efficiency and to reduce the carbon dioxide emission according to EU norm. The recent developments in the application of lightweight design concepts in the automotive industry were analysed from the viewpoint of hot forming of metallic materials. Firstly, basic principles of hot forming of steels, widely known as Press Hardening of Steels, were introduced and then hot forming and quenching of aluminium alloys were presented. PHS-steels are mainly boron alloyed manganese steels and among them, the 22MnB5 alloy is the most widely applied in automotive applications. Also, comparison has been made between the first generation AHSS (22MnB5) and the new generation high-strength Al-alloys (Ultralex®). The development of the latter unique process brought revolutionary change to the usability of aluminium and cost effectiveness along with the higher safety.

References

[1] Automotive Metal Market Analysis by Product (Aluminium, Steel, Magnesium) by Application (Body stucture) by End-use and Segment Forecasts 2018-2025 (2017), *Report ID: GVR-1-68038-560-1*

[2] Rick P, Reid W, Evan H, Akshay S (2017) 2017 Automotive Trends

[3] Ruth H, Nicolai M, Wolff van S, Anne S, Andreas T (2012) Lightweight heavy impact *McKinsey&Company* 2012

[4] Report from the commission to the European Parlament and the Council – *Two years after Paris: Progress Towards meeting the EU's climatec commitment* (2017) Brussels 7.11.2017

[5] Tisza M (2013) Recent development trends in sheet metal forming, *Int. J. Microstructure and Material Prop.* 8. No. 1-2. pp. 125-139.

[6] Tisza, M (2014) Advanced materials in sheet metal forming, *Key Engineering Materials*, 581 pp. 137-142. DOI:10.4028/www.scientific.net/KEM.581.137

[7] Ghassemieh E (2011) Materials in Automotive Application – State of the Art and Prospects, *in New Trends and Developments in Automotive Industry*, InTech Open Publications, DOI: 10.5772/1821

[8] Banik J *et. al* (2013) Warmumformung im Automobilbau, Süddeutsche Verlag, Munich, p. 84.

[9] Keeler S, Kimchi M and Mooney, P (2017) Advanced High Strength Steels – Application Guidelines 6.0, *World Auto Steel* April, pp. 314.

[10] Sam C (2017) Constellium gets it right with AutoForm-Thermosolver *AutoForm Blog-Customers Stories*

[11] Audi AG.: Historical background on use of aluminium at Audi, https://www.audiworld.com/news/02/aluminum/content1.shtml

[12] European Aluminium Association (2012) Aluminium in cars-unlocking – The light weighting potential http://www.alueurope.eu/publications-automotive/

[13] LoCoLite (2013), An industry system enabling the use of a patented materials processing technology for Low Cost forming of Light Weight structure for transportation industries EU FP7 NMP Project ID No. 604240

Acknowledgments

The authors express their grateful thanks to the European Commission for their support on the H2020 project entitled "Low Cost Materials Processing Technologies for Mass Production of Lightweight Vehicles (LoCoMaTech)", Grant No: H2020-NMBP-GV-2016 (723517). HFQ is a registered trademark of Impression Technologies Limited. Impression Technologies Limited is the sole licensee for the commercialization of the HFQ technology from Imperial College London.

IOP Conf. Series: Materials Science and Engineering **448** (2018) 012032 doi:10.1088/1757-899X/448/1/012032

Experimental Investigation on the Single Point Incremental Forming of AlMn1Mg1 Foils using Flat End Tools

Sherwan M. Najm[1,3, b)] and Imre Paniti[2,1, a)]

[1] *Department of Manufacturing Science and Engineering, Budapest University of Technology and Economics, Hungary, H-1111 Budapest, Műegyetem rkp. 3.,*
[2] *Institute for Computer Science and Control, Hungarian Academy of Sciences H-1111 Budapest, Kende u. 13-17., Hungary*
[3] *Northern Technical University, Iraq*

[a)] Corresponding author: imre.paniti@sztaki.mta.hu
[b)] sherwan@manuf.bme.hu

Abstract. Single Point Incremental Forming (SPIF) is one of the main and most used variants of Incremental Sheet Forming (ISF), a flexible rapid prototyping technique in the sheet metal sector. In this paper, the effect of flat end tools on SPIF with variable wall angle geometry was studied under different process parameters. Experiments on AlMn1Mg1 sheets with 0.22 mm initial thickness are carried out by a CNC Milling Machine. Geometry measurements of the formed parts carried out on a coordinate measuring machine (CMM) are also presented.

Keywords: Single Point Incremental Forming, Formability, AlMn1Mg1, flat end tools

1. Introduction

Single Point Incremental Forming (SPIF) is one of the most important types of Incremental Sheet Forming (ISF) due to the flexibility and capability to change the experimental parameters to achieve the desired shape. In fact, the main advantage of SPIF is the ability of rapid design by changing factors affecting the process, like sheet thickness, tool radius, tool speed, tool feed and type of lubricant used. This makes SPIF one of the most important methods of manufacturing prototypes before starting mass production. One of the most important factors is the tool geometry which is affecting the other related parameters especially the formability and the maximum forming angle in SPIF.

T. McAnulty et al, mentioned that seven articles indicate that increasing tool diameter led to increasing the formability and ten articles proved the opposite of that, and six articles agreed to the fact that the best tool diameter had adjusted for the highest formability since all the tools were solid hemispherical [1]. There are three different types of forming tools that are used in SPIF; spherical, ball bearing in the concave cavity with free movement on the forming sheet surface and the flat end tool [1]. These tool names are based on the end of the tool without relation to the shank of the tool.

There were many discussions on this topic, especially in earlier studies on the effect of shape and size of the forming tool. According to a study of Y.H. Kim and J.J. Park [2], a hemispherical-head tool is less efficient - in connection of formability - than a ball tool. They used an aluminium 1050 sheet with 0.3 mm in thickness in their experiments and obtained the best formability with ball tool of 10 mm in diameter. By applying a ball tool the decreased friction between the tool and formed sheet caused better formability than in the case of hemispherical tool [3]. In [4] the authors claimed that the geometrical accuracy of a product is better when a flat end tool is used instead of a hemispherical one. Furthermore, a flat end tool in some cases showed a better result in term of formability, and needed less force to form the sheet than the hemispherical end tool [4]. Failure occurs easily with the decrease of the contact area between the tool end and the formed sheet due to the high localized stress generated, which led to lack of required plasticity for the formability [5]. In SPIF, Sheet thickness, the diameter of forming tool and the interaction between them having a significant effect on the forming wall angle limit [6]. Best surface roughness was obtained with a 10 mm tool diameter which explicated to the biggest contact between the tool and the formed sheet [7]. Geometrical accuracy can be increased by decreasing the diameter of the tool and the step size [8].

Furthermore, it is claimed that formability in SPIF increases with friction and by applying smaller tool diameter [8].

In this paper, three tools with different corner radius are used to investigate experimentally the best flat tool corner radius in SPIF for achieving maximum forming depth and best accuracy of profile geometry. Profile measurements are carried out with a coordinate measuring machine (CMM), and thickness distribution of the plate along the wall of the product is examined with a Digital Micrometre. AlMn1Mg1 sheets with 0.22 mm are used to form a truncated cone with variable wall angle by a CNC milling machine. In all cases machine oil as lubricant was applied.

2. Material Properties

AlMn1Mg1 with the chemical composition in Table 1 had been used for the experiments. Tensile testing was used to measure the mechanical properties of the sheet in $(0, 45, 90)^0$ of the rolling direction, the mechanical properties are listed in the Table 2. Material properties are given by Széchenyi István University.

TABLE 1. Chemical Composition

Al	Si	Fe	Cu	Mn	Mg	Zn	Cr	Ni	Others
96,90	0,201	0,448	0,212	0,807	1,260	0,071	0,022	0,006	0,073

TABLE 2. Mechanical Properties

Direction	Rp0,2, MPa	Rm, MPa	Ag, %	A50, %	n5	r10
0^0	88,3	183,0	16,44	16,88	0,297	0,554
45^0	90,0	155,5	9,27	10,45	0,266	0,580
90^0	86,3	170,3	12,48	12,95	0,268	0,594

3. Experimental

Rieckhoff CNC milling machine with LinuxCNC real-time controller was used for the experimental part as shown in Fig. 1 (a). Flat steel tools with 2 mm end radius and different corner radius (0.1 mm, 0.3 mm and 0.5 mm) for the experiments are shown in Fig. 1 (b).

IOP Conf. Series: Materials Science and Engineering **448** (2018) 012032 doi:10.1088/1757-899X/448/1/012032

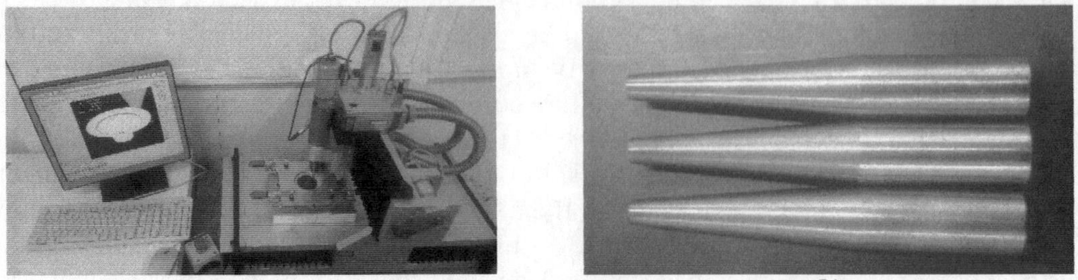

(a) (b)

FIGURE 1. (a) Set-up of the experiments, (b) SPIF Tools with different corner radius

Fig. 2 (a) shows the magnified image of the tool with corner radius 0.5 mm and the radius calculation from the obtained magnified area of the circle, while Fig. 2 (b) illustrate the designed shape of the truncated cone with an increasing wall angle. All parameters were fixed to identify the influence of the tool corner radius on the forming depth, geometry accuracy, and pillow effect (as defined in [9]). Inwards spiral strategy with analogical tool rotation were performed, the experiments were carried out with step size: 0.01 mm, feed rate: 1500 mm/min and spindle speed: 3000 Resolutions Per Minute.

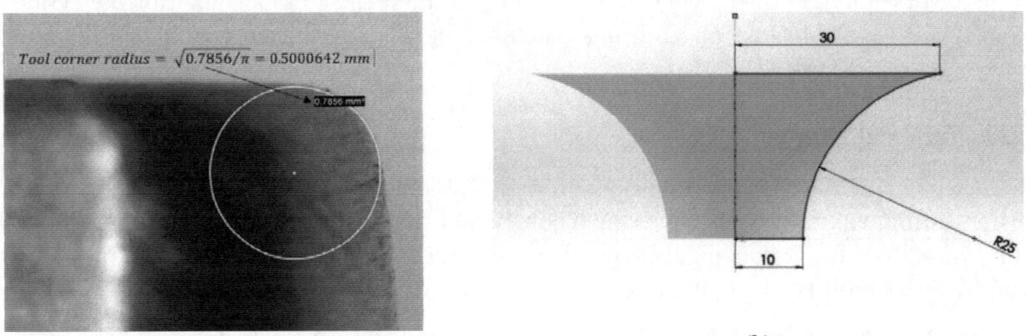

(a) (b)

FIGURE 2. (a) magnified image of the tool with 0.5 mm corner radius, (b) Section view of the test geometry

The result of the Pillow effect of the formed parts and the cone profile in the Z-direction were measured using a Mitutoyo CMM. Fig. 3 (a) shows the CMM measurement setup. A digital micrometer was used to measure the thickness distribution for each formed part along the cone depth, as shown in Fig. 3 (b).

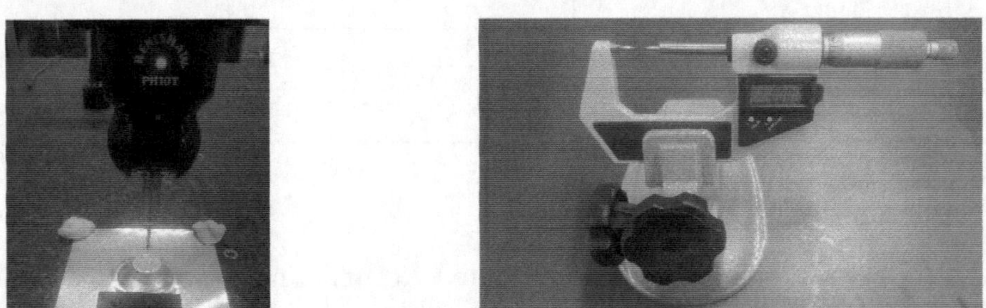

(a) (b)

FIGURE 3. (a) Measurement setup with Coordinate Measure Machine, (b) Digital Micrometre for thickness measurement

Three sets of experiments with the result showed in Fig. 4 (a) were carried out. Fig. 4 (b) shows the crack propagation appeared on the formed part.

IOP Conf. Series: Materials Science and Engineering **448** (2018) 012032 doi:10.1088/1757-899X/448/1/012032

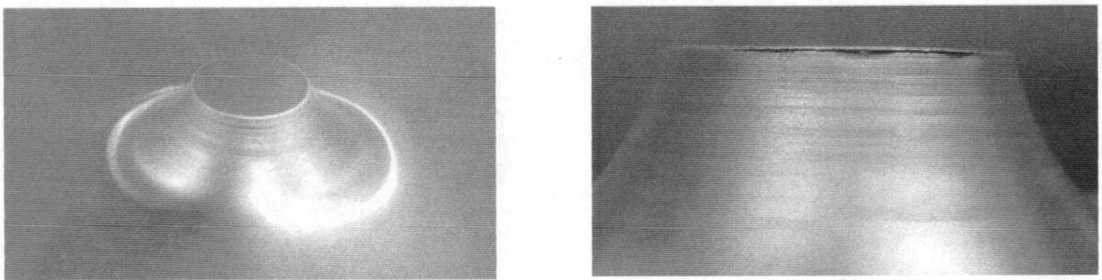

(a) (b)

FIGURE 4. (a) Formed product, (b) Crack appearance on the base of the cone

4. Result and Discussion

The maximum depth and formability was achieved with the 0.1 mm tool corner radius, so it can be concluded that forming depth increases with the decrease of the flat tool corner radius. The tool with 0.3 mm corner radius also showed good formability but slightly less than the smallest one. Fig. 5 (a) shows the relation between forming depth and tool corner radius. A new phenomenon occurred in the formed part using 0.5 mm corner radius since the sheet was neither formed nor downed from the inside. Forming depth decreased while pillow effect increased in this case. Due to the small step size (0.01 mm) the tool returned to the same formed zone several times. The previous mentioned deformation mechanism with the tool rotation in the same zone caused the flow and streching of the material and lead to an increase of the Pillow effect Fig. 5 (b) shows the undesired formed part with increased Pillow effect.

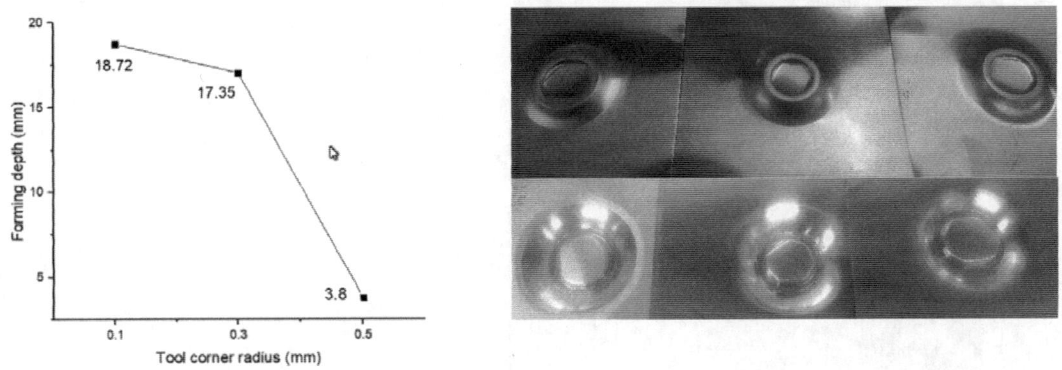

(a) (b)

FIGURE 5. (a) Forming Depth in relation to tool corner radius, (b) Undesired formed part

Fig. 6 (a) shows the thickness distribution which indicates that the thickness was continuously decreased from the beginning to the highest plasticity until the end of the process. The elastic-plastic behavior decreased and finally the material faced instability. Fig. 6 (b) shows the thickness measurement points on the part.

IOP Conf. Series: Materials Science and Engineering **448** (2018) 012032 doi:10.1088/1757-899X/448/1/012032

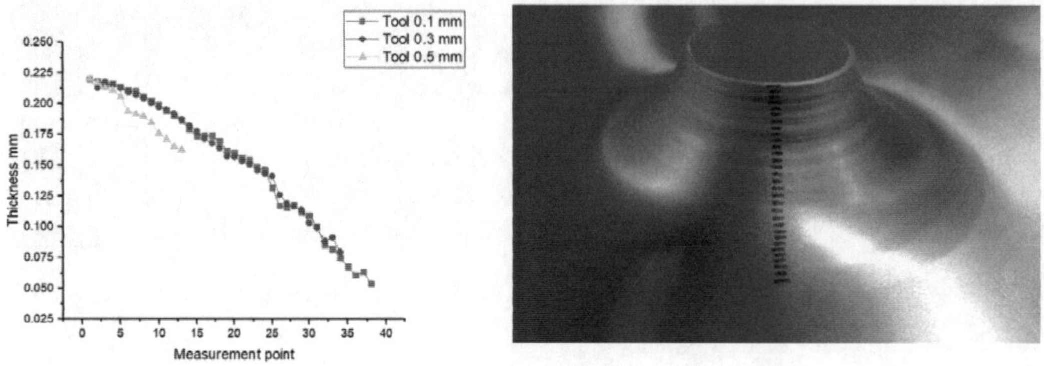

(a) (b)

FIGURE 6. (a) Thickness distribution, (b) Thickness measurement points on the part

The pillowed Surface at the bottom of the part was measured by a Digimatic dial indicator (gage) as shown in Fig. 7 (a). The effect of the tools realized in the deviation from the ideal geometry at the bottom (Pillow effect) is presented in Fig. 7 (b). The results are also summarized in Table 3. The results showed that the minimum deviation from the Pillow effect was obtained by the smallest corner radius of the tool, while the biggest corner radius leads to maximum deviation in connection with the Pillow effect. It follows that decrease of the tool corner radius is directly proportional to Pillow effect decrease.

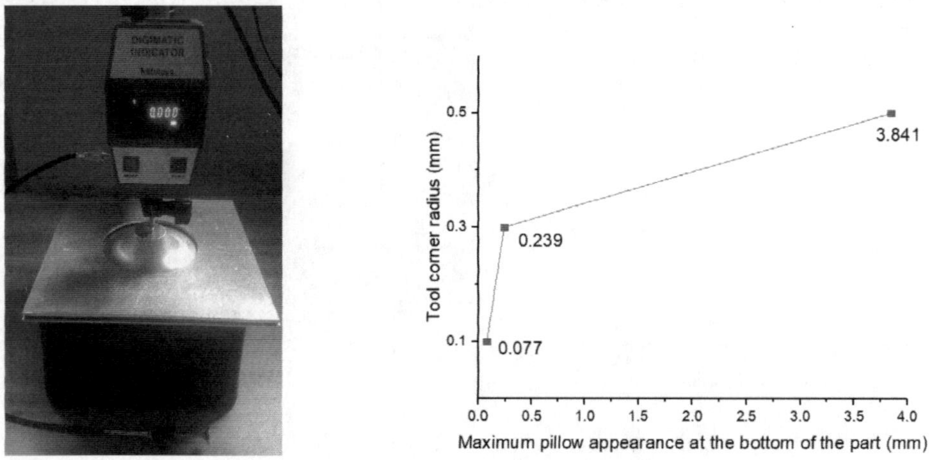

(a) (b)

FIGURE 7. (a) Digital gage (Digimatic Indicator), (b) Result of the Pillow effect

TABLE 3. Pillow effect at the bottom of the parts

Tool corner radius [mm]	Max. deviation caused by pillow effect [mm]
0.1	0.077
0.3	0.239
0.5	3.841

The geometry profile (circumferential) was measured with a CMM at the edge of the cone depth of the formed part downward to the edge of the base of the cone (see Fig. 8 (a)), deviations from the ideal profile are given in Table 4. Fig. 8 (b) also shows the effect of the tool regarding to the profile geometry. The results highlighted that the smallest corner radius of the tool gives the best geometrical accuracy, by decreasing spring-back.

TABLE 4. Profile deviation

Tool corner radius [mm]	Achieved profile radius [mm]	Deviation [mm]
0.1	50.638	0.638
0.3	45.75	-4.25
0.5	43.143	-6.857

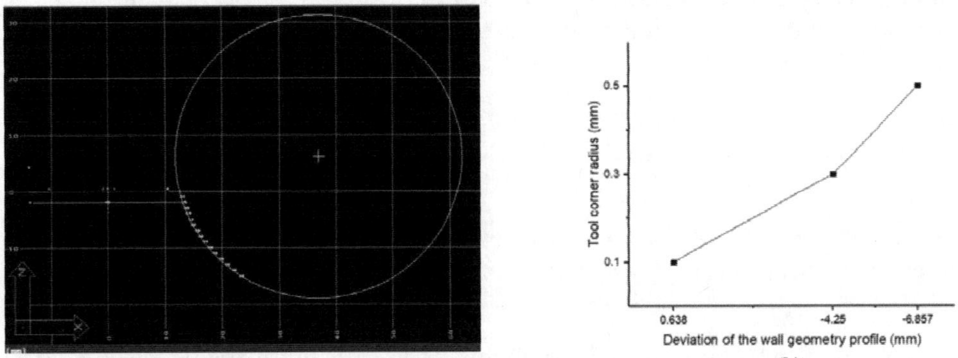

(a) (b)

FIGURE 8. (a) profile geometry points from CMM, (b) Deviation of the profile geometry

5. Conclusion

As a result of the experiments the flat tool with 0.1 mm corner radius showed the best results from all measured parameters of the formed part (forming depth, thickness distribution, pillowed surface and final geometric accuracy) and this result decreases by increasing the corner radius of the flat end tool. Finally, a new ratio can be defined in micro SPIF (where initial sheet thickness is below 0.5 mm) regarding corner radius of a flat tool and initial thickness.

6. Future work

More experiments are needed with different flat tool parameters to investigate other parameters of the formed part.

Acknowledgement

This research has been supported by the GINOP-2.3.2-15-2016-00002 grant on an "Industry 4.0 research and innovation center of excellence"; and by EPIC grant No. 739592. The authors would like to thank the staff of the Laboratories of Department of Manufacturing Science and Engineering, Budapest University of Technology and Economics, Hungary. The authors also thank Dr. Miklós Czampa Assistant Professor, Head of Laboratory for his help.

IOP Conf. Series: Materials Science and Engineering **448** (2018) 012032 doi:10.1088/1757-899X/448/1/012032

References:

[1] T. McAnulty, J. Jeswiet, and M. Doolan, "Formability in single point incremental forming: A comparative analysis of the state of the art", *CIRP J. Manuf. Sci. Technol.*, vol. 16, pp. 43–54, 2016.

[2] Y. . H. Kim and J. . J. Park, "Effect of process parameters on formability in incremental forming of sheet metal", *J. Mater. Process. Technol.*, vol. 130–131, no. 3, pp. 42–46, 2002.

[3] Y. Kumar and S. Kumar, "Advances in Material Forming and Joining", pp. 29–47, 2015.

[4] X. Ziran, L. Gao, G. Hussain, and Z. Cui, "The performance of flat end and hemispherical end tools in single-point incremental forming", *Int. J. Adv. Manuf. Technol.*, vol. 46, no. 9–12, pp. 1113–1118, 2010.

[5] B. Cawley, D. Adams, and J. Jeswiet, "Examining tool shapes in single point incremental forming", *Proc. NAMRI/SME*, vol. 41, no. August 2012, pp. 114–120, 2013.

[6] S. Gatea, H. Ou, and G. McCartney, "Review on the influence of process parameters in incremental sheet forming", *Int. J. Adv. Manuf. Technol.*, vol. 87, no. 1–4, pp. 479–499, 2016.

[7] S. A. Nama, N. S. M. Namer, and S. M. Najm, "The Effect of using Grease on the Surface Roughness of Aluminum 1100 Sheet during the Single Point Incremental Forming Process", vol. 1, no. 1, pp. 1–4, 2014.

[8] F. Maqbool and M. Bambach, "Dominant deformation mechanisms in single point incremental forming (SPIF) and their effect on geometrical accuracy", *Int. J. Mech. Sci.*, vol. 136, no. December 2017, pp. 279–292, 2018.

[9] Hussain, G., L. Gao, and N. Hayat. "Forming parameters and forming defects in incremental forming of an aluminum sheet: correlation, empirical modeling, and optimization: part A." Materials and Manufacturing Processes 26.12, pp. 1546-1553, 2011.

New multi-objective methods to solve resource-constraint multi-project scheduling problems in integrated ERP-APS systems

Krisztián Mihály[1], Mónika Kulcsárné Forrai[1] and Gyula Kulcsár[1]

[1] Department of Information Engineering, University of Miskolc, Miskolc-Egyetemváros, Miskolc 3515, Hungary

krisztian.mihaly@sap.com, aitkfm@uni-miskolc.hu, iitkgy@uni-miskolc.hu

Abstract. This paper presents a new scheduling model and methods to solve multi-project scheduling problems considering strict resource constraints and multiple objective functions simultaneously. The solution approach is based on an integration of enterprise resource planning system (ERP) and an advanced planning and scheduling application (APS). An ERP system software component defines a set of projects with their details (activities, due dates, processing times, precedence constraints, etc.) and the available resources with their capabilities and real constraints. The user can specify many objective functions and their priorities. The goal of this scheduling problem type is to create a detailed execution plan that meets all the hard constraints and gives a near optimal compromised solution from objective functions point of view. In real execution environments, more than one projects may run in parallel, and they may require the same resources. For example, projects may produce the standard type of products, and they use the same manufacturing resources. In different decision-making situations, dynamic priorities may be defined for projects or activities to express the relative importance of items. The paper focuses on modeling and solving of multi-objective multi-project scheduling problems, and we present the integration of ERP and APS systems.

1. Introduction

Field of operation research and project management has a significant history, with many proven results. Great books and articles show approaches to solve different problem types. The valid reason to invest further activities in this research area is the high variety of the real-life problems and the fact which they are proven as NP-Hard problems. Scientists and software engineers are looking for more efficient ways to deal with these problems or enhance the models with new, relevant aspects.

In this paper, we are focusing on the resource-constrained projects where the scheduling result shall complete multi-objective goal functions. Next, to the models, we are presenting a standard, state of the art market leader ERP solution to model and schedule projects and we target different alternatives to enrich the standard functionality.

A project management problem requires planning and scheduling decisions. The planning process focuses on distributing the total project into manageable tasks or activities. The project planner estimates the demand for different types of resources and processing time (duration) of all the tasks of the project. The tasks must be assigned to the resources, and the executions of tasks must be scheduled over time. The project scheduling deals with these decision-making problems. The role of the project scheduler covers the allocation of the available resources to the project tasks to determine the start time and finishing time of each task or activities [1].

2. Resource Constrained Project Scheduling

The resource-constrained project scheduling problem (RCPSP) is a well-known model describing a special version of project scheduling problems. The problem itself contains a various number of tasks (*activity, job*) and limited resources (*machine, people, executor*). The model is describing a single project, which consists of a set $T = \{1,2,3, \dots n\}$ of different tasks to be executed. Set T may be enhanced with virtual task 0 and $n + 1$ as "project start" and "project end", respectively. To execute the project a set of resource types $K = \{1,2,3, \dots, m\}$ are defined. Resource type k $(k \in K)$ has a limited capacity of R_k at any point in time. Resources are renewable, meaning execution of a task does not decrease the available capacity of the resource type k during the project execution. Execution of task j cannot be break, when execution on defined resources has been started. The capacity of a given resource type is constant for the time horizon.

Each task may define two kinds of constraints:

- If task j $(j \in T)$ has immediate predecessor tasks defined by set P_j, then all task i $(i \in P_j)$ shall be finished before execution of task j is possible.
- If task j $(j \in T)$ requires one or more resource they are defined by set of pairs $(r_{j,k})$.

Each task has a pre-determined execution time on resource type k, defined by $p_{j,k}$. It is not allowed to define a circle in the predecessor tasks chain. Further constraint for virtual tasks are, that task 0 cannot have predecessor tasks and task $n + 1$ cannot be predecessor for any task. Furthermore, virtual tasks do not require any resource and virtual task processing time is 0.

Let C_j denote the completion time of task j $(j \in T)$. A feasible solution can be described as a vector, where each completion time of a task is given by the corresponding C_j. Let $A(t) = \{j \in T | C_j - p_j \leq t \leq C_j\}$ the set of tasks, which are being processed at time t. The objective of the RCPSP is to find precedence and resource feasible completion times for all tasks such that the makespan of the project is minimized.

3. Schedule Generation Schemes and Priority Based Rules for RCPSP

One solution approach to solve an RCPSP problem is to generate a feasible schedule with predefined rules. These are called as schedule generation schemes (SGS), and they are the core of further specific implementations. The generation starts from an empty schedule and in each iteration, a feasible schedule is constructed until all tasks of the project are scheduled. The SGS implementations can be distinguished into activity- and time-incrementation based generations. In this paper, we are focusing on the activity-based generation.

Serial SGS is an activity-incrementation based generation scheme. It consists of $g = 1,2, \dots n$ stages. Each stage starts with a feasible solution and represents one task selection with its addition to the actual schedule at the earliest precedence- and resource-feasible completion time.

Two disjoint sets of tasks can be defined. Set S_g contains tasks, which have been already scheduled, set D_g contains tasks which are eligible to be scheduled. Let $R'_k(t) = R_k - \sum_{j \in A(t)} r_{j,k}$ be the remaining capacity of resource type k at time instant t and let $C_g = \{C_j \mid j \in S_g\}$ be the set of all finishing times. Let further $D_g = \{j \in T \setminus S_g \mid P_j \subseteq S_g\}$ be the set of eligible tasks. The serial SGS is given as follows [2]:

Initialization $C_0 = 0,\ S_0 = \{0\}$

For g = 1 to n do

Calculate $D_g, C_g, R'_k(t)\ (k \in K;\ t \in C_g)$
Select one $j \in D_g$
$EC_j = max_{h \in P_j}\{C_h\} + p_j$
$C_j = \min\{t \in [EC_j - p_j, LC_j - p_j] \cap C_g \mid r_{j,k} \le R'_k(\tau), k \in K, \tau \in [t, t + p_j[\cap C_g\} + p_j$
$S_g = S_{g-1} \cup \{j\}$

$C_{n+1} = max_{h \in P_{n+1}}\{C_h\}$

The algorithm starts from the initial feasible schedule when virtual project start task is scheduled with it completion time 0 and all resources are available with maximum capacity. At the beginning of each stage g, the decision set D_g, the set of completion times C_g are calculated. Then one activity j is selected from the decision set. Once j has been selected, the earliest precedence feasible completion time EC_j is calculated.

4. Multi-project Scheduling Problems

The allocation of resources and scheduling the tasks for multiple projects are non-polynomial (NP) hard problems and they are more difficult than a single project [3].

In the literature, different classical optimization methods have been used to solve the multi-project scheduling problems. Paper [4] proposed a zero-one programming approach and paper [5] introduced an integer programming for generating the schedule and a simulation method for testing the heuristic rules to choose the best schedule. In [6], Deckro et al. formulated the multi-project scheduling problem as a general integer programming model and presented a decomposition approach to solve large size problems. Jolayemi [7] proposed an integer programming approach for multi-project scheduling and considered a special penalty function. These outlined studies give good solutions for small size problems by using traditional optimization approaches. If the problem sizes are middle or large, then the classical methods cannot solve the problems within a reasonable time.

In the literature, we can found several heuristic and meta-heuristic methods for multi-project scheduling. Researchers developed many useful algorithms to generate feasible solutions for multi-project schedules [8]. The goal of this development was to increase the efficiency of the heuristic methods, to extend the scope of the problems with new methods. and reduced the computation time.

Many researchers have applied artificial intelligence methods to solve the multi-project resource-constrained project scheduling problems. For example in [9], Kim et al. proposed a combined genetic algorithm to create the schedule. Their goal was to minimize the total project completion time. Kumanan et al. applied a genetic algorithm based approach for generating the schedule of multi-project scheduling problems [10]. The objective function of the optimization was also the project completion time. Damak et al. also proposed a genetic algorithm variant with a local search strategy to solve the problem considering the project delay minimization as the optimization goal [11].

After reviewing the connected literature, we can conclude that there is a need to develop efficient and flexible methods to improve the quality and robustness of the solutions of resource constrained multi-project scheduling problems.

In real execution environments, multiple projects are running parallelly, and they are not independent of each other. They may have defined requirements of the same resource, or the execution of project tasks may be dependent on the completeness of a task in another project.

As circle in the tasks precedence path is still not allowed, technically it may be possible to restructure multi-project dependencies to independent projects, where connection points between projects are modeled as virtual sub-project start and sub-project end tasks. Each project has an own defined objective

IOP Conf. Series: Materials Science and Engineering **448** (2018) 012033 doi:10.1088/1757-899X/448/1/012033

to meet. The project management declares objective functions to be optimized varying in time, like as to achieve the earliest completion time, to minimize the maximum tardiness, to maximize the robustness, to minimize the sum of costs and to maximize the average utilization of resources. An extended, adaptive and effective scheduling software is needed to satisfy these flexible multi-objective requirements of the users.

The remaining part of this paper summarizes our approach, development, and implementation to realize an integrated multi-objective multi-project solver system.

5. Multi-objective Approach

5.1. Schedule generation scheme with multiple priority rules

Multi-objective optimization is an optimization problem when more than one objective function has to be optimized simultaneously.

Serial SGS implementation selects at stage g an arbitrary task from the decision set D_g. As enhancement of first RCPSP model all tasks may have a predefined priority. At stage g from decision set D_g the task with the highest (lowest) priority is selected. Definition of priorities may be calculated during runtime, depending on the actual applied schedule. These dynamic priorities are called as priority rules. For example, latest finishing time (LFT) or minimal slack (MS) based priority rules.

Let us denote priority rule of a task as $v_j(T)$. Depending on applied and supported priority rules, a set of priority rules can be combined to enforce tasks selection with multiple aspects.

5.1.1. Linear combination of multiple priority rules. Many priority rules can be used for calculating the priority of the task to be scheduled. The set of priorities can easily be combined into a single formalism. If task j $(j \in T)$ has a set $V_j = \{v_{j,1}, v_{j,2}, \dots, v_{j,L}\}$, then weighting factors can be introduced to each priority rule, respectively as $PrioV_j = \{a_1, a_2, \dots, a_L \mid \forall a_l \in \mathbb{R}_0^+\}$. Weighting factors are input parameters for the scheduling and the user can calibrate the desired combination.

At the priority selection step, the priority value is calculated for all priority rules, defined by $v_{j,l} \in V_j$ $(l = 1, 2, \dots, L)$ and then the linear combination of all priority members is calculated (1).

$$v_j = a_1 \cdot v_{j,1} + a_2 \cdot v_{j,2} + \dots + a_l \cdot v_{j,L} = \sum_{l=1}^{L} a_i \cdot v_{j,l} \tag{1}$$

The difficulty of this method is to give the appropriate values of the weighting factors. The choices of these values strongly influence the solution. It is not possible to distinguish the normalization and the expression of the importance sharply. Normalization is required to add priorities with different values and dimensions. The importance expression focuses on the role of the different priority rules in the actual decision-making process.

5.1.2. Multiple priorities with the relative comparison. We propose a new approach to choose the task to be scheduled from the decision set in the task selection steps of the schedule generation schemes.

A relative rating method has been developed to compare the task of the decision set according to multiple priorities. The priorities are given such that:

$$v_l : T \to \mathbb{R}_0^+, \forall l \in \{1, 2, \dots, L\} \tag{2}$$

Coefficients w_l $(l = 1, 2, .., L)$ as input parameters support that the user may calibrate the actual importance of each priority rule independently. Each w_l is an integer value $(w_l \geq 0)$ that expresses the importance of v_l.

Let $t_x, t_y \in T$ be two candidate tasks. The function F is defined by (3) to express the relative importance of t_y compared to t_x as a real number.

$$\mathrm{F} : T^2 \to \mathbb{R},\ F(t_x, t_y) := \sum_{l=1}^{L} (w_l \cdot D(v_l(t_x), v_l(t_y))) \tag{3}$$

The function D defined by (4) means the comparison of t_x and t_y according to v_l.

$$D : \mathbb{R}^2 \to \mathbb{R}, \; D(a, b) := \begin{cases} 0, \; if \max(a, b) = 0 \\ \frac{b - a}{\max(a, \; b)}, \; otherwise \end{cases} \tag{4}$$

The relational operators are overloaded with (5) using the definition (3) of the function F.

$$(t_y, t_x) := (F(t_x, t_y) \; ? \; 0) \tag{5}$$

Any of the relational operators can be used between two tasks to compare them like two real numbers. For example: t_y is less important than t_x, if $F(t_x, t_y)$ is less than zero.

The most important task can be selected in the actual step of the schedule generation scheme, comparing candidate tasks in an actual decision set. According to our proposal, the relative importance of a task is more expressive than the weighted linear combination of its priorities. The base of the approach is that the relative importance of the candidate task is measured by comparing it with another task in the decision set.

5.2. Multi-objective searching algorithm with relative qualification

Different scheduling goals can occur managing real-life processes or systems. In many cases, for example, it is very important that the task be completed as close as possible to its due date. A task that is tardy causes delays in the processes or projects, while early completion of a task can be damaging if the storage space is limited or this event has an extra cost. Typically, many other objective functions must be considered, and they are usually conflicting. The scheduling problems often involve more than one aspect, so it is required to use multiple criteria simultaneously. For this reason, a general method has been developed to measure the performance of the fine schedules by calculating management indices or performance indicators based on tasks, projects, and resources.

A generalized mathematical model has also been elaborated for a relative qualification that serves for comparing the created schedules according to multiple objectives [12], [13].

According to our proposal, the relative goodness (quality) of a solution is more important than the absolute goodness (quality) of one. The base of the approach is that the relative goodness of the candidate solution is measured by comparing it with another solution in the feasible solution space.

Let Ψ be the solution space. Suppose that the objective functions are given such that:

$$f_z : \Psi \to \mathbb{R}_0^+, \; f_z \to min. \forall z \in \{1, 2, \dots, H\} \tag{6}$$

Coefficients u_z $(z = 1, 2, .., H)$ as input parameters support that the user may calibrate the actual importance of each objective function independently. Each u_z is an integer value $(u_z \geq 0)$ that expresses the importance of f_z.

Let $s_x, s_y \in \Psi$ be two candidate solutions. The function F is defined by (7) similarly to (3), in order to express the relative goodness of s_y compared to s_x as a real number.

$$\mathrm{F} : \Psi^2 \to \mathbb{R}, \; F(s_x, s_y) := \sum_{z=1}^{H} (u_z \cdot D(f_z(s_x), f_z(s_y))) \tag{7}$$

The function D, which is defined before by (4), means the comparison of s_x and s_y according to f_z. Using the re-definition (7) of the function F, the relational operators can also be overloaded by (8).

$$(s_y, s_x) := (F(s_x, s_y) \; ? \; 0) \tag{8}$$

In this case, any of the relational operators can be used between two solutions to compare them like two real numbers. For examples: s_y is better than s_x, if $F(s_x, s_y)$ is less than zero.

These definitions of the relational operators are suitable for applying in meta-heuristics like taboo search, simulated annealing, evolutionary or genetic algorithms to solve multi-objective multi-project resource-constraint scheduling problems.

IOP Conf. Series: Materials Science and Engineering **448** (2018) 012033 doi:10.1088/1757-899X/448/1/012033

Our solver system solves the scheduling problem in two phases. In the first phase, combined heuristic schedule generation schemes and multiple priority rules create solutions. The best solution becomes the initial solution of the second phase. In this second phase, the solver performs iterative improvements by using multi-objective searching methods.

6. Integration of ERP and APS systems

ERP systems are mainly used to manage all resources in an enterprise company. They consist of set of different modules, responsible for coupled, separated and integrated application areas, like manufacturing, finance, project management, and so on. We were focusing on SAP ERP system in our work, but our conceptual model can be used by other ERP vendors as well, after adjustment of the implemented artifacts to the vendor-specific platform.

As all standard project management systems, SAP ERP PM supports management of project related entities, such as the creation of the project, the definition of tasks and dependencies, the definition of resources and their task assignments. In project execution phase the reporting of execution, collaboration with external parties are covered.

In SAP ERP this module is called as project management (PM) or project and portfolio management (PPM) [14] [15]. Both implementations follow the same architecture design. Data is persisted in a relational database which is Open SQL compatible; business logic is written in ABAP language; user interfaces are developed for various sets of modeled business roles in different UI technologies, like SAP Dynpro, ABAP WebDynpro or JavaScript-based SAP UI5.

The standard SAP solutions are supporting scheduling of projects. The state of the art implementation supports so-called up-down and bottom-up scheduling algorithms. These algorithms are mainly an extended CPM implementation. Two main directions have been identified to enrich project standard scheduling capabilities with advanced scheduling capabilities.

6.1. APS integration within ERP

SAP standard solution offers a set of extension points to influence system behavior and enhance the business logic. ABAP language and the SAP NetWeaver platform offers a rich set of features to implement a complete new APS module. The concept of the integrated APS is depicted in ***Figure 1***.

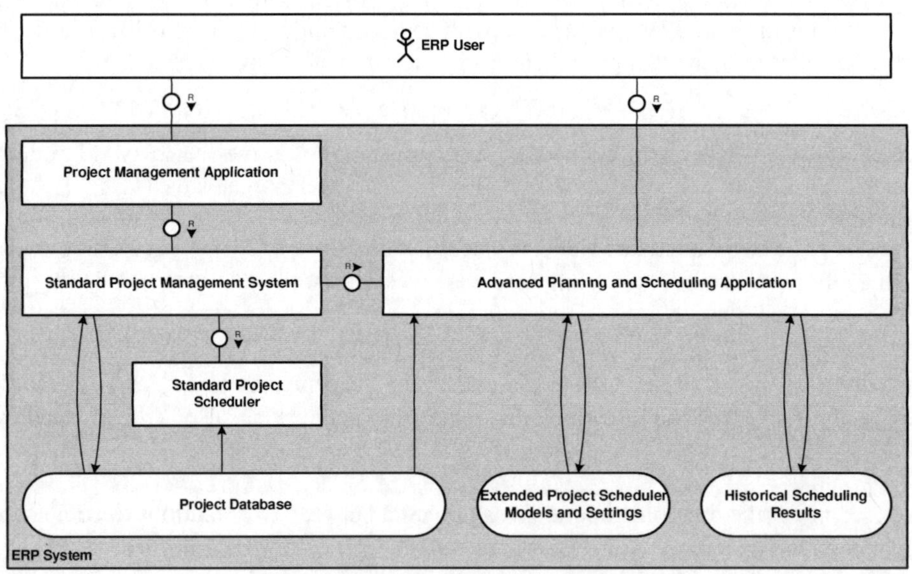

Figure 1. ERP integration with internal APS

The NetWeaver based APS implementation is using the standard project database as source and APS models and configuration used to enrich scheduling capabilities. The APS agent is triggered by the SAP PM or by ERP user. Scheduler reads actual project management and resource booking data, reads the scheduler configuration for calibrated scheduling parameters, may check the historical scheduling information and retrieves for the ERP user a feasible schedule or failure information in case of of non-feasible scheduling setup.

Advantages of this approach are to have tight integration with standard PM system, system-to-system communication is not required, and the NetWeaver platform supports all transactional requirements. The disadvantage of the approach is, that solution is SAP platform specific and requires a complete SAP NetWeaver stack.

6.2. Integration with external APS
Integration concept of external APS system usage depicted in ***Figure 2***.

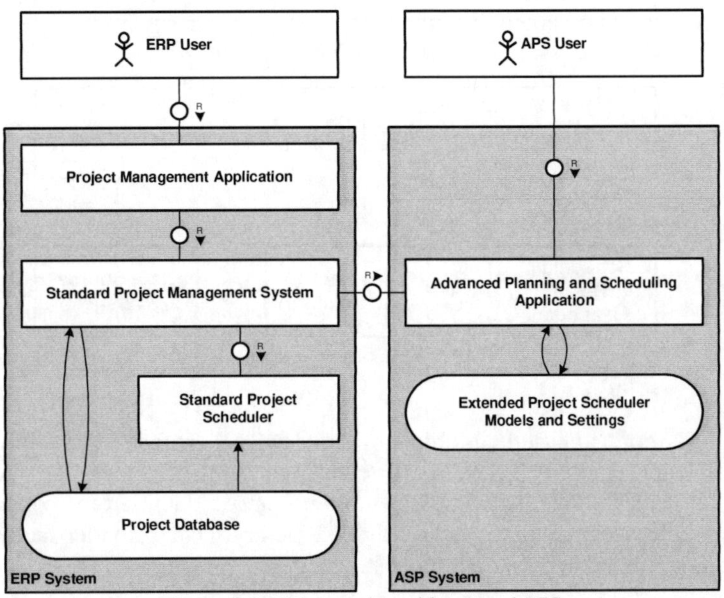

Figure 2. ERP integration with external APS

In this concept, APS is a separated system with own stack. ERP and APS system can communicate via predefined channels, like HTTP, RFC, IDoc and so on. The ERP user and the ASP user is modeled independently. The ERP system calls the APS system and the execution of an advanced project scheduling consists of the following steps:

- ERP system collects all relevant data required for the scheduling service and maps it to the expected structure. As an alternative, the APS system can call-back for required data by a reverse proxy.
- System-to-system communication is build up, and scheduling request is sent.
- APS system interprets the request, map it to APS models and execute the scheduling algorithms.
- In case of success task schedule is sent as a response, or in case of failure, a proper error code with detailed is replied.
- ERP system interprets the APS response.

The advantage of the approach is that the APS system is ERP system independent. APS and ERP system has a separated software lifecycle. The disadvantage is that scheduling requires system-to-system

IOP Conf. Series: Materials Science and Engineering **448** (2018) 012033 doi:10.1088/1757-899X/448/1/012033

communication, which is more resource- and time-consuming compared to an integrated solution. ERP system data is sent over to an external system what is in many cases not acceptable for an ERP customer.

7. Software Development Results

We implemented in SAP NetWeaver platform a standalone APS component. The high-level concept is presented in ***Figure 3***. The user is reaching the APS user interface via SAPGUI. In the user interface the user can select the data set, where the projects can be selected, and scheduling parameters can be calibrated. Once the project scheduler is started, the actual data is read from the data source, the detected scheduling algorithms are executed, and the scheduling results are stored via the scheduling result manager. Scheduling result is displayed in the APS user interface. In the subchapters, the main entities are detailed out on the conceptual compositional level.

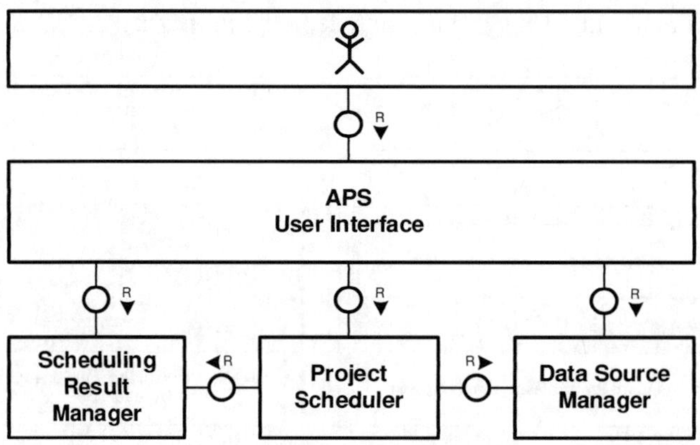

Figure 3. APS system agents

7.1. Data source

Access to project definition is unified via object-oriented ABAP artifacts. The data source package is separated into internal and external objects. For the consumer, the data source is an interface, which may have a various implementation. All implementations are hidden from consumers, and each implementation can be requested via the data provided factory. Data provider factory instantiates the hidden implementation, based on consumer request.

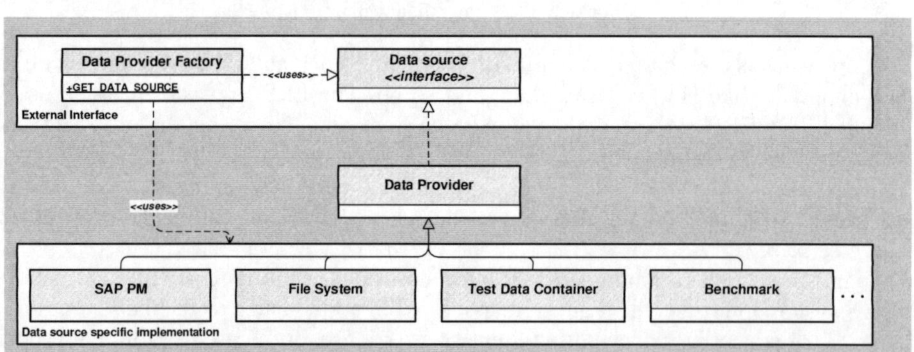

Figure 4. Data source modeling

As an example, the data source can be a test data container, used to evaluate scheduling algorithms functional correctness by comparing the project scheduling execution results with expected schedule result. A file system as input can serve a background task-based scheduling. Benchmark data sources can serve comparison of implementation by other solution in regard to quality, execution performance and correctness. SAP PM is aimed to be the access directly to SAP standard project management database tables or indirectly via SAP PM APIs.

7.2. Project Scheduler

Main development artifacts of the project scheduler are depicted in ***Figure 5***. APS scheduler consists of two disjoint sets of entities. One is visible for APS consumers and the second is hidden from consumers. Disjoint separation aims to offer a stable interface to code against it and be flexible to change the used implementation. The external interface consists of the supported project scheduling problem definition interfaces (CPM, RCPSP), the resource manager read interface and a solver factory.

RCPSP problems without the resource constraints can be solved with the critical path method (CPM) algorithm. CPM result may be an import parameter of a dynamic priority rule calculation, that is why RCPSP problem may require a CPM solver and CPM entity is an individual entry in the model. RCPSP problem solver has different implementation alternatives, and the concept is flexible to change at any time. The reason for this flexibility is to be able to analyze implementation alternatives without any effect on consuming system artifacts.

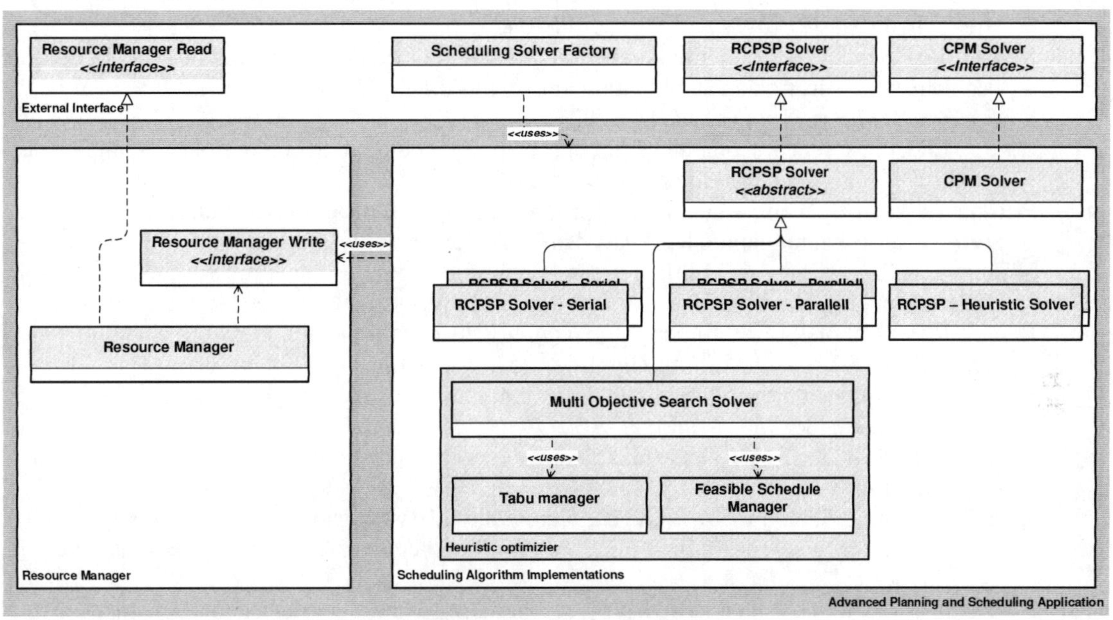

Figure 5. Project scheduler entities

The implemented solution realizes the generation-based solvers, like SGS serial, the extended generation based solvers as RCPSP heuristic solver and the multi-objective search based solver.

The resource manager is responsible for checking the feasibility of scheduling, provided by the scheduling algorithms. The resource manager has two interfaces:

- Write interface for management of resources and management of resource bookings
- Read interface for consumers to read and display scheduling in a user-friendly way

8. Conclusions

In this paper, a new approach is proposed to solve multi-objective multi-project scheduling problems considering strict resource constraints and multiple objective functions simultaneously. The description was planned from a practice-oriented point of view. The novelties of the solutions are based on the special combination of multiple priority rules in schedule generation schemes, and the flexible usage of multi-objective relational operators for comparing qualities of schedules in search algorithms. The proposed methods can consider exactly what the actual resource environment should perform in the

IOP Conf. Series: Materials Science and Engineering **448** (2018) 012033 doi:10.1088/1757-899X/448/1/012033

planned time horizon. Each schedule created for the extended problems is a feasible solution because all the hard constraints are considered. The proposed extended model and methods ensure to develop and implement adaptive scheduling software applications for the practice.

Our solver system prototype is based on an integration of enterprise resource planning system and an advanced planning and scheduling application. SAP technologies were used for implementation. The ERP functionalities can preferably be used for managing the projects and resources. In a conflicting decision-making situation, the integrated APS methods solve the actual problems by using the dynamic priorities and relative importance of items. The user can specify the solver settings.

Our research focused on modeling and solving of multi-objective multi-project scheduling problems. The proposed approach and the results are independent of the technical or management areas, so the results can effectively be deployed for multi-project scheduling problems in different fields, such as software development, manufacturing and assembly systems, supply chain management, human resource managements, and so on.

References

[1] Amol, S., 2014, Resource Constrained Multi-Project Scheduling with Priority Rules & Analytic Hierarchy Process, *Proc. Eng,, 69, pp*. 725–34.

[2] Kolisch R, Hartmann S, Heuristic Algorithms for the Resource-Constrained Project Scheduling Problem: Classification and Computational Analysis 1999, *Proj. Scheduling. Intern. Series in Op. Res. & Mgmg. Scn., vol 14.* 147-78.

[3] Garey M R, Johnson D S, Computers and intractability a guide to the theory of NP completeness 1979, *W. H. Freeman and Company*

[4] Pritsker A, Allan B, Watters L J, Wolfe P M, Multiproject scheduling with limited resources: A zero-one programmingapproach 1969, *Mngmt. Science, 16, pp*. 93–108.

[5] Mohanthy R P, Siddiq M K, Multiple projects multiple resources-constrained scheduling: Some studies 1989, *Int. Jrnl. of Prod. Res., 27, pp*. 261–80.

[6] Deckro R F, Winkofsky E P, Hebert J E, Gagnon R, A decomposition approach to multi-project scheduling 1991, *Europ. Jrnl. of Oper. Res., 51, pp*. 110–18.

[7] Jolayemi J K, Scheduling of projects under penalty and reward arrangements: A mixed integer programming model and its varients 2012, *Acad. of Inf. and Mngmt Science J., 15, pp*. 29–52.

[8] Gawiejnowicz S, Lin B M T, Scheduling time-dependent jobs under mixed deterioration 2010, *Appl. Math. and Comp., 216, pp*. 438-47.

[9] Kim K W, Yun Y S, Yoon J M, Gend M, Yamazaki G, Hybrid genetic algorithm with adaptive abilities for resource-constrained multiple project scheduling 2005, *Comp. & Ind. Eng., 56, pp*. 143–60.

[10] Kumanan S, Jegan J G, Raja K, Multi-project scheduling using a heuristic and a genetic algorithm 2006, *Int. Jour. of Adv. Manuf. Techn., 31, pp*. 360–6.

[11] Damak N J B, Siarry P, Loukil T, Differential evolution for solving multi-mode resource constraint scheduling problems 2009, *Comp. and Op. Res., 36, pp*. 2653–9.

[12] Kulcsár Gy, Erdélyi F, A New Approach to Solve Multi-Objective Scheduling and Rescheduling Tasks 2007, *Int. J. of Comp. Int. Res., Vol. 3, No. 4, pp*. 343-51.

[13] Kulcsár Gy, Kulcsárné F M, Detailed production scheduling based on multi-objective search and simulation 2013, *Prod. Syst. and Inf. Eng. Vol. 6, pp*. 41-56.

[14] SAP SE, SAP Project and Portfolio Management 2015, https://help.sap.com/doc/saphelp_ppm400/4.0.24/en-US/02/4b1a417c4bf423e10000000a155106/frameset.htm

[15] SAP SE, SAP S/4HANA 1709 FPS01, Project System (PS) documentation 2018, https://help.sap.com/viewer/4dd8cb7b1c484b4b93af84d00f60fdb8/1709%20001/en-US/1ad4b65334e6b54ce10000000a174cb4.html

Impact of Industry 4.0 on robotic welding

A Farkas[1,2]
[1] Flexman Robotics Kft.
[2] John von Neumann University, GAMF Faculty of Engineering and Computer Science

Abstract. In our compilation we have a look at what is meant by the notion 'Industry 4.0' and how we can place it in the history of technical progress. The article reviews the general characteristics of Industry 4.0 and it also mentions the expected benefits of its introduction, as well as the challenges and risks entailed by it. We present how the technical- technological elements of Industry 4.0 appear in robotisation of welding, in addition, we examine the Hungarian experiments concerning this.

1. Introduction

The term Industry 4.0 indicates that industry has globally moved to the fourth level of progress, which is often referred to as the fourth industrial revolution. Despite the fact that nowadays Industry 4.0 is discussed by many and even conferences are organized on this topic, experience has shown that not all people know what it means exactly. A lot of people expect a lot of results from this era of industrial progress: some experts prognosticate 30 % increase as a result of the technical-technological improvements of Industry 4.0. [1] It can be said that it is almost impossible to avoid this topic with regard to the robotisation of welding as well. But first of all, let us summarize what is meant by Industry 4.0.

2. About Industry 4.0 in general

Today manufacturing industry still has a significant impact on economy in the European Union, nevertheless compared to the EU's economy as a whole, the proportion of manufacturing industry has been decreasing in the last few decades. This is partly resulted from the relocation of production in part to countries outside the EU – mostly to China. EU leaders expect the situation to be changed by the overall transformation of industrial production that we can observe today, and the main characteristics of which is the fusion of traditional industry, digitalisation and the Internet. have the same style to the title page. This paragraph follows a section title so it should not be indented.

2.1. Where does the term Industry 4.0 originate from?

The term originates from Germany: first it was used as a name for a digital manufacturing project at the Hannover Messe in 2011. [2] At that time the participants of the project may not have thought that, a few years later, this would become such a wide-spread concept.

2.2. Defining the notion Industry 4.0

Industry 4.0 is used to indicate the present phase of technical progress, which has noticeably entered a new era mainly as a result of progress in information technology. We have not found a uniform definition yet; each field of expertise approaches and tries to define the term in different ways. At all events, together with progress, apparently the definition of Industry 4.0 itself also goes beyond its limits defined

IOP Conf. Series: Materials Science and Engineering **448** (2018) 012034 doi:10.1088/1757-899X/448/1/012034

earlier, which at the beginning was only applied to the so called digital manufacturing. It can help if we place the mentioned era on the historical timeline of technical progress, which we review in Table 1. [3]

Table 1. Industrial revolutions

Industrial revolution	Time peridos	Technology and capabilities
First	1784-mid 19th century	Water- and steam-powered mechanical manufacturing
Second	Late 19th century -1970s	Electric-powered mass production based on the division of labour (assembly line)
Third	1970s-Today	Electronics and information technology drives new levels of automation of complex tasks
Fourth	Today	Sensor technology, interconnectivity and data analysis allow mass customisation, integration of value chains and greater efficiency

Naturally, transition from the earlier time periods to the fourth one did not happen suddenly; for example, since the second half of the third period the progress of artificial intelligence (which is an integral part of the fourth period) has been largely contributing to, and it has even laid the foundations for entering the next era called the fourth industrial revolution.

Therefore, at the moment there is no such thing as a uniform definition based on unanimous consensus. Also, it is difficult to provide a definition as Industry 4.0 has become more complex, and its effect goes beyond the establishment of smart manufacturing, which is evidently included in it:

- The transformation of manufacturing industry exceeds the field of manufacturing itself; it is not limited to its earlier goal, the establishment of 'smart factory',
- Industry 4.0 is more than automation and data exchange in manufacturing technologies because it goes beyond manufacturing and it is oriented toward the chain between the end points; for example, it includes storage, logistics, recycling, energy systems, etc.
- Cyber-physical systems, which play a central role in Industry 4.0, as well as intelligent networks, intelligent logistics and even intelligent buildings promote the establishment of smart factories operating in the smart manufacturing of smart industry.

I-Scoop collected seven definitions, from which I would like to highlight the following statements:

- Most of these evaluate this era as a completely new step, which they identify assess as the fourth industrial revolution indicated in Table 1.
- A significant increase in productivity ca be expected as a result of the convergence, interconnection of the technical-technological improvements that are characteristic of Industry 4.0, in which the Internet, digitalisation, artificial intelligence, advanced data management and data analysis play an important role.

2.3. The technical-technological pillars of Industry 4.0

The expected increase and improvement of Industry 4.0 in manufacturing is based on 9 pillars: [4], [5]

1. **Autonomous robots** – autonomous, cooperative robots, high-end sensors, intelligent operation, standardized interfaces

IOP Conf. Series: Materials Science and Engineering **448** (2018) 012034 doi:10.1088/1757-899X/448/1/012034

2. **Additive manufacturing** – 3D printing, especially for producing prototypes and individual components, decentralized options: less transportation and stock
3. **Augmented reality** – for maintenance, logistics, various standardized operation processes, display, with additional information e.g. by using special glasses
4. **Simulation** – simulation of value networks, optimisation of real-time data originated from intelligent systems
5. **Horizontal and vertical system integration** – complete enterprise data-integration based on data transfer standards, preconditions for establishing a fully automated value chain (from the supplier to the customer, from management to the workshop)
6. **The industrial Internet of things** – the network of machines and products, multidirectional communication between the network members
7. **Cloud-based systems** – the management of large amount of data in open systems, real-time communication between manufacturing systems
8. **Cybersecurity** – operation in networks and open systems, organizing intelligent machines, products and system into a network
9. **Big Data and analytics** – the full evaluation of available data (e.g. from ERP[1], SCM[2], MES[3], CRM[4] and from data of machines), the support of real-time decision making and optimization[5]

In order to supplement the industrial Internet of things at point 6, we note that this one includes the cyber- physical systems, such as systems built from the building blocks – the so called IoT (Internet of Things) – of Industry 4.0.[6]
IoT devices are devices connected to the Internet with specific IP address; one of their functions is the capability of uploading data and information autonomously to the Internet[7], and in addition, it could allow distant interference via the Internet. [8]

Each component of the 9 pillars mentioned at the beginning of chapter 9 is already being applied separately in several areas (e.g. robotics, additive manufacturing etc.); however, we will only experience the advantages characteristic of Industry 4.0 when the convergence, the interconnected operation of all pillars will be achieved.

2.4. The change in the role of human resources in Industry 4.0
The introduction of Industry 4.0 also means a change in the tasks of professionals and in the requirements for them. At workplaces built on cyber-physical systems one has to find his way, make decision and interfere in more complex, more complicated systems. As opposed to the CIM manufacturing concept of the 80s, Industry 4.0 did not set the exile of human beings from production as an objective; instead, it aims to integrate humans into the structure of cyber-physical systems, which utilizes human experiences and skills. In order to achieve this, advanced HMI[8] devices that are capable of conforming to the operation of cyber-physical systems need to be invented. Typically, these are mobile devices with touchscreen, which are made according to industrial design, have indoor positioning systems and apply virtual and augmented reality devices. [9]

---

[1] ERP (Enterprise Resource Planning)

[2] SCM (Supply Chain Management)

[3] MES (Manufacture Executive System)

[4] CRM (Customer Relationship Management)

[5] It presents a solution provided for the purpose of e.g. service and maintenance [6].

[6] We note that the literature also distinguishes the industrial version of IoT: its abbreviation is IIoT (Industrial Internet of Things), More details under [7].

[7] A significant part of the huge amount of information available on the Internet has been uploaded/ is being uploaded by IoT devices.

[8] HMI: Human-Machine-Interface – means of communication between the operator and the appliance

2.5. The expected benefits of Industry 4.0

Benefits expected from the application of options provided by Industry 4.0 are the followings: [5]

- **Growth of productivity:** due to the higher level of manufacturing's automation, which reduces manufacturing time as a result of better utilization of assets and more efficient stockpile management
- **Improvement of flexibility:** mainly thanks to robotics, with the option of manufacturing wide range of products and variation of products.
- **Quality improvement:** with real-time, quick interference in case of errors with the help of advanced sensor technology and artificial intelligence
- **Improvement in speed:** the lead time between plans and their realization is reduced e.g. thanks to the advanced simulation devices and solutions efficiently fulfilling the customer's needs. [10]

Additional co-benefits:

- Improvement of safety at work as a result of higher level of automation
- Improvement of working conditions with a workplace designed more ergonomically
- More efficient education and cooperation as a consequence of consistent data available in the manufacturing network
- Higher level of environmental protection by utilising resources more efficiently e.g. by operating machines at a higher efficiency level while saving energy.

2.6. Challenges and risks of applying Industry 4.0

Besides the expected benefits, the introduction of Industry 4.0 also offers challenges for professionals; moreover, certain risks should be considered as well. [5]

In terms of challenges, first we have to consider the followings: determining the adequate strategy, rethinking organization and processes, changing the management, the effect of the company's culture, developing an efficient cooperation between different departments and a solution for supporting talents.

The introduction of Industry 4.0 will polarize the companies: based on the forecasts, those who apply it correctly (or those who apply it at all) will gain a considerable advantage compared to those who do not want or cannot move to this direction.

In terms of risks, experts primarily draw our attention to the fact that the IoT devices, such as interfaces between the cyberspace and physical world, used in Industry 4.0 mean a potential safety threat regarding cyber attacks; [11] also, they are capable of amplifying their effects by directly interfering in the physical world. [12] Therefore, information security is key in Industry 4.0. Information security can be interpreted in two ways; in both cases we need to find solutions in order to secure a reliable system operation and to provide protection against external, illegal intrusion and interference. [13

3. Applying Industry 4.0 in the robotisation of welding

When studying the field of welding, today it is too early to talk about the successful introduction of complex Industry 4.0 solutions; nevertheless, regarding the technical-technological solutions listed in chapter 2.3 progress has been made in some areas of the robotisation of welding.

One of the main pathways of development aimed at Industry 4.0 is **the robotisation of welding** itself; this is how developments that focus on this area and are firmly present in robotisation can get into the field of welding. Some examples for this: developments in sensor technology, application of artificial intelligence, [14] the development of standardized robot interfaces, interface for open source programming, remote diagnosis via the Internet, maintenances, intelligent forecast of the need for part replacement. [15] In the last few years, in the robotisation of welding we have observed a rapid development in the options of programming technology. With the help of these, the boundaries of the economical application of welding robots can be extended to smaller series and even to individual production. One option for this is simplifying the traditional direct on-line teaching, which would provide settings for the robot's welding torch through manual movement. [16] This method can make the teaching process of relatively short and simple welding programmes more efficient in smaller robot cells, as the programmer does not need to think in co-ordinate systems when moving the robot manually.

Bigger and more complex programmes can be efficiently made with off-line programming software so that the robot does not have to be withdrawn from production during most of the programming time. The data of robot cells connected to local networks can be remotely accessed and easily supervised. Welding process archiving systems, applied as part of quality management, are capable of storing huge amount of data, which can be efficiently processed by Big Data systems. Also, the field of **additive manufacturing**, mentioned in chapter 2.2 in relation to Industry 4.0, is primarily present in welding when robotisation takes place. In this respect we can give account of a Hungarian development as well. [17], [18]

The latest developments of welding equipment have also largely contributed to the improvements of the robotisation of welding. In this respect we can mention the solutions of **the industrial Internet of things** and **the cloud-based systems** for data transfer, which allows data transfer e.g. between welding machines (**M2M** – Machine to Machine). These solutions are suitable for e.g. identifying the welder, indicating the decrease in welding material, the remote observation of different status information, even from a smart phone. Therefore, these characteristics of welding equipment are not only available for robotic welding but for manual equipment as well.

- 11 point Times or Times New Roman.
- The text should be set to single line spacing.
- Paragraphs should be justified.
- The first paragraph after a section or subsection heading should not be indented; subsequent paragraphs should be indented by 5 mm.

4. Hungarian experiences

In order to get a realistic picture of this topic, we have collected information from welding technology trading companies, and primarily from the official Hungarian distributors of welding equipment manufacturers [19].

Essentially, developments characteristic of Industry 4.0 can be discovered at every national welding machine distributing company, which manufacture appliances suitable for robotisation, although to different degrees, in different areas and with different complexity.
In the first place this development process started with the improvement of communicational skills, remote diagnosis and options for process archiving and remote access. M2M communication, Big Data-based statistical methods are already applied and, in some cases, efforts to use cloud-based communicational solutions can be observed as well.

In Hungary there are great differences between users' demands that can be related to Industry 4.0. There are territories where everyday problems (usually caused by the chronic lack of labour force) distract people's attention from such developments. There are places where the introduction of "Industry 3.0's" developments could be the next step; at the same time, appliances integrating the newly available developments automatically include certain components of Industry 4.0, especially in the field of mechanization/automation. In this way solutions related to Industry 4.0 automatically become a part of the companies' lives.

As car industry is a significant field of use for welding in Hungary, solutions characteristic of Industry 4.0 and related demands "arrive" from Western Europe together with this industry.
At the same time, we have to note that certain developments – such as process supervision and data collecting/archiving that operate through networks – were already present as available solutions before Industry 4.0. These provided good basis for the next steps: introducing solutions now related to Industry 4.0.

Summary

Industry 4.0 stands for the fourth phase of industrial development, in which mainly digitalization and Internet solutions play a dominant role. Having reviewed the main fields of Industry 4.0 we concluded that nowadays, with regard to the robotisation of welding, we can observe such integral parts as advanced solutions in robotics, artificial intelligence, the additive manufacturing and cloud-based solutions. By all means, these are progressive solutions; still, according to the experts' prognosis, we will be able to really benefit from Industry 4.0 when these partial solutions are interconnected in almost every area and operate efficiently together. Having examined the Hungarian situation, we can say that solutions characteristic of the developments of Industry 4.0 are already available in our country. We are still in the introductory phase, which means that on the suppliers' side developments and certain employable solutions characteristic of Industry 4.0 are present, but related demand on the users' side still varies greatly.

References

[1] Otte C: What is Industry 4.0? BDI Internet: https://english.bdi.eu/article/news/what-is-industry-40/ 2018.03.15
[2] I-Scoop: Industry 4.0: the fourth industrial revolution – guide to Industrie 4.0 https://www.i-scoop.eu/industry-4-0/ 2018.03.16.
[3] Briefing September 2015 - European Parliament: Industry 4.0 - Digitalisation for productivity and growth Internet: http://www.europarl.europa.eu/RegData/etudes/BRIE/2015/568337/EPRS_BRI(2015)568337_EN.pdf 2018.03.16.
[4] Rüssmann M and co-autors: Industry 4.0: The Future of Productivity and Growth in Manufacturing Industries 2015. április 9. Internet: http://www.inovasyon.org/pdf/bcg.perspectives_Industry.4.0_2015.pdf 2018.03.25
[5] BCG: Sprinting to Value in industry 4.0. Internet: https://translate.googleusercontent.com/translate_c?depth=1&hl=hu&prev=search&rurl=translate.google.hu&sl=en&sp=nmt4&u=https://www.slideshare.net/TheBostonConsultingGroup/sprinting-to-value-in-industry-40&xid=17259,15700022,15700105,15700124,15700149,15700168,15700173,15700186,15700201&usg=ALkJrhjmzcbqbNJOJ42v0hYnkLkL_cSZuA 2018.03.21.
[6] Lee J, Kao H-A, Yang S: Service innovation and smart analytics for Industry 4.0 and big data environment. ScienceDirect, Procedia CIRP 16 (2014) pp. 3-8.
[7] i-scoop: The Industrial Internet of Things (IIoT): the business guide to Industrial IoT Internet: https://www.i-scoop.eu/internet-of-things-guide/industrial-internet-things-iiot-saving-costs-innovation/ 2018.03.21.
[8] Farkas B.: IoT alapú motorvezérlés. Szakdolgozat, 2017. BME Villamosmérnöki és Informatikai Kar, Villamos Energetika Tanszék pp. 9-14.
[9] Gorecky D, Schmitt M, Loskyl M, Zühlke D: Human-Machine-Interaction in the Industry 4.0 Era. Innovative Factory Systems, German Research Center for Artificial Intelligence (DFKI), Kaiserslautern, Germany 2014 12th IEEE International Conference on Industrial Informatics (INDIN) Porto Alegre, Brazil pp. 289 – 294.

[10] Zawadzki P, Zywicki Z: Smart Prodcut Design and Production Control for Effective Mass Customization in the Industry 4.0 Concept. Management and production Engineering Review. Volume 7. number 3 September 2016. pp. 105-112
[11] i-scoop: IoT security: smart business requires smarter Internet of Things security. Internet: https://www.i-scoop.eu/iot-security-smarter-internet-of-things-security/ 2018.03.21.
[12] Schneier B: The Internet of Things Will Turn Large-Scale Hacks into Real World Disasters. Internet: https://motherboard.vice.com/en_us/article/qkjzwp/the-internet-of-things-will-cause-the-first-ever-large-scale-internet-disaster 2018.03.21.
[13] Seacon Europe: Ipar 4.0 Internet: http://www.industry4.hu/hu/ipar4 2018.03.15
[14] Farkas A: A mesterséges intelligencia szerepe a hegesztés robotosításában. 26. Hegesztési Konferencia, Budapest, 2012. Május 10-12. ISBN: 978-615-5018-28-2 pp. 45-51.
[15] Garcia L A: Robot systems at YASKAWA: today and tomorrow 2018.02.26. Internet: https://www.yaskawa.eu.com/en/news-events/news/article/news/robot-systems-at-yaskawa-today-and-tomorrow-1/?tx_news_pi1%5Bcontroller%5D=News&tx_news_pi1%5Baction%5D=detail&cHash=4a98525248e3d1998a86f8ab23ea3681 2018.03.22.
[16] Yaskawa: Kinetiq teaching - Expanding robotics for agile manufacturing https://www.motoman.com/hubfs/Images/Kinetiq_Teaching.pdf?t=1480625354642 2018.05.29.
[17] Uzonyi S, Asztalos L, Farkas A, Dobránszky J: Az additív gyártás jelen és jövője. Hegesztéstechnika 28/3 (2017.) pp. 89-92.
[18] Uzonyi S, Asztalos L.: Additív gyártás robotosított, huzalelektródás, védőgázos hegesztéssel. 28. Hegesztési konferencia Dunaújváros, 2016. május 26-28.
[19] Farkas A, Palotás B: Az Ipar 4.0 hatása a hegesztés szakterületére. 29. Nemzetközi Hegesztési konferencia, Miskolc, 2018. május 24-26. pp 11-25.

IOP Conf. Series: Materials Science and Engineering **448** (2018) 012035 doi:10.1088/1757-899X/448/1/012035

New technology for manufacturing quasi-globoid worm gearings

L Dudás

Department of Information Engineering, University of Miskolc, Miskolc, Hungary
E-mail: iitdl@uni-miskolc.hu

Abstract. The advantage of the globoid worm gearing, that all teeth of the worm are in mesh in every moment, is well-known. The main advantage of the helical worm gearing, the easy production is also known. The paper presents a new gearing construction that tries to combine these two characteristics in one novel worm gearing. This solution, similarly to the manufacturing of helical worm, applies turning machine instead of the special teething machine of globoid worm, but the path of the cutting edge is not parallel to the axis of the worm but has an angle in the vertical plane. The resulted in form is a hyperbolic surface of revolution that is very close to the hourglass-form of a globoid worm. The worm wheel then generated by this quasi-globoid worm. The paper introduces the geometric arrangements of this new worm generating method then investigates the meshing characteristics of such gearings for different worm profiles. The considered profiles are circular and elliptic. The meshing curves are generated and compared. For the modelling of the new gearing and performing the meshing analysis the Surface Constructor 3D surface generator and motion simulator software application was used.

1. Introduction

One of the most loadable types of gearings is the globoid gearing that has hourglass-like worm. This gearing is famous from its precision and durability. These advantages originate from the very long summated meshing line acting simultaneously [1]. But one of the disadvantages also comes from this property: the high sensitivity of manufacturing and assembly errors [2]. Because of this the production of these gearing requires extra precision which makes the manufacturing expensive. One of the components of the expensive production is the special teething machine that applies a circular motion of the cutting tool for manufacturing the worm as demonstrated in figure 1. Additionally the original Cone-type gearing suffers from an undercut region in the middle of the worm wheel teeth and parallel to surface-surface mesh edge-surface contact also exists [1]. These drawbacks induced many researches that aimed to eliminate the undercut zone. Zhao at al. suggested concave worm profile with grinded surface to achieve better meshing characteristics and to avoid the undercut region [3]. Zhao and Zhang modified the mentioned gearing type with increasing the height of the tool relatively to the unmodified worm and improved such a way the meshing pattern [4]. Utilizing the gathered experience Zhao and Wu developed a universal modification method for the globoid drives [5]. In this modifying effort they performed a strict meshing analysis with numerical computer simulation to exclude the most dangerous contact point from the meshing [6]. Team of Lagutin analysed the load capacity of the new modified globoid drives [7]. The patented shaping method of Wildhaber [8] also eliminated the undercut region.

IOP Conf. Series: Materials Science and Engineering **448** (2018) 012035 doi:10.1088/1757-899X/448/1/012035

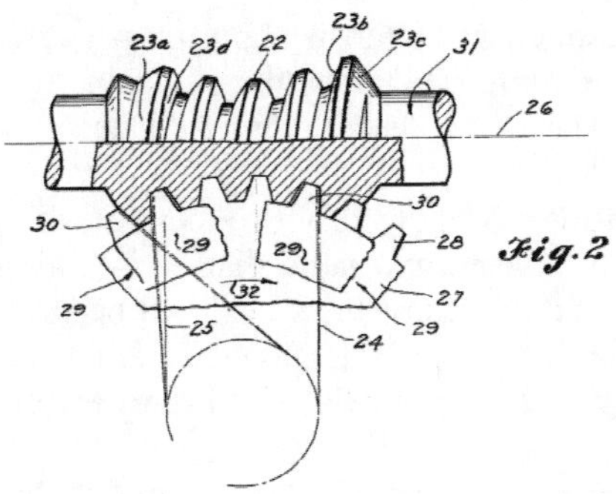

Figure 1. Teething of the Cone-type globoid worm (Figure with the original numbering as given in the patent.) [9].

Some of the improvements like the method patented by Kobayashi (see figure 2) suggested modified technology that applies rotating tool instead of discrete edge resulting in such a way the grindability of the worm [10]. The additional advantage of this solution is that the grinded worm can be more precise and has better sliding features.

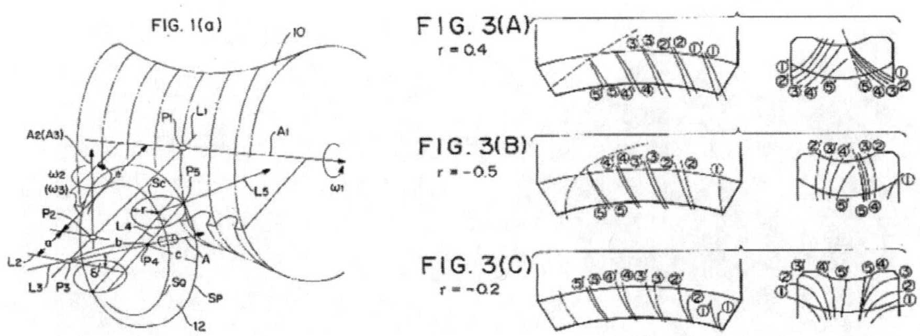

Figure 2. The generation of the worm suggested by Kobayashi applies rotary cutter or grinding wheel and results in advantageous meshing lines (Figures with the original numbering as in the patent documentation.) [10].

Similar investigation was published by Zhao et al. which presented a double-enveloping hourglass worm characterised by point-like meshing. Unfortunately such a gearing has not constant transfer ratio as it applies modified worm wheel, but the error insensitivity is remarkably improved and it works with less noise and vibration [11]. See figure 3.

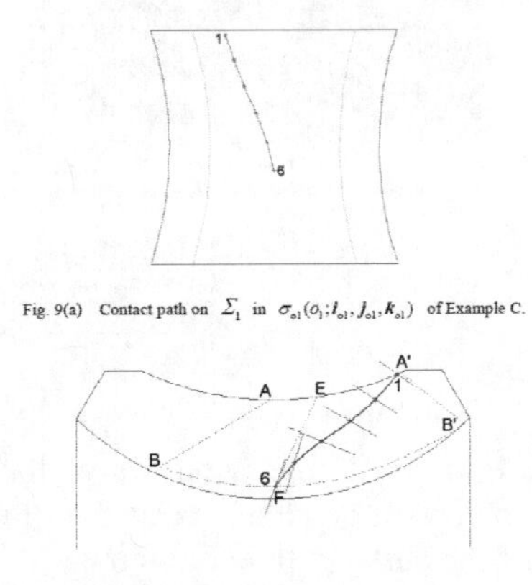

Figure 3. Contact path on the worm and on the gear (Figures are shown with the original captions.) [11].

Though all these investigated gearing types resulted in better meshing properties eliminating the undercut region and the edge-surface contact, they require special teething machine for manufacturing the worm providing circular motion of the tool. The worm wheel can be produced with usual gear hobbing machine.

The need for the expensive teething machine of the worm of the above mentioned gearings induced a new generation method for producing the worm on a normal lathe that has a tilting appliance for setting the linear motion path of the cutting tool. As the form of the worm will be similar to a surface of revolution shape having hyperbolic profile in the axial section, this worm is named quasi-globoid worm. Instead of single edged tool this technology makes possible of using rotary cutting tools like milling tool or grinding tool.

In the following sections the paper will describe the inexpensive generating of the worm and the generating of the worm wheel surface using the worm. The mating characteristics of this new gearing type will be analysed for different cases of the shape of the generating edge. All the surface generations and the analysis of meshing properties through the shape and wandering of the meshing lines will be accomplished by the Surface Constructor (SC) software application [12]. This tool applies the original meshing theory named Reaching Theory of the author [13]. The structure and some of the windows of SC are shown in figure 4 and in figure 5.

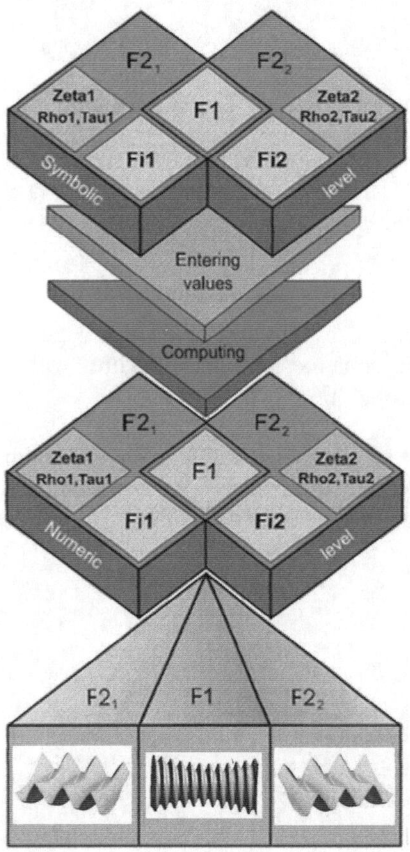

Figure 4. The structure of the Surface Constructor surface generator and kinematic modelling software.

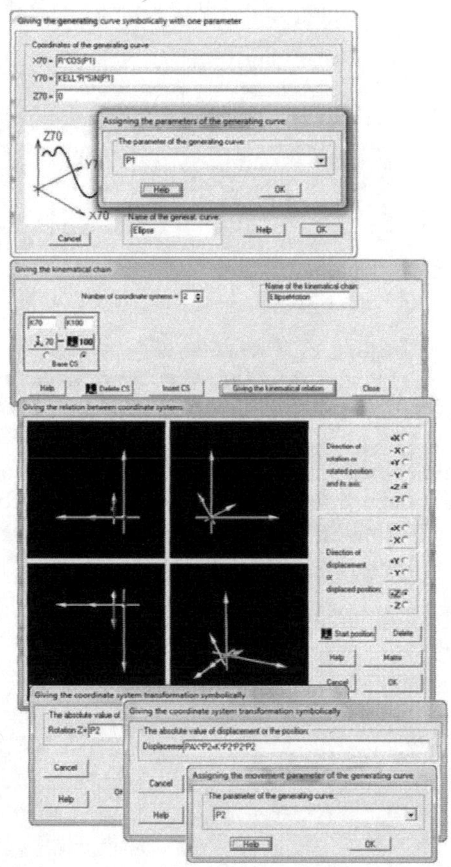

Figure 5. Some windows of the tool for entering a generator curve and the motion of the curve with expressions.

IOP Conf. Series: Materials Science and Engineering **448** (2018) 012035 doi:10.1088/1757-899X/448/1/012035

The application is intended for innovation of machine parts having enveloping surfaces, like gear improvement. SC proved its power many times, e.g. for modelling a patented internal combustion engine having only rotary parts [14]. In the closing part along with the summation of the advantages and disadvantages of the new gearing type a short comparison to globoid and cylindrical gearings also will be given before the future research plans.

2. The generation process of the gearing elements

The generation of the swept by the discrete edge worm surface and the enveloping process of the worm wheel will be presented in this section with the description of the applied coordinate systems.

2.1. Worm surface generation

The advantage of this gearing type is the easy manufacturing of the worm. The manufacturing uses a normal lathe that has some additional parts to provide the relation of the workpiece and the tool as shown in figure 6. The applied co-ordinate systems for the modelling of the worm thread generation are as follows: CS100 co-ordinate system is fixed to the worm, the tool edge is given in the CS70 frame. While the tool moves in z70 direction the edge sweeps the surface of the thread. The hyperbolic head surface is made similarly.

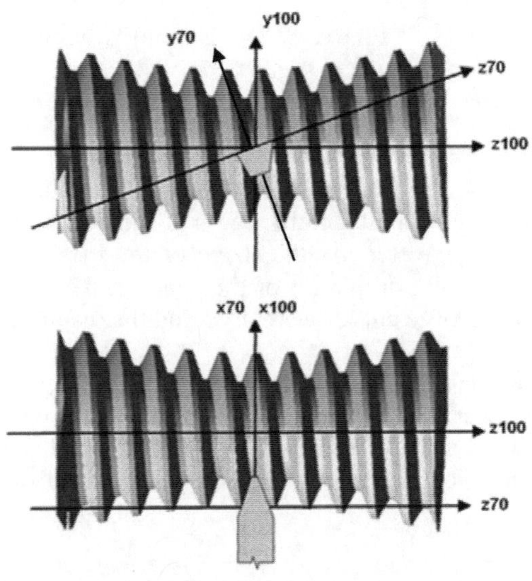

Figure 6. Co-ordinate systems for the worm generation.

The tilting angle can be realised in the practice by shifting the living centre of the tailstock and applying a Cardano-like or CV joint at the driving side or using a tilted tool post slide with a fixture that positions the tool above the workpiece.

2.2. Worm wheel surface generation

The worm wheel surface is made as usual, using a gear hobbing machine and applying radial feed of the hob. For the modelling of the worm wheel tooth surface the SC software is used. In the generating process the worm surface envelopes the tooth surface. All the teeth have same surface, so it is enough to envelope one and then it can be rotated creating the other teeth. The arrangement of the worm and the worm wheel for the surface enveloping is the same as for modelling the working of the gearing. This relation and the used co-ordinate systems are shown in figure 7. The co-ordinate system CS100 fixed to the worm while CS230 frame holds the worm gear surface. The CS201 co-ordinate system is connected to the gear housing. Using the SC software the tooth surface of the worm wheel can be generated. As every worm wheel teeth are similar only a segment having three teeth was generated. See figure 8.

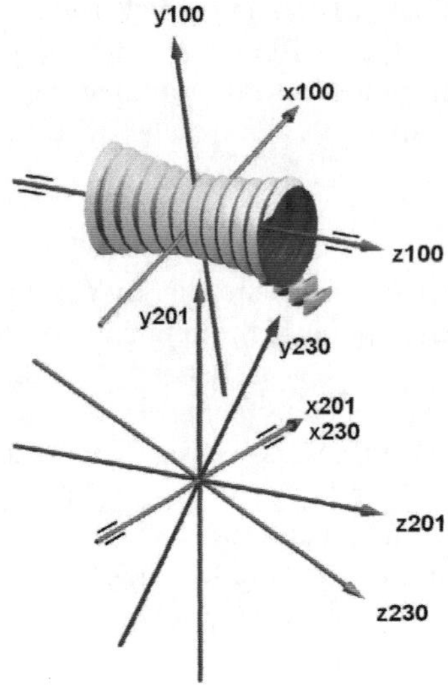

Figure 7. Arrangement of the coordinate systems for the surface generation and for the meshing analysis.

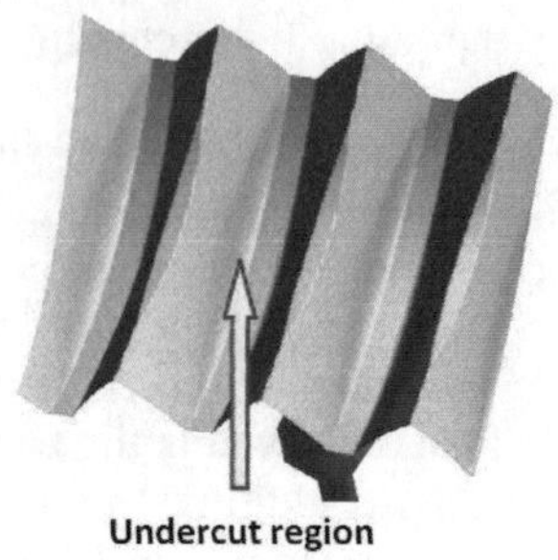

Figure 8. The generated worm wheel segment.

The figure shows that the worm wheel teeth have an undercut region at the dedendum of the gear. It can be eliminated decreasing the base circle diameter. As the shape of the meshing lines is very informative and determines the quality of the mating properties of the gearing, the next step was the analysis of them. For this, three different worm profile curves were used and the results are compared.

3. Analysis of the contact pattern in case of different worm profiles

The results for linear profile revealed that the worm wheel teeth have large undercut region at the foot of the tooth. Moreover the meshing lines were also disadvantageous, almost parallel to the sliding velocity direction, as can be checked in figure 9. In such case the formation of the loadable oil film is almost impossible.

Figure 9. Disadvantageous meshing lines on the gear segment.

The previous practice of the author proved that the circular and elliptical profile curves of the worm result in better meshing line shapes in case of cylindrical worm gearings. Because of this in the following circle and ellipse profile curves will be used for worm generation. After enveloping the mating worm wheel surfaces the contact patterns will be evaluated and the gearings will be compared.

3.1. Circle profile curve in the axial plane of the worm

Similarly to the case that is shown in figure 6 the theoretical circular edge is placed in the x70-z70 co-ordinate plane. As the calculation time for the enveloping process depends on the number of surfaces that used in it, to save time only one side of the teeth was used. Figure 10 shows such a model with the used co-ordinate systems.

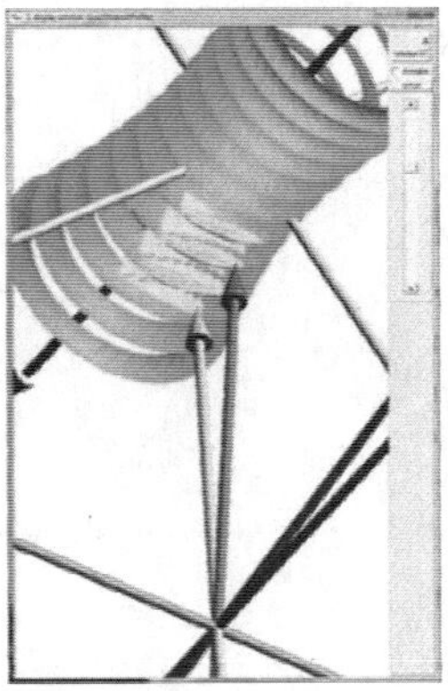

Figure 10. Meshing of the working tooth side surfaces.

In contrast to the expectation the worm wheel tooth side was only generated by the positive half of the worm measured in z100 direction. This happens because of the asymmetric location of the generating motion path to the centrally symmetric helicoids of the threads of the worm. This asymmetric position which bear resemblance to the globoid drive shown in Figure 3 originates from the ALPHA angle measured between z100 worm axis and z70 generating curve translation direction. The axial relocation of the generating edge doesn't modify this characteristic. Figure 11 shows successive phases of the meshing.

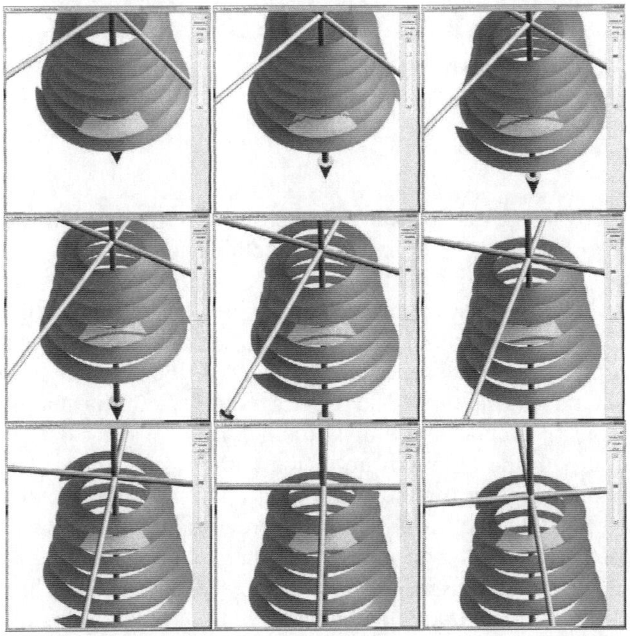

Figure 11. Meshing lines of the mating worm and worm wheel tooth surfaces.

IOP Conf. Series: Materials Science and Engineering **448** (2018) 012035 doi:10.1088/1757-899X/448/1/012035

Only the effective positive half of the worm was used. It can be checked that in case of circle profile the enveloped worm wheel, the enveloping worm and the characteristic curve between them show similarity to the helical worm having linear profile form.

3.2. Ellipse profile curve in the axial plane of the worm
The pictures of the mating surfaces of the gearing generated using ellipse profile curve are demonstrated in figure 12. The characteristic lines slightly better curved than the meshing lines of the circular gearing.

Figure 12. The elliptical profile results in slightly more curved contact arcs.

4. Summary
The paper analysed the advantages and disadvantages of today modern globoid worm drives. Every unmodified or modified solution uses the expensive teething machine for cutting the threads of the worm with circular motion of the tool. The paper suggests a novel drive construction that is very similar to the globoid gearings, but uses a normal lathe with a tilted worm holder to cut the threads of the worm. Really this worm has hyperbolic form which is very close to the hourglass-like form of the globoid worms. As the production of such gearings seems to be cheaper it worst to investigate the meshing properties of such a gearing type. The paper accomplished this analysis for three drive types that have worms with linear, circular and ellipse tooth profile in the axial plane of the worm. The modelling was performed with the SC software of the author (see figure 13). The upper three input windows are used for entering data for the generating curve, for the generated worm and for the generated worm wheel segment. The lower seven different visualisation windows are opened at the same time to analyse the motion of the gears and to visualise the wandering of meshing lines during motion. The variations showed that the meshing properties become better if the worm profile is more curved. It was a not expected finding that only the half length of the worm is in contact with the wheel in the meshing. This feature is similar to the meshing pattern of one of the modified globoid drives. The future improvements have to deal with the searching for suitable parameter set and worm profile shape that results in balanced localisation of the meshing lines. Other investigations may apply surface

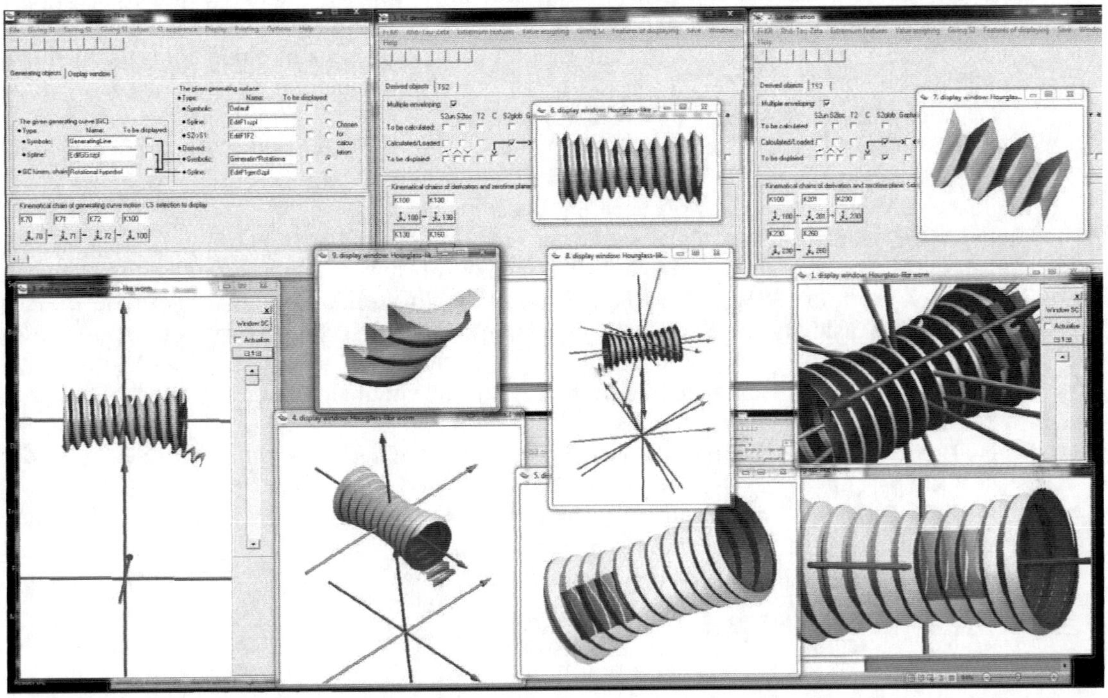

Figure 13. The Surface Constructor tool is in the midst of development.

of revolution tool for the generation of the worm. As this type of worm gearing can be manufactured more easily than the normal globoid types and all the applied profile modifications that can be used for producing cylindrical worms can be used in this worm generation also, it is worth to continue this research and improve the meshing pattern.

Acknowledgments
This research was partially carried out in the framework of the Center of Excellence of Mechatronics and Logistics at the University of Miskolc. The described study additionally was carried out as part of the EFOP-3.6.1-16-00011 "Younger and Renewing University – Innovative Knowledge City – institutional development of the University of Miskolc aiming at intelligent specialization" project implemented in the framework of the Széchenyi 2020 program. The realization of this project is supported by the European Union, co-financed by the European Social Fund. The financial support is acknowledged.

References

[1] Mohan L V and Shunmugam M 2009 Geometrical aspects of double enveloping worm gear drive *Mech. Mach. Theory* **44**(11) 2053–65
[2] Shi W, Qin D and Xu W 2004 Meshing control of the double-enveloping hourglass worm gearing under the conditions of existing the errors and the load *Mech. Mach. Theory* **39**(1) 61–74
[3] Zhao Y P, Su D and Zhang Z 2010 Meshing analysis and technological parameters selection of dual tori double-enveloping toroidal worm drive *Mech. Mach. Theory* **45**(9) 1269–85.
[4] Zhao Y P and Zhang Z 2010 Computer aided analysis on the meshing behavior of a height-modified dual-torus double-enveloping toroidal worm drive *Comput. Aided Des.* **42**(12) 1232–40

IOP Conf. Series: Materials Science and Engineering **448** (2018) 012035 doi:10.1088/1757-899X/448/1/012035

[5] Zhao Y P and Wu TC 2011 A general modification theory for double-enveloping hourglass worm drives *Key Eng. Mater.* **486** 279–82
[6] Zhao Y P and Wu T C 2011 Determination of the most dangerous meshing point for modified hourglass worm drives *Proc. 11th ASME Int. Power Transmission and Gearing Conf. PTG 11.* August 28–31 Washington DC USA pp 1–7
[7] Lagutin S, Gudov E and Fedotov B 2011 Manufacturing and load rating of modified globoid gears *Balkan J. Mech. Transm.* **1**(2) 45–53
[8] Wildhaber E 1960 Enveloping worm gearing US 2935886 A patent
[9] Skog G 1956 Gear worm thread generating machine US 2756641 A patent
[10] Kobayashi T 1994 Hourglass worm gear US 5325634 A patent
[11] Zhao Y P, Kong J, Li G, Wu T C and Shi S Y 2012 Computerized simulation of tooth contact and error sensitivity investigation for ease-off hourglass worm drives *Comput. Aided Des.* **44**(8) 778–90
[12] Dudás L 2010 Advanced software tool for modelling and simulation of new gearings *Int. J. Des. Eng.* **3**(3) 289–310
[13] Dudás L 2010 New way for the innovation of gear types *Engineering the Future* ed. L Dudas (InTech) pp 111–40
[14] Dudás L 2018 Optimal manufacturing technology determination for the main parts of a rotary internal combustion engine *Proc Vehicle and Automotive Engineering Conf.* vol 2, ed K Jármai and B Bolló (Springer International Publishing AG) pp 1–15

IOP Conf. Series: Materials Science and Engineering **448** (2018) 012036 doi:10.1088/1757-899X/448/1/012036

The role of strategy in business network

M Magyar [1, 2]

[1] Faculty of Economics and Business, John von Neumann University, Kecskemét, Hungary
[2] Institute of Business Economics, Eötvös Loránd University, Budapest, Hungary

Abstract. The focus of this paper is the conceptualisation of strategy and strategy development in the philosophy of IMP (Industrial Marketing and Purchasing) Group - a research group which examines business organizations from the network theory perspective. The most important factor of the IMP strategic approach is that it talks about actors embedded in business networks through their business relationships and therefore they're in mutual dependence. The main purpose of the paper is to better understand the strategy, namely, 'strategizing' activity through using the mentioned theoretical approach and analysing two customer relationships of a Hungarian contract manufacturer by using case study research. After the theory interpreting the paper is presenting a research framework which is adapted to the empirical findings. In the strategy of contract manufacturer it is important to realize how partners are able to synchronize key activities, resources and actors while taking the network of the companies into consideration.

1. Introduction

The conceptualisation of strategy and strategy development in the philosophy of IMP (Industrial Marketing and Purchasing) Group - a research group which examines business organizations from the network theory perspective – is the focus of the paper. The basic question is that the strategy process of companies is best described by the approach of traditional strategic management or by the IMP network approach. Although in strategic management it is recognized that the resources between organizations play important role in strategy [1], the fundamental difference compared to the IMP concept is that in strategic management (1) the network consists of the relationships between actors within the industry and (2) the relationships and so the network is managed by the managers. In contrast, according to the IMP approach, the network means the network of relationships, it is not an industry-related one, and the relationships and network can not be controlled by a single actor. The inter-relationships within the network are important resources, but none of the companies own them. [2] Consequently, business activities - see strategy creation or strategizing - are not limited either to one or the other party in a diad [3] or in the network. The process of strategizing in business networks can be interpreted as attempting to select, develop, control and manage the interdependencies of oneself and others, the relationships in which they work and the resources they derive from. [4] Therefore the most important difference between the strategic management and IMP network approaches is that while the former assumes independent company [5], the later talks about actors embedded in business networks through their business relationships and therefore they're in mutual dependence. [2] [5] [6] [7] [8]

The main purpose of the paper is to better understand the strategy, namely, 'strategizing' activity through using the mentioned theoretical approach and analysing two customer relationships of a Hungarian contract manufacturer by using case study research. After the theory interpreting the paper is presenting a research framework which is adapted to the empirical findings. The conclusions presented in this article are largely related to researches in a similar topic in the Hungarian IMP Research Center (hIMP). [9] [10] [11] [12] [13] [14]

2. Strategic approach of business networks

To understand the strategic approach of the IMP (Industrial Marketing and Purchasing) group, it is important to clarify the following. The network approach focuses on business relationships. Business relationships are embedded in a network, where nodes of the network are companies, and relationships between them are bonds, which ties to companies with opportunities but also constraints. [2] The network does not have a well-defined boundary in their view, and the network is continuously shaped by the relationships and interdependencies between the actors involved. [4] The processes of the network are both conscious and non-conscious interactions between actors. [4] Accordingly, interaction and interdependence are the decisive starting points for IMP strategic approach. The strategy focuses on the ability of business organizations to build and maintain a relationship with each other. [15] It is important to note that strategy and strategy creation in the IMP approach has not yet resulted in similar definitions as Chikan [16], well-known from strategic management saying that "the strategic planning is an integrated unit of strategy implementation and feedback based company management "or Hamel's statement. [17] The latter describes strategizing as a discovery, a work of art, and planning for programming to technocrats. Design is thus a rigid, repetitive process, as opposed to the flexible, plastic work. It is no coincidence that the IMP group never uses the word design, but uses terms of creation and development in the context of the strategy.

In recent years, however, IMP researchers are increasingly addressing the issue of strategy building. [3] [4] [18]

Table 1. A matrix for analysing strategy in business networks

	Existing process	**Evolving process**
Existing structure	1. review	2. change relationships
Evolving structure	3. establish new relationships	4. change network positions

Source: [4]

The network structure, network process (consists of interactions), network position and the network picture are essential terms to understand Ford and Mouzas [4] approach to strategy development. "The network picture is a picture of the network in the minds of the network's managers" and the network position "shows the network situation of a particular company". [19] Based on all these, Table 1 illustrates a system of context, in which Ford and Mouzas [4] investigated the change of relationships in a network. They compare the process and structure dimensions by which they distinguish between (1) review (existing process and structure), (2) change relationships (evolving process and existing structure), (3) establish new relationships (existing process and evolving structure) and finally (4) change network positions (evolving process and structure) strategy. It is important to emphasize that the strategy is not connected to the company, but according to the changes in network structure and processes it is understood on the network level.

Summarizing in Table 2 the differences between strategy and strategy creation, namely, 'strategizing' in strategic management and IMP network approaches, the following can be observed:

Table 2. Comparison of strategy and strategizing with strategic management and IMP network approach

Subject of comparison	*Strategic management*	*IMP network approach*
Unit of analysis	market (company)	network
Participants	seller, buyer etc.	participants (actors)
Center of strategy	seller's, buyer's etc. relationship with the market	participants (actors) relationship with the network
The company's dependency	independence	dependency
The nature of the process of creating a strategy	prescriptive	evolving

Strategy creating	analysis of external and internal company factors, evaluation of strategic alternatives	it develops from the interactions of each business relationships at a network level

Source: based on [2] [3] [4] [18] [20]

The process of strategic planning, its individual steps, environmental analysis (e.g. STEP, STEEPLE, etc.) or SWOT analysis related to a particular company can not be interpreted in IMP's understanding. [21] This is due to the fact that "interaction with others limits the freedom of management and the ability to create strategies independent of others". [3]

3. Strategy and strategizing according to contract manufacturers

Since the analysis of the strategy creation in this article is done in the network approach of the Hungarian automotive suppliers, it is indispensable to take into account specific characteristics of the relations in automotive industry. One of the key pillars is contract manufacturing (CM), whose conceptual clarification is primarily sought in the supply chain management literature. Numerous studies on contract manufacturers (including [22] [23]) start from the strategic capabilities of Original Equipment Manufacturers (OEMs). Consequently, one of the main objectives of contract manufacturing is to determine the reduction of OEMs' costs.

Sodhi and Tang [24] treat the strategy of contract manufacturers and OEMs separately/compared, and conclude that strategies lead to 'cooperation', 'competition' and mixture 'co-opetition', where companies cooperate and compete simultaneously. [22] Using the game theory in supply chain management, the authors have created the value network of contract manufacturers and OEMs, where each connection between each player shows the potential presence of 'co-opetition'.

Obviously, looking at the strategy of the contract manufacturers, partners can not be considered as independent companies. [14] Regardless of the fact that we are analyzing the original part manufacturers or the contract manufacturer's point of view, relationships between partners play key role in strategy building. At the same time, in the strategic management approach the strategy puts the company in the center, as opposed to the IMP network approach, according to which the strategy can only be created when the company takes into account its business relationships. However, in practice, business executives are well aware of - and apply - the mainstream interpretation of the concept of strategy. Accordingly, the main aim of this research is to understand the strategy of companies (contract manufacturers, suppliers) especially in terms of IMP network approach.

4. Empirical research

In the following section the framework and method of the empirical research is described, as well as the case of the Hungarian contract manufacturer (CM).

Figure 1. Research framework

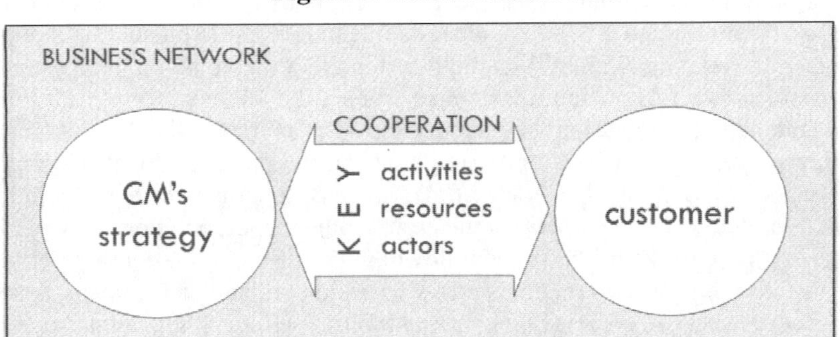

Source: based on [7] and [14]

Figure 1 illustrates the research framework which incorporates the CM with its strategy, customer(s). The framework also contained a certain degree of cooperation through key activities, resources and actors. [7] Key activities, key resources and key actors themselves are also common elements in the strategic management approach of the strategy and IMP theory. [14] The research framework also includes the features of network, because relationships are embedded in networks. [14] Customers are indicated with a dashed line because business models can be modified through the acquisition of new customers, hence this represents the dynamics of the business model. Through comparing theory the research framework is adapted to the empirical findings. In the strategy of contract manufacturers it is important to realize how partners are able to synchronize key activities, resources and actors while taking the network of the companies into consideration.

The next step is to present the case of the Hungarian contract manufcaturer (CM) called Videoton as the primary information of the empirical research. The case study research method was chosen to reveal the role the strategy in business network of the CM. In IMP industrial marketing studies the case study research is widely used [25], and the theoretical contributions of case studies are generally accepted in the literature. [26] [27] [28]

The results presented here are parts of the results of the comprehensive research carried out about Videoton Holding and VT Automotive Electronics Ltd. (the research was about the company's historical development and business relationships). In the framework of qualitative research, deep interviews were conducted with 23 mid- and top managers of Videoton in 2012 and with the two CEOs of the company in 2013. For the analysis primary qualitative datas are used from five deep interviews, including the two company CEOs, the vice president of business development, the director of international projects at Videoton Holding and the managing director of Videoton Automotive Electronics Ltd. In addition to the primary informations, the company's datas, internal publications, and the company's website are also used as important secondary informations for the results. These interviews were chosen because the respondents are responsible for the strategy-related tasks in the company.

4.1. About Videoton

The Videoton was established in 1938 in Székesfehérvár in Hungary as a private company in the military mechanics industry. After the Second World War it has been nationalised. Over a forty years period Videoton has been operated as one of the most important and largest Hungarian state-run company. It has changed its profile and became a huge electronic consumer goods company covering the big part of the Hungarian market and making an important export activity to the neighbouring socialist countries and to the USSR. Just after the political transition of 1989, the state-run Videoton underwent a major crisis and was ultimately privatized in 1995. The new owners had immediately stopped all the money loosing activities, it means most of the manufacturing activities and carefully analysed and used Videoton’s old capabilities and began to rebuild a completely new industrial complex.

Today the vertically integrated Videoton is the largest Hungarian industrial group in local private ownership offering manufacturing and related services for industrial firms. The company has nine locations in Hungary, one in Bulgaria (Stara Zagora) and one in Ukraine (Mukachevo). Employing 7600 people, Videoton’s turnover was 326 million EUR in 2011. Videoton is a professional, regional, integrated supplier and contract manufacturing company being the 4th most significant European EMS provider and the 26th largest globally at the same time. The company is also a competent multi-commodity supplier of parts, assemblies and modules, a professional regional EMS provider with extended engineering services, a turn-key contract manufacturing partner for outsourcing and transfer projects and also a complex service provider for the establishment and operation of industrial parks.

According to the purpose of the paper, in the following two different relationships of Videoton will be introduced. In the case of these two relationships with customers named Company A and Company B (the names of the companies are not relevant to the research), the interesting point is that both of them are milestones in the development of Videoton's story. The process of developing from an Original Equipment Manufacturer (OEM) to a CM is illustrated through Videoton (more precisely through the presentation of Videoton's most important subsidiary, Videoton Automotive Electronics Ltd.) and its two important relationships which commenced in the 1990s. Figure 2 illustrates the relationships of Videoton as a contract manufacturer where the Tier1 mark (used in the automotive industry) means direct supplier to the manufacturer (OEM), Tier2 is the key supplier of Tier 1 suppliers, Tier3 companies are suppliers of Tier2 companies.

Figure 2. Business relationships of Videoton with Company A and Company B

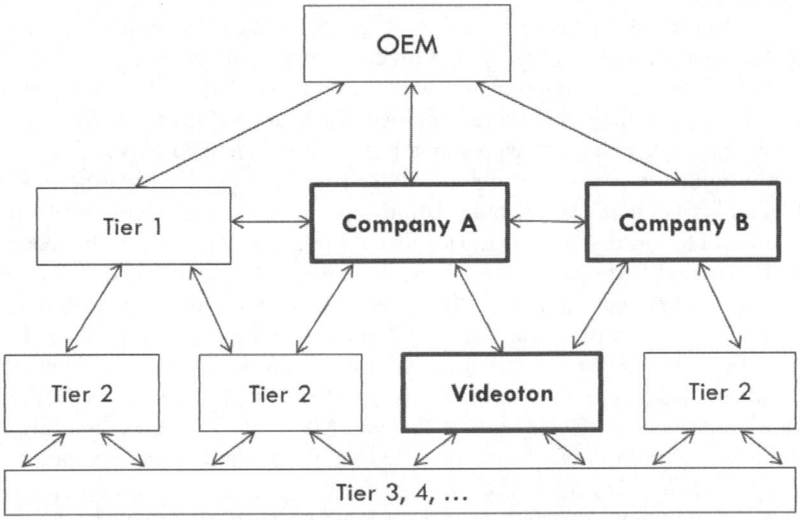

Source: author's own construction

4.2. Videoton and Company A

Videoton's 25 specialized medium sized subsidiaries provide a complete solution to its business partners from Automotive Industry (39%), Household Appliances (22%) and Industrial Applications (29%). One of the major activities of Videoton is supplying for automotive companies which is provided by VT Automotive Electronics. The subsidiary's main cooperation began in 1998 with the French company, Company A which is a Tier 1 supplier in the automotive industry and still the most important partner for VT Automotive Electronics and Videoton Holding as well. At that time Videoton already had several years of experience in supplying to the automotive industry. It means they were aware of the strict quality requirements of the industry and they had the expertise, the manufacturing facilities and the human resource as well. The whole business of VT Automotive Electronics was based only on one customer. But in 1998, this relationship stalled, they could not develop further, so VT had to look for new customers. The truth is that the relationship with Company A wasn't created from a natural step in the company's life. In 1998, the company still did not have adequate capital, liquidity; the financial risk was on the highest level. The managing director was careful and restrained with the establishment of the French relationship; however, the two CEOs of Videoton have supported it. In fact, the Company A found VT and not inversely. VT sent approximately 180 proposals to Company A to different locations. The Company A roughly defined the size of potential partner and thus actually reached VT. But after proposals there was no answer. A twist in the relationship has occurred when the CEO's at Videoton persuaded a manager at upper level at Company A to visit Videoton. The risk bearing and "go ahead" was paid off. They have been taken steps without specific customer needs because they seriously believed in the potential for development and finally, therefore could the French relationship developed. As the director of the automotive electronics company nicely formulated that situation: "This French manager has been arrived, he went to the machine-shop ... we said him look there was the machine-shop and there were thousand workers, we had been very nicely able to supply for automotive electronics industry since six-seven years and believe we would be able also to delivery to him. He gave credence to us. I always say that marketing and sales are about that I have to sell my capability which doesn't yet exist but I'm surely able to develop it when there will be demand for it." First of all they negotiated about the switches and contacts, than electronic components. Since 2002 the VT Automotive Electronics has been delivering to 22 of Company A's sites (e.g. Argentina, Brazil, Mexico, China, Turkey, Portugal, Spain, Poland, Czechoslovakia, Czech Republic, and Romania).

4.3. Videoton and Company B

The story of this relationship started in 1995. The antecedents of the relationship go back to 1988, to the Videoton before the transition in Hungary. At that time one of the subsidiaries of Videoton, dealing with audiotechnics, was in for a fruitful relationship with Sony Vega. The responsible person at Videoton was the same manager who is the business development director of Videoton today. They were already near to sign the contract about manufacturing sound box for radio cassette players, but in 1990 when the political situation in Hungary became troublous, Sony finally stepped out of the business. After the transition the director started to reveal the old relationship. Sony showed interest and the sales leader of the component manufacturing division in Japan visited Videoton. He wanted to sell components for manufacturing CD players. Videoton had business in this field but they had commitments to other suppliers. The Japanese sales leader was so enthusiastic that he had frequent visits – twice in a year – to Videoton to check the opportunities. The last visit occurred in the beginning of 1995. The CEO's of Videoton did not support these visits, but the business development director kept open house. He remembers: „A manager from Sony used to visit us regularly he wanted to sell us CD pick-ups. We have said him that we were producing CD players but that was the client who determined what kind of pick-ups must be used, thus we couldn't buy from him. Never mind he came every year and checked how we had developed. A day he played golf together with the European boss of Company B, an other Japanese who asked him whether he knew an Eastern-European company to manufacture car radios. This manager gave him my visit card." In September 1995, the business development director at Videoton got a phone call from the European boss of Company B. He explained that he was looking for a partner in Eastern Europe for manufacturing car radio. He also promised to send products by mail and they arrived within 2 weeks. The relationship with Company B emerged, the first personal meeting occurred in October 1995, and manufacturing started in March 1996. In that year out of the 12.000 employees of Videoton 300 worked in this project and this number increased up to 550 employees (out of a total of 19.000). This relationship was significant in terms of added value for Videoton and lasted until 2002 (Company B had to finish the project because its European audio division closed).

5. Findings and conclusions

The purpose of the paper was to analyse two different but determinative relationships of Videoton by which the process of developing from an OEM to a CM is illustrated.

The Videoton's strategy through the relationships with Company A and Company B is also illustrated. The time dimension is so important in the company's strategy development because what Videoton learned from their relationship with Company A was capitalized on in its relationship with Company B. As an interesting point of the research, according to the deep interviews with the top management the word strategy often appeared in connection with opportunity. The opportunity in this context appears in the attempts to balance key activities, key resources and key actors – which elements should be investigating when the relationship between the partners is analysed – between the CM and its customers, taking the business network into consideration.

Let's see the two relationships as milestones in the Videoton's strategy development. As the first milestone Videoton's transformation occurred when a subcontracting relationship with the Company A began in 1994. This relationship ultimately led to the foundation of Videoton Automotive Electronics Ltd. The company shipped automotive electronic controls, wiring harnesses, relays and distribution boxes as a subcontractor (these correspond to key activities of the Videoton). One year later, production was moved from Ireland to Hungary together with supply, meaning that the former suppliers were replaced by Hungarians. Company A also transferred technical documentation (corresponding to key activity of the Company A) and technology (corresponding to key resource of the Company A), from Ireland to the factory installation in Hungary. The raw material, storage and continued delivery or return of processed products was guaranteed by the partner (also corresponding to key activities of the Company A). On the other side the experience in military electronics and professional leadership are in line with key resources of the Videoton. As key actors in the relationship the ongoing learning played an important role at Videoton, as well as the managing processes at Company A.

As the second milestone the Videoton took action without specific customer demand because it believed in the potential for development (corresponding to key actor from the side of Videoton, on the other side the ongoing learning is in line with key actor of the Company B). This was not a common habit at the company, because the risk management at the company is always associated with conscious and systematic decision making. The relationship with Company B emerged. First of all, negotiations were held about switches and contacts, than about other electronics (corresponding to key activities of the Videoton). As key activities in the relationship the production and feedback played roles at Company B. The management of Company B was constantly changing throughout those times, while Company A's leadership remained intact (corresponding to key resource of the Videoton, on the other side the technology corresponded to key resource of the Company B) during the entire period of the development of the relationship. This period of cooperation with Company B clearly helped to transform Videoton, formerly a product manufacturer into a highly developed contract manufacturer in the electronic manufacturing service (EMS) industry. This involved a major change not only in its mode of operating, but primarily in the culture and the identity of the company. Today, Company B is one of the 10 most important customers of Videoton.

After examining the two relationships, it can be stated that there is a clear difference between pure subcontracting and EMS activity. At the pure subcontracting, Videoton as a supplier is responsible for the following issues: facility, regional know-how and contacts, production and management team, quality control, maintenance and logistic services. The customer (Company A) is responsible for quality assurance, technology, new product introduction, procurement, and sourcing. In its EMS activity Videoton is responsible for all of these issues, except for logistical services, the responsibility for which varies according to each relationship. It can be said that the Videoton's transformation involved a shift from classic product-oriented industrial activity to customer-oriented, service-based activity [14] and this case illustrates that the company's success came from its strategizing [18] activity, and not as outcomes from a planned and deliberate strategy.

References

[1] Madhavan R, Koka B R and Prescott J E 1998 Networks in transition: How industry events (re)shape interfirm relationships *Strategic Management Journal* **19** 439 – 59

[2] Håkansson H and Ford D 2002 How should companies interact in business networks? *Journal of Business Research* **55** 133 – 9

[3] Ford D 2011 IMP and service-dominant logic: Divergence, convergence and development *Industrial Marketing Management* **40** 231 – 9

[4] Ford D and Mouzas S 2008 Is there any hope? The idea of strategy in business networks *Australasian Marketing Journal* **16** (1) 64 – 78

[5] Achtenhagen L, Melin L and Naldi L 2013 Dynamics of business models – strategizing, critical capabilities and activities for sustained value creation *Long Range Planning* **46** 427 – 442

[6] Håkansson H. 1982 *International Marketing and Purchasing of Industrial Goods: An Interaction Approach* (Chichester: Wiley)

[7] Håkansson H and Snehota I 1995 *Developing Relationships in Business Networks* (London: Routledge)

[8] Håkansson H, Ford D, Gadde L-E, Snehota I and Waluszewski A 2009 *Business in Networks* (United Kingdom: Wiley & Sons)

[9] Mandják T, Szalkai Zs, Neumann-Bódi E, Magyar M and Simon J 2014 Phoenix in the network: The genesis of a Hungarian industrial company *The IMP Journal* **8** (3) 107 – 19

[10] Mandják T, Szalkai Zs, Neumann-Bódi E, Magyar M and Simon J 2014 *Emerging relationships: Where are they coming from?* 30th IMP Conference, Bordeaux

[11] Mandják T, Szalkai Zs, Neumann-Bódi E, Magyar M, and Simon J 2015 Emerging relationships: How are they born? *Industrial Marketing Management* **49** *32 – 41*

[12] Mandják T, Szalkai Zs, Neumann-Bódi E, Magyar M and Simon J 2016 Trigger issues in emerging relationships *Industrial Marketing Management* **58** 137 – 47

[13] Szalkai Zs, Magyar M, Mandják T and Neumann-Bódi E 2015 *Videoton – a vadásztölténytől a színes televízión át az autó hátsó lámpájáig* (Budapests: Budapesti Corvinus University, Műhelytanulmány)

[14] Szalkai Zs and Magyar M 2017 Strategy from the perspective of contract manufacturers *IMP Journal* **11** 150 – 72

[15] Gadde L E, Huemer L and Håkansson H 2003 Strategizing in industrial networks *Industrial Marketing Management* **32** 357 – 364

[16] Chikán A 2000 *Vállalatgazdaságtan* (Budapest: Aula Kiadó)

[17] Hamel G 1996 Strategy as revolution *Harvard Business Review* **July-August** 69 – 82

[18] Baraldi E, Brennan R, Harrison D, Tunisini A and Zolkiewski J. 2007 Strategic thinking and the IMP approach: A comparative analysis *Industrial Marketing Management* **36** 879 – 894

[19] Gelei A, Mandják T (eds.) 2011 *Dzsungel vagy esőerdő? Az üzleti kapcsolatok hálózata* (Budapest: Akadémiai Kiadó)

[20] Lynch R 2009 *Strategic management* (Harlow: Pearson)

[21] Magyar M and Szalkai Zs 2015 „Strategizing", avagy a stratégiaalkotás szerepe a hálózatban egy magyarországi autóipari beszállító példáján keresztül. *TAYLOR. Gazdálkodás- és szervezéstudományi folyóirat* **VII** (1-2.) 119 – 26

[22] Han Ch, Porterfield T and Li X 2012 Impact of industry competition on contract manufacturing: An empirical study of U.S. manufacturers *International Journal of Production Economics* **138** 159 – 69

[23] Hu G 2011 Strategies for Outsourcing Peptide Contract Manufacturing *BioProcessing Journal* 10 (2) 14 – 16

[24] Sodhi M S and Tang Ch S 2013 Strategies and tactics of Chinese contract manufacturers and western OEMs (2001-2011) *International Journal of Production Economics* **146** (1) 14 - 24

[25] Dubois A and Araujo L 2004 Research methods in industrial marketing studies *in* Håkansson H, Harrison D and Waluszewski A. (ed.) *Rethinking marketing – Developing a new understanding of markets* (Chichester: John Wiley and Sons) 207 – 27

[26] Eisenhardt K M 1989 Building theories from case study research *Academy of Management Review* **14** (4) 532 – 50

[27] Eisenhardt K M 1991 Better stories and better constructs: The case for rigor and comparative logic *Academy of Management Review* **16** (3) 620 – 7

[28] Eisenhardt K M and Graebner M E 2007 Theory building from cases: opportunities and challenges *Academy of Management Journa* **50** (1) 25 – 32

IOP Conf. Series: Materials Science and Engineering **448** (2018) 012037 doi:10.1088/1757-899X/448/1/012037

MONITORING OF PRODUCTION LINE SUPPLY SYSTEMS WITH DRONES

B Mészáros[1], K Bóna[1], M Bertalan[1]

[1:] Department of Material Handling and Logistics Systems, at
Faculty of Transportation Engineering and Vehicle Engineering
Budapest University of Technology and Economics
Budapest, Műegyetem rkp. 3, 1111

meszaros_balint@hotmail.com

Abstract. There is extensive literature about drones in a logistics context, and numerous applications have been implemented, but there is one unique use case that hasn't gotten momentum so far, yet offers significant potential. Our paper focuses on this opportunity, which is deploying drones to support the monitoring of intralogistical processes, i.e. Multi Moment Analysis (MMA). This way the observation and measurement of manufacturing processes could be automated to a large extent, making it faster, more reliable and cheaper, therefore offering benefits to both the logistics company performing the MMA and the customer. Our paper describes the architecture of a system needed to perform MMAs using drones, focusing on two key components: the indoor localization sub-system and a real-time closed-loop control algorithm, that enables the drone to track the monitored object. In order to test our algorithm, we built a simulator in MS Excel, where a drone is tracking an object moving along a straight line and a curve. The results of our experiments indicate that the drone was able to stay well within 1 meter of the object, despite the introduced uncertainty in their motion, therefore our algorithm appears to be validated and ready to be tested in a physical environment.

1. Introduction
This paper describes our research efforts at the Budapest University of Technology and Economics, that is focusing on a special use case for Unmanned Aerial Vehicles (UAVs, or drones) in a manufacturing context. We propose that UAVs could be utilized for Multi Moment Measurements, instead of a human workforce, offering significant financial and other benefits. In this paper we will describe why we think this use case offers substantial potential thru a Gap analysis, then the two core results that we accomplished so far, specifically a suitable system architecture, and a closed loop algorithm that is capable to control the drone. In the end we will describe the next steps we are working on and plan to incorporate in our research.

2. Gap analysis

2.1. UAVs outdoors
Drones are getting more and more utilized in various industries, logistics among them. We see that applications receiving the biggest attention are outdoors applications, specifically home delivery. This is a futuristic scenario where people imagine ordering packages from Amazon which are then promptly

IOP Conf. Series: Materials Science and Engineering **448** (2018) 012037 doi:10.1088/1757-899X/448/1/012037

dropped at their drone from the company's much advertised Prime Air drones [1]. More sophisticated systems for deliveries are the Flying Sidekick Travelling Salesman Problem (FSTSP), and the Parallel Drone Scheduling Travelling Salesman Problem (PDSTSP), see figure 1. In the former, a drone could be supplied with packages on the go, would supply a few customers, return back to the vehicle, charged, supplied and deployed again, thereby extending a delivery vehicle's route. The latter is simply a form of segregation of duties, where customers close the depot would be supplied with drones, the rest with trucks. [2] However, according to our understanding, these delivery scenarios are not overly competitive, as UAVs' technological constraints limit their potential. Drones nowadays are operating from batteries, which means limited range, and weight carrying capacity, which, furthermore, present tradeoffs – the larger the battery, the less weight they can carry. This tradeoff can be mitigated using hydrogen fuel cells, but it is a very new technology, and still only offers a payload around 5 kgs. [3] Also, it must be noted, that when flying outdoors, certain regulations apply, that differ from country to country.

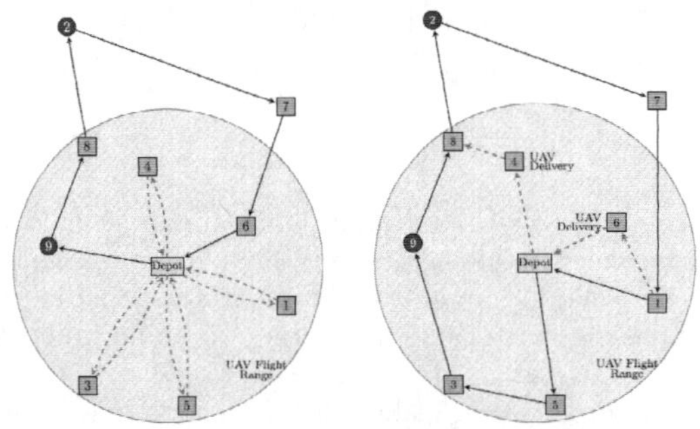

Figure 1: PDSTSP and FSTSP

2.2. *UAVs indoors*

Drones appear to be much more capable indoors. One use case is picking, where they could substitute human labor, however the limited payload again limits the available savings potential. We also argue that if a process can only be replaced partially, then the added coordination efforts will subtract as well. [4]

An area where UAVs can be used very well is the inventory taking processes. This seems to be a use case where painstaking, extensive labor efforts can be significantly relieved when using drones. The main reason for this is that the required tasks of the UAVs fall in line with the capabilities they already process, which are high manoeuvrability, agility and very cheap movement in general. The only reason we decided not to focus on inventory taking is that there are already a couple existing solutions on the market, such as Aeriu, Eyesee and DroneScan.[5][6][7]

2.3. *Traditional Multi Moment Analysis [8]*

Multi Moment Analysis is a measurement, where objects serving in the production supply processes, such as people, forklifts, etc. are observed with the intention of determining their utilization. During the measurement people walk around with clipboards and printed out forms, and at random times observe a certain object writing down what activity it was performing at the moment. This is a useful and popular measurement, even though it is fairly expensive. According to Krisztián Bóna PhD's expert opinion, an average measurement requires 10 people plus a project lead, lasting 5 days, therefore requiring travel, accommodation and also training for the participants. Other drawbacks of the current method is that people have different judgements therefore increasing uncertainty in the measurement and reducing data integrity, and that collecting and digitalizing the forms take rather long.

2.4. Gap analysis outcome

Bringing together the above topics we concluded that Multi Moment Analysis is a perfect use case where – similarly to inventory taking – the natural advantages of UAVs can be leveraged as they are not required to carry a payload. Also, this is a concept that has not been researched outside of our department.

In our concept, the setup costs of a measurement would slightly increase, as first an indoor localization system had to be installed, but afterwards only one person monitoring the measurement would be needed, as drones would be able to autonomously find an object and take a few seconds long video of it. This video could be livestreamed into (or recorded and then played in) a control room, where a single person could record the activity of the object (already digitally), therefore homogenizing and speeding up the measurement.

3. System Architecture

The task at hand is clear: we need to design a system that is capable to monitor the objects that take part in production supply processes in a specified area, with the help of drones. The former move on the surface level, on paths that are unknown, while the latter are able to fly above them in a designated area, therefore the available space can be physically separated to an observed airspace and an observer airspace.

The static architecture of the system can be grouped according to 3 main categories: the above two (observed and observer airspace) plus IT&HR, as visualized on figure 2. The more important sub-systems are the following:

- The observed objects are part of the overall system, because they carry some device (a tag, beacon, etc.) that enables the system to collect their real-time location data.
- The number of required drones can be determined depending on the number of objects that need to be monitored. According to the methodology of Multi Moment Analysis, measurements of objects need to be taken at random time intervals, which means that a specific drone takes off, tracks down the object it gets assigned to, captures a few seconds long video of it, then either gets assigned to another object or returns home.
- The indoor localization sub-system, which collects and transmits real time 3D localization information of the observed as well as the observer objects is crucial and will be discussed in detail in section 3.1.
- A central control hub with the necessary interfaces would be responsible for processing all the localization and other data and performing the assignment tasks mentioned above. Also, part of the IT landscape is the closed-loop control algorithm, responsible for the flight paths of the drones. This autonomous control could run on the flying objects' onboard computer or in the control hub as well.
- The HR sub-system would be responsible for ensuring the safety and reliability of the measurement, therefore would be able to interject at any time.

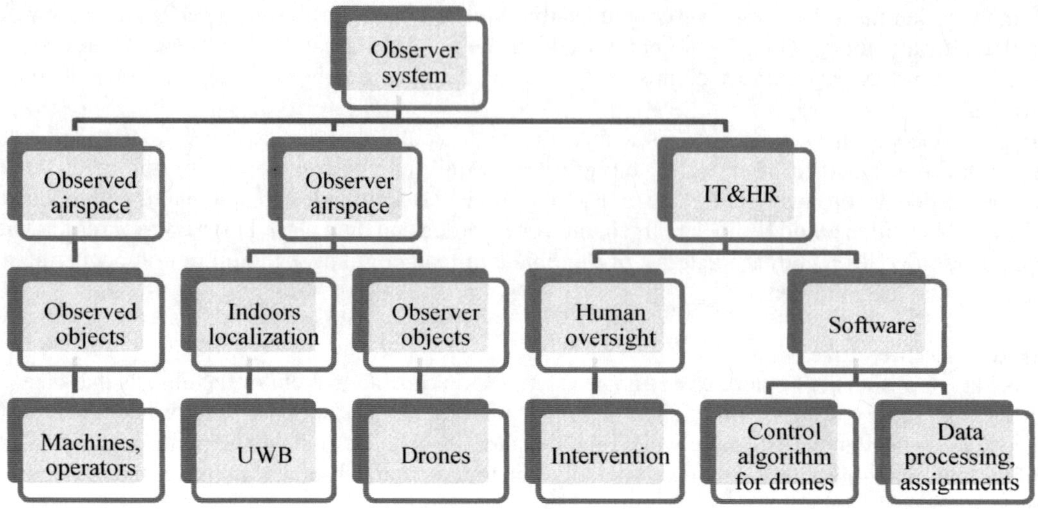

Figure 2. System architecture

3.1. Indoors localization sub-system

Our concept is based on available location information, which allow to control the drones, therefore the types of systems that can be used is a vital question. Because of the indoors environment, and the proximity of objects to one another GPS is out of the question due to its lacking accuracy and signal reliability.

3.1.1. Requirements

Fundamental requirements when considering the applicable systems are described in table 1.

Table 1. Requirements toward the localization sub-system

Category	Requirement
Accuracy	In order to ensure sound flight for the drones, a few centimetres accuracy is needed.
No. of objects	Naturally, there would be numerous objects (up to 100) that need to be tracked simultaneously, with unique IDs.
Frequency	Drones are significantly faster than humans, or other objects in the production supply systems, which need to be taken into account.
Range	Observed objects are as much as 50 meters apart from each other.
Simple tags	Because of the limited capacity of the drones, the physical tags should be as small and light as possible.
Visibility	The localization system would need to "see thru" obstacles such as concrete columns, walls, etc., therefore camera-based approaches are not applicable.
Misc.	Other aspects such as cost efficiency, simple installation, and processability of the provided information are also important.

3.1.2. Applicable technologies
Most plants possess their own Wi-Fi networks which can be used for indoors localization, however this offers an accuracy of only 5-15 meters, which removes it from the list of options. [9]

Another alternative is Bluetooth, specifically to use *Bluetooth LE (Low Energy) „beacons"*. The most popular products are iBeacon and Eddystone, which are used for example in shopping malls, or onboard on flights, because they don't disturb other communication channels. These are completely unrelated fields, and although the accuracy is higher than Wi-Fi's – around 1 meter – it is still not in the desired range. [10]

The next option is the Ultra Wide Band technology (UWB), which is a radio technology characterized by low energy level, short-range, high-bandwidth communications, and is capable of sharing its spectrum. [11] The wide bandwidth allows for high data transmission speed, and the narrow impulses can be differentiated by the receiver. Moe Z. Win and Robert A Scholtz were able to prove in an indoor office setting that barrier objects are not a concern, as they put it, *"an UWB signal does not suffer multipath fading"*. [12] Furthermore, Anton Ledergerber, Michael Hamer and Raffaello D'Andrea were able to demonstrate the accuracy of an UWB localization system while navigating an Unmanned Aerial Vehicle (UAV, i.e.: drone) as published in their work titled *„A Robot Self-Localization System using One-Way Ultra-Wideband Communication"* [13] According to them, UWB *"is suitable to be used in a feedback control system, and enables the robot to track and perform high-speed, dynamic motions"*, which is very promising to our current use case.

Figure 3. DecaWave UWB Active RFID modul

The last alternative is an Infrared-based Real Time Location System (Irid-RTLS), which in theory checks all the boxes. It offers centimetre accuracy, high number of objects, RF immunity and easy operation and maintenance. The only concern is the range which is said to be 0-20 meters. However this technology seems to be in an R&D phase, therefore definitive conclusions are difficult to make regarding its usability. [14]

All in all, we conclude that given the requirements of our use case, currently Ultra Wide Band looks to be the most suitable from the available indoor localization technologies.

3.2. Closed-loop control algorithm
The other focus area of our work has been to create an appropriate closed-loop control algorithm that is capable of giving flight commands to the drones based on their and their objective's position. In the first step in our research, we consider a 1:1 setup, where 1 drone is tracking and then following 1 object and both are moving in a 2D plane.

3.2.1. Principle, behaviour
It is a linear discreet algorithm that approximates movements during a time interval (dt) as linear motion, therefore equation (1), (2) are true.

$$x_{t+1} = x_t + dt * v * \cos(\alpha) \quad (1)$$

$$y_{t+1} = y_t + dt * v * \sin(\alpha) \quad (2)$$

The very first step is to input variables into the algorithm. For this we assume that coordinates of the tracked object, and the drone are readily available from the indoor localization sub-system, and in the followings will be denoted by x_{i_obj}, y_{i_obj} for the tracked object and x_{i_dro}, y_{i_dro} for the drone, where i is the identifier of the time interval. These are the only inputs, and the aim is to determine the speed and angle for the drone that will be used during the next time interval.

After determining the inputs, the next step is to forecast the object's position at the end of the *next* time interval $(i + 1)$. In order to do this, we need to calculate its speed and angle, which are based on its location at i and $i - 1$:

$$v_{i\_obj} = \frac{\sqrt{\left(x_{i\_obj} - x_{i-1\_obj}\right)^2 + \left(y_{i\_obj} - y_{i-1\_obj}\right)^2}}{dt} \tag{3}$$

$$\alpha_{i\_obj} = \tan^{-1}\left(\frac{x_{i\_obj} - x_{i-1\_obj}}{y_{i\_obj} - y_{i-1\_obj}}\right) * \left(\frac{180}{\pi}\right) \tag{4}$$

Substituting these values into (1) and (2) we get the forecasted coordinates of the object which form the objective for the drone at *i.*

Afterwards, we are able to calculate the angle of the vector pointing from the drone to the object and also, the absolute distance between them:

$$\alpha_{i\_obj-dro} = \tan^{-1}\left(\frac{\Delta y_{i\_obj\_fc-dro}}{\Delta x_{i\_obj\_fc-dro}}\right) * \left(\frac{180}{\pi}\right) \tag{5}$$

$$s_{i\_obj-dro} = \sqrt{\left(\Delta x_{i\_obj\_fc-dro}\right)^2 + \left(\Delta y_{i\_obj\_fc-dro}\right)^2} \tag{6}$$

The distance is important, because along with the maximum speed of the drone it determines the desired speed for the next time interval, according to equations (7). Here, ε is a range up to which the drone is tracking the object, and within which it is able to perform the monitoring. In practice we expect ε to be cca. 2 meters.

$$v_{i\_dro} = \begin{cases} \dfrac{v_{max} - v_{max} * e^{-2*\left(v_{max}^{-1,2}\right)*s_{i\_obj-dro}}}{v_{max}} & ha\ \Delta s > \varepsilon \\ \dfrac{average\left(v_{0\_obj} \dots v_{i\_obj}\right)}{v_{max}} & ha\ \Delta s < \varepsilon \end{cases} \tag{7}$$

The function in case of tracking ensures that the drone is flying close to its maximum capability when it's far from the object and slows down gradually as it is approaching. Figure 4 shows the function's main characteristics in case of maximum speeds of 5,10 and 20 m/s, where the x axis is the distance differential in meters, and y axis is the speed in m/s.

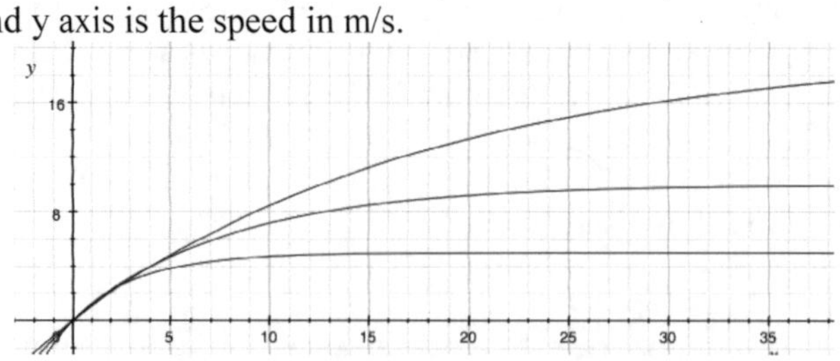

Figure 4. Speed function characteristics

4. Validating the algorithm

4.1. Excel simulator

In order to validate our algorithm, we built a simulator in Excel, where we were able to perform experiments with different scenarios. The reason we used Excel, is that it is a simple, yet capable tool that almost everybody has access to, therefore our experiments may be repeated or checked easily. In the simulator, each step (time interval) was represented with 1 row of data, and visualized using simple x-y scatter plots.

4.2. Experiments

We designed two different experiment scenarios, where the path of the object was simulated with a) linear, and b) arched paths. We'd like to emphasize that we added a 1°-5° (a), and 10°-20° (b) noise to the motion of the object between each time interval to better simulate real-world objects.
As described in section 3.2.1, the outputs from the algorithm are the speed and the angle of the drone, determined by equations (5) and (7), which allowed us to simulate the drone's position at each time interval:

$$x_{i+1} = x_i + dt * v_{max} * v_{i\_dro} * cos\left(\alpha_{i\_obj\_fc-dro} * \frac{\pi}{180}\right) + \Delta E \quad (8)$$

$$y_{i+1} = y_i + dt * v_{max} * v_{i\_dro} * sin\left(\alpha_{i\_obj\_fc-dro} * \frac{\pi}{180}\right) + \Delta E \quad (9)$$

Here, ΔE is the simulated noise in the drone's motion, which we set to plus/minus 1 meter, which we consider to be significantly large (remember, the required accuracy of the localization technology is just a few centimetres).

4.3. Results

After performing these two sets of experiments we are confident to say that the algorithm is viable, and performs according our expectations. After performing 50 experiments for both a) and b), with the length of the time intervals (dt) at 1 sec, ε at 3 meters, the maximum speed of the drone at 10 m/sec, and the other constants as described above, we were able to achieve 75,4 cm (a), and 65,7 cm (b) average distances between the drone and the object (std dev: 7,7 and 11,8 cm). A couple experiments are visualized in figures 5.

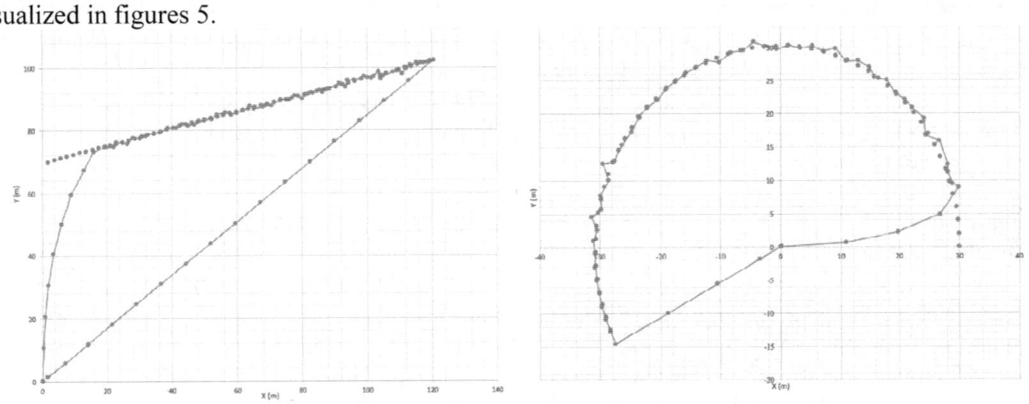

Figure 5. Experiments visualized

It is apparent that the drone (orange, starting from 0,0), finds the object (blue), follows it for a specified period of time with just a few hiccups, and then returns home. Using an arched path was important because it demonstrates that the algorithm does not get confused from an object that constantly is changing its direction. Also, a significant experience was that the accuracy depends solely on the length of the time intervals, ergo on the technological limitations and not the logic itself. At 20 ms, it can achieve 4-5 cm, and on the other hand, is still serviceable at 2 seconds.

5. Future research

5.1. Testing in a physical environment

Our department purchased a Parrot Ar.Drone 2.0, which is a surprisingly capable quadcopter with a large number of sensors. It is programmable in a Javascript environment thru an SDK made by a third party. Currently, we are implementing the described control algorithm, assuming an ideal physical environment with no obstacles. The first step was successful, we were able to send commands to the drone which it followed. Moreover, we could get the drone to perform certain cycles that were controlled by its own live location data (e.g.: altitude).

The next step will be to install a localization system, process the data that it transfers and control the drone based on that. The physical installation of an UWB system is ready, we are working on implementing communication channels. That will allow us to test tracking a stationary object first to get information on the UWB-PC connection speed, accuracy and usability of the location data. If that is successful then the first phase of the physical testing would conclude with tracking an object in motion.

5.2. Extending the logic

When monitoring multiple objects with multiple drones, the problem of assignment comes into the picture. This is basically a VRP problem with customers constantly in motion. This is a rather hard problem, where multiple heuristics, or metaheuristics would need to be tested in order to determine what logic could be most suitable. In our opinion this could be carried out initially by building an appropriate simulation, and not necessarily in a physical environment.

It is also an exciting topic to find a method to calculate the number of required drones depending on a certain parameter set. These parameters would most likely include the floor size of the plant, the number of observed objects, the number of required measurements depending on the reliability and the physical environment (light, temperature, humidity).

5.3. Image recognition

We forecast that in the future the tracking of the objects could be done in a hybrid way. By this we mean that the tracking phase (as mentioned in section 3.2.1) would be done as described in this paper, while to following phase could be done thru image recognition. We are already using the drone's camera, and suitable image recognition technologies are more or less already available.

A step further would be to recognize and record the activity of the object using purely image recognition, thereby further reducing the required human manpower, and improving speed and reliability.

6. Conclusions

We conclude that the utilizing UAVs to assist the MMA measurement is a unique research topic that holds significant practical value that appears to be realizable in the rather near future. The measurement can be made cheaper, faster and more reliable at the same time.

We described a system architecture that is needed for an MMA measurement with drones, focusing on the indoor localization, and the closed loop control algorithm sub-systems. Also, we described the measures we took to validate this algorithm, and the results we were able to achieve. At the end of the paper we presented our current and planned future research efforts.

References

[1] Amazon Prime Air. https://www.amazon.com/Amazon-Prime-Air/b?ie=UTF8&node=8037720011
[2] Chase C M, Amanda G C 2015 The flying sidekick traveling salesman problem: Optimization of drone-assisted parcel delivery *Transportation research* **C** 86-109
[3] Hydrogen cell drones. http://www.mmcuav.com/drones/hydrone1550/#spec
[4] Portal robots http://www.photoneo.com/product-detail/bin-picking-solution/
[5] Aeriu.co – inventory with drones: http://www.aeriu.co
[6] Eyesee – inventory with drones: https://eyesee-drone.com
[7] DroneScan – inventroy with drones: https:// http://www.dronescan.co
[8] Dr. Bóna K: Az RST folyamatok vizsgálata (lecture material for Material handling and warehousing proceses) https://www.logisztika.bme.hu/hu/hallgatoknak/tananyagok.html
[9] Insoft- indoor positioning accuracy WiFi https://www.infsoft.com/indoor-positioning

[10] Insoft- indoor positioning accuracy Bluetooth https://www.infsoft.com/indoor-positioning
[11] The Code of Federal Regulations, Title 47, Section 15, SubPart F – Ultra-Wideband Operation https://www.gpo.gov/fdsys/pkg/CFR-2005-title47-vol1/pdf/CFR-2005-title47-vol1.pdf
[12] Moe Z W, Scholtz R A 1998 On the Robustness of Ultra-Wide Bandwidth Signals in Dense Multipath Environments. *IEEE Comm. Letters* 51-53
[13] Anton L, Michael H, Raffaello D 2015 A robot self-localization system using one-way ultra-wideband communication *Int Robots and Syts 2015 IEEE/RSJ Int Conf*
[14] Infrared Real Time Location System Recosense https://irid-rtls.com/technology/

IOP Conf. Series: Materials Science and Engineering **448** (2018) 012038 doi:10.1088/1757-899X/448/1/012038

Research results and future orientations for the development of a pre-estimation system for the measure of the quantitative determination of milled granite surfaces

E Cserta[1], I G Gyurika[2]

[1] Budapest University of Technology and Economics, H-1111, 3. Műegyetem rkp, Budapest, Hungary

[2] Department of Mechanics, Institute of Mechanical Engineering, University of Pannonia, H-8200, 3. Egyetem Street, Veszprém, Hungary, mobile: +36 20 476 1909; gyurika@almos.uni-pannon.hu

Abstract. On the manufacturing of different CNC machine-tools, the tables of ultraprecision machine-tools and CNC coordinate measuring machines are made of granite. Despite of the continuous industrial investigations, an accurate estimation system has not been developed so far. Although, an estimation program is necessary to predict the rate of the average edge chipping of different granite types. In this article, a summary of a research work is given connecting to this topic. Furthermore, the trendency of our future research is defined.

1. Introduction

Granite is a favoured raw material used in the construction industry, as it has excellent abrasion resistance. However, due to this feature, it is often applied in other engineering fields as well. For example, in machinery engineering, granite is used as the primary raw material for tables and machine beds of ultra-precision machine tools. Major problem is that pre-estimation system is not developed to the present day in this topic. In this paper the researchers have summarized the achievements of research on the subject and they have determined the future research trends.

2. Literature review

For machining, the shaping of the granite surface is required. The primary forming process is the milling. During this treatment, unexpected surface damage can occur on the milled surface or its edge. The granite surface loses its quality and market reach due to these surface deformations and irregularities. In order to minimize this problem, estimation systems have been developed in the recent years that can efficiently determine the degree of force during the machining process varied according to the applied technological parameters and the properties of the stone materials. Turchetta et al. [1, 2, 3, 4] developed a system to describe the cutting conditions in the immediate vicinity of the stone-manufacturing area, whereby the process elements of the material scaling were defined. Tanovic et al. [5] investigated the properties of diamond segment tools applied in micro-cutting process to define the optimal parameters of the treatment. Jiang et al. [6] measured the effect of the pressure and cooling intensity of mining drill equipment on the breaking and fracture parameters of natural stones. Using finite element simulation, Wang et al. [7] analysed the damage processes of manufactured stone in mining works as well. Furthermore, interesting results have been published by Zhai et al. [8] about

IOP Conf. Series: Materials Science and Engineering **448** (2018) 012038 doi:10.1088/1757-899X/448/1/012038

surface damage and breaking in granite materials caused by mechanical vibration with different frequencies.

A further important aspect of the international research is the determination of the technological parameters (cutting speed, feed rate, cutting depth) that minimize specific energy in the milling process of granite [9, 10]. Specific energy is a fundamental parameter for manufacturing of natural stone [10, 11]. The effectiveness and efficiency of a given process can be measured efficiently with the specific energy. Xu et al. [12] studied the specific energy consumption of stone manufacturing processes. In their work, the shape of the segments of the applied tools and the properties of the grains in the segments were taken into account. The influence of technological parameters on the product quality was examined by Aydin et al. [13] as well. Based on their microscopic results, they conclude that the average surface roughness increases under increasing the cutting speed and the cutting depth, but it decreases due to the increase of the intensity of cooling. These models provide effective help in selecting optimal technology parameters and manufacturing tools.

The discrete elements method is a new technology in the impact analysis of grain structures of different stone materials. According to the definition given by Cundall and Hart [14], the fundamental assumption of the method is that the material consists of separate, discrete particles. These particles may have different shapes and properties. Each element has its own freedom and the model is able to follow their shift and turn. The effects of elements on each other may change. Thanks to the method, the relation between the elements as well as their material and geometrical characteristics can be determined. In this way, the manufacturing deformation, surface properties, and surface loads can be simulated. Using this method, Van Wyk et al. [15] studied the change of geometry, forces and force components (such as shear, friction, friction) on the surface of the stone.

3. Experimental processes

Quantitative and qualitative analysis of surface defects has already been monitored by our research team. Our long-term aim in this research is to develop a model system that provides accurate results that is capable of predicting surface defects and thereby optimizing selected materials and tools. After defining the materials and methods, the results of our studies are presented that were obtained so far.

3.1. Used stone materials

Based on the new standard, rocks can be divided into 3 groups according to their particle size. In our measurements, one material was selected from each group and the milling was carried out using tools containing diamond grains of different sizes. 300 x 300 mm granite blocks were used. When selecting arrays (blocks), we took into account the particle size classes defined in the Building Material Standards (BMS) standard and assigned to each class an experimental array (Table 1).

Table 1. Classification of natural stones according to their grain size [16]

Average size of the stone	Description
$d > 10\ mm$	Large grained
$5\ mm < d < 10\ mm$	Coarse-grained
$1\ mm < d < 5\ mm$	Middle grained
$d < 1\ mm$ grains are separated with naked eye	Small-grained
$d < 1\ mm$ grains are not separated with naked eye	Fine-grained

4-6 different surfaces were prepared for blocks with different grain sizes so that the currently selected parameter of the surface manufacturing treatment was changed per measurement.

3.2 Evaluation options
During our previous research, we have evaluated the extent of the edge damage by several methods.

One of the excellent solutions is the use of optical microscopes. The edge damage ratio of the shaped surface as well as the degree of the edge chipping are determined by optical microscopy. During the evaluation, the length of the edge chipping is measured manually. Thus, the distance between the starting and end points of the edge chipping is measured and the maximal edge chipping depth is determined as well (Figure 1). Handling these data, the volume of the edge chipping can be determined.

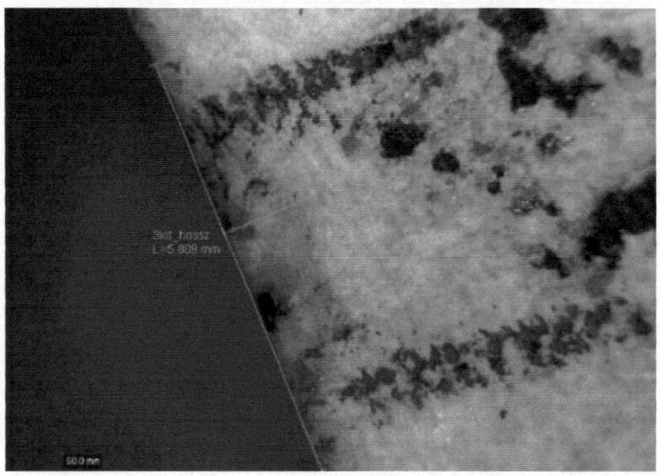

Figure 1. Determination of the typical dimensions of an edge chipping [16]

Our other device was the profile scanner (Figure 2). The device uses a scanner mounted on a movable slide that digitizes the surface in small contour sections. The test surface is fitted to the resulting data cloud. Using this 3D scanner, the evaluation is time-consuming but it provides accurate measurement results. The evaluation of these data is done by the MiniTab 15 statistical software. The software gives result for the standard deviation of the measurement sequence, the expected function of normal distribution, the curve prediction band of the set of statistical results, and the confidence interval.

Figure 2. Measurement with profile scanner [16]

3.3 Examined parameters

Gyurika et al [17] investigated the effect of cutting speed on the edge damage. In this measurements they have examined granite blocks of 5 different particle sizes. Each set consisted of 5 sample surfaces with 5 different cutting speeds. The other technological parameters that relate to the cutting, such as feed rate, cutting depth, cutting width, tool and track were kept constant. The results were evaluated by means of an optical microscope. In this way the degradation on the surface of the leaving range.

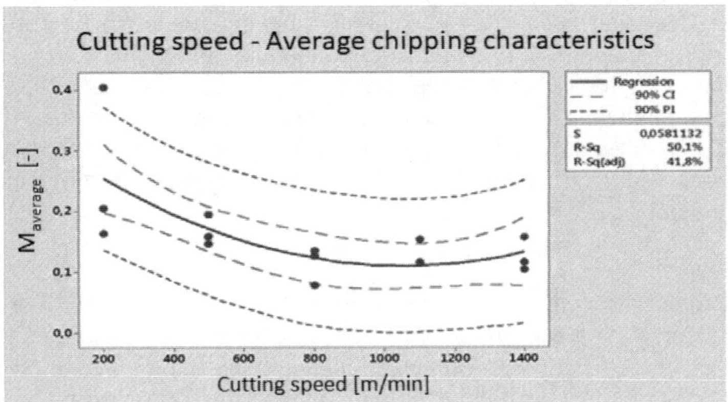

Figure 3. Alteration of the cutting speed in case of the Rosa Betha type granite [17]

They observed in the range between 200 m/min and 1400 m/min that for some granite types the degree of edge chipping decreases up to a given value (1100 m/min), after that, the alteration trend reverses and starts to increase. We can see the result for one of the stones in Figure 3. A second degree regression curve was fitted to the obtained data. The evaluation was done using the ANOVA statistical method.

During other research Gyurika [18] have investigated the effect of grain sizes of the diamond tool and the effect of feed rate in this theme. Considering the previous experiment, 4 different feed rates (200 mm/min, 300 mm/min, 400 mm/min, 500 mm/min) were adjusted to the 3 different stone materials. The minimum chipping value can be seen even better if the diamond's grain size is increased at fixed feed rate. Changing the feed rate influenced the degree of edge chipping less than the size of the diamond grains. At the same time, a local minimum is observed for each stone sample depending on the particle size, where the degree of the edge chipping is the smallest. The position of the minimum point is affected by the grain size of the diamond and that of the machined rock as well (Figure 4). The larger the particle size of the rock, the lower the feed rate is where the local minimum appears, while this local minimum is observed at high feed rates if the grain size is smaller.

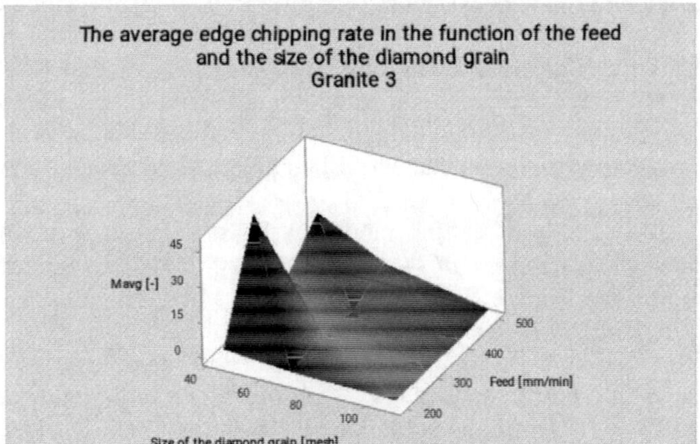

Figure 4. Test results by the experiment [18]

IOP Conf. Series: Materials Science and Engineering **448** (2018) 012038 doi:10.1088/1757-899X/448/1/012038

4. Research conceptions

In case of the milling of granite surfaces, the rupture properties of different types of stone materials have not been analysed yet considering the edge chipping and rupture properties of the grains. Therefore, the next step in our research work is to study the rupture properties of the stone samples, since, these properties have a main influence on the average surface quality and the edge quality as well. In case of a manufacturing process, some grains break apart as a result of various forces, while other stone components dislodge from the surface and from the edge. These deformations can significantly alter surface and edge quality. That is way, the properties of the binding, adhesive components of the stone samples will also be the subject of the further study. Here, the qualitative and quantitative evolution of the forces between the particles will be discussed as well.

In the literature, no publication is found about the impact of the tool paths on the level of surface and edge damage if automated stone processing is used. For that reason, the study of the influence of the applied tool paths and motility on the surface quality of stone material offers an important industrial topic of research.

5. Summary

There is little data available to develop a proper estimating model of the effect of milling on the quality of the stone sample and the existing data is not accurate enough. Therefore, the influence of the technological parameters (cutting speed, feed rate, cutting depth) and the size of the diamond particles in the machining tool is an important field of research. Its results are proposed to be implemented in industrial application as well.

6. References

[1] Turchetta S 2009 Cutting force on a diamond grit in stone machining. *International Journal of Advanced Manuf. Technol.* **44** 854-861
[2] Turchetta S 2010 Cutting force in stone machining by diamond disk. *Advances in Materials Science and Engineering*, Article ID 631437
[3] Turchetta S Polini W 2011 Cutting force in stone lapping. *International Journal of Advanced Manuf. Technol.* **57** 533-539
[4] Carrino L Polini W Turchetta S 2011 Stone chipping volume using a diamond mill. *Proceedings of 16th International Conference on Production Research* (ICPR-16)
[5] Tanovic L Bojanic P Puzovic R Milutinovic M 2011 Experimental investigation of microcutting mechanisms in granite grinding. *Journal of Manufacturing Science and Engineering* **133** Article number: 024501
[6] Jiang H Du C Zheng K Liu S 2015 Experimental Research on the rock fragmentation loads of a water jet-assisted cutting head. *Technical Gazette* **22:5** 1277-1285
[7] Wang S Li N Li C Cao C 2015 Distribution characteristics analysis on pressure-arch in horizontal stratified rocks under coal mining conditions. *Technical Gazette* **22:4** 997-1004
[8] Zhai Y Li N Wang S Zhang Y 2015 Space wave equation considering damage-induced weakening and strain rate dependency of rock. *Technical Gazette* **22:4** 1035-1042
[9] Ertingshausen W 1985 Wear processes in sawing hard stone *International Diamond Review* **5** 254-258
[10] Ersoy A Atici U 2004 Performance characteristics of circular diamond saws in cutting different types of rocks *Diamond Relat. Mater.* **13** 22-37
[11] Buyuksagis I S 2007 Effect of cutting mode on the sawability of granites using segmented circular diamond sawblade *Journal of Materials Processing Technology* **183** 399-406
[12] Xu X P Huang H 2004 High-efficiency cutting of granite blocks with multi-blade diamond saws. *Advances in Abrasive Technology* 297-302
[13] Aydin G Karakurt I Aydiner K 2013 Investigation of the surface roughness of rocks sawn by diamond sawblades *International Journal of Rock Mechanics and Mining Sciences* **61** 171-182
[14] Kovács Á 2015 Diszkrét elemes módszer (DEM) alkalmazásának lehetősége a mezőgazdasági géptervezésben *National Scientific Students' Associations Conference* Budapest University of Technology and Economics

IOP Conf. Series: Materials Science and Engineering **448** (2018) 012038 doi:10.1088/1757-899X/448/1/012038

[15] van Wyk G Els D N J Akdogan G Bradshaw S M Sacks N 2014 Discrete element simulation of tribological interactions in rock cutting. *International Journal of Rock Mechanics and Mining Sciences* **65** 8-19.
[16] Gyurika I G 2015 Mart gránitfelületek-élek minősítése és kitöredezéseinek vizsgálata technológiai optimalizálás céljából *PhD dissertation* Budapest University of Technology and Economics
[17] Gyurika I G Szalay T 2015 Examination of the average chipping rate on the edges of milled granite surface *Measurement Science Review* **15:1** 1-8
[18] Gyurika I G 2018 Researching the Effects of Feedrate and Diamond Grain Size on Edge Chipping of Milled Granites *Technical Gazette* **25** 49-55

IOP Conf. Series: Materials Science and Engineering **448** (2018) 012039 doi:10.1088/1757-899X/448/1/012039

Measurement technique analysis of shaped surfaces

R Monostori[1]

[1]Institute of Manufacturing Science, University of Miskolc, 3515 Hungary

Abstract. The article describes the measurement technology of various complexity surfaces and the reconstruction of the measurement task using CAD software.
For example, I present a wave-shaped wheel, a crown gear of spiroid worm gear drive, and a hip joint prosthesis measurement with a coordinate measuring machine. In the case of the prosthesis I have also reconstructed the measurement tasks for further analysis and testing. These results are also presented in the article.

1. Three coordinate measurement

Basic principle of the coordinate measuring technique is simple extremely. The co-ordinate is the physical base a measuring machine (figure 1.) he makes sure, which is at right angles standing with axie of the three (the letter x, the letter y, the letter Z) he is embodying a spatial frame of reference, and hereupon along axies moving possibility they assure a basic system for the measure probe. He means the relating to matters of principle base, that any shape is can be mapped with points, at the same time you the point with time value of the letter x, the letter y and the letter Z with one voice representable in a frame of reference. Measurement has it cultivated it is lying in during the coordinate measuring technique so that we decide upon time values of the letter x, the letter y and the letter Z to different points, then we are figuring the characteristic wished for out with the help of points hereupon. To this on the other hand a perform also need the following conditions is:

- he has to measure taking place moving along up the three a forming a right angle to axie to each other need is in all 3 directions some kind of lenght measured for system,
- the wished for points he has to be able to record, need was a probe system,
- the registered he has to size points up , there is need for computational background.

IOP Conf. Series: Materials Science and Engineering **448** (2018) 012039 doi:10.1088/1757-899X/448/1/012039

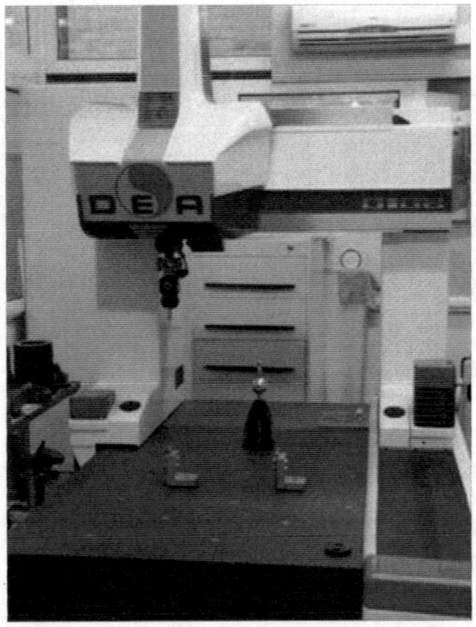

Figure 1 DEA type 3D measuring machine

Main point: the workpiece. is scannable with a touching element. For the workpiece as you wish put measured points, and these taking up of his is possible with in a short while large punctuality. Theoretically a co-ordinate measuring machine separately a lenght measured system 3 conquer it in its seminal, an orthogonal measured axes, what a right-angled co-ordinate system create. In this coordinate system known 3 coordinate of the central point of the measured probe [1].

With this measurement, in principle any workpiece of geometry can be measured and verified. Some practical applications are shown in the following diagrams (figures 2 - 5).

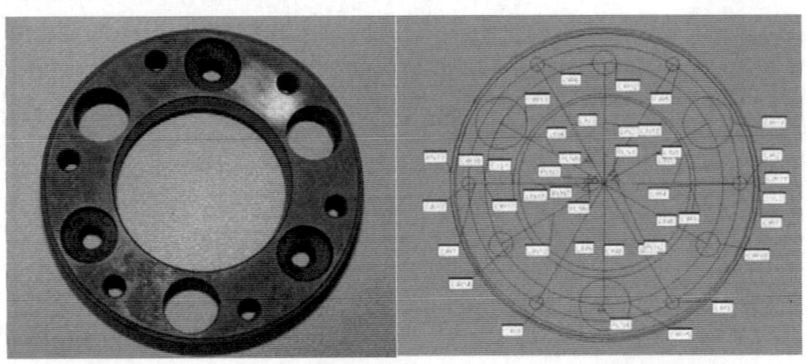

(a) the part to be measured (b) graphical display

Figure 2. Dimension options in case of an engineering part

IOP Conf. Series: Materials Science and Engineering **448** (2018) 012039 doi:10.1088/1757-899X/448/1/012039

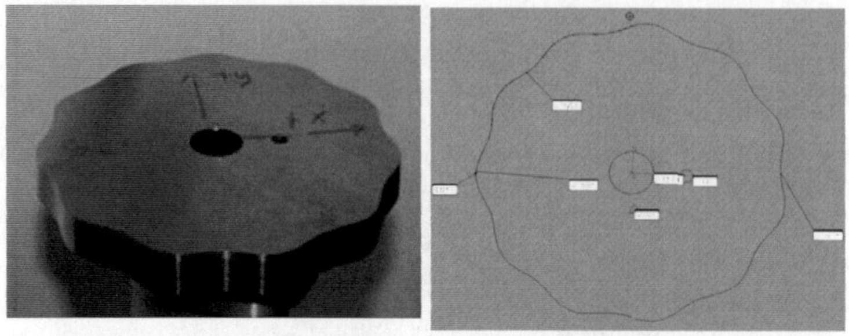

(a) the part to be measured (b) graphical display
Figure 3. Profile measurement of a wave-shaped wheel

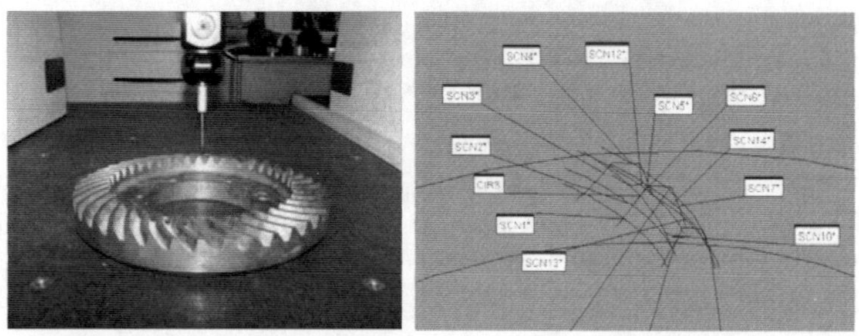

(a) the part to be measured (b) graphical display
Figure 4. Profile measurement of a crown gear of spiroid worm gear drive

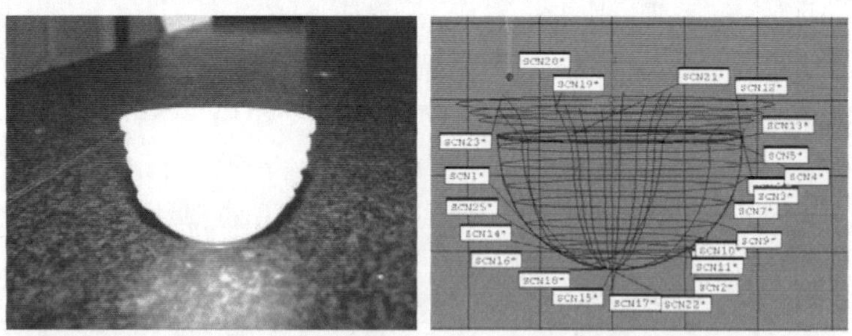

(a) the part to be measured (b) graphical display
Figure 5. Profile measurement on an orthopaedic joint prosthesis

2. Complicated shaped surface – hip joint prothesis

Hip prosthesis is one of the most complicated prosthetic prostheses, as it must allow provide three-dimensional movement. The hip prosthesis (the implant) is made of either stainless steel or other metal, such as a stem and ball head made of titanium, and a polyethylene "counter". The metal ball head spins in the plastic counterbalance, which is implanted in the hip bone.

2.1. Acetabular measurement

To measure acetabular, as it is a hemispherical body, it was necessary to have a clamping device (figure 6). To make the measurements, I wrote a measuring program using the PC-Dmis software.

Figure 6. Measure of acetabular cup

After the program, we also received the model of the vase wireframe and extracted the coordinate values in the IGES file. I was able to draw the already weighed, defective surface, using this file and a design software.

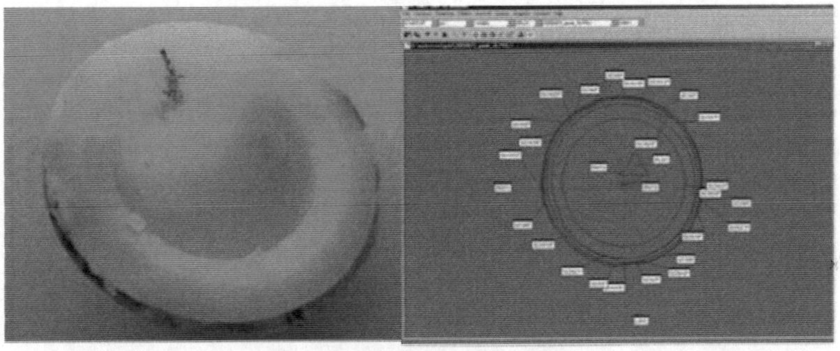

Figure 7. The measured socket the representation of his picture and a graphic designer

2.2. Reconstruction of the measurement points

The reconstruction of the measurement points carried out with the help of several types of 3D design software, at which I applied the bases of computer geometry and graphics. For instance I used Hermite – interpolation [2] for the composition of the curves.

2.2.1. Hermite – interpolation

Tertiary Hermite – interpolation used not only with interpolated points, but we presuppose the tangent vectors for known one in these [3].

These are sketched with the help of Hermite – arcs visible on the figure below (figure 8).

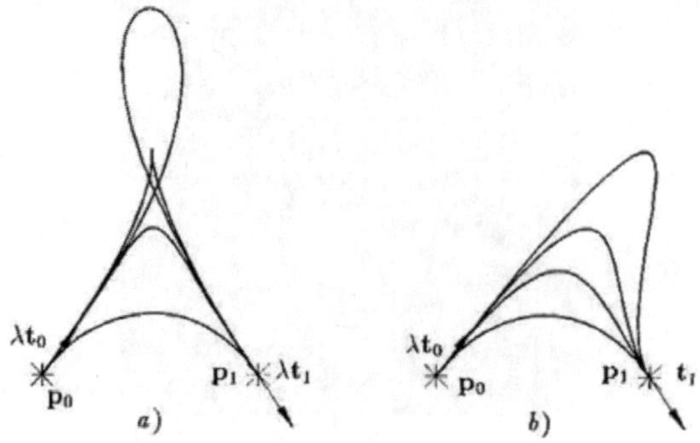

Figure 8 Hermite arc shape

The surface of the worn cup was reconstructed with the help of the measured coordinate values (figure 9). After it the differences in wear were determined as it can be seen on figure 10.

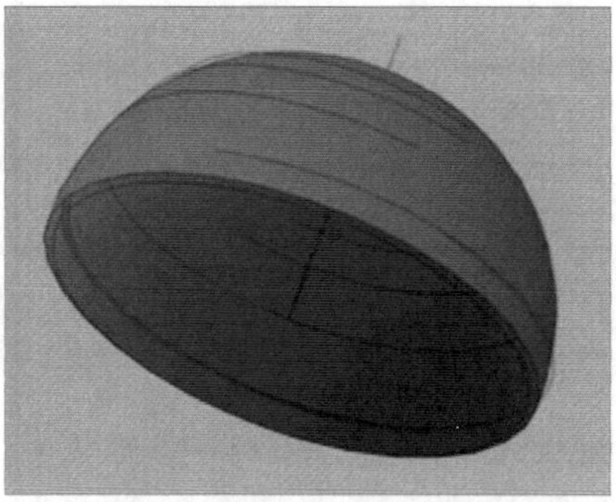

Figure 9 Reconstructed acetabular cup

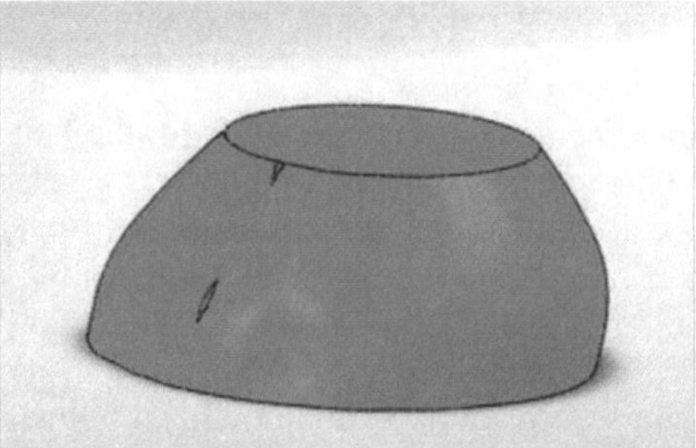

Figure 10 Real (worn) acetabular cup with 3D software

3. Determine a coordinate system that is needed for modelling

As I have already mentioned, the femur head moves in the vault. To create the motion model, it is necessary to determine the coordinate systems that will make the movement of the vagina and the femur head more transparent. This requires the coordinate system of the vase as a standing coordinate system, as the hive is recorded in the pool bone. The femur head can be moved by a coordinate system and a rotating coordinate system. The former is the resting position, the latter when the patient moves. These coordinate systems were taken from the Dudás mathematical model (figure 11).

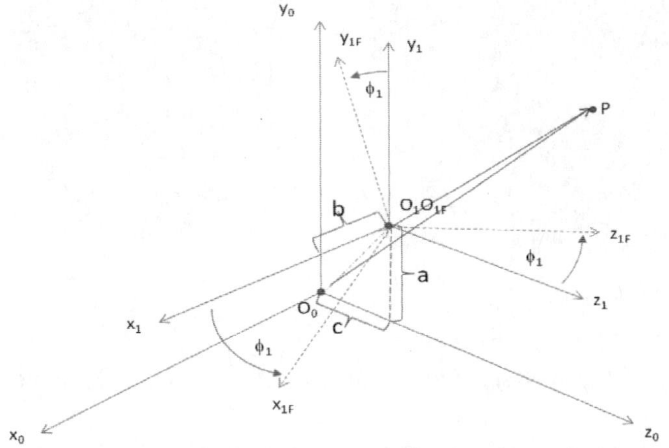

Figure 11 Coordinate systems

These coordinate systems were used in the simulation.

3.1. Simulation of the deviations

A FEM simulation was performed in order to get information about the deviation of wear values. The ANSYS finite element software [4] has been used for the investigation. Construction of the 3D model of the socket and ball was necessary to the investigation. It was designed with the software Pro Engineer.

The analysis was done in static state.

Simulation of dynamic for reaching dynamic state simulation, input parameters were set up as changing in time.

3.2. Steps of the analysis

3D models of the acetabular cup and the ball was designed as two parts. Before simulation they have to be assembled (figure 12).

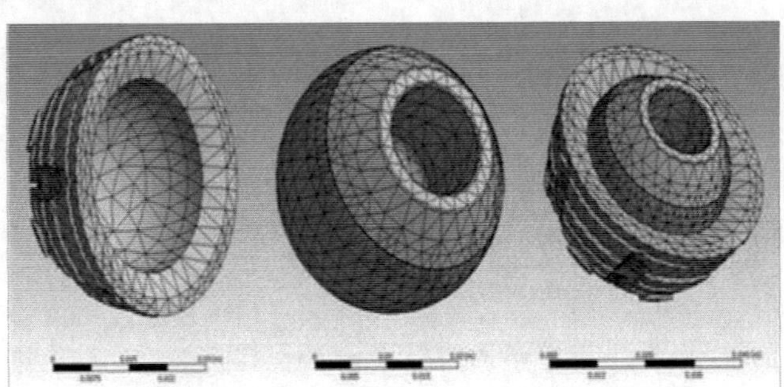

Figure 12. Finite element mesh

The test was made in two different ways:

First we put the sphere head and the socket in each other and some force was applied towards the centre (figure 13). After the test we analysed the findings separately for the sphere head and the socket (figures 14 and 15).

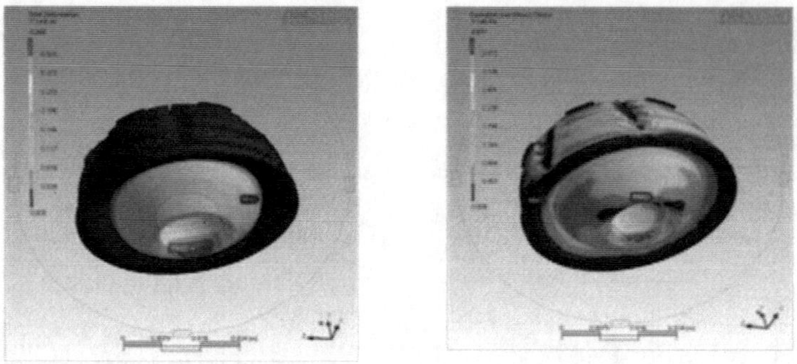

Figure 13 The analysis of tenseness and deformation, in a complex state

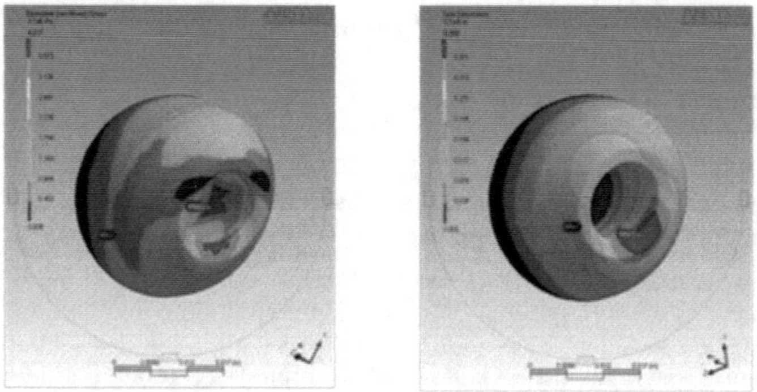

Figure 14 The analysis of tenseness and deformation on the head

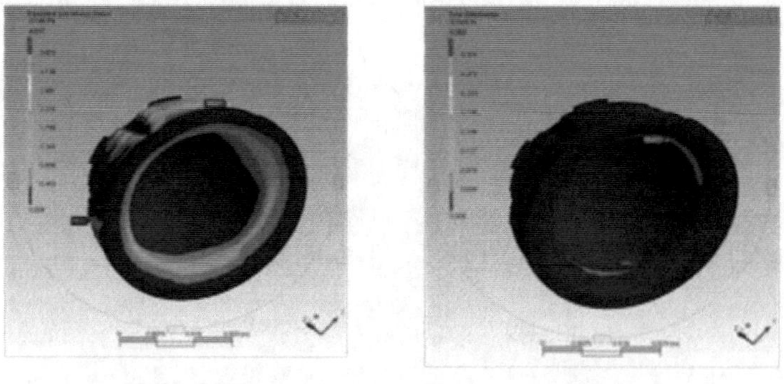

Figure 15 The analysis of tenseness and deformation on the socket

In the second stage we applied the torque to the system and we examined tension and values of deformation (figures 16).

IOP Conf. Series: Materials Science and Engineering **448** (2018) 012039 doi:10.1088/1757-899X/448/1/012039

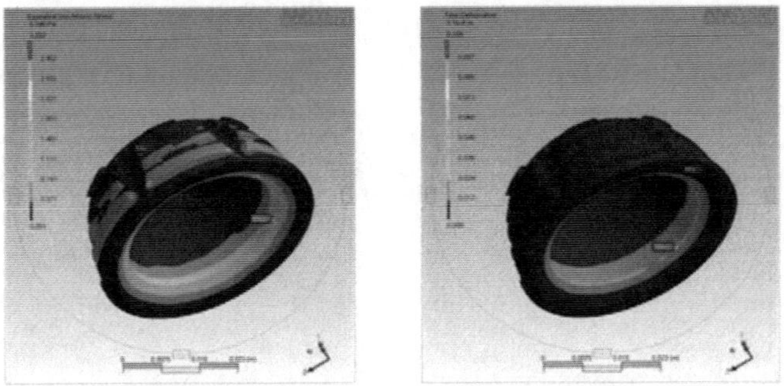

Figure 16 The analysis of tenseness and deformation on the socket

4. Summary

3D Measurement Techniques, as we have seen, can be measured with a very wide variety of complex shapes and complexes. Depending on the purpose of using high-precision gauges, we can use it to audit production processes or to identify defects in the quantities, both in volume and graphically. Using the coordinate values defined by the coordinate measuring machine, we can create 3D models which can be further analyzed, simulated, and examined for its function and the effect of influencing factors.

By modelling or simulating, we can change geometric and other parameters even in the design phase.

5. Acknowledgement

"The described article was carried out as part of the EFOP-3.6.1-16-2016-00011 "Younger and Renewing University – Innovative Knowledge City – institutional development of the University of Miskolc aiming at intelligent specialisation" project implemented in the framework of the Szechenyi 2020 program. The realization of this project is supported by the European Union, co-financed by the European Social Fund."

6. References

[1] Hörcsik Renáta: 3D measurement with two different software, microCAD 2005 Proceedings of International Scientific Conference, Section M1: Production Engineering and Manufacturing Systems, March 10-11. 2005., University of Miskolc, Hungary

[2] Juhász Imre: Számítógépi geometria és grafika. Miskolci Egyetemi Kiadó, 1993.

[3] Ábel József – Balajti Zsuzsanna: Fogazatok szerszámkopás vizsgálata az ábrázoló geometria segítségével, OGÉT, 2018

[4] Dr. F. Szabó – Z. Bihari – F. Sarka: Finite element modeling and optimization of the products, structures, machinery parts, Miskolc 2006.

IOP Conf. Series: Materials Science and Engineering **448** (2018) 012040 doi:10.1088/1757-899X/448/1/012040

Elaborating Industry 4.0 compatible DSS for enhancing production system effectiveness

N Tóth[1], **R Ladányi**[1] **and E Garamvölgyi**[1]
[1] Department of Logistic Systems, Bay Zoltán Nonprofit Ltd. for Applied Research, Miskolc, Hungary

Abstract. Developments on the base of the Industry 4.0 philosophy are the most important drivers of the recent production efficiency increasing activities. For meeting the requirements of the production facility operators, integration of digitalization-, virtualization-, simulation- and info-communication technologies are essential. Automatic detection and elimination of the errors in the actual production processes is one the most promising outcome of utilization of the data resulted by the online data acquisition and monitoring of the production system. Having these thoughts in mind, an innovative application of the discrete event driven process simulation on actual sensor data will be presented in this paper. Minimizing the idle times and maximizing the capacity utilization are the required consequences of this application.

Keywords. production process, discrete event driven process simulation, industry 4.0, production monitoring, data acquisition sensors

1. Introduction

Impacts of the recent developments in info communication technologies became part of our everyday life. Widespread use of mobile phones rapidly contributed to the development and use of online marketplaces and e-commerce solutions. As a result, production and service companies shortly faced the challenge of processing and fulfilling on-line orders and requests on a level that they were not ready to handle. Despite such customer orders are usually random considering time and amount of purchase, fast, prompt and high quality fulfilment has to be ensured. Complying such requirements with traditional production systems optimized for mass production is hard and costly, as medium or small quantities has to be produced with widening and continuously changing product scale. Frequent change in production results in losses both in production and logistics.

Thus, processing online orders call for a flexible production system where even special customer requests can be fulfilled without losses. To supply such production systems with parts and materials improved logistics are needed: JIT/JIS supplies shall be managed even in production-stock relations. Such goals can be achieved by strong co-operation between production and logistics: prompt and adequate data interchange using up-to-date communication channels. The need for increased flexibility and custom manufacturing induced the development of production systems where universal and autonomous equipment communicate and co-operate to reach a common goal: to produce a product item. As equipment status can be monitored by sensors in such a system, production effectiveness can be improved through decision making based on collected sensor data. Industry 4.0 focuses on this approach.

2. Digitalisation

As huge number of parameters is typical for complex production-logistic systems, high level computerised representation can be a suitable tool for their monitoring, analysis and characterisation. Tools with dynamic process modelling capability can be used to determine optimal parameter values. Primary goal of this modelling is to enable experiments in the system without interfering with the processes in the real system [1]. Data acquired from the real production systems have to be processed displayed and transferred by means of modern IT technologies for the process simulation [2][3]. Two main categories of production system models are as follows:

- Models based on static data preloaded from a database: In case of these passive models time dependency of indicators in the system can be described using the simulation; and results can be used to predict the performance of the real system using predefined KPIs. Thus, such models are suitable not only for optimisation and KPI improvement, but also for intervening Figure 1 a).

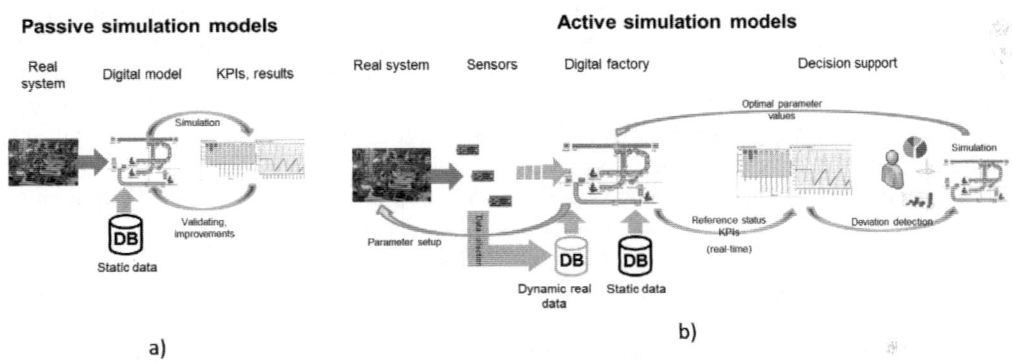

Figure 1. Simulation model types

- Models based partly on sensors collecting data from the real production system Figure 1 b): In case of such active models a Cyber Physical System (CPS) [4][5] is generated according to the Industry 4.0 philosophy that integrated the five architectural levels of CPS. Data access that is needed for the simulation is carried out either via direct communication and data interchange or through common databases (Smart Connection-, Data Conversion Level). Measured sensor data is continuously evaluated, thus comparing system KPIs with pre-calculated values are executed even real time. Mitigation and intervention measures can be evaluated immediately if significant difference is emerged between real and target KPI values (Cyber and Cognition Level). Changing parameters in the real system can either be carried out by direct access through the available communication channels or by indirect human interactions (Configuration Level). Such models and simulation techniques are suitable for real time, parallel evaluation of real system data and detection of problems in the real system through the changes in parameter values and for decision support for mitigation and intervention.

3. Concept of a decision support tool based on active simulation models

A digital twin can be developed according to the digital factory concept based on the data collected in the real system real-time [6]. The digital twin can be used not only for monitoring parameters of the real system, but also for defining and managing intervention measures to improve system performance in case of insufficient KPI values (Figure 2).

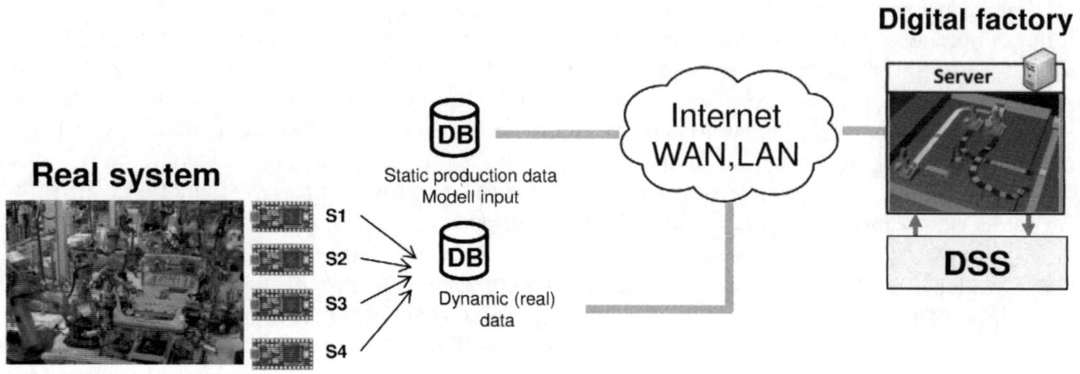

Figure 2. System concept

Data that originated from the real production systems through the acquisition mechanisms of the different types of sensors is essential for the appropriate functions of the digital factories. The data transfer between the sensors and the digital factory can be realized by IoT solutions directly or indirectly through database tables either. The digital factory receives, processes and evaluates the sensor data continuously. Establishment of a digital factory could be carried out in the Siemens Plant Simulation software that is developed for modeling and for discrete event driven simulation of production and logistic processes [7]. It supplies with the object classes and with the communication interfaces that are necessary for the process modeling:

- socket communication for sending and receiving data through the specified IP address and port;
- ODBC related objects for accessing databases that are located on shop floor computers or on computer networks or in remote clouds;
- interfaces for accessing SQLite- or Oracle databases.

Requirements against the Decision Support System to be developed can be summarized as follow:

- digitalization of a real production system has to be supported;
- support for communication and for data transport among the system elements (e.g. sensors and databases) is necessary;
- ability of near-real-time monitoring of the production system parameters and ability for detection of the deviances arisen between the actual values of the system parameters and their targeted values;
- support for performing the necessary interventions automatically (e.g. setting the system parameters to the targeted values).

4. Demonstrative setup of the Decision Support System

A DSS prototype was elaborated to demonstrate the features and advantages of the digital factory application (Figure 3).

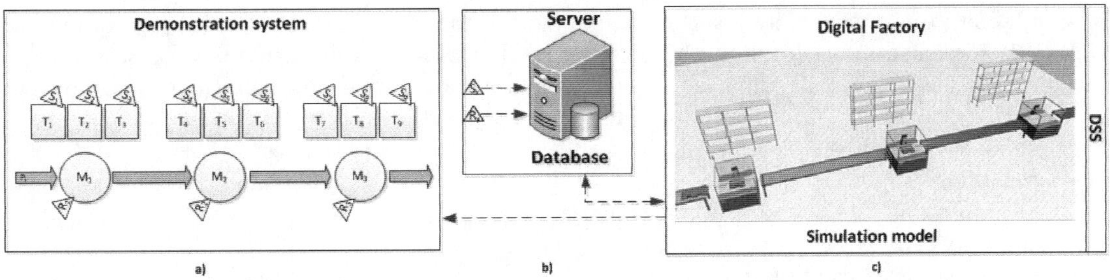

Figure 3. Real system illustration and DSS Components

Figure 3 a) shows the flow-shop model [8] based illustration of a real production system which assembles different kind of products (P_i). The sequence of the assembly operations is predetermined which means that the same types of products are follow the same path among the production steps. A production machine ($M_1,...,M_3$) performs one assembly operation at once and there is no operation interruption permitted. The assembly operations cannot overtake each other. Every operation has its own time consumption which is necessary for assembling all the parts into the products at that specified production machine. There are 3pcs. of enumerated part storages ($T_1,...,T_9$) allocated to each of the machines. Storage ($S_1,...,S_9$) is dedicated to a specific part therefore a part can be placed only into its dedicated storage at the machines. The part consumption of an assembly operation depends on the type of the product to be assembled. The semi-finished product transports between the operating machines are characterized by the transport times. There are two types of data collection sensors applied in the system. Weight measuring sensors were applied for monitoring the part consumption at the production steps. These sensors supply with all the necessary information to calculate the intensities of the part consumptions. These intensities can be calculated by part types, and by part storages etc. RFID sensors ($R_1,...,R_3$) were applied for identifying the type of the actually assembled product and for identifying the parts that are assembled into them. There is a possibility to identify the parts or products individually which enables the integration of advanced production track&trace functions into the DSS. The central component of the system is a database - Figure 3 b). It comprises on one hand the sensor supplied dynamic data and on the other hand the static data which is related with the construction and the set-up of the digital model of the production system:

- bill of materials (BOM) by product types;
- characterization of assembly steps by product types (including machines, time consumption, sequence, etc.);
- semi-finished product transport times between the production machines;
- lists of production machines, of part storages, of part types and of sensors.

However, all the data which describe the preliminary conditions of the simulation examinations on the digital model of the production plant are stored at the same database either. For example the production plan which determines the quantities, the types and the sequence of the products to be assembled can be stored there. There are data management mechanisms integrated with the central database also. Continuous comparison of the sensor bound data and the results of the simulation examinations are the main function of these mechanisms. Detection of the deviances and errors that may arise during the production process is the reason behind this comparison. Another element of the DSS system is the digital model of the production plant - Figure 3 c). This model is built within the Siemens Plant Simulation software. Giving possibility for simulation examination of the production processes is the main function of this system component. The production plan and all the production related preliminary conditions including the production supply processes are taken into account during this simulation examination. The result of a simulation examination is called reference status time series (RS) for the examined future production period. Predicted values of the necessary process indicators and their changes during the examined future time period are included in the reference status. The following Key Performance Indicators of the production system are determined [9]:

- regarding the production processes:
  - utilization of the production machines;
  - productivity;
  - production lead times;
  - overall equipment effectivenesses (OEE) [10].

- regarding the production supply processes:
  - capacity utilization of the part storages;
  - indicators of the part supply processe;
  - effectiveness of the warehousing and commissioning processes.

The reference status time series describes the progress of the production from the viewpoint of the actually assembled products and the quantity of the products that are already completed besides the already mentioned data. The DSS continuously compares these results to the sensor bound actual data. Stochastic effects that might be arisen at the real production can cause deviances (K) between the actual data and the reference status time series. The actual production process can be qualified to be in normal status (NS) depending the measures of this deviance. There are several indicators that have to be taken into account during the qualification of the actual status of the real production system. The indicator (I) of the part consumption (κ_{S_i}) and the indicator of the production intensity (κ_{R_j}) are among these indicators. The deviance regarding these indicators can be calculated with the following formulas:

$$\kappa_{S_i} = | I_{S_i}^{RS}(t) - I_{S_i}^{*}(t) | \leq d_{S_i} \tag{1}$$

$$\kappa_{R_j} = | I_{R_j}^{RS}(t) - I_{R_j}^{*}(t) | \leq d_{R_j} \tag{2}$$

The members of these formulas are the following:

κ_{S_i}: deviance between the value ($I_{S_i}^{*}$) that supplied by the i. weight sensor and its reference value ($I_{S_i}^{RS}$) at t. time [%];

κ_{R_j}: deviance between the start time ($I_{R_j}^{*}$) of an assembly operation that registered by j. RFID sensor and the reference value ($I_{R_j}^{RS}$) of this event [%];

d_{S_i}, d_{R_j}: the predetermined maximum of the deviations between the sensor bound data and their reference values [%].

Average deviance [%] of aggregated indicators regarding the data that supplied by the weight and the RFID sensors:

$$K = \frac{\sum_{n}^{i=1} \kappa_{S_i}}{n} + \frac{\sum_{k}^{i=1} \kappa_{R_j}}{k} \tag{3}$$

where:
K : means the average deviance [%] between the real production system and the reference value which is calculated on the base of the simulation examinations.

Behaviors of the production capacities may influenced by stochastic errors. These stochastic features have effects on the production procedures continuously. Some of them strengthen each other's effect and some of them weaken it. Overall influence of these effects on the production system can be qualified to be negative because it often manifested in efficiency reduction (e.g. daily production plan cannot be fulfilled). The reference status is calculated with having ideal production circumstances in mind. Therefore the reference values of the indicators can be the base of the comparison for detecting the deviances. The reference status (RS) means the forecasted time series of the production indicators, sensor bound data represents the actual values of these indicators. Probably there will be always a deviation among these values. This overall deviation is characterized by the value of the K which is calculated continuously during the real production process. The production process state can be qualified to be normal while the K value does not exceed a predetermined maximal value ε :

$$NS = RS \pm K \qquad (K \leq \varepsilon) \tag{4}$$

A failure status (FS) developed if the value exceeds its limitation. This failure can be caused by an external impact which has significant influence on the production system. (e.g. raw material shortage):

$$FS = RS \pm K \qquad (K > \varepsilon) \tag{5}$$

There is a necessity for intervention into the real production procedure in case of such failure status. The aim of this intervention is to mitigate the effects of the errors and to ensure sustainable normal production circumstances. A new reference status has to be calculated after the interventions as production circumstances are characterized by the preliminary production conditions. The following Table 1 contains sample values of the sensor bound data and of their reference values for explaining the relation between the deviations and their possible appearance in the production system.

Table 1. Sensor bound data and their reference values

	Reference status (RS) (between 3h – 4h)	Real production system ($*$) (between 3h – 4h)	Deviances
$R_1 : M_1$	P3, P2, ***P1, P3, P4, P5***, P1, P2, P2, P4, P3	**P3, P2, P1, P3**, *P4, P5, P1, P2*, P2, P4, P3	$t_{(3h)}^{RS}$: P1 $t_{(3h)}^{*}$: P3
M_1 production quantity	20-25pcs	18-23 pcs	$t_{(3h)}^{RS}$: 20 $t_{(3h)}^{*}$: 18
$S_1 : T_1$	180 pcs	192 pcs	$t_{(3h)}^{RS}$: 180 $t_{(3h)}^{*}$: 192
$S_2 : T_2$	80 pcs	95 pcs	$t_{(3h)}^{RS}$: 80 $t_{(3h)}^{*}$: 95

The data that collected by the R_1 sensor confirm the production delay compared to the Reference Status. M_1 production machine accomplish 2pcs fewer assembly than it was forecasted to the specified time interval. This delay is the reason for the greater amount of parts left in the part storage at this machine. Increasing delay will have impacts on the part supply processes as well. There will be more parts in the part storages than it was expected at the time of the part delivery. Parts that exceed the storage capacity of the storage place have to be returned to the warehouse. This part return process means unnecessary work and it decreases the efficiency of the part supply system. Early detection of the errors and determining their types is necessary to avoid situations with failures. A possible intervention to solve the situation that is described in the Table 1 is to increase the assembly speed at the M_1 machine or to apply a redundant unit in parallel with M_1. Synchronizing the production and the production supply systems is marked to be the further aim of the DSS. By determining the decreasing intensity of the parts in the storages next to the machines the time of the shortage can be calculated precisely. Therefore the warehouse subsystem can schedule accurately the preparation procedures of the parts and punctual preparation increases the efficiency of the part supply procedure. The DSS system continuously evaluates the operation of the production system as it was described above. It can propose intervention in case of significant deviances. Its activity facilitates the application of JIS methods in the part supply process [11][12]. Furthermore, only the necessary quantities of the parts have to be located next to the machines by means of the DSS therefore logistic related areas could be transformed into productive places in the plants.

5. Decision support system

The system concept of an industry 4.0 compatible DSS was described in the previous chapters. Its main function is the continuous comparison of the forecasted and the real values of the process indicators to preemptively detect the production errors:

- determines the location and the type of the errors;
- supports the decision regarding the most effective way of the intervention if it is necessary;

Hierarchical levels of the system are represented in Figure 4. based on the concept introduced in Figure 2.

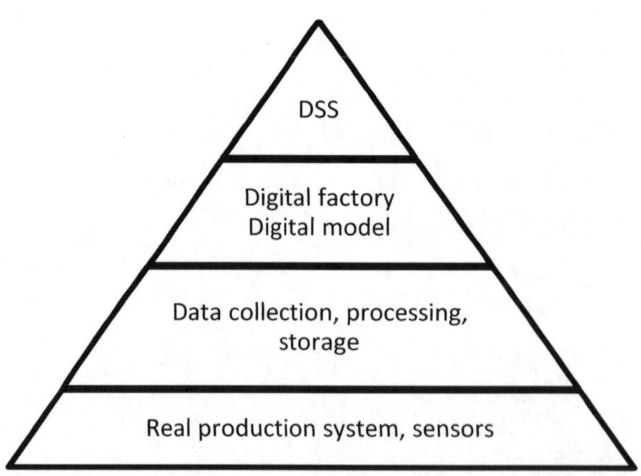

Figure 4. System concept of an industry 4.0 compatible DSS

The real system itself is situated on the lowest level of the hierarchy, with the sensors – the data collection units – connected to them. The next level covers data collection, processing and storage, which processes and stores data on collected by sensors and calculated for reference status characterisation into an appropriate database. While third level contains the digital model of the system, the top level represents the decision support system that makes decisions on intervention based on the previous levels results. Generalised operation process of a DSS is represented in Figure 5 [13].

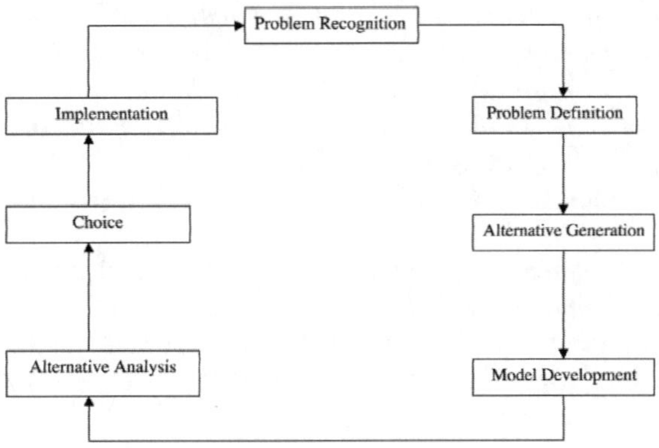

Figure 5. DSS [13]

While problem identification is the first step towards a DSS, categorising such problems for identifying mitigation measures are next. As several ways may be available to mitigate impacts or eliminate errors, these possible ways have to be generated in the model and their side-effects have to be investigated. Intervening may be carried out either manually by operators or automatically directly by the DSS according to the rules defined for the error type. Despite operation of the DSS being simple, types of errors can be numerous due to the complexity of the real system. As a result, number of mitigation measures or error elimination procedures may be high. Next step of the development will be the development of a powerful decision support algorithm based on the aforementioned.

6. Conclusions

A possible use of digital factory approach, a central element in Industry 4.0 philosophy has been introduced in this paper: stochastic impacts causing errors in a real system can be calculated using digital factory models. Earliest possible identification of errors occur is crucial to determine the best way to intervene the system operation. Defining and categorisation of errors may lead to an effective decision support system suitable for error avoidance.

References

[1] VDI-Richtlinie 3633-1 1993 *Simulation von Logistik-, Materialfluss- und Produktionssystemen-Grundlagen* (Düsseldorf: VDI-Verlag)

[2] Russmann M, Lorenz P, Gerbert P, Waldner M, Jastuss J, Hengel P, Harnisch M, 2015 *Industry 4.0: the future of productivity and growth in manufacturing industries* The Boston Consultin Group report

[3] https://inform.tmforum.org/internet-of-everything/2016/10/industry-4-0-digitization-sensorization-optimization/

[4] Lee J, Bagheri B, Kao H-A, 2015 *A Cyber-Physical Systems Architecture for Industry 4.0-based Manufacturing Systems (Manufacturing Letters* 3) pp 18-23

[5] Monostori L, Kádár B, Bauernhansl T, Kondoh S et al. 2016 *Cyber-physical systems in manufacturing* (*CIRP Annals – Manufacturing Technology* Volume 65 Issue 2) pp 621-641

[6] http://www.techmonitor.hu/termek-megoldas/az-ipar-4-0-es-a-siemens-plm-digitalizacio-20161001

[7] Siemens AG 2017 Tecnomatix Plant Simulation Help

[8] Miloš Š 2007 *Mathematical Models of Flow Shop and Job Shop Scheduling Problems* (*International Science Index Physical and Mathematical Sciences* vol 1 no 7) waset.org/Publication/14079

[9] Amrina E, Yusof S 2011 *Key performance indicators for sustainable manufacturing evaluation in automotive companies (IEEE International Conference on Industrial Engineering and Engineering Management* Article number 6118084) pp 1093-1097

[10] Gamberini R, Galloni L, Lolli F, Rimini B, 2017 *On the Analysis of Effectiveness in a Manufacturing Cell: A Critical Implementation of Existing Approaches* (*Procedia Manufacturing* 11) FAIM2017 pp 1882-1891

[11] Bányai T 2013 *Just-In-Sequence ellátási stratégiák* (*Multidiszciplináris tudományok 3. kötet*) 1. sz. pp 141-152

[12] Aleksandra A, 2017 *Implementation of Just-in-sequence concept in automotive industry – comparation of Austrian and Serbian model* (*Industrija* vol 45 No 3) pp 83-99

[13] Shim J, Warkentin M, Courtney J, et al. 2002 Past, present, and future of decision support technology (Decision Support System 33) pp 111-126

IOP Conf. Series: Materials Science and Engineering **448** (2018) 012041 doi:10.1088/1757-899X/448/1/012041

Isothermal crystallization kinetics of flame retardant PET containing melamine based FR

Balázs Ádám

Department of Materials Technology, Faculty of Engineering and Computer Science, John von Neumann University, 10 Izsáki Street, Kecskemét, 6000, Hungary

adam.balazs@gamf.uni-neumann.hu

Abstract. Isothermal crystallization behavior of poly(ethylene terephthalate) (PET) was investigated by differential scanning calorimetry. 1 to 5 wt% halogen-free flame retardant materials were mixed into PET and these samples were examined. Avrami equation was used to analyze the crystallization process. Arrhenius equation was applied to calculate the activation energy of crystallization. It was proven that flame retardant material affects the crystallization.

1. Introduction

The poly(ethylene-terephthalate) (PET) is a thermoplastic semi-crystalline polyester having many excellent properties including dimensional stability, chemical resistance, high strength, and stiffness. Therefore, PET is widely used in areas of packaging industry for producing fiber, film, beverage bottles and as engineering plastics in electronics, automotive and construction. [1-3]

However, PET is quite flammable, hence its fire behaviors should be improved to extend its potential application where fire retardant properties are required. [2] Flame retardant (FR) additives are developed to stop or slow down the possibility of a fire. In the past, FR additives, which contained halogen atoms, were the most effective, therefore they were the most popular. Spreading of environmental regulations caused that halogen-free additives were developed to protect the environment. [2, 4]

Another disadvantage of PET that its crystallization is slow. [1, 3, 5, 6] Hence it is difficult to reach high crystallinity using pure material and conventional processing techniques respectively technological settings. Nevertheless, the rate of the crystallization can be improved by nucleating agents. Nucleation is the process when new phases appear. Heterogeneous material with appropriate surface energy can act as a nucleating agent. Among others, nucleation can be controlled by the concentration of the additive. [6]

There are overlaps between the flame retardant and the nucleating additives, such as montmorillonite (MMT), which is often used as a nucleating agent for PET [3, 5] but is also used as a flame retardant for PLA. [7]

In this study, isothermal crystallization of PET containing halogen-free flame retardant additive material has been investigated. Avrami equation was applied to analyze the overall crystallization kinetics and Arrhenius equation was used to calculate the activation energy of the phase transformation.

IOP Conf. Series: Materials Science and Engineering **448** (2018) 012041 doi:10.1088/1757-899X/448/1/012041

2. Experimental

2.1. Materials

Matrix material was SKYPET BL8050 PET copolymer. Its density is 1.4 g/cm3, glass transition temperature is 95-98°C and Vicat softening temperature is 78-82°C. [8]

Halogen-free flame retardant was melamine-based Melapur MC25, developed mainly for polyamides and thermoplastic polyurethanes by CIBA. The empirical formula is 1,3,5-triazine, 2,4,6 (1H, 3H, 5H)-trione, and 1,3,5-triazine. Its decomposition temperature is 350°C and the minimum nitrogen content is 65%. [4, 8]

The samples were prepared by the Department of Polymers of the University of Miskolc. The blends were made by a Haake Rheomix 3000p heated chamber mixer. The lowest oxygen index was measured by Polymer-Laboratories unit and the rheological properties of the mixtures were calculated from the mixing torques. [8]

In this study, three samples were tested containing 1, 2, 3, 4 and 5% Melapur. The pure material was not examined in the absence of a sample.

2.2. Method

The thermal behavior of PET was measured by TA Q200 heat-flux DSC instrument, which was calibrated by Indium. The sample weights were about 5 mg. The applied gas during the DSC scan was nitrogen 50 mL/min flowing rate. The temperature range was from 30°C to 300°C.

At first, the sample was heated to 300°C at a heating rate of 20°C/min to eliminate any thermal history. Then the sample was cooled (by the maximum rate of the instrument) to the crystallization temperatures, which were 217, 219, 221, 223°C at 1%, 216, 218, 220, 222°C at 2 and 3%, 219, 221, 223, 225°C at 4%, and 222, 224, 226, 228°C at 5% Melapur content. These temperatures were kept until the end of the crystallization. [10-12]

3. Result and discussion

Determination of the temperatures and the holding times of the isothermal crystallization can be quite difficult, therefore it was necessary to carry out several pre-tests to find the correct data. Due to the different Melapur content, the applied crystallization temperatures were different.

Figure 1 illustrates the DSC curve of the 3% Melapur containing sample during 220°C isothermal crystallization temperature. It has two Y-axes, the thick blue line shows the heat flow, and the thin orange line shows the area percent versus time. The area percent is the relative crystallinity. The time zero (t_0) is that point when the isothermal crystallization starts.

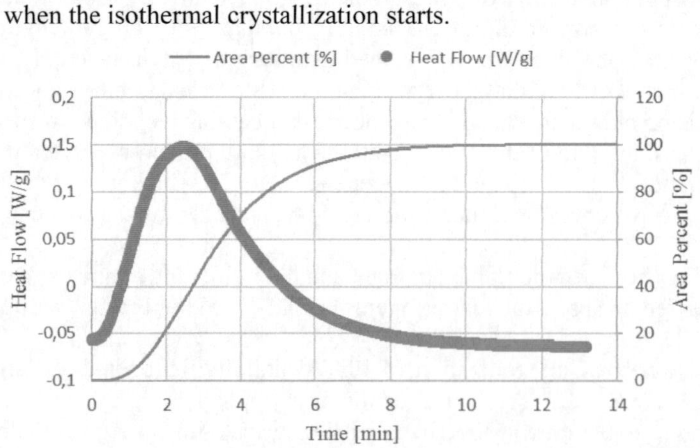

Figure 1. DSC curve of 3% Melapur containing sample, $T_{c\ iso}$: 220°C

The asymmetrical shapes of the heat-flow peak suggest that the growing crystals collide therefore they blockade each other. When the isothermal temperature is increased, the maximum of the heat flow is lower and the time of the crystallization is longer. The change of relative crystallinity at a time can show the rate of the crystallization. The slope of the curve is reduced when the isothermal temperature increases, so it means that the rate of the crystallization becomes slower.

The Avrami equation (1) was used to describe the change of the relative crystallinity with time:

$$x(t) = 1 - \exp(-Kt^n) \quad (1)$$

where x is the relative crystallinity, K is the rate constant of crystallization, t is the time and n is the Avrami exponent. The values K and n are considered to be characteristic to the mechanism of crystallization. [10-13]

Figure 2 depicts the relative crystallinity after transformation by Avrami equation. The slope of the linear trend line can show the Avrami exponent, and the rate constant of the crystallization is the intersection of the axis.

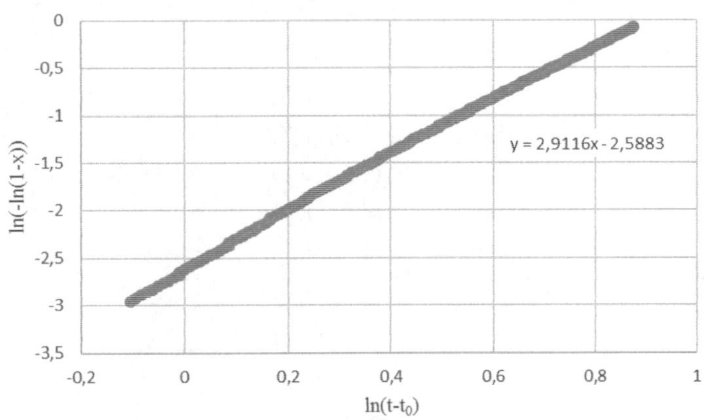

Figure 2. Transformed 2% Melapur containing sample, $T_{c\ iso}$: 216°C

Table 1 [13] summarizes the meaning of the Avrami exponent, and Table 2 summarizes the results derived from Avrami analysis after the isothermal crystallization on every sample.

Table 1. Avrami exponents

Dimension of growth	Geometry	n	
		athermal nucleation	thermal nucleation
1	Fibrillar	1	2
2	Lamellar	2	3
3	Spherulitic	3	4

IOP Conf. Series: Materials Science and Engineering **448** (2018) 012041 doi:10.1088/1757-899X/448/1/012041

Table 2. The result of Avrami analysis

Melapur [%]	Tc iso [°C]	1/Tc iso [K]	n [-]	ln K [-]
1	217	0,002040816	2,2018	-0,2724
	219	0,00203252	2,4824	-1,5506
	221	0,002024291	2,4716	-2,5898
	223	0,002016129	2,6108	-4,0716
2	216	0,00204499	2,9116	-2,5883
	218	0,00203666	2,8038	-3,4537
	220	0,002028398	2,8465	-4,5441
	222	0,002020202	2,9335	-5,8772
3	216	0,00204499	2,6754	-1,3356
	218	0,00203666	2,624	-2,1421
	220	0,002028398	2,5564	-3,0631
	222	0,002020202	2,6553	-4,3383
4	219	0,00203252	2,9534	-1,746
	221	0,002024291	2,7758	-2,7534
	223	0,002016129	2,6555	-3,735
	225	0,002008032	2,5944	-4,6402
5	222	0,002020202	2,9848	-2,2447
	224	0,002012072	2,767	-3,169
	226	0,002004008	2,8404	-4,5505
	228	0,001996008	2,797	-5,6602

Figure 3 shows the changing of the Avrami exponent. Despite the fact that the crystallization temperature was changed, there was no significant difference between the n values of the measurements, this suggests that the nucleation is probable athermic.

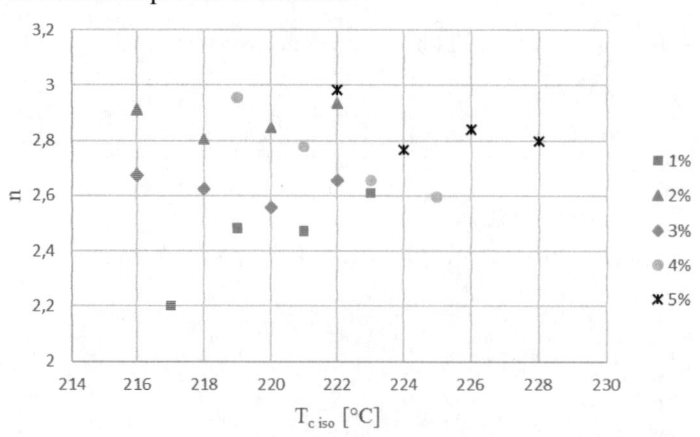

Figure 3. Avrami exponent as a function of $T_{c\ iso}$ and Melapur content

All n values are between 2 and 3, suggest that the geometry can be changed from 2 dimensions to 3, nevertheless the transformation is not complete. The 1% Melapur containing sample is different from the others, its points are mainly below than 2.6 while the points of 2, 3, 4 and 5% Melapur containing samples are preferably larger than 2.6. [9-13]

Figure 4 shows the altering of the rate constant of the crystallization as a function of reciprocal isothermal crystallization temperature, in Kelvin.

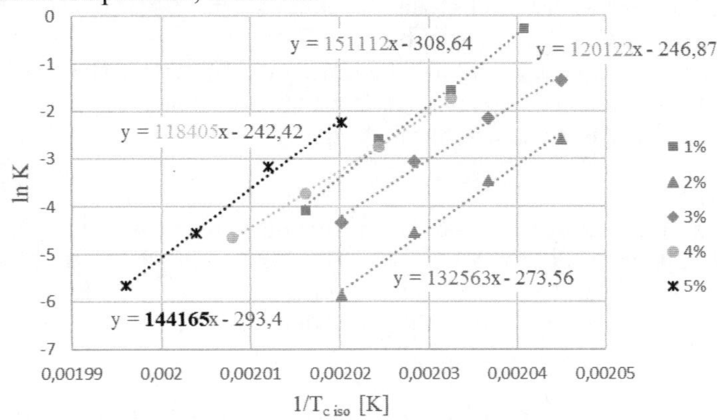

Figure 4. The rate constant as a function of reciprocal $T_{c\,iso}$ and Melapur content

The activation energy of the crystallization process was determined from the rate constant by the Arrhenius equation (2):

$$K = A\exp(-E_a/RT) \tag{2}$$

where K is the rate of crystallization, A is a pre-exponential constant, T is the crystallization temperature, R is the universal gas constant and E_a is the activation energy of crystallization. The activation energy can be calculated from the slope of the rate constants curves. [10-13]

Figure 5 shows the varying of activation energy as a function of Melapur content. It can be seen, that increasing Melapur content from 1% to 4% decreases the activation energy of crystallization, as a consequence, the crystallization becomes easier. However, the activation energy of the 5% sample began to increase again. At present, the reason for the difference is not known, therefore further investigation is needed.

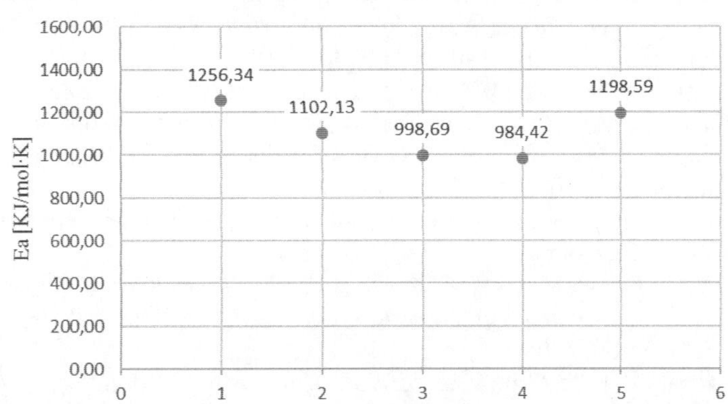

Figure 5. Activation energy as a function of Melapur content

4. Conclusion

1, 2, 3, 4 and 5 wt% halogen-free flame retardant materials were mixed into PET then isothermal crystallization behavior was investigated by DSC. Avrami equation was used to describe the crystallization process, and Arrhenius equation was applied to calculate the activation energy of crystallization. Increasing the amount of the flame retardant Melapur material changes the crystallization process. All values of Avrami exponents are between 2 and 3, suggesting that the geometry can be changed from 2 dimensions to 3, but the transformation is not complete. The sample containing 1% Melapur was the lowest, separated a bit from the rest. The values of 2, 3, 4 and 5% were similar. From the Avrami exponents, it became apparent that nucleation was athermic. The activation energy between 1% and 4% has steadily decreased, resulting that the crystallization has become easier. The activation energy of 5% was larger and this requires further testing.

5. Acknowledgment

I would like to thank Prof. Károly Belina for his invaluable help in acquiring thermal analysis.

This research is supported by EFOP-3.6.1-16-2016-00006 "The development and enhancement of the research potential at John von Neumann University" project. The Project is supported by the Hungarian Government and co-financed by the European Social Fund.

References

[1] Gao Y, Wang Y, Shi J, Bai H W, Song B 2008 *Functionalized multi-walled carbon nanotubes improve nonisothermal crystallization of poly(ethylene terephthalate)* Polymer Testing, 27, 179-188

[2] Si M M, Feng J, Hao J W, Xu L S, Du J X 2014 *Synergistic flame retardant effects and mechanisms of nano-Sb2O3 in combination with aluminum phosphinate in poly (ethylene terephthalate)* Polymer Degradation and Stability, 100, 70-78

[3] Wang Y M, Gao J P, Ma Y Q, Agarwal U S 2006 *Study on mechanical properties, thermal stability and crystallization behavior of PET/MMT nanocomposites* Composites Part B-Engineering, 37, 399-407

[4] Nagy D, Ádám B, Czél Gy 2016 *Thermal Investigation of flame retarder containing PET* Gradus, Vol 3, No 2

[5] Ke Y C, Wu T B, Xia Y F 2007 *The nucleation, crystallization and dispersion behavior of PET-monodisperse SiO2 composites* Polymer, 48, 3324-3336

[6] Belina G, Menyhárd A, Juhász P, Belina K 2001 *Heterogén gócképzők hatékonysága PET hulladékban* Műanyag és Gumi, 38 évf 12 szám, 441-446

[7] Wang X, Wang D-Y 2017 *Fire-retardant polylactic acid-based materials: preparation, properties, and mechanism. Novel Fire Retardant* Polymers and Composite Materials, http://dx.doi.org/10.1016/B978-0-08-100136-3.00004-2

[8] Czél Gy, Kelemen A, Sólyom D 2010 *Flame retardant polyethylene terephthalate fibers added with melamine based FR additive* Materials Science Forum Vol 659, pp 135-140

[9] Saeidlou S, Huneault M A, Li H, Park C B 2012 *Poly (lactic acid) Crystallization* Progress in Polymer Science, 37, 16657-1677

[10] Belina K, Halász L, Vorster O C, Juhász P 2006 *α-Olefin Homo- és Kopolimerek Kristályosodási Tulajdonságai* Műanyag és Gumi, 43(8), 347-356

[11] Ádámné Major A, Belina K, Pósa M, Kecskés B 2008 *Investigation of mechanical and thermal properties of oriented polyamide copolymer* DOI 10.1007/s12289-008-0324-5, Springer/ESAFORM 2008

[12] Lee S W, Cakmak M 1998 *Growth Habits and Kinetics of Crystallization of Poly (Ethylene 2.6-Naphthalate) under Isothermal and Nonisothermal Conditions* Journal of Macromolecular Science, Part B, 37:4, 501-526

[13] Pukánszky B, Varga J. 2012 *Bevezetés a műanyagok fizikájába* Kiadja a BME Fizikai Kémia és Anyagtudományi Tanszék, Műanyag és Gumiipari Laboratórium, Budapest

The special capabilities of Surface Constructor kinematic modelling and simulation software

L Dudás

Department of Information Engineering, University of Miskolc, Miskolc, Hungary
E-mail: iitdl@uni-miskolc.hu

Abstract. The paper introduces the innovative and unique capabilities of the Surface Constructor software application. This tool is intended for modelling and optimizing mating enveloping surfaces that can be used e.g. as tooth surface of gears. The software applies the original Reaching Model theory that can provide surface enveloped by a given surface. The theory handles local and global methods, generating not only the enveloped surface but all the types of local undercuts and global cut such a way. Surfaces and kinematic relations are represented by symbolic algebraic expressions giving the maximal freedom in modelling. Special capability of the tool is showing the $R = R(\Phi)$ functions that can explore and qualify the abnormalities in mating of surfaces. Moreover it provides special velocity-space features that can reveal the nature and characteristics of the space where motions happen. The tool can apply intermediary generating surface making also possible generating mating surfaces characterized by point-like contact. All these unique features integrated in a complex design system that uses symbolic, numerical and visual representations of the modelled objects. The paper demonstrates every capability of the software tool with concrete applications, in some cases with patented solutions.

1. Introduction

The paper gives an overall overview and demonstration of the capabilities of the *Surface Constructor* (*SC*) enveloped surface generator and kinematical modelling software application. In the last decades this software got name as a versatile and powerful tool proving its capabilities in solving different innovation tasks. Its success origins from the fact that one of the requirements of the machine industry is to resolve the general task of conjugate surface pairs, so generate surface by other moving surface and this task is not solvable easily by usual 3D CAD applications. The most important area is designing and machining gearings and their tools. For the resolving such tasks the *SC* includes a novel enveloped surface determination theory, the *Reaching Model*. This theory solves the general task of meshing surfaces fully including the detection of all the possible types of undercut situations. This fundamental theory underlies the *SC* tool and will be detailed in the second section.

SC works similarly to the well-known 3D CAD applications, gives a universal and task type independent environment for modelling enveloping surfaces and analysing their mesh properties and animating their motion. The three-level structure of the application that provides this flexibility and versatility will be described shortly in the third section.

Then, in the fourth section the unique capabilities will be introduced and demonstrated through example tasks. The first speciality will describe the visualization and use of the $R = R(\Phi)$ functions. The second speciality will be the visualization of the velocity-space properties, while the capability of using intermediary theoretical surface and creating such a way surface pairs that mate point-like manner will be demonstrated as the third special feature.

2. The core theory: the *Reaching Model*

There are two well-known existing methods to determine the enveloped surface [1]:

- the differential-geometric method developed by *Gochman*, and
- the kinematical method, which applies the $\boldsymbol{n}\cdot\boldsymbol{v}^{1,2} = 0$ product where $\boldsymbol{n}$ is the normal vector and $\boldsymbol{v}^{1,2}$ is the relative velocity at the meshing point.

These quick analytical methods are local methods, so they can work in infinitely small space and time domain and they can produce the local undercuts or the boundary of required enveloped surface. As a consequence, these methods are suitable for detecting all types of local undercuts, which can appear in an infinite small domain but fail in determination of the real surface if it has swept region which accompanies to local undercut almost every time. The determination of such not enveloped part of surfaces need global surface determination method, like the more slow discrete digital simulation of the surface generation.

In the last decades researchers have been applied computers for enveloped surface modelling expansively. Among the pioneers *Holler* produced worm-gear and hypoid-gear surfaces using computer model of meshing and local method [2]. Later *Weck* and his colleague applied computer simulation of cutting process and global method to produce different types of gears [3].

Other researchers used both local and global methods to determine quantity of the interference. *Seveleva* computed a *Gleason* bevel gear tooth surface using theory of enveloping surfaces and numerical simulation of cutting at the same time [4]. In our days the most famous gear investigator in addition to the others, who uses discrete digital simulation is *Lunin* in its gear development institute *ZAKGEAR* [5].

It is necessary to emphasize that the above-mentioned researchers and their papers did not use the local-global terminology. The original model of the author proposed in the next section gives a consistent and easy to understand resolution of enveloping surfaces and all the types of local undercuts and global cut. This theory, that gives the core of the algorithm of *SC* software application, proved its fullness, power and versatility in solving many concrete tasks.

2.1. The general task

Given the generating surface $F1$ in coordinate system $K1$ by vector $\boldsymbol{r1} = \boldsymbol{r1}(x1,y1,z1)$ as shown in figure 1, and given the relative motion between $K1$ and $K2$ coordinate systems. Surface $F2$ has to be generated by surface $F1$ in the coordinate system $K2$ as vector $\boldsymbol{r2} = \boldsymbol{r2}(x2,y2,z2)$. The figure shows a general contacting point P and its trajectory d in the $K1$ coordinate system.

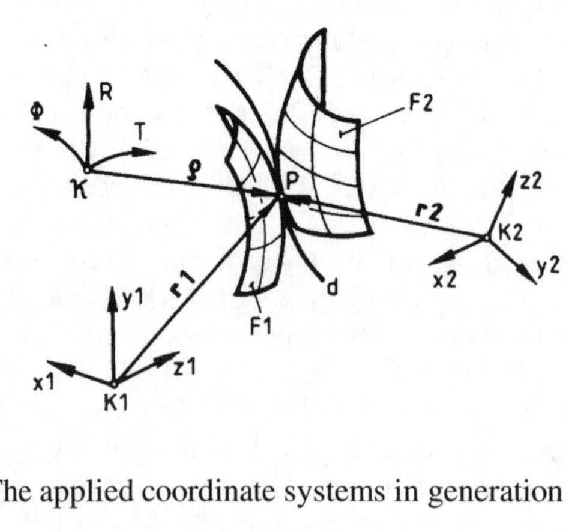

Figure 1. The applied coordinate systems in generation $F2$ by $F1$.

The *Reaching Model* applies a special non-Descartes coordinate system κ. The Φ coordinate lines as well as the coordinate axis Φ itself coincide with the trajectory of the moving points of a surface that is in the coordinate system *K*2 so Φ has two roles: 1. motion (time) parameter, 2. one of the three space coordinates of coordinate system κ. When the motion of *K*2 starts this surface coincides with the *R–T* coordinate surface of coordinate system κ. The vector $\boldsymbol{\rho} = \boldsymbol{\rho}(\Phi,R,T)$ gives the generating surface *F*1 in the coordinate system κ with its general point *P*. The *R* and *T* coordinate directions are set up using suitable shape and relation to Φ coordinate lines. The *K*1 and κ coordinate systems are fixed to each other.

*2.2. The reaching process to generate one point of the generated surface F*2

The reaching process is demonstrated in figure 2.a. In the reaching process we choose a Φ coordinate line that does not intersect *T*1. Stepping from one Φ coordinate line to another going *R* direction, the generating Φ coordinate line will be that, which can reach *F*1 first. This Φ line is the trajectory of point P_k' that will be one surface point of generated surface *F*2. Point P_k will be the meshing point. Using the κ_l coordinate system we can determine one curve of the generated surface *F*2 like a set of points. Applying a series of different κ coordinate systems from κ_0 to κ_s using a new *Z* parameter the full generated surface *F*2 can be got as a series of curves. Looking at figure 2.b it can be recognized that to find the contacting point P_k using the reaching process is equal to resolve a minimum-value problem. This easy task becomes evident if the curved Φ–R surface has been transformed into a plane with straight Φ–R coordinate axes. Due to the aforementioned the complex task of determining surface *F*2 and recognizing all the types of local undercuts can be replaced with resolution and discussion of a minimum-value problem. Detailed description and examples of local undercuts and global cuts are given in [6].

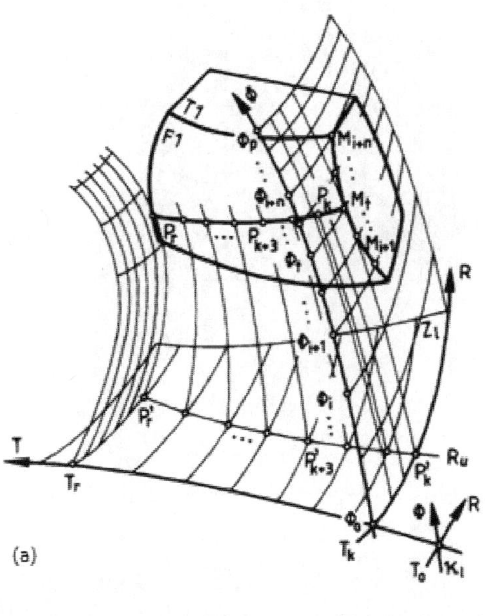

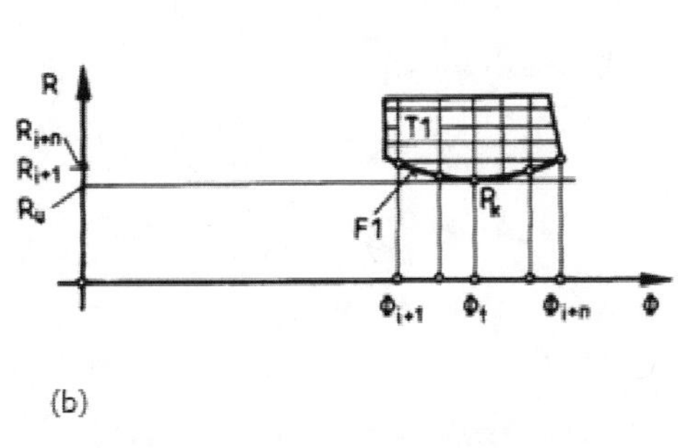

Figure 2. Determining one meshing point of *F*2 is equivalent to solve a minimum value task.

3. The *Surface Constructor* software

The *Surface Constructor* computer program implements the theory. The structure of a newer improved version is shown in figure 3. This version can utilize more generating surfaces one after another and can also produce point-like contact between more generated surface pairs. The software has three levels:

- Symbolic representation of input objects
- Numerical representation of input and computed objects
- Visualization and simulation possibilities.

Among the unique capabilities there is the inbuilt symbolic algebraic computation [7] for manipulation of input expressions and for multiplying the automatically created symbolic transformation matrices, see figure 4. The *SC* program is able to demonstrate the motion of the elements in maximum eleven independent opened visualization windows and can visualize the different features of the analysed kinematical problem e.g. wandering of meshing line dynamically. The selection and visualization options are as follows:

- *F2glob*: global computational method and the surface calculated with it
- *F2lok*: local computational method and the resulted surface points and meshing lines
- *F2al*: computation and visualization of the variations of local undercuts
- Φ: computation and visualization of trajectories of selected points
- *R*–Φ : computation and visualization of $R = R(\Phi)$ functions as a special feature of this software
- *v_a*: computation and visualization of the space of relative velocity and acceleration
- *PT* : computation and visualization of axoids and features of velocity screw.

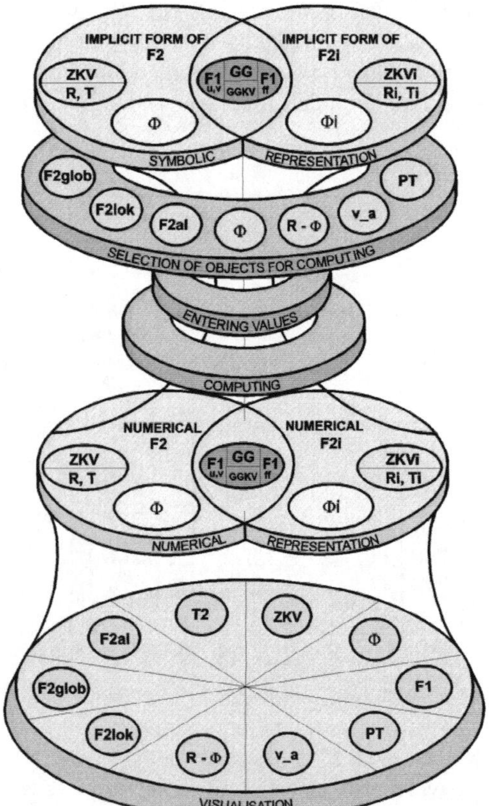

Figure 3. The symbolic, the numerical and the visual representation levels in *SC*. Before calculations and for visualization different objects can be preselected.

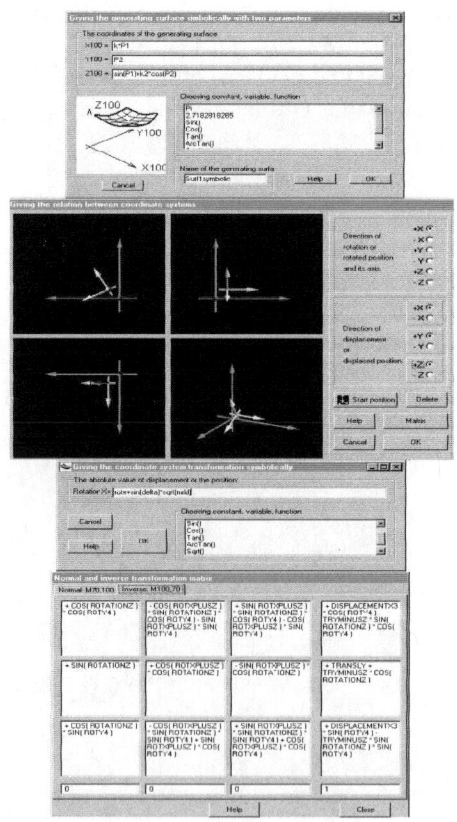

Figure 4. Curves and surfaces can be entered with symbolic expressions, symbolic transformation matrices are generated automatically.

In case of analysis of point-like mesh the generated members of a gearing are surfaces *F2i*, e.g. $F2_1$ and $F2_2$. The intermediary theoretical generating surface between them is *F*1. *F*1 can be give with *GG* generating curve as a swept surface or as a two-parametric surface $F1 = F1(u,v)$. The third planned variant is giving *F*1 as free form (*ff*) surface with spline. The kinematic relations that define the generating motions are given by automatically created symbolic transformation matrices as functions of *Φi*, e.g. *Φ*1 and *Φ*2 motion parameters. The *ZKVi* kinematic relations and the *Ri* and *Ti* parameters are used in the algorithm realising the *Reaching Model.*

4. Introducing some of the additional unique capabilities

These capabilities will be presented through examples. First the $R = R(\Phi)$ functions will be introduced that can explore and qualify the abnormalities of the generated surface. Then the special velocity-space features of the velocity screw will be demonstrated that can reveal the nature and characteristics of the space where motions happen. Finally the possible use of intermediary generating surface will be exemplified making possible generating mating surfaces with point-like contact.

4.1. Visualization of $R = R(\Phi)$ functions

The power of $R = R(\Phi)$ functions comes from the main role of them in *Reaching Theory*. As figure 2(b) shows the minimum point of this function generates a meshing point in the enveloping that corresponds to one of the generated point of the *F*2 generated surface. To exploit this power *SC* includes a special window that mappings the $P_k = [\ T_k;\ Z_l\ ]$ points of the generated *F*2 surface into the points of this window. Figure 5 shows a surface generation by a parabolic surface (a) and the corresponding $R = R(\Phi)$ functions visualizing window that corresponds to the $F2 = F2(T,K)$ parametric representation (b). The window can visualize the $R = R(\Phi)$ function for every point of *F*2 as a small function curve as shown in figure 5(b).

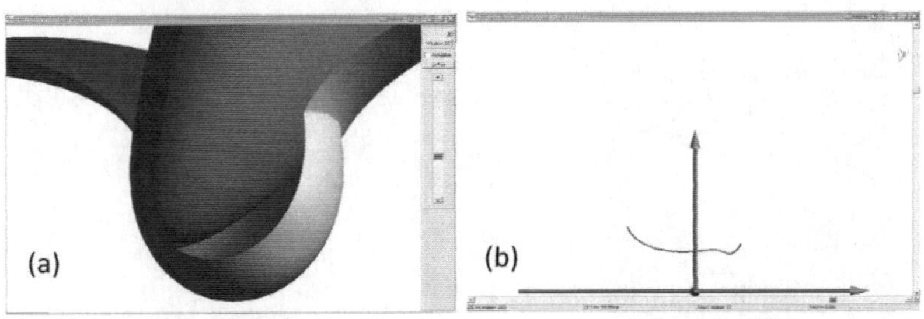

Figure 5. Global cut type interference on the generated surface (a) and detecting it as special shape of the $R = R(\Phi)$ function (b).

Moving the mouse cursor in the window, the $R = R(\Phi)$ function curve that corresponds to the $[T,K]$ parameters of *F*2 appears instantly. Moving the mouse cursor in the window the $R = R(\Phi)$ function for any point of *F*2 can be visualized. This is a comfortable way for analysing these function curves. As the shape of this function curves can reveal the undercut situations, this visualization capability of *SC* can be used advantageously. The local *minimum* point means healthy generated point of *F*2. A *maximum* point means a generated point which is a local undercut point because its neighbour points are globally cut points. The typical form of the local undercut is an *inflexion* shape having horizontal tangent on the $R = R(\Phi)$ function curve.

The concrete example shown in figure 5 introduces both local undercut and global cut. As the undercut that can be determined by local mathematical methods is a point-like phenomenon and in this example – and usually – the motion is not limited to a point, the continued motion causes global cut that removes the previously enveloped surface region, such way decreasing the area of the connecting surface. Figure 5a shows when the parabolic generating body destroys the previously enveloped part of the lower surface in the continued motion causing global cut and resulting in an edge on the surface.

The right window shows this situation with the corresponding $R = R(\Phi)$ function curve where R is the vertical reaching direction and Φ is the motion parameter measured along the horizontal axis. The left minimum point of the curve results in an enveloped point that will be destroyed later by the right lower minimum point, which is the generator point of one of the points of the enveloped undercut surface. The detailed explanation of the *Reaching Model* with detailed explanation of undercut phenomenon can be found in [8].

The $R = R(\Phi)$ function curves visualization window has a more comfortable visualization possibility for detecting undercuts. In this possibility all the function curves having the same value of Z or T $F2$ surface parameter can be shown as a surface. In this method changing the other parameter the full surface of $F2$ can be checked against undercut trough the set of appearing checking surfaces. The demonstrative example solves the next task: given a special quasi-helical surface as a compressor rotor characterised by changing pitch. The grinding of such special helicoids needs a special grinding machine invented by the author [9]. The grinding wheel of this machine has a working surface created enveloping the grinding wheel with the compressor rotor. Because the generated wheel surface is fairly complex, the checking against undercut is necessary. Figure 6 demonstrates the use of the mentioned checking method in case of this grinding wheel surface.

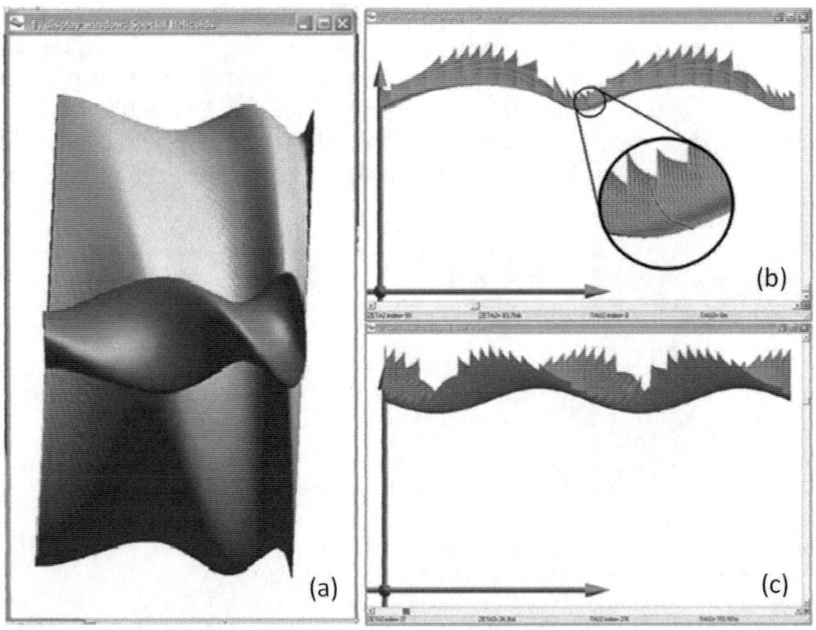

Figure 6. The rotor and the special grinding wheel (a), inflexion means undercut (b) and the perfect valley form of checking surface (c).

4.2. Velocity space features

One of the unique capabilities of *SC* is handling the velocity screw of the moment of the motion. The trajectories, the velocity vectors and the acceleration vectors can be visualized. The axis of the screw can be determined by *SC* and the set of axes can be displayed parallel to the motion. So the well known theoretical formation, the axoid of the motion can be visualized and analysed dynamic manner. Figure 7 expounds the velocity screw while figure 8 shows an example on moving point trajectories, on the axoid of the motion and on velocity and acceleration vectors of the motion along the trajectories.

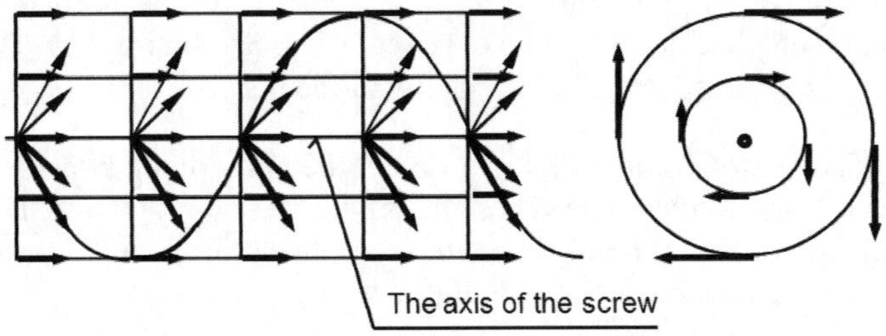

Figure 7. The velocity screw formed by the tangential to helical lines velocity vectors in a moment of the motion. Shown in two views.

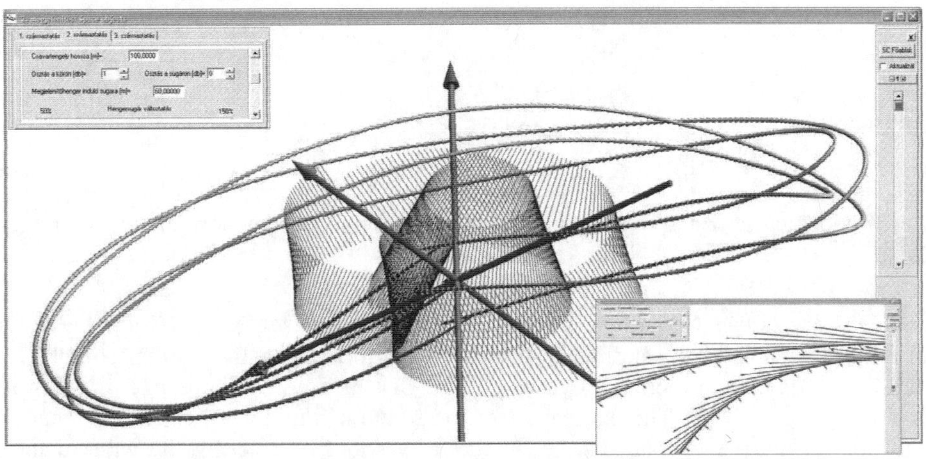

Figure 8. Trajectories of selected moving points and the set of velocity screw axes as an axoid. The small window shows velocity vectors that are tangential to the trajectories and acceleration vectors of the moving points.

Higher level space objects of the velocity space will be discussed in the next subsections. The base of the calculation was the determination of the axis of the velocity screw. It is known, that the spatial instantaneous motion of a rigid body can be represented by a velocity screw [10]. The calculation details are given in [6].

4.2.1. Visualizing the Velocity Screw. The goal of the visualization of the instantaneous velocity screw is to provide an imagination on the structure of the velocity space and to discover special moments when the space changes to simple rotation or translation momentarily, or the motion stops for a moment. The axis of the screw is a special loci itself where the points characterized by simple translation or standing, in case of instantaneous rotation. The space of velocity is demonstrated by spirals that give the direction of the tangential velocity vectors. The visualization mode gives the maximal freedom to analyze the space: scrolling the time-parameter scroll bar the motion can be animated with the continuously changing spirals. *SC* needs symbolic input of motions. After entering the numerical values the selection of the world coordinate system and the turning on the velocity space visualization follows. Eleven display windows can be used in the same time with different kinematical and visualization parameters providing maximum freedom in visual analysis. Here only some example window is given to make the possibilities felt. Figure 9(a) shows an axoid having elliptical form and the momentary screw.

4.2.2. Visualizing the Normal Set. The null-system is pictured by one normal set. The radius, the density and the position along one outer spiral can be modified. See figure 9(b). The set of normal sets forms the normal-complex of the screw. The cardinality of the normals in the normal-complex is ∞^3.

4.2.3. Visualizing the Reciprocal Polars. Giving a line in the null-complex that is not a normal in the screw determines a second non-normal line. These lines determine each other unambiguously and play central role in the geometry of connection because they are the holder of a ∞^2 cardinality of normals in the normal-complex. See the reciprocal polars in figure 9(c).

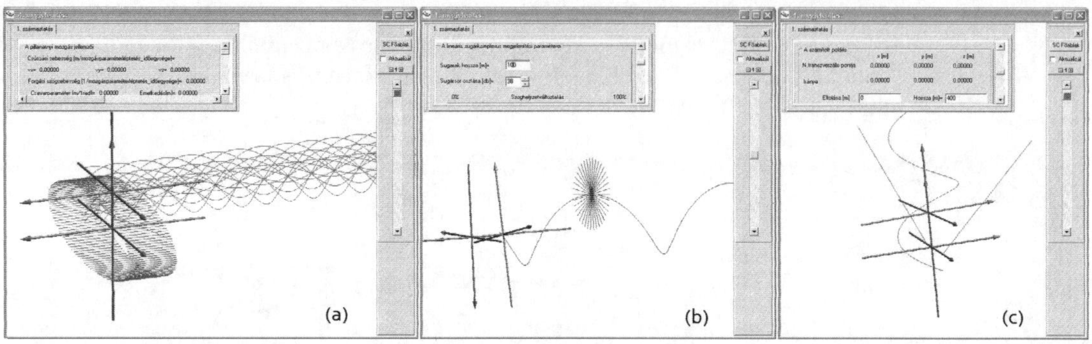

Figure 9. Elliptical axoid of the motion and the momentary screw (a), normal set (b) and reciprocal polars (c).

4.2.4. Visualizing the Linear Congruent. The set of normals that intersect the reciprocal polars forms the linear congruent. It is a ∞^2 cardinality formation of normals in the normal-complex. For the visualization of the congruent there are two ways. The first method can draw one or two normal sets on one or on two linear polars. The parameters of the figure of the normal set can be changed freely and the location on the linear polar can be modified by slider or entering the value manually. See figure 10(a). The second method draws all normals of the normal-complex that intersects the reciprocal polars in given number points. The density of this formation can be modified by up-down controls. Normal sets can be turned on and off additionally. See figure 10(b).

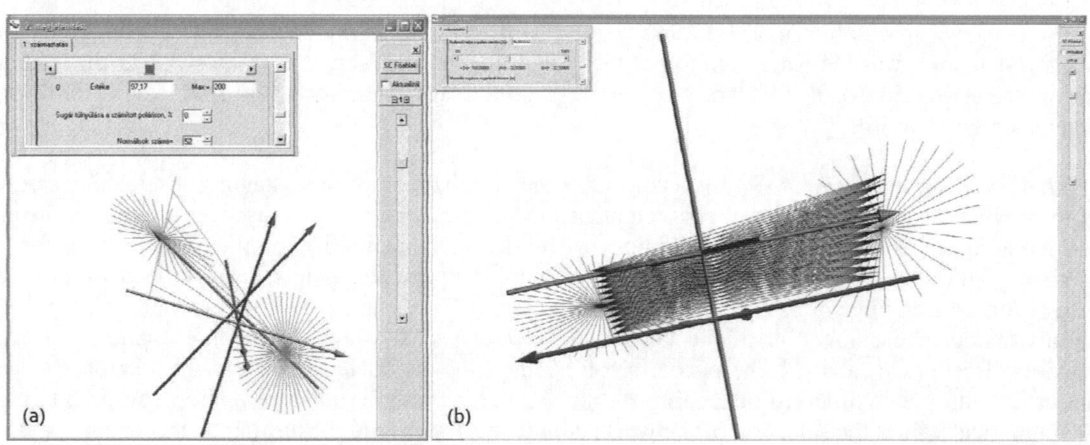

Figure 10. Normal sets on the reciprocal polars (a) and the linear congruent and two normal sets on the reciprocal polars (b).

4.3. Modeling point-like contact with intermediary generating surface

This capability simple repeats the enveloping process with the same generating $F1$ surface with different relative kinematics. See figure 11. The $F1$ generating surface is a swept by the $VG(p1)$ generating curve with $p2$ as the parameter of the sweeping motion. Using $F1$ as a common theoretical generating surface the $F2_1$ and $F2_2$ generated surfaces can be calculated. In the enveloping process of $F2_1$ the Φ_1 motion parameter determines the working of the generating kinematic between $K1$ and $K2_1$ coordinate systems while in the enveloping of $F2_2$ the Φ_2 plays a similar role activating the motion between $K1$ and $K2_2$ coordinate systems. The two generating processes result in conjugate to $F1$ generated surfaces that mesh with $F1$ in lines. Removing the theoretical $F1$ surface out of the two generated surfaces, the intersection point of the two momentarily meshing curves will remain as the contact point of $F2_1$ and $F2_2$. Such gearings having point-like contact are famous for good machining error and assembling error toleration and work with less noise and vibration.

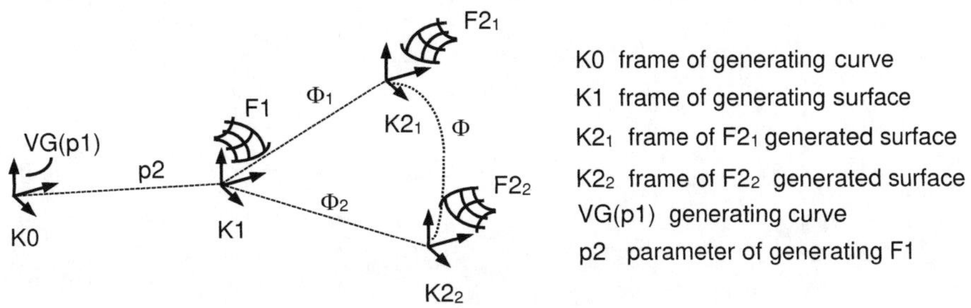

Figure 11. Meshing pair generation with intermediary generating surface $F1$.

The improvement of worm gearings and avoiding noise that can originate from backlash [11] and from not constant transfer ratio is important [5, 12]. The application example solves the problem of modified worm gearings [13, 14]. The profile or tooth modification applied to achieve noiseless and misalignment-tolerant worm gearings results in error in transmission generally. The use of intermediary generating surface for generation of worm gearing members solves this problem because produces transmission error free continuous point-like contact between the worm and the worm gear. Figure 12 demonstrates the working of the mentioned method in case of cylindrical worm gearing. Subfigure (a) shows the worm, the segment of the worm gear, the intermediary generating helicoids segment and the contact arc between the worm gear and the generating helicoid. Subfigure (b) shows three moments of the contact patch wandering. The dark patch on the arc is the momentary point-like contact pattern. The method is given in detailed form in [8].

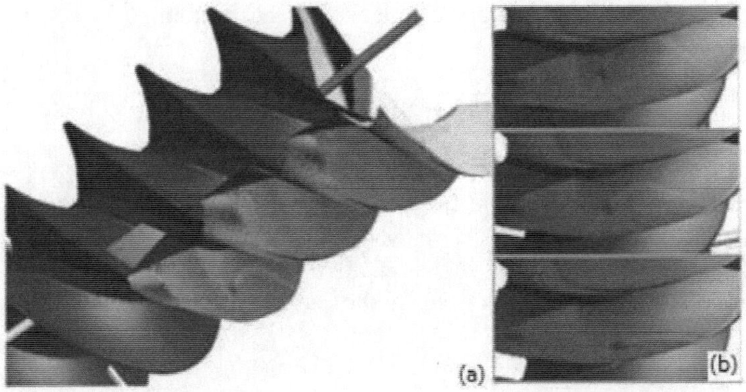

Figure 12. Visualization of worm gear surfaces having point-like meshing (a) and the animation of the contact patch wandering (b).

IOP Conf. Series: Materials Science and Engineering **448** (2018) 012042 doi:10.1088/1757-899X/448/1/012042

5. Summary

The paper focused on the unique capabilities of the *SC* software application. The modelling flexibility and fullness originate from the *Reaching Model* as the core theory and from the symbolic algebraic representation ability of the software realisation. A short description refreshed the structure and working of the software. In the main part of the paper, some solved task emphasised the unique capabilities of *SC*. The visualization of $R = R(\Phi)$ functions help detecting and avoiding local undercuts and global cut. The peerless spatial velocity space object visualisation capability helps to reveal the best regions of the velocity space for optimization of gearings. Then the method of producing cylindrical worm gearings having point-like connection was demonstrated. The advantage of this master technique is that the point-like connection was realised without the ruining of the exact motion transfer that was made by the previously published suggestions. These examples prove the applicability of *SC* for gear investigation and innovation. Figure 13 demonstrates the profuseness and the ease of use of this tool in enveloped surfaces development and improvement.

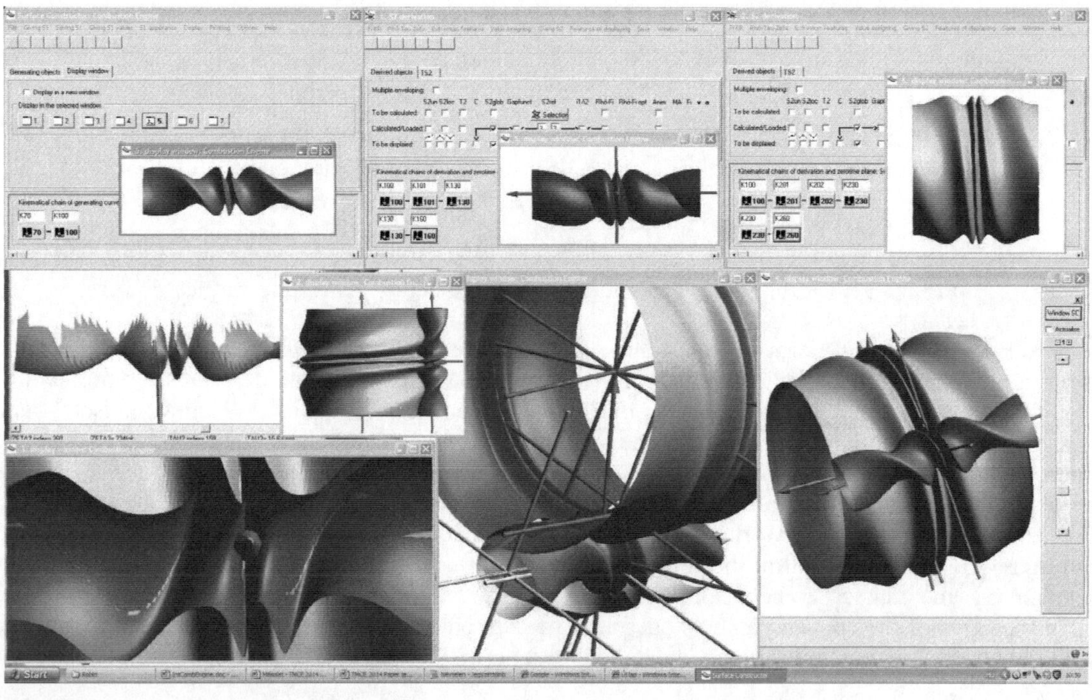

Figure 13. Designing the rotor, the rotary chamber and the grinding wheel surface for a patented rotary internal combustion engine [15].

Acknowledgments

This research was partially carried out in the framework of the Center of Excellence of Mechatronics and Logistics at the University of Miskolc. The described study additionally was carried out as part of the EFOP-3.6.1-16-00011 "Younger and Renewing University – Innovative Knowledge City – institutional development of the University of Miskolc aiming at intelligent specialization" project implemented in the framework of the Széchenyi 2020 program. The realization of this project is supported by the European Union, co-financed by the European Social Fund. The financial support is acknowledged.

References

[1] Litvin F L 1994 *Gear Geometry and Applied Theory* Prentice Hall, Englewood Cliffs, NJ.

[2] Holler R 1976 *Rechnersimulation der Kinematik und 3D-Messung der Flankengeometrie von Schneckengetrieben und Kegelradern*, PhD Thesis, TH Aachen

[3] Weck M and Reuter W 1989 Optimiertes Verzahnungs-scleifen durch rechnergestützte Verfahrenssimulation *Industrie-Anzeiger* **95** 28–31.

[4] Seveleva G I, Gundaev S A and Pogorelov V S 1989 Numerical modelling of cutting tapered gears having teeth with circular arc form (in Russian) *Vectnik Masinoctroenia* **3** 44–7

[5] Lunin S 2009 *Design of Worm Gears for Low Noise Applications Using Worm Gear Software from ZAKGEAR,* http://www.zakgear.com/Worm.html, Accessed: 05.20.2018.

[6] Dudás L 2010 New way for the innovation of gear types *Engineering the Future* ed. L Dudás (InTech) pp 111–40

[7] Lazard D 1979 Systems of algebraic equations *Symbolic and Algebraic Computation, Lecture Notes in Computer Science* Eurosam '79, An International Symposium on Symbolic and Algebraic Manipulation, Marseille, France, ed E W Ng, pp 88–94

[8] Dudás L 2010 Modelling and simulation of a novel worm gear drive having point-like contact, *Proceedings of TMCE 2010 Symposium,* Horváth, I.; Mandorli, F. & Rusák, Z. (Ed.), pp. 685–98, ISBN 978-90-5155-060-3, Ancona, Italy, April 2010.

[9] Dudás L 1992 *Grinding machine, for grinding non-surface of revolution surfaces, especially conical and globoid worms*, Hungarian patent HU P9003803

[10] Xiao D Z and Yang A T 1989 Kinematics of three dimensional gearing, *Mech. Mach. Theory* **24** 245–55

[11] Kacalak W, Majewski M and Budniak, Z 2016 Worm gear drives with adjustable backlash. *Journal of Mechanisms and Robotics* 8/014504-1

[12] Bercsey T and Horák P 1999 Error analysis of worm gear pairs *Proc. 4th World Congress on Gearing and Power Transmissions. CNIT-PARIS France* 419–24

[13] Ohshima F and Yoshino H, 2001 Ideal Tooth Surface Modification Method For Cylindrical Worm Gears *Proc. JSME International Conference on Motion and Power Transmissions VT:MPT I I*(01-202), pp 460–5.

[14] Seol I and Chung S 2000 Design and Simulation of Meshing of New Type of Worm-Gear Drive with Localized Contacts *KSME International Journal* **14** 408–17.

[15] Dudás L 2017 Developing a rotary internal combustion engine characterised by high speed operation *Vehicle and Automotive Engineering. Lecture Notes in Mechanical Engineering*, ed K Jármai, B Bolló (Springer) pp 79–89

IOP Conf. Series: Materials Science and Engineering **448** (2018) 012043 doi:10.1088/1757-899X/448/1/012043

A new method for analysing the efficiency of cutting inserts

Sándor Sipos
Óbuda University, Institute of Material and Manufacturing Science, 1081. Budapest, Népszínház u. 8., Hungary
sipos.sandor@bgk.uni-obuda.hu

Abstract – The operational productivity and the economic efficiency of turning, drilling and milling inserts can be expressed by cutting performance. This complex definition is composed of several main and supplementary features, therefore it is not possible to characterise it by only one indicator. In the present article a new method will be introduced, being able to measure and compare the productive efficiency of turning inserts: the four most important features will be summarised in one indicator. The procedure has a low time requirement, it is built on instrumental investigations, so it does not require great financial and material expenses, it makes possible to use the tools in an optimal way, according to the four aspects, investigated by us, and to qualify the inserts in many ways. The so called complex efficiency indicator serves to measure the effectiveness of innovation, too. The procedure, developed by us, has already been used successfully in case of tools, produced by leading companies and having different functions, design and materials.

1. Introduction

Nowadays, in the area of cutting operations there are appearing new versions of mounted tools. The cutting elements, the inserts, optimised to the given task, are fixed in the tool holders, drill and milling cutter bodies by the manufacturing companies. Although the amount of operationally applicable shape variations is quite limited, the number of the chipbreakers and edge designs, usable under different circumstances, is considerably great. The operational productivity and economical efficiency of turning, drilling and milling inserts can be mostly described by the cutting performance. This term is a complex feature of the tool (its edge geometry and/or material), serving to judge and qualify its behaviour during cutting operation (wear, tool life, force effects, the shape of the developed chips etc.) and the achieved surface quality (waviness, roughness) of the machined workpiece. The *judgement* is always built on practical experiences, it provides a characterisation, describable with words, i. e. it has a quantitative character. For example, an insert is considered to be appropriate if it breaks the chip well and its wear process is low. The *qualification* is mainly built on instrumental measurements, it characterises the productive efficiency of the tested tool with the results of measurements, depending from the adjusted data combinations, i. e. in a qualitative way [1][2].

The following article is going to introduce a method for measuring and comparing the cutting performance of cutting inserts. The procedure is built on short-time examinations; therefore it does not require great material, instrumental and financial expenses. The developed method has been successfully applied in case of tools, having different tasks, constructions, cutting materials and produced by world-famous tool manufacturing companies [3][4].

2. Machinability and cutting perfomance

Among the specialists, being familiar with the technology, the pair of terms „machinability" and „cutting performance" is well-known. The first mentioned characterises the workpiece material based on three different sets of parameters. The first set of parameters provides information on the *chemical composition* of the workpiece: it refers to the percentage ratio of main and auxiliary alloy components, furthermore to the rate of the non-metallic alloys (for example, S, Ca, Se, Te etc.), having a positive influence on the machinability. The chemical composition determines the possible microstructure of

IOP Conf. Series: Materials Science and Engineering **448** (2018) 012043 doi:10.1088/1757-899X/448/1/012043

the given material. It includes also undesired alloy materials, getting into the material as pollution (for example, S, O, Si). The second set of parameters refers to the *condition* of the workpiece, including the method of the production (preproduction), the (preliminary) heat treatment and structure of the material etc. To this belongs the different machinability of the cast or forged pre-production products (compared to the basic structure). The third set of parameters is formed by the *material properties* of the workpiece. These cover the mechanical, the thermo-physical and the metallographycal properties of the given grade. The first group of features includes the strength and density at low and high temperature, etc., the thermo-physical contains, for example, the specific heat, thermal conductivity and absorption properties, while the last one involves the material structure, the ratio and microhardness of hard phases, macrohardness, tendency of the self-hardening, critical temperature, etc. [5].

From the material properties of the workpiece, the listed ones have the greatest influence on the tool:

- *basic hardness* of the workpiece material (having a close correlation with the strength) has a significant influence on the cutting force and the temperature, to be developed during the machining, and it determines the mechanical and thermo-physical load of the direct environment of the tool edge.
- good ductility (appropriate elongation) has a correlation with the toughness. Some alloy materials (Cr, Mo, Ni, W) increases the toughness, while in case of other ones (Cr, Ni) it may be accompanied by the *adhesion* tendency (to sticking) and it may be a precondition to the development of built-up edge.
- hard phases of multiphase materials (for example, the presence of the primary silicon crystals with a great hardness in the hypereutectic aluminum alloys etc.) have an increased *abrasive* influence, therefore, they can cause an intensive tool wear.
- presence of some alloy materials (Cr, Ni etc.) in the workpiece causes *self-hardening* during the machining. If the allowance is detached with some cuts, then there is an increase in the hardness of the upper layers, already machined.
- thermo-physical properties of the workpiece can be influenced by alloy materials. These features determine the ratio of the (thermal) energy, developed during the machining and getting into the workpiece and chip, and also the percentage, causing tool damage.

In the last decades, a range of novel materials has appeared (ADI – Austempered Ductile Iron, CGI – Compated Graphite Iron, duplex corrosion resistant steels, heat resistant (HRSA) nickel- and cobalt-based alloys, used in gas turbine manufacturing, AMC – aluminum-based composits, AMFS – aluminum-based, syntactic composite metal foams etc.), where the economic machining causes a real challenge for tool manufacturing companies. It means that the very poor workpiece machinability should be offset by the increased cutting perfomance of applied tools.

The cutting perfomance is a measure of the efficiency of tools, having different designs, geometries and materials (including substratum and coatings, too), its main feature is the tool life (and its wear curve and wear model, describing this). Without that knowledge it is not possible to calculate, analyse or compare the manufacturing (operational) costs, the productivity (expressed, for example, in piece/hour). It requires quite great material and financial expenses to determine its value for a given task, while its validity depends to a great extent on the specific conditions of production (the composition, the condition and hardness of the workpiece, the condition of the machine etc.).

The *supplementary features* of cutting performance include several factors: material flow velocity (with other words, the material removal rate), cutting temperature, cutting force components and power requirement, the shape and size of the chips, the microgeometry (wavieness, roughness) and the layer properties of the machined surface. Some of the listed properties can only be measured with complicated instruments (for example, the temperature in case of dry-machining) or the measurement can cause damage in the produced workpiece (for example, the material structure, the change in the microhardness, the distribution of the residual stress etc.).

IOP Conf. Series: Materials Science and Engineering **448** (2018) 012043 doi:10.1088/1757-899X/448/1/012043

3. Examination of the efficiency of cutting inserts

In the indicator, developed to qualify cutting inserts, research and practical experiences of two decades have been summarised. The quintessence of the developed method is to determine the cutting performance of a single insert with four selected supplementary features. Important factors will be considered that first can be easily calculated and observed, and, second, they can be quite easily measured. An advantage of this procedure is that the test of the tool occurs with material-, instrument- and cost-friendly way, i. e. with short-time use. This requires only nine different setups and it means turning a length of just 150-200 mm.

The cutting performance is a complex definition and it is impossible to express it with a single indicator. If we asked a practicing specialist what are the characteristics of fine working tools than his reply probably would be the following: „productive and easy chip detachment, furthermore, favourable, foreseeable surface roughness, resulting from the low cutting force and favourably broken chips".

Accordingly, the selected supplementary features are: the material removal rate (MMR), the chip shape, the active cutting force and the Rz parameter of surface roughness. The specific feature of this method is that all the four factors are described with a single indicator and in this figure the most important requirements, concerning the tool, are summarised.

In case of any setting, the formula of efficiency indicator, describing the cutting performance of a turning insert, is the following:

$$H = \frac{1000 \cdot V' \cdot g}{F_a \cdot Rz} = \frac{1000 \cdot a \cdot f \cdot v_c \cdot g}{F_a \cdot Rz} \left[\frac{\text{mm}^3}{\text{min} \cdot N \cdot \mu m} \right] \qquad (1),$$

where there are: MMR (V´, mm^3/min), the figure, referring to the chip shape (g), the active cutting force (F_a , N) and the achieved surface roughness (Rz, μm). The formula (1) suggests that the greater material flow rate and more favourable chip shape is, the better the efficiency is. On the other hand, the smaller power requirement of the chip detachment and break is, and the smaller roughness on the machined surface is, the more favourable the use of the insert is.

The selection of the four „critical" features (beside their simple measurability and exact determination) can be explained by the following considerations.

1. ***Material removal rate*** refers to the productivity of the cutting process. In case of turning operation it is the product of depth of cut, the feed and the cutting speed, and – after substitution – it has the unit of measurement: cm^3/min. By reason of the scale of „H" indicator it is recommended to use the unit mm^3/min in formula (1).

   During the tests, the depth of cut has a constant value, while – according to the function of tested insert – the feed rate and cutting speed values were varied on three-three levels (i. e. altogether 9 settings).

   The cutting perfomance features can be really shown when there is a several-fold increase in the intensity of the chip detachment during the tests. During the tests the tool wear should not affect the results of the measurements, therefore all the three technological data have to be adjusted to the workpiece material (cutting speed), and to the chip breaking range of the insert (depth of cut and feed).

2. ***Favourable chip shape*** decisively affects the process-sure operation of machines with limited superivision (CNC-lathe centres, multitasking machines etc.), but it has an effect on the aesthetic appearance of machined products, too. The rebound movement of chip to the machined surface may cause faulty products, as the chip may cause damage on other clearly visible surfaces or on the coated surface.

   The chip samples, detached during the tests, will be arranged in a chart form (in a so-called „chip chart") and after that the value of „g", reffering to the chip shape, can be determined. It is a mark (is a whole number) and its value may be between g= 1...5. The most favourably broken chips were awarded the mark „5", while the dangerously tangled, flowing chips receive the mark „1". We would like to note that this classification matrix – to our knowledge – was applied first in Hungary by the authors [6][7][8].

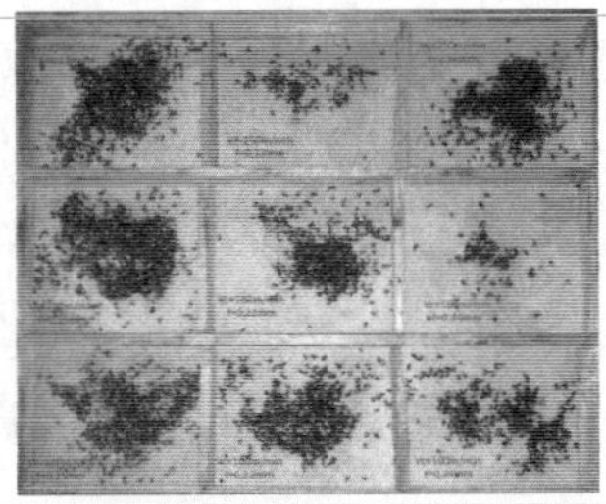

Efficiency: 89%	f=0,15 mm	f=0,22 mm	f=0,34 mm
v_c=250m/min	4	5	5
v_c=160m/min	4	4	5
v_c=100m/min	4	5	4

Figure 1 Chip chart and marks of the detached chips
Insert: CNMG120408 MP3 WPP20S
Machining conditions: C50 unalloyed steel (HB220±5); a=2 mm; dry machining

The ***Figure 1*** shows the chip chart of an insert (together with the evalution, „mark"). In the heading of the table the general chipbreaking efficiency is shown, it is a number, exactly the percent ratio of the sums of the marks, awarded to the inserts, and the sum of the maximum available marks (in this case it is 9x5 = 45).

3. ***The active force*** is the summarised value of two force components, determined by instrument measurement. Although the force measuring equipment is able to register 3 force components, the method, introduced earlier, is built on the simultaneous registration of cutting (F_c) and feed (F_f) force components. The passive force is important from the point of view of the workpiece accuracy, therefore it will not be considered in case of our method. The active force (F_a) is the vector resultant of F_c and F_f forces therefore

$$F_a = \sqrt{F_c^{\ 2} + F_f^{\ 2}} \qquad (2).$$

The cutting force determines – beside its effects, carried out on the tool – the necessary power requirement. The feed force depends – in case of tools, having the same lead angle values, and inserts, having the same corner radius value – mainly on the tool edge design (edge rounding off, edge preparation etc.) and on the design of the chipbreaker.

This F_a force component is able to characterise the chipbreaking ability: the greater the F_f force component is, the greater the plus force is, being necessary to cut (break) the detached chips into pieces of sufficient size. Therefore the qualification of inserts happens based on the active F_a force and important conclusions can be drawn from the ratio of F_f and F_c.

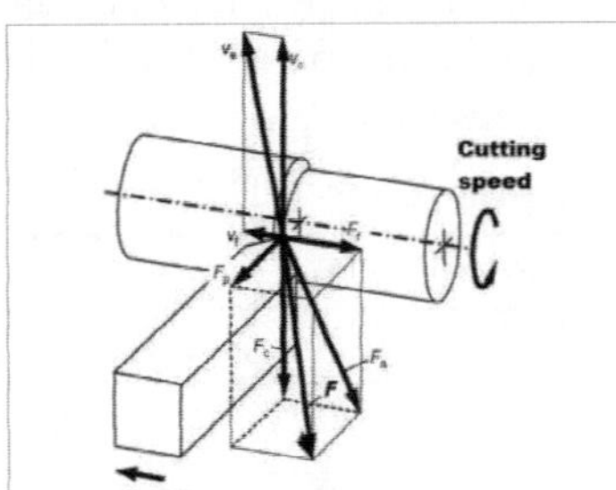

Two-factor force models
(a=constant ; v_c=100…200 m/min)

$$F_c \cong 3185 \cdot f^{0,79} \cdot v_c^{\ -0,03} \ \text{[N]}$$

deviation : ±20 N ; R^2>0,95

$$F_f \cong 1200 \cdot f^{0,41} \cdot v_c^{\ -0,07} \ \text{[N]}$$

deviation: ±15 N ; R^2>0,95

Figure 2 The cutting force components and their models
Machining conditions: Ko36 stainless steel (HB170±5); a=2 mm; dry machining

On **Figure 2** the force components and the two-factor model of force components of an insert, having PP chipbreaking geometry and modern coating, and produced by a well-known company, are summarised and shown. As it can be seen well the feed rate has a determining effect on the development of both force components, while the cutting speed does not affect significantly the forces, developing during the machining.

4. The fourth feature is the ***surface finish*** producing capability of the tool, it clearly belongs to the qualitative requirements. This term means that the tools are able to produce surfaces, having nearly the same roughness parameters, under determined circumstances, *almost* independently

IOP Conf. Series: Materials Science and Engineering **448** (2018) 012043 doi:10.1088/1757-899X/448/1/012043

from their measured (or detected) wear condition in a way, where it can be calculated/predicted planned in advance [2].
The roughness of the turned surfaces can be characterised with different measures. On **Figure 3** the development of the roughness (and R_e theoretical roughness) is shown in function of feed rate in case of different settings.

Although the use of the average roughness (Ra) is quite widespread in the practice in Hungary, in this case the unevenness height (Rz) was used to compare the different inserts. This can be explained by the fact that in the supplier segment almost only Rz parameter is applied, and the Rz shows an *acceptable agreement* with Re theoretical roughness in case of some feed rate ranges. In case if we know the value of corner radius (r_ε , mm), then the lastly mentioned can be calculated [2] with the following formula (Bauer-formula, 1937):

$$R_e \approx 125 \cdot \frac{f^2}{r_\varepsilon} \ [\mu m] \qquad (3).$$

As it can be seen well on Figure 3, the theoretical roughness follows the development of Rz value in the range of $0{,}25 \le f$, mm $\le 0{,}35$ mm; at the same time, the „popular" average roughness shows a moderate change in the same range. We note that in case of the tested insert the general rule Rz=(5…6)×Ra prevails, it means that the approach $Rz \cong 4 \cdot Ra$, commonly used in the technical practice, is not correct. The denial of this myth can be found in [9][10].

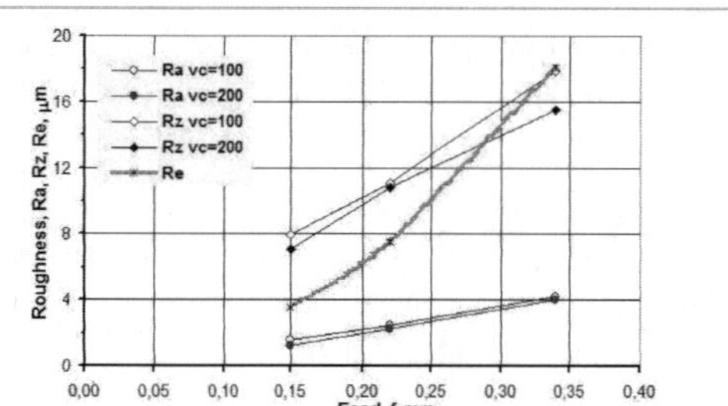

Two-factor Rz model:

(a=constant; v_c=100…200 m/min)

$$Rz \approx 103 \cdot f^{0,99} \cdot v_c^{-0,15} \ [\mu m]$$

Deviation: ±0,34 µm; R^2>0,99

Figure 3 The roughness parameters of the machined surface
Insert: CNMG120408 MR WP15CT
Machining conditions: C50 unalloyed steel (HB220±5); a=2 mm; dry machining

With the following example it will be presented how to substitute the factors, mentioned earlier, into the (1) relationship. In the **Table 1** the efficiency indicators of an inserts, having the marking EM/YBN and produced by a well-known company, are presented.

Table 1. The efficiency indicators of insert EM/YBN

Efficiency of the cutting insert (H)									
v_c	a	f	V'	F_c	F_f	F_a	Rz	g	H
100	1,5	0,1	15	405	253	478	4,98	3	18,9
100	1,5	0,15	22,5	539	330	632	4,33	4	32,9
100	1,5	0,25	37,5	870	300	920	5,67	4	28,7
160	1,5	0,1	24	382	230	446	5,61	3	28,8
160	1,5	0,15	36	518	300	599	5,67	4	*42,4*
160	1,5	0,25	60	826	423	928	8,98	4	28,8
200	1,5	0,1	30	355	220	418	4,68	2	30,7
200	1,5	0,15	45	498	282	572	5,53	4	**56,9**
200	1,5	0,25	75	762	357	841	8,41	4	*42,4*

Machining conditions: dry machining
Insert: CNMG120408 EM YBN253
Workpiece: C50 unalloyed steel (HB 220±5)

Analysing the data, gained from the nine different settings (v_c = 100…200 m/min ; f = 0,1...0,25 mm) and in case of constant (a=1,5 mm) depth of cut, the following conclusions can be drawn:

- in case of combination of the feed and cutting speed values, the insert has a greatly varying efficiency ($H_{1\text{-}9}$ = 18…56). The simultaneous increase of both cutting speed and feed values has a clearly favourable effect on the machining productivity and on the tool efficiency, too. In case of this combination of chipbreaker/material, the fivefold increase of the material removal was followed only by a moderate increase in the efficiency;
- using the insert under the most favourable conditions, the value of its efficiency nearly *tripled*, compared with the lowest value (H_8 = 56,9). It is mainly due to the reason that the material removal, already mentioned, increases, at the same time, the chip shape and the roughness, achieved during the process, have a very favourable values;
- as it can be seen well, the insert efficiency is extremely good in case of feed value of f = 0,15 mm (almost independently from the settled cutting speed value).

Based on the method, introduced earlier, it is easy to select the best efficiency of application and the settings, enabling the favourable results (see the coloured columns). With these, the optimal conditions of application can be selected for any insert. With this method it is possible (moreover, from the aspect of the operational practice it is desirable) to qualify the inserts in a form of tests (carried out under the ***same*** conditions), in a versatile and objective way (taking four factors into consideration at the same time), to compare them (or to discover the differences between same insert types).

4. The complex efficiency of cutting inserts

It is not usually to compare inserts, having different functions (finishing, semi-finishing and roughing), material grades, edge constructions (ISO or wiper) and chipbreaking geometries. The method, proposed in the present article, makes it possible to realistically compare them with introducing a new term, the so-called complex efficiency.

The basis of the complex comparison of tools, intended for different fields of use, is the behaviour of the tested inserts: this will be determined based on the analysed cutting data (v_c, a, f, V') and based on the measured and calculated features (F_a, Rz, g, H). The complex efficiency (H_K) of any insert builds on H_1 … H_9 values, determined for individual settings: it expresses the efficiency of the given insert geometry and/or edge construction in percentage, taking the examined four features into consideration. Its value represents the average cutting perfomance of an insert, with reference to all settings, it means it is a complex, percentage indicator. When evaluating the inserts, the main statistical features (average, deviation, variance ratio) will be used, at the same time, the ***percentages*** of some measured/calculated values will be considered.

The complex efficiency can be calculated with the following formula:

$$H_K = 100 \cdot \frac{V'_{\%} \cdot g_{\%}}{F_{a\%} \cdot Rz_{\%}} \ [\%] \qquad (4).$$

The interpretation and the calculation method of this formula will be introduced with a specific example.

Table 2. The complex efficiency of an insert, having a shortened edge

Value	v_c	a	f	V'	F_c	F_f	F_a	Rz	g	H
Average	128,3	1,5	0,17	32,1	634	339	721	9,0	3,7	18,2
Deviation	26,1	0	0,07	14,5	196	42	192	2,2	0,5	5,1
Var,%	20,3	0	39,7	45,2	31	12	27	24,6	13,6	28,2
In %				**100**	**100**	**53,5**	**114**	**206**	**73**	
Complex efficiency			**Insert: TT9080/09 PC**				**H_K = 31,2%**			

Machining conditions: dry machining
Insert: CNMG090408 PC TT9080
Workpiece: Ko36 austenistic stainless steel (HB 170±5)

In the **Table 2** the turning results of a hard-to-machine, austenitic corrosion-resistant steel will be summarised, in case of a fixed depth of cut a = 1,5 mm, (limit values: v_c = 100...160 m/min; f = 0,1 … 0,25 mm). The data, shown in Table 2, have been calculated with formula (4). From the results, the following conclusions can be drawn:

- if the range of the test settings is the same, the average of the material flow ($V'_{\%}$) will agree, too, in the evaluation it means 100 percent. However, it there is any change in the test conditions (for example, the comparison of semi-finishing and roughing chipbreaking geometries) then this change has to be mentioned, $V'_{\%} = 100 \times V'_{new\%}/V'_{base\%}$ [%];
- the percentage ratio of the active force ($F_{a\%}$) can be calculated in the following way: the average active force (F_{aave}), measured during the tests, will be compared to the average cutting force (F_{cave}), it means: $F_{a\%}=100 \times F_{aave}/F_{cave}$ [%]. Accordingly, the $F_{a\%}$ is always greather than 100%, and, the greater the ratio of the feed force is, the greater this deviation is. We note that this proportion depends crucially on the tool holder construction (PCLNR or PSSNR) and on the corner radius of the insert (r_ε). The average ratio of F_f/F_c is a subsidiary, but very useful information, in this case it was 53,5%. In the present case of corrosion-resistant steels this value is in the range of 48-62% (in case of PCLNR tool holder, corner radius is 0,8 mm, without edge preparation, depending on the chipbreaker construction);
- the percentage value ($Rz_{\%}$) of the surface finish producing capability compares the measured and theoretical roughness values. The average of the measured unevenness height was compared with the average theoretical roughness, its value was calculated with formula (3) for the average feed. With the percentage data, indicated in the table, it is expressed that the measured values are approx. 2 times greater than the theoretical roughness values. It is common in case of corrosion-resistant steels ($Rz_{measured}/Rz_{calculated} \approx$ 200...250%) as at low cutting speed and feed values there is a development of built-up edge. Some types of corrosion-resistant steels, having a „sticky" (tensile) character, result in an undetachable material layer (so-called „Spanzipfeltheorie"), and its appearance requires rather to apply the Brammertz roughness model [11]. The tested austenitic steel is extraordinary inclined to self-hardening. The initial wear phase of the coating is not able to prevent the developing material adhesion on the rake face, close to the tool nose, and it makes the roughness even worse;
- the percentage value of chip shape ($g_{\%}$) expresses the average chipbreaking efficiency of an insert (see Figure 1). The chipbreaking efficiency of inserts, tested till now, varied in the range of 40-90%. It has been confirmed by other tests that the applied PC chipbreaking geometry has an outstanding performance in case of turning corrosion-resistant steels [12][13];
- the value of complex efficiency (i.e. 31,2%), calculated according to formula (3), is - in our experience - markedly acceptable in case of hard-to-machine materials [10][13].

With introduction of the term „complex efficiency" and with the formula (4) it became possible to compare the efficiency of tool materials and /or chipbreakers, formed on inserts, having the same shape, but modified sizes, to analyse the tools, having different edge constructions (for example, ISO or wiper) and to qualify the innovation effectiveness of certain insert developments. The experiences of the tests, carried out till now, have been summarised in **Table 3**.

Table 3. Evaluation of the complex efficiency

Conditions: Insert/workpiece material		Complex efficiency, H_K, %			
		poor	proper	good	very good
ISO	unalloyed steel	≤20	≤40	≤60	≈80
ISO	stainless steel	≤18	≤36	≤50	≈65
wiper	unalloyed steel	≤40	≤80	≤120	≈160
wiper	stainless steel	≤35	≤70	≤100	≈130

The complex efficiency of inserts can be divided into four categories according to the material grade of workpieces and the edge constructions of inserts. The outstanding performance of the wiper edge

IOP Conf. Series: Materials Science and Engineering **448** (2018) 012043 doi:10.1088/1757-899X/448/1/012043

construction can be explained mainly by the lower active force and with the very low roughness values.

5. The weigthed efficiency of cutting inserts

The requirements for the inserts are obviously different in case of different machining conditions or tasks, like roughing, semi-finishing and finishing. International surveys have been published in this topic (for example, the CIRP-study, published in the eighties), however, to our knowledge there was not carried out such an assessment about the situation in Hungary in the second decade of the 21st century. Due to this reason, a questionnaire has been sent out to tool sales companies and some manufacturing companies to collect their opinions: what ratios are represented by the four supplementary features - selected *systematically* and considered to be the *most important* - during the turning operation, according to their experiences.

Three companies, having a close partnership with our Faculty, summarised and sent us their practical experiences, while the representation of tool manufacturing companies was the following: 4 companies from Sandvik Group, 2 from IMC Group, 1 from Kennametal Group. The ten replies, received from the companies, have been averaged. The weighted numbers of requirements for the inserts, calculated from the data of questionnaires, are shown in **Table 4**.

Table 4. Weighting of the requirements for the inserts

Supplementary features	The weighted numbers of machining conditions			Weighting ratios
	Roughing	Semi-finishing	Finishing	
Removal rate	**0,4**	**0,3**	0,15	λ_1
Chip shape	0,3	**0,3**	**0,35**	λ_2
Force effects	0,2	**0,25**	0,1	λ_3
Roughness	0,1	0,15	**0,4**	λ_4
Sum total	1,0	1,0	1,0	

On **Figure 4** it is expressively shown what kind of priority principles are necessary to select and to apply the appropriate insert. The data, gained from the questionnaires, have confirmed our expectations as regards the material removal rate, the developing force effects and the produced surface roughness. Only the chipbreaking requirement had the same weight in case of all the three machining ranges. It is understandable in case if the machines, used during the turning proccess, are able to operate under conditions of production with poor monitoring (with other words, „unmanned machining"). Comparing the data from the Table (1) and Figure (4), it is clearly seen that in case of roughing there is one, in case of finishing there are two, while in case of semi-finishing there are three features, having the greatest weight and playing a determining role.

It means that the complex efficiency of inserts, involved in the test series, has to be modified with weight numbers, expressing the usability based on the tool distribution trends and the operational practical experiences.

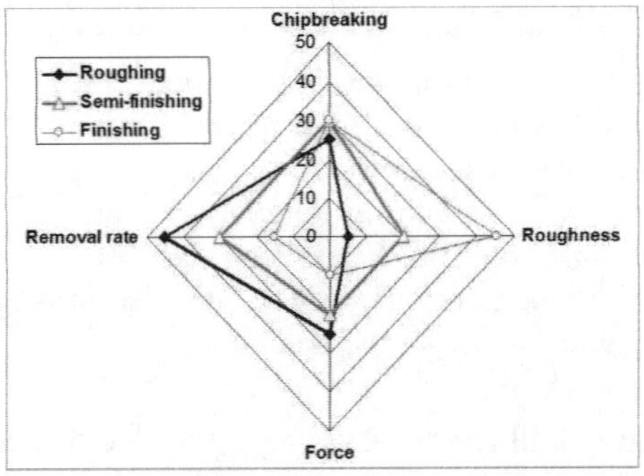

Figure 4 The percentage weight ratios of supplementary features

IOP Conf. Series: Materials Science and Engineering **448** (2018) 012043 doi:10.1088/1757-899X/448/1/012043

The individual measurement results and the chipbreaking behaviour of inserts can be gained with consideration of Table (1) and with the correction of formula (4). Concerning the suitability of inserts, the formula, to calculate the *complex efficiency, corrected by weight numbers* and expressing the demands of the users, is the following:

$$H_{SK} = \frac{1000 \cdot (1+\lambda_1) \cdot V'}{(1-\lambda_3) \cdot F_a} \cdot \frac{(1+\lambda_2) \cdot g}{(1-\lambda_4) \cdot Rz} \; [\%] \qquad (5).$$

The interpretation of the published formula and the calculation method will be introduced with a help of an example. In **Table 5** the results of semi-finishing turning operations are shown, carried out on C60 structural steel grade, with wiper edge insert (limit values: v_c=100…250 m/min; f=0,15…0,34 mm).

Based on the weighted, complex efficiency indicators the following conclusions can be drawn:

- as a result of weighting, the average indicators have changed only slightly, in case of some of them there is an increase (material flow, chipbreaking efficiency), while in case of others, there is a decrease (force effects, roughness). It can be considered that in the category of semi-finishing, the *weighting* has *not* basically *changed* the percentage ratios, created by the complex efficiency indicators. The reason for this is that three from the considered four supplementary features had *almost* the **same** weight;
- the efficiency of inserts, having a wiper edge construction, means a special category, as it has already been introduced [14][15]. It is contributed by the fact that all tested features are extremely favourable, in particular, the surface finish producing capability and chipbreaking efficiency are outstanding. We note that similar experiences have been gained in case of tools with this (wiper) edge design, produced by other companies: the weighted complex efficiency is nearly two times greater, compared to the ISO edge tools;

Table 5. The weighted complex efficiency indicator of a tested insert

Evaluation	v_c	f	V_s'	F_c	F_f	F_{as}	Rz_s	g_s	H_s
Average	170	0,24	78,5	1065	591	914	3,8	5,1	150,9
Deviation	65,4	0,08	42,1	301	116	238	1,4	1,0	118,6
Var, %	38,5	35,2	53,7	28	20	26	36,5	20,1	78,6
in %			**100**	**100**	**55,5**	**85,9**	**43,2**	**101,1**	
Correction factors	λ_1	***0,3***	λ_2	***0,3***	λ_3	***-0,25***	λ_4	***-0,15***	**2,651**
Corrected complex efficiency		**Insert type: NF chip breaking geometry**							**H_{SK}=273%**
Machining conditions: dry machining **Insert:** CNMG090408 NF (wiper) WPP20S **Workpiece:** C60 unalloyed structural steel grade (HB 215±5)									

- In this connection we note that based on our measurement results the passive force component is extraordinary great in case of wiper insert; due to this reason, the characteristic features of the edge desing have to be considered! For the purposes of productivity, the longitudinal turning of components, having a small ratio of cutting length/diameter, carried out on lathes with a great stiffness, could be a favourable solution. In case of machining surfaces with tapering and other profile (for example, torus) shapes, the favourable decrease in roughness does not occur; on the other hand, the accuracy, achieved during the machining, develops unfavourable, due to the wiper edge geometry [14][15].

In order to increase the reliability of all results, it is recommended to select the inserts, to be tested, by thorough examinations (carried out with microscope) and to test only inserts of appropriate (well pressed) condition, intact coatings and correct edges.

6. Long-term efficiency of cutting inserts

The long-term efficiency of inserts is closely related to the deterioration of the tool (or with other words, the degradation of cutting performance of the tool): by this term is understood the sum of *unfavourable* influences, expressly resulting from the tool wear or – maybe even indirectly – they can play a role in it. As a result of the wear, there is a clear increase in the cutting temperature, in the force, torque and power demand, in the vibration tendency and noise impact, and there is a change in the roughness and/or in the waviness, too [16]. It is obviously, a tool life, belonging to an appropriately determined wear criterion, can be considered as the main feature for the purposes of the workpiece machinability and the cutting performance of the tool.

It would be a self-evident thought that the time, spent in machining, had to be considered when calculating the efficiency, based on the formula (2) and containing four selected features. To monitor the wear process of a tool from the beginning till the end, would make the examinations very time-consuming (and cost-intensive) as it would be necessary to collect chip samples, to measure the values of force components and the development of the Rz roughness parameter at the same times of individual wear measurements. Although the constancy of the cutting data (a, f, v_c) and material removal rate (V') can be ensured during the lifetime measurements, the other three features (chip shape, force components and the Rz roughness values) change to a great extent during the wear period of tool.

The modification of the chip shape is illustratively shown on **Figure 5**. The change, created by crater wear, may cause even 3-class difference in the value of „g" as the crater, developing gradually on the rake face, impairs the initially favourable chipbreaking efficiency. In the first phase of the persistent wear process, there are developing chips, having C and G shape and divided into small pieces, later - depending from the place, width and depth of the crater - there are cylindrical, spiral chips, having a length of 40-90 mm, even later, there are unmanageable, dangerously coiled, flowing chips. When achieving the wear criterion, there are again well manageable chips, having a C shape and/or chips, coiled up like a snake and spreading dangerously.

The influence of the tool wear on the cutting force components is well-known. The change in the F_c cutting force, resulting from the wear, means an increase of 25-30 percent in the force formula of Kienzle and Victor [2]. This experience has been confirmed by our measurements, too (see Table 6). An even greater change (i. e. increase) is shown in case of feed and passive force components (F_f and F_p), thus it is understandable that the condition monitoring systems are based, among others, on the change of these two force components (or their ratio).

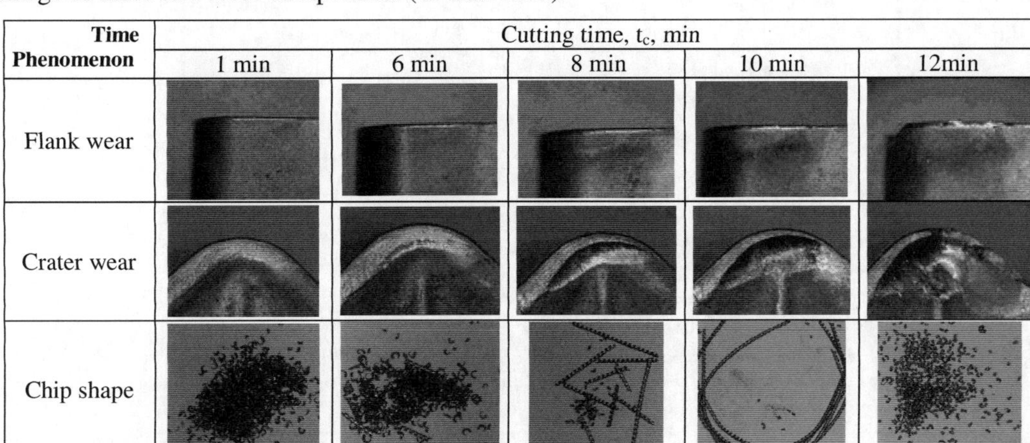

Figure 5 The change in chip shape due to the tool wear
Insert: CNMG120408 PM K10+TiAlN-mono (PVD); Workpiece: C60 unalloyed steel (HB220±5);
Machining conditions: a=1,5 mm; f=0,25 mm; v_c=200 m/min; dry machining

The roughness data of machined surface are also in relation to the wear condition of the tool. On **Figure 6** the change of the roughness features (Ra, Rz) is shown in function of the cutting time in case of a hard-to-machine, austenitic, corrosion-resistant steel. As it can be seen well on the diagram, in the phase of uniform wear (t_c = 2...6 min) the value of Rz suddenly doubles, after that - in function of the current condition and ironing effect of the edge - there is a sharp fall in the value of unevenness height.

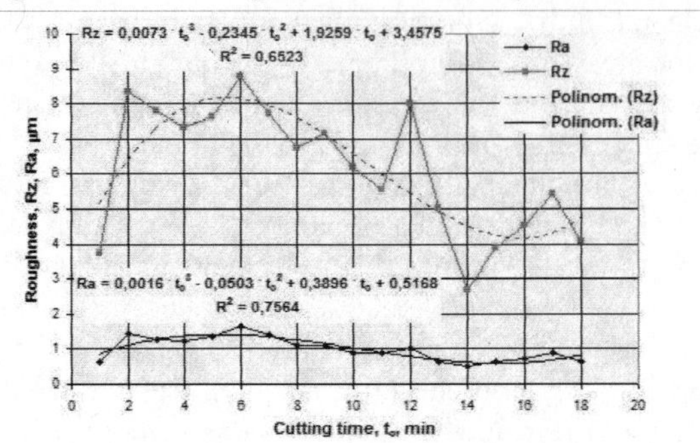

Figure 6 The change in roughness of machined surface in function of time [10][17]
Insert: CNMG120408 MT TT5100; Workpiece: Ko36 stainless steel (HB170±5)
Machining conditions: a=1 mm; f=0,1 mm; v_c=160 m/min; dry machining

Towards to the end of the tool life (t_c > 4 min), there is again an increase in both measured roughness features. The change of the surface finish producing capability, shown on the mentioned figure, is not allowed, for example, in case of series production of components, having a small diameter and length measurement or in case of production conditions, when the specified quality has to be ensured.

The protective effect of PVD-coatings of different materials, structural compositions, thicknesses and nano-hardnesses, has been examined by long-term (lasting) test series. In the **Table 6** the results of wear test of a single-layer coating are summarised.

Table 6. The effect of tool wear on the efficiency of insert

Long-term efficiency of inserts							
Wear data		Force values, N			Roughness	Chip	Efficiency
Time, min	VB, mm	F_c	F_f	F_a	Rz, μm	g	H
1	0,08	840	361	914	8	4	41,0
2	0,09	830	390	917	9	4	36,3
4	0,1	830	407	924	10	4	32,5
6	0,11	835	434	941	11	3	21,7
8	0,12	725	355	807	13	1	7,1
10	0,14	820	426	924	12	1	6,8
12	0,2	1100	605	1255	5	4	47,8

Machining conditions: dry machining
a=1,5 mm; f=0,25 mm; v_c=200 m/min; V'=75 cm³/min
Insert: CNMG0120408 PM K10 + single-layer of TiAlN (PVD)
Workpiece: C60 steel (HB 225±5)

As long as the wear criterion (VB_{meg} = 0,2 mm) has not been achieved, there was a 3-class difference in the classification of chip shape, the active force (F_a, N) increased by approx. 25 percent, the feed force (F_f) grew nearly by 70 percent, while Rz, the unevenness height, changed rhapsodically (almost following the development, shown on Figure 6). In function of the cutting time, the efficiency first decreased gradually, then - unexpected - fell to a third. In the last phase of the wear process, the efficiency indicator (H) increased to such a great extent that it exceeded largely the value, measured at the starting phase. This can be explained by the fact that the Rz roughness of the machined (turned) surface reduced (due to the ironing effect of the tool edge) to a small proportion, the classification of chip shape has became better by three marks. The wear measurement results and the efficiency indicators of multilayer coating, having the same material, are summarised in **Figure 7**.

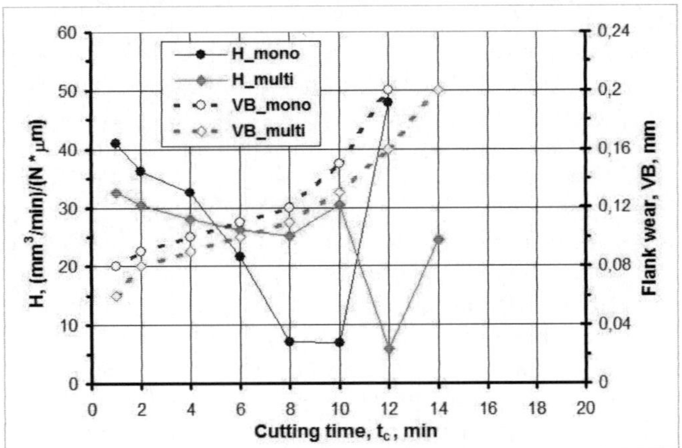

Figure 7 Wear and insert efficiency in function of the time, in case of different coatings
Insert: CNMG0120408 PM K10; Coating: TiAlN single-layer and TiAlN multilayer coating (PVD)
Machining conditions: C60 steel (HB225±5); a=1,5 mm; f=0,25 mm; v_c=200 m/min; dry machining

The wear curve of TiAlN single-layer coating can be described by the formula

$$VB(t_c) = t_c \cdot EXP\left(4{,}7 - 0{,}46 \cdot t_c + 0{,}03 \cdot t_c^2\right) \ [\text{mm}] \quad (6)$$

$$s = \pm 0{,}007 \ ; \quad R^2 = 0{,}972$$

while the TiAlN multilayer-coating by the following:

$$VB(t_c) = t_c \cdot EXP\left(4{,}3 - 0{,}33 \cdot t_c + 0{,}02 \cdot t_c^2\right) \ [\text{mm}] \quad (7)$$

$$s = \pm 0{,}01 \ ; \quad R^2 = 0{,}945$$

Comparing the equations (6) and (7), it can be seen that the greater number of layers - reducing the wear on the flank land - improves the insert efficiency and increases the tool life by 20 percent. In function of the cutting time, the efficiency has a dynamic decreasing tendency throughout the process in case of single-layer coating, and it has a favourable value only shortly before the failure. At the same time, in case of multilayer coating, the efficiency first decreases slightly, then it starts to increase and at the next wear measurement there is a drastical reduction. In the final phase of the wear process, according to the formula (1) we have again a great value. This can be explained by the fact that the increasing value of developing force effects can be compensated by the edge, ironing the surface (leaving a very low roughness values behind) and by the very favourable (snake/spiral-like) chip shape.

Our opinion has been confirmed by the results of durable tests that it is not recommended to apply the efficiency, calculated by the formula (1) to describe the long-term productive efficiency of inserts, due to the following reasons:

- the wear period of the tool has an inconsistent effect on the most important four supplementary features. With other words, the value of „H" indicator has a sudden and drastic reduction close to the inflexion points of wear curves (t_c = 8...10 min);
- the flank wear, used as criterion in case of lifetime measurement, has a clear effect only on the developing force effects from the supplementary features (in case of tools, having a great lead angle value, it affects mainly the F_f component), it has an indirect effect on the developing roughness, while it has no influence on the chipbreaking effiviency;

- the wear on the front of the insert (mainly the crater wear formation) has a hectic effect on the chip development;
- the actual condition and real shape of the tool edge, depending from the wear status, may make the roughness measurement on the workpiece unsure. An instabile cutting procces, the current shape of tool edge, having been copied and/or the ironing effect at the same time, may lead to fluctuations, shown on Figure 7.

Due to the reasons, mentioned above, it is not recommended to make any connection between the efficiency, as interpreted by us and calculated according to formula (1), and the long-term productive efficiency of inserts. It does not offer an appropriate solution to average the efficiency values, belonging to the dates of wear measurements or even to weight these values.

7. Result and discussion

In the present article a new qualification method has been introduced, it can be beneficially applied when determining the cutting ability of inserts. The method is able to compare inserts, having a different shapes, edge constructions, sizes, chipbreakers, material grades and coatings, in an objective way; furthermore, it makes possible to make efficiency rank orders and to measure the of success the innovations, too. It defines three important efficiency elements which enable to select the optimal data settings, using the formula (1) and, defining the complex efficiency (see the formula (4)), there is only a single percentage ratio to describe the tested insert. The applicability, modified with weighted numbers based on the manufacturing trends and operational practical statistics, can be calculated with formula (5). With the weighted complex efficiency it is possible to compare inserts, having different functions, in an objective way.

The advantage of the elaborated method is that it is based on four, simple features (to be exact: the material removal rate, chip shape, active force and surface roughness): these can be easily calculated, evaluated and measured; furthermore, this process is sufficiently cost-effective as it needs a low time-, material and financial expense.

The disadvantage is that it is possible to make only short-time tests with this. It is well-known that the main feature of the cutting performance of an insert is the long-term productive efficiency, with other words, the wear behaviour and the edge durability of the tool. The connection between the wear process development (see formulas (6) and (7)) and the efficiency is rather accidental. The example, introduced in the present article, illustrates that the efficiency, calculated from four main parameters of the supplementary features, should not be confused with the productive efficiency of inserts.

Three world-leading tool manufacturing company groups provided us with inserts to these tests; the usability and the adequacy of method has already been confirmed by test results, carried out on 85 different inserts. In the future we are going to test newly developed inserts with the method, described earlier: these inserts, to be launched on the market, can almost revolutionise the turning operation.

Acknowledgements
The author wishes to thank to his colleagues, taking part in the measurements and in the preparation of the previous articles. Special thanks must go to Dr. Zoltán Pálmai, for his valuable professional comments.

References

[1] Sipos S, Palásti-Kovács B 2007 A forgácsoláskutatás néhány eredménye a szerszámminősítés területén *Nemz. Gép. & Biztech. Szimp. (Budapest)* p 28
[2] Sipos S, Palásti-K B, Horváth R 2015: Forgácsoló technológiák és szerszámai *(Budapest, https://elearning.uni-obuda.hu)* p 410
[3] Biró Sz, Sipos S, Kisa Zs, Uracs P 2009 *microCAD Int. Scie. Conf. (Miskolc)* M pp 27-33
[4] Biró Sz, Cselle T, Hájek V, Sipos S 2010 *microCAD Int. Scie. Conf.* (Miskolc), N pp 25-32
[5] Pálmai Z 1980 Fémek forgácsolhatósága Budapest, MK p 320
[6] Sipos S 1990 *Korszerű technológiák*, vol 4 No 5 pp 9-19.

[7] Sipos S, Repáruk I 1990 *Gépgyártástechnológia* vol XXX No 7 pp 317-20.
[8] Palásti-Kovács B, Dajcs L, Sipos S 1990 *Gépgyártástechnológia* vol XXX No 8 pp. 346-51.
[9] Palásti-Kovács B, Sipos S, Czifra Á 2012 *Proc. Intl. Conf. on Tools (Miskolc)* pp 237-44
[10] Palásti-Kovács B, Sipos S, Biró Sz 2014 *Acta Polytechnica Hungarica* vol 11 pp 5-24
[11] Brammertz P H 1961 *Industrie Anzeiger* vol 83 No 2 pp 25-32
[12] Biró Sz, Nagy A, Sipos S 2010 *Int. Conf. on Manufacturing (Budapest)* on CD ISBN 978-963-9158-30-6
[13] Biró Sz, Nagy A, Sipos S 2011 *microCAD Intl. Scie. Conf. (Miskolc)* L pp 17-22
[14] Sipos S, Biró Sz, Biró J 2005 *Int. Scie. Conf. on Met. Cut. DMC2005 (Kosice)* on CD ISBN 978-80-8073-858-7
[15] Sipos S, Biró Sz, Tomoga I 2006 *Gépgyártás* vol XLVI No 4 pp 17-24
[16] Sipos S 1997: A forgácsolás minőségbiztosítása *(PHARE)*, TDQM-HU9305- 1337/B, Budapest, pp 256-318
[17] Palásti-Kovács B, Sipos S, Biró Sz 2014 *Gyártóeszközök* vol XX pp 3-9

IOP Conf. Series: Materials Science and Engineering **448** (2018) 012044 doi:10.1088/1757-899X/448/1/012044

Mechanical Properties Study of High Impact Polystyrene Under Impact and Static Tests

L Hýlová[1], A Mizera[1], M Mizera[1], R Grund[2] and M Ovsík[1]

[1] Tomas Bata University in Zlín, nám. T.G. Masaryka 5555, 760 01 Zlín, Czech Republic
[2] TREVOS a.s., Mašov 34, 511 01 Trutnov, Czech Republic

hylova@utb.cz

Abstract. This article deals with the study of high impact polystyrene (HIPS) mechanical properties. Injection moulded samples were subjected to the tensile impact test, tensile test, impact test – Charpy and three-point bending test at the ambient temperature. Tensile test was measured at two crosshead speeds 50 and 500 mm/min and bending test at 2 and 50 mm/min. The results show the great adaptability of HIPS which classify this material to the group of very well adaptable polymers to different loadings.

1. Introduction

High impact polystyrene (HIPS) is generally produced by the radical polymerization of styrene monomer with polybutadiene rubber. It is a thermoplastic polymer, which micro-structure is composed of an amorphous polystyrene matrix containing spheroid domains which are responsible for HIPS´s opacity, elongation, craze plasticity and energy absorption. The rubbery phase of HIPS is dispersed in the rigid polystyrene matrix which causes the excellent elongation at break and the impact resistance. The thermal and oxidative degradation of this polymer is caused by double carbon-carbon bonds (C=C). HIPS has very good properties in comparison to common thermoplastics such as the impact resistance, flexibility, low cost, shatterproof and easy process-ability. This material is commonly used to the production of the toys, housewares, packaging, bottles light-duty industrial components and electronic appliances. [1-2]

From HIPS it is possible to create a super-hydrophobic and wear abrasion resistance HIPS/silica nanocomposite coatings for metal surfaces. The best concentration of silica is 50 wt. %. [1]

The scientists from Brazil dealt with the HIPS derived from electronic devices and its subsequent recycling. They created virgin and recycled HIPS samples and exposured these to an accelerated aging process. Subsequently they studied the consequences of the photo-oxidation on the properties. They found out that the recycled samples did not show big changes in the properties of process-ability compared to virgin samples. [2]

Luca Andena et al. studied the influence of processing conditions on the impact resistance of two types of HIPS. First one was „standard" and second was specifically resistant to environmental stress cracking. They found out that higher degree of orientation increased the impact resistance in the air but had no effect in the active sunflower oil. [3]

Recycled HIPS is able to be used to the production of low-pressure membranes. It was found that in the comparison to the membranes produced from pure HIPS, recycled membranes with additives made the membranes more hydrophilic. In comparison to the membranes from pure HIPS, the thermal properties were similar and also they had similar asymmetric membrane structures. [4]

Our scientists in cooperation with Slovakia´s scientist investigated the influence of fall height on HIPS deformation. It was found out that the optimal energy for penetration of tested samples is 50 J. Higher doses are not needed because of the higher power consumption of the testing device. [5]

This study deals with the study of mechanical properties of HIPS and shows the adaptability of this material on different loadings with minimal differences in the results.

2. Experimental

2.1. Sample preparation

HIPS (EDISTIR SR 550 07) was used as the material for the production of tested samples. An ARBURG Allrounder 470 H injection moulding machine was used for the test samples preparation, with the processing conditional to comply with HIPS producer's recommendations, as can be seen in Table 1. The samples for the impact test – Charpy and three-point bending test were in the shape of blocks with the thickness of 4 mm, width 10 mm and length 80 mm. For the tensile impact test and the tensile test, the samples were in the shape of "dog-bone" with the thickness of 4mm, width 10 mm and length 100 mm. Conditioning was taken for 5 days in the ambient temperature and relative humidity of 50 %. All subsequent measurements were measured 10 times to the creation of the arithmetic mean (values displayed in Figures) and standard deviation.

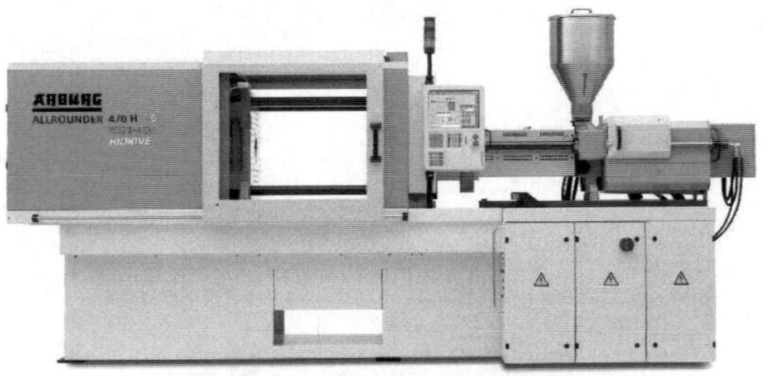

Figure 1. Arburg Allrounder 470 H injection moulding machine.

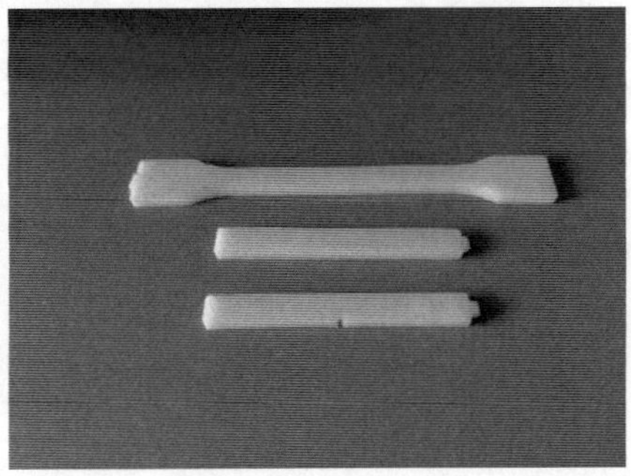

Figure 2. Test samples.

IOP Conf. Series: Materials Science and Engineering **448** (2018) 012044 doi:10.1088/1757-899X/448/1/012044

Table 1. Injection moulding parameters.

Injection Parameters	HIPS
Injection Pressure [MPa]	70
Injection velocity [mm.s^{-1}]	50
Holding Pressure [MPa]	60
Cooling Time [s]	20
Mould Temperature [°C]	30
Melt Temperature [°C]	225

2.2. Tensile Impact test

The tensile impact test was carried out on Zwick HIT50P according to ISO 8256 standard. In this test impact pendulum with potential energy 15 J was used. Samples were tested and values were evaluated in program TestXpert II.

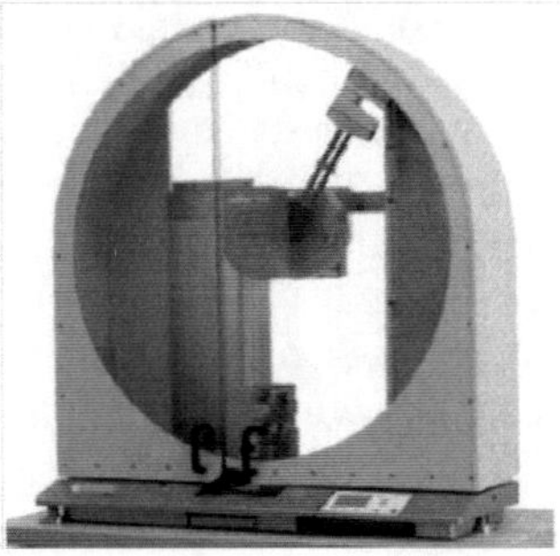

Figure 3. Zwick HIT50P machine.

2.3. Tensile test

A ZWICK 1456 tensile machine was used for the estimation of the tensile behaviour. Measurements were carried out according to the ISO 527 standard at ambient temperature, the crosshead speeds were 50 and 500 mm/min.

Figure 4. Zwick 1456 machine.

IOP Conf. Series: Materials Science and Engineering **448** (2018) 012044 doi:10.1088/1757-899X/448/1/012044

2.4. *Impact test - Charpy*
The Charpy impact test was carried out on Zwick HIT50P device at ambient temperature according to the ISO 179 standard. There are two types of samples, with the notch of width 2 mm and without notch. In this impact hammer test, 50 J of potential energy was used for research purposes. Samples were tested and their impact toughness values were evaluated in the program TestXpert II.

2.5. *Three-point bending test*
Three-point bending test was carried out on ZWICK 1456 device according to the CSN EN ISO 178 standard. The crosshead speed was 2 and 50 mm/min.

3. Results and Discussion

3.1. *Tensile Impact test*
In Table 2 the arithmetic mean and standard deviation of the maximal force and impact toughness calculated from 10 measurements are displayed. The maximal force of HIPS samples is around 628 N and impact toughness is around 548 kJ/m^2.

Table 2. Tensile impact test results.

	Maximal Force (N)	Impact Toughness (kJ/m^2)
Arithmetic Mean	627,91	548,13
Standard deviation	8,92	7,65

3.2. *Tensile test*

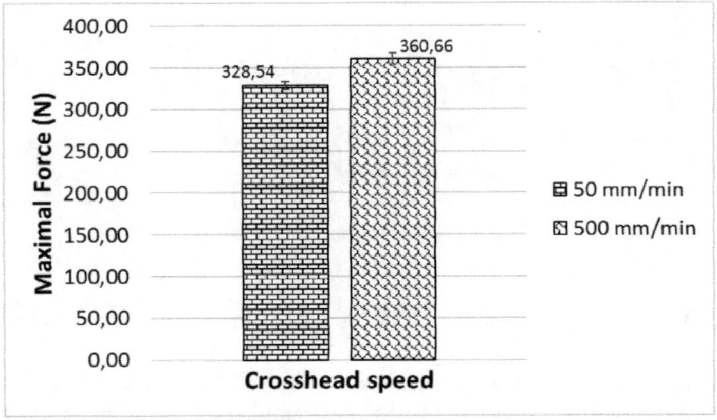

Figure 5. The comparison of maximal force with crosshead speed 50 and 500 mm/min.

As can be seen in Figure 5, maximal force of tensile test was measured at two crosshead speeds. There is a small difference between crosshead speed of 50 and 500 mm/min. The maximal force at the speed of 50 mm/min is 328.54 N and at speed of 500 mm/min is 360.66 N. The small difference can be caused by the polybutadiene particles dispersed in the polystyrene matrix, which are able to absorb the energy inside the material during the test.

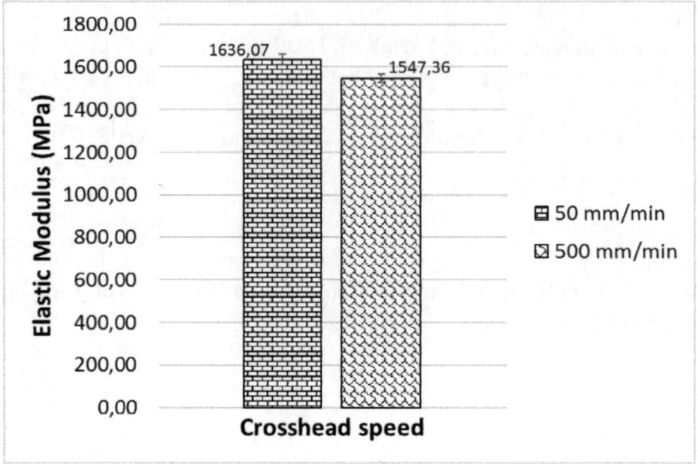

Figure 6. The comparison of elastic modulus with crosshead speed 50 and 500 mm/min.

The elastic modulus of HIPS samples is displayed in Figure 6. There is also only small difference between crosshead speed of 50 and 500 mm/min. The elastic modulus at the speed of 50 mm/min is 1636.07 MPa and at speed 500 mm/min 1547.36 MPa. The inner structure of HIPS is able to adapt during the tensile strength test because of the rubbery phase.

3.3. Impact test - Charpy

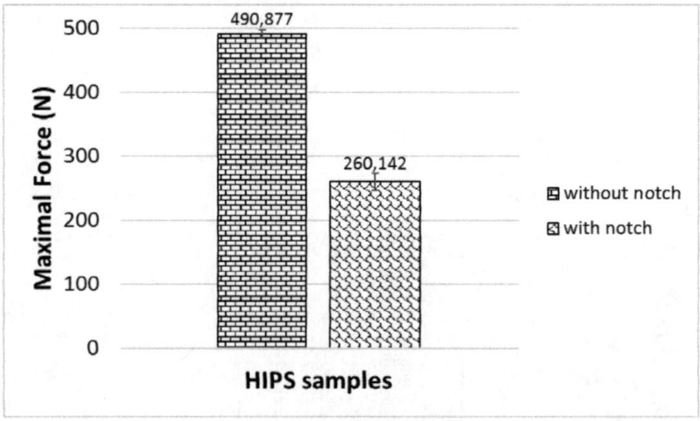

Figure 7. The comparison of maximal force of the samples without notch and with 2 mm width notch.

The Charpy test was carried out at samples with (width 2 mm) and without the notch. From the results in Figure 7 it is clearly visible that HIPS samples without notch has significantly higher maximal force – 490.877 N, while the sample with the notch only 260.142 N. This phenomenon can be explained by the width of the samples. In the case of the sample with the notch, there is about one-fifth narrower width and thus the easier way of the crack growth. And as was already mentioned above, the measurement conditions were the same for all samples.

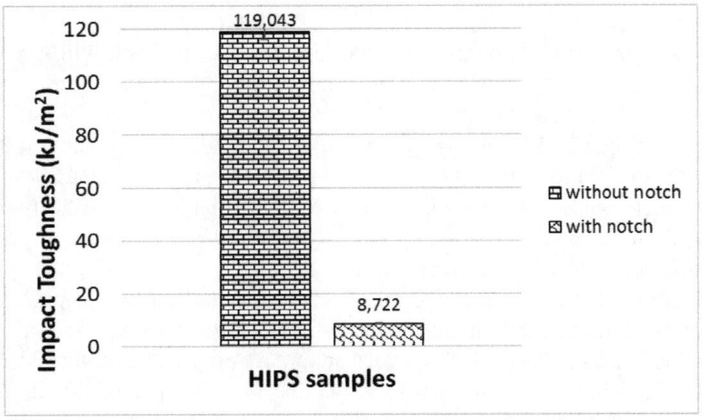

Figure 8. The comparison of impact toughness of the samples without notch and with 2 mm width notch.

The impact toughness is displayed in Figure 8. The comparison of HIPS samples without notch and with notch shows a big difference. The impact toughness of the sample without notch is 119.043 kJ/m^2, while at the sample with the notch, it is only 8.722 kJ/m^2. These results can be also connected with the width of the sample. In the case of the samples with 2 mm notch there is easier way of the crack growth caused by the notch cut into the sample.

3.4. Three-point bending test

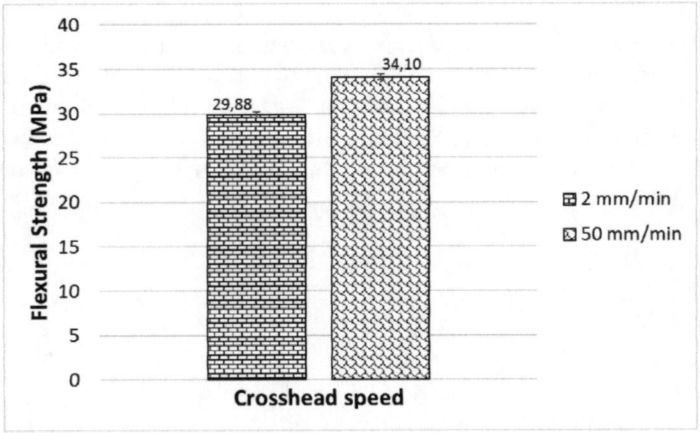

Figure 9. The comparison of flexural strength with crosshead speed 2 and 50 mm/min.

In Figure 9 it is possible to find the results of three-point bending test in the form of the flexural strength. HIPS samples were subjected to two crosshead speeds of 2 and 50 mm/min. In this case the difference is very slight. It can be as in above displayed cases caused by the rubbery phase of the HIPS, which causes a very good adaptability of this material.

IOP Conf. Series: Materials Science and Engineering **448** (2018) 012044 doi:10.1088/1757-899X/448/1/012044

4. Conclusion

In this article, HIPS samples were injection moulded in the shape of block with the thickness of 4 mm, width 10 mm and length 80 mm. These samples were conditioned for 5 days at the ambient temperature and 50 % of humidity and subsequent subjected to the tensile impact test, tensile test, impact test – Charpy and three-point bending test. In case of the tensile test, there were two crosshead speeds – 50 and 500 mm/min and in the case of three-point bending test 2 and 50 mm/min.

The biggest difference was found out in the case of the impact test – Charpy compared to HIPS samples with and without the notch. The notch causes the increase in the crack growth, which shows that the sample without notch has a greater impact toughness.

During measurement of the tensile test and the three-point bending test, it was found out, that there is a minimal difference in different crosshead speeds, which can be caused by the rubbery phase in HIPS.

As a conclusion it can be said that HIPS is a very adaptable polymeric material which is able to react quickly during big changes in the conditions of the loading and because of this it is a suitable material for the application in the practice where the great adaptability on the different loading is needed.

References

[1] Masood T M, Heredia-Guerrero J A, Ceseracciu L, Palazon F, Athanassiou A and Bayer I S 2017 *Italy. Chem. Eng. J.* **322**, p. 10-21.
[2] Dilly J, Renan D and Alves C L H 2016 *Brazil. IJAET* **9**, p. 522-531.
[3] Andena L, Rink M, Marano C and F Briatico-Vangosa 2016 *Italy. Polym. Test.* **54**, p. 40-47.
[4] Garcia-Ivars J, Wang-Xu X and Iborra-Clar M I 2017 *Spain. Sep. Purif. Technol.* **125**, p. 340-351.
[5] Mizera A, Chalupa P and Hudec I 2017 *Czech Republic, Slovakia. MATEC web conf.* **125**, p. 1-4.

Acknowledgment

This work was supported by the European Regional Development Fund under the project CEBIA-Tech Instrumentation no. CZ.1.05/2.1.00/19.0376 and by the Czech Ministry of Industry and Trade in Program "Aplikace" No. CZ 01.1.02/0.0/0.0/16_084/0009949 and by the Ministry of Education, Youth and Sports of the Czech Republic within the National Sustainability Program project no. LO1303 /MSMT-7778/2014). Moreover, it was supported by the Internal Grant Agency of TBU in Zlin: no. IGA/FT/2018/012.

IOP Conf. Series: Materials Science and Engineering **448** (2018) 012045 doi:10.1088/1757-899X/448/1/012045

Implementing Education Vision in the Context of Industry 4.0

Numan Durakbasa, Gökcen Bas, Jorge Bauer

TU-Wien, Vienna University of Technology, Austria

numan.durakbasa@tuwien.ac.at, goekcen.bas@tuwien.ac.at

Abstract. Transformation of the traditional industry through digital technologies shapes the education system, as comprehensive knowledge in the interdisciplinary areas of technologies is required. Amid the challenges of competitive and digital technologies in the advanced manufacturing industry, the universities and industry cooperation is imperative. This study presents the planning and implementation of the educational projects in metrology to create a vision of building the bridge between educational institutes and the industry. The implementation of tele-presence and tele-operation case studies focus on flexibility, efficiency and quality of the processes that does not limit the resources to a location, time difference or facility. Vision in education is of extreme importance in this present time of worldwide international competition in Industry 4.0 that will enable to achieve sustainable environmental friendly goals with contribution to both industry and society.

1. Introduction

Digitalization and deployment of growing technologies in ICT are transforming the machining and the manufacturing industry worldwide. The term "Industry 4.0" was first emerged in 2011 to refer to a new era in industry following a research agenda and recommendations for implementation in 2013 by Acatech - the German Academy of Engineering Sciences [1]. Industry 4.0 includes a higher priority on individualized products, which requires transformations in the designing, manufacturing, and service of industrial systems and products as well as their business models [2]. A long term strategy to dovetail the demand in today's manufacturing industry assures a sustainable development [3]. Besides, the roadmaps and plans be routinely reviewed and updated until a formal iterative process occurs to better explore and understand certain technologies [4].

The industrial and technological key concepts need to be specified and comprehended not only by industry but also, even more importantly by educators and scholars. This goal will be achieved by establishing guidelines, roadmaps, and scientific literature [5]. Moreover, the strategic work will enable to address the guidance of transitioning to new forms of engineering and manufacturing address [6].

The cooperation of international partners in research focusing on providing borderless collective experience, expertise and use of laboratory facilities sets the background of a roadmap towards efficient integration of applications both in academy and in industry [7]. When we talk about the education, sustainability is the key term to overcome the challenges of competitive and digital technologies in the advanced manufacturing industry. The sustainable model of educational improvement is basically based on a methodology of Life Long Learning easy to adapt moreover to other topics with adequate constraints [8].

2. The Scope of Metrology in Advanced Manufacturing Systems

In advanced manufacturing systems there exist close interactions between intelligent design, intelligent metrology and industrial and technological developments. Special points of interest are integrated intelligent management, digitalization, automation, high accuracy, high efficiency, next generation manufacturing and nano/pico metrology.

The production of very precise components goes hand in hand with the development of the necessary metrology, and a wide range of measuring instruments has been devised to cater for the evaluation of surfaces and structures down to the 0.1 nm level. This powerful array of instruments provides a measuring capability in nano- and pico-metrology. In modern metrology it is possible to use instruments capable of creating atomic resolution images of the surfaces of different specimens and 3D measurements as well as scanning of high precision parts. SCM, AFM, STM, Scanning Probe Microscopy (SPM) and high precision CMMs are such advanced measurement technologies.

The needs of the industry for ultra-high precision engineering and workpieces with a surface roughness less than few nanometers call for measurement instrumentation that can be applied reliably in modern production processes, together with international standards defining parameters and tolerances in the nanometer scale. The requirements on the measurement systems and the measurement strategy to determine suitable parameters, time, costing and the guarantee of predetermined process stability by means of measurable and correlated parameters come into focus.

On basis of industry needs, the demands on industrial nanometrology can be subdivided into three major scientific attributes:

- Reliability: Measurement results have to mirror the real surface structure and statistic and systematic errors may be reduced to an absolute minimum.
- Comparability: Measurement results must be comparable when they are measured with different measurement systems of the same kind. Ideally measurement results taken with different systems should be comparable too.
- Reproducibility: Several measurements of the same sample under the same conditions must result in the same results. Changes in measurement conditions must result in comprehensible changes in the measured parameters.

Increasing demands on precision and accuracy of measurement results in manufacturing and metrology is a particular important challenge in industry. Adequate competencies, knowledge and experience are required for sustainable development. As a part this vision, the High Precision Measurement and Nano Metrology Laboratory at the AuM-TU Wien (The Department of Interchangeable Manufacturing and Industrial Metrology / Production Metrology and Quality, Vienna University of Technology) is chosen as a node to create an inter-university network.

As the technology advances rapidly, the qualification of required education has to catch up providing the required knowledge. Otherwise, the classical system of education will be irrelevant for soon future. In order to establish required engineering and technology educational models, one has to consider a methodology based on continuous improvement. This study performs a model and strategically implementation of continuous improvement in education with modern methods of quality management.

2.1. Quality Management in Education

Implementation of quality management in the vision of education offers an advantage in fast developing technological era with defined research and development processes at the universities and institutes. The applicable standards for the quality management in education are used as guidelines as they support three pillars of sustainable development namely economic, social and environmental. They can be listed mainly [9-16]:

- ISO/IEC 19796-1:2005:Information technology - Learning, education and training - Quality management, assurance and metrics - Part 1: General approach. It is a framework to describe, compare, analyze, and implement quality management and quality assurance approaches with the main aspect of Reference Framework for the Description of Quality Approaches (RFDQ).
- ISO/IEC 19796-3:2009: Information technology - Learning, education and training -- Quality management, assurance and metrics - Part 3: Reference methods and metrics. The reference framework for the description of quality approaches (RFDQ) by providing a harmonized description of the methods and metrics required to implement quality management and quality assurance systems for stakeholders designing, developing, or utilizing information technology systems used for learning, education, and training.
- ISO/IEC 40180:2017:Information technology - Quality for learning, education and training - Fundamentals and reference framework. It provides the fundamentals and the reference framework for quality assurance, quality management and quality improvement in IT-enhanced learning, education and training (E-Learning) harmonizing existing approaches, concepts, specifications, terms and definitions related to quality for E-Learning, education and training.
- ISO 29993:2017:Learning services outside formal education - Service requirements. It specifies requirements for learning services outside formal education, including all types of life-long learning (e.g. vocational training and in-company training, either outsourced or in-house) including any learning services provided by a learning service provider (LSP) that are addressed to learners themselves, as well as to sponsors who are acquiring the services on behalf of the learners. The key features of these kinds of services are that the goals of learning are defined and the services are evaluated, and that they involve interaction with the learner. The learning can be face-to-face, mediated by technology, or a combination of both.
- ISO 29990:2010:Learning services for non-formal education and training - Basic requirements for service providers. It specifies basic requirements for providers of learning services in non-formal education and training.
- ISO 21001:2018:Educational organizations - Management systems for educational organizations - Requirements with guidance for use. It specifies requirements for a management system for educational organizations that uses a curriculum to support the development of competence through teaching, learning or research, regardless of the type, size or method of delivery such as professional training departments.
- PAS 1069:2009-07 Learning, Education and Training with special consideration of e-learning - Guideline for the Reference Process Model for Quality Management and Quality Assurance - Planning, Development, Realization and Evaluation of Processes and Offers in Learning, Education and Training.
- PAS 1032-1:2004-02 Learning, education and training focusing on e-learning - Part 1: Reference model for quality management and quality assurance - Planning, development, realization and evaluation of processes and offers in learning, education and training.

The quality system to approach contributes to understanding and managing interrelated processes as a whole and moreover enhances effectiveness and efficiency in education models fulfilling its targets. The quality system approach principle address the requirements through:

- Defining the system objectives
- Establishing, implementing and maintaining the management system as a set of processes
- Describing the system
- Continual improvement
- Interconnection, interrelation and sequence of processes
- Maintaining the integrity of the system

Establishing educational models will not only enable adequate qualification but also enable processes to overcome the challenges for organizations that proceed towards the modern manufacturing environment in such as the hidden costs leading to potential savings. Therefore, the documentation requirement covers the reports to be implemented in the education strategy.

2.2. Applications in Education

The time, cost and expertise for establishing metrology and mechanical laboratories is very high for many small and medium size enterprises. Besides, required expertise and knowledge can't be found in every country. The High Precision Micro and Nano Metrology Laboratory at the AuM-TuWien (The Department of Interchangeable Manufacturing and Industrial Metrology / Production Metrology and Quality, Vienna University of Technology) is chosen as a node in this study to create an inter-university network. A developed platform serves as a practical experience of an intelligent metrology process model, which is also available as educational models to the students on a global scale. An educational concept for the current students taking part in high precision measurement activities and intelligent metrology cycle as the target, research and development process is established at the cooperating universities and institutes. Based on continuous improvement approach from the referenced standards, the experience obtained by integration of the processes will further create a roadmap for future developments.

The cooperation in this study between the local laboratories in Argentina enables the practical applications to develop and create a resource for future connection nodes that is already in the near future plans. The current status of the project focuses on automatic control of the high precision measurement instruments through the network established using a server. The operation requirements of remote control results an increase of the workload of the channel in the network. Hence, the bandwidth limitations generate a reduction speed of the process. To resolve this difficulty; a client server computing technology is used for the control of all the equipment in the laboratory. Establishing a paradigm in the Linux server enabled us to connect to the equipment that is required for further applications. The measurement applications are represented in the Figure 1.

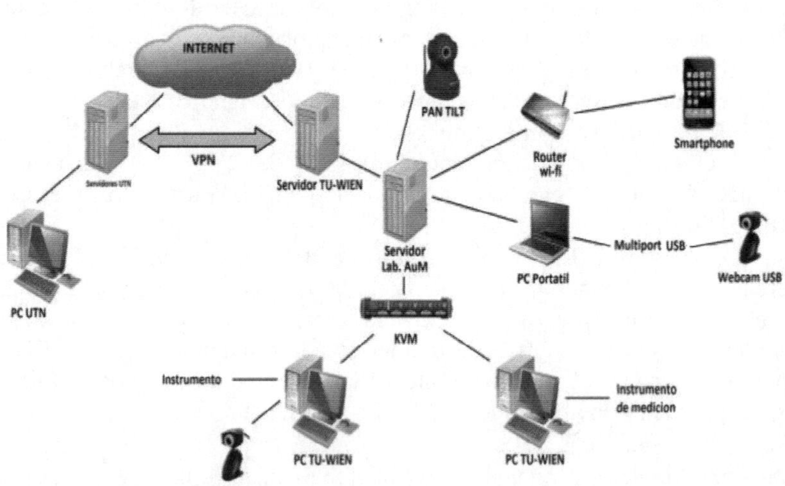

Figure 1. The overview of working environment

The existing status of the inter-university network exhibits good results using also mobile phone technology. The mobile phone technology enables currently to connect to the other nodes (e.g. Buenos Aires) using the Wi-Fi network established in the laboratory environment. The international network for telepresence and teleoperation educational models is represented in Figure 2.

Figure 2. The international network for tele-presence and tele-operation educational models

3. Conclusion and Future Work

Establishing and implementation of educational vision to serve adequate knowledge and expertise throughout generations sustainably will also enable the methods to overcome the challenges of the new era Industry 4.0. The educational models based on flexibility, efficiency and quality will provide the advanced manufacturing industry an advantage in competitive and digital technologies.

The practical approach of this inter-university network project offers collaboration to any industrial or university organization without boundaries. Learning and continuous improvement are the basics of each module of the system. Once the targets are achieved, it is possible to implement the modules in another node using the systematic knowledge obtained throughout this project. Moreover, as the technology develops, improvement will be supported further by means of documentation as a part the quality management system integration.

References

IOP Conf. Series: Materials Science and Engineering **448** (2018) 012045 doi:10.1088/1757-899X/448/1/012045

[1] https://www.bmbf.de/de/zukunftsprojekt-industrie-4-0-848.html

[2] Schuh, G., Potente, T., Wesch-Potente, C., Weber, A. and Prote, J. (2014). Collaboration Mechanisms to Increase Productivity in the Context of Industrie 4.0. *Procedia CIRP*, 19, pp.51-56.

[3] Westkämper, E. (2014). Towards the re-industrialization of Europe: a concept for manufacturing for 2030. ISBN 9783642385025, *Springer*.

[4] Garcia, M. L. and Bray, O. H. (1997). Fundamentals of Technology Roadmapping. *Sandia National Laboratories*.

[5] Santos, C., Mehrsai, A., Barros, A., Araújo, M. and Ares, E. (2017). Towards Industry 4.0: an overview of European strategic roadmaps. *Procedia Manufacturing*, 13, pp.972-979.

[6] Erol, S., Schumacher, A. and Sihn, W. (2016). Strategic guidance towards Industry 4.0 – a three-stage process model. *Proc. of International Conference on Competitive Manufacturing 2016 (COMA16)*, Stellenbosch, South Africa.

[7] Durakbasa, M.N., Bas, G., Bauer, J.M. and Poszvek, G. (2014). Trends in Precision Manufacturing Based On Intelligent Design and Advanced Metrology. *Key Engineering Materials*, Trans Tech Publications, Switzerland, Vol.581; pp.417-422.

[8] Weckenmann A., Werner T. (2011) Sustainability in Education. In: Seliger G., Khraisheh M., Jawahir I. (eds) Advances in Sustainable Manufacturing. *Springer*, Berlin, Heidelberg

[9] ISO/IEC 19796-1:2005: Information technology - Learning, education and training - Quality management, assurance and metrics - Part 1: General approach.

[10] ISO/IEC 19796-3:2009: Information technology - Learning, education and training - Quality management, assurance and metrics - Part 3: Reference methods and metrics.

[11] ISO/IEC 40180:2017: Information technology - Quality for learning, education and training - Fundamentals and reference framework.

[12] ISO 29993:2017: Learning services outside formal education - Service requirements.

[13] ISO 29990:2010:Learning services for non-formal education and training - Basic requirements for service providers.

[14] ISO 21001:2018:Educational organizations - Management systems for educational organizations - Requirements with guidance for use.

[15] PAS 1069:2009-07: Aus- und Weiterbildung unter besonderer Berücksichtigung von E-Learning - Leitfaden zum Referenzprozessmodell für Qualitätsmanagement und Qualitätssicherung - Planung, Entwicklung, Durchführung und Evaluation von Bildungsprozessen und Bildungsangeboten (Learning, Education and Training with special consideration of e-learning - Guideline for the Reference Process Model for Quality Management and Quality Assurance - Planning, Development, Realisation and Evaluation of Processes and Offers in Learning, Education and Training).

[16] PAS 1032-1:2004-02: Aus- und Weiterbildung unter besonderer Berücksichtigung von e-Learning - Teil 1: Referenzmodell für Qualitätsmanagement und Qualitätssicherung; Planung, Entwicklung, Durchführung und Evaluation von Bildungsprozessen und Bildungsangeboten (Learning , education and training focusing on e-learning - Part 1: Reference model for quality management and quality assurance - Planning, development, realization and evaluation of processes and offers in learning, education and training).

IOP Conf. Series: Materials Science and Engineering **448** (2018) 012046 doi:10.1088/1757-899X/448/1/012046

Horizontal traffic signs anomalies and their classification

Henrietta LENGYEL, Dr. Zsolt SZALAY
Budapest University of Technology and Economics

henrietta.lengyel@gjt.bme.hu

Abstract: In addition to the traffic signs, the road markings are those that help to control the traffic with of paint and other materials in the road surface. The general requirements and standards contained in this specification apply to daytime, night time visibility, distance, quality and usability. The significance of road markings is essential in terms of driving and helping traffic participants in the possible directions of travel. The traffic lanes give a sense of security to the drivers and passengers, as they are less visible for certain reasons (light, quality, and visibility) instantly appear insecurity, which in turn increases the risk of an accident. These signs are not only of great importance but also provide a basis for the development of both traffic trends and vehicle technology. Everywhere they are present in traffic, they are already used in the early stages of autonomy (parking, track tracking), and likely the road markings will be used and on the road in the future. But it often happens that traffic lanes and road markings quality is deteriorating and less visible, making it difficult to recognize and interpret human drivers and autonomous vehicles. That's why need to classify the mistakes which can use to the test later.

1. INTRODUCTION

This is only possible if autonomous cars can perceive the environment with the correct accuracy. In many cases, human drivers cause accidents because they cannot see the signs or they cannot properly interpret them. The recognition problem can be solved by autonomous systems.

This article draws attention to the anomalies of traffic lanes and road markings currently used. Weather, environmental effects, and traffic situations all have an impact on recognizability, these factors cause difficult recognition ability on autonomous systems. To improve the errors, need to gather situations that are causing problems in detecting the environment in your systems. Horizontal and vertical signals are provided on roads with many information, and the article pay attention to the horizontal signs.

2. IMPORTANCE OF HORIZONTAL SIGNS

Horizontal signals are always needed as long as the old vehicles are present in traffic. Horizontal signs include traffic lanes and other road markings. The lanes help drivers and later autonomous vehicles stay on track and follow the trajectory. The traffic lanes and the road markings used by them provide a wealth of information on the road. They have different colors, materials and shapes with special information content. For their correct use a standard book is considered, which the *Road Technical Specifications* for the road is. (Fig.1.)

IOP Conf. Series: Materials Science and Engineering **448** (2018) 012046 doi:10.1088/1757-899X/448/1/012046

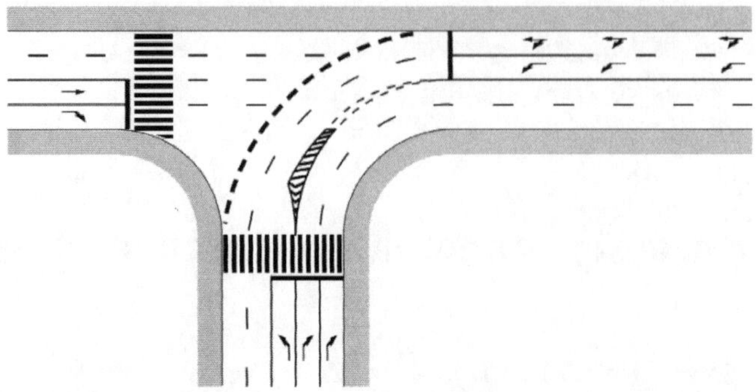

Fig. 1. Complete crossroads with Different type of road markings.

Road Technical Specifications e-UT2-1.113-2001-Design of road markings

It contains all the sizes, colors and materials which need to use on the roads. Completed crossroads can also be found in finds that help later designers work and show how need to plan a crossroad or construction area. (Fig.2.)

Traffic participants receive a lot of information and external stimulation while traveling, so be sure to place as many traffic signs in order in which the drivers are able to interpret and reaction.

If the traffic lanes with their length, shape, colours help to define the track what autonomous vehicle can follow. Use different type of the lane and road markings to the road construction. (Fig.2.)

These factors can be of great help to the autonomous systems in the perception of the traffic signs later.

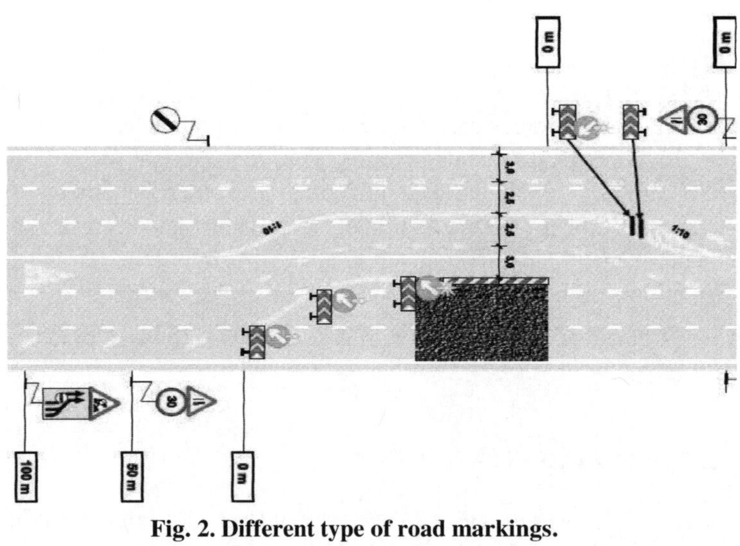

Fig. 2. Different type of road markings.

Road Technical Specifications e-UT2-1.113-2001-Design of road markings

IOP Conf. Series: Materials Science and Engineering **448** (2018) 012046 doi:10.1088/1757-899X/448/1/012046

3. DIFFICULTIES OF TRAFFIC LANES AND ROAD MARKINGS

In general, circumstances in the environment significantly influence the visibility of the signs. In addition to this, the natural wear, and fault of the signs makes it more difficult to recognize by the recognition systems and drivers. However, recognition systems can only solve the situations for which they are prepared and taught.

In addition to common errors, the most critical are traffic restrictions. The shape of the temporary road markings must be the same as the shape, color, dimensions and, in some cases, the intended arrangement of the permanent signals. Temporarily used cover road markings can show the track with two types. Depending on the type, budget, length, and duration of the work, which ones are used. Signs can be applied to the road surface by painting or glutted pre-fabricated elements. It is permissible to deviate from the individual symbols (arrows, inscriptions) in case they can be solved with fastenable tapes for faster and more economical execution.

In vain, it is expensive to buy, but in the case of long-lasting work, it is a more durable and reliable solution. The advantages of glued symbols are that they are more economical, faster and easier to apply to the surface, but have a significant disadvantage that they are move away from the original place more frequently due to the road surface and weather conditions. In many cases, it may cause an accident if all lanes and road markings are "slip" from the original place. (Fig.3.)

Fig. 3. Traffic lanes and road markings are "slip" from the original place

It is important that car drivers notice the traffic lanes and road markings changes in time. The sooner the drivers are placed in the right lanes, the less chance of sudden moves in the place. Often found on the roads that these were not started early enough or the signs were unclear and visible, so the information about the traffic situation was unexpected for the drivers.

Removing temporary signs when painting will also cause problems. The natural wear is not expected by the workers handling the restrictions, but they are blown in a gray color that does not completely delete the provisionally used signal. This often causes misunderstandings for the drivers and other traffic participants, which can disturb the band sensors. (Fig.4.)

Fig. 4. Use two types of traffic lanes and cause misunderstandings

These many reasons why we started to collect cases from various traffic signs anomalies that can be integrated into the simulation phase of the test, autonomous systems could be prepared. For the teaching, the first step is classifying the mistakes, the visibility, clarity, recognizability, position and complexity of the boards at the permitted speed.

The visibility means when a traffic lanes and road markings cover by other one. The clarity when you can see a clean traffic lanes, or a very dirty one. Recognizability is when you watch a traffic markings and the size, form, faults and worn is acceptable. The position means that the sign is not at the right place. Simplicity when too much or opposite information should be observed.

In daytime conditions, there were several examples of the fact that the temporary burglary was difficult to notice due to the sun. At limited visibility (rain, fog) and at night, even greater difficulty in detecting the new lane created by them.

Autonomous vehicles have to recognize the signs when it comes to camera recognition, and to other support systems. They must be helped to secure access to the area.

Fig. 5. Use glued traffic markings on the road but can see the old ones

The interpretation of traffic signs for people and systems is not an easy task, so there will be continuous improvements in the future so that no further accidents occur.

4. POSSIBILITIES OF TRAFFIC SIGNS BY RECOGNITION SYSTEMS

Many factors and effects make it difficult to track traffic lanes and recognize road signs. All of these make it difficult for the detection systems to function properly and may lead to an accident in case of erroneous interpretations. Road markings are part of the currently used road signaling systems. Recognizing their travels can provide great help and security for the drivers. The most common cause of an accident is that they lose their bands from fatigue or inattention. [14]

The essence of the system is to keep track of the signals on the road. The **Lane Keeping Assist** warns the driver of the steering wheel vibration when the vehicle accidentally slides off its tracks to prevent accidents. The road markings are scanned with a "camera" on the inside of the front windscreen. An electronic control unit can detect boundary lines with contrast differences. It determines the position of the vehicle during the unit, so it can even more accurately determine whether to leave the track. In the Mercedes vehicles, lane departure warning is shown in (Fig.6.) [18] [17]

Fig. 6. Lane Keeping Assist show the traffic lanes in the Mercedes

Unlike conventional systems, this system also has the ability to monitor driver behavior and activity while driving. With this feature, you can distinguish between intentionally or accidentally leaving the track. If it feel that the driver wants to go over the midline for stopping or overtaking, it does not intervene and does not give a warning vibration. You can sense the unit even if a vehicle is moving on a continuous or dashed line.

The lane system has additional built-in ancillary assistants in the vehicle. This is the **Active Steering Assist**, which helps the steering torque, so that the driver can keep the vehicle in the track on a straight or bendy road. The system uses orientation for the vehicles in front and the traffic lanes. The signals are progressively evaluated by senses of a stereo camera, which is located behind the windshield, near the rear-view mirror. The assistant switches on when the driver starts lane change with a direction indicator and the system does not detect another vehicle.

In the event that the system detects that the lane change is unsafe, the driver does not react or act in the correct manner and does not seize the steering wheel despite the warning, the system will slow down the vehicle. The active steering system itself can be activated up to 210 $^{km}/_{h}$. The system is already included in several car brands in new generation driving packages. This option is already at the lower level of autonomy, which allows for comfort and safety. [19]

The system works perfectly when the road markings are correctly positioned on the road. Particular attention should be paid to what happens if the individual signs change their color or shape.

From a road safety point of view pedestrian crossings are the most vulnerable places. In order to reduce the number of accidents, the route of pedestrian crossing has to be drawn with a variety of indications for road users. Already doing tests to help them better draw drivers attention to seeing the pedestrians as soon as possible. This is a great example of the use of multiple markers, since they use markers for signaling, which can be used for further travel, with reflective properties. There was an opportunity to consider what happens when manipulated by the shape of the crossing point to force the driver to slow down to avoid collision. Earlier, they had been trying in China, and recently two Indian artisans first came up with the idea of what would happen when 3D-acting crossings were painted. (Fig.7.) Their effect is from the perspective of the motorists as if they were a physical obstacle to them, so they would certainly slow down or stand in front of the pedestrian crossing. But other signaling can be applied to the autonomous vehicles, for example LEDs. [21]

Fig. 7. Lane Keeping Assist show the traffic lanes in the Mercedes

There is a need to prepare for this, so there is a need for combined systems. When testing autonomous systems, it is important to address them in such a way as to detect shortcomings and malfunctions as soon as possible.

5. EXPLANATION OF CLASSIFICATION PROCEDURE BY EXAMPLES

The classes of the mistakes: ***the quality, status, quantity, visibility, perception, recognizability, clarity, and interpretability of the boards at the permitted speed.***

In order to prevent accidents, the classification system would help to evaluate traffic signals based on different aspects. Each attribute gets a grade and based on the scale created in the developed system,. This evaluation system would be applicable to any signal. The next table include the attributes, values and grades.

Table 1. Traffic signs anomaly classification

	Visibility 1= not visible 5= visible	Clarity 1= dirty 5= clear	Recognizability 1= not recognizable 5= recognizable	Position 1= Improper 5= standard	Simplicity 1= complex 5= simply
Figure 3. 	2	2	1	1	3
Figure 4. 	2	2	1	4	1
Figure 5. 	5	5	3	5	5
Figure 7. 	5	5	2	4	3

This procedure can be developed to configure anomalous traffic lanes and road markings installation into an autonomous vehicle test environment. This system help to find the weaknesses of recognition systems.

6. CONCLUSION

The difficulties of recognizing traffic signs and road markings will probably still be up for a while. The solution is that the recognition systems must be taught and prepared for possible errors. The other, which promotes the development of systems, is testing that specifically points to existing errors.

For this reason, the tests can show to errors in systems that can still be eliminated, thus reducing accident accumulation on the roads. And the classification method can help to find the weaknesses of recognition systems.

Zalaegerszeg is currently building a test track for the highly automated and autonomous vehicles (ZalaZONE). Our anomaly classification about the horizontal traffic signs are contribute to this implementing of such test environment. Where traffic sign recognition systems can be effectively tested and safely validated. After the protocols the systems will allow safe traffic in the future. [10]

The project has been supported by the European Union, co-financed by the European Social Fund. EFOP-3.6.2-16-2017-00002

References

1. J.STALLKAMP, M.SCHLIPSING, J.SALMEN, C.IGEL 2016 *Man vs. computer: Benchmarking machine learning algorithms for traffic sign recognition.* Neural Networks Volume 32 Pages 323-332.
2. ÁDÁM NYERGES, ZSOLT SZALAY, 2017, *A new approach for the testing and validation of connected and automated vehicles,* 34th International Colloquium on Advanced Manufacturing and Repairing Technologies in Vehicle Industry.
3. MONIKA LASOTA, MARCIN SKOCZYLAS, 2016, *Recognition of Multiple Traffic Signs using Keypoints Feature Detectors,* 2016 International Conference and Exposition on Electrical and Power Engineering (EPE)
4. HIGINIO GONZALEZ, B. RIVEIRO,GEOMETRIC, JULIA ARMESTO, PEDRO ARIAS, 2011, *Evaluation of road signs using radiometric information from laser scanning data,* Research Gate, 2011
5. KRISZTIÁN PINTÉR, ZSOLT SZALAY, GÁBOR VIDA, 2017, *Autonomous Vehicles - novel types and causes of traffic accident, responsibility*, 34th International Colloquium on Advanced Manufacturing and Repairing Technologies in Vehicle Industry
6. TOM SIMONITE, *Even artificial neural networks can have exploitable backdoors,* Last access time: 2018.04.03.
   URL: https://www.wired.com/story/machine-learning-backdoors
7. J. A. Barria, S. Thajchayapong, 2011, *Detection and Classification of Traffic Anomalies using Microscopic Traffic Variables,* IEEE Transactions on Intelligent Transportation Systems (Volume: 12, Issue: 3)
8. LEONARDO BRUNO, GIUSEPPE PARLA, CLARA CELAURO, 2012, *Improved Traffic Signal Detection and Classification via Image Processing Algorithms,* Procedia - Social and Behavioral Sciences Volume 53

9. POTÓ VIVIEN, SOMOGYI ÁRPÁD, LOVAS TAMÁS, BARSI ÁRPÁD, TIHANYI VIKTOR, SZALAY ZSOLT, 2017, *Creating HD map for autonomous vehicles - a pilot study,* 34th International Colloquium on Advanced Manufacturing and Repairing Technologies in Vehicle Industry
10. SZALAY ZSOLT, VIKTOR TIHANYI, 2017, *Zalaegerszegi tesztpályához kapcsolódó kutatás-fejlesztési területek,* IFFK 2017: XI. Innováció és fenntartható felszíni közlekedés
11. IVAN EVTIMOV, KEVIN EYKHOLT, EARLENCE FERNANDES, TADAYOSHI KOHNO, BO LI, ATUL PRAKASH, AMIR RAHMATI, AND DAWN SONG, 2017 *Robust Physical-World Attacks on Deep Learning Models,* Computer Vision and Pattern Recognition (CVPR 2018) (supersedes arXiv preprint 1707.08945, August 2017)
12. ZSOLT SZALAY, TAMÁS TETTAMANTI, DOMOKOS ESZTERGÁR-KISS, ISTVÁN VARGA, CESARE BARTOLINI, 2017, *Development of a Test Track for Driverless Cars Vehicle Design, Track Configuration, and Liability Considerations,* Periodica Polytechnica Transportation Engineering Vol 46 No 1 (2018)
13. D. FÖLDES, Cs. CSISZÁR, *Az autonóm városi személyközlekedés hatásai,* 2016, Magyar Mérnökakadémia (MMA)
14. Cs. CSISZÁR, A. ZARKESHEV, *Demand-capacity coordination method in autonomous public transportation,* 2017, Transportation Research Procedia 27, 784-790
15. A. MUNAWAR, C. CREUSOT, *Structural Inpainting of Road Patches for Anomaly Detection, 2015,* MVA2015 IAPR International Conference on Machine Vision Applications
16. DPA/MUENCHEN.DE, *Schilder mit geometrischen Figuren auf schwarzem Kreis,* 2016
17. A. HECHRI, *Lanes and Road Signs Recognition for Driver Assistance System, 2011,* IJCSI International Journal of Computer Science Issues, Vol. 8, Issue 6, No 1
18. Z. FAZEKAS, P. GÁSPÁR , *Computerized Recognition of Traffic Signs Setting Out Lane Arrangements,* 2015, Acta Polytechnica Hungarica
19. MERDECES'S ENGINNER, *On the right track: the way to autonomous driving,* Last access time: 2018.01.03.
URL: https://www.mercedes-benz.com/en/
20. HVG, *A britek kifejlesztették a gyalogátkelőhelyet, ami ezrek életét mentheti meg évente,* Last access time: 2018.06.03.
URL: http://hvg.hu/cegauto/20171101_direct_line_okos_gyalogatkelohely_video

IOP Conf. Series: Materials Science and Engineering **448** (2018) 012047 doi:10.1088/1757-899X/448/1/012047

Structural Changes in Polymethyl Methacrylate during Laser Machining

L Sýkorová[1], J Knedlová[1], V Pata[1], M Kubišová[1*]
[1]Faculty of Technology, Tomas Bata Univerzity in Zlin, Vavrečkova 275, 46001 Zlín. Czech Republic.
E-mail: sykorova@utb.cz, knedlova@utb.cz, pata@utb.cz, *mkubisova@utb.cz

Abstract: The focus of the article is an issue of laser machining of polymethyl methacrylate materials. Specifically, it deals with evaluation of structural changes in polymers as due to the effect of concentrated energy there appears vaporization of the material in the place of cut as well as heat transfer into the material around the place of cut and thus to the heating. As a result of the raised temperature, different structural changes may occur in the material; some high-molecular substances begin to break down into low-molecular substances, and depolymerization or degradation of the material may occur. All these changes may result in a change of the final properties of the product. For experimental machining and subsequent study using the "Fourier Transform Infrared" vibration spectroscopy, PMMA was chosen as it is one of the most feasible polymeric materials in this technology.

Keywords: Laser Technology, Structural Changes, Polymeric Material, Vibration Spectroscopy.

1 Introduction

At material machining, a laser is commonly used as a source of an intense electromagnetic radiation, which is directed to a particular object. The effect of the laser depends on the characteristics of the workpiece material (its physical properties, structure, etc.), the parameters of the laser itself; next, it is also affected by the interaction time between the material and the laser beam. In terms of machining, surface quality assessment and possible changes around the cut, the part of the laser beam that is absorbed by the material is a key aspect with respect to the material. [1].

Polymeric materials are brittle at normal temperatures; with more load and stress they can crack. With increasing temperature, the polymers become more ductile and elastic. Therefore, the properties of these materials are temperature-dependent. In some temperature areas, changes occur faster, or they may occur in fits and starts. These areas are characterized by the transition temperature Tg (glass transition temperature), Tf (flow temperature), which gives us the flow temperature, shows a large increase in deformation and is irreversible (see in Fig. 1). However, changes in the chemical structure of the polymer occur only behind the temperature of Tf, in the areas before the evaporation temperature when the material evaporates. [2].

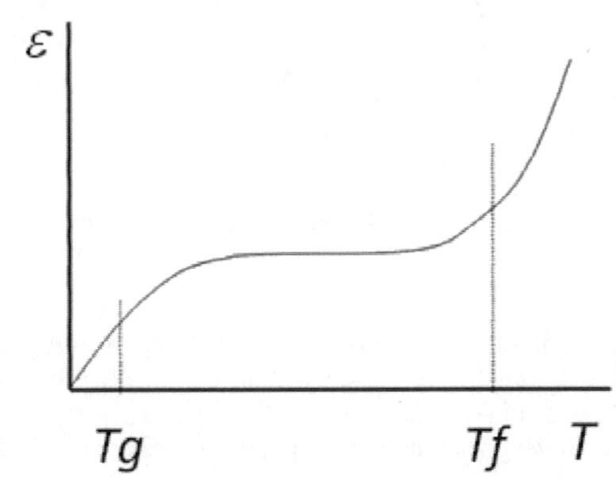

Figure 1 The thermomechanical curve of the amorphous polymer

IOP Conf. Series: Materials Science and Engineering **448** (2018) 012047 doi:10.1088/1757-899X/448/1/012047

Because of the elevated temperature, the polymers begin to soften and, eventually, the kinetic energy of the macromolecule chain can increase until it overcomes the intermolecular forces and the polymer obtains properties that are very close to the properties of the liquid. There may also appear a change in the structure of the polymers by starting some high-molecular substances to break down into low-molecular products. Sometimes monomer cleavage can occur. However, this may happen without changing the chemical composition of a particular polymer. The tendency to depolymerize up to the monomer without changing the chemical composition has also PMMA. The depolymerization process takes place by a chain reaction in which the monomer is released from the ends of the macromolecular chains. [2].

2 Experimental part

Because of the elevated temperature, the polymers begin to soften and, eventually, the kinetic energy of the macromolecule chain can increase until it overcomes the intermolecular forces and the polymer obtains properties that are very close to the properties of the liquid. There may also appear a change in the structure of the polymers by starting some high-molecular substances to break down into low-molecular products. Sometimes monomer cleavage can occur.

However, this may happen without changing the chemical composition of a particular polymer. The tendency to depolymerize up to the monomer without changing the chemical composition has also PMMA. The depolymerization process takes place by a chain reaction in which the monomer is released from the ends of the macromolecular chains. [2].

2.1 Polymethylmethacrylate – PMMA

Acrylic PMMA is a transparent, synthetic, highly transparent thermoplastic polymer that is mainly obtained by polymerization of methyl methacrylate. Its chemical formula and is evident from Figure 2.

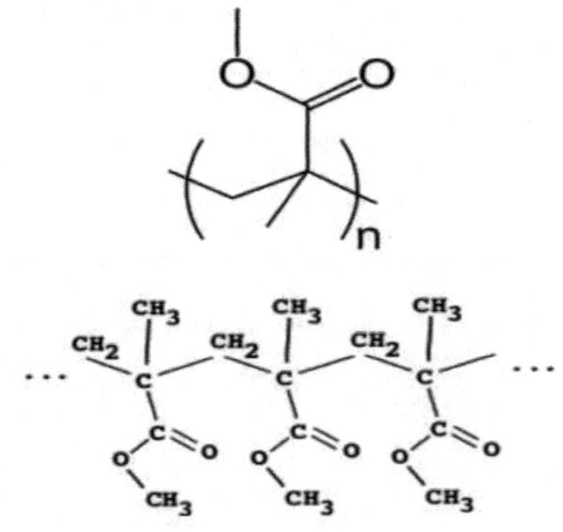

Figure 2 – PMMA formula

PMMA transmits light in 92% across the spectrum, reaching up to the UV range. It has excellent mechanical properties, such as water resistance and electrical insulation qualities and is also wholesome. A small surface resistance is considered to be its deficiency as it gets scratched easily.

It is applied in more complex machine parts, often in replacing glass - "glazing" of transport means, protective covers of instruments and machines, glasses for watches, office supplies. It is also used in medicine, for example in dentures. [3].

2.2 Sample Preparation

In PMMA samples, the grooves were cut by the EUROLASER LCS XL – 1600 (CO_2 Laser, wavelength 9,3μm) under various technological conditions (varying power and feed rate, see Table I), marked with digits 1 to 5 (see in Fig. 3). These grooves were then cut through by a band saw to allow access to the place of cut. FTIR measurements were performed in individual grooves. To compare the FTIR results, a base material untreated by the laser beam was set.

Table I Technological parameters

Cut	Power [W]	Feed rate [mm/s]
1st groove	255	12
2nd groove	255	17
3rd groove	255	35
4th groove	255	25
5th groove	225	45

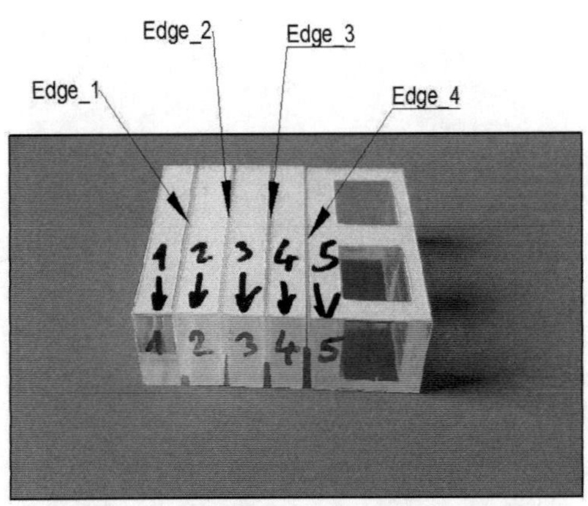

Figure 3 Sample with individual cuts

2.3 Surface Evaluation Using FTIR Infrared Spectroscopy

This is a non-destructive analytical method that allows, based on different interaction of infrared radiation, to identify and structurally characterize organic and inorganic compounds in a given substance. It is generally possible to analyse gaseous, solid and liquid substances. The principle of this method consists in examining the absorption of IR radiation when passing through the material; it changes the rotational vibrational energy state of the molecules, depending on the transformation in the dipole moments of the molecules of the substance. Each atom present in the molecule exhibits vibration around its equilibrium position; the frequency of this vibration is determined not only by the mass of the particles involved, but also by the strength of the bonds.

The frequency of radiation can resonate on certain frequencies with molecular vibrations, so the energy of this radiation can thus be absorbed. Substances that do not absorb IR radiation (for example oxygen, nitrogen, hydrogen, graphite or sulphur) cannot be analysed by this method.

The result of the measurement is a graphical representation of the energy dependence expressed as absorbance at the wavelength of the incident radiation. The absorption intensity is usually described as a percentage, with IR unit being the so-called reciprocal centimetre (cm^{-1}). It is true that each substance under

IOP Conf. Series: Materials Science and Engineering **448** (2018) 012047 doi:10.1088/1757-899X/448/1/012047

investigation has a unique spectrum. [4].

FTIR spectroscopy enables to distinguish the molecular structure of the materials based on their various absorption in the infrared range of electromagnetic radiation. For FTIR, a Germanium crystal was used in the form of a tip touching the measured surface under constant pressure. The measured area was 100 μm x 100 μm.

Measurements were performed at 25 ° C, atmospheric pressure 1016,4 hPa and huminitydy 45 %.

3 Results and Evaluation

The curves have a characteristic geometry of PMMA. The dark blue colour shows the curve for the groove 1 (Edge_1), the green curve belongs to the groove No. 2 (Edge_2), the orange is the groove No. 3 (Edge_3), the light blue includes the groove No. 4 (Edge_4) and the red is the 5th groove. The X axis shows the wavelength in cm^{-1} while the Y axis represents the absorption (see in Fig. 4).

When comparing the geometry of the curves 1-5, it can be stated that in terms of qualitative material changes of the machined surface, these changes were not proven by FTIR. All samples (grooves 1 to 5) have almost the same infrared spectra. Thus, based on the results, it can be concluded that changes of the technological parameters (the power and feed rate), do not affect the molecular bond in the examined material.

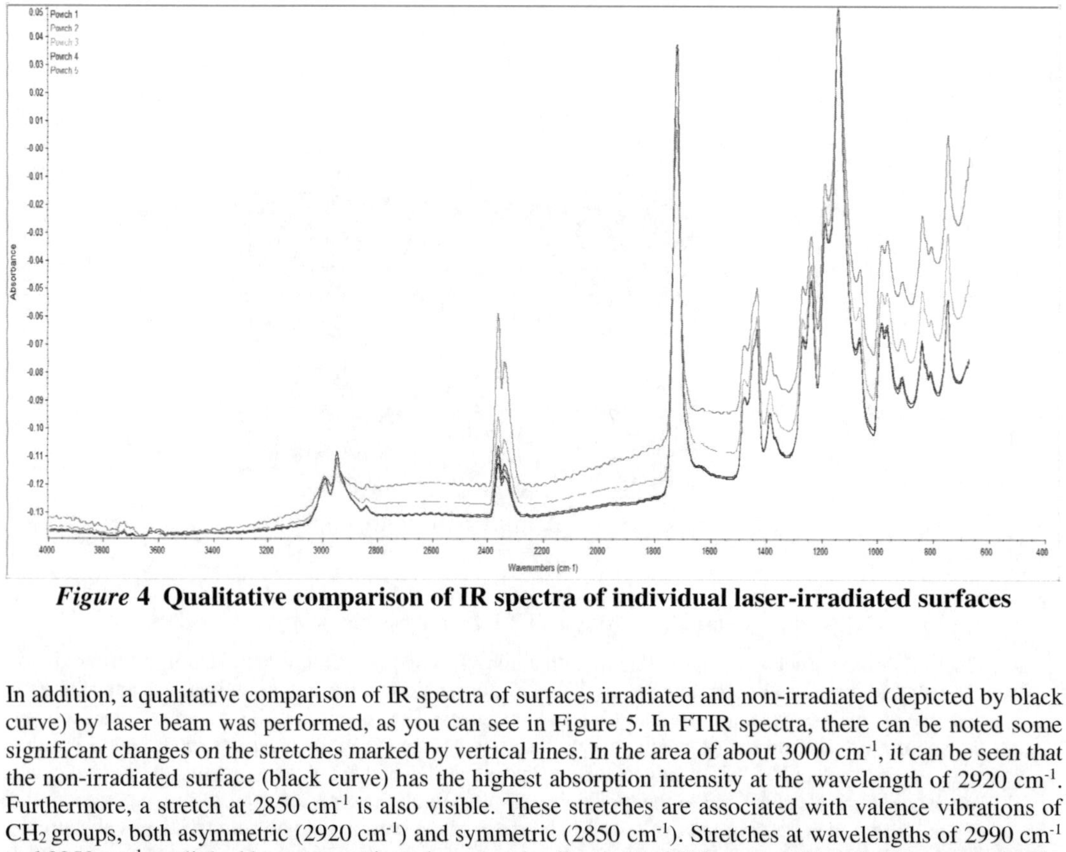

***Figure* 4 Qualitative comparison of IR spectra of individual laser-irradiated surfaces**

In addition, a qualitative comparison of IR spectra of surfaces irradiated and non-irradiated (depicted by black curve) by laser beam was performed, as you can see in Figure 5. In FTIR spectra, there can be noted some significant changes on the stretches marked by vertical lines. In the area of about 3000 cm^{-1}, it can be seen that the non-irradiated surface (black curve) has the highest absorption intensity at the wavelength of 2920 cm^{-1}. Furthermore, a stretch at 2850 cm^{-1} is also visible. These stretches are associated with valence vibrations of CH_2 groups, both asymmetric (2920 cm^{-1}) and symmetric (2850 cm^{-1}). Stretches at wavelengths of 2990 cm^{-1} and 2950 cm^{-1} are linked by asymmetric and symmetric vibrations of CH_3 groups. It can be seen that with the irradiation of the surface, the stretch intensity decreases to 2920 cm^{-1}, almost disappears, and then only appears as a stretch edge of 2950 cm^{-1}. The 2850 cm^{-1} stretch almost completely vanishes with the surface irradiation.

This phenomenon can be attributed to the material degradation and chain shortening with surface irradiation, resulting in a decrease of CH_2 groups in the spectrum and an increase of terminal CH_3 groups. Another observable phenomenon is that the 1475 cm^{-1} wavelength, which belongs to the deformation vibrations of CH_3 groups, has only a very little noticeable edge in the non-irradiated surface spectrum. On the other hand, irradiated surfaces show a totally differentiated (though intensely small) stretch. This could confirm the consideration of the CH_3 moieties in the material and the possible degradation of the -CH2- groups. The 1720 cm^{-1} stretch is typical for esters (bond - (CO) -O-). It is a valency vibration. Furthermore, in the region of about 1100 cm^{-1} there is a stretch for the deformation vibration of this bond.

Thus, it can be stated that structural changes in the material caused by laser radiation have been demonstrated using vibrational spectroscopy - FTIR spectroscopy.

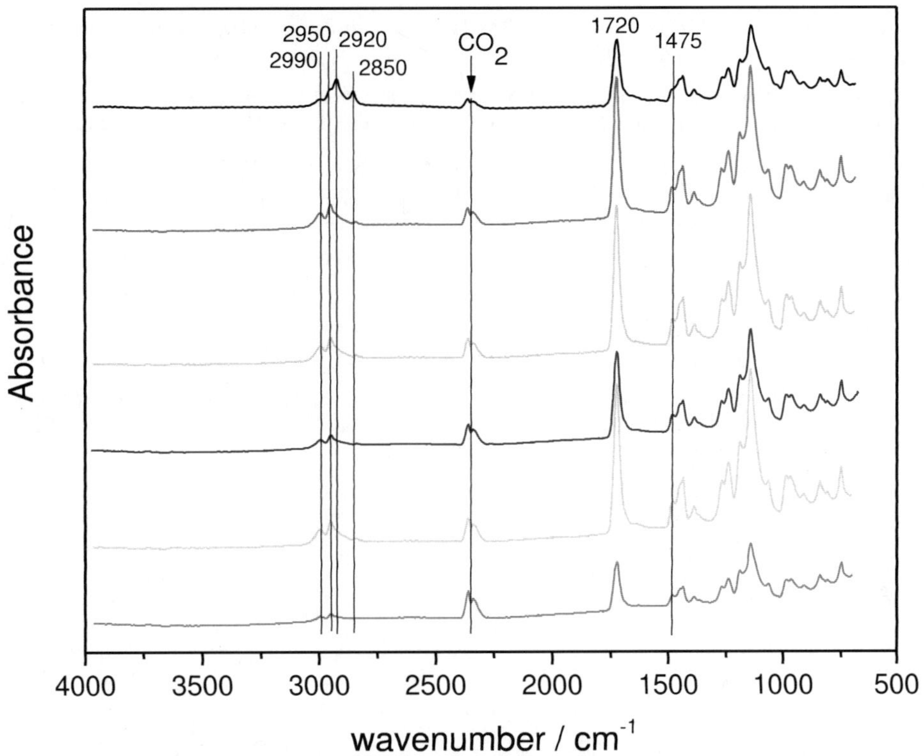

Figure 5 Qualitative comparison of IR spectra of individual laser cut surfaces with non-irradiated surface

4 Conclusion

Nowadays, the laser technology is an indispensable part in a wide range of industries, including engineering, medicine, research and others. In spite of all its advantages, however, it concerns a high-energy source of radiation; thus, the heat-affected zone at the cutting area occurs during machining. The size of this area varies, depending on many factors, such as laser wavelength, technological parameters, and the thermal conductivity of the workpiece plays a big role here. Although plastics belong among materials with a low thermal conductivity, there is also a very high temperature at the cutting point for a short time, and the heat spreads into the internal volumes of the material. As mentioned above, due to the raised temperature, various undesirable changes may happen in the material, even if they occur only in close proximity to the cutting point. [4], [5], [6].

The PMMA was selected for experimental machining and subsequent study using the Fourier Transform Infra Red Vibration Spectroscopy. A qualitative comparison of the IR spectra of surfaces was performed, on which a laser beam with a non-irradiated surface was applied. There are some significant changes in the FTIR spectra of the irradiated sample on the vertical lines, which suggests.

Acknowledgement
This work was supported by an internal grant of UTB in Zlín IGA/FT/2018/004 financed from Funds of specific academic research.

References

[1] VASILE, C., KULSHRESHTHA, A. K. (2003). Handbook of Polymer Blends and Composites, pp. 389 – 460. Rapha Technology. UK.

[2] CONNES, J.: CONNES, P. (2016). Near-Infrared Planetary Spectra by Fourier Spectroscopy. In: Journal of the Optical Society of America, Vol. 56, No. 7, pp. 896-210, OSA Publishing. America.

[3] VAJDOVÁ, A., MIČETOVÁ, A., NESLUŠAN, M., KOLAŘÍK, K. (2014). Analysis of Surface Integrity of Parts after Non-conventional Methods of Machining. In: Manufacturing Technology, Vol. 14, No. 3, pp. 470 - 474, UJEP, Czech Republic.

[4] Felho, C., Kundrák, J. (2014). Characterization of Topography of Cut Surface Based on Theoretical Roughness. In: Key Engineering Materials, Vol. 581, pp. 470 - 474, Switzerland.

[5] PATA V., SÝKOROVÁ, L., ŠUBA, O., Knedlová, J. (2018). The Influence of Laser Beam technological paramters on the Polymethyl methacrylate Surface Quality. In: NTPDS IV, Vol. 919, pp. 190 - 198.

[6] Major, A. A., & Belina, K. (2013). Investigation of the structure of multiwall carbon nanotubes in polymer matrix. Paper presented at the IOP Conference Series: Materials Science and Engineering, , 47(1) doi:10.1088/1757-899X/47/1/012031

IOP Conf. Series: Materials Science and Engineering **448** (2018) 012048 doi:10.1088/1757-899X/448/1/012048

Utilization of cluster analysis in the multi-dimensional evaluation of mould surface quality in product correlation

V Pata[1], M Kubišová[1]*, L Sýkorová[1], O Šuba[1], L Hýlová[1]
*mkubisova@utb.cz

[1] Tomas Bata University in Zlín, Faculty of Technology, Department of Production Engineering, Vavrečkova 275 760 01 Zlín, Czech Republic

Abstract. Quality assessment of mould surfaces, of the products they produce is often done in both scientific and industrial practice separately and without considering the links of the individual descriptive parameters.
The present article will describe how to find amplitude, frequency parameters that will be grouped into so-called parametric sets. Each parameter set will be characterised by its values for both the mould and the product it produces.
The final step is to apply cluster analysis and to determine the similarity of the individual sets of parameters between the form and its product.
Therefore, the presented methodology will provide complex correlation evaluation of products and mould not by individual parameters but within the whole set of parametric sets.

1. Introduction

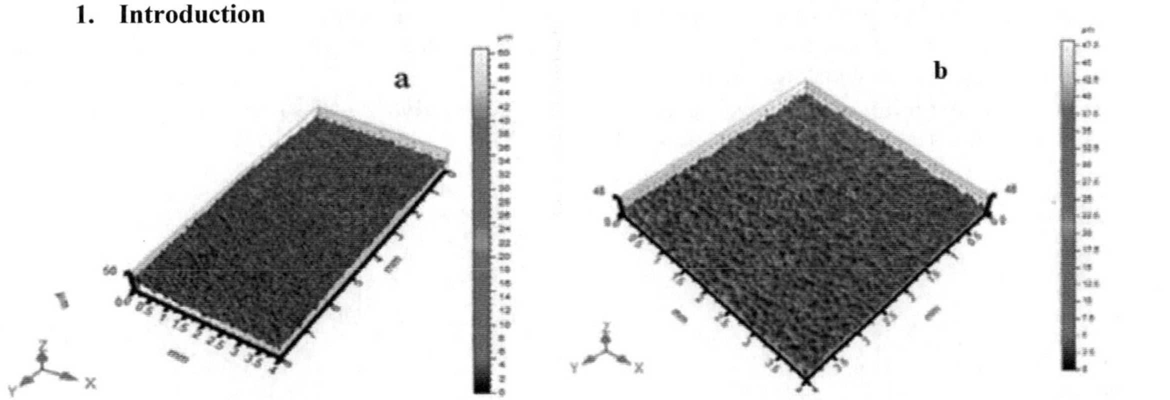

Figure 1 3D surface a) form, b) product

A classic approach to assessing compliance is to use hypothesis testing, individual parameters independent of each other. This, however, leads to the following problem, for example, I do not ignore the match of the Rp and Rv parameter between the form and the product, but I will reject the match between Parameters Rz, which is the sum of Rp and Rv. The same problem is even when using 3D parameters, such as SP, Sv and Sz [1].

However, scientific practice requires a much more complex approach, such as solving the question of the consistency between the set of Ra, Rz, Rp and Rv parameters between form and product.

Here is the possibility to directly visualise the form (the form is small, and the weight is low) and it is rarely possible to capture this form. After the mould and product have been developed at an equivalent site, it can be said that the parameter has very different results [1,3].
Looking at the statistical results obtained directly through the TalyMap program, we see differences in diameters and direction. The deviations generally differ in tenths of microns to units of micrometres. However, we are not able to determine with what probability the results have been determined, whether they contain gross errors, whether the skewness or spike of the individual parameters differs significantly from one another and above all the data come from the normal distribution [1,4].
This type of results, which is often presented at scientific conferences, can only be considered as indicative in view, similar to 3D graphics scenes.
If we evaluate the parameters precisely in the sense of mathematical and statistical rules, we find that the parameters of skew and spikes fluctuate at a 95% confidence level between insignificant and significant differences from the normal distribution. Normality is often not observed at all. Such types of heterogeneous data cannot be solved using classical averages and direction. Deviation (as is often the case), but it is necessary to use data transformations that give much more accurate results [1,8].

2. Material and method

If we want to evaluate complex amplitude and frequency parameters, between form and its product, it is appropriate to use multidimensional statues, methods, in particular, factor analysis and cluster analysis.

Subsequently, the use of factor analysis, in particular, the diagram of the footnote of own numbers. On this diagram, it is quite clear that when using the Kaiser criterion, only two factors will be sufficient to describe all parameters, as is evident from the graph of factor loads. Thus, using factor analysis, namely the Varimax method, we can state that the parameters describing the surface quality of the mould and the product form 2 separate clusters that differ in their external correlation.
Both clusters are factor-net - they have a high projection value in the factor axis and the minimum projection value in the axis of the opposite factor [1,2].

Finally, after factor analysis, we can assert that for the description of 8 surface quality parameters it is sufficient to describe only two factors [1,5].

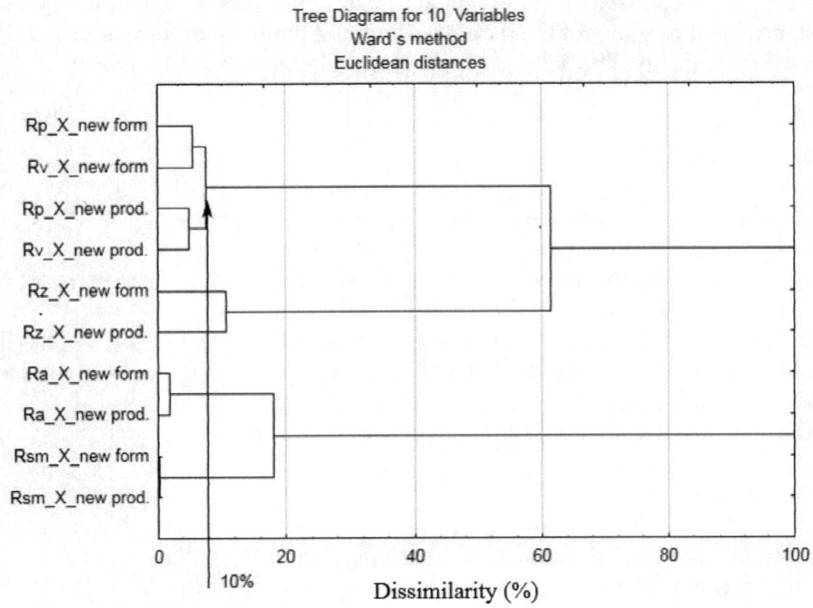

Figure 2 cluster analysis for the new form and its product

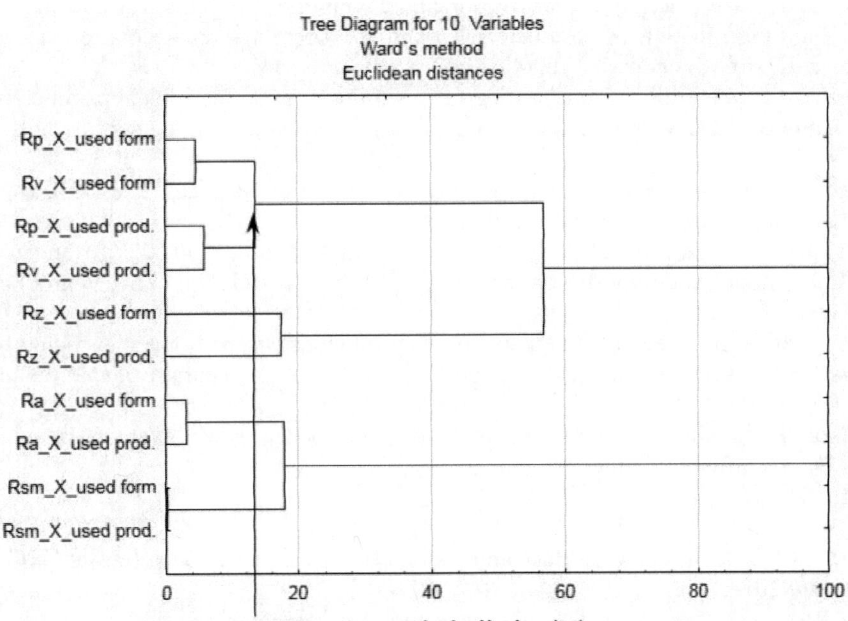

Figure 3 cluster analysis for the form used and its product

This problem helps us solve cluster analysis, specifically Ward's method, using the so-called Euclidean distance calculation.

Clusters were found between the parameters describing the surface quality of the new mould and the product of the mould.

It is possible to state, based on the distance matrix, the degree of similarity of the individual parameters describing the quality of the mould surface and the product formed from that place.

As a measure of quality, the increasing or decreasing value of Euclidean distances, which is dimensionless, can be used between individual parameter aggregates.

It is also possible to see here for a better understanding of the binding of individual parameters merged into corresponding clusters.

For example, it can be seen that the cluster of parameters Rp and Rv measured on the product of the new form is similar to the cluster of parameters Rp and Rv in the new form.

This disparity is expressed either by dimensionless numbers or by percentage. Looking at our dendrogram, we can say that the unity of the first and second clusters of the new form and its product is approximately 10%. (90% similarity for Rp and Rv) (see fig.2).

In the case of a product of the mould used, this same cluster is dissimilar to 15% (see fig.3), and it can be stated that the parameters Rp and Rv of the mould and product used are less similar to those of Rp and RV in the new form of product

Similarly, the percentage disparity is expressed for the other clusters.

3. Conclusion

Using cluster analysis, specifically Ward's method, using the so-called Euclidean distance calculation, clusters were found between the parameters describing the surface quality of the new mould and the mould used and the products formed in these forms.

Within individual clusters, we can say that the parameters Rp, Rv and Ra respectively form similar clusters based on the calculation of distance by the Euclidean method.

The similarity of parameters based on arithmetic averages and parameters generated by summaries, expressed by the distance in the assessment of the worn form and the product from this worn form.

Parameters Rz, which are the function of parameters Rp and Rv, form a separate cluster that is unlike the parameters Ra and Rsm, which are the function of arithmetic means.

Based on the distance matrix, it is possible to state the degree of similarity of the individual parameters describing the surface quality of the new mould and the resulting product (4.3) and the surface of the mould and product used (4.4).

As a measure of quality, the increasing or decreasing value of Euclidean distances, expressed as a percentage, between individual parameter aggregates can be used.

Factor analysis has shown that the product varies from one form to another, but the differences have not been proven. It has only been demonstrated that there is a higher correlation between product and form related parameters.

This means that the methodology recommended by prof. Whitehouse is no longer sufficiently prominent and only applies to surfaces that are scanned in the form of separate cuts (contact profilometer).

For spatial (non-contact) scanning, this method is inadequate because we evaluate several parameters that have arisen in a different mathematical way.

References

[1] WHITEHOUSE, D. J., 2011. Handbook of surface and nanometrology. 2nd ed. Boca Raton: CRC Press. ISBN 978-1-4200-8201-2.

[2] BILODEAU, M. A BRENNER D. Theory of multivariate statistics. New York: Springer, c1999. Springer texts in statistics.

[3] REISS, ROLF-DIETER AND M. THOMAS. Statistical analysis of extreme values: with applications to insurance, finance, hydrology and other fields. 2nd ed. Basel: Birkhäuser Verlag, 2001. ISBN 3-7643-6487-4.

[4] BÍLEK, O. Cutting tool performance in end milling of glass fibre-reinforced polymer composites. Manufacturing Technology - Journal of Science Research and Production [online]. 2016, 16(1), 12-16 [cit. 2017-03-21]. ISSN 12132489.

[5] JINFENG ZHANG, CHAO FENG, YUNHUI MA, WEI TANG, SHUAI WANG, XIN ZHONG. Non-destructive analysis of surface integrity in turning and grinding operations (2017) Manufacturing Technology, 17/3, pp. 412-418.
[6] HANZL P., ZETKOVÁ I., MACH J. Optimization of the Pressure Porous Sample and Its Manufacturability by Selective Laser Melting (2017) Manufacturing Technology, 17/1, pp. 34-38.
[7] E. HNATKOVA, D. SANETRNIK, V. PATA, B. HAUSNEROVA, ZDENEK DVORAK. Mold Surface Analysis after Injection Molding of Highly Filled Polymeric Compounds. (2016) Manufacturing Technology, 16/1, pp. 86-90.
[8] JAN PODANÝ, ALEXEY MOLOTOVNIK. 3D Measurement of Surface Texture Parameters (2014) Manufacturing Technology, 14/4, pp. 596-600.

Acknowledgment
This work was supported by an internal grant of UTB in Zlín IGA/FT/2018/004 and IGA/FT/2018/012 financed from Funds of specific academic research.

IOP Conf. Series: Materials Science and Engineering **448** (2018) 012049 doi:10.1088/1757-899X/448/1/012049

The possibility of statistically eliminating systematic errors of replicated mould surfaces in evaluating their quality.

Abstract. When assessing the quality of the mould surfaces, the basic problem arises that, for reasons of dimensional form, it is often impossible to carry out a quality assessment of its surface. A surface impression of mould, called a replica, is then performed, which is then evaluated instead of the original. However, the question remains, how exactly the replica is made and, above all, how it transmits information about the mother surface. This article describes a method of statistical evaluation and, above all, a method of assessing the correlation accuracy of the mould surface and its replica at the individual parameter level and by ISO 4287 and ISO 4288.

V Pata[1], M Kubišová[1]*, L Sýkorová[1], O Šuba[1], L Hýlová[1]
*mkubisova@utb.cz,

[1] Tomas Bata University in Zlín, Faculty of Technology, Department of Production Engineering, Vavrečkova 275 760 01 Zlín, Czech Republic

1. Introduction

The article brings new directions and methods of evaluation. For the production of replicas are used impression compounds used mainly in dentistry. Therefore, in the field of dentistry, we are looking not only for the necessary inspiration for the choice of materials suitable for surface sensing but also for knowledge in the field of preparation and the application of impression materials themselves.
The Talysurf CLI 500 is used to evaluate the basic parameters of roughness (Ra, Rz, Rp, Rv and Rt) both on the original surface and on the surface replicated. Subsequently, these parameters are compared using statistical methods. Due to the size of the mould, it was possible to carry out complete testing and thus proof of complete data. The replication of the moulds was based on the burning need to devise a way of testing large forms that cannot be easily removed and compared.
For replication, it is very important to ensure stable temperature conditions and qualified service to ensure repeatability.
If it is a large articulated form, it is necessary to get a form of surface etalon from the manufacturer so that it is possible to perform the necessary control through replication [1,4,7].

IOP Conf. Series: Materials Science and Engineering **448** (2018) 012049 doi:10.1088/1757-899X/448/1/012049

2. Materials and methods

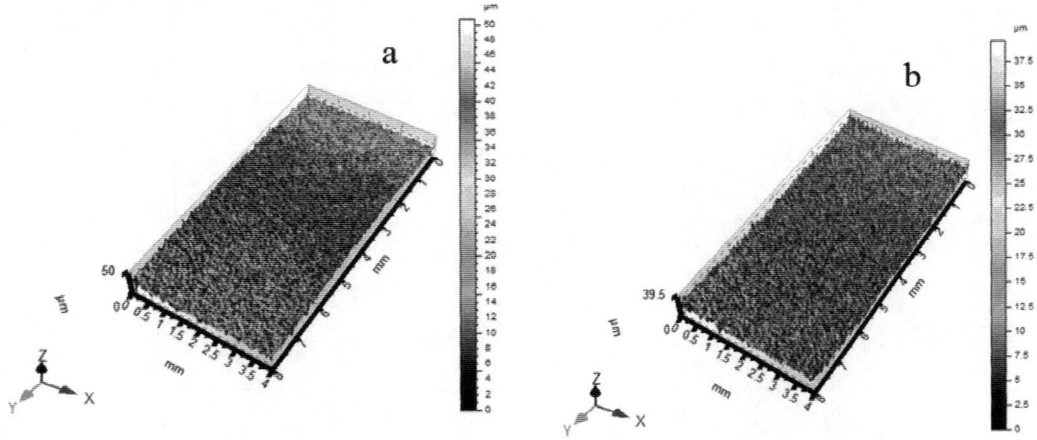

Figure 1 a) Original surface, b) replicated surfac*e*

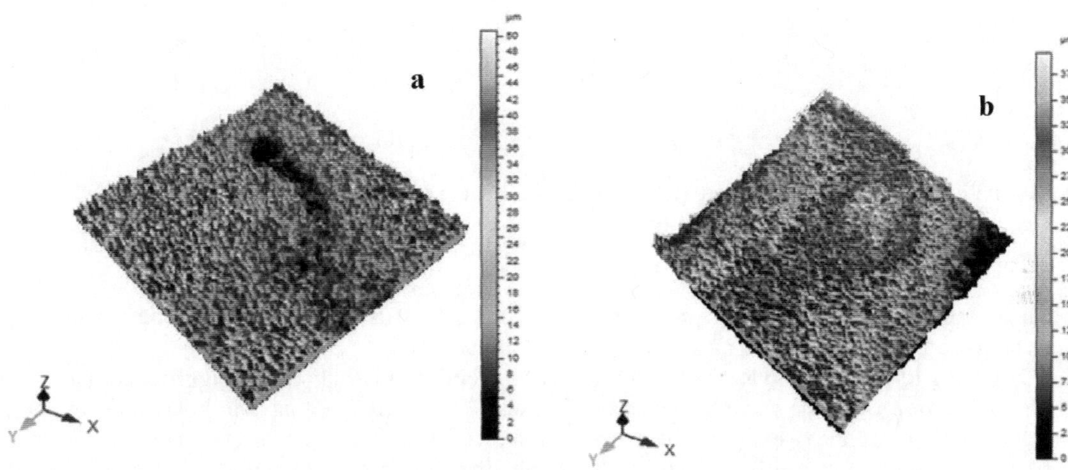

Figure 2 3D scan replica a) new form, b) old form

From the measured surface profile, the shape and subsequently the waviness was first removed, resulting in a roughness profile according to ISO 4287 and ISO 4288, which was then drawn in sections in the Y-axis (North-South) and especially in the X-axis (East-West). In both axes, 252 slices were created. Based on ISO 25178, the 3D surface profile was calculated. In conclusion, roughness parameter values were generated according to ISO 4287 [1,3,5]

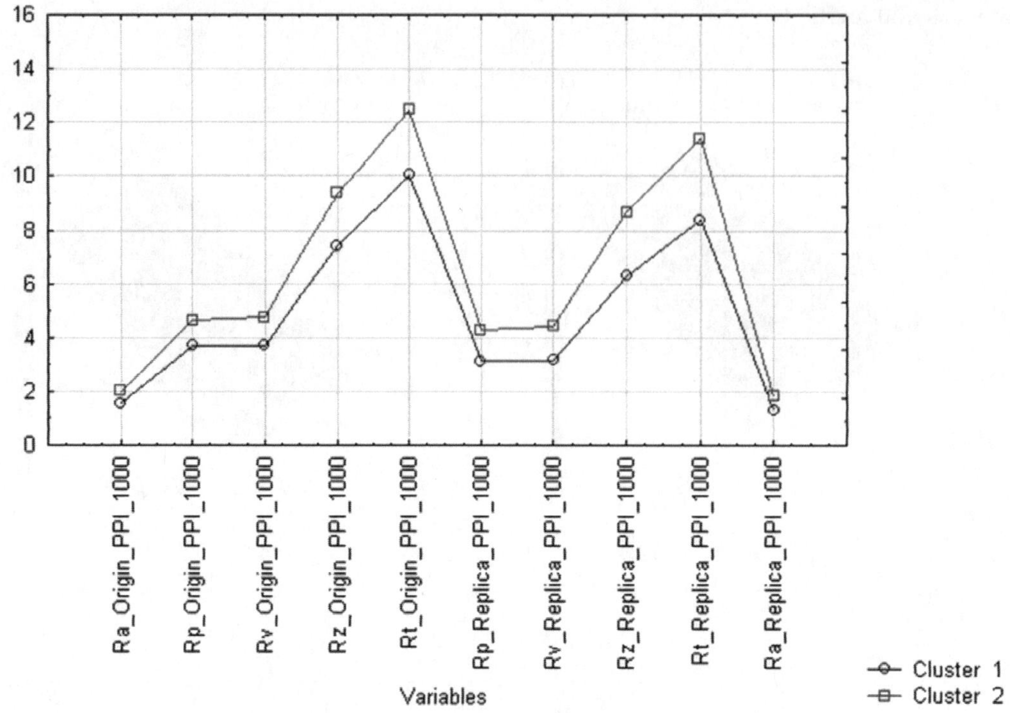

Figure 3 Plot of means for each cluster

All parameters (Ra, Rz, Rp, Rv and Rt), were divided into two basic clusters, where the number 1 denotes the original surface parameter cluster, on the other hand, the parameter number 2 for the surface replicated (see Fig. 3).

As can be seen from the above clustering graph (see Fig. 4), it is possible to state the closeness of individual parameters within own clusters number 1 and number 2.
Instead of using Rp, Rv, Ra and Rz for the surface and its replica, it is statistically significant to use only two factors that represent the original eight parameters to 72%. We do not have to judge one parameter against another (Rz replica versus Rz form). We create two clusters using factor analysis, which we describe using factors. As can be seen from the graph "the footnote of the numbers", it is possible to state that the variability of all parameters for the form and its replica can be expressed using only two factors, which express 72.13 % variability.
Therefore, we call factor 1 as a factor describing the surface properties of the new form expressed by amplitude parameters and "factor 2" by a factor describing the surface properties of a replica of a new form expressed by amplitude parameters [1,2,9]

2.1 Factor analysis Quartimax normalised for the new form and its replica

Factor 1 - describing the amplitude characteristics of the new form.
Factor 2 - describing the amplitude characteristics of the replica of the new form.

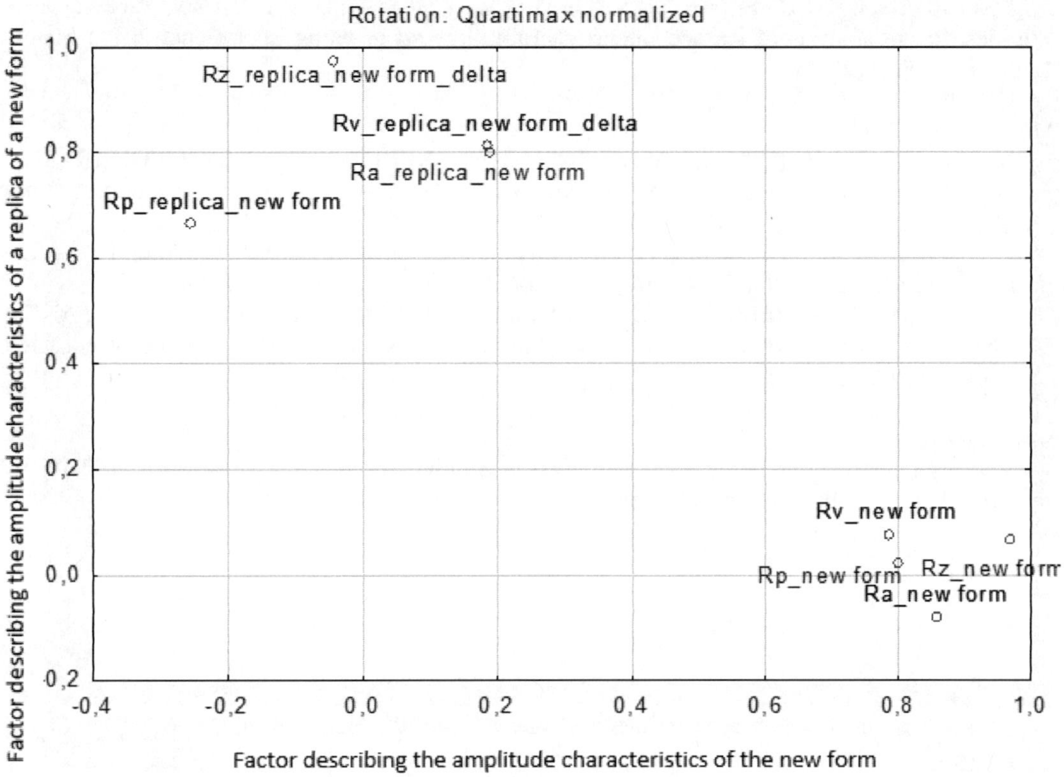

Figure 4 Factor analysis

3. Conclusion

By evaluating the amplitude parameters: Ra, Rz, Rp, and Rv in new forms and replicas of the new form using factor analysis and 3 types of rotation, namely Varimax normalized, Quartimax normalized, and Equamax normalized, we can state that in all these cases, which are factor-net and belong to individual factor axes. Thus, we can claim that by the individual clusters' affinity to different factors (axes), the global surface of the form (expressed by amplitude parameters) of the statistics is significantly different from the surface of the replica of this form. In this case, the rotation is meant to find the greatest covariance between the individual clusters.

References

[1] WHITEHOUSE, D. J., 2011. Handbook of surface and nanometrology. 2nd ed. Boca Raton: CRC Press. ISBN 978-1-4200-8201-2.
[2] BILODEAU, M. A BRENNER D. Theory of multivariate statistics. New York: Springer, c1999. Springer texts in statistics.
[3] REISS, ROLF-DIETER AND M. THOMAS. Statistical analysis of extreme values: with applications to insurance, finance, hydrology and other fields. 2nd ed. Basel: Birkhäuser Verlag, 2001. ISBN 3-7643-6487-4.
[4] BÍLEK, O. Cutting tool performance in end milling of glass fibre-reinforced polymer composites. Manuf. Tech. - Journal of Science Research and Production [online]. 2016, 16(1), 12-16 [cit. 2017-03-21]. ISSN 12132489.

[5] JINFENG ZHANG, CHAO FENG, YUNHUI MA, WEI TANG, SHUAI WANG, XIN ZHONG. Non-destructive analysis of surface integrity in turning and grinding operations (2017) Manuf. Tech., 17/3, pp. 412-418.
[6] HANZL P., ZETKOVÁ I., MACH J. Optimization of the Pressure Porous Sample and Its Manufacturability by Selective Laser Melting (2017) Manufacturing Technology, 17/1, pp. 34-38.
[7] E. HNATKOVA, D. SANETRNIK, V. PATA, B. HAUSNEROVA, ZDENEK DVORAK. Mold Surface Analysis after Injection Molding of Highly Filled Polymeric Compounds. (2016) Manufacturing Technology, 16/1, pp. 86-90.
[8] JAN PODANÝ, ALEXEY MOLOTOVNIK. 3D Measurement of Surface Texture Parameters (2014) Manufacturing Technology, 14/4, pp. 596-600.
[9] MANAS, D., OVSIK, M., MIZERA, A., MANAS, M., HYLOVA, L., BEDNARIK, M., STANEK, M. The effect of irradiation on mechanical and thermal properties of selected types of polymers (2018) Polymers, 10 (2), art. no. 158.

Acknowledgements
This work was supported by an internal grant of UTB in Zlín IGA/FT/2018/004 and IGA/FT/2018/012 financed from Funds of specific academic research.

IOP Conf. Series: Materials Science and Engineering **448** (2018) 012050 doi:10.1088/1757-899X/448/1/012050

Improvement of adhesive joining of hybrid aluminum – GFRP using surface modifications

M Berczeli[1], Z Weltsch[1]

[1]John von Neumann University Department of Material Science,

berczeli.miklos@gamf.uni-neumann.hu

Abstract. One of the biggest research discipline todays is the weight reduction of the vehicles. The manufacturers are focusing on making new hybrid materials and innovative technologies that satisfy the newest requirements, such as light weighting. There exist many solutions to reach lower vehicle weight, reduce the air pollution and CO2 emission. One of them is the hybrid materials pairing in the different part of the vehicle body. These joining technologies have to complete the new requirements for hybrid joints for this the weakest part of the bonding have to increased. The strength of the adhesive joining can be increased with different surface preparation methods. In this research five different surface treatment were investigated to increase the tensile stress of the overlapped Glass Fiber Reinforced Plastic and Aluminum sheets. The macro geometry of the connecting materials, the tension stress and the breaking method were investigated during the analysis. The research investigates the connection between the surface properties and the adhesive technology using hybrid materials connection.

1. Introduction

Nowadays the vehicle industry has certainly begun to show signs of rebounding from the economic downturn, however, companies needs to "do more with less" as production volumes approach the levels of several years ago. Over the decades, these numbers of the new vehicles show a negative impact on the environment and human health. The increasing air pollution from the vehicle sector has led many government agencies to lay strict regulations on the automobile manufacturers to curb the harmful emissions under permissible limits [1].

Over years, the old, standard vehicles get more and more new features, new luxury elements, higher power and higher safety, these buyer requirements cause the increasing of the car weight.

Figure 1 shows the increasing of the weight for example on a Suzuki swift from 785 kg to 1200 kg over the years. The EU agency has set mandatory emission reduction targets for the automaker companies and this cause a breakpoint in the increasing weight. According to EU rules, the fleet average by cars to be achieved by 2020 must be 95°gramms of Co2/kilometer, which works out to a fuel consumption rate of around 4.1 liter/100 km of petrol or 3.6 liter/100 kilometer of diesel [2].

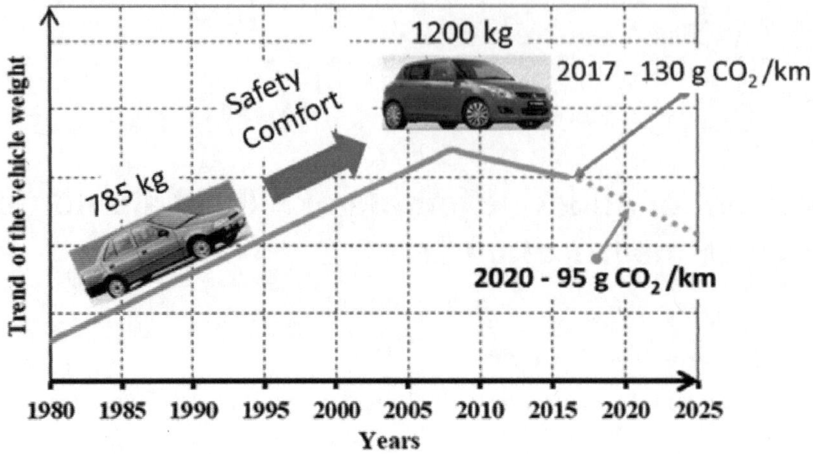

Figure 1. Trend of the vehicles weight over years [2].

The weight reduction can be reached with new, innovative, hybrid material usage in the vehicles. The manufacturers are using different types of steels in the body, depending on what the function of the structural element. Where they need higher strength and stiffness, because of the safety, they use high ultimate strength steels. In case of accident, where the deformation and the energy absorption is the important, they use lower strength, generally used non-alloyed DC steels (figure 2) [3]. The most of the body contains aluminum sheets, with this the weight can be reduced, with the same strength of the body.

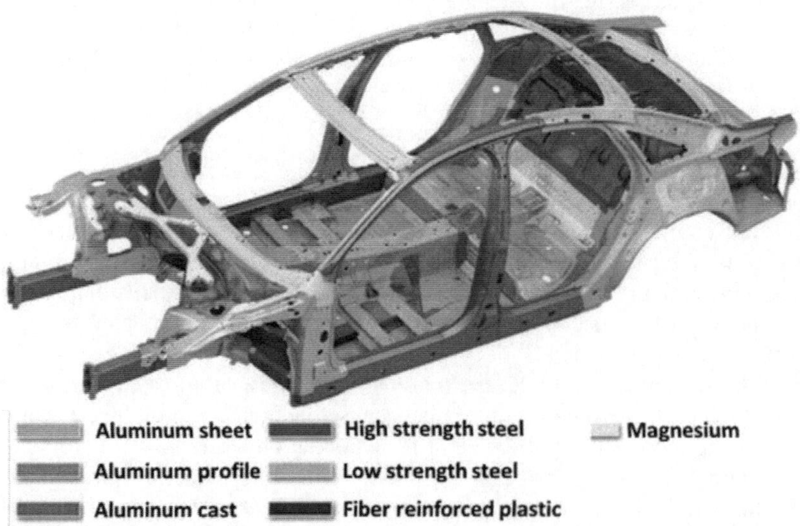

Figure 2. Different hybrid material usage in a new 2018 vehicle [4].

The typical joining methods on the vehicle body are the riveting, welding and bolts. The hybrid combination of these materials would not be possible without the researches of new bonding technologies.

The adhesive bonding technologies are a rapidly growing disciplines. Applications of adhesives for joining elements made of dissimilar, hybrid materials. Joining of fiber reinforced plastics with aluminum alloys using adhesive bonding is by far the most conventional method with both advantages and limitations [5]. The adhesives as the main part in the adhesive bonding should have good wettability with respect to joining components [6]. Using mechanical or chemical surface preparation on the materials the surface energy can be increased and it causes the increasing of the joining quality. The surface treatment modifies the surface roughness of the joining materials, reduce the oxide layer thickness and increase the wettability [7]. There are some treatments which has negative effect on the surface energy, these are the super hydrophobic surfaces [8] There is also manufacturing methods which can decrease the smooth of the surface of the GFRP [9,10]. To increase the adhesive strength between the materials and the adhesive, the optimal surface energy status is the hydrophilic.

Our goal is the optimization of making a hybrid adhesive joint between aluminum sheets and Glass Fiber Reinforced Plastic using different surface treatments. For this it has to be investigated the effect of the different surface treatment on the hybrid adhesive bonding.

2. Experimental procedure

The materials for the present investigation are non-coated AlSiMg1 sheet of 2 mm and Glass Fiber Reinforced Plastic (GFRP) of 2.5 mm thickness. The GFRP were vacuum laminated during 24 hours using 10 layer of 1 x 1 mm weave glass fiber and epoxy glue. The aluminum plates were mechanically sectioned to avoid heat input, into 70 x 70 mm, the GFRP into 70 x 25 mm. Before the joining method, the surface off the GFRP were grinded to remove the form separating layer. The GFRP has not modified with any treatment, only the grinding and the degreasing method by cleaning spray. The aluminum and GFRP surfaces were all cleaned and degreased with cleaning spray. 5 different surface preparation method were used during the research: the un-treated and un-cleaned surface, only cleaned surface, grinded surface with a P10 grinding stone, sand blasted surface using 0.3-0.7 mm abrasive and chemical etched with hydrochloride acid mixed with distillated water in 1:5 ratios for 5 minutes.

The adhesive joining method of the overlapped joints was performed using Loctite 9466 2 component epoxy adhesive, with this adhesive high strength of hybrid bonding can be created. First step of the adhesive method was the cleaning of all the contact surfaces (expect the uncleaned type of the materials), and glue the materials on 20 x 20 mm area overlapped surface (figure 3). The hybrid joints were pressed over 24 hours in a bending machine, this time needs to reach the 100% strength of the adhesive.

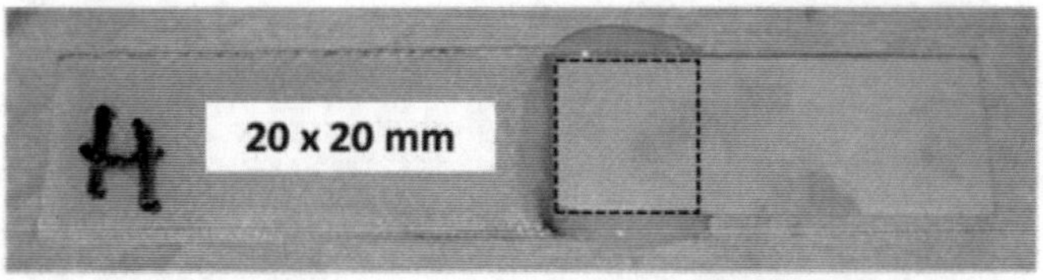

Figure 3. Adhesive procedure setup and connecting area.

In order to analyze the effect of the different surface preparations to the tensile strength of the hybrid CFRP and aluminum joints surface roughness measurement was performed. The macroscopic structure of the surfaces was investigated. Shear tensile test were carried out in a Instron 5900R testing machine at a cross-head speed of 2 mm/min (figure 4). Five samples were tested for each surface modification method. The breaking method was investigated to find out which component has lower adhesion force to the adhesive.

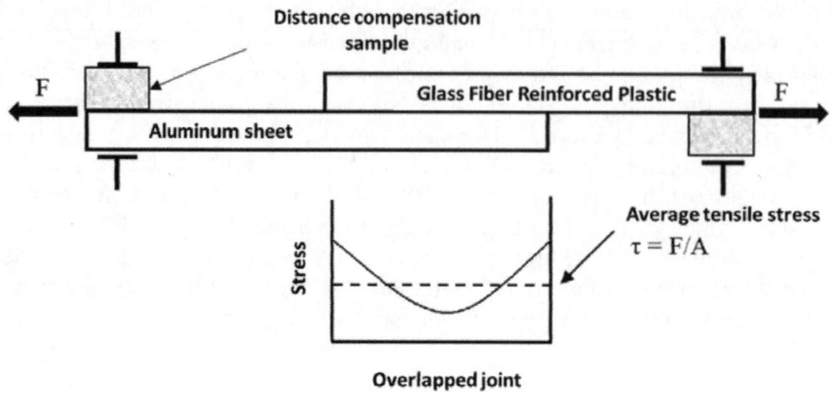

Figure 4. Test setup of tensile strength measurements.

3. Results

3.1. Effect of the different surface modification for the macroscopic topography and surface roughness

After the surface preparation of the aluminum sheets the surface roughness measurement were performed. In figure 5 the macroscopic topography and the roughness of the surface can be seen

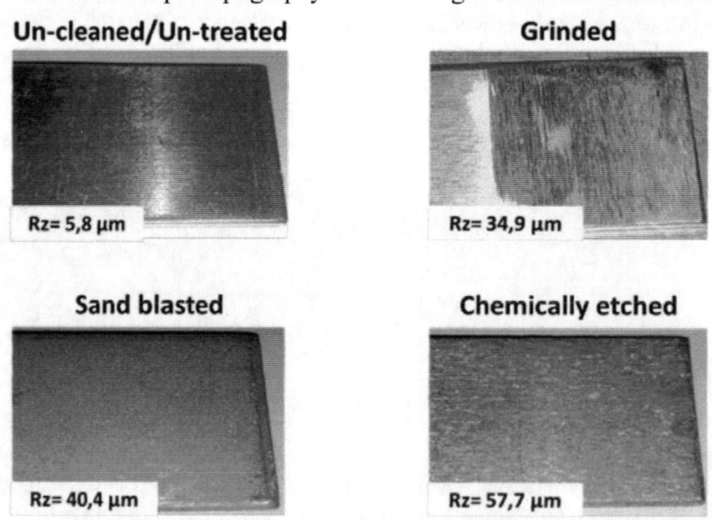

Figure 5. Surface macro topography and roughness using different preparations.

The original, as received surface is a very smooth, mirrored surface with a Rz= 5.8 µm, on this surface preparation the effect of the un-treated and un-cleaned strength of adhesive joints were investigated. The first mechanical prepared was the grinding with Rz= 34.9 µm, unfortunately the grinding is not enough homogenous surface preparation and only can be used on flat surfaces properly. A homogenous preparation could be the diameter of 0.3 – 0.7 mm sand blasting on the aluminum surface, it causes Rz= 40.4 µm. It both take the surface homogenous and can be used on any type of surface. With these methods the oxide layer can be removed mechanically. The chemical etching in thin hydrochloride acid cause homogenous surface with Rz= 57.7 µm, also the etching has effect on the oxide removal.

3.2. Effect of the different surface modification for the tensile strength

In figure 6 the effect of the surface cleaning and degreasing to tensile strength of the hybrid joints can be seen, using aluminum surface of un-cleaned/ un-treated and cleaned/un-treated. The diagrams contain the error bars of each type of surface treated joints.

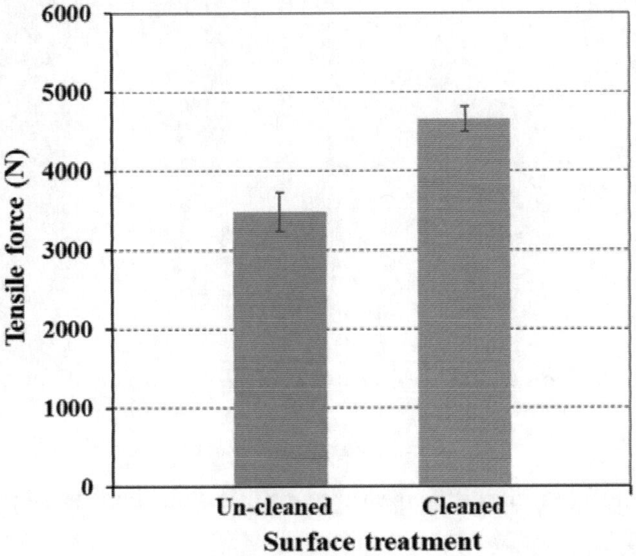

Figure 6. Effect of the surface chemical cleaning on the tensile strength.

As the results show, the un-cleaned joints has tensile strength of 3485 N and 8.7 MPa. The strength of cleaned samples was 4662 N and 11,7 MPa. The chemical spray cleaning both of the GFRP and aluminum surface cause 34% increasing of the strength. The first tensile results predict, that the cleaning should be used every time before the adhesive joining.

In figure 7 the effect of the surface treatment can be seen for the tensile strength of the hybrid GFRP and aluminum joints. The diagram contains the error bars and the Rz values of each surfaces. In this case all the connecting surfaces were chemically cleaned, only the effect of the surface topology can be seen.

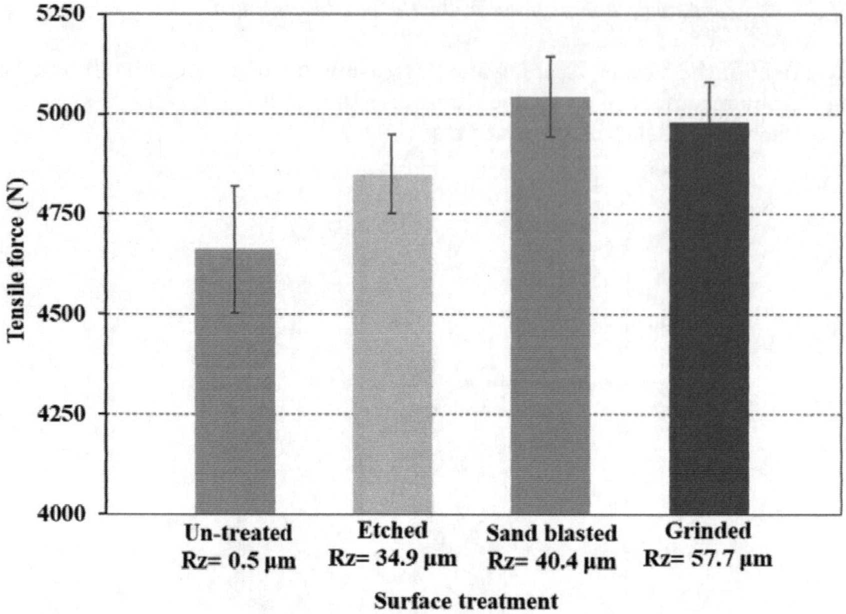

Figure 7. Effect of the different surface modification on the tensile strength.

The tensile strength of the adhesive joints changed using different surface preparation. The un-treated has 4662 N, 11.7 MPa, the chemically etched has 4850 N, 12.1 MPa, the grinded has 4985 N, 12.5 MPa and the sand blasted 5044 N, 12.6MPa of tensile strength. The error in case of the treated surfaces shows lower value because of the homogenous surface topography compared to the un-treated aluminum. The highest strength can be reached using sand blasted surface with these technical setups. The connection between the surface roughness and the tensile strength shows parabolic trend, so there is an optimal roughness value for this adhesive. The sand blasting preparation can increase the strength of the hybrid joints more than 10%.

3.3. Effect of the different surface modification for the breaking method

In addition to the changing of the tensile strength, the surface modification procedures resulted in different breaking methods for the specimens. In figure 8 the connection surfaces of the un-cleaned/un-treated samples can be seen after the breaking.

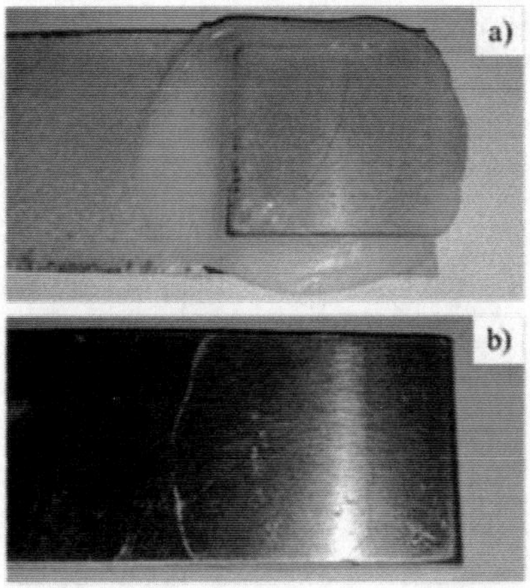

Figure 8. Separation method of un-treated/un-cleaned specimens.

The adhesion between the adhesive and the aluminum is totally chased, the adhesive separated from the aluminum surface. This could be because of the dirt or remained grease, these cause the lower adhesion forces and the bad wettability properties. In contrast, the figure 9 shows the cleaned/un-treated joints.

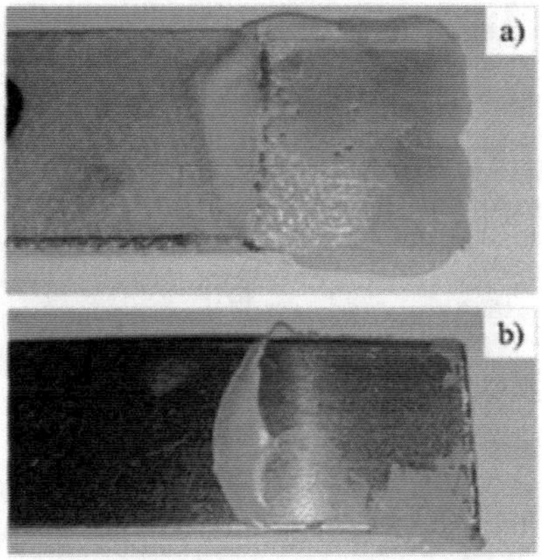

Figure 9. Effect of the surface cleaning and degreasing on the separation method.

In this case the adhesive shows lower adhesion force to the aluminum too, but the is remaining adhesive on the aluminum surface. The adhesion force is higher on the aluminum surface locally, than on the GFRP.

The figure 10 shows the hybrid joints using etched aluminum specimens.

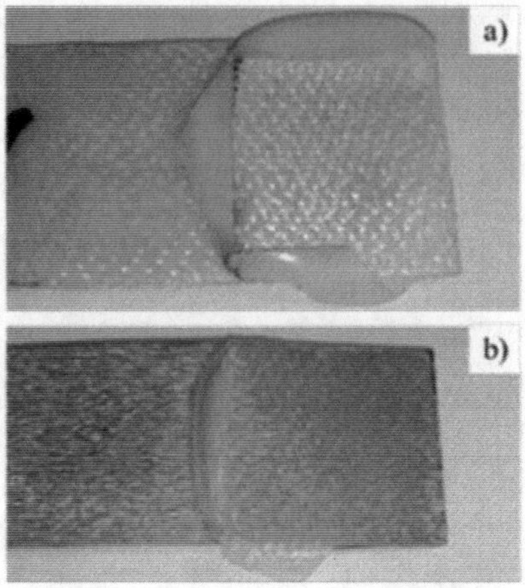

Figure 10. Breaking method using etched aluminum samples.

The adhesion force is higher on the aluminum, than the GFRP. The adhesive almost full remained on the etched surface aluminum sheet, then the GFRP.

Figure 11 shows the breaking method using grinded treatment.

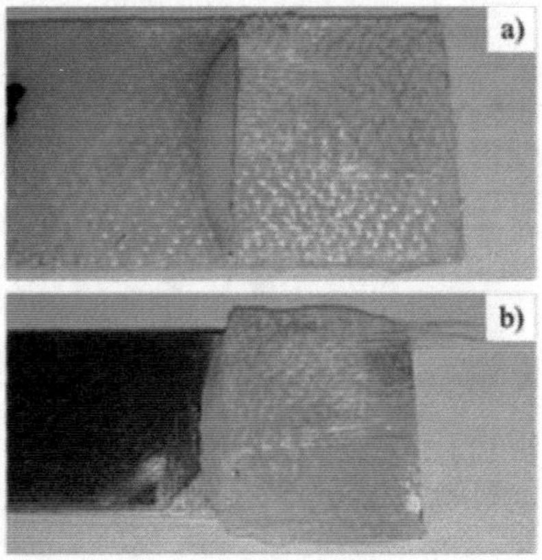

Figure 11. Breaking method using grinded aluminum specimens.

The adhesion force is acceptable on the aluminum surface in case of grinded surfaces. All of the adhesive is separated from the glass fiber reinforced plastic. The breaking method shows that from this value of strength the weakest part of the hybrid joints is the plastic.

The breaking method of hybrid joints with the highest strength can be seen in figure 12.

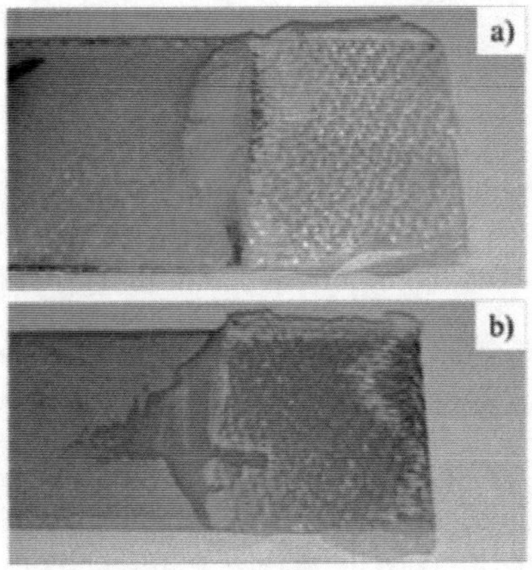

Figure 12. GFRP ripping out from the structure during the breaking.

The adhesive remained on every part of the aluminum, the sand blasting increased the wetting properties between the adhesive and the aluminum sheets. During the strain one layer of the GFRP ripped out from the plastic in the 20 x 20 mm connection area, with the blasting the strength limit of the GFRP has been reached.

4. Conclusion

In this research, the optimization of hybrid GFRP and aluminum sheets joints were investigated using high strength epoxy adhesive and different surface modifications only on the aluminum surface. The following conclusions can be drawn from this study:

- Different surface treatments can be used to create homogenous surfaces for adhesive procedures.
- The wetting properties of the connection surfaces can be changed with surface treatments for adhesive technologies.
- The chemically cleaning increased the tensile stress of GFRP and aluminum with 34 percent, compared to the un-cleaned situation.
- In case of bad wetting condition there is no remaining adhesive on the aluminum surface after the separation.
- The connection between the surface roughness and tensile strength shows parabolic trend using high strength 2 component epoxy adhesive, GFRP and aluminum.
- The highest tensile strength can be created with sand blasting surface preparation with this technical parameter. The increasing of the strength is more than 10%.

5. References

[1] Malte M, Nicolai M, William H, Sarah R, Kajta F, Reto K, David J F, Myles R A.2009. Greenhouse-gas emission targets for limiting global warming to 2°C. Nature. 459/909. pp. 1158-1162.

[2] Joeri R, David L M, Brian C O, Keywan R. 2013. 2020 emissions levels required to limit warmint to below 2°C. Nature Climate Change. 3. pp. 405-412.

[3] Budai D, Tiszai M. 2015. Development trends in aluminum car body production. Material Engineering, 148/5, pp. 29-36.

[4] http://automotive.arcelormittal.com/News/2791/AHSSforAudiA8 [Downloaded: 2018.05.26.]

[5] Boutar Y, Naimi S, Mezlini S, Carbas R J C, Silva L F M, Ben M. 2018. Fatigue resistance of an aluminium one-component polyurethane adhesive joint for the automotive industry: Effect of surface roughness and adhesive thickness. International Journal of Adhesion and Adhesives.
[6] Lekatou A, Gkikas N, Karantzalis AE, Kaptay G, Gacsi Z, Baumli P, Simon A. 2017. Effect of Wetting Agent and Carbide Volume Fraction on the Wear Response of Aluminum Matrix Composites Reinforced by WC Nanoparticles and Aluminide Particles. Archives of Metallurgy and Materials. 62/2. pp. 1235-1242.
[7] Sergio C, Vitor A, Luis R, 2018. Effect of surface treatment on adhesively bonded aluminium-aluminium joints regarding aeronautical structures. Engineering Failure Analysis. 84. pp. 34-45.
[8] Deepak M, Sebastian A, Yang Z, Jesper H, Jon S. 2018. A study of laser surface modification of polymers: A comparison in air and water, Journal of Manufacturing Processes.32.
[9] Liska J , Kodacsy J. 2012 Drilling of glass fibre reinforced plastic, Advanced Materials Research 472-475: pp. 958-961.
[10] K Liska , J Kodacsy , J Liska. 2012. Investigation of the microgeometry after hard turning and diamond burnishing, Advanced Materials Research, 472-475: pp. 902-907.

Acknowledgments
This research is supported by EFOP-3.6.1-16-2016-00014 project. The Project is supported by the Hungarian Government and co-financed by the European Social Fund.

IOP Conf. Series: Materials Science and Engineering **448** (2018) 012051 doi:10.1088/1757-899X/448/1/012051

Relationship between the keyhole laser welding and the plasma

M Berczeli[1], G Buza[2]

[1] John von Neumann University Department of Material Science,
[2] Bay Zoltán Nonprofit Ltd. for Applied Research, Budapest, Hungary

berczeli.miklos@gamf.uni-neumann.hu

Abstract. In this paper we investigated the behaviours of the laser beam keyhole welding with different laser beam sources, CO2 and a solid state Nd: YAG. We showed the differences between the types of laser welding, the geometry and method. Today the most used laser welding technology is the laser beam keyhole welding. In our research we investigated the heating effect of photons and the absorption method on the substrate surface. The photons can significantly absorbed better in case of 10^6 W/cm2 or more because of the plasma state in the welding material. The quality of the keyhole plasma depend on the free electrons of the materials and working medium. A theory of the creation of plasma phase and cloud were investigated. There is a difference working inside the metal vapour and working gas plasma cloud using CO2 and Nd: YAG. The solid state laser beam source shows better quality of the joint in keyhole welding.

1. Introduction

1.1. The behaviours of laser keyhole welding

The opportunity of optimizing bonds done with laser beam welding has become a decisive research area in today's developing industry, since it is a prevailing and frequent type of bonds both in machine and vehicle industry. Welded bonds must suit the increasingly strict quality requirements, all the welding structures must be specificated. There are two types of heating "modes" used to describe the resulting melting of the metal during laser welding (figure 1).

These are called "conduction mode" and "keyhole mode" heating. These modes of heating are created by different power densities and produce different results. Power density is defined as the power of the laser beam divided by the area the focused laser spot. In conduction mode heating, the power density is great enough to cause the metal to melt. Weld penetration is achieved by the heat of the laser conducting down into the metal from the surface. An example would be pulsed laser welding in the millisecond pulse length range. The depth of the weld penetration is controlled in part by the length of the pulse. The longer the pulse the more time heat has to "conduct" into the part. In conduction mode heating the welds are typically wider than they are deep [1-3].

Keyhole mode heating achieves its weld penetration in a different way. During keyhole mode welding the power density is great enough that the metal goes beyond just melting. It vaporizes. The vaporizing metal creates expanding gas that pushes outward. This creates a keyhole or tunnel from the surface down to the depths of the weld. As the laser beam is moved across the surface, the keyhole

IOP Conf. Series: Materials Science and Engineering **448** (2018) 012051 doi:10.1088/1757-899X/448/1/012051

follows and creates a typically deep and narrow weld. As long as the laser power is great enough and the travel speed is not excessive, this keyhole will remain open [2-4].

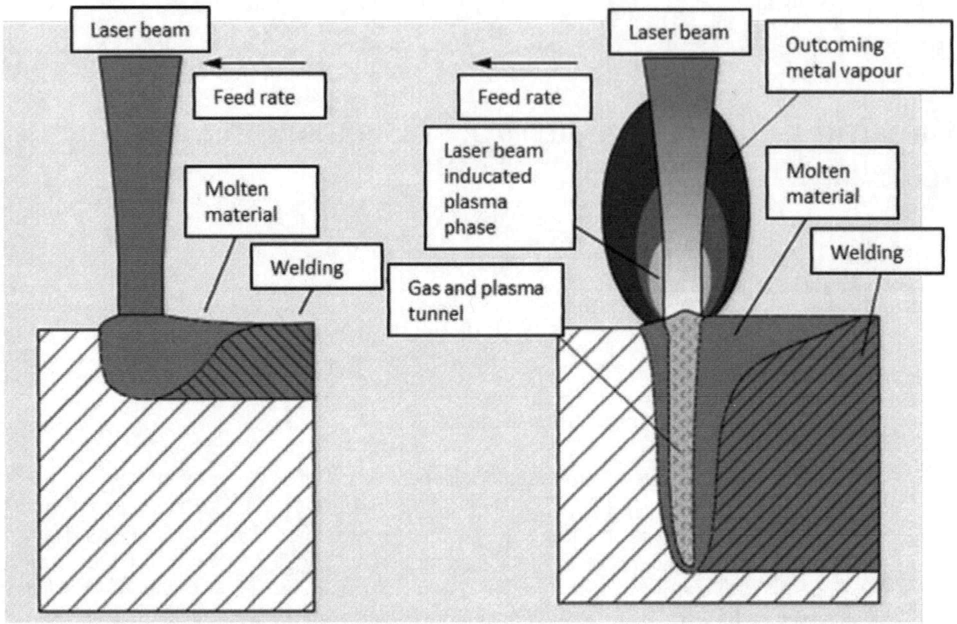

Figure 1. Conduction and keyhole laser welding [2].

Beyond this limit –due to another physical process- the technological procedure is controlled by the conditions of keyhole welding. As an aspect of technology and quality assurance it is important how different welding parameters have effects on different joint geometry. These parameters are the power of the laser beam, the velocity of welding, the position of focus spot, use of working gas and the excitation mode of the laser beam (CW, pulsating). All changes in these parameters have an essential effect on the joint geometry and the behavior of the plasma. Despite the diversity we must stick to the technical requirements and geometric specifications. Typical parameters of joint geometry are the following: depth of penetration, weld width, weld height, root weld with and the height of reinforcement [5].

1.2. Difference between traditional(MIG) and laser welding

Technically laser welding is similar to MIG welding only without the shielding gas. Laser welding fuses the metal by pulses of intense light and heat melting the members together while TIG welding fuses the members by heat created from an electric arc. MIG welding and TIG welding are similar in that they both use electricity to melt the material but MIG welding uses a filler material (usually a wire of the same type of material fed through the electrode).

New advancements in laser welding called "metal deposition laser welding" are similar to MIG in that they introduce a filler material by concentrating a cone of powdered metal into the laser's focus point [6]. This will help to solve one of laser welding's biggest obstacles: part fit up. Since the laser can only create enough heat to melt the metal in a small focus point it doesn't weld consistently well when part fit up varies. Any gap between the 2 members will cause an inconsistent weld. For this reason the most common form of welding (at least in a production environment) is MIG, because MIG welding is currently the most consistent and easiest at welding inconsistent fit up of parts because of the filler material. The geometry differentiation can be seen in figure 2.

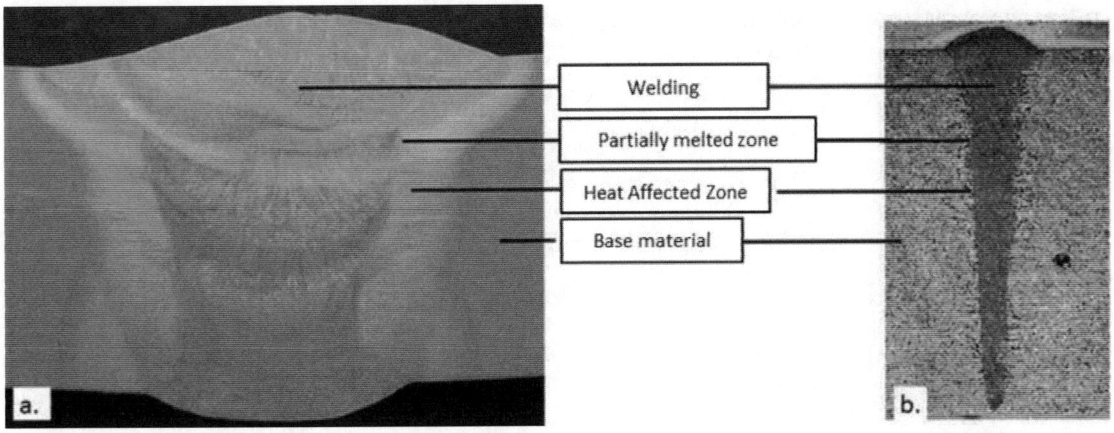

Figure 2. Geometry differentiation between traditional welding (a.) and laser keyhole welding (b.)

In case of laser welding the heat only absorbed where the joint will be so the heat affected zone is much smaller than the traditional welding method. During the laser keyhole welding a plasma tunnel created so the slenderness of the welded joint could be high. The aspect ratio (depth of welding divide with the width) can be 10 or more. The base material has lower inner stress because of the lower and focused heat input.

2. The method of photon heating

The energy of the photon (E) depends on the Planck constants (h), the speed of light (c) and the wavelength of the photon (λ).

$$E = h * \frac{c}{\lambda} \tag{1}$$

, where h= $6.626*10^{-34}$ Js, c= 299 792 458 m/s and the λ depends on the type and the material of the laser beam source.

From the 1. equation we can calculate how many photons we need to generate 1 kW laser beam power per second. In case of Nd: YAG (neodymium-doped yttrium aluminium garnet) laser beam source $5.35*10^{18}$ photon need, using Yb: YAG (Ytterbium-doped yttrium aluminium garnet) $5.18*10^{18}$ and in case of CO_2 laser beam source $5.33*10^{19}$ photon required.

These photons have high energy, when they collision with the atoms in the welding base material, the material's atoms energy will increase and the heat will grow. But the photons are only capable of transmitting their energies and interacting to free electrons of the base materials' atoms. So the photon - material interact is more precise if we call it to photon – free electron interact. The electrons can transmit energy to each other, but according only to the thermodynamics II.

When laser irradiation (in the form of electromagnetic waves or photons) impinges on the surface of a workpiece, a fraction of the energy is reflected at the surface, and a part penetrates of the energy is reflected at the surface, and a part penetrates into the substrate. A surface cannot absorb or emit photons. Attenuation takes place inside the solid, as does emission of radiative energy (and some of the emitted energy escapes trough the surface into the adjacent medium).

In practical system the thickness of the surface layer over which absorption of irradiation from a laser beam occurs is very small compared with the overall dimensions of a workpiece – usually a s few nanometres for metals and a few micrometres for most non-metals. The same may be said about emission from within the solid that escapes into the adjacent medium. In case of opaque materials it is customary to speak of absorption by and emission from "surface", although a thin surface layer is implied. Consider thermal radiation impinging on a medium of finite thickness, some of the irradiation will be reflected away from the medium, a fraction will be absorbed inside the layer, and the rest will

IOP Conf. Series: Materials Science and Engineering **448** (2018) 012051 doi:10.1088/1757-899X/448/1/012051

be transmitted trough the slab. Not that all three of these properties are nondimensional and may vary in magnitude between the values 0 and 1.

3. Investigation of plasma phase during the keyhole welding with different wavelength of laser beams

Transition to keyhole laser welding occurs at about 10^6 W/cm2 due to the enhancement of metal vaporization. In this case the absorption on the surface of the workpiece is drastically change, the absorption of laser beam is nearly 0.9 (Figure 3).

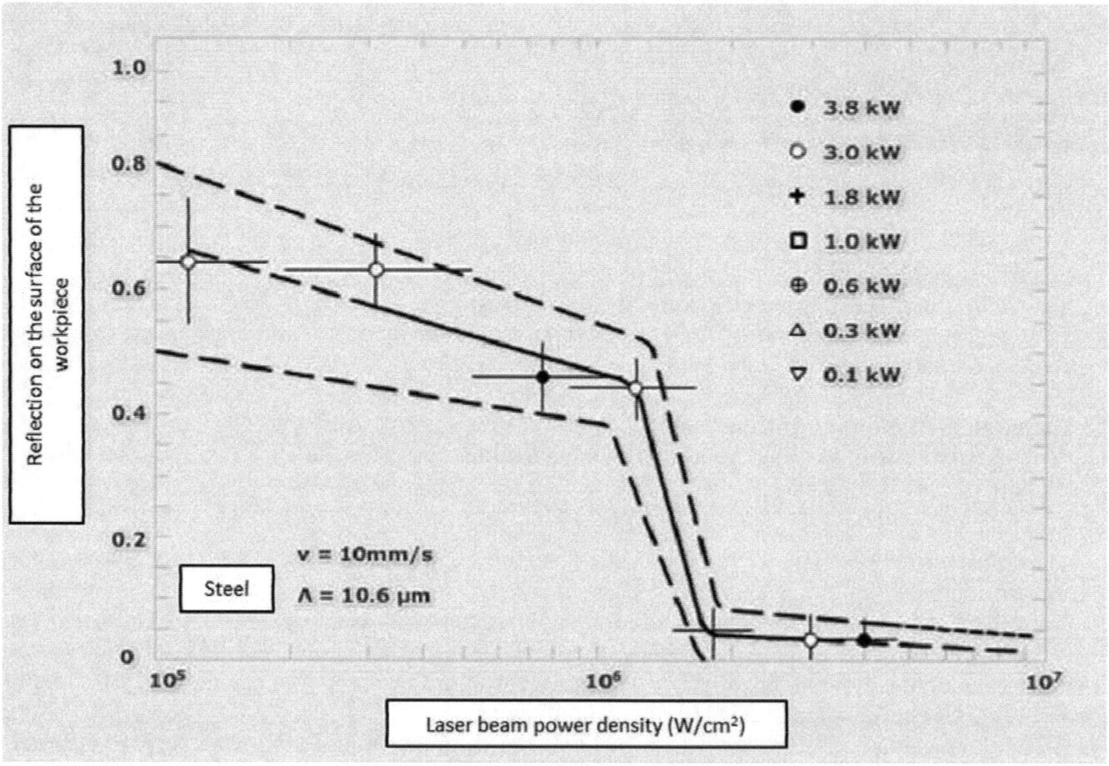

Figure 3. Absorption level changing depending on the laser beam power density [2].

The intense vaporization distinguishes keyhole laser welding from other conventional joining methods. It causes a large increase in vapour pressure (recoil pressure) that drills a depression in the melted metal, forming a long and narrow cavity or keyhole. The laser beam can then penetrate deeper into the metal through the cavity and be refracted and damped while traveling through the vapour. As the beam rays reach the keyhole surface, beam energy is partially absorbed on the surface and partially reflected towards a new point of interaction. The free electrons absorbe a part of the photons. This succession of absorption/reflection (or multiple Fresnel absorption/reflection) increases the overall energy absorption (up to 90% in the present example). The keyhole cavity is filled with metal vapour that partly absorbs the incoming laser light. The charged particles present in the vapour can gain kinetic energy from the beam photons. When the gain is significant it results in beam damping, further ionization of the metal vapour (plasma) by the highly energetic electrons, increase of the plasma temperature. There is a high electron density and temperature in the plasma phase compared to the wavelength of the laser beam, also there is a rise in the plasma cloud's radiation energy loss. The optical densities of the plasma is changing, the reflective index is increasing. With these effects less photons can reach the plasma tunnel.

IOP Conf. Series: Materials Science and Engineering **448** (2018) 012051 doi:10.1088/1757-899X/448/1/012051

Figure 4 and 5 shows when the laser beam source is CO_2 during the keyhole welding. In this case the laser beam heat the plasma cloud too much, and it can be overheated. The plasma reflection rate increase and, the keyhole plasma tunnel shutdown and it goes to heat convection welding. The unstable plasma can cause cracks in the welding because of the high rate of density of the heat input [10].

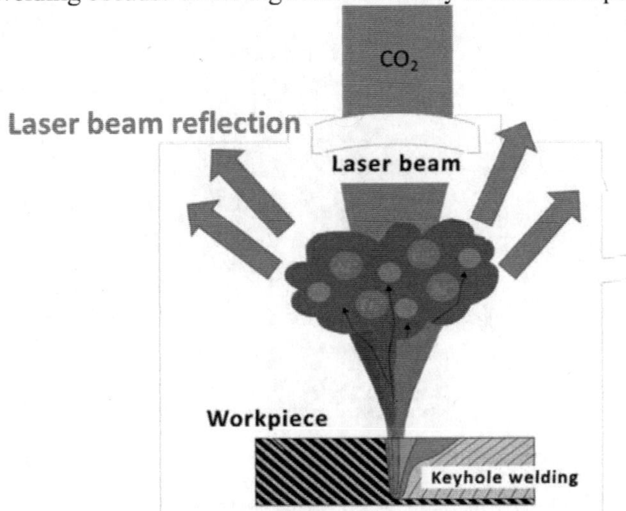

Figure 4. Behaviour of the plasma cloud in case of using CO2 laser beam source.

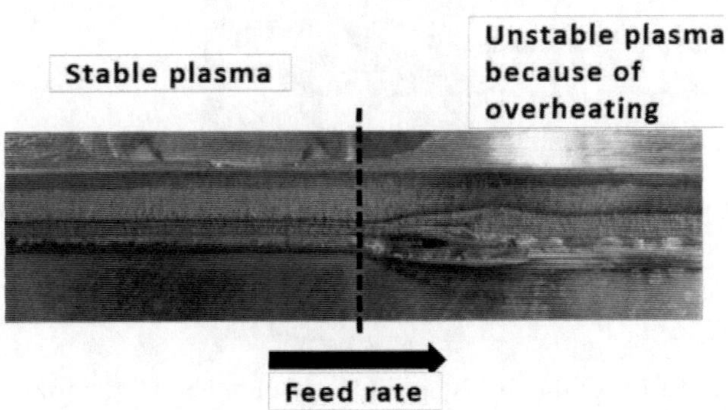

Figure 5. : Stable and unstable plasma using CO2 laser beam during the keyhole welding.

Using a solid state laser beam source, for example Nd: YAG the laser beam wavelength is much shorter, and it has different effect when interact with the plasma cloud during the keyhole welding process.

Figure 6. and 7. shows the effect of a solid state laser beam source to the plasma cloud. Because of the different wavelength just a few part of photons absorbe in the plasma cloud, the most part of it reach the keyhole welding. This type of plasma cloud can be overheated too, in this case the metal vapour with the working gas overheat and has high rate of infrared irradiation. This cloud goes trough on the welding and the top of the keyhole welding geometry will be wider because of the overheating of infrared irradiation.

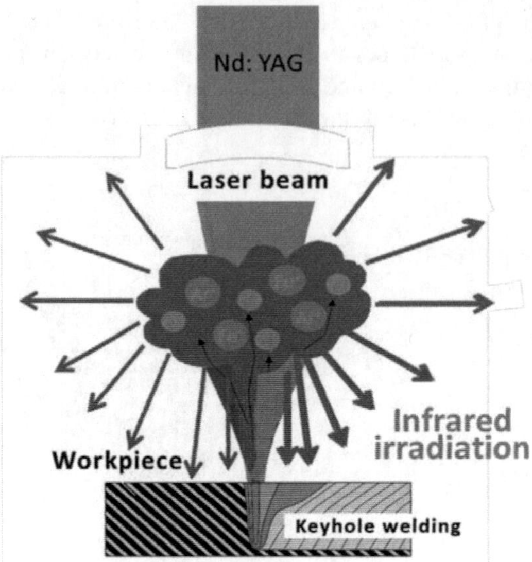

Figure 6. : Behaviour of the plasma cloud using a solid state laser beam heating.

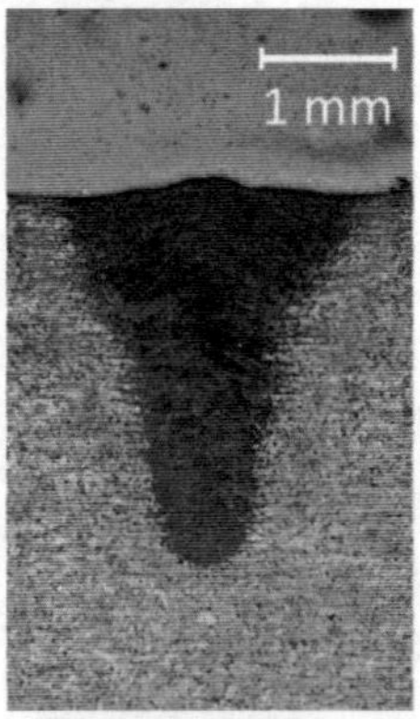

Figure 7. : Effect of the overheated plasma cloud and the infrared irradiation.

Summary Figure 8 shows the connection between the free electron density and the wavelength in the plasma cloud. The dotted lines shows the change between the reflection or transmission behaviors of the plasma cloud.

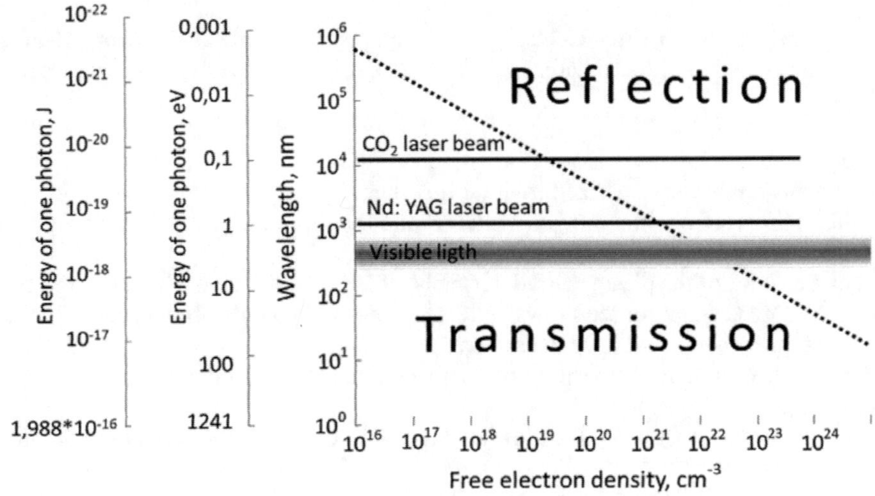

Figure 8. Effect of the laser beam wavelength, photon energy and the electron density in the plasma cloud [2].

The figure shows the wavelength of the CO_2 laser beam 10600 nm and the Nd: YAG 1064 nm. In case of CO_2 laser beam welding the plasma cloud will reflect the incoming photons because the free electron density level is smaller during the overheating. In case of solid state Nd: YAG laser beam the plasma cloud can reach about 3 magnitude higher electron density, and the photons can reach the keyhole welding.

The plasma has then a positive role: it can protect the keyhole cavity from cooling by the surrounding atmosphere and plasma radiation can strengthen vaporization at the keyhole surface This effect also can support the plasma state using laser beam cutting [11]. Above the critical temperature damping becomes dominant and vaporization vanishes.

There is different type of plasma mediums and the plasma state quality depend on the quality and the free electron density of the mediums (Figure 9.).

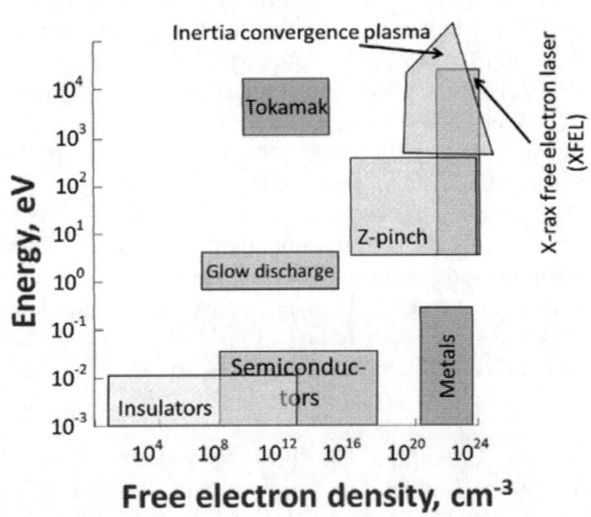

Figure 9. Different plasma mediums depending on the free electron density [8].

Metals have high density of free electrons but the energy of the plasma "lower". Z-pinch, Tokamak, X-ray free electron laser and Inertia convergence plasma is an innovative research areas in the energy industry [7-9].

4. Summary

The major conclusions are summarized in the followings:

- The accelerated photons can only interact and capable of transmitting energy to the medium's free electrons.
- The behavior of the plasma cloud during laser keyhole welding depend on the type of the laser beam source. Nd: YAG beam source shows better tolerance with overheating because of the free electron density can be higher than using CO_2 laser beam.
- The quality and the composition of the plasma medium define the quality of the plasma cloud and the welding geometry.
- In case of CO_2 laser beam the plasma cloud be overheated, the reflection rate increase and the plasma became unstable.
- In case of solid state laser beam the plasma cloud has infrared irradiation during the overheating and it causes the wider top of the welding geometry.

References

[1] W A Ayoola, W J Suder and S W Williams 2017 Parameters controlling weld bead profile in conduction laser welding *J. of Mat. Proc.* Tech. 249 pp 522-530

[2] G. Buza 2012 Laser technology I-II. (Hungary)

[3] M Berczeli and G Búza 2017 The effect of laser beam welding paramters onto the evolving joints geometry *Mat. Sci. F.* 885 pp 178-183

[4] Y Ai, P Jiang, X Shao, P Li, C Wang, G Mi, S Geng, Y Liu and W Liu 2016 The prediction of the whole weld in fiber laser keyhole welding based on numerical simulation *App. Therm. Eng.* 113 pp 980-993

[5] J. F. Ready and D. F. Farson 2001 LIA Handbook of laser materials processing (USA, Laser Instituta of America)

[6] X Zhan, Y Li, W Ou, F Yu, J Chen and Y Wei 2016 Comparison between hybrid laser-MIG welding and MIG welding for the invar36 alloy *Opt. & Laser Tech.* 85 pp 75-84

[7] M G Haines and T W L Sandord and V P Smirnov 2005 Wire-array z-pinch: a powerful x-ray source for ICF *Plasma Phys. & Control. Fusion* 47 B1

[8] B Koncar, M Draksler, O C Garrido and B Meszaros 2017 Thermal radiation analysis of DEMO tokamak *Fusion Eng. & Design* 124 pp 567-571

[9] E Part, S Bettoni and S Reiche 2016 Enhanced X-ray free-electron-laser performance from tilted electron beams *Nuc. Instr. & Meth. in Phys. Rsch.* 865 pp 1-8

[10] G Gergely, D K Horvath, Z Weltsch, Z Gacsi. 2017. Development of crack detection method with 2 dimensionally generated 3 dimensionally reconstructed images in THT solder joints, Archives of Metallurgy and Materials 62:(2) pp. 1033-1038.

[11] Z. Weltsch, K Klam, A. Lovas. 2017. The Comparison of Hardness and Coercivity Evolution in Various Fe–B Based Glasses (Including FINEMET Precursor) during Relaxation and Crystallization, Acta Physica Polonica 131:(4) pp. 669-671.

Acknowledgement

The publication of the work reported herein has been supported by NTP-NFTÖ-17-B-0514, National Talent Programme, Ministry of Human Capacities, Human Capacities Grant Management Office (EMET).

IOP Conf. Series: Materials Science and Engineering **448** (2018) 012052 doi:10.1088/1757-899X/448/1/012052

Investigation of inserts surface structures on injection moulded parts

Krisztián Kun[1], **Pál Boza**[1]

1 John von Neumann University, GAMF Faculty of Engineering and Computer Science, Department of Vehicle Technology, Kecskemét H-6000, Hungary

kun.krisztian@gamf.uni-neumann.hu

Abstract. Nowadays, the surface quality of injection moulded products is becoming increasingly important to users. Due to increasing demands, engineers (especially in the automotive industry) creates structures on the surface of the parts that are close to the appearance of natural materials (e.g. leather, wood, etc.). These aspirations are not only aesthetically but functionally important. In order to create quality product, besides processing parameters, it is also crucial how to design the mould. However, it is important to know how the processing parameters affect the mould surface mapping. The preliminary objective is to investigate the effect of different macro-geometric structures created by cutting technology on the flow of polymer melt.

1. Introduction

The production of a quality finished part requires perfect knowledge of the filling and compression phase. One of the main tasks of this research is to study the relationship between polymer melt and the manufactured surface structure. The preliminary objective is to investigate the effect of different macro-geometric structures created by cutting technology on the flow of polymer melt. The backbone of the research is provided by a special, experimental injection molding tool, consisting of two symmetrically positioned cavities which used for producing bending specimens (Figure 1.).

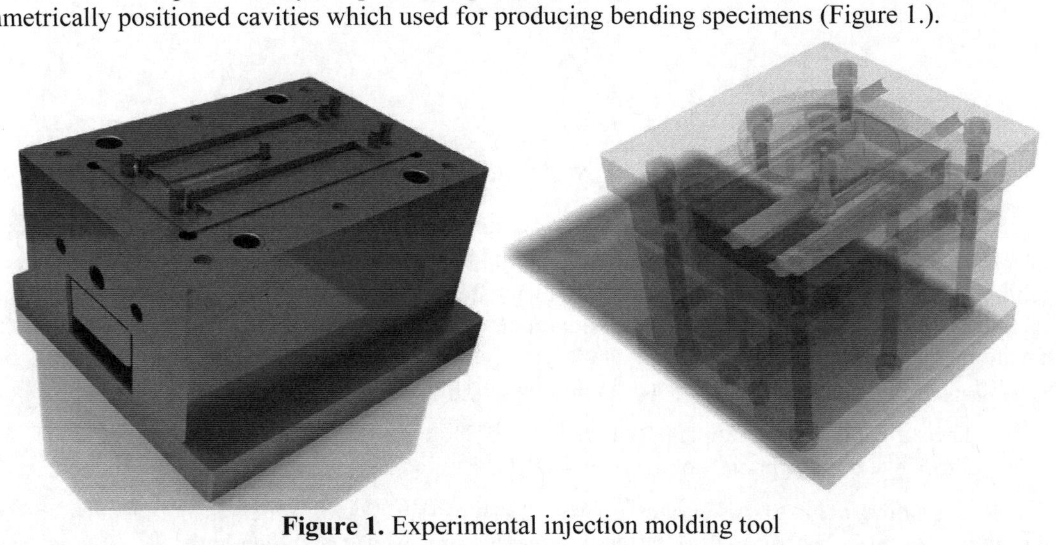

Figure 1. Experimental injection molding tool

2. Description of the research and experimental conditions

The premise was to create two paths on the insert's surface that result significantly different orientation during injection molding. We designed a special mold insert, kept the dimensions of the former active insert, and then created grooves on its surface as shown on Figure 2.

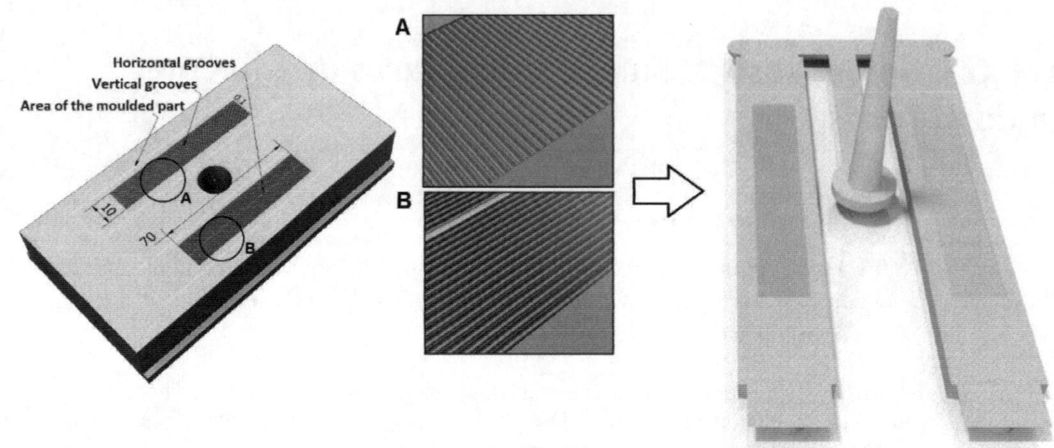

Figure 2. The designed insert and the moulded part.

In terms of flow direction, we designed perpendicular and parallel grooves which depths were 0.1 mm. As for the manufacturing technology, the selected tool was a TiN coated, carbide engrave tool, with 30° tip angle and 0.2 mm corner radius. It is shown in the Figure 3.

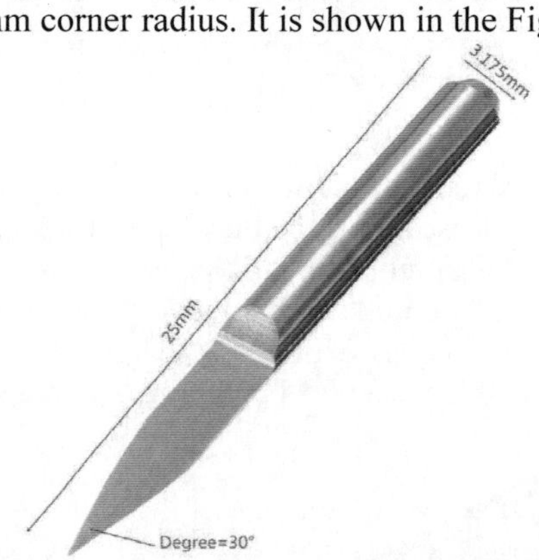

Figure 3. TiN coated, carbide engrave tool

After the geometrical boundary conditions have been recorded, material selection has taken place. Our aim was to use an amorphous and semi-crystalline polymer. In the semi-crystalline polymers, the melting of the crystalline phase has a higher heat demand during heating and, when cooled, the crystallization need a higher heat dissipation compared to the amorphous polymers. During the phase transition, the density also varies: the crystalline phase is of greater density than the amorphous phase or the melt, so the shrinkage of the semi-crystalline polymers is greater:

- The shrinkage of crystalline polymers is 1.5-2.5%.
- Shrinkage of amorphous polymers: 0.4-0.8%.

In crystalline polymers, a further approx. 1% post-shrinkage, which is completely absent in amorphous polymers. The properties of the two selected materials are shown in the Table 1.

1. Lexan Resin 141R – Polycarbonate (amorphous)
2. Dowlex 2027G – LLDPE (semi-crystalline)

Table 1. Examined polymers and its properties

Properties	**Lexan 141R**	**Dowlex 2027G**	unit
Recommended processing temperatures			
Tool surface temperature	100	16	°C
Melt temperature	320	230	°C
pvT properties			
Density (melt)	1,0534	0,72883	g/cm³
Density (solid)	1,1983	0,92819	g/cm³
MFI (Melt flow index)	11	4	g/10 min

3. Experimental results

3.1 Microscopic recordings

The Figure 4 shows the engrave tool, captured with a Dino-Lite Pro HR AM7000 microscope. That more or less confirmed the properties of the tool's datasheet. The tip angle was close to the description, however the corner radius was not 0.2 mm, but rather a parallel edge to the surface. The tools were exposed to increased wear due to the long tool path. Since the horizontal and vertical grooves were made with different engraving tools, the quality was less affected by the tool wear.

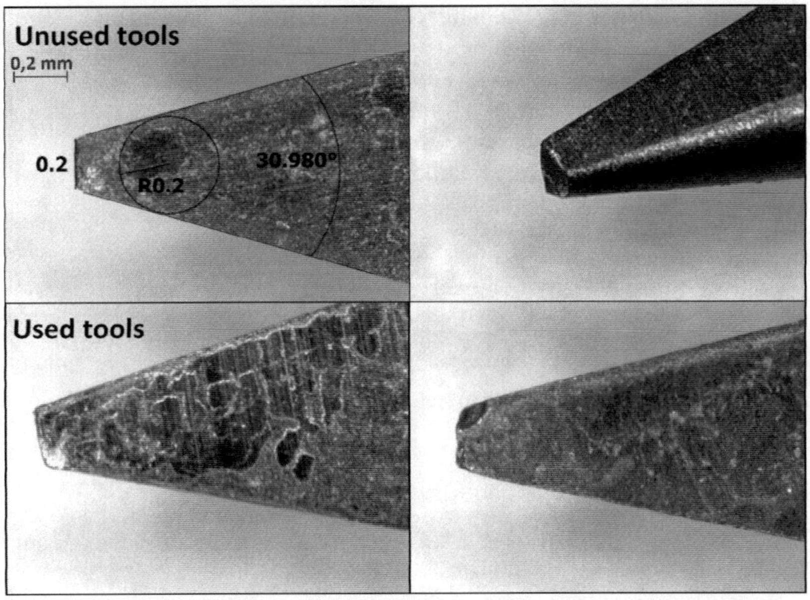

Figure 4. Microscopic recordings about the TiN coated, carbide engrave tool

To check the quality of the engraved grooves a Mitutoyo QuickVision Elf Pro microscope was used. The macro and micro-geometric characteristics of the grooves are well visible on the Figure 5.

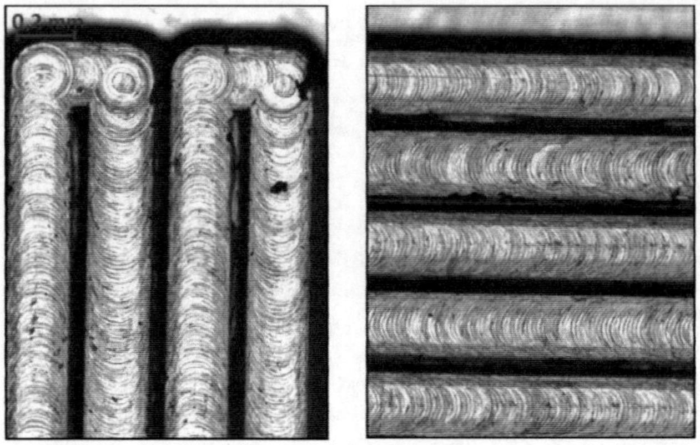

Figure 5. The vertical (left) and the horizontal (right) paths recorded with Mitutoyo Quick Vision microscope.

3.2 Examination in contour measuring instrument

The cavity filling quality was measured using a Mitutoyo Formtracer Sv-C3000 contour measuring instrument. The primary purposes of contour measurement were:

- Find the differences between the actually created and the planned geometries.
- Investigating the filling of injection products
  - → in case of Dowlex 2027G LLDPE
    - with back pressure (500 bar)
    - without back pressure
  - → in case of Lexan 141R polycarbonate
    - with back pressure (500 bar)
    - without back pressure

3.2.1 Contour measuring of the insert

Both the horizontal and the vertical paths were divided into 3 to 3 sections. In these areas we measured the surface contour on a 3.5 mm long distance. The sections were recorded at the beginning, middle and end of the path. The difference in height due to tool wear was also investigated (Figure 6).

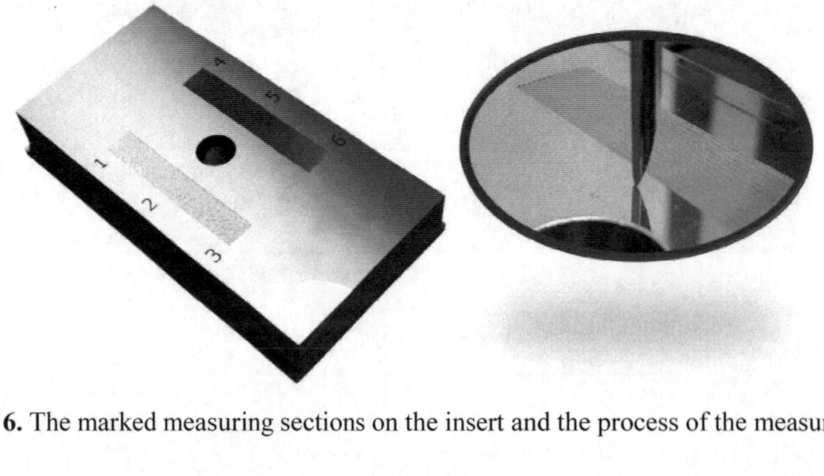

Figure 6. The marked measuring sections on the insert and the process of the measurement.

IOP Conf. Series: Materials Science and Engineering **448** (2018) 012052 doi:10.1088/1757-899X/448/1/012052

The differences between the engraved grooves and the planned geometries were clearly visible in the microscopic images. After the contour measurement, the facts found there are proved. However, there is a regular arrangement between the engraved grooves, which is shown in Figure 7. The geometric differences of the engraving tool and wear after manufacturing were also noticeable on the paths. The differences between the depths of the grooves in the measuring sections were not significant, so it can be stated that the tool suffered a most part of the wear during the plunge. The wear was 0.02 mm on average. The figure also shows that the distance between the tool paths was in accordance with the programmed value (0.346 mm)

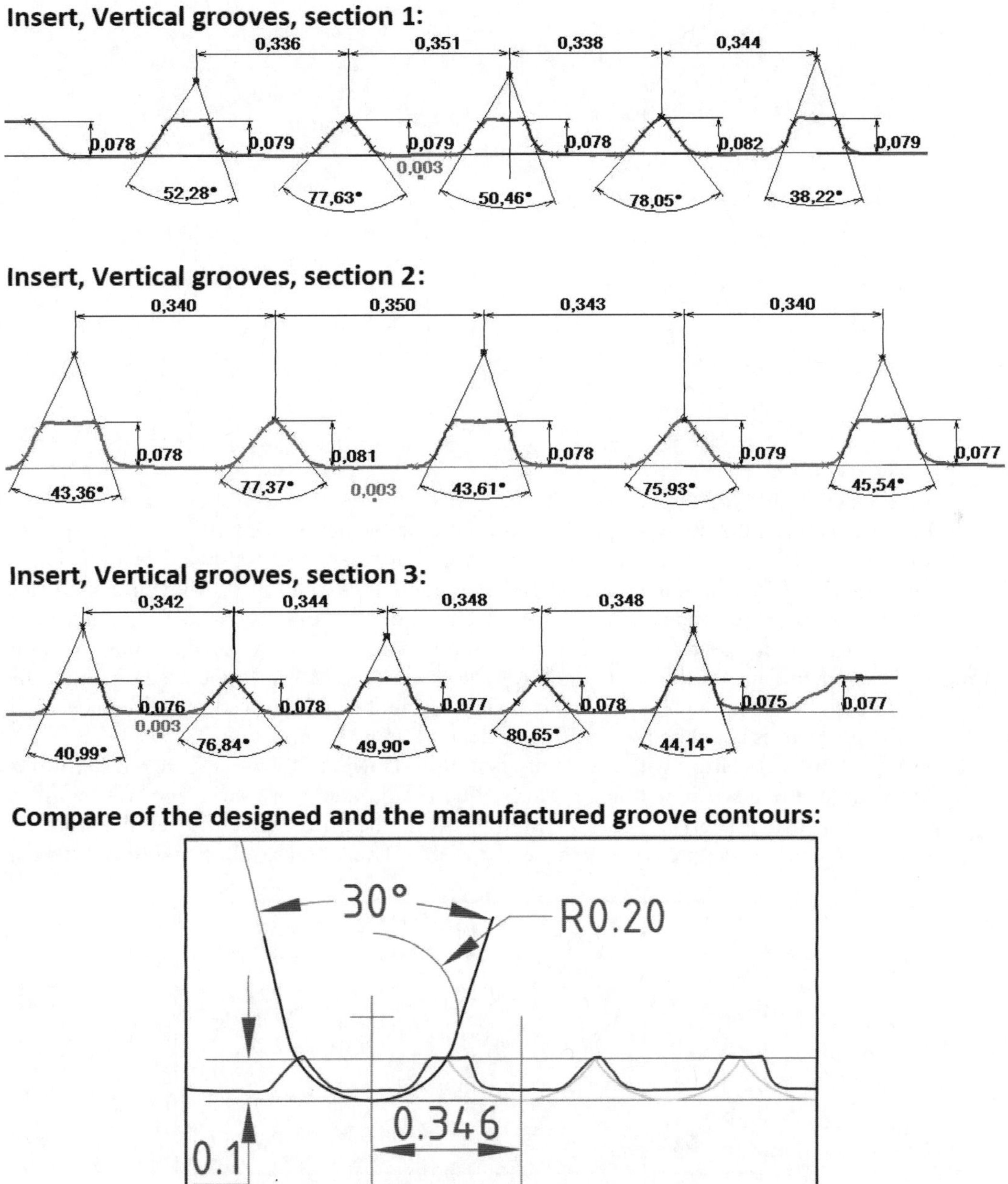

Figure 7. Measured geometries on the insert's surface.

3.2.2 Contour measuring of the moulded parts

The comparative examination of the structures on the moulded products was carried out similar to the insert (Figure 8). We compared products with backpressure, and without backpressure, which had a value of 500 bar. The examined section of the moulded product was combined with the same contour section of the insert. The objective was to analyze the geometric mapping.

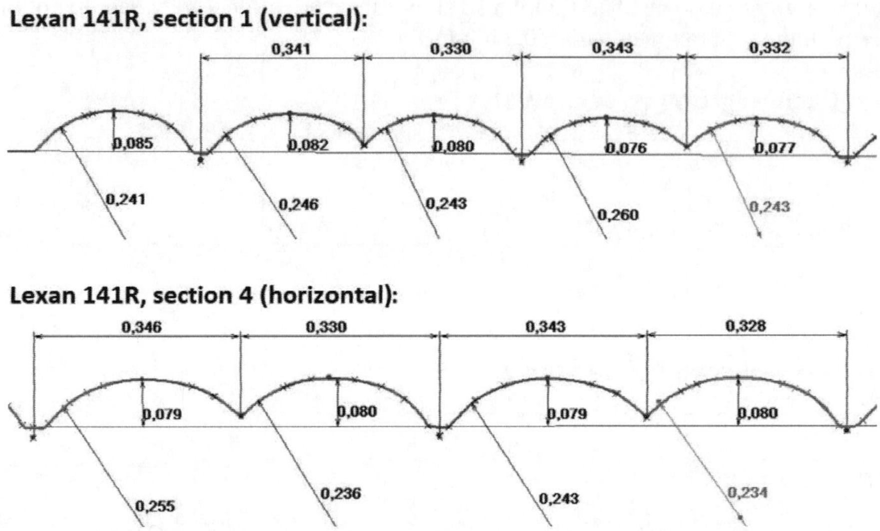

Figure 8. Measured geometries on the insert's surface.

The figure shows that, the contours of Lexan 141R horizontal and vertical paths did not show significant differences in the moulded products. In the case of the Dowlex 2027G, it also got similar results, so I did not differentiate the structures that were engraved. The following sections have been studied on vertical paths. The points measured by the contour measuring instrument can be saved in ".dxf" format. The software can automatically align these points with polynomials. The number of the points were a magnitude greater than the desired limit of measurement, so the approximation was sufficient. The comparison was done in AutoCAD using the ".dxf" files. In the case of height, the flat surfaces were used, while in the direction of the length the first groove served as a reference. With this technique, the geometries could be fitted with the accuracy of <0.01 mm.

The examination was started with the semi-crystalline Dowlex 2027G (Figure 9). It can be seen that the structure of the insert is not nearly covered by the product. The shrinkage was significant in longitudinal parts and sharp corners. In addition to its depth, the differences between the products with backpressure and without backpressure were clearly visible. This often resulted two-fold differences.

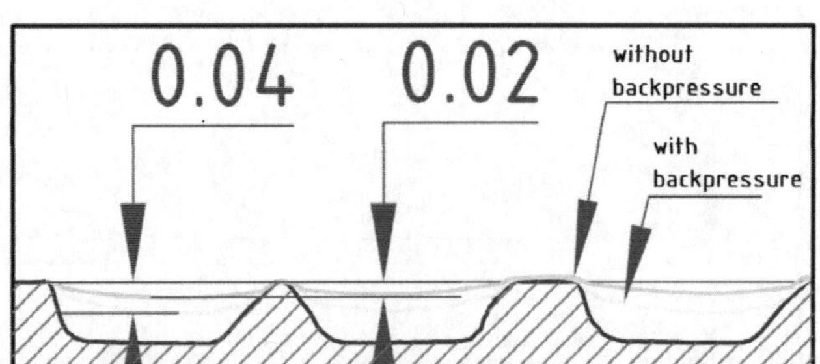

Figure 9. The measured contours of the semi-crystalline Dowlex 2027G with and without backpressure.

Then we examined the products made from Lexan 141R (Figure 10.). Since it is an amorphous material, the expected shrinkage was less. The grooves were better filled, and its depths were far better than on the LLDPE parts. The sharp corners and contour lines were better returned by product geometry. In this case the presence of the backpressure was not significant.

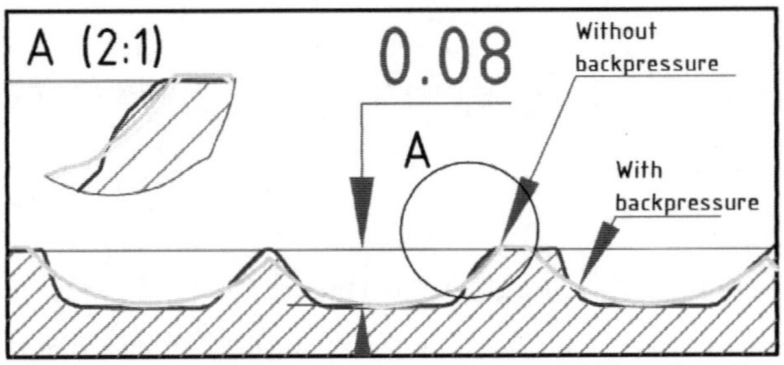

Figure 10. The measured contours of the amorphous Lexan 141R polycarbonate, with and without backpressure.

4. Results and discussion

The moulded contours did not show significant differences in the horizontal and vertical paths. The orientation of the paths less influenced the inserts filling. The filling of the structures were not perfect either. The moulded parts had a positive effect on shrinkage after backpressure. The Lexan 141R polycarbonate (amorphous and higher MFI) had a significant shrinkage and its surface contour did not followed well the insert's geometry. In spite of all these, it had a better mapping ability with the applied technologies compared to the LLDPE.

5. Conclusions

Examination of geometric mapping with contour measurements gave clear results. In order to have the designed insert surface structure on the manufactured parts, it is appropriate to have a careful geometric design and further experiments. The thickness of the moulded parts and the depth of the grooves can also be definitive perspectives and steps forward in the near future. As a long-term goal, it is possible to design a surface structure that guides the molecule orientation of the melt to different needs. However, the structure of the surface which affecting the molecular orientation on boundary layer, and the influence it has on the mechanical properties requires a series of further research and experiments. The measurements in this article made may serve as a basis for these investigations.

References

[1] M Pósa, K Belina, P Boza, A Szűcs *Examination of the effect of machining technology on the surface of injection moulding parts,* (Inter-Ing 2005, Targu Mures 2005.) pp 19-25.

[2] K M B Jansen, D J van Dijk, M H Husselman *Effect of processing conditions on shrinkage in injection molding*, (Polymer Engineering and Science, 1998.) pp 838-846

[3] Bánhegyi. Gy. *Kristályos és amorf polimerek termikus vizsgálata differenciális pásztázó kalorimetriával (DSC)*, In: Műanyagipari szemle 2014.

[4] Deligio, T.: Thinner thin-wall. Modern Plastics Worldwide, 83. k. 4. sz. 2006. p. 78–81. Yokoi, H.; Han, X.: *Effects of molding conditions on transcription molding of microscale prism patterns using ultra-highspeed injection molding*. Polymer Engineering and Science, 46. k. 9. sz. 2006. p. 1140–1146.

[5] C. Liu, L. T. Manzione *Process studies in precision injection molding*, (I: Process parameters and precision, Polymer Engineering and Science, 36, 1996.) pp 1-9.

IOP Conf. Series: Materials Science and Engineering **448** (2018) 012053 doi:10.1088/1757-899X/448/1/012053

Machining of rotationally symmetric parts with abrasive waterjet

K Kun-Bodnár[1], J Kundrák[1], Zs Maros[1]

[1]Institute of Manufacturing Science, University of Miskolc, H-3515 Miskolc, Egyetemváros, Hungary

krisztina.bodnar@uni-miskolc.hu

Abstract. Recently, water jet cutting is widespread, because almost any material can be machined by this machining method. The technique and technology of cutting has also improved a lot. Further utilising of waterjet cutting advantages could Application of waterjet cutting for machining surfaces with different geometries (rotationally symmetric parts) beyond simple cutting tasks, could be a further utilisation of this method. This paper demonstrates a solution for machining cylindrical surfaces by abrasive waterjet. For this examination of kinematics at machining of cylindrical parts could be the starting point Appropriate movement conditions can be created by considering the machine as a basic machine and complementing it with a water jet cutting head or a device which can be installed on the water-jet cutting machine. We have developed an equipment for this latter solution. Design aspects and the device are presented in the paper.

1. Introduction

The main advantage of waterjet cutting is that it is possible to cut so complicated planar curve which with other technology is not workable. During the application of abrasive waterjet cutting technology, heat effect does not affect the material so its' structure does not suffer during the machining. The great advantage of the waterjet cutting in contrast with other technologies, that this is a cold cutting process [1]. This technology does not cause fracturing in the material, and suited for cut different materials, therefore the application field is wide.

This technology is applied on cutting in the food and automotive industry, in manufacturing of medical appliances. In addition it is applied on work of arts and mosaics. Almost all materials can be machined: non-metallic materials (simple and fiber reinforced plastics, rubber, leather, paper, textiles), brittle materials (glassware, ceramics, rocks, and concrete, building materials), tough materials (non-ferrous metals, iron and nonferrous metals, in any heat treatment condition), soft, flammable materials, plastic foams.

In general terms, abrasive waterjet cutting machining offers a set of unique capabilities: cuts any materials regardless of their mechanical, wear-resistant properties which makes the process attractive for part generation in difficult to cut materials, generates negligible heat during the material removal process which is an advantage when machining heat sensitive alloys [2].

Aim of this work is to create a device that allows the machining of rotational symmetrical parts from the above mentioned materials with help of waterjet machining. This machining from external point of view is very similar to turning operation that is why this method is called waterjet turning.

IOP Conf. Series: Materials Science and Engineering **448** (2018) 012053 doi:10.1088/1757-899X/448/1/012053

2. Essence and characteristics of waterjet turning

Because of the good characteristics of abrasive waterjet cutting there are a lot of researches oriented on investigation of application waterjet in other machining methods. Most often, the kinematic relationships of the chip removal processes are applied for the machining with abrasive waterjet. From this point of view, we can say about waterjet milling, drilling or turning. In the following sections the abrasive waterjet turning will be discussed.

2.1. Kinetics of waterjet turning

Waterjet turning from external point of view is very similar to the normal turning process accomplished with a single point tool. Workpiece has a rotational motion while the tool has a linear motion, usually parallel with the axis of the workpiece. The depth of cut determined by the relative displacement of the waterjet and the workpiece [3]. Some elements of the waterjet turning can be seen on Fig. 1. [4] Forces effecting on the workpiece are neglectable.

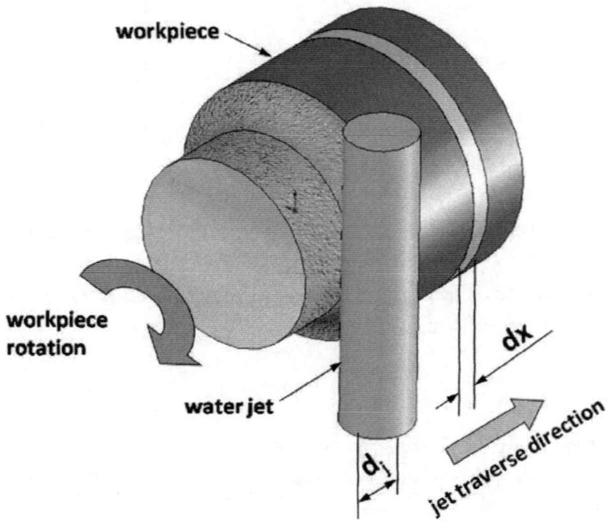

Figure 1. Elements of waterjet turning

The depth of cut is a result of more process parameters and defined by the machinability of the material, the displacement of the waterjet and the real speed of the grains, which defined among others by the water pressure and the abrasive mass flow rate. [4]

According to the kinematics of the turning the main cutting speed is the speed of the jet, ie. the speed of the abrasive grains in the jet. There are two auxiliary motions here, a linear feedrate of the jet along the contour of the workpiece and a circular feedrate of the workpiece (rotation).

2.2. Technological parameters of abrasive waterjet turning

There are a lot of parameters affecting on the geometry of the workpiece at waterjet cutting. The depth of cut determined not only by the displacement of the jet, but the efficiency of the jet and the machinability of the workpiece as well. The diameter of the jet is a technological characteristic too. The formed geometry of the workpiece depends on the cutting speed (v_c), linear feedrate (v_f), circular feedrate (v_{fcirc}), the displacement and direction of the jet and the energy of the waterjet

The efficiency of the jet is determined by the water pressure (p), the abrasive mass flow rate (m), and the geometry of the primary (water) and secondary (abrasive) nozzle of the cutting head, the extent,

IOP Conf. Series: Materials Science and Engineering **448** (2018) 012053 doi:10.1088/1757-899X/448/1/012053

and the material of the abrasive grains. From these parameters usually, the pressure and the abrasive mass flow rate are easy to change.

Depending on the energy of the jet, the actual machining diameter can be formed in two ways. The jet either completely cuts the material or just at a certain depth. (Fig. 2) When machining rotationally symmetric parts, extent of the removed material, the diameter of the workpiece depend on the energy of the jet [5][6]. When planning the machined diameter of the workpiece results of kerfing tests (not cot through tests) can be a very good base information to define technological parameters (abrasive mass flow rate, pressure, and feedrate).

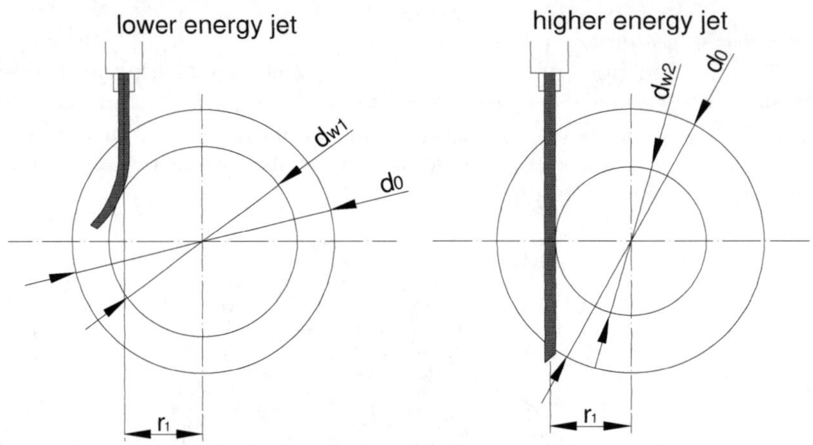

Figure 2. Formation of the workpiece diameter depending on the energy of the jet

Formation of the machined diameter of the workpiece is shown on Fig. 2. Extent of the final diameter depends on the energy of the jet, and the displacement of the jet according to the workpiece axis.

3. Characteristics of the planned device for waterjet turning

For turning with abrasive waterjet, a special device should be planned. This device should be meet the following requirements:

- the waterjet machine tool assures the waterjet and the linear feedrate, the device assures the circular feedrate
- the device should be stable fixed on the machine tool
- workpiece should be fixed in the device
- electrical and corrosive parts of the device are protected from the water

During the design, the workpiece rotating device was divided into three main units: a power unit, one waterproofing structure and one supporting structure.

The power unit is an infinitely variable drive ensuring the fixing and rotation of the workpiece. This task is solved by a three-jaw chuck and pulleys. For the infinitely adjustment of the number of revolutions a toroid-type transformer with speed meter was applied. The device is able to change of the direction of the rotation motion. The speed meter is shown in Fig. 3.

Figure 3. The speed meter

For protect the device from the coming back water a special waterproofing structure was planned. All these parts should be place on a special supporting structure with help of which the device can be fixed on the waterjet machine-tool. This part should ensure the parallelism of the axis of the workpiece end the linear motion of the cutting head. Considering the supporting structure, the dimensions of the machine-tool and the range of motions, the place of the waterproofing element can be defined.

Figure 4. Power unit of the planned waterjet turning device

After planning and manufacturing of the parts, the waterjet turning device was assembled (Fig. 4). First, dry motion tests have been accomplished for today. The power of the motor, the stability of the rotation parallelism of the motions were tested. During these preliminary tests all the appeared mistakes were eliminated.

For the waterproofing the device has a special protective cover made of polymer, soft PVC. Fig. 5 shows the waterproofing structure of the waterjet turning device. The test part are under pre-machining from steel and aluminium-alloy material.[7] Different fixing mandrels and other fixing elements are available for the tests as well. For today the device can work for real cutting tests.

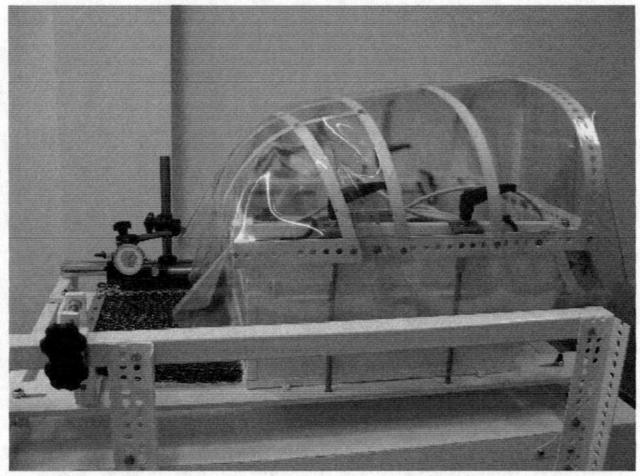

Figure 5. Waterproofing structure of the waterjet turning device

For planning and implementation of cutting experiments technological parameter should be determined. These main parameters are as follows: cutting speed (v_c), which is defined by the pressure of the water, the abrasive mass flow rate and the geometry of the nozzles; feedrate speed of the cutting head (v_f) and the circular feedrate speed (v_{fcirc}) of the workpiece defined by the rotation of the workpiece.

On base of previously accomplished cutting tests oriented to the determination of depth of kerf on different these turning parameters can be estimated. In the first waterjet turning test we will estimate the needed cutting parameters by this way, with help of previously made kerfing nomograms.[8]

4. Summary

Planning of a special waterjet device was shown I this paper. For today the device has been produced. With help of this device waterjet turning can be accomplished on a two-dimensional waterjet machine tool. This gives the possibility to machine cylindrical part (or rotationally symmetric parts) from special materials difficult to cut by chip removal. Future tasks of the research work to investigate the connection between the very high numbers of technological parameters and the geometry of the machined parts, the analysis of the efficiency and accuracy of the cut.

5. Acknowledgement

"The described study was carried out as part of the EFOP-3.6.1-16-00011 "Younger and Renewing University – Innovative Knowledge City – institutional development of the University of Miskolc aiming at intelligent specialisation" project implemented in the framework of the Szechenyi 2020 program. The realization of this project is supported by the European Union, co-financed by the European Social Fund."

References

[1] Hashish M 1998 The Waterjet as a Tool *14th International Conference on Jetting Technology* Bruges Belgium pp iXX-iXLIV

[2] Axinte D A, Stepanian J P, Kong M C, McGourlay J 2009 Abrasive waterjet turning – An efficient method to profile and dress grinding wheels *International Journal of Machine Tools & Manufacture* **49** pp 351-356

[3] Momber A W and Kovacevic R 1998 Principles of Abrasive Water Jet Machining *Springer-Verlag London Limited* p 394

[4] Manu R, Babu N R, 2009 An erosion-based model for abrasive waterjet turning of ductile material *Wear* **266** pp 1091-1097

IOP Conf. Series: Materials Science and Engineering **448** (2018) 012053 doi:10.1088/1757-899X/448/1/012053

[5] Srivastava A K, Nag A, Dixit A R, Tiwari S, Scucka J, Zelenak M, Hloch S, Hlavacek P 2017 Surface integrity in tangential turning of hybrid MMC A359/B_4C/Al_2O_3 by abrasive waterjet *Journal of Manufacturing Processes* **28** pp 11-20

[6] Li W, Zhu H, Wang J, Ali Y M, Huang C 2013 An investigation into the radial-mode abrasive waterjet turning process on high tensile steel *International Journal of Mechanical Science* **77** pp 365-376

[7] Liu D, Huang C, Wang J, Zhu H, Yao P, Liu Z W, 2014 Modeling and optimization of operating parameters for abrasive waterjet turning alumina ceramics using response surface methodology combined with Box-Behnken design *Ceramics International* **40** pp 7899-7908

[8] Maros Zs 2014 Effect of load energy on the form of the gap at waterjet cutting *Key Engineering Materials* **581** pp. 304-309

IOP Conf. Series: Materials Science and Engineering **448** (2018) 012054 doi:10.1088/1757-899X/448/1/012054

Application and modelling of polygon joints torque transmissions in the power transmission of electromobiles

Ottó Szabó

Institute of Manufacturing Science, University of Miskolc, Miskolc, Hungary

e-mail: otto.szabo@uni-miskolc.hu

Abstract. The fourth industrial revolution has started, and one of its determining areas is electromobility. The author advises the exchange of the traditional torque transmission joints for polygon joints, which have proven to be more favourable. This recommendation is confirmed by many supporting facts. The parts of polygon joints can be machined with simpler and cheaper tools and target machines. The technical advantages mean economical benefits as well. This paper shows digital model equipment which assists in the graphical design of polygon joints. This tool makes the engineering and aesthetic work easier, too.

1. Introduction

The fourth industrial revolution has started, and one of its determining areas is the electromobility. E-mobility is a pillar of the 21st century. Electromobility relies on the earlier development phases of the machine and electronics industry (automatization, intelligent technologies, computers, CNC techniques, super-hard tools, novel technologies, etc.) and the results of the past hundred years of the automotive industry. The substitution of vehicles driven by environmental polluting internal combustion engines has already started. Significant results are found in both international and Hungarian studies in this area [1-3].

The relatively simple and easy to control powertrain is one of the greatest advantages of electrical driven vehicles. Because there is no need to apply complex internal combustion engines and transmissions, which are expensive to produce, this can lead to cheaper vehicle prices.

It is advisable to analyse and modernise the machine design parts of vehicles, for example: power transmission mechanisms and shaft-hub joints. This is a good time for re-evaluation of the traditional solutions. One of these issues is the expensive joints used in current designs.

2. Comparison of traditional and polygon joints

Shaft-hub joints ensure the connection between the electrical power source (motor) and the driven shaft in the power train of electromobiles. The traditional shaft-hub joints (key, taper key, rib, serrated, kerb) show many disadvantageous properties (Figure 1a). The sharp edges on these are the starting points of stress peaks, fatigue and/or shaft breakage [4-5]. This is taken into consideration by the safety factors in the strength dimensioning. The results are the following: greater shaft diameter and higher material cost. The strength dimensioning with safety factors increases the weight of the vehicle.

Furthermore, the machining is performed on more expensive, special machine tools (serrated milling machine, serrated grinding machines, broaching machines) and with expensive cutting tools with complex geometry. The traditional joints have eccentricity error.

The technical errors and the manufacturing difficulties of traditional joints are eliminated by the polygon shafts, which are non-circular, and these can be given with epicycloid and hypocycloid curves. These polygon joints can provide reliable shaft-hub connections (Figure 1b).

Polygon joints show numerous favourable properties. Among these are that stress collecting points do not form; there is no need to increase the cross-section of the shafts; their life time is higher, and they ensure self-centring [5].

The manufacturing of the polygon and cylindrical surfaces of polygon joint shafts is done in one operation by an up-to-date, target machine-like designed polygon lathe and polygon grinding machine. Furthermore, hubs with polygon bores are machined on a polygon lathe and their finishing operation is done by a small, short broaching tool. After the roughing operation, the allowance for the broaching is minimal [6-7].

(a)

(b)

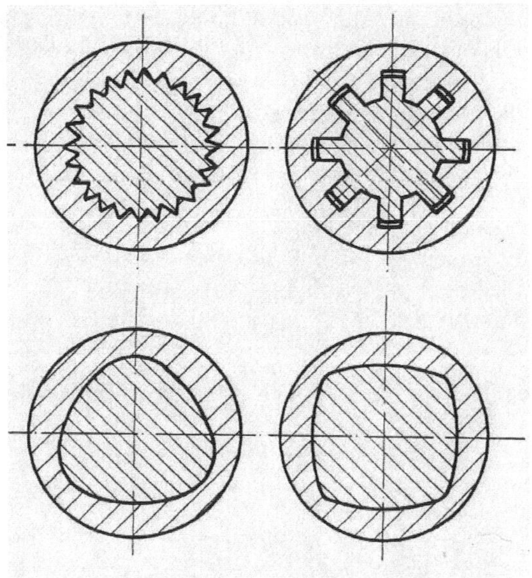

Figure 1. (a) Traditional shaft-hub joint: Kerb-toothed and serrated surface (b) Polygon joints with three- and four-sided hypocycloid cross-sections

Precise accuracy can be achieved for hardened polygon bores and shaft by super hard cutting tools on polygon lathe [10], with dry machining [16] and by quasi-honing with super hard grains in fine finish (quasi honing) [11]. The hardening of the surface layer can be ensured with diamond burnishing [17-18]. The pre-machining of the torque transmission parts (shaft and hub with bore) does not change compared to the traditional technology. The production of the short broaching tool is done on a polygon lathe in roughing and on a polygon grinding machine after hardening. The manufacturing system can be assembled by target machines.

The industrial suitability of the polygon joints is justified by the polygon grinding machines manufactured by the Fortuna Works and their widespread usage [5]. The Sandvik Coromant produced the Coromant Capto fast tool-changing system (in three sizes) for machine tools with high precision and high accuracy, which apply polygon joints. These are the tools of the modern machine industry [8-9].

Development taking place in the Department of Production Engineering (nowadays: Institute of Manufacturing Science) and Department of Machine Tools (nowadays: Institute of Machine Tools and Mechatronics) of the Faculty of Mechanical Engineering and Informatics, University of Miskolc has been used e.g. in the machine industry, tool manufacturing for the mining industry, and the production of forms for glass industry [6-7]. The manufacturing of many workpiece-actuated polygon lathes and grinding machines was carried out in the two departments. These results can provide a prospective future for the automotive industry. The novel results of this research led to the publication of many papers, dissertations and patents. The enumeration of those exceeds the extent of this paper. Polygon joints can be designed and manufactured with different tools and shapes according to the German standard [12]. The draft for a Hungarian standard has been completed [13].

3. Parametrical equations and graphical representation of polygon joints

3.1 Epi- and hypocycloids

In normal sectional view the polygon joints are epi- and hypocycloids or curves related to those. Numerous international and national polygon machining methods are known. Most of the procedures can be traced back to one base case: the generation of the cycloid. Figure 2c models the generation of polygon curves. The basis of this is the following: the epi- or hypocycloid is generated by the Point P assigned on a circle rotating around Point O_1, which rolls down without slip on the base circle rotating around Point O (which is the generating plane).

In the practice, the normal sectional view for most of the polygon joints is a stretched hypocycloid (stretching factor: $\lambda < 1$). The parametrical equation of the hypocycloids with the notation of Figure 2 is:

$$\left.\begin{aligned} x = x(\delta) = L_t \cos\delta + e\cos C_2\delta \\ y = y(\delta) = L_t \sin\delta + e\sin C_2\delta \end{aligned}\right\}, \quad (1)$$

where $L_t = R - r\ ;\ e = \lambda r$ and $C_2 = \frac{R-r}{-r}$

R: radius of base circle; r: radius of pitch circle; δ: rolling angle; e: excentricity.

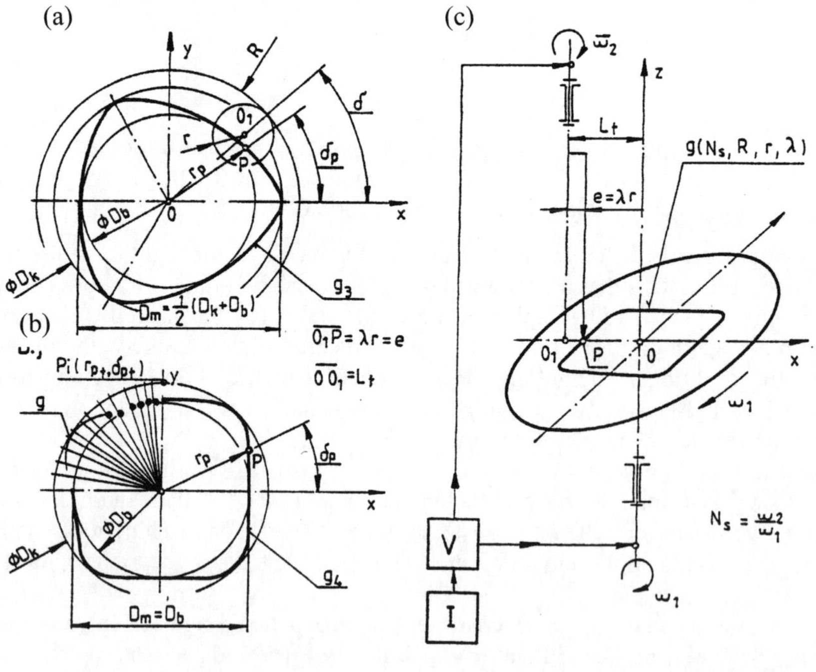

Figure 2. (a) Geometry of the hypocycloid profile (b) g-marked plane curve given by external program carrier (c) Kinematic model of profile mapping

The following equations can be written based on the geometric and kinematic relations which must be fulfilled:

$$\omega_2 = \omega_1 \cdot \frac{R}{r} = \omega_1 \cdot N_S \;;\; N_S = \text{integer}\,\frac{R}{r} = \text{integer}\,\frac{\omega_2}{\omega_1} \quad , \tag{2}$$

where the used notations are the radius of the base circle (R), the radius of the pitch circle (r); and their angular speed (ω_1, ω_2). It is a requirement that the number of the polygon sides must be an integer: N_s = 1, 2, 3, 4, …. If the rotation direction of the two axes is the same, a hypocycloid is generated. If these directions are the opposite, the result is an epicycloid. In the working model equipment, the adjustable parameters are: ω_1, ω_2, R, r, L_t, r, N_s, e (Figure 3). The kinematic mapping system (Figure 2c), which solves Equation (1) and (2) and which is based on the superposition of two rotary movements, is capable of mapping different hypo- and epicycloids, ellipses, Pascal spirals and cardioids. The criterion is that the executing unit V should operate based on Equations (1) and (2). The realisation of the g-curve in Figure 2b can be done by external data input depending on the need.

3.2 Model equipment to generate digital epi- and hypocycloids

The visual fast supervision of the designed polygon profile is necessary in the design of the polygon joint and in the adjustment of the polygon machining equipment. Therefore, the author planned and built model equipment capable of generating of digitally controlled polygon-curves for the visualisation of epi-, hypocycloids and other different curves.

The theoretical structure of the equipment is shown in Figure 2c, where there is a writing tool in the P point (which symbolises one point of the cutting edge) and the plane of registration is the x-y plane (symbolises the normal cross-section of the workpiece). The two revolving axles (with ω_1 and ω_2 angular speed) of the digital model equipment are rotated by controlled stepper motors (V in Figure 2/c), which are connected to the transmission, according to the orders of control unit I.

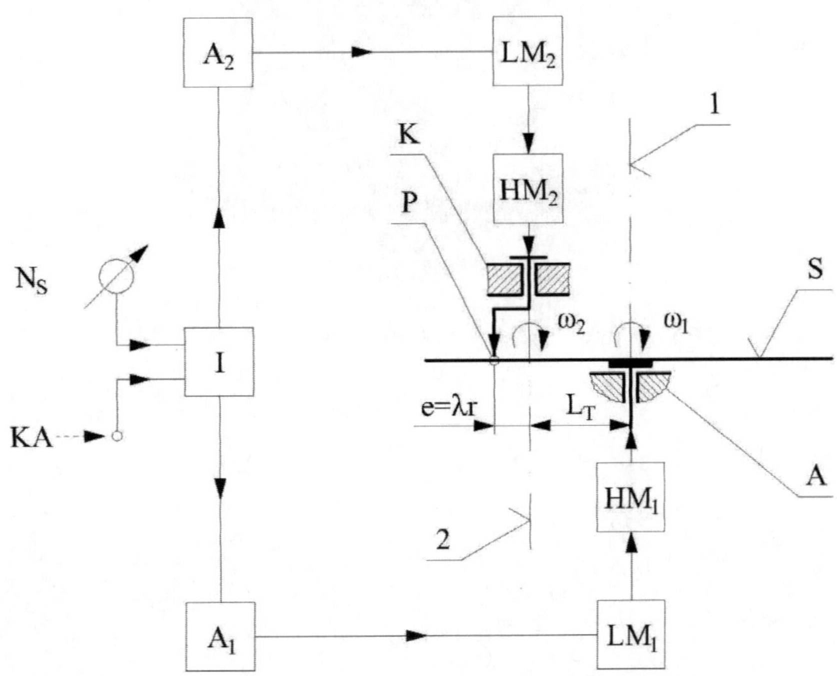

Figure 3. System structure of the digital model equipment

The schematic of the equipment is shown in Figure 3. The symbols are: base plate with the bearing of axle 1 (A), console with the bearing of axle 2 (K), plane of registration (S), writing tool (P), adapter

IOP Conf. Series: Materials Science and Engineering **448** (2018) 012054 doi:10.1088/1757-899X/448/1/012054

transmissions (HM_1, HM_2), electric stepper motors (LM_1, LM_2), the distance of P point from axle 2 (e), polygon number selector (Ns) and external data input (KA). Figure 4 shows a photograph of the graphical unit of the model equipment.

The inspection of the system is provided by the generation of a Cardano line, whose basis is: the model equipment works correctly if the registrate of two controlled, given rotary movements is a line. The controlled, correct function check of control unit I and the two stepper motors in unit V was done by this method.

The shape of the cycloid can be changed with the help of the polygon number (N_s) and the other adjustable parameters shown in Equations (1) and (2) (Figure 2a and 2c). There is an option for external data input for special path mapping tasks. These tasks occur for example in the machining of cam plates and decorative surfaces (Figure 2b). Recordings generated by the model equipment are shown in Figures 5-8.

In modern torque transmission polygon joints, the usage of convex or nearly linear curved hypocycloids is advisable. The two outer curves in Figure 7, which are extended hypocycloids, are good example for this. The two inner curves are peaked hypocycloids.

Figure 4. Graphical unit of the modelling equipment

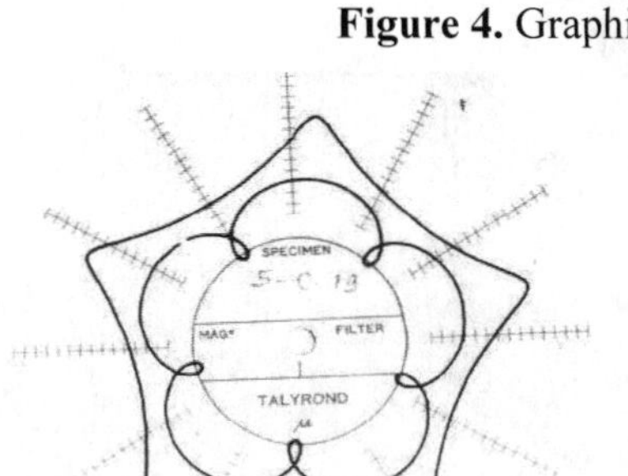

Figure 5. Five-sided peaked and looped epicycloid

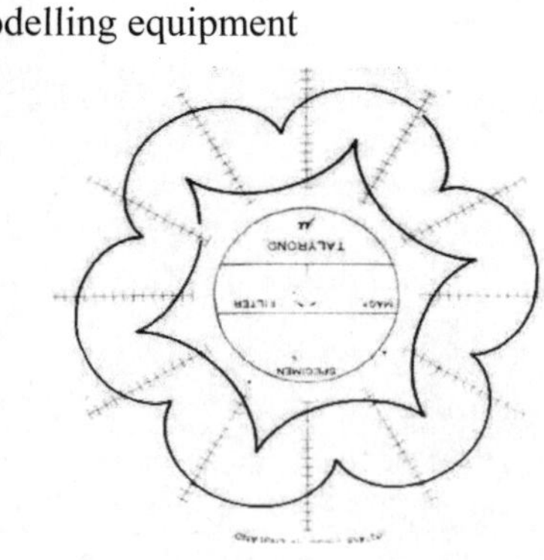

Figure 6. Six-sided peaked epi- and hypocycloid

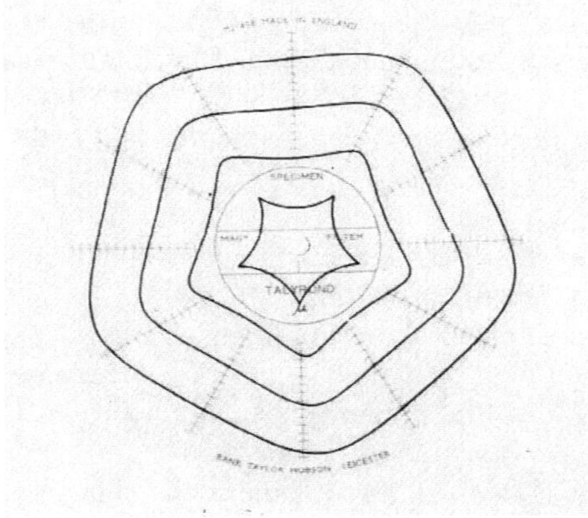

Figure 7. Five-sided stretched hypocycloid (excluding the inner curve)

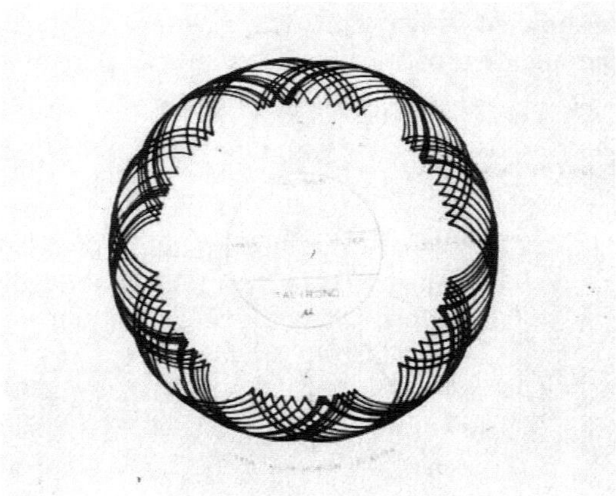

Figure 8. Family of curves for decorative purpose

The normal cross-sectional shape, characteristic dimensions, fit and surface quality (roughness) requirements of the polygon joints (fitted shaft-hub connections) can be determined by the German standard. The optimal strength design and check is done by the method of Mechnik furthermore Citarella and Pertella [14-15].

4. Conclusion, proposal and vision

The internal combustion engine and the needed transmission are very complex constructions from the points of view of both design and production. These are made with numerous parts, which need certified materials and blanks, special, expensive machine tools for machining, and complex tools and technologies. The service and repair of internal combustion engines are more expensive than the electric power supply of the electromobiles.

In the last century, the automotive industry settled on the traditional shaft and hub joints. It is time to change to polygon joints in the starting century of electromobility: they are better in quality, advantageous in characteristics and more economical. It has been proven that better quality parameters, easier and cheaper tooling, lower number of machining operations and target machines, smaller workspace, lower energy consumption can be achieved. Polygon joints, machining target machines and technologies are promising for the automotive industry based on the results of the research and development work so far.

The engineering tasks (analysis and design of polygon profiles, drafting of cam plates) often need fast graphical visualisation, which is helped by the digital model equipment presented in this paper.

Many problems need to be answered in Hungary as well, for example the building of a suitable electrical charger network, the research and development of fast recharge and energy storage, the installation of technical bases and services, the teaching of specialists and the reduction the expenses. An electrical driven vehicle could be significantly cheaper if the costs of electrical power storage could be lowered. This is still to be expected, because lithium batteries are used nowadays. Sadly, the world market price of lithium is increasing.

Hungary can become one of the leading countries in electromobility with the application of the earlier and current novel research results. This work aims to contribute to this objective.

Motto

Henry Ford dreamed of and manufactured a car which could be bought by the workers of the Ford Industries. This was achieved!
This thought must act as a model for the present-day automotive industry.

IOP Conf. Series: Materials Science and Engineering **448** (2018) 012054 doi:10.1088/1757-899X/448/1/012054

Acknowledgement
The author greatly appreciates the support of the Hungarian National Research, Development and Innovation Office (OTKA, Budapest) T048760.

References

[1] 2013. Why Electromobility and What is It? In: System Perspectives in Electromobility (B. Sandén, ed.), Chalmers University, Gothenburg (Sweden), pp. 10-21 http://publications.lib.chalmers.se/records/fulltext/211430/local_211430.pdf (14 May 2018)

[2] Klüser R (IFA/DGUV) 2018 Elektromobility https://oshwihi.eu/wuju/elektromobility (downloaded 14 May 2018)

[3] Conference on E-mobility 2018 Az e-mobilitás a 21. század egyik pillére (in Hungarian). https://jedlikanyosklaszter.hu/e-mobilitáskonferencia (downloaded 14 May 2018)

[4] Takacs J. 2004 *Advanced technologies in forming surface properties*. Budapest University of Technology Budapest 346 p. (in Hungarian)

[5] Fortuna-Werke Maschinenfabrik AG 1968 *Wissenwertes über Polygon* Stuttgart-Bad Cannstatt Anhang 20 p. Nr: 854 1069 (300E). 87 p.

[6] Lierath F, Tajnafői J, Menz P, Demeter P, Hammer A 2002 Polygonbearbeitung an Drehmaschinen. *Gép* **LIII** (2-3) Budapest pp. 22-25.

[7] Tajnafői J, Gellért K, Gribovszki L, et al. 1978 *Procedure and grinding machine for machining polygonal workpieces* Patent Hungarian Intellectual Property Office (OTH) 167.643. Budapest (in Hungarian)

[8] https://sandvik.coromant.com/en-us/products/coromant_capto (downloaded 14 May 2018)

[9] Rivin E 2000 Tooling Structure: Interface between Cutting Edge and Machine Tool *Annals of the CIRP* **49** 2 pp. 591-633.

[10] Kundrák J 1996 Cutting with Superhard Tools under Periodically Changing Conditions (in Russian) *Rezanie pri Periodicheskikh nagruzkakh rezcami iz sverkhtvjordydh materialov* Kharkov-Miskolc 90 p.

[11] Szabó O 2001 Optimisation of Technology and "Quasi Honing" of Polygon Bores *Journal of Materials Processing and Technology* **119** pp.117-121.

[12] DIN 32711 Polygonprofile P3G; DIN 32712 Polygonprofile P4C. https://www.inkoma-albert.com/wp-content/uploads/2015/07/inkoma-shaft-hub-connections.pdf (downloaded 14 May 2018)

[13] MI-05-24.0035-1982. *Metal cutting machine tools. Polygonal sectional shaft couplings* 27 p. (in Hungarian)

[14] Mechnik R P 1991 Beitrag zur Festigkeitsberechnung von Polygon-Welle-Nabe- Verbindungen unter reiner Torsion *Konstruktion* **43**

[15] Citarella R, Pertella M 2015 Robust Design of a Polygonal Shaft-Hub Coupling *Frattura ed Integritá Strutturale* **34** pp.554-563.

[16] Mamalis A G, Kundrak J, Gyani K 2002 On the Dry Machining of Steel Surfaces using Superhard Tools *International Journal of Advanced Manufacturing Technology* **19** 3 pp. 157-162.

[17] Varga G 2016 Possibility to Increase the Life Time of Surfaces on Parts by the Use of Diamond Burnishing Process, *Key Engineering Materials*, **686** pp. 100-107.

[18] Varga G 2014 Effects of Technological Parameters on the Surface Texture of Burnished Surfaces, *Key Engineering Materials* **581** pp. 403-408.

IOP Conf. Series: Materials Science and Engineering **448** (2018) 012055 doi:10.1088/1757-899X/448/1/012055

Oriented Isotactic Polypropylene Structure and Morphology

M.Kaszonyiova [1*], F.Rybnikar [1], J.Sadilek [2] J.Kucera [2], M.Kubisova[3]
[1]Department of Polymer Engineering, Faculty of Technology, Tomas Bata University in Zlin, Czech Republic
[2] UNIPETROL RPA, s.r.o. – POLYMER INSTITUTE BRNO, odštěpný závod, Czech Republic
[3]Department of Production Engineering, Faculty of Technology, Tomas Bata University in Zlin, Czech Republic

mhribova@utb.cz

Abstract. Structure, morphology and mechanical properties of uniaxially oriented in the solid state and subsequently annealed isotactic polypropylene sheets were examined by X-ray diffraction, electron microscopy, differential scanning calorimetry (DSC) and mechanical testing. The sheet structure and morphology showed parallel fibrils with orientation degree, crystal size, crystallinity, melting temperature and mechanical properties depending on draw ratio and annealing temperature.

1. Introduction

Isotactic polypropylene (iPP) is a widely applied technical material mainly due to its versatility and relatively low price. The low mechanical strength is often improved by sample preferred orientation [1 - 3]. There are several types of orientation, one, two or more directional drawing, with or without subsequent annealing at temperatures near but below the melting point. [7, 8]. For instance, unidirectional drawing leads finally to a microfibrillar morphology with increased tensile strength and Young's modulus [1 - 3, 8] and different melting behaviour [12].

According to the classical model proposed by Peterlin [11], drawn films and fibres consist of aligned small lamellar blocks mutually connected through non-crystalline molecules and/or crystalline bridges. Several reports do not support his model and the proposed transition scheme of original crystal lamellae into microfibrils and a longitudinal sliding motion of these microfibrils [10,11]. Those studies describe the melting of crystalline lamellae under tension followed by the recrystallisation into the oriented fibrillar structure. At present, there is no accepted general model for the drawing behaviour of crystalline polymer samples.

In this contribution, we describe the preparation and characterisation of iPP sheets unidirectionally oriented and subsequently annealed taut at various temperatures below the melting temperature. The relations between structural changes and mechanical properties owing to preferred orientation and annealing are discussed.

Experimental

1.1. Samples

Isotactic polypropylene, Mosten TB002, was obtained as pellets from Unipetrol RPA, Czech Republic, with isotacticity 98.3%, number molecular weight 67 935 g/mol and weight molecular average 428

800 g/mol. Sheets of this iPP were prepared in the Polymer Institute Brno (Czech Republic). In the first step the pellets were extruded at 125°C as sheets. The sheets were cut to 10 mm wide bands and extended uniformly at 145°C to extension ratios 5, 8 and 11. Annealing, was performed either 30 min in an oven while bands were held taut, or moved slowly (10 min in contact with plates and a roller heated to annealing temperature (Ta). The sample speed and thickness were continuously monitored. The sheets with various elongation ratio, λ, and annealing treatments, with the samples held taut, are listed in Table I. All further measurements were done at room temperature.

1.2. X-ray diffraction

The wide angle diffractogram scans were taken in the longitudinal and radial directions in the 2Θ range 5 - 30° in transmission and reflection modes. Scans were collected on a horizontal goniometer HZG-6 (Freiberg Prazisionsmechanik Holding GmbH, Germany) and the diffractometer A1P1 (AREM PRO, s.r.o., Czech Republic) using CuK_α Ni filtered line shape radiation. The measuring conditions were: 0,05° step size, hold time 5s at room temperature.

Crystal size - The sample crystal size, L110, was calculated from the 110 crystal diffraction peak profiles according to the Scherrer equation [12].

Crystallinity - The sample crystallinity (X) was calculated from the X-ray scans as the ratio of areas of crystalline peaks and total scan area over background.

Orientation degree - The most important characteristic of preferentially oriented samples is the type and, primarily, the degree of orientation. The orientation degree was characterized by the azimuthal 2Θ angle where the original 110 peak intensity on the equator decreased to one half (Or).

1.3. Morphology

Transmission electron microscopy (TEM) was performed using a Tesla 500 (TESLA ELMI, a.s., Czech Republic) at 90kV using replicas of selectively etched sample surfaces. After selective etching the samples were vacuum shadowed with gold or platinum and vertically coated with an evaporated carbon film. The magnification of the micrographs is indicated by 1μm bars and the extension direction by an arrow.

1.4. DSC-measurements

Disc shaped specimens, 1 to 5 mg in weight, were cut from the sheets and measured in a DSC (TA Instruments Inc., Q100, USA). The scan conditions were heating rate 10 °C/min in nitrogen atmosphere. From the scans the fusion enthalpy (ΔH), peak temperature (Tm), melting range and crystallinity (X) were evaluated. For crystallinity calculation as a standard for 100% iPP the $\Delta H0$ = 209 J/g was taken according to Mezghani and Phillips [13].

1.5. Mechanical properties measurements

The tensile properties of the original and annealed polypropylene film strips were tested using an Instron 4302 (Illinois Tool Works Inc., USA).

2. Results and Discussion

2.1. X-ray diffraction – crystal structure

Figure 1 shows the transmission X–ray diffraction scans of the non-oriented sample, TB, after annealing at various Ta. The scans show 7 reflections corresponding to the (110), (040), (130), (111), (031) and (041) superimposed, (060) and (220) crystallographic planes, respectively. The diffraction patterns belong to the stable α monoclinic unit cell of isotactic polypropylene with cell dimension a = 6.65, b = 20.96 and c= 6.50 Å [10]. Annealing at higher temperature increased the peak intensity, and decreased the peak width so that the two peaks (111 and superimposed 131 and 041) in the range 2Θ = 21 - 22° could be better resolved. Annealing the TB samples (Fig. 1) at higher T_a tended to increase the (040) and (130) intensities relative to that of (110). In non-annealed, oriented samples (T_a = 23°C) the ratio of I (040) / I (110) decreased with growing extension ratio λ from 0.55 of the non-oriented sample (λ = 1) to 0.34 for the temperature and sample 11/10 with λ = 11. This is in accord with increasing sample orientation.

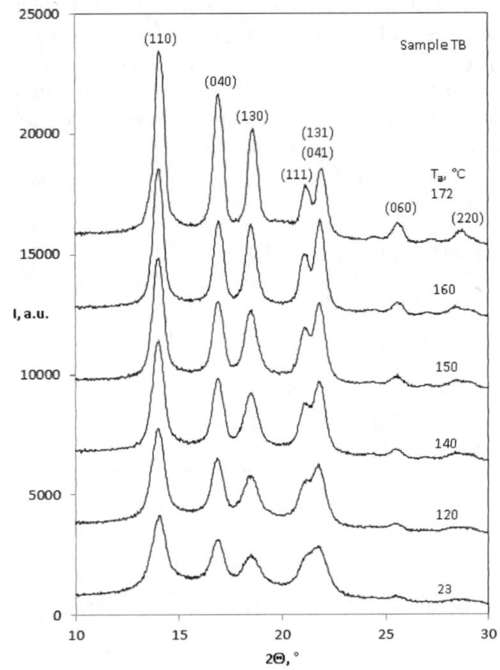

Figure 1. Typical equatorial WAXS diffractograms of the TB samples non-oriented and annealed at various Ta.

Increasing T_a had an opposite effect, the ratio $I_{(040)}/I_{(110)}$ increased with growing T_a. Annealing, though in a fixed position, slightly decreased the sample orientation degree. A hypothetical explanation of this effect (higher 040 peak intensity) would be the formation of a small amount of the γ – phase in the α-phase matrix, however, this is improbable because the main γ-phase characteristic peak (117) at ~ 21° 2Θ was not detected in any of our samples.
The crystallinity increased with elongation degree λ and annealing temperature Ta as expected.

2.2. Crystal size

The tendency for the L110 values to increase with increasing Ta was apparent for sample TB annealed for a relatively long time (30 min), while in oriented samples, annealed for a shorter time (10 min), the L110 values growth was less pronounced. The L110 value at Ta = 23° C for the non-oriented sample, TB, was slightly higher (107 Å) than for the oriented samples (101-104 Å) indicating that the extension process resulted in decrease of the original crystal size or changed the tilt of molecules in the crystals. With increasing Ta the L110 values increased for the non-oriented sample TB up to 143 Å. In the oriented samples the L110 values also increased with Ta, but to a lesser extent. Although the L110 values might also be affected by crystal defects, it is often accepted that they reflect mainly the crystal size. Thus, we suggest that the crystal size's growth during annealing depended on two factors: the original sample extension and the annealing temperature.

2.3. Orientation degree

The orientation degree values O_r were close to 4° and increased with increasing T_a only slightly. Thus annealing decreased the orientation degree (increase of the O_r or of I_{040}/I_{110}) a little with growing T_a. Sample annealing, mainly above 140°C, led to structure changes, including increased crystallinity and crystal size but on the other side, to lower orientation. Annealing of the drawn samples at higher T_a brought two types of effects: 1. Increasing crystallinity and crystal size. 2. Decreasing of the orientation degree and sample volume which, the latter in samples in taut position increased the internal tension and resulted in perpendicular crack formation, leading to the tensile strength decreasing in samples annealed at T_a = 150°C and higher.

2.4. Melting

For the non-oriented and non-annealed sample TB the melting scan (Fig. 2) shows only one broad melting peak (Tm = 165,4°C). Besides of the main peak in the TB samples, annealed at higher Ta, a second lower melting peak (or shoulder in sample with Ta = 160°C) appeared. The oriented and annealed samples had more complicated melting scans, showing 2-3 lower melting peaks or shoulders. The melting peaks and shoulders also had the tendency to shift to higher Tm with growing □ and Ta. The melting behavior was influenced by two factors: extension and annealing. The sample extension and annealing increased the melting temperature of the sample. Also lower melting peaks or shoulders shifted to higher temperatures.

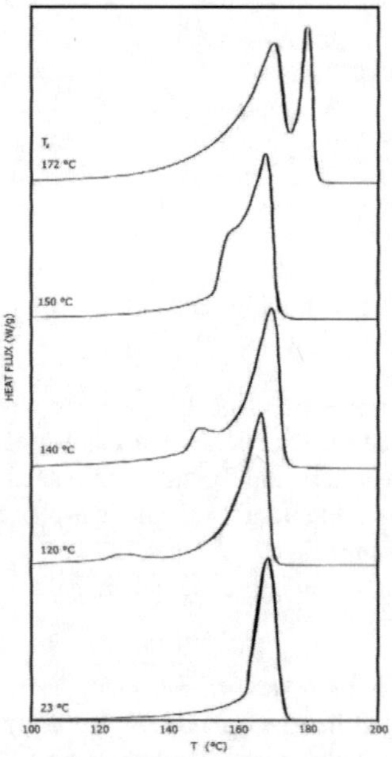

Figure 2. DSC melting curves of non-oriented (TB) and oriented samples iPP annealed at various Ta. a – non-oriented sample, b – sample 5/10, c – sample 8/10, d – sample 11/10.

2.5. Morphology

The examination of oriented iPP sheets by SEM and TEM has shown that the spherulitic structure was transformed by orientation and annealing to a system of parallel fibrils oriented along the orientation direction.

In Fig. 7a of the sample 5/10, the fibrils form 10 ~ 100μm long pointed elongated sections of probably emerging and sinking fibrils parts. It appears that the fibrils are not straight but slightly waved up and down in extension direction. The fibril diameter lies in the range 2 - 6 μm. Similar picture shows of the sample 8/10 Fig. 7b, where the emerging parts are longer (50 – 150 μm) and thicker (4 – 8 μm).

Fig 3c of the sample 11/10 has a different appearance. There are two features: the elongated fibrils in orientation direction such as in Fig. 3a and b and a system of perpendicular lines or cracks (10 – 20 μm apart) that are visible as a silvering effect. The silvering effect (opacity) was observed only in the sample 11/10 not annealed or annealed at Ta below 170°C. In the sample 11/10 annealed at 170°C the opacity decreased.

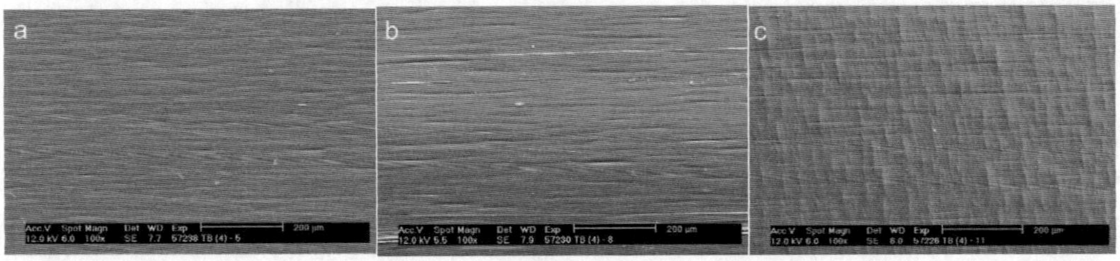

Figure 3. SEM micrographs of iPP parallel oriented fibrils : a – sample 5/10 annealed at Ta = 140°C, b –sample 8/10 annealed at Ta = 150°C, c – parallel fibrils and perpendicular cracks in sample 11/10 annealed at 120°C. Orientation direction horizontal.

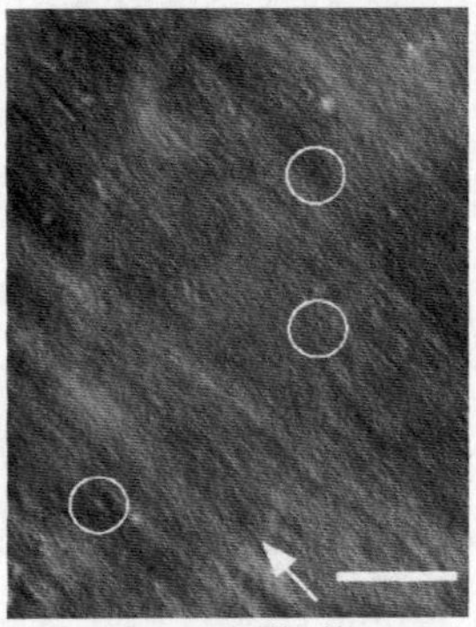

Figure 4. SEM micrographs of iPP parallel oriented fibrils : a – sample 5/10 annealed at Ta = 140°C, b –sample 8/10 annealed at Ta = 150°C, c – parallel fibrils and perpendicular cracks in sample 11/10 annealed at 120°C. Orientation direction horizontal.

The figure 4 of the sample 8/10 annealed at 120°C shows that the fibrils consist of thin parallel (120-160Å) lamellae oriented in the extension direction and perpendicular or slightly inclined to the sheet surface (marked by circle). Fig. 5a, b of the highest oriented not – annealed sample 11/10 disclosed details of iPP oriented structure.

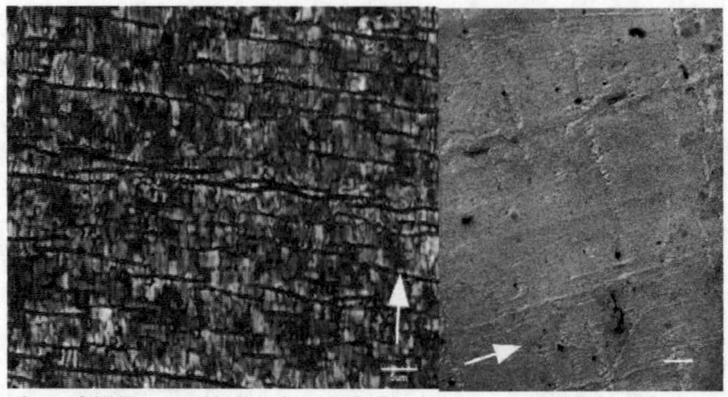

Figure 5. SEM micrographs of iPP parallel oriented fibrils : a – sample 5/10 annealed at Ta = 140°C, b –sample 8/10 annealed at Ta = 150°C, c – parallel fibrils and perpendicular cracks in sample 11/10 annealed at 120°C. Orientation direction horizontal.

Polymer chain breaking releases the local superheating and partial melting of broken chains ends parts (Fig. 10a). Annealing of highly oriented samples disclosed details of the cracked surface with lamellar parts extended to fibrils, melted fragment or voids and crazes (Fig. 10b). After annealing at Ta = 140 or 160°C the amount of such material increased. In those samples the individual fibrils and thin lamellae were clearly resolved. In some locations was evident that from their longitudinal fibrils or thin lamellae short branches grow in a perpendicular direction (marked R in Fig.10b).

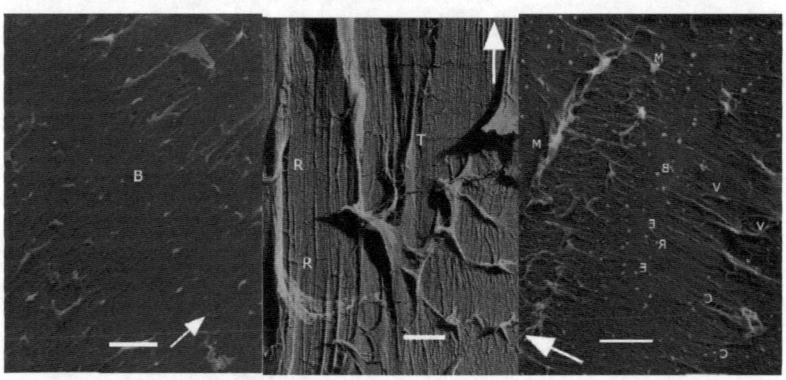

Figure 6. TEM micrographs of the 11/10 not annealed iPP sample.

Perpendicular bands (B) to extension direction formed by lamellar twisting and traces of chain breaking and local surface melting (M) are visible. 6b Details of the 11/10 surface after annealing at 140°C with twisted lamellae (T) and lamellae with short branches (R).

6c TEM micrograph shows details of the sample surface of a iPP sample 11 /10 after annealing at 160°C. In band parts with lamellae perpendicular to sample surface cracks and voids appeared (marked C and V). Perpendicular bands are marked (B), short branches (R), lamellar parts extended to fibrils (E), melted fragments (M).

2.6. Mechanical properties - *Tensile strength (TS)*
The tensile strength of the starting, not-annealed sample (35 MPa) increased with the elongation degree λ. In the direction perpendicular to the sample elongation the TS decreased an order of magnitude as shown for the sample with λ = 5 in Fig. 2. The substantial increase of the tensile strength in oriented samples was due mainly to the molecular orientation (Or) in the elongation direction. Sample annealing at Ta up to 140°C had only a negligible effect; the tensile strength decreased only slightly. Increasing the Ta to 170°C resulted in a sharp decrease of tensile strength and final sample failure. The main reason for the relative sharp TS decrease is the presence of growing number or depth of the perpendicular crazes formed initially in the extension process and further enlarged by annealing at Ta above 140°C.

3. Conclusion

The samples consisted of 63-83% crystalline, highly oriented, iPP α-modification. The structure and property changes, compared to the initial nonoriented spherulitic material, were due to 1. the elongation and 2. the annealing process.

In the elongation process the original lamellar spherulites transformed to a system of parallel fibrils (ca 2 – 6 μm diameter) oriented in the extension direction. During this conversion also elongated parallel cracks and voids and finally also perpendicular cracks formed in high oriented samples.

The crystallinity and crystal size values of oriented samples were slightly lower than in the nonoriented samples (TB) annealed at the same Ta. The void and cracks formed in highly oriented samples resulted in an anomalously low sample density. Annealing of the extended samples in taut position brought further structure changes mainly at Ta above 140° C. Annealing increased the sample crystallinity, crystal size, melting temperature and mechanical strenght. Annealing tended to decrease the crack and void content (crack healing).

The elongation process obviously represents a relatively drastic breaking up of the original lamellar spherulitic structure to smaller crystalline blocks of different size and lower Tm. During this transformation a small number of voids and cracks formed. After extending further structural changes occurred and their extent depended on Ta.

The sample fixation during annealing obviously partially hindered the chain segments mobility at Ta values below 150°C. At higher Ta the segmental mobility increased and the conformational changes took place similar as in the non-oriented sample. The enhanced segmental mobility at Ta above 160°C was confirmed by the decrease of silvering, which was observed in highly oriented samples 11/10. The silvering was due to cracks oriented perpendicular to the extension direction

References

[1] Bigg, D. M.: *Mechanical Properties Enhancement of Semicrystalline Polymers – A Review.* Polym Eng Sci, 1988, 28, 830 – 841.
[2] Turetskii A.A., Baranov, A.O., Chvalun, S.N., Yerina, N.A., Zubov,Yu.A., Prut, E.V., Bakeyev, N.F., Yenikolopyan, N.S.: The structure of higly oriented polypropolene and its effect on physico-mechanical properties. *Polym Sci U.S.S.R.,* 1986, vol. 28, no. 10, p. 2380-2386
[3] Karger-Kocsis, J. Polypropylene: structure, blends and composites: Structure and morphology, London,: Chapman and Hall, 1995, 392 p.
[4] Karacan, I., Taraiya, A.K., Bower, D.I., Ward, I.M.: Characterization of orientation of one-way and two-way drawn isotactic polypropylene films, Polymer, 1993, vol.34, no. 13, 2691-2701.
[8] Porter, R.S. Wang, L.-H., Uniaxial Extension and Order Development in Flexinble Chain Polymers, *JMacromol Sci – Review of Macromolecular Chemistry and Physics,* 1995, vol. 35, p. 63-115
[9] Alcock, B., Cabrera, N. O., Barkoula, N.-M., Reynolds, C. T., Govaert, L. E., Prius, T.: The effect of temperature and strain rate on the mechanical properties of highly oriented polypropylene

IOP Conf. Series: Materials Science and Engineering **448** (2018) 012055 doi:10.1088/1757-899X/448/1/012055

tapes and all-polypropylene composites. Composites Science and Technology, 2007, 67, 2061 – 2070.

[10] Peterlin A. Molecular model of drawing polyethylene and polypropylene. 1971, *J Mater Sci*, vol. 6, p.490-508.

[11] T. Sakurai, at all, Structural deformation behavior of isotactic polypropylene with different molecular characteristics during hot drawing proces, 2005, *Polymer*, vol. 46, p. 8846–8858

[12] Scherrer, P. (1918) Bestimmung der Grosse und der inneren Struktur von Kolloidteilchen mittels Rontgenstrahlen. *In:* Alexander, L. E.; X-Ray Diffraction Methods in Polymer Science. New York : Wiley-Interscience, **1969**. 582 p.

[13] Mezghani K.; Phillips P.J., The γ phase of high molecular weight isotactic polypropylene: III. The equilibrium melting point and the phase diagram, 1998, *Polymer*, vol. 39, no. 16, p. 3735-3744

Acknowledgment

This work was supported by an internal grant of UTB in Zlín IGA/FT/2018/004 financed from Funds of specific academic research.

IOP Conf. Series: Materials Science and Engineering **448** (2018) 012056 doi:10.1088/1757-899X/448/1/012056

Effects of the tool edge design on the roughness of face milled surfaces

B Karpuschewski[1], J Kundrák[2], C Felhő[2], G Varga[2], D Borysenko[3]

[1] Leibniz Institute for Materials Engineering IWT, Badgasteiner Str. 3, 28359 Bremen, Germany

[2] Institute of Manufacturing Science, University of Miskolc, Hungary

[3] Institute of Manufacturing Technology and Quality Management, Otto von Guericke University Magdeburg, Germany

csaba.felho@uni-miskolc.hu

Abstract. Increasing the cutting speed and/or the feed rate is often chosen to increase the efficiency of the material removal. Of course, there are limits to the selection of specific values in both cases, such as, the rigidity of the machining system, the loadability and lifetime of the cutting edges, the prescribed roughness values of the surfaces, etc. When increasing the feedrate in face milling, the conditions of the chip removal change considerably, because at low feed values, the material forming effect of the side edge (i.e. the edge on the lateral surface of the tool) is the most determinative and the chip deforms perpendicular to it, while increasing the feed gradually moves the determinative part of the chip removal to the edge which lies perpendicular to the tool axis. Therefore, both of the two edges have a distinct role in chip removal due to both their position and geometry. The two edges are located on one cutting insert; therefore, their geometry needs to be examined individually and collectively in the search for the most suitable for effective chip removal. The study investigates how the different tool geometrical characteristics affect the roughness parameters of surfaces in normal and in high-feed face milling.

1. Introduction

In face milling, the extent of roughness values on the different points and surface elements of the milled surface differs due to the movement (trajectory) of the cutting edge. This roughness inhomogeneity characteristic of the surface means that the roughness values in planes measured in different directions are different [1]. In the path of the rotation axis of the tool, in the direction of feed, in relation to the symmetry line, theoretically, only those planes have the same roughness values which are at the same distance, and the greatest roughness of the milled surface is in the symmetry plane. In addition, the magnitude of roughness is influenced by the settings and conditions of the surface generation. These topics are undergoing extensive research, which is also very diverse in methods and testing purposes.

In the modelling introduced in [2], which is based on the geometric analysis of the trails left behind on the milled surface, particular attention has been paid to the consideration of tool set-up errors (axial and radial). The procedure used not only determines the theoretical roughness characteristics, but also generates two-dimensional theoretical roughness profiles. The effect of different technological parameters such as speed, feed rate and depth of cut on the roughness, flatness and shape errors of machined surfaces was investigated experimentally in [3] when face milling of wrought cast steel (WCB) workpieces, using ANOVA to analyse the results. Krolczyk et al. [4] investigated the surface texture (two- and three-dimensional surface roughness) in the turning of austenitic stainless steel by coated carbide tools. The effects of the runout errors of milling cutter inserts on surface topography, surface location error and stability were investigated in [5]. The authors concluded that runouts in milling remain an important issue in milling, because they significantly affect the machining process and can lead to premature failure of the cutting edges. In the surface roughness estimation method described in [6], not only the cutting parameters and the axial and radial setting errors were taken into account, but also the dynamic phenomena resulting from the cutting tool errors in end milling. Miko et al. [7] investigated the effects of the feed per tooth f_z, depth of cut a_p, and width of cut a_e on different surface roughness parameters in ball-end milling. The measured surface roughness was compared with the analytically calculated theoretical roughness and an estimation method was presented. Ref. [8] also analyses the flatness deviations in addition to 2D and 3D roughness parameters in the milling of aluminium alloys. Miko and Farkas [9] also investigated the flatness as well as 2D and 3D surface roughness parameters in face milling and face turning with different parameters. The effect of the coolant and lubricant fluid, technological data (spindle speed, feed rate and depth of cut) and the milling method (down- and up-milling) on residual stresses, cutting forces and roughness of the machined surface was investigated in [10] by the RSM method. Theoretical description of the surface pattern in face milling was studied in [11], where both mathematical calculations (MATLAB) and CAD simulations were used to study the effects of different workpiece and tool designs and positions, the feed rate and the angle of the milling axis to the surface on the quality (roughness) of the machined surface. A so-called gray-fuzzy modelling procedure was used in [12] to determine the optimal process parameters for end milling of an aluminium alloy. The examined process parameters were the arithmetical average roughness (Ra) of the root mean square roughness (Rq) and the material removal rate (MRR). A hybrid approach for the modelling of surface roughness in slot milling is described in [13], where the analytical calculation of the specific energy consumption (SCEC) and the empirical relationship between SCEC and surface roughness were combined in a model. It was found that there is a direct relation between the specific energy consumption and the roughness parameter Ra and a new model was suggested to estimate the expected roughness. Ref. [14] focuses on the determination of the optimal cutting parameters that result in minimal roughness characteristics in plain peripheral down milling, where theoretical parameters were estimated using a model using an artificial neural network (ANN). RSM and ANOVA methods were used to determine input and output parameters. Experiments were conducted for the training and verification of the model by machining Ti-6Al-4V ELI titanium alloys. A formula for estimating surface roughness was reported in z-level milling of steep surfaces in [15] when using a 2.5D milling strategy. The technological parameters considered in the estimation are the depth of cut (a_p), feed per tooth (f_z), and the milling strategy (climb or zig-zag milling). The analytical modelling of theoretical roughness was described for face milling in [16], where the effect of technological parameters (feed rate, cutting speed) was investigated on the roughness of the machined surface in case of two different cutting insert geometries. A new procedure was described in [17], where the determination of the theoretical 2D and 3D roughness characteristics is based on the CAD modelling of the machined surface. The article describes the application of the method in face milling. The great advantage of the process is that it can be used to determine the theoretical roughness of practically any surface machined by a tool having defined edge geometry. The efficiency of the milling can be increased by increasing the cutting speed and feedrate, but changing their value also modifies the surface finish. The effect of the technological parameters on the three-dimensional roughness parameters of the machined surface was investigated in [18] for high speed face milling. It was found that the feedrate has the greatest impact on the surface topography.

In this paper, the effect of the feed on the roughness of the milled surface was examined. More specifically, the study analyses how the values of roughness parameters vary in the rotation axis (symmetry plane) in symmetrical face milling depending on the edge angles and chip ratios (a_p/f_z).

IOP Conf. Series: Materials Science and Engineering **448** (2018) 012056 doi:10.1088/1757-899X/448/1/012056

2. Characteristics of the tool edge geometry in face milling

The goal of the investigation was to determine the effect of different a_p/f_z (b/h) ratios on the surface roughness when face milling with different insert geometries and axial rake angles. It is important to examine the ratio of the depth of cut and feed because by increasing the feed value, the shape of the chip changes, and the role of the cutting edges in chip removal also changes. When the a_p/f_z ratio is much greater than one ($a_p/f_z >> 1$), the chip deformation is predominantly perpendicular to the edge on the outer surface, while in case of $a_p/f_z << 1$ is perpendicular to the face edge. This change takes place not only for a tool with $\kappa_r = 90°$ (Fig. 1), but also with $\kappa_r = 45°$ (Fig. 2).

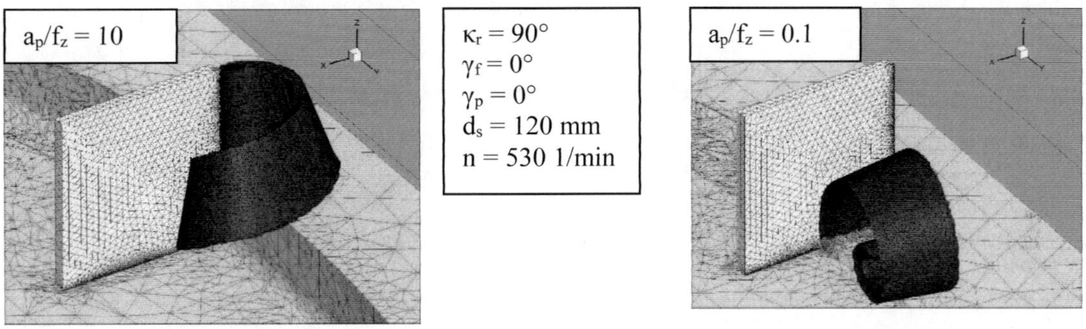

Figure 1 Chip formation for different a_p/f_z ratios when $\kappa_r = 90°$

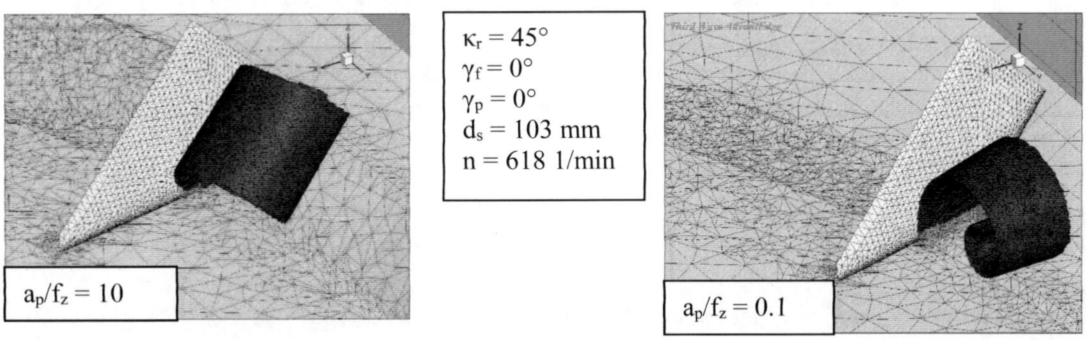

Figure 2 Chip formation for different a_p/f_z ratios when $\kappa_r = 45°$

In face milling, in addition to the placement angles, the chip removal ability of the insert is predominantly determined by the rake and clearance angle as well as the inclination angle, similar to other machining operations such as turning. The markings used in Figure 1 show the change in the role of working edges on the rake and peripheral surfaces of the insert. Depending on the ratio of the depth of cut (a_p) to the feed per tooth (f_z), the working edges and angles also change (Table 1). If this ratio is greater than one, the chip deformation occurs along the peripheral edge of the tool; when it is less than one, it takes place along its face edge.

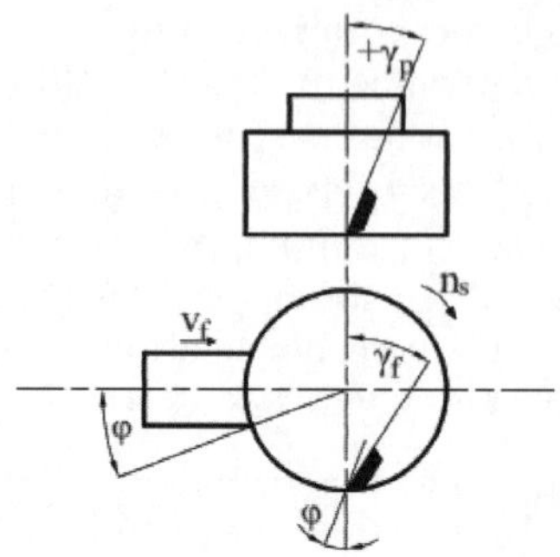

Figure 3. Interpretation of axial and radial rake angles in face milling

Table 1. Working angles at different a_p/f_z ratios

a)	b)
$a_p/f_z > 1 \rightarrow$	$a_p/f_z < 1$
$\lambda_s = \gamma_p, \gamma = \gamma_f$	$\gamma = \gamma_p =, \lambda_s = \gamma_f$

It has been established by analyzing the work of the cutting-edge in face milling that due to the changing of cutting conditions (technological data) not only the distributed load of the cutting edge varies, but the ratio of the load interchanges as well between the edges that are parallel to and perpendicular to the lateral surface of the tool [19]. This is illustrated in Figure 3. This means that the major and minor cutting edges and thus the inclination and rake angles are also interchanged. Because of the changed conditions, the method of generating the machined surface also changes. This justifies the examination of the effects of altered (active) angles on both the chip removal and on the machined surface topography.

3. Experimental conditions

3.1. Cutting tool

The applied milling tool is a unique tool design was developed at Otto von Guericke University in Magdeburg [17]. The essence of the tool design is that the cutting inserts are fixed by means of fixing cylindrical shafts in the tool body, and the insert seats in these shafts are individually formed, adapted to the desired insert geometries. This construction allows the usage of a variety of edge geometries and different cutting edge orientation angles in one milling head. For example, for examining different radial and axial rake angles, it is enough to produce the fixing shafts for each setting rather than to modify the whole milling head.

Two different milling insert geometries were used during the experiments:

- Garant XOEW 120508 PDER-W insert ("Garant"). This insert has a major cutting edge angle of 90°. The other characteristics are summarized in Table 3.
- Sandvik Coromant R24512 T3 EW K15W ("Wiper"). This insert has a major cutting edge angle of 45°. The other characteristics are summarized in Table 4.

Table 2. Characteristics of the "Garant" insert

a_e [mm]	κ_r [°]	r_ε [mm]	r_β [mm]	D [mm]	v_c [m/min]	γ_p [°]	γ_f [°]	α [°]
70*	90	0.8	0.015	120	200	0	0	12

IOP Conf. Series: Materials Science and Engineering **448** (2018) 012056 doi:10.1088/1757-899X/448/1/012056

Table 3. Characteristics of the "Wiper" insert

a_e [mm]	κ_r [°]	r_ε [mm]	r_β [mm]	D [mm]	v_c [m/min]	γ_p [°]	γ_f [°]	α [°]
70*	45	1,5	0.015	103	200	0	0	30

3.2. Machine tool
A Heller FT 2000 five-axis machining center was used in the milling tests.

3.3. Workpiece
The milling samples were designed as 70×70×170 mm³ blocks. The workpiece material was C45 medium carbon steel (WNr 1.0503, Ck45), which is widely used for general mechanical engineering and for automotive components.

3.4. Measurement of surface roughness
2D and 3D surface roughness measurements were performed using an Altisurf 520 surface roughness tester device using a CL2 confocal chromatic distance measuring sensor and an MG140 magnifier. The measurement head which has been assembled in this way has an approximate axial precision of 0.01 µm. The 2D measurements were measured and evaluated in accordance with ISO 4287: 1997 and ISO 4288-1996 standards, while for 3D surfaces the recommendations of ISO standard 25178 were followed.

3.5. Evaluation method
To evaluate the measured results of the 2D and 3D surface roughness parameters, we use the Full Factorial Experiment Design method. According to the literature, experimental designs can be labelled full factorial when all experiments are performed at each level combination of all factors to be tested. We perform the measurements on two levels, although only linear effects can be considered at this time. Within the examined range of factors, the results are shown visually and the empirical (experimental) formulas describing the correspondence between the examined parameter and the factors are determined. When applying the method, from among factors shown in "Experimental conditions" the radial inclination angle (γ_f) was chosen to be $\gamma_f = 0$ and is kept constant, while the values of the additional three factors (κ_r, a_p/f_z, and γ_p) are determined on two different levels (bottom and top). Table 4 contains the data for the experiments required for the factorial analysis.

Table 4. Experiments necessary for the factorial analysis

No.	k_r, [°]	a_p/f_z	γ_f [°]	γ_p [°]
1	45	0.1	0	0
2	90	0.1	0	0
3	45	10	0	0
4	90	10	0	0
5	45	0.1	0	12
6	90	0.1	0	12
7	45	10	0	12
8	90	10	0	12

4. Experimental results
After completing the planned experiments, the results shown in Figure 4 were obtained. The figure shows the experimental settings and the 2D and 3D roughness profiles.

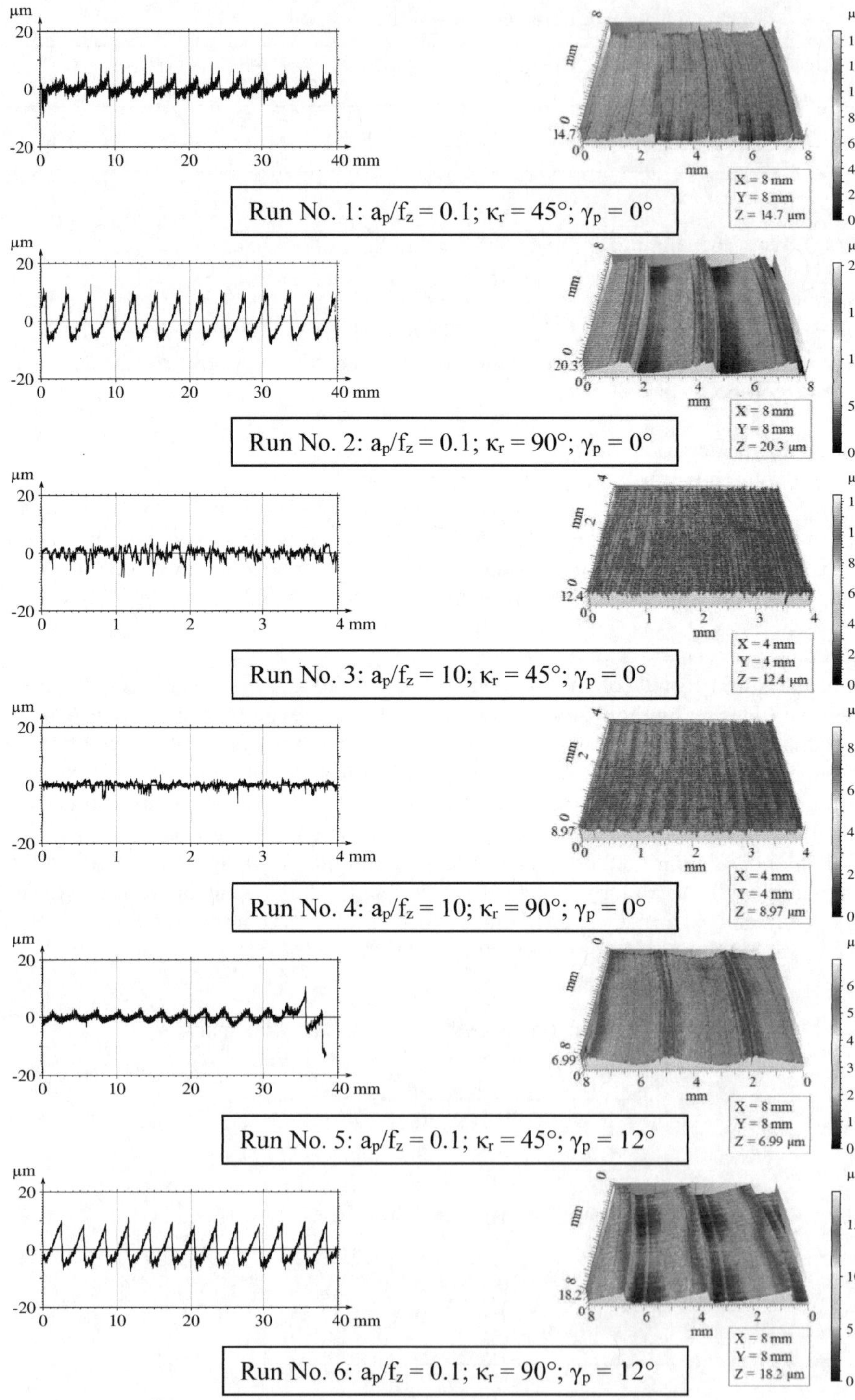

Figure 4. Measured 2D profiles and 3D surfaces

IOP Conf. Series: Materials Science and Engineering **448** (2018) 012056 doi:10.1088/1757-899X/448/1/012056

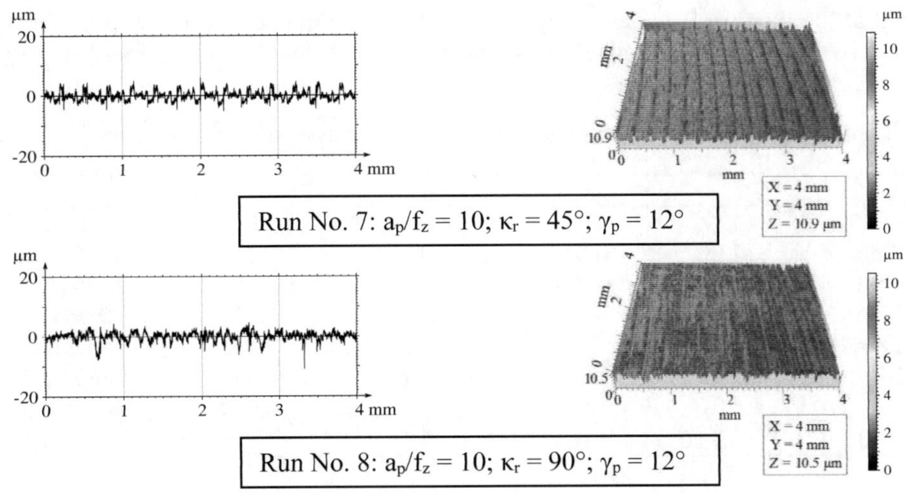

Figure 4 (continued). Measured 2D profiles and 3D surfaces

The roughness data measured on surfaces of the specimens are shown in Table 5.

Table 5. Measured surface roughness values of the milling experiments

No.	κ_r, [°]	a_p/f_z, [-]	γ_p, [°]	Ra, [µm]	Rz [µm]	Rq [µm]	Sa [µm]	Sz [µm]	Sq [µm]
1	45	0.1	0	1.520	14.2	1.93	1.440	14.7	1.89
2	90	0.1	0	4.010	19.2	4.69	4.150	20.3	4.85
3	45	10	0	1.380	11.2	1.71	1.350	12.4	1.71
4	90	10	0	0.798	7.18	1.04	0.813	8.97	1.07
5	45	0.1	12	1.170	9.79	1.46	0.848	6.99	1.04
6	90	0.1	12	3.540	19.0	4.07	3.580	18.2	4.15
7	45	10	12	1.310	9.73	1.63	1.260	10.9	1.62
8	90	10	12	1.310	11.0	1.68	1.250	10.5	1.58

5. Evaluation and discussion

Analyzing the profile graphs, the following statements can be made: In the first two figures (Runs 1 and 2) it can be seen that at $\kappa_r = 90$ ° the roughness indexes are lower and a more uniform profile was obtained than for $\kappa_r = 45°$. If the cutting ratio changes (Runs 3 and 4), the effect of κ_r also changes.

The response function was approximated by a 3-parameter first-degree (linear) polynomial. The individual and joint effects of the change of the three parameters were investigated to determine the polynomials of the response function of 2D roughness parameters (Ra), (Rz) and (Rq) and 3D roughness parameters (Sa), (Sz) and (Sq). All the lower and upper levels of the 3 factors resulted in a total of $n = 2^3 = 8$ total experiments. The coefficients of the first polynomial of the three variables were determined by the coefficients of the roughness measurements of the 8 different settings, so that the coefficients of 8-8 of the linear three-factor equation for the approximation of the three 2D and three 3D roughness parameters were determined. For the evaluation of the factorial experiment, the response function (y) was approximated with the linear polynomial for the surface roughness parameters of 2D as (Ra), (Rz) and (Rq) and of the 3D as (Sa), (S) and (Sq).

$$y = k_0 + k_1 \cdot \kappa_r + \frac{k_2 \cdot a_p}{f_z} + k_3 \cdot \gamma_p + k_{12} \cdot \kappa_r \cdot \frac{a_p}{f_z} + k_{13} \cdot \kappa_r \cdot \gamma_p + k_{23} \cdot a_p/f_z \cdot \gamma_p + k_{123} \cdot \kappa_r \cdot a_p/f_z \cdot \gamma_p \quad (1)$$

The results of the measured roughness values shown in Table 1 were processed by the factorial experimental design method to determine the response functions, the constants of which are given in Table 6.

Table 6. The constants of the response function for every dependent variables

	Ra	Rz	Rq	Sa	Sz	Sq
k0	-0.9996	9.1392	-0.8624	-1.3019	9.032	-1.1045
k1	0.056	0.113	0.062	0.061	0.126	0.067
k2	0.296	0.296	0.324	0.319	0.68	0.345
k3	-0.019	0.608	-0.026	-0.051	1.117	-0.084
k12	$-6.896 \cdot 10^{-3}$	-0.02	$-7.699 \cdot 10^{-3}$	$-7.288 \cdot 10^{-3}$	-0.02	$-8.081 \cdot 10^{-3}$
k13	$-2.354 \cdot 10^{-4}$	$7.773 \cdot 10^{-3}$	$-2.941 \cdot 10^{-4}$	$3.129 \cdot 10^{-5}$	0.01	$2.694 \cdot 10^{-4}$
k23	$-3.552 \cdot 10^{-3}$	0.016	$-4.04 \cdot 10^{-3}$	$-2.525 \cdot 10^{-5}$	0.074	$2.609 \cdot 10^{-3}$
k123	$1.313 \cdot 10^{-4}$	$2.02 \cdot 10^{-4}$	$1.627 \cdot 10^{-4}$	$9.446 \cdot 10^{-5}$	$4.826 \cdot 10^{-4}$	$8.418 \cdot 10^{-5}$

The roughness values calculated by the equations of the response function are shown in axonometric representation in Fig. 5. To evaluate the effect of the change of the edge geometry, to make the evaluation simpler and to make the representation more apparent, we showed the measured roughness values in the form of three-dimensional response surfaces as the function of κ_r and a_p/f_z with constant values of γ_f. To make the evaluation more representative, we have depicted the surfaces of $\gamma_p = 0°$ and $\gamma_p = 12°$ in one diagram (Fig. 5).

After studying Fig. 5, it can be stated that both the 2D and 3D drawings are similar for Ra and Sa; Rz and Sz; as well as for Rq and Sq. In the 2D case, the $\kappa_r = 90°$, $a_p/f_z = 10$ and $\gamma_p = 0°$ factor levels provide the lowest roughness values for all three parameters (Ra, Rz and Rq). For 3D roughness parameters, the situation is not so homogeneous. For Sa, the set values ($\kappa_r = 90°$, $a_p/f_z = 10$ and $\gamma_p = 0°$) gave the lowest roughness value (Sa = 0.813 μm). However, for the roughness parameters Sz and Sq, the minimum roughness values were obtained when setting the values $\kappa_r = 45°$, $a_p/f_z = 0.1$ and $\gamma_p = 12°$. When analyzing the two-dimensional roughness, it can be stated that the a_p/f_z parameter has the maximum effect on the roughness parameters. For Ra its value reduces from Ra = 4.01 μm to Ra = 0.798 μm, if the $\kappa_r = 90°$ and $\gamma_p = 0°$ are left unchanged, the $a_p/f_z = 0.1$ is increased to $a_p/f_z = 10$. So, the improvement rate is 80%. In this technology and geometric parameter setting, Rz is reduced from Rz = 19.2 μm to Rz = 7.18 μm (63% improvement), while for Rq these two characteristic values are Rq = 4.69 μm and Rq = 1.04 μm (78% improvement). In case of Sa the situation is similar: with the increase of $a_p/f_z = 0.1$ to $a_p/f_z = 10$ the Sa is reduced from Sa = 4.15 μm to Sa = 0.813 μm, which is also a 80% reduction. In case of Sz, the change in the angle κ_r has the greatest effect on roughness. If the values $a_p/f_z = 0.1$ and $\gamma_p = 12°$ are left unchanged and only the side cutting edge angle is reduced from $\kappa_r = 90°$ to $\kappa_r = 45°$, the roughness value of Sz will decrease from Sz = 18.2 μm to Sz = 6.99 μm, i.e. the improvement rate is 62%. By a similar adjustment, Sq decreases from Sq = 4.15 μm to Sq = 1.04 μm with a 75% improvement.

6. Summary

In face milling, the edges of the inserts in the milling head - by altering the a_p/f_z ratio - affect the chip removal differently, resulting in the roughness of the milled surface. If the deviation of the a_p/f_z ratio is significantly different from the value of 1 in any direction, the characteristics of the chip deformation also change, i.e. the deformation is performed either by the peripheral or the face edge. As a result, the loads on the edges of the inserts, the force components, the magnitude and direction of the thermal effects are changed, thus the roughness of the milled surface also changes, and consequently the position of the edges of the insert are also of decisive importance.

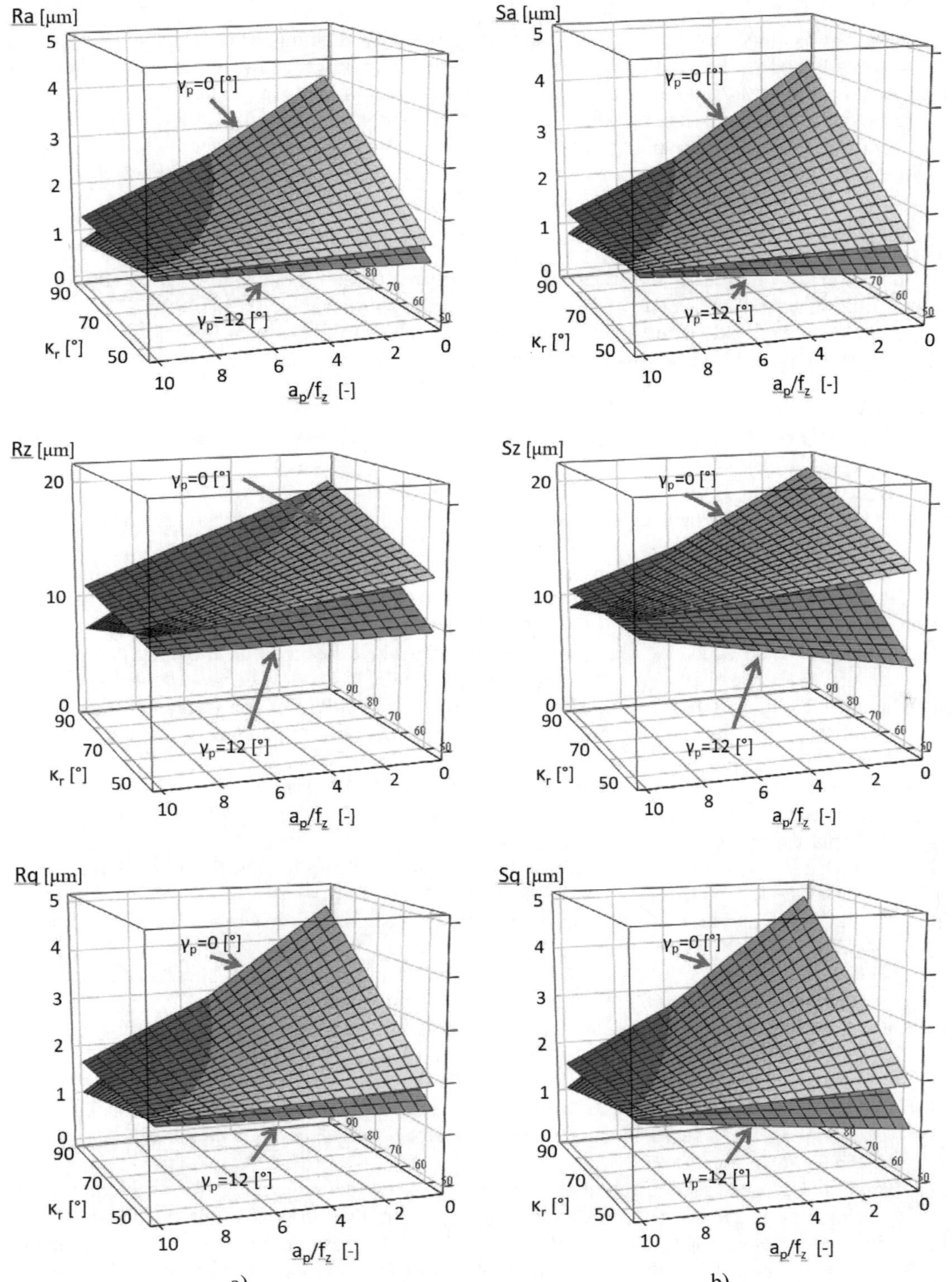

Figure 5. Relationship between surface roughness parameters and factors. In case of a) 2D roughness parameters; b) 2D roughness parameters

Overall, it was found that the roughness can be minimized by selecting the κ_r value of 90°, the higher a_p/f_z with $\gamma_f = 0°$ and $\gamma_p = 0°$. At the same time, in the case of $\kappa_r = 45°$, the lowest roughness value can be observed for the lower a_p/f_z ratio and for $\gamma_p = 12°$, but its value is slightly larger than the previous value. Favorable surface roughness can be achieved for both 45° and 90° κ_r when $a_p/f_z = 10$ and $\gamma_p = 12°$.

In our further research, we would like to frame our research task in order to identify how the required roughness requirements for the production can be provided at higher feed rates and with which variations of the different cutting edge angles.

Acknowledgements
The authors greatly appreciate the support of the German Research Foundation (DFG) (Project title: "Inverse cutting technology - a new strategy in face milling", Project No. 316141494).

The authors also greatly appreciate the support of the Hungarian National Research, Development and Innovation Office – NKFIH (No. of Agreement: OTKA K 116876).

References
[1] Kundrak J and Felho C 2016 3D roughness parameters of surfaces face milled by special tools *Manufacturing Technology* **16:3** 532-53
[2] Munoz-Escalona P and Maropoulos P G 2015 A geometrical model for surface roughness prediction when face milling Al 7075-T7351 with square insert tools *J. Manufacturing Systems* **36** 216–223
[3] Sheth S and George P M 2016 Experimental Investigation and Prediction of Flatness and Surface Roughness during Face Milling Operation of WCB Material *Procedia Technology* **23** 344 – 351
[4] Krolczyk G M, Legutko S, Maruda R W 2014 Influence of technological cutting parameters on surface texture of austenitic stainless steel *Appl. Mechanics & Materials* **693** 430-435.
[5] Schmitz T L, Couey J, Marsh E, Mauntler N, Hughes D 2007 Runout effects in milling: Surface finish, surface location error, and stability *Int. J. of Machine Tools & Manufacture* **47** 841–851
[6] Wojciechowski S, Twardowski P, Pelic M, Maruda R W, Barrans S, Krolczyk G M 2016 Precision surface characterization for finish cylindrical milling with dynamic tool displacements model *Precision Engineering* **46** 158–165
[7] Mikó B, Tóth B, Varga B 2017 Comparison of Theoretical and Real Surface Roughness in Case of Ball-End Milling *Solid State Phenomena* **261** 299-304
[8] Varga G and Kundrak J 2017 Effects of technological parameters on surface charateristics in face milling *Solid State Phenomena* **261** 285-292
[9] Mikó B and Farkas G 2017 Comparison of Flatness and Surface Roughness Parameters when Face Milling and Turning *Development in Machining Technology* **7** 18-27
[10] Masmiati N, Sarhan A A D, Hassan M A N, Hamdi M 2016 Optimization of cutting conditions for minimum residual stress, cutting force and surface roughness in end milling of S50C medium carbon steel *Measurement* **86** 253–265
[11] Hadad M and Ramezani M 2016 Modeling and analysis of a novel approach in machining and structuring of flat surfaces using face milling process *Int. J. of Machine Tools & Manufacture* **105** 32–44.
[12] Tamiloli N, Venkatesan J, Ramnath B V 2016 A grey-fuzzy modeling for evaluating surface roughness and material removal rate of coated end milling insert *Measurement* **84** 68–82
[13] Liu N, Wang S B, Zhang Y F, Lu W F 2016 A novel approach to predicting surface roughness based on specific cutting energy consumption when slot milling Al-7075 *Int. J. of Mechanical Sciences* **118** 13–20
[14] Karkalos N E, Galanis N I, Markopoulos A P 2016 Surface roughness prediction for the milling of Ti–6Al–4V ELI alloy with the use of statistical and soft computing techniques *Measurement*

90 25–35

[15] Mikó B 2016 Surface quality prediction in case of steep free form surface milling *Key Engineering Materials* **686** 119-124

[16] Felho C and Kundrak J 2012 Characterization of topography of cut surface based on theoretical roughness indexes *Key Engineering Materials* **496** 194-199

[17] Felho C, Karpuschewski B, Kundrak J 2015 Surface roughness modelling in face milling *Procedia CIRP* **31** 136-141

[18] Logins A and Torims T 2015 The Influence of High - Speed Milling Strategies on 3D Surface Roughness Parameters *Procedia Engineering* **100** 1253 – 1261

[19] Karpuschewski B, Kundrak J, Felho C, Varga G, Sztankovics I, Makkai T, Borysenko D 2018 Preliminary Investigations for the Effect of Cutting Tool Edge Geometry in High-Feed Face Milling *Lecture Notes in Mechanical Engineering: Vehicle and Automotive Engineering 2* eds Jarmai K and Bollo B pp 241-254

[20] Karpuschewski B, Kundrák J, Emmer T, Borysenko D 2017. A New Strategy in Face Milling - Inverse Cutting Technology. *Solid State Phenomena* **261** 331–38.

IOP Conf. Series: Materials Science and Engineering **448** (2018) 012057 doi:10.1088/1757-899X/448/1/012057

What Can Cosmic Radiation Cause in Polymers?

A Adamne Major[1], D Boja[2]

[1]Professor, John von Neumann University, Kecskemet, Hungary
[2] Test engineer, ACPS-Automotive, Kecskemet, Hungary

E-mail: major.andrea@gamf.uni-neumann.hu

Abstract. In our research the influence of the cosmic radiation on different types of polymers was investigated. Main focus was on the mechanical properties, which includes elongation at break, tensile strength and impact strength. Polypropylene, polyamide 6.6 and polycarbonate were used as materials. All of the test pieces were injection molded with the same technological settings. These specimens were placed at ground laboratory and different flying height airplanes (on Earth, at 5500 m, at 12000 m). Mechanical properties were investigated. We found that cosmic radiation may induce change in some of the mechanical properties of polypropylene, polyamide 6.6 and polycarbonate.

1. Introduction

Both passenger and commercial airplanes are being bombarded by high-energy particles. These particles come from source called cosmic rays. Most of them interacts the Earth's magnetic field but a lot of them can penetrate it and reach the atmosphere. The radiation rate increases by higher altitudes. There were three types of widely used polymers in our focus. These polymers were polyamide 6.6, polycarbonate and polypropylene. The main question was how these polymers behave, how their chemical, physical and other properties change during the radiation, how the specific polymers mechanical properties will change after a relatively short exposure to this type of radiation [1-9]. Finally, these ongoing degradations how seriously affect their lifetime. Standard specimens were made by injection molding method, later they were placed on two types of planes. Then the returned specimens were investigated with different mechanical analytic techniques.

1.1. Cosmic radiation

There are various types of particles which can be found in cosmic radiations; mainly protons, accelerated atoms, electrons [10-16]. These particles come from deep space and from our Solar system. Cosmic radiation levels increase due to the low activity of the Sun. These particles firstly collide with Earth's magnetic field and if they have enough energy they can reach the atmosphere where they hit atoms or particles which results new but weaker particles.

There are three types of cosmic radiation: galactic cosmic rays (weakest), anomalous cosmic rays (weak) and particles from the Sun (strong).

Galactic cosmic rays are the weakest of all because of the distance between Earth and the source. Supernova events, eruptions, neutron stars black holes are the sources. They contain protons and alpha particles.

The other two types come from the Solar system. Anomalous cosmic rays are ionized particles which are accelerated by the solar wind meeting interstellar space associated with magnetic field. After reaching the heliopause UV radiation ionizes these particles, the solar wind moves toward interstellar space colliding with the magnetic field where these particles can freely travel back into the solar system. The strongest radiation comes from the solar system's central star in forms of solar wind and solar eruptions. Radiation can be soft X-ray and hard X-ray. The solar wind contains mostly protons.

2. Experimental

Tensile test pieces were made by injection molding method with a Sumitomo DEMAG IntElect 100-500 injection molding machine using basic, additive-free polypropylene, polyamide 6.6 and polycarbonate. The technology settings were the same during the production, each polymer had its own technological settings, including temperature, pressure. Later these test samples were placed in a ground laboratory (above sea level ~110 meter) as reference and on two different flying altitude airplanes (5500 and 12000 meter).

The airplanes were an An-26 (a) and a Boeing 737 Cargo (b) which can be seen in figure 1. An-26 had an average flying altitude of 5500 meters. On this altitude the specimens were exposed for a total of 2.97 μSv radiation dose. The Boeing 737 had an average flying altitude of 12000 meters, the specimens were absorbed a higher dose of radiation of 51.8 μSv. Radiation doses were calculated after the flights with meteorological ballon data. The reference samples were in a special box, with the help of it the influence of the cosmic rays could be increase to ~ 0 μSv.

After the flying the samples were investigated. 10 samples were used for each test. Charpy impact strength was measured with a CEAST IMPACTOR II with a 5 J hammer. Elongation break and tensile strength were measured with an INSTRON 3366, 10 mm/min was used as preload speed, measurement speed was 200 mm/min. The tests were executed in room temperature.

(a) (b)

Figure 1. Airplanes used during the experiment

3. Results and discussion

3.1. Results of polyamide 6.6

3.1.1. Charpy impact strength

Charpy impact strength test results are shown in figure 2. The test samples were shaped in the standard test specimen size for the testing. It's clearly visible that PA 6.6's impact strength had no significant change on the airplane samples. Compared to the ground referential samples both airplane samples impact strengths were greater by ~12%.

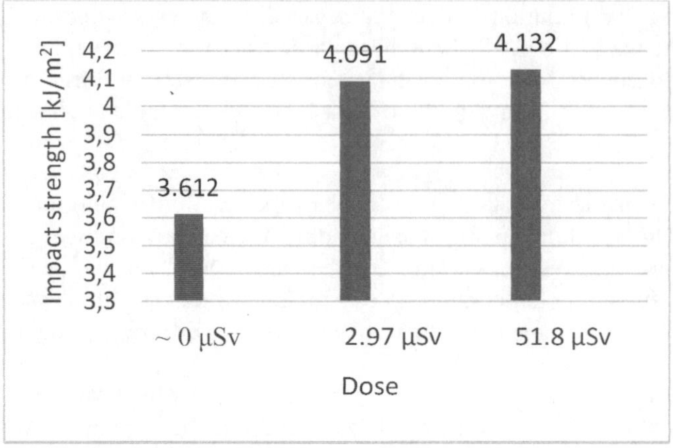

Figure 2. Charpy impact strength of polyamide 6.6

3.1.2. Tensile strength and elongation at break

Polyamide 6.6's tensile strength results can be seen in figure 3. Compared to the ground referential samples the 5500-meter flying altitude specimens had almost negligible difference. While the 12000-meter flying altitude samples had almost the same results as the ground samples, the 5500-meter samples had ~5% lower tensile strength compared to the two other samples.

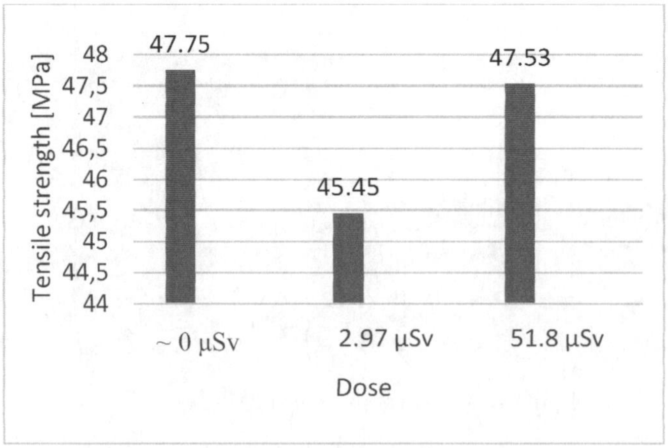

Figure 3. Tensile strength of polyamide 6.6

Elongation at break results can be seen in figure 4. It is shown that the elongation increasing depending on the altitudes. Both airplane samples' elongation break increased. 5500-meter specimens increased by 4.2%, while 12000-meter samples increased by 10.8% in elongation.

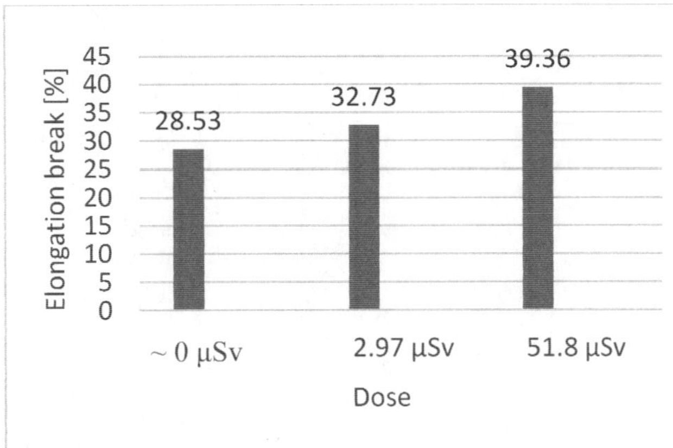

Figure 4. Elongation at break of polyamide 6.6

3.1.3. Brinell hardness

Brinell hardness is shown in figure 5. Both the ground and the 5500-meter specimens values stayed the same. The 12000-meter samples hardness value decreased by 2.5%.

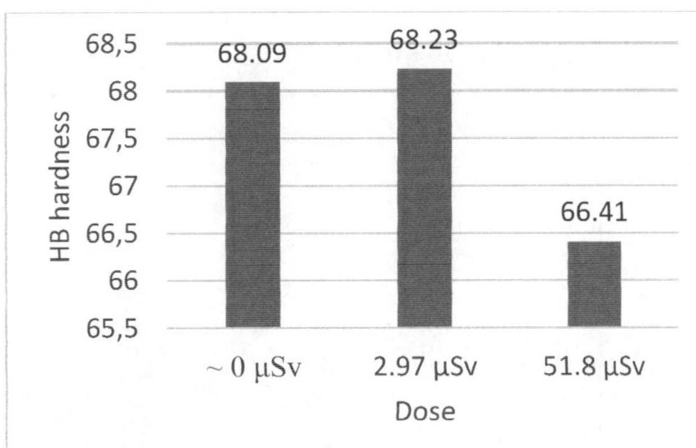

Figure 5. Brinell hardness of polyamide 6.6

3.2. Results of polycarbonate

3.2.1. Charpy impact strength

Polycarbonate impact strength results can be seen in figure 6. All the samples impact strength values stayed close to the ground referential specimens' values. While 5500-meter airplane samples had a slightly increase in the impact strength (~7%), the 12000-meter airplane specimens impact strength decreased (~3%) but not significantly.

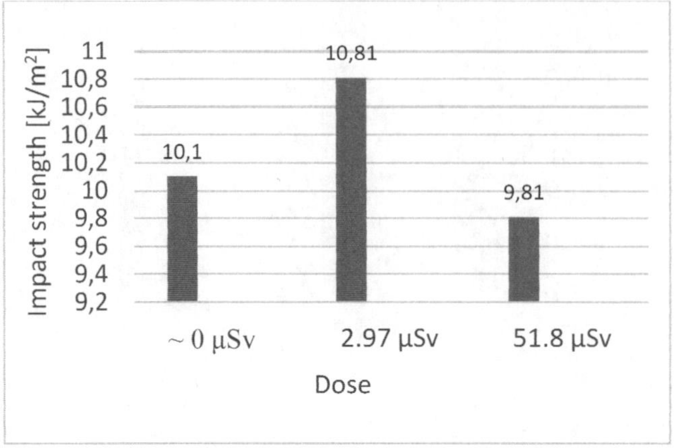

Figure 6. Impact strength of polycarbonate

3.2.2. Tensile strength and elongation at break

Tensile strength results can be seen in figure 7. Both airplane samples had an increase in tensile strength compared to the ground specimens. Both airplane samples had an increase in tensile strength by 7-9%.

Elongation at break results can be found in figure 8. In this case the results were interesting. It is known that cosmic radiation is more likely localized not effecting the whole system. This can be a sign to this effect, more high energy particles reached the 5500-meter airplane samples. Compared to the ground specimens, the 5500-meter airplane samples elongation break increased by 14.4%. The 12000-meter airplane samples increased only by 2.5% and remained the closest to the original referential values.

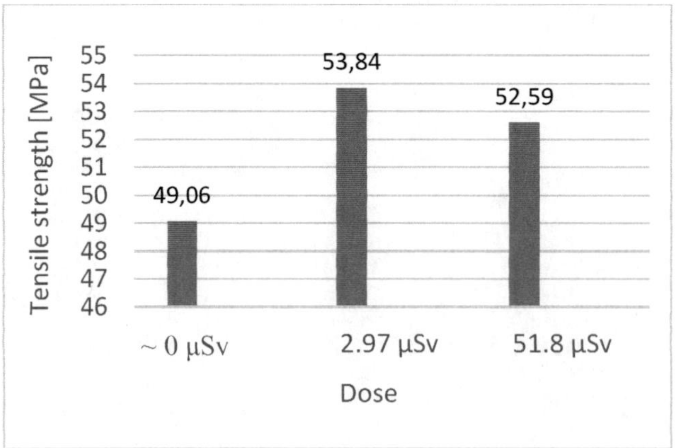

Figure 7. Ttensile strength of polycarbonate

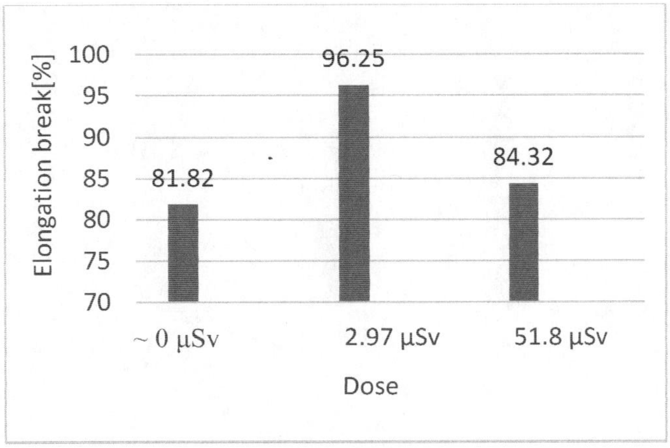

Figure 8. Elongation at break of polycarbonate

3.2.3. Brinell hardness

The results of Brinell hardness can be found in figure 9. These are very interesting data. Compared to the ground samples both airplane specimens' hardness increased. While the 5500-meter sample with „only" 3.1%, the 12000-meter sample increased by 10.6%. This can be the clearest indication of a degradation event which ended up in increased hardness.

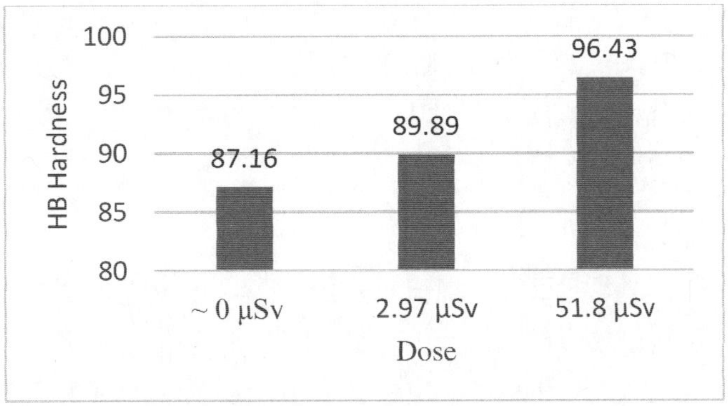

Figure 9. Brinell hardness of polycarbonate

3.3. Results of polypropylene

3.3.1. Charpy impact strength

Polypropylene had the least significant change in impact strength. There is only ~0.1 kJ/m^2 difference between the samples. While 12000-meter airplane samples kept the same values as the ground reference, 5500-meter airplane sample had 0.1 kJ/m^2 difference. Cosmic radiation had almost no effect on this property of polypropylene. The results can be seen in figure 10.

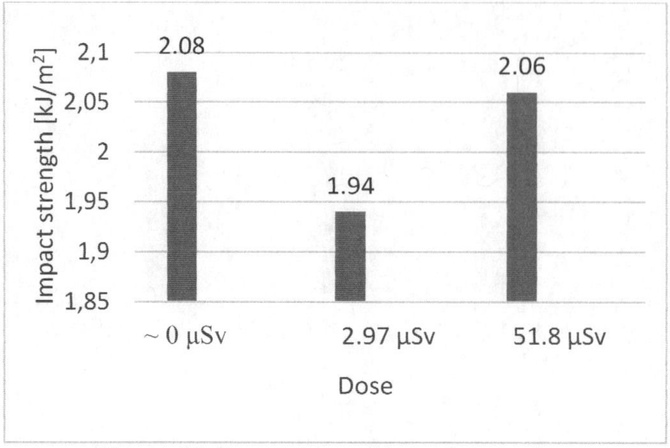

Figure 10. Impact strength of polypropylene

3.3.2. Tensile strength and elongation at break

While it has a really slow and steady decreasing tendency it can be the effect of the increasing radiation dose on the samples. Compared to the ground samples all of the airplane specimens' tensile strength decreased but with an almost negligible value. The highest decrease is only 1.2 %. Tensile strength results can be seen in figure 11.

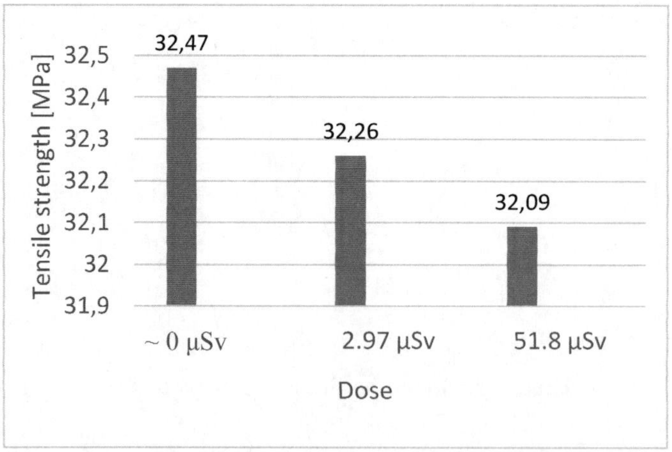

Figure 11. Tensile strength of polypropylene

Elongation at break had an interesting result (figure 12). The reason can be the same: the cosmic radiation localized degrading effect. The 5500-meter airplane samples' values had the same results as the ground reference, while the 12000-meter specimens had an increase but only about 0.2%.

IOP Conf. Series: Materials Science and Engineering **448** (2018) 012057 doi:10.1088/1757-899X/448/1/012057

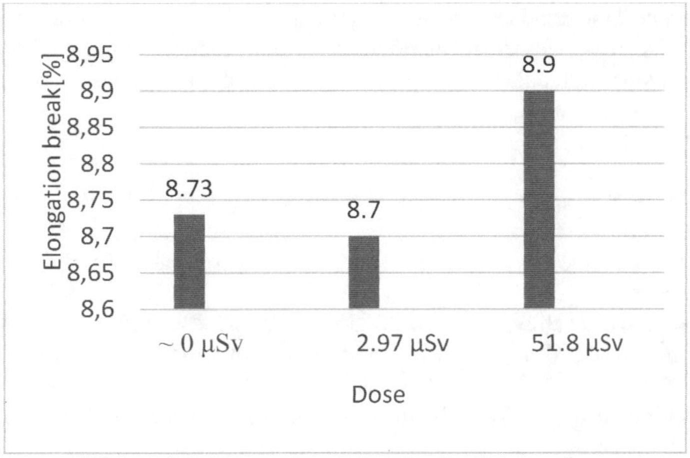

Figure 12. Elongation at break of polypropylene

3.3.3. Brinell hardness

There is not a big difference between the sample results, most of these remains in a limit; no significant changes can be seen. The 5500-meter sample's values had the lowest of all; compared to the ground it has 3.6% lower hardness.

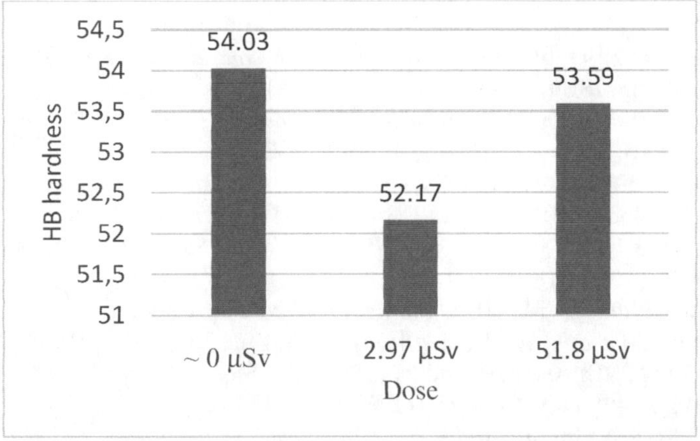

Figure 13. Brinell hardness of polypropylene

4. Summary

Polymers can exhibit various and large scale of radiation effects. Chemical reaction can occur during an irradiation. For example new chemical bonds can occur, some chains can break up, which leads to irreversible events. These events can be investigated by changes of appearance, chemical, physical, mechanical and thermal properties. Radiation induced degradation shows up in the form of elongation break increase and decrease, tensile strength, hardness changes. In this case, our aim was to find any possible mechanical alteration that could be induced by cosmic radiation. It is clear in the results that, a very low change was showed up. On the other hand, these properties shifts are visible, compared to the ground reference, because they were made by the same technological settings. The results show that this short period of flight time and a low radiation dose can be enough for induce a radiation degradation which shows up on the interestingly increasing hardness in polycarbonate. These materials are still useable and the radiation had no fatal effect on them but foreseeably these changes by the time can lead to risk these materials and the products made from them. More experiments are needed with these and other types of materials. It is planned to use bigger surface, longer radiation doses and the ability to examine these materials on molecular level searching for visible radiation induced degradation.

References

[1] Epstein L M 1959 *Radiation Degradation of Polymethylene and Polypropylene* Rept. P 5992-2 Westinghouse Electric Co.
[2] NASA 1970 *Nuclear and Space Radiation Effects on Materials - Space Vehicle Design Criteria* NASA Techn. Rep. https://ntrs.nasa.gov/search.jsp?R=197100155582018-10-20T10:40:10+00:00Z
[3] Le Clair H G and Cobbs W H 1958 *Effects of Radiation on Plastic Packaging Films* Ind. & Eng. Chem. **50** (3) 323
[4] Sears W C and Parkinson W W 1956 *Post-Irradiation Oxidation in Rubber and Plastics* J. Pol. Sci. **21** (98) 325
[5] Bopp C D and Sisman O 1952 *Radiation Stability of Plastics and Elastomers* Rept. ORNL-1373 Oak Ridge National Lab.
[6] Sisman O and Bopp C D 1951 *Radiation Stability of Irradiated Polymers* Rept. ORNL-928 Oak Ridge National Lab.
[7] Bruce M B and Davis M V 1981 *Radiation effects on organic materials in nuclear plants* Georgia Inst. of Tech. Atlanta (USA). Nuclear Eng. Dept.
[8] Shulman H and Ginell W S 1970 *Nuclear and space radiation effects on materials* NASA Techn. Rep.
[9] Jiri George Drobny 2013 Ion*izing Radiation and Polymers, Principles, Technology and Applications* Elsevier ISBN 978-1-4557-7881-2
[10] Sarangi A et al 2018 *Dust in Supernovae and Supernova Remnants I: Formation* Scenarios Living Rev. Sol. Phys. **15:**1 doi.org/10.1007/s41116-017-0011-z
[11] Plane J M C et al 2018 *Impacts of Cosmic Dust on Planetary Atmospheres and Surfaces* Space Sci. Rev. **214**:23, doi.org/10.1007/s11214-017-0458-12017
[12] Bykov A M et al 2018 *Cosmic Ray Production in Supernovae* Space Sci. Rev. **214**:41 doi.org/10.1007/s11214-018-0479-4
[31] Grbino M and Lattanzi M 2018 *Status of Neutrino Properties and Future Prospects—Cosmological and Astrophysical Constraints* Front. Phys. **5**:70. doi: 10.3389/fphy.2017.00070
[14] Benghim V V et al 2018 A*n Experiment in Radiation Measurement Using the Depron Instrumen*t Space Sci. Rev. **214**:9, DOI 10.1007/s11214-017-0445-6
[15] Reames D V 2018 Ab*undances, Ionization States, Temperatures and FIP in Solar Energetic Particles* Space Sci. Rev. **214**:61 doi.org/10.1007/s11214-018-0495-4
[16] Richardson I G 2018 *Solar wind stream interaction regions throughout the heliosphere* Living Rev. Sol. Phys. **15** doi.org/10.1007/s41116-017-0011-z

Acknowledgement

This research is supported by EFOP-3.6.1-16-2016-00006 "The development and enhancement of the research potential at John von Neumann University" project. The Project is supported by the Hungarian Government and co-financed by the European Social Fund.

IOP Conf. Series: Materials Science and Engineering **448** (2018) 012058 doi:10.1088/1757-899X/448/1/012058

Design and manufacture of orthopedic corset using 3D digitization and additive manufacturing

I Molnár [1] , L Morovič [2]
[1,2] Slovak University of Technology, Faculty of Materials Science and Technology, Institute of Production Technologies, Department of Machining and Forming, Jána Bottu 2781/25, 917 24 Trnava, Slovakia

[1] ivan.molnar@stuba.sk, [2] ladislav.morovic@stuba.sk

Abstract. The paper discusses the use of 3D digitization and additive manufacturing technologies in the field of medicine. The paper focuses on design and manufacture of an orthopedic corset. Described are the methods of designing and manufacturing of orthopedic corsets according to the individual requirements of the patient at present, and a more detailed description of the design and manufacturing process using 3D digitization and additive manufacturing. The paper introduces three variants of corset design and manufacture, with the description of the materials selected for use. The main purpose of the paper is to describe the process of the effective design and manufacture of orthopedic corset using 3D digitization and additive manufacturing technologies.

1. Introduction

At present, the methods of 3D digitization and additive manufacturing are increasingly used in various sectors of industry, with increasingly widespread use being made in the field of medicine. In the medical field, the development and use of 3D digitization and additive manufacturing influences several factors, which are advances in technology, improving health infrastructure or investing in science and research.

The purpose of this paper is to describe the design and manufacture of the individual orthopedic corset using 3D digitization and additive manufacturing. This orthopedic aid was chosen for the dissertation thesis with the title „Research of design and manufacture of orthopedic aid by 3D digitization and Rapid Prototyping". The orthopedic corset was chosen in collaboration with the Neoprot company, which designs and manufactures individual orthopedic and prosthetic aids in Slovakia. The choice of orthopedic corset was made on the basis of requirements for saving manufacturing material, shortening of manufacturing time, introducing a new design and creating an effective methodology of design and manufacture of individual orthopedic corset [1].

2. 3D digitization and additive manufacturing

The purpose of 3D digitization is to convert real physical objects into their digital form. This fact is used in medicine to obtain computerized patient data that serves, for example, to determine the exact diagnosis or to make computer adjustments for the manufacture of various medical aids and models [2].

The following part of the paper describes the 3D scanning of the human body with the method of active triangulation (structured light), which subsequently serves as a basis for making an individual orthopedic corset [3]. This method of 3D digitization is based on obtaining the coordinates of points represents the surface of the scanned object (cloud of points) or component by transmitting light patterns from the light projector to the scanned object´s surface. The patterns on the surface are captured by the sensor´s of the cameras (CCD (Charge Coupled Device) or CMOS (Complementary Metal-Oxid Semiconductor) sensor) and the exact position of the scanned points is then calculated on the principle of triangulation 'figure 1' (a) [1,3].

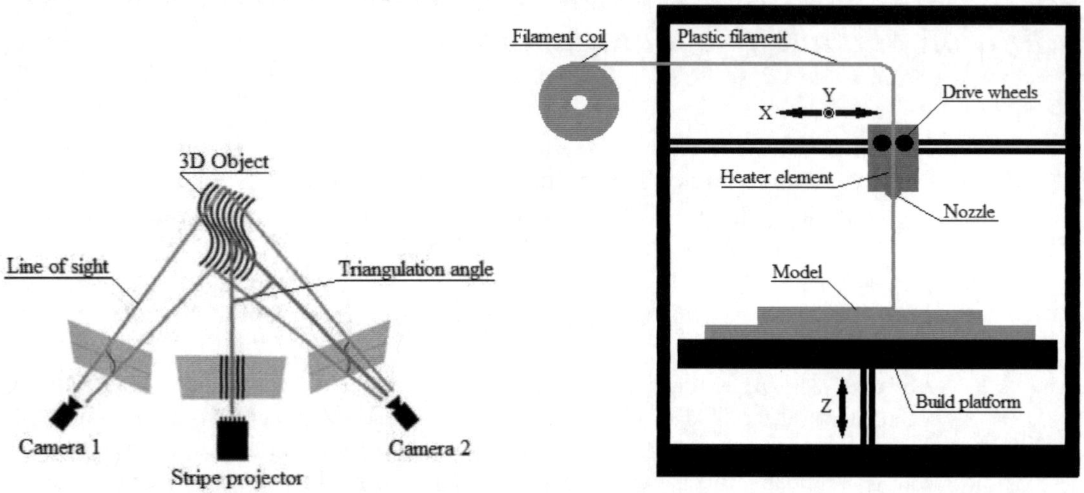

(a) (b)

Figure 1. Principle of active triangulation (a) and Fused Deposition Modeling (b) [1].

At present, the several methods of additive manufacturing (AM) are used in the field of medicine. These are mainly Selective Laser Sintering (SLS) [4], Selective Laser Melting (SLM) [5], Poly Jet Modeling (PJM) [6], Stereolitography (SLA) [7] and Fused Deposition Modeling (FDM) [6,7].

The following part of the paper describes the principle of AM method FDM 'figure 1' (b) used for the manufacture of individual orthopedic corset. FDM is one of the most common AM in practice.

Fused thermoplastic fibers are extruded from the tip of a heated nozzle that moves in the X and Y axes. Thin fibers are deposited on a platform that has a significantly lower temperature, which ensures fast cooling of the melted thermoplastic. After the platform has been reduced to a precisely defined thickness of one layer, another layer is deposited. In this way, a component is created.

The height of individual layers ranges from 0.05 to 0.3 mm, depending on the quality requirements of the resulting surface of the component. In this method, depending on the complexity of the component, it is also necessary to use the support material [8].

3. Current methods of design and manufacture of an individual orthopedic aids

At present, there are several methods used for designing and manufacturing of individual orthopedic aids, the use of which depends on the patient's state of health or, respectively, the state of his / her injury, which is the consequence of the need to use an individual orthopedic aid. The orthopedic aids are determined by a specialist after a medical examination and the extent of the injury.

At present, the following three methods of designing and manufacturing of the individual orthopedic aids are mainly used. Corsets as well as other orthopedic aids can be produced by the next methods:

- design and manufacture of gypsum / "light gypsum" orthopedic aid,
- design and manufacture using 3D digitization and CNC milling,
- design and manufacture using 3D digitization and additive manufacturing [9].

The use of these methods of designing and manufacturing of an individual orthopedic aids has been found based on a literature review in the field of medicine. Currently, the above mentioned methods are used, the choice of which of them depends on the case, respectively, type and extent of injury to the patient [10].

The gypsum and "light gypsum" based methods are entirely manual methods of designing and manufacturing of individual orthopedic aids. The other two methods work with the use of computer aided systems, which are computer aided design and computer aided manufacture software.

The gypsum individual orthopedic aids are made of a plaster, a handy and relatively „dirty" way of applying of plaster bandages on the injured place on patient body. The „light gypsum" individual orthopedic aids are made of a resin that hardens similar as gypsum after contact with water [1]. The 'figure 2' describes the above mentioned methods of designing and manufacturing of individual orthopedic aids. The highlighted is the method by which the article is dealt with more closely.

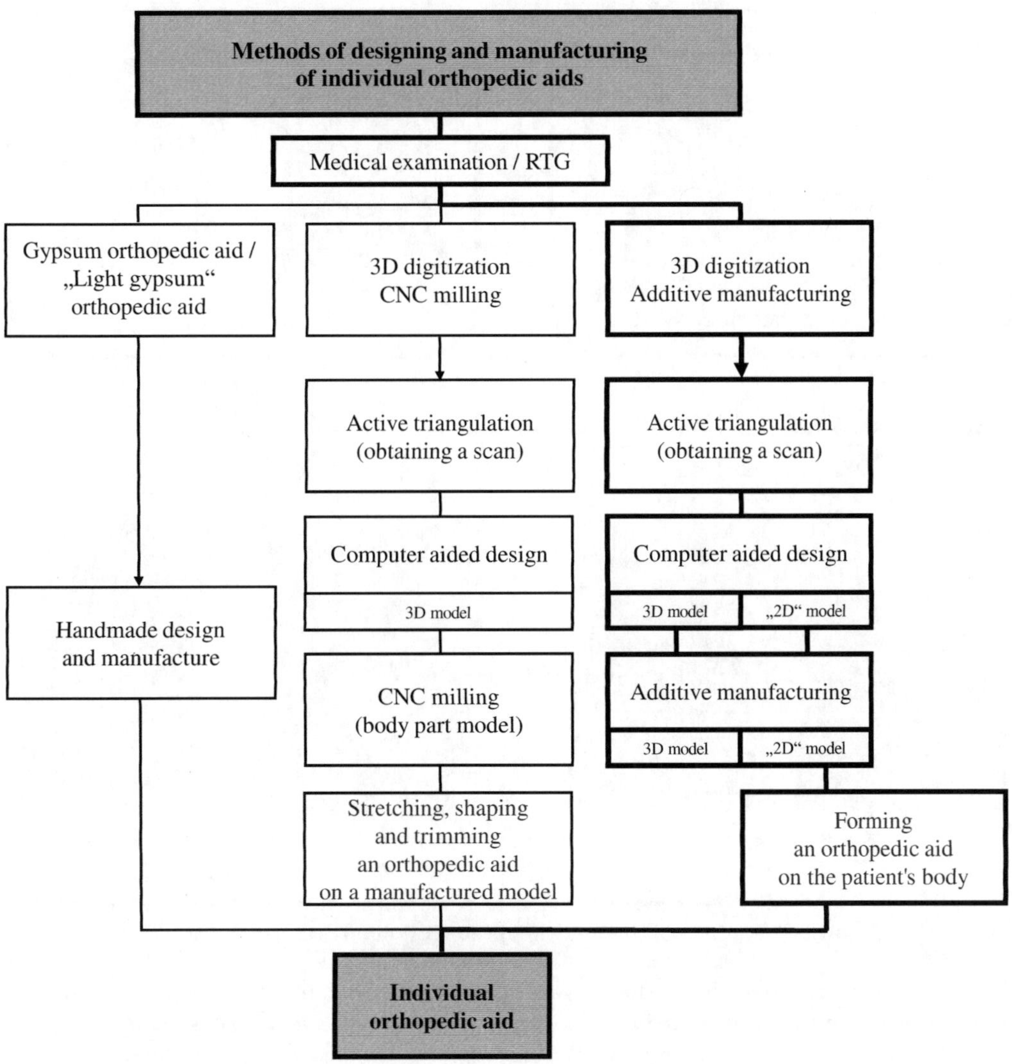

Figure 2. Methods of designing and manufacturing of individual orthopedic aids [2].

IOP Conf. Series: Materials Science and Engineering **448** (2018) 012058 doi:10.1088/1757-899X/448/1/012058

3.1. Design and manufacture using active triangulation and Fused Deposition Modeling

In the design and manufacture of individual orthopedic corset two ways of designing and manufacturing can be made:

1. design and manufacture of a 3D model of the individual orthopedic corset,
2. design and manufacture of a „2D" model of the individual orthopedic corset.

3.1.1. 3D model of individual orthopedic corset. 3D model of the individual orthopedic corset is created using CAD software according to the data obtained by 3D scanning a model that will be designed and manufactured in the shape and size according to the particular patient, and then it is possible to use after manufacture.

Variants created on the basis of the 3D scans were designed as 3D models of corsets. These two variants were created, based on their own pattern and on the basis of the polygonal mesh pattern. The above mentioned two variants of corsets are seen in the following 'figure 3' [1].

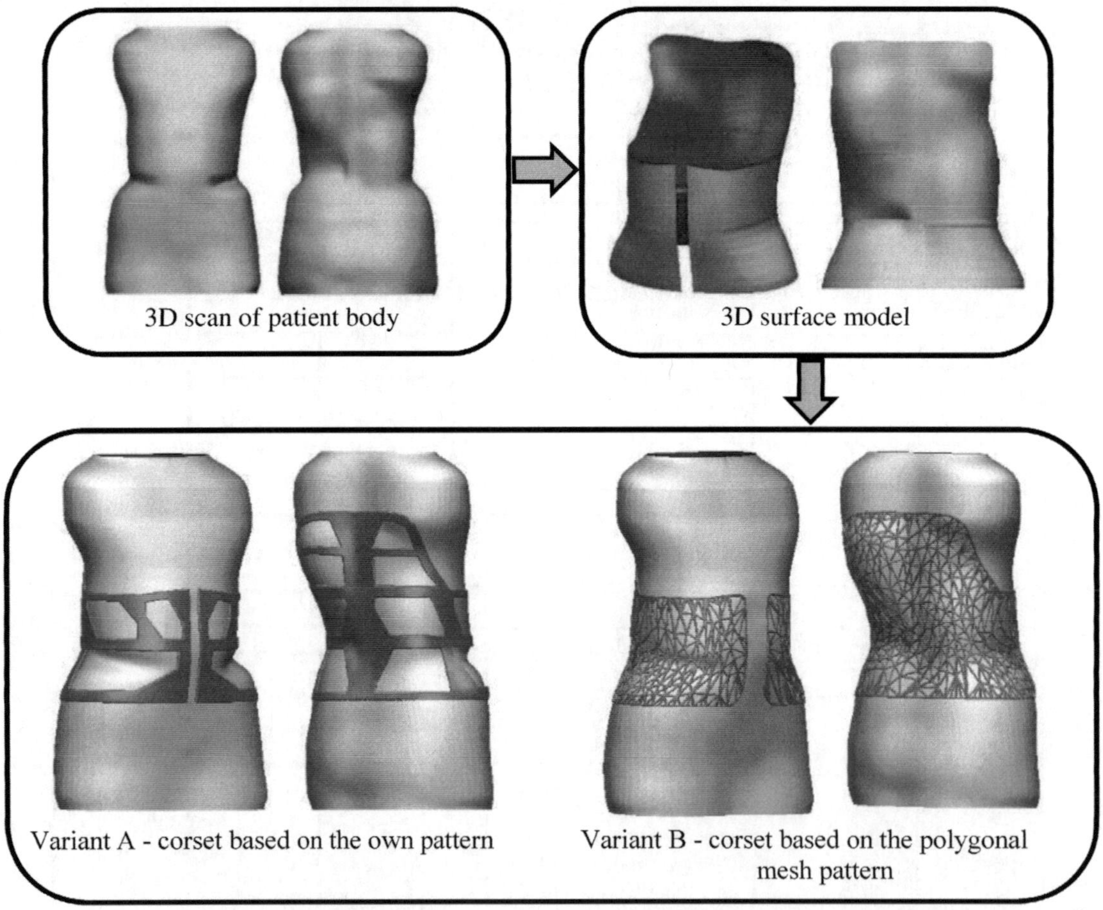

Figure 3. Variants of 3D models of the individual orthopedic corsets.

Variant A was created in software Autodesk PowerSHAPE. In the first step, it was necessary to create curves on scan model by means of a dynamic section. Subsequently, it was possible to create surfaces between the curves. Therefore, reverse engineering was used to create the corset 3D model. Because it was a rather complicated area in the back, it was necessary to create a larger number of curves so that the software could be counted and packed between the curves. The next step was the cropping of the created surfaces. The digital surfaces had to be cut to the desired shape of the corset, which was possible by creating further curves copying these surfaces. From these curves, additional areas were created, this time perpendicular to the corset surfaces. Using the cutting tool, the separation of the areas between them at the points of contact was achieved.

In this design and manufacture variant, emphasis was placed on minimizing material consumption, saving manufacturing time and resulting in lower production costs, while maintaining the strength and hence the functionality of the device.

In variant B, the corset was created using software Autodesk Meshmixer. The surface model from software Autodesk PowerSHAPE has to be re-introduced into the polygonal mesh. Subsequently, this mesh was reduced from 128,000 triangles to 6,300. The reduction was performed to reduce the size of the model file and thereby increase the speed of the modeling itself. Mesh reduction generally reduces the shape and dimensional accuracy of the model, but in this case it will not affect the future functionality of the model (corset).

After the reduction, the mesh was exported in the STL format into software Autodesk Meshmixer in which was possible to further reduce the number of triangles of the mesh and to work exactly to the required number and size of the given triangles. These processes were performed using the crossover functions, as in the previous case, and then by the triangle number reduction function. In the last step, it was possible to create a STL corset model, which is automatically generated according to the number of triangles in the mesh [1].

As has already been mentioned, the corset manufacturing will be made by FDM method. In case of the printed elements require improve surface properties, finishing operations are recommended [11,12].

3.1.2. „2D" model of individual orthopedic corset. On the basis of the outputs and information (time for design of the corsets, manufacturing time, quantity of building and support material) obtained from the previous two variants of 3D models of the corsets, the „2D" model of the corset will be designed and manufactured in the next step of dissertation thesis. The variant of „2D" model expect greater efficiency in the design and manufacture processes of the individual orthopedic corset. The benefits of this variant compared to previous two variants of 3D models are as follows:

- lower material consumption,
- shorter production time,
- higher accuracy of shapes and dimensions after shaping on the patient's body.

The „2D model" of the individual orthopedic corset is designed and manufactured in the plane, i.e. as an unfolded 3D model, is created using CAD software according to data obtained by 3D digitization (active triangulation). This model will be formed (i.e. shaped) on the body of the patient after model production.

The process consists in performing the following main steps:

1. obtaining the 3D digitized model of the desired part of the body using active triangulation,
2. edit and convert to STL format (input format to AM),
3. automated production using FDM,
4. individual orthopedic corset [1].

4. Materials of individual orthopedic corsets

Two types of materials were selected to make an orthopedic corset, namely PLA (Polylactic Acid) and PET-G (Polyethylene Terephthalate Glycol).

The reasons for choosing these materials are as follows:

- health-conscious materials that do not cause medical complications in contact with the skin,
- both materials belong to the cheapest polymers that can be purchased on a regular basis, so they are not difficult to produce,

- during manufacture itself, by the chosen technology, there is no complication and no special conditions need to be created during the additive manufacture of medical devices by the chosen technology [13],
- PLA material has features that are particularly suited to the application. This material is a biocompatible polymer which, in some forms, is used for the manufacture of various medical devices, whether for external or internal use (i.e. placement in a human body). Another important feature is its forming ability, under certain conditions. This material is so soft when it is heated to a certain temperature that it is possible to shape / form it, in this case directly on the body of the patient. This shaping can be done many times in succession without breaking the coherence of the material or its degradation (variants of „2D" model) [14].
- PET-G material is a health-conscious material, has good mechanical properties (strength, elasticity) among the plastic materials used in the additive manufacturing technology. It can not be shaped to the same extent as PLA material, so the desired shape of the model can be manufactured without its further modification by molding (variants of 3D models) [15].
- The detail description of material properties and shrinking of both materials (PLA and PET-G) has been researching and described in the article [16]. On the basis of the results of the article, it is possible to assume the minimum values of the shape and dimensional deviation of these materials.
- There are currently a number of materials designed specifically for medical purposes, but only a few of them can be used for FDM technology and also their price is higher [6,7,13].

5. Conclusion

The paper discusses the processes of design and manufacture of individual orthopedic aids, mainly the method which uses 3D digitization and AM. The use of 3D digitization and AM for the design and manufacture of individual orthopedic aids is an effective way of designing and manufacturing them. With these technologies, it is possible to design and manufacture quickly and efficiently orthopedic aids, which have many advantages for the patients themselves as well as the doctors.

Using these technologies for the design and manufacturing processes has many advantages over both doctors and patients. They are:

- low material consumption,
- automated production,
- easy, comfortable, airy and visually appealing solution for the patient,
- cost savings,
- higher accuracy of orthopedic aids (by shaping the PLA material)..

This paper represents part of the dissertation thesis which is being elaborated in collaboration with Neoprot company. This company is engaged into the design and manufacture of individual orthopedic aids. The result is a simple, fast, and therefore efficient process of designing and manufacturing of individual orthopedic corset with regard to its functionality, suitability for use and practical application [1].

The dissertation thesis will be dealt with already mentioned variant of design and manufacture of „2D" model of corset. It is necessary to choose the appropriate software functions and their sequence for efficient editing of the scanned data and then select the optimal parameters of manufacturing process. The aim is to create a methodology for the design and manufacture of orthopedic corset using 3D digitization and AM, with a view to the practical use in Neoprot company.

References

[1] Molnár I 2018 Research of design and manufacture of orthopedic appliance by 3D digitizing and Rapid Prototyping. (Written project for dissertation examination). Slovak University of Technology in Bratislava, Faculty of Materials Science and Technology in Trnava, Institute of Production Technologies. Supervisor: doc. Ing. Ladislav Morovič, PhD.

[2] Milde J 2016 Use of 3D digitalization and additive technologies in the medicine (Written project for dissertation examination). Slovak University of Technology in Bratislava, Faculty of Materials Science and Technology in Trnava, Institute of Production Technologies. Supervisor: doc. Ing. Ladislav Morovič, PhD.

[3] Morovič L 2013 Design, machining, measuring and scanning of complex shapes (Habilitation thesis). Trnava. Slovak University of Technology in Bratislava, Faculty of Materials Science and Technology in Trnava, Institute of Production Technologies.

[4] Borzan C 2017 Post-processing of customized implants made by laser beam melting from pure Titanium. In *Journal of Optoelectronics and Advanced Materials*. Vol. 19, Iss. 11-12, pp. 738-747. ISSN 1454-4164.

[5] Cosma C, Balc N, Moldovan M, Morovič L, Gogola P, Borzan C, Berce P, Leordean V D, Luca A, Miron A V 2013 Study of a tridimensional model of a custom implant in cranio-maxillofacial surgery In *Academic Journal of Manufacturing Engineering*. Vol. 11, Iss. 2, pp. 38-43. ISSN 1583-7904.

[6] Tappa K, Jammalamadaka U 2018 Novel biomaterials used in medical 3D printing techniques. In *Journal of Functional biomaterials*. Vol. 9, Iss. 17

[7] Dodziuk H 2016 Applications of 3D printing in healtcare. In *Kardiochirurgia i Torakochirurgia Polska*, 13(3). pp. 283-293.

[8] Miltiadis A B 2010 CAD - CAM & Rapid Prototyping Application evaluation. 174 p. ISBN 978-87-7681-676-6.

[9] Neoprot 2017 Orthopedic and prosthetic medical devices [cit. 2017-11-16] Available at: http://www.neoprot.sk/

[10] Mok S W, Nizak R, Fu S CH, Kevin Ho K W, Qin L, Saris D B F, Chan K M, Malda J 2016 From 3D printer: Potential of three-dimensional printing for orthopaedic application. In: *Journal of orthopaedic translation* 6. pp. 42-49.

[11] Galantucci L M, Lavecchia F, Percoco G 2009 Experimental study aiming to enhance the surface finish of fused deposition modeled parts. In *CIRP Annals*. Vol. 58, Iss. 1, pp. 189-192.

[12] Siketová K, Lipa Z, Baránek I, Líska J 2009 Position of the superfinishing on the finish machining. In *Annals of DAAAM for 2009 & Proceedings of the 20th International DAAAM Symposium "Intelligent Manufacturing & Automation: Focus on Theory, Practice and Education"*. pp. 1299-1300. ISBN 978-3-901509-70-4.

[13] Mihov D, Katerska B 2010 Some biocompatible materials used in medical practice. Faculty of Medicine, Medical University of Sofia, Bulgaria. [cit. 2017-11-06] Available at: http://www.uni-sz.bg/tsj/vol8,Suppl.2,2010/D.Mihov.pdf

[14] Simonovsky F 2004. Poly(lactic acid) (PLA) and Poly(glycolic acid) (PGA). University of Washington Engineered Biomaterials. [cit. 2017-11-27] Available at: https://www.uweb.engr.washington.edu/research/tutorials/plagla.html

[15] Harper A CH 1999 Modern plastics handbook. McGraw-Hill Companies, Inc. ISBN 0-07-026714-6. Available at: http://allaboutmetallurgy.com/wp/wpcontent/uploads/2016/10/020552.pdf

[16] Molnár I, Hrušecký R, Görög A, Morovič L 2018 Observation of shape and dimensional accuracy changing of parts in time intervals manufactured by additive method Fused Deposition Modeling. In *Materials Science Forum. Novel Trends in Production Devices and Systems IV.* Vol. 919, pp. 182-189. ISBN 978-3-0357-1265-0

Acknowledgement

The article was written with the support of the institutional project 1307 „Utilization of Rapid Prototyping technology on CNC machines" (project supervisor: Ing. Ivan Buranský, PhD.) of the Faculty of Materials Science and Technology in Trnava, Slovak University of Technology in Bratislava.

IOP Conf. Series: Materials Science and Engineering **448** (2018) 012059 doi:10.1088/1757-899X/448/1/012059

Investigation of constitutive equation parameters on dual phase steel sheets

G Béres[1] and M Tisza[2]

[1]John von Neumann University, Faculty of GAMF Engineering and Computer Science, Department of Materials Technology, 10. Izsáki road Kecskemét 6000 Hungary.
[2]University of Miskolc, Faculty of Mechanical Engineering and Informatics, Institute of Materials Sciences & Technology, Miskolc-Egyetemváros 3515 Hungary.

beres.gabor@gamf.uni-neumann.hu

Abstract. This study is about using of the least squares' method for parameter defining and model accuracy investigation of mechanical-based work-hardening laws of automotive dual phase steels. The experimental true stress true strain curves were determined by standard, quasi-static tensile tests at room temperature. As most of the commercial finite element codes are using mechanical-based constitutive equations for describing the stress-strain relationship of the materials, the coefficients of these equations are one of the most important input parameters of the simulations. The results show that most of the models are able for data estimation, but the physical meaning of each model parameters is not taking into account by this mathematical technique. The highest of the stress values are observed at Ludwik's solution, while the lowest are given by the Voce model at each steel grades. Although the differences are negligible in the strain range of uniform elongation, but significant discrepancies are displayed at higher strains, which typically occur during sheet forming processes in the press shop. The model parameters' developing in the function of the martensite volume fraction was also considered.

1. Introduction

Finite element (FE) analysis is a useful technique for evaluation of metal forming processes. The success of these simulations is significantly depends on the accuracy of the input parameters. One of the most important input parameter of FE simulations is the flow curve of the material. Current FE codes contain more types of integrated equations in which the accuracy of each parameters are generally depending on the approximation method of the measured stress-strain values. The best-known mechanical approximations models are originated from such famous scientists like Nádai [1], Zener and Hollomon [2], Ludwik, Swift [3], Voce [4], Johnson and Cook [5], as well as Hockett and Sherby [6]. These researches elaborated their equations in different mathematical concepts, due to the complexity of the deformation process of metals. Nevertheless, this complexity is the reason why the understanding of exact processes which happen in the microstructure during plastic deformation is still actual topic at metallic materials.

The applied automotive dual phase (DP) steels are widely used in the automotive industry due to their beneficial attributions, not only in the field of forming but in the field of joining technologies like welding [7] or soldering [8,9] also. These steels contain soft ferrite and hard martensite in the microstructure, thus the mechanical properties are mainly depending on the amount, the morphology and the distribution of each phases [10]. Consequently, the microstructure-based modelling of these steels is also a popular field, so new hardening model strategies are continuously spreading with the development of computer performance and microscopic techniques. This modelling means the mechanical simulation of each phases of the microstructure, taking into account the microscopic properties like grain size, dislocation density, length of the burgers vector, etc. The method is called as representative volume element (RVE) technique [11,12,13].

Although, these materials are particularly appropriate for RVE simulations (as the mechanical properties can be defined well by the amount, the morphology and the distribution of each phases), we primarily focused on the accuracy of the classic, mechanical-based approximation methods.

2. Materials and experiments

Three types of conventional DP steels as DP600, DP800 and DP1000 with the chemical composition and phase proportions (i.e. ferrite and martensite volume fractions) shown in table 1 were applied in this study. The chemical compositions were defined by optical spectrometry. For the microstructure analysing and the calculation of the phase dispersions, small pieces were grinded and polished by using standard metallographic sample preparation techniques and then etched in 2% nital. The calculation of the area ratios in percent of each phases were performed by the usage of Zeiss Imager M2m program wizard software, in at least five different cross sections. Figure 1 shows some specific microscopic pictures of each sample. It is worthy to note that DP800 has a bit higher average carbon content than DP1000. It is contrary to the general experiences, that higher martensite volume fraction (MVF) belongs to higher carbon content.

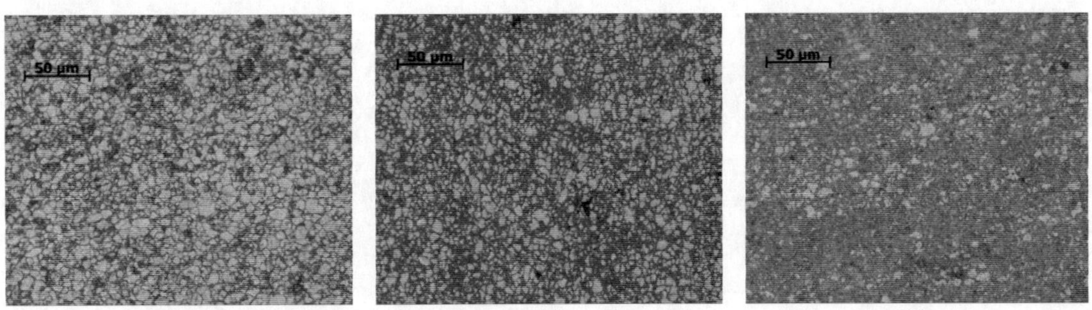

(a) (b) (c)

Figure 1. Microstructure of (a) DP600, (b) DP800 and (c) DP1000 materials, etched in 2% nital.

Table 1. Chemical composition of the applied DP steels.

	Fe (wt%)	C (wt%)	Si (wt%)	Mn (wt%)	P (wt%)	S (wt%)	Ferrite vol.%	Martensite vol.%
DP600	98.6	0.085	0.171	0.87	0.013	0.005	0.734	0.266
DP800	97.9	0.161	0.187	1.52	0.012	0.003	0.579	0.421
DP1000	97.8	0.148	0.181	1.50	0.012	0.004	0.350	0.650

The engineering stress engineering strain (e) curves were determined by tensile tests according to MSZ EN ISO 6892-1:2010 European standard, with 30 mm/min frame rate at room temperature. The tensile specimens with gauge length of 80 mm and 1mm thickness were manufactured based on the mentioned prescription in three directions: parallel, perpendicular and 45° to the rolling direction. The measurements in all different specimen conditions were repeated five times. For the strain measuring we used contact proof AVE video extensometer. The calculation method from engineering data (across tensile force – F_i and initial cross section – A_0) to true data is well-known and given by equation (1) and (2). As these materials show near isotropic behaviour and constant thickness-transverse strain development during outer loading (figure 2 (d)), only the results of specimens parallel to the rolling direction can be seen in details, in the coordinate system of true and equivalent quantities (figure 2 (a) – (c)). The isotropic behaviour is also demonstrated by the $\bar{r}$ value. The main mechanical properties are summarized in table 2.

$$\sigma_{eq} = \frac{F_i}{A_0}(1+e) \tag{1}$$

$$\varepsilon_{eq} = \ln(1+e) \tag{2}$$

Table 2. The main mechanical properties of the applied DP steels.

	UTS N/mm²	YS N/mm²	A_{80} (%)	Ag (%)	ε_{total}	ε_{crit}	$\bar{r}$
DP600	656	445	20.6	13.6	0.187	0.128	0.92
DP800	879	571	16.0	10.8	0.148	0.102	0.75
DP1000	1099	767	10.6	7.0	0.101	0.067	0.76

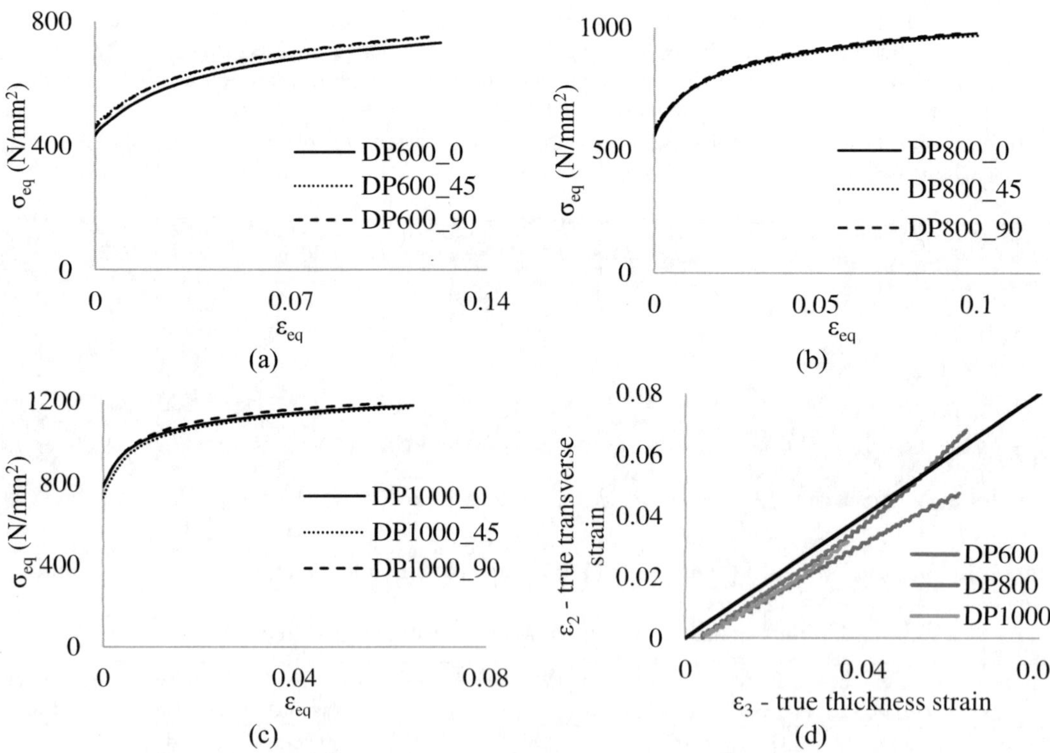

Figure 2. Flow curves of DP600 (a), DP800 (b) and DP1000 (c) materials and the true thickness – true transverse strains developing during the tensile test (d).

In table 2, Ag belongs to engineering uniform strain, while ε_{total} is the total true and ε_{crit} is the uniform true strains. In the same place, $\bar{r}$ means the Lankford coefficient.

3. Parameter fitting by the least squares' method

The parameter fitting of each flow curve model equations were performed by the least squares' method. The least squares' method is a mathematical method to calculate the differences of the estimated ($\breve{y}_i$)

and the measured values (y_i) of a given function. The object is the minimizing of the sum of the squared errors (χ^2), which is the square of the differences of the values mentioned above (3).

$$\chi^2 = \sum(\breve{y}_i - y_i)^2 \tag{3}$$

With the appropriate changing of the parameters between the iteration steps in the mathematic solver program, the most accurate estimation can be available. The estimation accuracy is generally defined by the quadratic regression (4), which already contains the value of the sum of the squared totals (SST in equation (5)) next to the χ^2. It is worthy to mention, that one of the main disadvantage of this method is that the results are slightly depending on the user defined initial value of the predicted function. Fortunately, now it does not mean serious problem, as the start points of the estimated curves are known, as the initial yield strength of the materials.

$$R^2 = \frac{SST-\chi^2}{SST} \tag{4}$$

$$SST = \sum(y_i - \bar{y})^2 \tag{5}$$

The fitted stress-strain curves according to five different theories as well as the experimental data with different colors can be seen in figure 3 respectively, for DP600, DP800 and DP1000 materials.

It is observed that all the theories can approach the experimental results with sufficient accuracy in the strain range of uniaxial tension (table 3). Nevertheless, the models can show definite discrepancies at higher strains. It can be made visible with the extension of the strain axis up to the possible strain values at equi-biaxial tension. So as the figures below point it out, the solution of Ludwik calculates the highest stresses, while the Voce and Hockett-Sherby models calculate the lowest stresses for each strain values, for all investigated steels.

Table 3. The estimation accuracy as the quadratic regression values for the different models.

	R^2				
	Nádai	**Ludwik**	**Johnson-Cook**	**Voce**	**Hockett-Sherby**
DP600	0.8949	0.9860	0.9930	0.9936	0.9995
DP800	0.9332	0.9758	0.9894	0.9881	0.9991
DP1000	0.8790	0.9495	0.9790	0.9857	0.9969

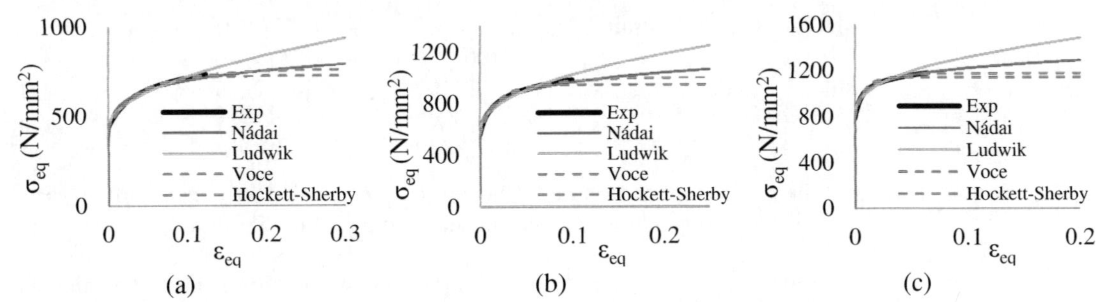

Figure 3. Experimental ("Exp") data and the fitted flow curves' offset for DP600 (a), DP800 (b) and DP1000 (c) steels.

4. Discussion

One classic constitutive model (6) is originated from Nádai [1], which was expanded later with a constant value (A in equation (7)) refers to the yield strength by Ludwik [3]. Similar expanding with an added constant but not refers to the yield strength is originated from Swift [3] (not published here). Former equation is often attributed to the names of Zener and Hollomon also. They reported more studies [2,14,15] about the stress-strain relationship of steel materials and about the effect of the strain rate and temperature, still in the negative degrees also. The genesis of these papers are nearly the same as the paper of Nádai.

$$\sigma_{eq} = K\varepsilon_{eq}^{n} \quad (6)$$

$$\sigma_{eq} = A + K\varepsilon_{eq}^{n} \quad (7)$$

The physical meaning of the n exponent in the Nádai expression can be described as the value of the uniform equivalent strain belonging to the onset of local necking (ε_{crit}) and as the steepness of the linear relationship between the stresses and strains in the logarithmic coordinate system (8). Therefore, it should remain a constant value during the uniform deformation process. Nevertheless, in the case of DP steels, due to the non-proportional deformation of the ferrite and martensite phases [16], it cannot be realized in the practice. Figure 4 visualizes the changing of Nádai and Ludwik exponents respectively in the function of the MVF fraction. The figures indicated the logarithmic axial strain at the onset of neck formation also (ε_{crit}).

$$\ln(\sigma_{eq}) = \ln K + n \cdot \ln(\varepsilon_{eq}) \quad (8)$$

$$\ln(\sigma_{eq} - A) = \ln K + n \cdot \ln(\varepsilon_{eq}) \quad (9)$$

In the Ludwik approximation, this exponent receives a quiet new meaning, and mostly differs from the limit uniform axial strain. Besides, the n exponent also should hold the linear relationship not between the logarithm of stresses and strains, but between the differentiation of the logarithm of tensile stress and of A parameter, and the logarithmic tensile strain (9). The higher discrepancies in figure 5 appear at the lower strains, as only the ferrite phases deform in the early progress. Later, the strain dispersion increasingly homogenised, as the low carbon content martensite also start to deform slightly.

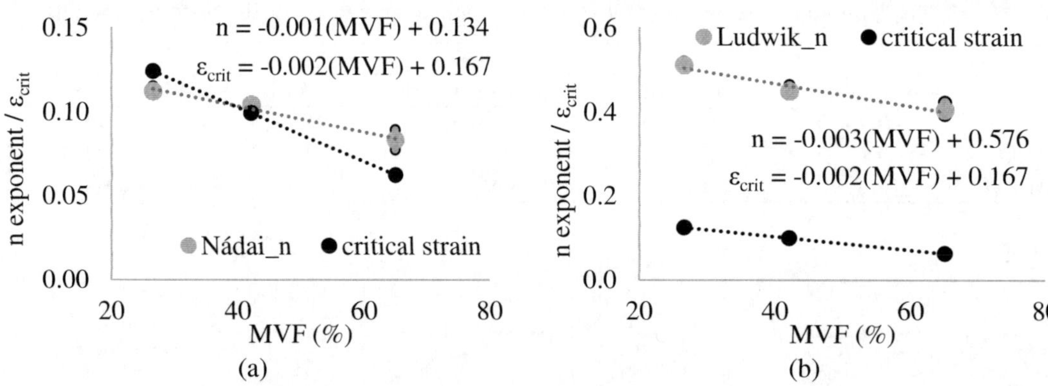

Figure 4. Changing of the n exponent according to the expressions of Nádai (a) and Ludwik (b) moreover the critical axial true strains depending on the MVF.

The shown examples refer to specimens manufactured in the rolling direction again. The calculated straights are presented as continuous lines - black line for Nádai and grey line for Ludwik approximation - while the measured values are symbolized by colourful dot series - red dots for Nádai and green dots for Ludwik approximation. It can be seen, that with the increasing of the MVF, the inhomogeneous deformation, thus the deviation of the measurements from the calculated straights are increasing.

Even let us here to mention, that the values of the n exponent show similar decreasing in both equations, with the growing of the MVF of the material. Regarding the mean values, with the changing of the MVF with 0.58% and 0.54% from DP600 to DP800 and from DP800 to DP1000 respectively, the changing of the exponents produces ~10% and ~20% decreasing in case of the Nádai, as well as ~12% and ~10% decreasing in case of the Luwik-type equations.

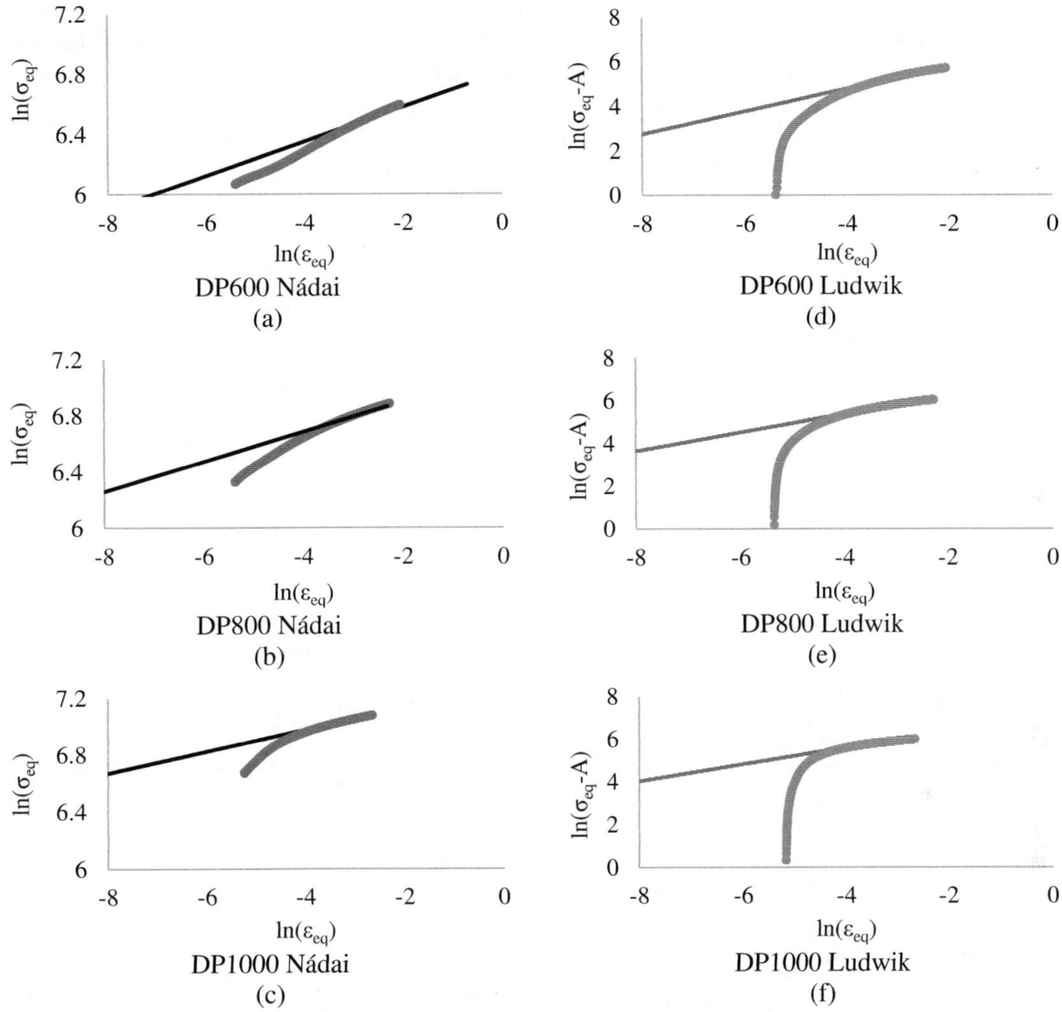

Figure 5. The calculated (black and grey straights) and the measured (colorful dots) stress-strain pairs in the log-log coordinate systems.

Next to conventional power laws, we investigated the compliance of strain-rate sensitive models also. The Ludwik solution gives the basis of two rate-dependent models, one from Jonson-Cook [5] and the other from Cowper-Symonds [17]. Former one attaches the strain rate to the plastic deformation by a logarithmic relationship (10). They performed high number of high strain rate dynamic tests subjected to large strains in twelve different grades of materials, during the experimental validation. Their constitutive model contains the effect of the thermal softening also with exponent of 'm'. Since the effect of the temperature is ignored in this study, only isothermal condition, in this way only the effect of the strain rate multiplier (C) in the function of the MVF is considered.

$$\sigma_{eq} = \left[A + K\varepsilon_{eq}^{n}\right]\left[1 + C \cdot \ln(\dot{\varepsilon}_{eq})\right] \tag{10}$$

$$\sigma_{eq} = \left[A + K\varepsilon_{eq}^{n}\right]\left[1 + \left(\frac{\dot{\varepsilon}_{eq}}{D}\right)^{1/p}\right] \tag{11}$$

Although, Johnson and Cook used much higher strain rates ($\dot{\varepsilon}_{eq}$) than we did, the calculated values of the multiplier (C) fell enough far from the experimental results in different strain rates. In their paper,

IOP Conf. Series: Materials Science and Engineering **448** (2018) 012059 doi:10.1088/1757-899X/448/1/012059

this multiplier connected the stresses to the logarithm of the strain rates, up to 1000 1/s. We used 0.4 mm/min, 4 mm/min, 40 mm/min and 400 mm/min cross head speeds to make some observation on the behaviour of this parameter. Considering the 80 mm gauge length of the specimens, it means that all of the applied strain rates were well below the unity 1/s, which logarithm is below than zero. It could be the reason, why the measured values of DP600 steels were two orders of magnitude higher, than the values were accepted by the least squares' method (figure 6).

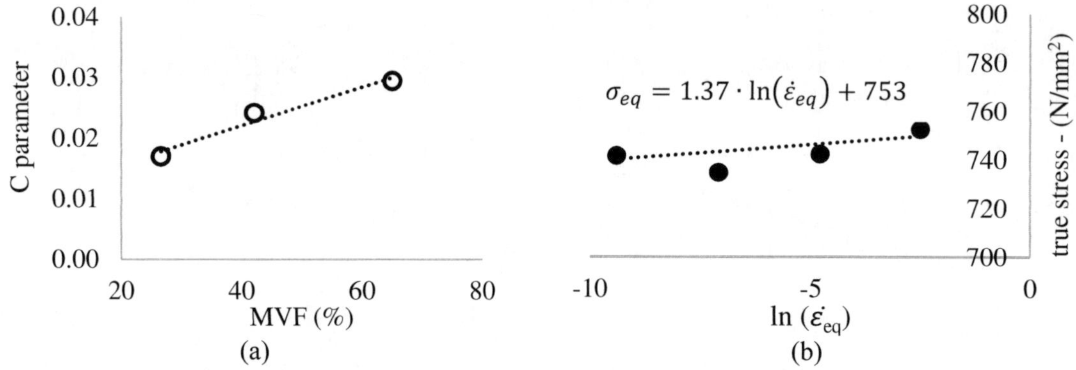

Figure 6. The calculated C parameters in the Johnson-Cook expression in the function of the MVF (a), and the experimental result of C with different cross head speeds (b) for DP600.

Cowper and Symonds calculated the effect of the strain rate through the pth root form equation (see equation (11)). This form of strain rate dependency does not change significantly the approximated results, since the D parameter came to very close to unity and the p exponent came to around 0.25. For as much the term in the second bracket is actually so close to unity, which means that no noticeable different was born compared to Ludwik's expression.

Voce [4] examined stress-strain curves for Cu and its alloys to true strain not beyond -1.0 (compression). Although, he firstly introduced an exponential relationship between the stress-strain values maybe, but such limited strain data made his correlation difficult to establish reasonable parameters for large strains. Voce's model is described by equation (12).

$$\sigma_{eq} = A + q[1 - \exp(-b\varepsilon_{eq})] \quad (12)$$

$$\sigma_{eq} = \sigma_s - \exp(-N\varepsilon_{eq})^p(\sigma_s - A) \quad (13)$$

Some decades later, Hockett and Sherby [6] proposed that cell formation dominates the plastic flow process, and got similar equation this way than Voce. They introduced the concept of steady-state flow stress for large strains (σ_s in (13)). They assumed that a steady-state flow stress (and consequently a steady-state microstructure) can be approached asymptotically, so that is why the true stress is approximated by a pth exponential form. Note that if the value of p in equation (13) is taken as unity, the Voce and Hockett-Sherby models are identical.

For these DP steels, p parameter is nor got as unity neither as 0.58, as it was supposed by Hockett and Sherby for α-Fe. The similarity of the varying of the strain multiplying factors (b and N) for Voce as well as Hockett-Sherby relationships is well visible in figure 7 (a). Both can be fitted quite well by exponential approximation with the growth of MVF. The same figure depicts the p exponent and the q multiplier (b) as well as the steady-state flow stress value (d) likewise in the function of the MVF. σ_s is able to well fitted by a power law, just like the K parameter of the former equations (c). In turn, there is a local minimum on the curve of the varying of the p exponent, and there is a local maximum on q, but no reasonable relationships are possible to determine in the function of MVF.

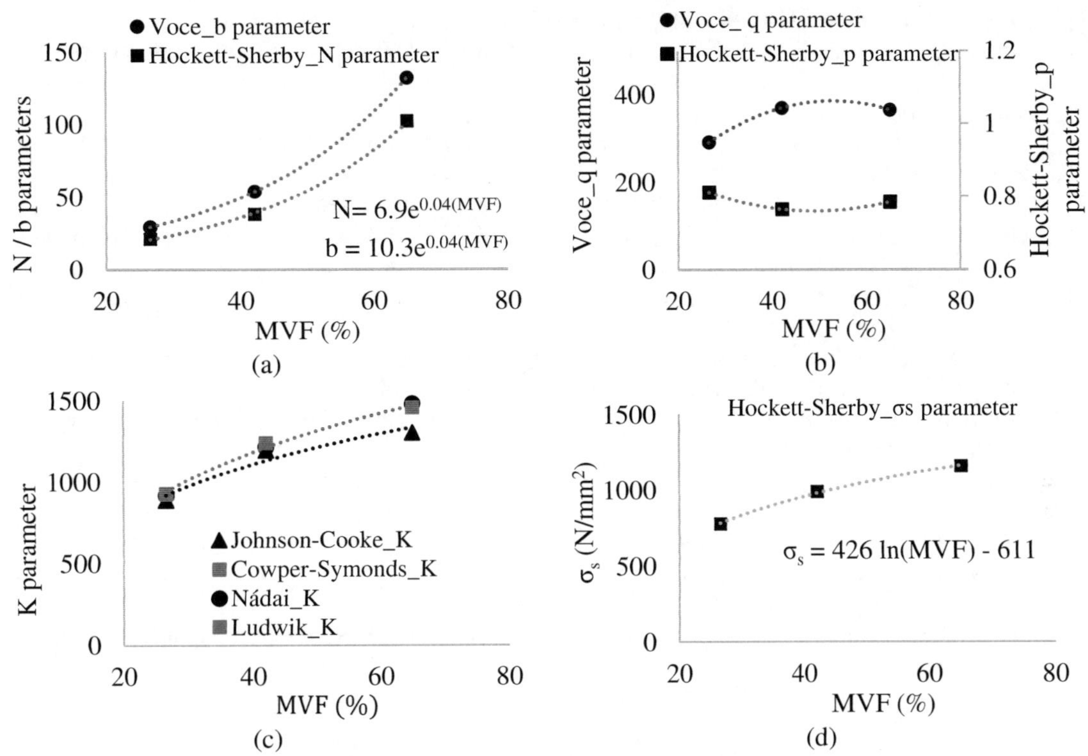

Figure 7. Predicted model parameters in Voce's and Hockett and Sherby's equations (a, b, d) and the K parameter (c) in the function of MVF.

5. Summary

This paper presents the experimentally determined and mathematically fitted flow curves of three types of dual phase steels. It can be observed that all the published constitutive laws are able to properly describe the stress-strain relationship using of least squares parameter fitting, besides, there are no significant discrepancies in the strain-range of uniaxial tension. Nevertheless, considerable disparities may occur between the models at larger strains, in which the Voce model estimate the lowest, and the Ludwik model estimates the higher stresses for the same strain data, in case of all steels. The differences will be well visible with the strain data offsetting up to such an equivalent strain value, which can typically occurs during biaxial tension.

Furthermore, the parameter sensitivity for the martensite volume fraction (MVF) was also displayed. It can be stated that all of the strain multipliers like K, b, N as well as the strain rate multiplier (C) and the one stress featured parameter (σ_s) of the models were always increasing with the increasing of the MVF. Only for the q and p parameters in the Voce and Hockett-Sherby models could not be found any functional relationship with the MVF.

Acknowledgement

This research is supported by EFOP-3.6.1-16-2016-00006 "The development and enhancement of the research potential at John von Neumann University" project. The Project is supported by the Hungarian Government and co-financed by the European Social Fund.

References

[1] Nádai A L 1937 *J. Appl. Phys.* **8** pp 205–213.

[2] Zener C, Hollomon J H 1944 *J. Appl. Phys.* **15** pp 22
[3] Kim S, Lee J, Barlat F, Lee M-G 2013 *Journal of Materials Processing Technology* **213** pp 1929–1942
[4] Voce E *J. Inst. Metals* **74** pp 537
[5] Johnson G R, Cook W H 1985 *Engineering Fracture Mechanics* **21** pp 31-48
[6] Hockett J E, Sherby O D 1975 *J. Mech. Phys. Solids* **23** pp 87-98
[7] Berczeli M, Buza G 2017 *Materials Science Forum* **885** pp 178-183
[8] Gergely G, Koncz-Horvath D, Weltsch Z, Gacsi Z 2017 *Archives of Metallurgy and Materials* **62**:(2) pp 1033-1038
[9] Berczeli M, Weltsch Z 2018 *Periodica Polytechnica-Transportation Engineering* **46**:(2) pp 63-68
[10] Uthaisangsuk V, Prahl U, Bleck W 2011 *Engineering Fracture Mechanics* **78** pp 469–486
[11] Perzynski K, Madej L 2017 *Arch Computat Methods Eng* **24** pp 869–890
[12] Perzynski K, Wrożyna A, Kuziak R, Legwand A, Madej L 2017 *Finite Elements in Analysis and Design* **124** pp 7–21
[13] Madej L 2017 *Archives of Civil and Mechanical Engineering* **17** pp 839-854
[14] Zener C, Hollomon J H High speed testing of mild steel *Am. Soc. Metals, in print*
[15] Zener C, Hollomon J H Plastic flow and rupture in metals *Am. Soc. Metals, in print*
[16] Kalashami G, Kermanpur A, Ghassemali E, Najafizadeh A, Mazaheri Y 2016 *Materials Science & Engineering A* **678** pp 215–226.
[17] ANSYS Inc. Release Notes 18.0, Structures with Material Nonlinearities, Rate-independent Plasticity

IOP Conf. Series: Materials Science and Engineering **448** (2018) 012060 doi:10.1088/1757-899X/448/1/012060

Tailor-made 3D model design of human implants for additive technologies

Laura Georgina Varga[1], János Dr. Takács[1], Ferenc Dr. Dömötör[1]

[1] BME KJK Department of Automotive Technologies

Abstract. This paper deals with the development of a model of individualized implants. It is presented how to produce an implant (bone model) based on digital medical information (CT, MRI). The biocompatibility criteria and human bone properties were taken into consideration during the studies. When creating models, the geometric design possibilities are focused, from the solid 3D body to the creation of hollow, bone-forming structures. The steps for creating models as well as the transitions between each work phase that are needed as input for additive production are presented.

1. Introduction

The new procedures of automated technology using new materials shall be capable to produce implants in high quality even among tailor made conditions. Tailor made manufacturing procedures of the so-called "health industry" reached the level of application of spatial additive manufacturing in a large number of cases. Tailor made implants can be manufactured from properly selected materials (e.g. metal powders), using laser beam methods. Designed implants shall be built from layer to layer by these methods. There are two methods, belonging to this procedure: the selective laser sintering (SLS), and the selective laser melting (SLM). Another important requirement is that tailor made implants shall be made of biocompatible materials in order for them to be capable to be built in the living human body and to cooperate with it. These materials are typically 316L and Ti-6Al-4V, respectively. Before the production a lot of tests shall be carried out. A doctor-engineer consultation and several steps of design procedure are necessary. [1]

2. Pre-modelling steps

The first step is the investigation of the bone to be implanted by using MRI or CT. MRI is an abbreviation for magnetic resonance investigation, while CT stands for computer tomography enabling X-ray tests. Both methods can be used successfully. However, CT has a wide range priority when modelling. The prepared photos then shall be digitalized. The most important task is the analysis of the gray values. This can efficiently characterize the density of the bones, which depends on the age, the sex and the various diseases. Density is an important piece of information, which has to be taken into account, because the goal of modelling is to manufacture implants, having nearly the same features as the original bones to be replaced. Only after this can the computer modelling be started. First, a simple 3D model shall be manufactured. All the necessary changes have to be made on this. [2]

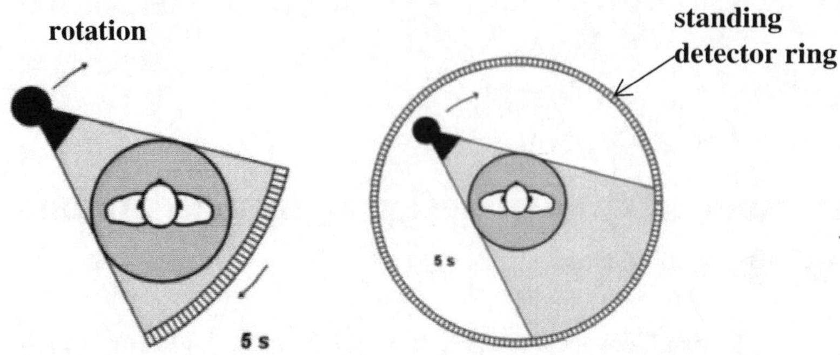

Figure 1: Operation of CT [2]

3. Changing the CT-pictures

CT-pictures are available in DICOM format. This type of file enables the storage of all photo-slices in one file only, making the processing and the transfer easier. These files contain text information about the patient. However, it has to be mentioned that only special types of software can open, process and transform them. The final purpose is that files of STL type shall be available for the manufacturing unit. An important aspect was that commercially available software has to be selected for processing. And this is what happened, so the CT-photos were transformed by InVesalius. Processing of the photos consists of several steps. After importing the data, surfaces to be tested shall be selected. Then, after generating 3D model surface, data shall be exported in a special file format. Shaping the bone models can be done only after this. This process shall be demonstrated via two examples. One of them shall be the middle bone of a big toe, while the other one shall be the middle bone of the hand. [3]

4. Modelling of the middle bone of a big-toe

As the first step CT photos were taken of the foot. Then a starting model was shaped about the complete foot using InVesalius.

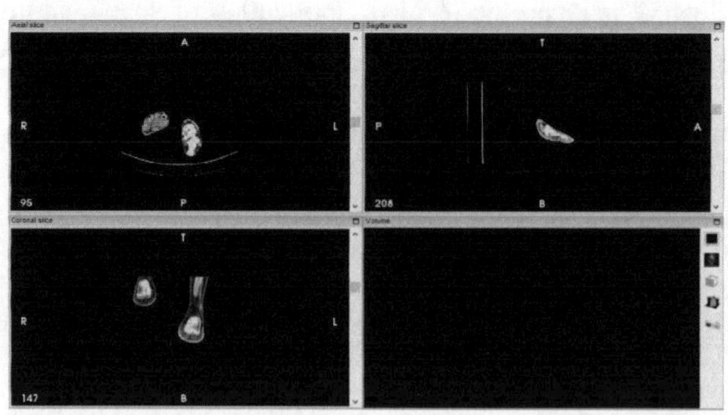

Figure 2: CT-photos by InVesaliusban

Further transforms were carried out by using the freely available Autodesk Meshmixer. First, the middle bone of the big toe had to be "cut out" from the foot. This means a cutting by planes (encountered). After this step the tight bone, as a starting model, was created. This has to be transformed in order to get the features of the natural bone. The goal is to create a model lighter, with holes, but having proper load carrying capacity. Apart from the tight version, several other versions were constructed. There are holes both on the outer surface and the internal parts of this first model. This kind of structure enables the adaptation of the bone, so the implant can be adopted into the human body. The rest of the powder unused during the manufacturing can be removed via these holes. Due to the connection of planes created during the process a repair afterwards might be necessary. E.g. to connect the open edges.

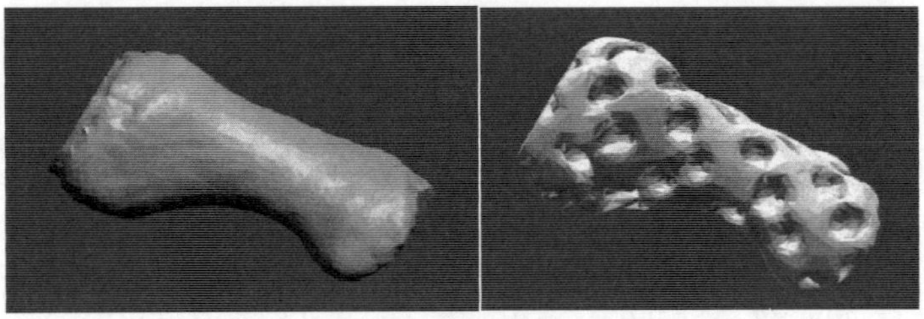

Figure 3: Models, tight and with holes

The second type is a model with tight outer surface, but inside having a grid structure. This means a thin (having determined thickness) outer surface, and an internal structure with perfect grid structure. There is a problem to be solved yet, namely that due to the closed outer surfaces the unused (not melted) powder cannot be removed from the internal parts of the model. This is a problem because the powder cannot be left in the human body.

As another version a model with grid structure was manufactured. The thin outer surface has a grid structure, as well. The wall thickness can be changed within a certain range. The load carrying capacity of the model can be increased in order to be applicable for building into the human body.

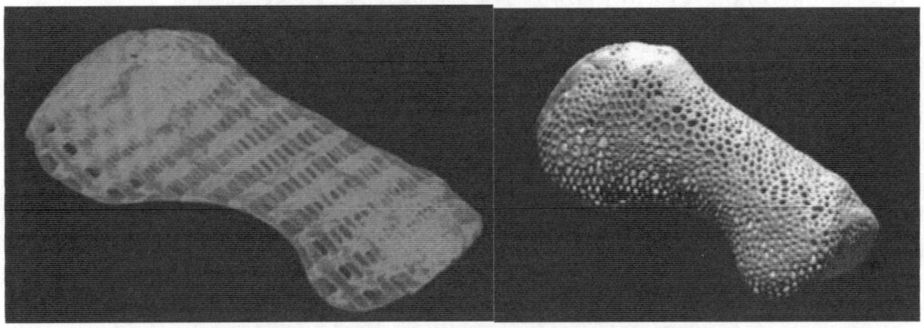

Figure 4: Model with tight outer surface and grid structure inside, and another model with grid structure outside

The next two models have similar structures. Both of them have full grid structure, but one of them with while the other one without internal stiffening. Both of the mentioned bone models are compatible with the human body, because the natural bones can emulate the implants. It is also important for the implants to be capable of bearing the forces acting on the bones. In order to prove this a simulation by finite element method shall be carried out. The question, whether an internal strengthening is necessary or not, can be answered by a finite element simulation only. In both cases it has to be taken into account that due to the non-contacting surfaces a repair afterwards might be necessary.

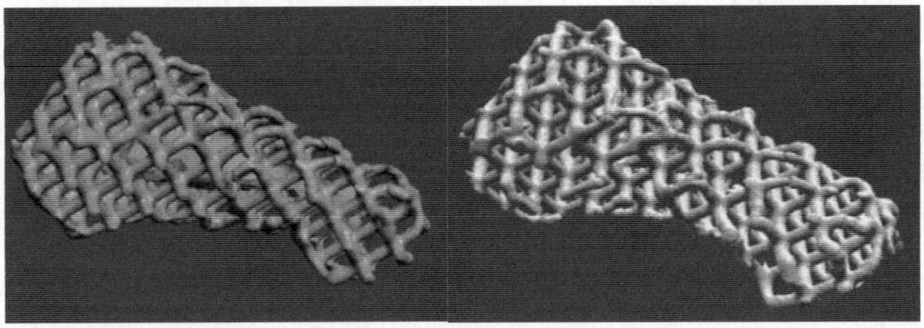

Figure 5: Models with fully grid structure, with and without internal stiffening

IOP Conf. Series: Materials Science and Engineering **448** (2018) 012060 doi:10.1088/1757-899X/448/1/012060

5. Modelling middle bone of hand

Apart from modelling bones of the foot, a model for the middle bone of the hand was created. The mechanical load on this bone is quite complicated. The middle bone of the hand plays an important role in the movement of the hand and the fingers. If this bone is damaged, and it cannot be regenerated, or replaced, some functions of the hand are lost. Based on this, it seems to be obvious that implants shall be modelled and manufactured individually.

That is why an X-ray picture of the hand is necessary. The first step in this case was again the creation of the starting model using InVesalius. After this, three further models have been created.

The fully tight model in this case served as the starting point. For the final construction a transformation into hollow structure was necessary. This model provides better mechanical features, similar to the natural bones.

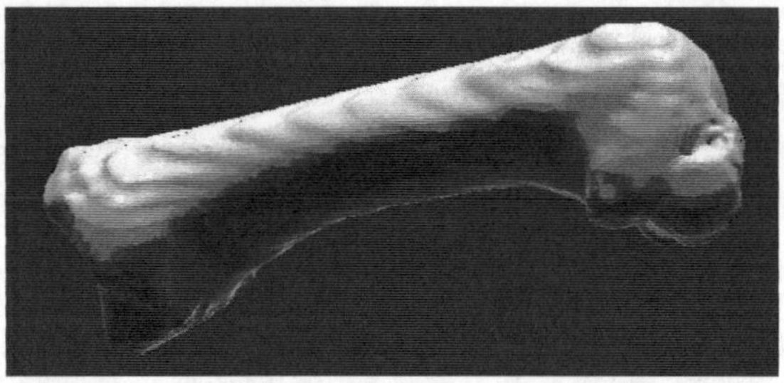

Figure 6: Model of a tight middle bone of a hand

The second version was again a model with internal holes. On the surface again hollows and indentations were created. They promote the adaptation of the bone structure. Moreover, the powder created during manufacturing can be removed through the hollows. If the edges of the hollows cannot be closed, damages shall be repaired afterwards. Apart from this there might be surface damages as well.

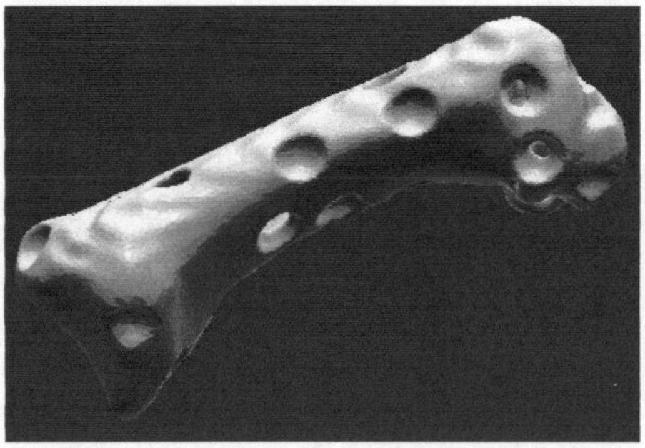

Figure 7: Model with inside holes and surface indentations

The following two models have a complete grid structure, which enables the implants to be compatible when being built into the human body. The bone structure has the capability of adopting the built-in implant. It can happen that only a part of the bone has to be replaced, not the full bone. Due to this, a connection surface to the bones and to the joints has to be provided. According to the medical needs either tight or hollow type structures can be created.

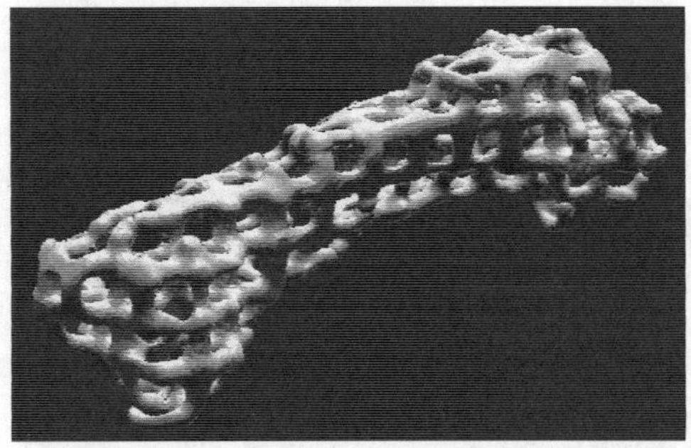

Figure 8: Model of the middle bone of the hand, having fully grid structure

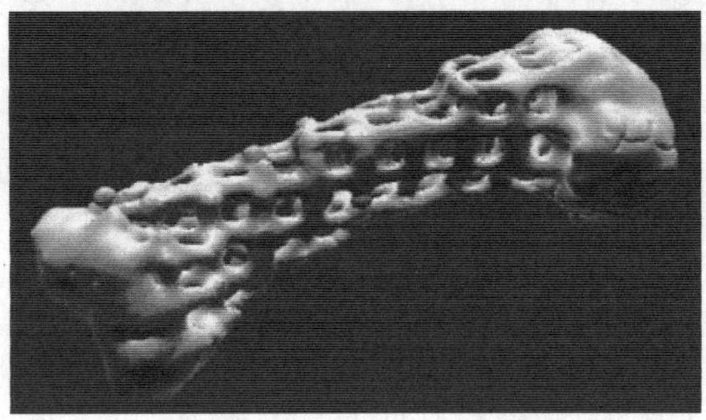

Figure 9: A fully grid structured model, at both ends with tight material

If all the versions are available, the analysis by Finite Element Method shall be used. [4]

The individual constructions can be compared from the load carrying capacity point of view by using these results. The final decision can be made upon the results of these tests. The decision criteria are based on the comparison of the features to the natural bones. The best one shall be declared to be the version to be manufactured.

6. System plan

In order for the modelling and manufacturing process to be easily controlled, the creation of a system plans is necessary. This plan shall have the following steps. (Fig. 10.)

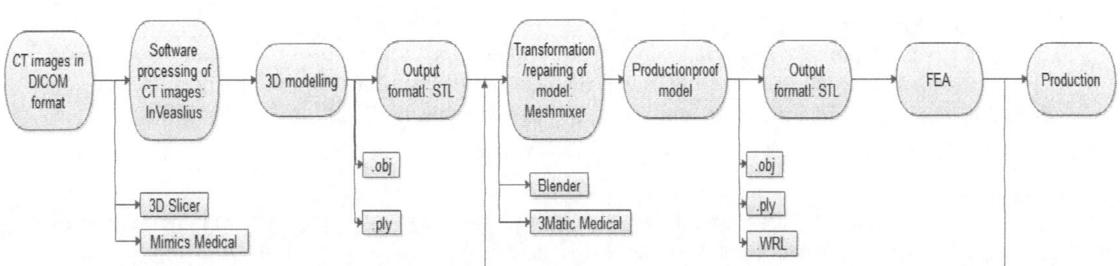

Figure 10: The final plan of the system, with consultations and feedback

Figure 11: Photos of the firstly manufactured models

Steps of the system plan:

1. Creating CT-pictures in DICOM formats;
2. Processing the photos by InVesalius software;
3. Creation of the starting 3D model in STL file formats;
4. Creation and repair of the model in various forms of construction;
5. Creation of the final version suitable for manufacturing in STL format;
6. Analysis using Finite Element Methods;
7. Manufacturing of implants; design of implants to be placed in powder-beds;
8. Testing of trial versions;
9. Fine tuning according to feedback.

In case the created model does not meet the requirements (features of the natural bones) according to the Finite Element Method testing, a step back is necessary. A new model shall be created. This procedure has to be repeated as long as the proper model of implant meeting all the requirements shall be available.
The first trial pieces after the models shall be manufactured using the material Ti-6Al-4V. (See the attached *Fig. 11.)*.

7. Acknowledgements

This project was supported by the National Institute for Research and Innovation. Title of the project: „Manufacturing new generations of individual medical-biological implants and auxiliary tools using additive technologies"; identification number of the support project: NVKP_16-1- 2016-0022.
Authors are grateful for the support.

References

[1] Bose, S., Ke, D., Sahasrabudhe, H., Bandyopadhyay, A., Additive Manufacturing of Biomaterials, *Progress in Materials Science* (2017), doi: http://dx.doi.org/10.1016/j.pmatsci.2017.08.003

[2] Bogner P.: Physics of creating pictures for the medical students Laboratory and Diagnostics courses (2014, In Hungarian) ISBN 978 963 226 450 9

[3] Features of DICOM formats: https://www.lifewire.com/dicom-file-2620657

[4] Dr. Molnár László: Practice of Finite Element Method, Ministry of Employment and Labour (In Hungarian): Program for Operation for Development of Human Resources *„Development of higher education both in structure and contents (in Hungarian)* "HEFOP-3.3.1-P.-2004-06-0039/1.0

IOP Conf. Series: Materials Science and Engineering **448** (2018) 012061 doi:10.1088/1757-899X/448/1/012061

Comparative study of the joining technologies of vehicle bodywork sheets

Z Weltsch[1]

[1]John von Neumann University, Faculty of GAMF Engineering and Computer Science, Department of Materials Technology, Hungary, 10. Izsáki road Kecskemét 6000.
weltsch.zoltan@gamf.uni-neumann.hu

Abstract. In the automotive industry lighter vehicles are needed with more fuel-efficient engines, due to reduced fleet consumption. Many development deals with new materials, but steel will remain probably the primary material for vehicle structures. Other materials, like new high-strength steel grades are finding their way into vehicle structures. Aluminium and magnesium with low-density and fiber-reinforced composites are increasing in demand in the vehicle industry. For the new materials new bonding technologies are needed. Bonding these new materials are the most important in this research.

1. Introduction

Many automotive designers and engineers have been particularly concerned about the "reduction of curb weight/curbside body weight" for the past three decades because vehicle weight reduction has been very important, as well as engine combustion improvement and friction reductions [1]. However, the efforts were unrewarding for the automobile industry. In the past two decades, the average curb weight of light-duty vehicles has risen by 20 % in European and 25 % in the US despite their best efforts [2, 3].

Fuel consumption can be reduced by a number of methods in motor vehicles, but the most commonly used method is lightweight construction. Lightweight construction immediately contributes to a reduction in consumption, and thus a reduction in the emission of pollutants. In addition to the vehicle body, the weight reduction of the transmission elements results in further secondary effects. Various studies have shown that a 100 kg weight reduction results in a reduction in CO_2 emissions between 5 and 12 g. As 40% of the total mass of vehicles is the body, this is one of the most important areas of weight reduction. If we make a change in the vehicle body, it works very effectively. As security for increasing acoustics demands are generally opposed to lightweight construction, they show the complexity and difficulty of the challenges inherent in it [2, 5].

The mass reduction of cars, automakers worldwide, are strategically implemented to achieve regulatory and market goals. Reducing the weight of a vehicle is mostly achieved by optimizing, reducing or reducing design, and by combining lower density materials with a greater weight and / or greater rigidity and weight ratios. New generations of automobiles are expected to contain an increasingly larger quantity and diversity of innovative material in their components (Figure 1.). The leading light weighting material candidates for vehicle body structures include advanced high strength steels, aluminium, magnesium, and plastic and polymer composites. Innovative combinations of these materials are available in product forms such as sheet, plate, moldings/castings, and extrusions.

Different composite body parts have a particular difficulty in choosing the right technologies, such as cutting technologies [2-4].

Figure 1. New generations of automobile with several innovative materials [6].

Thanks to the advanced materials, mixed materials are needed together to create cost effective solutions for them. Manufacturers are widely used in modern bonding technologies against conventional steel-steel welded joints. As supply chains are gradually evolving, more and more lightweight materials are becoming increasingly important in safety critical components, requiring extensive control over production quality.

Designing joining technologies is an extremely important engineering process that takes into account the materials used and their properties such as energy absorption, surface quality and shapability. In order for the latest materials to be properly connected, provide engineering features for different models. In order to establish such binding, it is necessary to examine the chemical and thermodynamic interaction of the adjacent components. [2].

In order to increase the confidence of car manufacturers, safety critical parts are often tested in both destructive and non-destructive ways. For the successful use of compound materials, life cycles such as end-of-life use should also be taken into account. These complex challenges can be solved in cooperation with research institutes, car manufacturers, suppliers and universities.

This paper investigates some of the popular methods for mixed material joining and identifies gaps and challenges and opportunities in their implementation for a mass-produced vehicle.

2. Experimental conditions

2.1. Applied materials and testing method

Two types of conventional DP and DC steels as DP600, DC01 with the basic mechanical properties, were applied in this study. The Figure 2. shows the geometric dimensions of the test specimens.

IOP Conf. Series: Materials Science and Engineering **448** (2018) 012061 doi:10.1088/1757-899X/448/1/012061

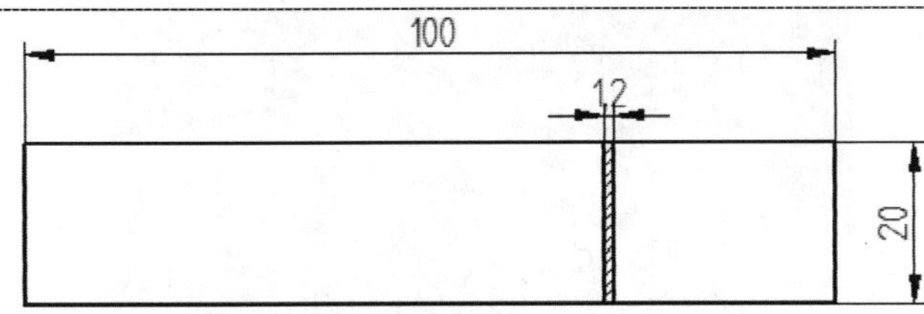

Figure 2. Dimension of the investigated sheet steels (unit of measure mm).

The Table 1. summarizes the chemical composition of the two different DC01 and DP600 sheets.

Table 1. Chemical composition of the investigated steels

	Fe %	C %	Si %	Mn %	P %	S %	Cr %	Mo %	Ni %	Al %	Cu %	Nb %	Ti %	W %	Pb %	N %
DC01	99.4	0.0291	0.0091	0.251	0.0063	0.0091	0.0394	0.0069	0.0238	0.0579	0.0513	0.0085	0.0019	0.0186	0.0065	0.0182
DP600	98.5	0.0873	0.197	0.924	0.0124	0.0026	0.0305	0.0029	0.0333	0.0479	0.0077	0.025	0.0027	0.0205	0.0075	0.0101

There are three combination of the investigated sheets for the experiments: DC01-DC01, DC01-DP600, DP600-DP600.

The plates are overlaid by 20 mm were coupled to the various joining modes. After this shear stress has been investigated by mechanical tensile test to determine the breaking force of the bonds.

2.2. Adhesive bonding technologies

Prior to gluing, the adhesive surfaces were thoroughly cleaned and then, as described in the catalog, the adhesive was allowed to bind.

TEROSON EP 5055 is a 2-component high-strength epoxy-based structural bonding adhesive with excellent corrosion protection properties for bonding metal (coated and uncoated) e.g. steel, aluminium or sheet moulding compounds. It is ideal for roof and panel bonding. The product exhibits excellent corrosion prevention and spot-weldable properties.

2.3. Brazing technology

The MIG brazing technology was carried out with a Migatronic Flex 3000 DUO in pulse mode with 5.0 purity argon gas and CuSi3Mn1 brazing materials wire 1.2 mm thick. The used fix current - voltage parsers are: current: I = 95 A, voltage: U = 19.4 V. The brazed bond was formed along the overlapping (Figure 3.)

Figure 3. Spot welded DP600-DP600 sheet.

2.4. Welding technologies

For spot welding were used SILA Schweissmaschinen NKLP 26. The machine in pulse and in traditional operation, with several parameter settings. The maximum current at which the machine can weld 14 kA. Figure 4. shows an example of a spot welded bond, the microscopic image shows the cross section of welded DP600-DP600 plates with the enriched martensit texture.

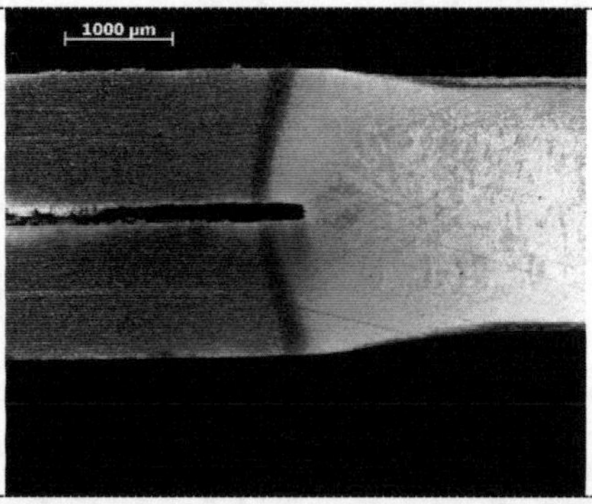

Figure 4. Spot welded DP600-DP600 sheet.

For the CO_2 arc welding used a Migatronic Sigma 300 C Pulse. It is also suitable for conventional and pulse welding, and I used only the conventional welding method. The used current and voltage parameters are: current: I = 98 A, voltage: U = 18.8 V.

2.5. Riveting technology

Clinching is a common joining technology that does not require consumables or pre-drilled holes. It is performed in a single step where stacked, ductile materials are pressed into a die with a punch. The punch forces the materials down and radially out into the die which creates a strong mechanical bond.

For the TOX Clinching use a CEC 008.100 type squeamish press with hydropneumatic cylinder. The Figure 5. shows the cross section of the investigated type of Clinching.

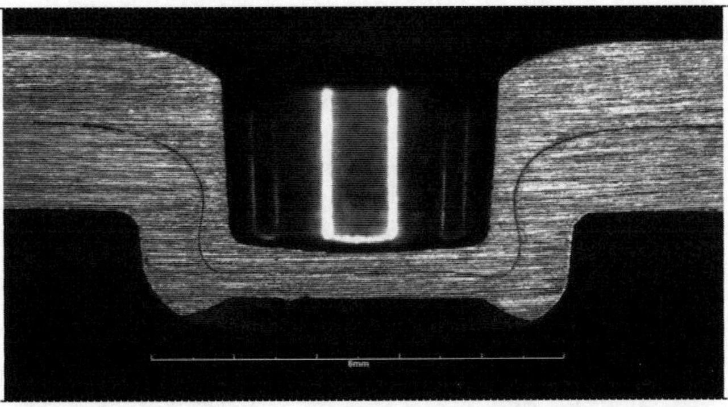

Figure 5. Cross section of a DC01-DC01 Clinched joint.

3. Results and discussions

Figure 6 shows the result of the shear test of the various joints. The DC01-DC01 specimen result shows that the welding of the MAG welding and the MIG welding line resulted in a more uniform stress distribution, thus providing these two technologies with the highest bonding force. At spot welding, the material is torn apart around the point due to increased stress. The bond strength of the TOX clinching technology could be closer to the bonding force of a larger bonding diameter or using thicker disks as the thickness of the neck and the area to be absorbed would increase.

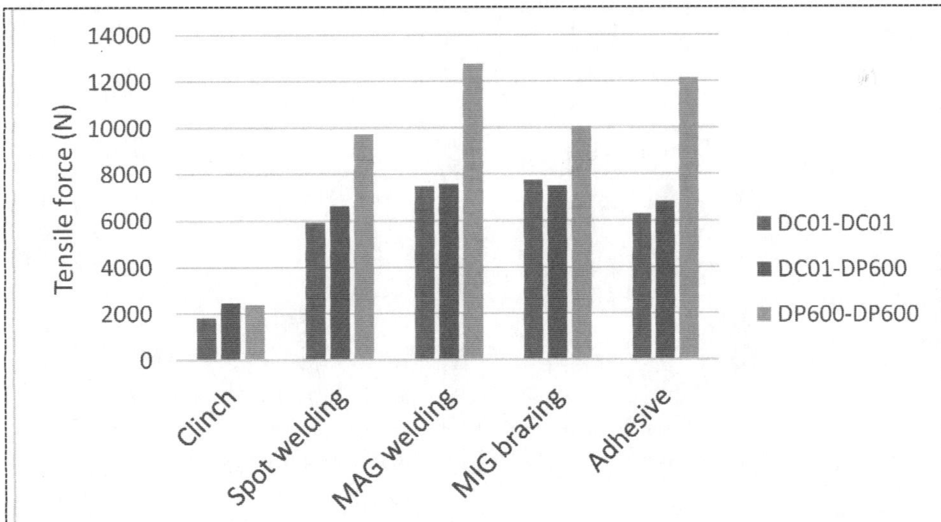

Figure 6. Shear test of the various joints.

In DC01-DP600 there was no significant change in DC01-DC01 for tensile forces, which means that the strength of the DC01 plate determined the maximum breakage value. This also means that the strength of the bonds formed is determined by the strength of the sheet in the vicinity of the bond.

Compared to this, the DP600-DP600 has a significant change in tensile forces. Except for clinch binding, there is a significant increase in tensile force for each type of joint. Observing the tensile specimens, it seemed clear that in these cases the sheet was not the weak point, but the lost bond in every case. Thus, in the case of plate pairing, the particular types of bonds are comparable.

IOP Conf. Series: Materials Science and Engineering **448** (2018) 012061 doi:10.1088/1757-899X/448/1/012061

If we introduce a specific index number, then we can refer to the measurement results on a unit surface. this is shown in the Table 2.

Table 2. Proportionate tensile force

	Joint surface (mm^2)	Proportionate tensile force (N/mm^2)
Chlinch	28,27	84,1
Spot welding	19,63	495,6
MAG welding	80	159,5
MIG brazing	100	100,7
Adhesive	400	30,4

The results shown in Table 2 are graph graphically shown in Figure 7. It can be seen that the spot of welded bonding can provide the best bond strength on the surface. From this point of view, the adhesive bond will remain below the same type of bond.

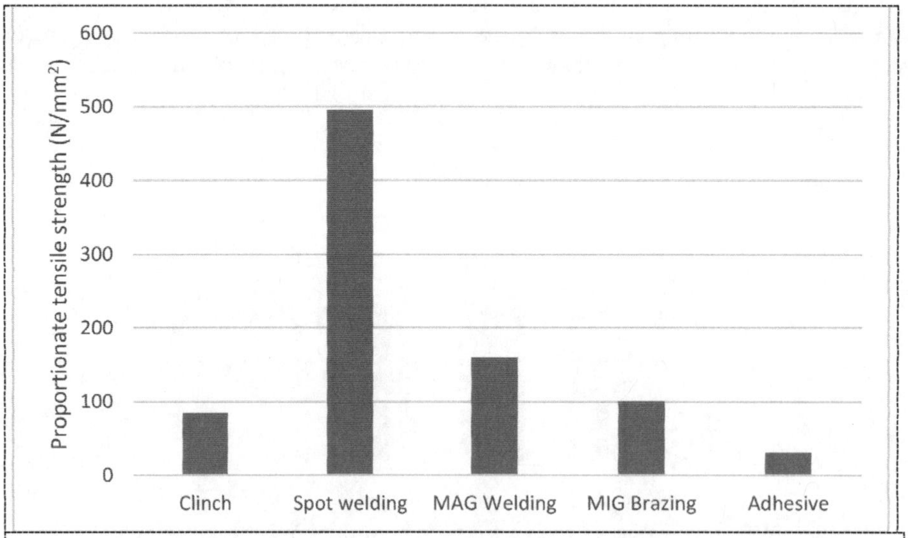

Figure 7. Proportionate tensile strength of different joining technology.

4. Conclusion

In this article, the joining technology of different vehicle body plates has been matched, depending on the tensile forces formed by the joining. This alone can not be used to classify bindings in all respects, but in this article the goal to get a comprehensive view of the evolved binding forces. Based on the results, the following conclusions may be drawn:

In the case of base-strength steel, all bonding types, with the exception of clinch, could provide the bond strength for low-strength steel.

IOP Conf. Series: Materials Science and Engineering **448** (2018) 012061 doi:10.1088/1757-899X/448/1/012061

In the case of matching the high and the base strength steel, the properties of low strength steel mechanics determine the bond strength. In the case of cohesion bonds significant structural changes are taking place on the boundaries.

When using high strength steel, the bond strength is determined by the bonding environment. Considering a specific bonding surface, there are significant differences between the various joining pairs. We can achieve the smallest specific bond strength with spot welding, but in this form the adhesive bond is the weakest.

5. References

[1] – Lankford W I, Snyder S C, Bauscher J A. 1950. *Trans. ASM* **42** pp 1196-1232

[2] – Modi, S., Stevens, M., & Chess, M. Mixed Material Joining Advancements and Challenges. Center for Automotive Research, Ann Arbor, MI. May2017

[2] – K Liska , J Kodacsy , J Liska: Investigation of the microgeometry after hard turning and diamond burnishing Advanced Materials Research 472-475: pp. 902-907. (2012)

[3] – János Kodácsy , János Líska , Zsolt Kovács: Surface finishing by roller burnishing using apparatus of permanent magnet Tehnicki Vjesnik-Technical Gazette 24:(Suppl. 1) pp. 49-54. (2017)

[4] – J Kodacsy , J Liska: Magnetic Assisted Roller Burnishing and deburring of flat metal surfaces, Advanced Materials Research 472-475: pp. 908-911. (2012)

[5] – Maghrour Zefreh, M., Török, Ádám and Mészáros, F. (2017) "Average Vehicles Length in Two-lane Urban Roads: A Case Study in Budapest", Periodica Polytechnica Transportation Engineering, 45(4), pp. 218-222.

[6] – https://www.topspeed.com/cars/porsche/2010-porsche-panamera-ar67301.html

Acknowledgments

The project has been supported by the European Union, co-financed by the European Social Fund. EFOP-3.6.1-16-2016-00014.

Vision 2030 on Manufacturing by the HLG of the European ManuFuture Technology Platform

G. Haidegger - & members of the ManuFuture HLG

Senior CIM research engineer, MTA SZTAKI, Hungary
Honorary prof at AMITY University

E-mail: haidegger@sztaki.hu

Abstract. The ManuFuture European Technology Platform (ETP) among other ETPs is probably the strongest, most highly reputed entity within the EU; representing the value-generation sectors in the manufacturing domain of the economy. The High-Level Group (HLG)– with experience in the last 15 years in harmonizing Visons, Strategic Research Agendas and Road-Maps – has taken up the task and role to prepare a vision paper for the European manufacturing scene for 2030. The presentation will first detail the scenario of the ETP's output processes, and will show the experiences gained in the Hungarian ManuFuture NTP. It will give a nut-shell overview of the presently publicised, 'consultation-version' of the VISION paper on European Manufacturing in 2030. The summarized findings was authored by the large number of participants in the HLG group. The details of the ideas are to be fed to the decision makers in the EC and EP, while forming the base for Strategic Research Agenda and Road-Maps for the FP9.

1. Introduction

More and more high-level politicians are requesting usable, harmonized, consensus-based working documents to help their responsible tasks, e.g. to generate reliable options for valid decisions. Technology Platforms – by definition – are experts groups that have basically the right partnership of all involved players or actors in a given scientific-technologic field, to cope with the preparation of such interdisciplinary set of issues. The members of the EP (European Parliament) and EC (European Commission) are requesting verified, harmonized ideas generated by European Technology Platforms (ETPs).

2. The wide range of Technology Platforms

Since TP-s are self-organizing fora, during the past 2 decades, many formulations had emerged on various industrial topics. Naturally, there are very relevant and less dominant economic sectors, but each has its own field of essential importance as given in Table 1. [1]

Bioeconomy	Energy	Envir	ICT	Prod@Process	Transport
ETIP	Biofuels	WssTP	ARTEMIS	ECTP	ACARE
ETPGAH	EU PV TP		ENIAC	ESTEP	ALICE
FABRE TP	TP OCEAN		EPoSS	EuMaT	ERRAC
Food fo Life	RHC		FTP 4 HPC	FTC	ERTRAC
Forest-bnased	SmartGrids		euROBOTICS	ManuFuture	Waterborn
Plants	SNETP		NEM & NESSI	NanoMedicine	
TP ORganics	ETIPWind		NETWORLD2020	SMR	
	ZEP		Photonics 21	SusChem	

Table. 1. The structured table of the EUROPEAN Technology Platforms

IOP Conf. Series: Materials Science and Engineering **448** (2018) 012062 doi:10.1088/1757-899X/448/1/012062

The following list highlights those ETPs, that are relevant for the Manufacturing industry:

- ETPS related to Manufacturing;
- Advanced Engineering Materials EuMaT – A4M;
- Embedded computing ARTEMIS
- European Nanoelectronics ENIAC
- European Techn Platform on Smart System Integration (EPoSS)
- Construction Techn Platform
- European Steel Technology Platforms (ESTEP)
- European Textile and Clothing –ManuTex
- Food for Life – Foodmanufuture
- Robotics (EUROP)
- Industrial Safety
- ManoFuture
- Photonics
- ConXEPT
- Sustainable Chemistry(SusChem)
- Waterborme

The ManuFuture ETP recognized several involving small platforms, to serve as subdomains within the manufacturing, and these are called the SUB-PLATFORMS of the ManuFuture ETP.

- Additive Manufacturing- AM
- Agriculture Engineering and Technologies –AET
- Joining
- Zero Defect Manufacturing
- Micro-Nano Manufacturing –MINAM

When the expert group of ManuFuture ETP prepares harmonized summaries, all sub-platforms are requested to feed in their topic-related relevant requirements and specifics.

3. What activities are required from ETPs?

European Technology Platforms are industry-led stakeholder fora recognized by the EC as key actors in driving innovation, knowledge transfer and European competitiveness. [2]

ETPs develop vision paper(s), research and innovation agendas (SRAs) and roadmaps for actions at EU and national level to be supported by both private and public funding. They mobilize stakeholders to deliver on agreed priorities and share information across the EU. ETPs are independent and self-financing entities. They conduct their activities in a transparent manner and are open to new members.

ETPs are bottom-up driven expert groups, their roles are set in their constitution. Nevertheless, the EP and the EC declared a common set of requirements for ETP-s, to define a reliable source for information and Decision Support.

Three major documents are required from each ETP, to be fed to the EP, EC decision makers. They must be prepared in a straightforward sequence, in harmony with entities involved within the area in a very broad sense.

Preparing, developing a **VISION document**, with a time-domain of 10 or 10+ years, so that the members of that Platform can share the same view for the future, explaining it in a harmonized, detailed view.

Preparing, developing a **Strategic Research and Innovation (SRA) Agenda**; listing what are the missing knowledge and practice solutions building blocks, that are not yet ready, but are in need to apply rather soon in order to reach the Vision, the visional future environment by the end of the next 10 years.

Preparing, developing a **ROADMAP**, that has scenarios for possible best or worst estimates or most probable scenarios, by allocating financial support, manpower and infrastructure parameters together with the time-duration estimates.

ETPS are also responsible to establish a close, industry – academia- and educational partnership, to have balanced view from many domains.

4. Experiences at ManuFuture ETP and Hungarian National TP

From 2003 to 2004, the European ManuFuture Technology Platform [3] has been involved in the preparation and development of the Vision 2020document, followed by the SRA in 2006 [4], and finally deployed the ManuFuture Roadmap in 2013 together with the EC's relevant DGs. These docs had been the fundamental stepping stones for the EC and EP to launch the 7th Framework Program and the HORIZON 2020 for a pan-European joint research-development and innovation program.

As other countries, the Hungarian Government has also been pushed by the EC to help the establishment of National TPs. The Scientific Society for Mechanical Engineers (GTE) [5] had the privilege to be hosting the ManuFuture-HU. The NTP in Hungary had not just prepared the translations of the working documents, like the Vision and the SRA [6], but had also gone beyond that EU level, and had generated the possible National allocation of funds, resources, etc. as shown in Figure 1.

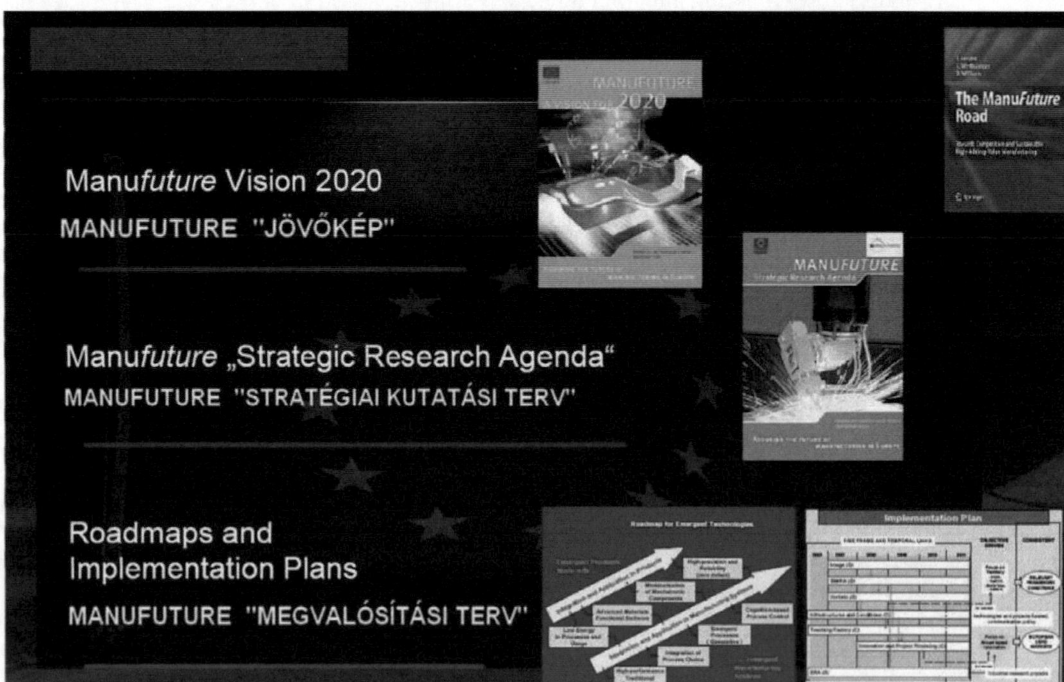

Fig. 1. European and National (Hungarian) ManuFuture main documents.

IOP Conf. Series: Materials Science and Engineering **448** (2018) 012062 doi:10.1088/1757-899X/448/1/012062

5. Beyond Horizon 2020

The ManuFuture ETP [3] has continuously been very active even after the deployment of the ROADMAP in 2008, a large group of its members has joined the High Level Group, the Industrial Support Group, the Industrial Advisory Group, the Mirror Group, for strengthening the ETP, for establishing PPPs [7] as EFFRA [8] and for monitoring the progress along the implementations.

During the past 2 years, the new Version for a new Vision document could be developed, that has a foresight timeframe for 2030.

The first public presentation, just as a summary of the work, had been given at the Tallin ManuFuture conference in 2017 October [9]. Prof H.Flegel president [10] and J.C.Caldeira [11] had given briefing on the new Vision doc.

The document reached its present version, referenced as 'Consultation Version' ManuFuture Vision2030, after the April 2018 HLG Working Conference in Graz. The following sections give a short overview of the 38 pages long document.

6. The structure and content of the of the Consultation Version of Vision2030

The first chapter shows the Manufacturing Industry Today, while Chapter 2 details the megatrends and drivers for manufacturing. Chapter 3 gives vision for scenarios and models for the future manufacturing processes. Vision and strategy are detailed in Chapter 4, and the Vision Building Blocks are described in Chapter 5. A short terminating Chapter deals with Manufacturing &Society relevancies in the vision. The document is a contribution for political, economic, ecologic, and social orientation from a European perspective.

7. European Manufacturing -global challenges

This is the backbone of the European economy: with 2.1 million enterprises, employing 30 million workers. Industry 4.0 will impact at global level. Presently Europe is the global leader for the technology and also for its implementation. At international dimensions, Europe is the world's biggest exporter, but the decline of added value produced by the manufacturing sector is disturbing, so there is a need to give priority to European manufacturing. Regarding other global areas, most of them give high priority to digitization and Key Enabling Technologies, like USA, China, Japan, South Korea, but even South-East Asia.

Competition and cooperation both increase at a global level, and thus the complete manufacturing innovation ecosystem needs to be involved for a change.

Today, society and the world economy are undergoing major changes, driving a social transformation as important as the first industrial revolution. These changes are a global phenomenon, affecting the way we live, work and behave. An unprecedented increase in the speed of development in science and technology, a fast diffusion of knowledge, the scarcity of resources and new generation of consumers will pose challenges and opportunities for Manufacturing. This will lead to a new paradigm shift at a global level.

The following are the most relevant trends and drivers for the future of European Manufacturing:

Fig. 2 Megatrends and drivers by importance and certainty.

Fig. 3 Megatrends and drivers for future Manufacturing.

Each block is detailed in the Vision paper.

8. Future Manufacturing Scenarios and Models

In this part, only short comments are extracted from the vision documents:

Understanding the challenges and the foreseen opportunities, European manufacturing will have to evolve to exceed the customer's expectations in design, quality and service, and become even more flexible and adaptable. It must be user-centric, the customers will have a central role in the value creation. Bionic manufacturing will enhance and augment relevant human capabilities. Nature inspired manufacturing can lead to new frontiers.

Circular economy is a large collaborative endeavor and manufacturing is at its core. Education and life-long learning will become a critical functions, and the new concept of learning factories will offer new challenges.

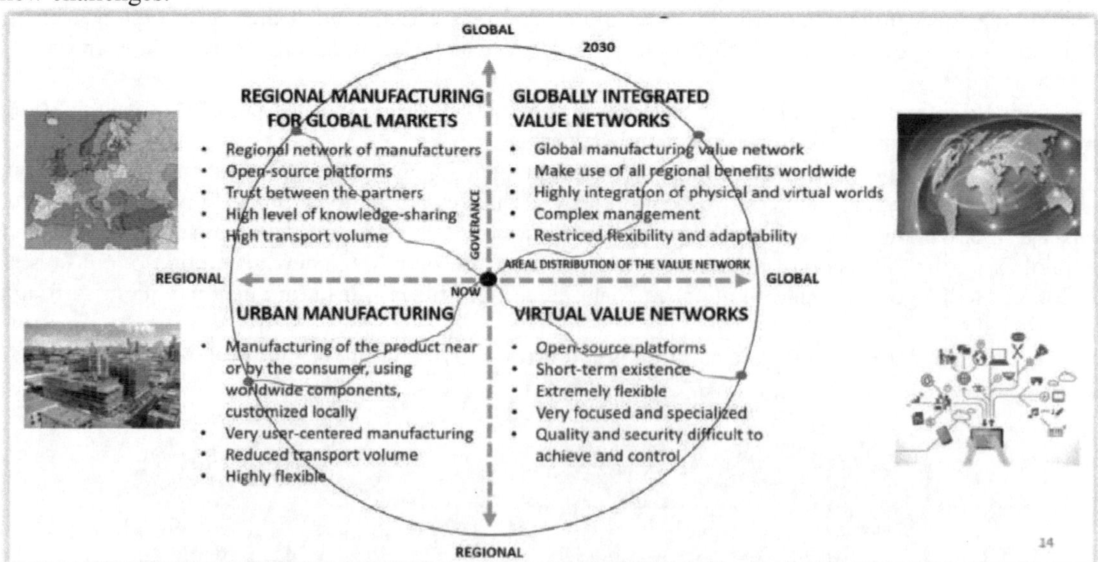

Fig.4. Scenarios for future manufacturing

Aligned with the foreseen evolution of manufacturing towards 2030, MANUFUTURE ETP developed four models for the development of value networks and manufacturing systems: Globally

IOP Conf. Series: Materials Science and Engineering **448** (2018) 012062 doi:10.1088/1757-899X/448/1/012062

Integrated Value Networks (Model 1); Regional Value Creation for Global Markets (Model 2); Regional Value Creation for Regional Markets (Model 3); Regionally Regulated Virtual Value Networks (Model 4). Companies will adapt and combine these generic models to optimally address specific market requirements, product characteristics and manufacturing resources availability.

Model 1: Globally Integrated Value Networks

This Model encompasses a further development from today's global supply networks. It will comprise a globally harmonized regulation and governance and will exploit the use of technologies. This is supported by systems predominantly based in artificial intelligence (AI), within or external to the value network, which monitor the potential customers, identify needs and wishes and interact with product design and a high and real-time integration of physical and virtual worlds.

In this context, globally acting companies such as vehicle OEMs operate and orchestrate a global supplier network, including highly automated assembly factories in the most relevant markets, with high cost, speed and flexibility pressures. Companies have full control of market access of their products.

Challenges for the development of Model 1 are:

- Complex management – governance, coordination, collaboration – possibly supported by AI-based systems;
- Development of novel digital/physical architectures from the factory to the networks (digital twin, factory as a product, etc.) to allow for vertical and horizontal integration and more flexibility and responsivity;
- Optimization of cost and quality: standardization at the highest technological level, zero failure, minimization of the use of resources.

Model 2: Regional Value Creation for Global Markets

As Model 1, this Model targets global markets. Dominating the production side are regional networks of design and manufacturing companies, including SMEs. Manufacturing systems are built according to frugal principles, cost and effort optimised and conformant to all relevant regulations. Digital platforms support product design and development, possibly including the end consumer, and operations management. These networks are unique and world leading and are based on the appropriate norms and standards, taking into account the regionally available manufacturing capabilities and services. According to the principle of the Regional Smart Specialization, the regions specialized on specific successful products, deliver them to the global world. These regional networks are highly flexible and dynamic.

In the described context, highly specialized companies such as factory equipment and automation suppliers have a high ratio of in-house production while controlling complete customer-specific value chains.

Challenges for the development of Model 2 are:

- Regional intelligent specializations

- Regions with many leading protagonists, in relevant technology fields;
- Flexible structure of medium-sized enterprises with high specialization (hidden champions);
- Focus of efficiency and intelligence;
- Capabilities for customer-specific system integration.

- Regional technological centres, including digital service centres (cooperative society models)

- Safety and security mechanisms supported by mutual trust in the region;
- Regional innovation hubs.

- Policy tasks: to create supportive conditions, infrastructure and a regulation framework
- Focus on highly specialized and complex products

Model 3: Regional Value Creation for Regional Markets

This prototypic Model has an even stronger regional focus than Model 2, by encompassing regional manufacturing for local customers. Due to the Urbanization Megatrend, manufacturing will be increasingly realized in urban areas, as cities have strong economic and social assets. An example of this paradigm is Urban Manufacturing, but manufacturing in more rural areas is alike. Sourcing of elements and components is done at a global level, while the final assembly or personalization is mainly performed near or even by the consumer.

In a context of urban manufacturing, the complete production, the final assembly or personalization takes place on demand for products such as shoes, clothes and food, as well as furniture or household equipment. Urban manufacturing is also applied to craftsmanship oriented small industries that produce specialized products for niche markets, including maintenance, personalization and product lifecycle services industries. The close-to-service manufacturing model is also managed under these circumstances.

Challenges for the development of Model 3 are:

- Emission-free factories (zero emission, zero waste);
- Flexible work organization and flexible automation;
- Local service centres;
- Compact design of factories and availability of affordable estate in urban areas;
- Customer driven and individual personalization
- Responsive environment and cost effective goods transportation in urban areas.

Model 4: Dynamic Virtual Value Networks

Complementing the centrally controlled value networks of Model 1, Model 4 is completely decentralized. Platform-based ad-hoc value networks emerge spontaneously for the production of specific lots of specific products. Manufacturing companies of different sizes and service providers offer their competences and capacities on the platform. Each person with a business idea can set up a manufacturing network to produce and sell a defined lot of a specific product. Professional platforms support the management of the ad-hoc manufacturing value network of regional manufacturing service providers, including the legal and financial transaction. The network maybe global.

An example of an application within this context are temporal companies, which operate in a platform economy model. Platform providers assemble temporal limited process chains, while each authorized user can have access to the platform.

Challenges for the development of Model 4 are:

- Logistics and sales platforms for supply, manufacturing services and end market products, supporting interoperability and open standards;
- Highly flexible, adaptive automation throughout highly networked value network partners, including legal contracting and money transfer;
- Demanding, complex and unique products;
- New business logics, business intelligence and advanced decision-making;
- Total quality and safety assurance.

9. Detailing the value of RESILIENT manufacturing

From the 2010-s the main priority of manufacturing was set to have **competitiveness**. During the 2015-to 2020, the focus was set to have **sustainability**. In our 2030 vision, we estimate to achieve **resilience** in manufacturing, having the following qualifications:

•GLOBAL: changes in markets, technology, politics, etc.,
•FLEXIBLE: adjustable to changes (small or disruptive),
•FAST: information and proactivity,
•MODULAR: matching demand (geography and capacity),

IOP Conf. Series: Materials Science and Engineering **448** (2018) 012062 doi:10.1088/1757-899X/448/1/012062

•NETWORKED & COLLABORATIVE: integrated value chains, shared risks,
•INDEPENDENT: resources; knowledge; circularity,
•SUSTAINABLE: use less resources with less impact,
•COMPETITIVE: last but not the least!!!

10. Summary

This conference paper gave a summary of the European Technology Platforms, highlighting the very active role of the ManuFuture ETP. Present activity is to produce the VISION, the SRA – and later on the ROADMAP for the VISION of European Manufacturing in 2030. The paper could only highlight some messages from the VISION document, e.g. the 4 scenario models, so interested readers are suggested to have a look at the full VISION document downloadable from the web.

Acknowledgements

The author declares, that the VISION for European Manufacturing 2030 document is written and edited by the ManuFuture ETP HLG group members, and this paper is just an appetizer to read the full document. The research in this paper was (partially) supported by the European Commission through the H2020 project EPIC (https://www.centre-epic.eu/) under grant No. 739592. This research has been supported by the GINOP-2.3.2-15-2016-00002 grant on an "Industry 4.0 research and innovation center of excellence".

References

[1] A,ArunJunai, Co-ordination of ManuFuture ETP with Sub-Platform and other related ETPs, 30th ManuFuture HLG Meeting,Graz, April2016.
[2] ETP; https://ec.europa.eu/research/innovation-union/index.cfm?pg=etp
[3] ManuFuture ETP; http://www.manufuture.org/
[4] SRA; http://www.manufuture.org/strategic-research-agenda/
[5] GTE; www.gteportal.eu
[6] English and Hungarian doc. repository; http://www.gteportal.eu/manufuture.php
[7] PPP for ManuFuture; http://www.manufuture.org/ppp-fof/
[8] EFFRA, European Factories of the Future Research Association; https://www.effra.eu/
[9] Manufuture conference2017; http://manufuture2017.eu/.
[10] H.Flegel; http://manufuture2017.eu/wp-content/uploads/2017/10/pdf-Heinrich-Flegel.pdf
[11] J.C.Caldeira; http://manufuture2017.eu/wp-content/uploads/2017/10/pdf-Jos%C3%A9-Carlos-Caldeira.pdf

IOP Conf. Series: Materials Science and Engineering **448** (2018) 012063 doi:10.1088/1757-899X/448/1/012063

Materials and measurements for additive manufactured customized human implants

M Fehér[1], J Takács[1] and F Dömötör[1]

[1] Department of Automotive Technologies, Budapest University of Technology and Economics, Faculty of Transportation Engineering and Vehicle Engineering, Hungary Budapest, Sztoczek street. 6. Building J, Floor V. PostCode: 1111

Abstract. Designing and manufacturing medical implants are complex tasks. These implants have to fit numerous requirements in their geometries and materials for a proper behavior. To determine the exact composition of their materials is important, and by the use of GDOES we can get reliable and quick results about that. This presentation demonstrates the results of a GDOES measurement taking the implants' quality requirements into account.

1. Introduction

Designing and manufacturing medical implants are complex tasks. In most cases we have to create customized, irregular geometric geometries made of different expensive materials which have to fulfill various requirements to be able to behave naturally in vivo. Taking these into account one can easily tell that the traditional manufacturing of these implants could cost a lot of money and time. These obstacles, however, can be eliminated by additive manufacturing technologies. This presentation demonstrates the Selective Laser Melting (SLM) procedure, which uses metal powder to build parts, reviews the most critical factors effecting the results of the SLM, the made geometries. It also mentions the important requirements regarding the contact surfaces for in vitro application. Materialistic adequacy is a key component. To quickly and accurately determine the material composition the use of „Glow-discharge optical emission spectroscopy (GDOES)" can offer a proper solution.

2. Selective laser melting

The Selective Laser Melting (SLM) is an additive manufacturing technology. The term „additive manufacturing" includes the layer-by-layer building technologies, which were mostly used in the Rapid Prototyping (RP), but today, by the fast development of these technologies we have become able to manufacture medical implants fulfilling most of the quality requirements.

SLM uses metal powder (Ti6Al4V, 316L stainless steel, etc.) and laser beam to build layers. The laser beam's absorbed energy melts the top of the powder bed in a preset layer thickness by scanning it in the cross sections of the built part. The melted pool's dimensions have to secure a total melted connection between the connecting layers and between the adjacent scanned lines so a perfect cohesion connection can be created [1]. The mechanism of action of the scanning is shown on figure 1. After melting a layer, the powder bed is sinked by the amount of a layer thickness, then a mechanism (blade, roller, etc.) covers the top of the powder bed with new powder layer taken from a powder container, and the overflowing powder is collected by a different powder container [2]. The basic schmeatics of this kind of equipment can be seen on figure 2. These steps continue until the part is totally manufactured.

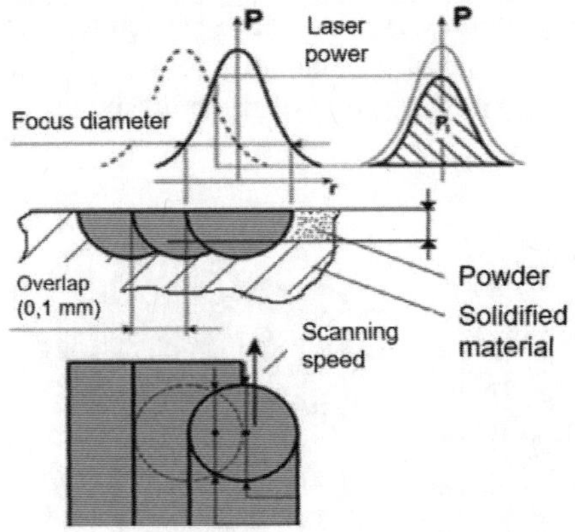

Figure 1. The mechanism of SLM [1]

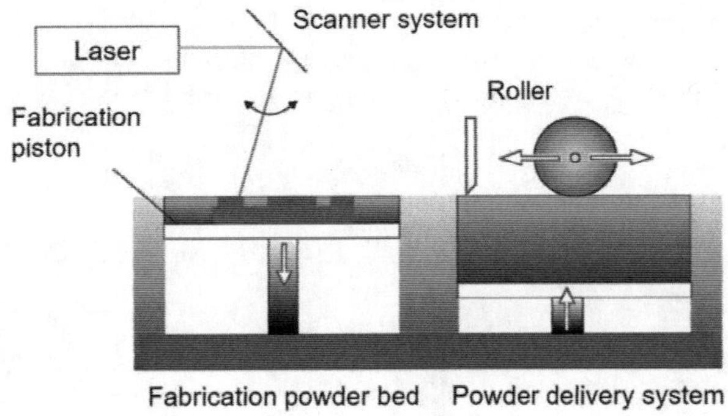

Figure 2. Schematics of an SLM equipment [2]

3. Designing and manufacturing implants

3.1. The bone

The bone contains organic (39%) materials (95% collagen, 5% proteoglykan), inorganic (49%) materials (calcium hydroxiapatite crystals) and fluids (12%). Bone has two types: cortical (solid) and canecllous (spongy) bone. The cortical bone is hard, rigid, this wears most of the stress, but it has low breaking elongation (~2%). The cancellous bone is more porous, can maintain a higher elastic deformation (~75%) but can not bear high loads. The proportion of these two types, age, sex, way of life (e.g. sportsman or not), the function and placement of the bone (craniofacial or tibia) form the bones' mechanical attributes, so it is hard to exactly determine that [3]. However, it can be declared, that bone and implant metals represent a way different mechanical behaviour, as shown in Table 1.

Table 1. Mechanical properties of implant metals and bone [4]

Material	E modulus [GPa]	Yield strength [MPa]	Ultimate strength [GPa]
316L SS	190	221-1213	586-1351
Co-Cr alloys	210-253	448-1606	655-1896
Titan	110	485	760
Ti6Al4V	116	896-1034	965-1103
Cortical bone	15-30	30-70	70-150

3.2. Implants

Implants are medical devices made of biologically satisfying materials, built into the human body by purpose and are partially or totally covered by tissue. In the designing phase it is important to pay attention to the effects when operating or maintaining the implants, it must not ransom invasive reactions in the organism [5].

IOP Conf. Series: Materials Science and Engineering **448** (2018) 012063 doi:10.1088/1757-899X/448/1/012063

The requirements of implants [6, 7] are biocompatibility (the organism should not reject it), bioactivity (cells can stick to and grow on its surfaces), good surface quality, appropriate porosity, pore shape and size (proliferation can start and tissue can grow) and proper mechanical properties (so it will not damage the ambient bone tissues. Latter is defined by the Wolff's law, which describes the stress shielding, where stress is worn by the more rigid implant instead of the more flexible bone (figure 3). This can lead to osteoporosis in the ambient area of the implant, that can elad to the losing of the built in part.

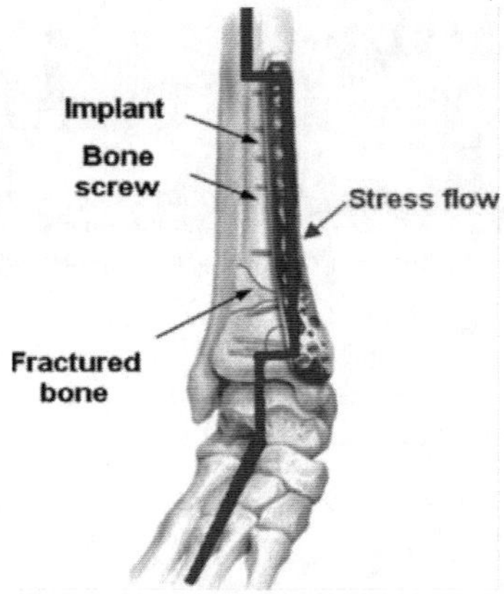

Figure 3. Bone to implant stress shielding [8]

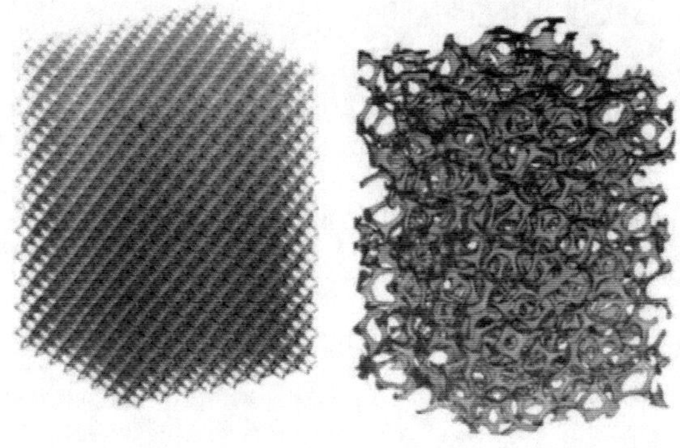

Figure 4. Open cellular structures for implants [9]

4. Main factors effecting the SLM-built parts' quality

The results of the SLM are dependent on many parameters of the technology, these can be categorized into 4 groups [10]:

- attributes of the laser beam (wave lenght, beam diameter ont he surface, energy density etc.)
- material attributes (powder size, melting point, flow-rate etc.)
- scanning parameters (scanning speed, hatch distance, ratio of overlapping, scanning strategy etc.)
- ambient factors (temperature, atmosphere, residual O_2 level etc.)

However, it is impossible to vary all paramteres in most cases. The usually variable and most important parameters are described in the followings.

4.1. Energy density

It has the biggest impact on the manufacturing process, and it's formula is:

$$\boldsymbol{E} = \frac{\boldsymbol{P}}{\boldsymbol{v} \cdot \boldsymbol{h} \cdot \boldsymbol{t}}$$

where E: energy density [J/mm^3], P: power of laser beam [W], v: scanning speed [mm/s], h: hatch distance [mm], t: layer thickness [mm] [11]. Energy density means the energy implied into a unit of space. In general it can be stated that by the right adjustment of these parameters – and all in all the energy density - any material can be manufactured to the required quality [12]. An example of the effect of energy density can be seen on figure 5 through the porosity of the created parts.

4.2. Scanning strategies

Scanning strategy (figure 6) means the way the powder layer is scanned. It greatly affects the developing heat profiles, textures and grain growth direction during the solidification. Choosing the strategy influences the manufacturing time and the product mechanical (residual stresses and surface quality). For a long time, the uni-direction (x or y) scanning was used (figure 6 (a)), which caused different heat

IOP Conf. Series: Materials Science and Engineering **448** (2018) 012063 doi:10.1088/1757-899X/448/1/012063

gradients, poor quality and high surface roughness. Using alternating directions while scanning can lead to better results (figure 6 (b)). The best quality can be reached if the whole cross section is divided into smaller areas (e.g. 5x5 mm squares), and these are scanned individually using the zig-zag strategy rotated by 90° after each other (figure 6 (c)), then the strategy applied to the layers are rotated by 45-90°.

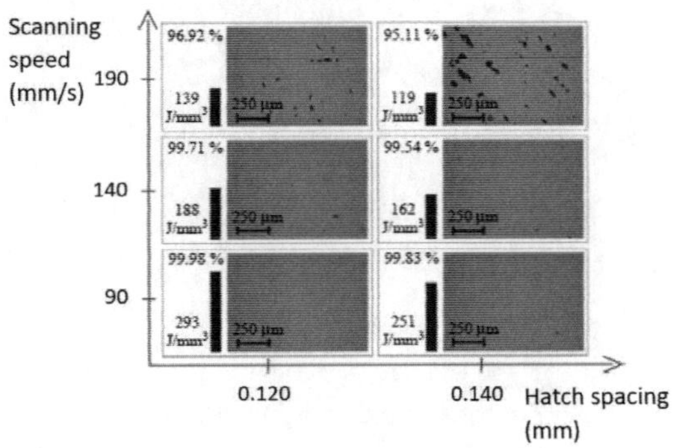

Figure 5. Effect of energy density on porostiy [11]

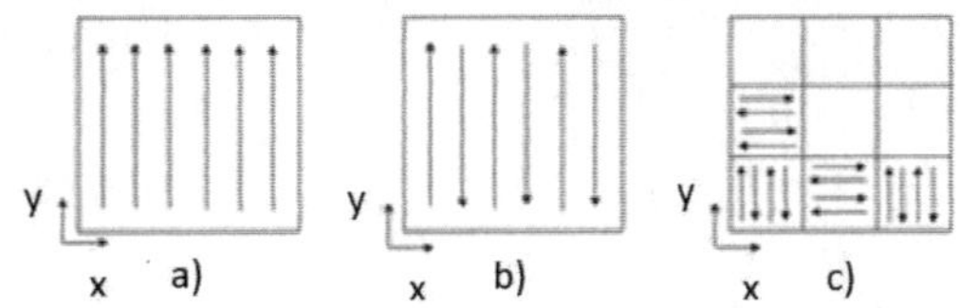

Figure 6. Scanning strategies. (a) uni-directional, (b) alternating directional (zig-zag), (c) chesstable-like

5. Material composition measurements

Implants have strict biological and healthcare requirements, so their materials have to maintain a composition fulfilling these. Regarding the laser technology any kind of disturbance during the manufacturing process can change the proper composition. For example, ambient oxygen or nitrogen may diffuse into the part, or overheating of the powder can cause a partial vaporization of a component in the alloy despite of the well-balanced process control of the machine.

This is the reason, why it is neccessary to use a fast and accurate material analysing measurement to quickly check the process quality through inspecting samples made in the same process along with prototype samples, and can analyse the presence of polluting elements, gas molecules (H, O, N) and carbon besides of the main alloy elements.

5.1. GDOES optical emission spectroscopy

The Glow Discharge Optical Emission Spectroscopy, GDOES is capable of carrying through quality and quantity analysis of metallic and non-metallic materials. The Glow Discharge Source is a Grimm type discharge tube characterized by a special arrangement of the electrodes: The two electrodes of the DC Current Source are set up of a cylindrical hollow anode and a cathode. The sample is placed directly onto the cathode and hence, is switched as cathode itself. The sample must seal the anode tightly so a vacuum can be generated (figure 7). Thus, a flat and preferably smooth sample surface is required.

The sample is placed into the Glow Discharge Source and switched as cathode. The Glow Discharge Source is filled with argon gas under low pressure. Direct Voltage is applied between a hollow anode and the cathode (≙ sample). Due to the energy input of the DC voltage the argon atoms are ionized resulting in the formation of a plasma. Argon cations are accelerated towards the negative sample surface and knock out some sample atoms ("sputtering process"). The knocked out sample atoms diffuse into the plasma where they collide with high-energy electrons. During these collisions energy is transferred to the sample atoms promoting them to excited energy states. Returning to the ground state the atoms emit light with a characteristic wavelength spectrum (figure 8). In the spectrometer the light is dispersed into its spectral components, which are registered by the detection system. The intensity of the lines is proportional to the concentration of the corresponding element in the plasma.

Figure 7. Schematic presentation of the Glow Discharge Source. The sample seals the anode tightly. [13]

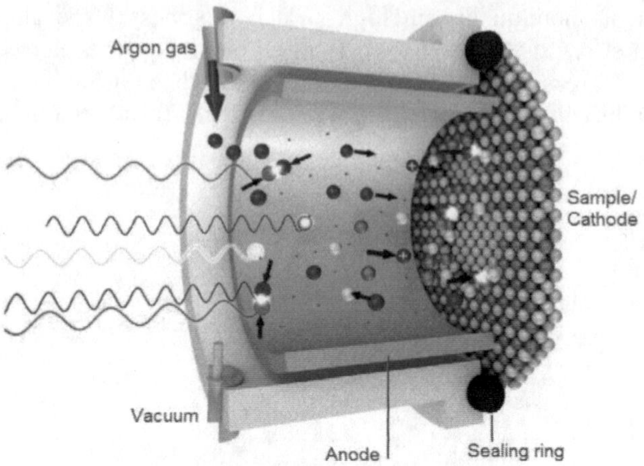

Figure 8. Sputtering process in the Glow Discharge Source. Yellow/red spheres: sample atoms; blue spheres: argon atoms or cations [13]

5.2. Results

We used GDOES to analyse an SLM-made sample regarding its depth-dependant compostition. The graph containing the results can be observed on Table 2. and figure 10, where the lines with different colours represent different elements (in weight %) in 1-10-100-1000x magnification. The sample is a Ti6Al4V part made by SLM, which can be seen on figure 9 along with the analysed spots. It was stored in an open air environment for a long time (>> 72 hours) before the measurement.

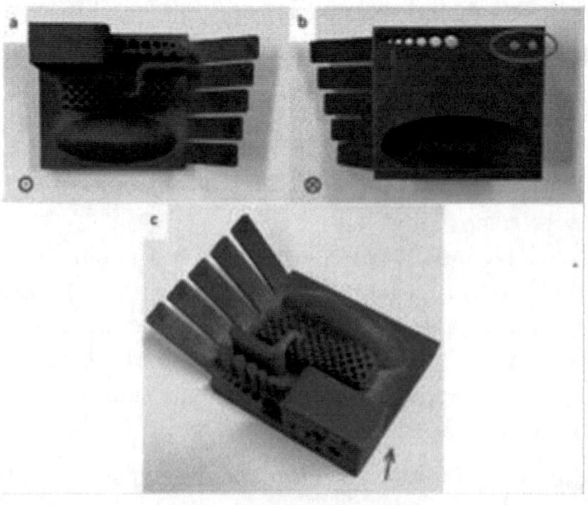

Figure 9. The sample. (a) top view, (b) bottom view with the analysation spots, (c) axonometric view. The z direction of the building is shown on every picture

The deviation of the composition (figure 10.) is stopped after 4-5 μm depth (refer to Table 2, column 1). This is where the equilibrium composition of the elements are stable. It can be observed, that in the top layers (0-4 μm) the oxygen and nitrogen are present in a way higher ratio than they should have been according to the theoretical composition. This may be caused by the storing method of the sample.

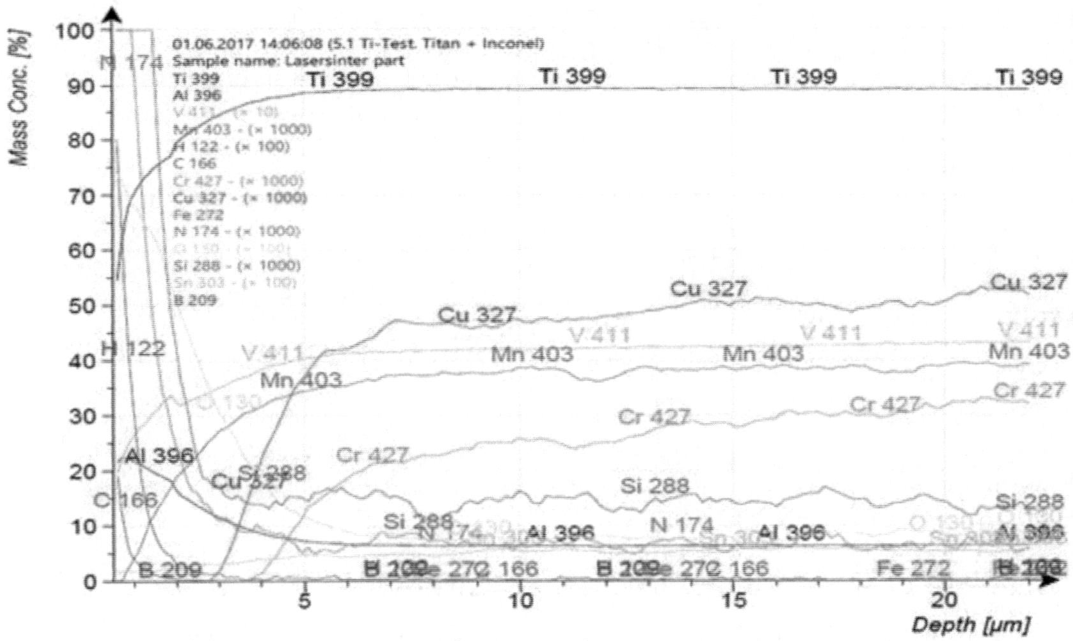

Figure 10. GDOES results of the material composition according to sampling depth [13]

Table 2. Material composition of specified elements in different depths

Depth, µm	Ti %	Al %	V %	C %	H %	N %	O %
2,6119	71,746	19,263	2,849	4,956	0,231	0,072	0,637
4,7782	85,942	9,972	3,692	0,064	0,008	0,01	0,216
6,9616	88,764	6,85	4,105	0,023	0,005	0,006	0,091
9,1781	89,15	6,404	4,179	0,018	0,003	0,007	0,075

The structural differences can also be seen by Table 2 at the depth of 2,6119 µm. The amount of Ti is almost 20% lower than the equilibrium level (≈90%), while Al represents a 19,263% quantity instead of 5,5-6,5% and vanadium is present with a 2,849% amount instead of 3,5-4,5%. This ratio of Al and Ti can tell, that a non-equilibrium structure or alloy is present in this depth that can be the cause of different events, e.g. the rapid heat abstraction occuring during the building phase. These effects should be investigated with further targeted examinations, especially because if aluminium contacts or diffuses into bone, it delays, slows or prevents the growth of bone cells and tissue. Aluminium salts are also capable of causing chondrogenesis and osteogenesis [14]. The non-equilibrium ratio of Al can mean that most of it is not stored safely in a compound phase, so as an implants surface this composition may be easily diffuse into the ambient organs and damage them.

In deeper layers (deeper than 5 µm) the elements' ratio stabilizes to a certain level. Comparing this to EOS's standardized material composition regarding this powder material, it can be seen that after 9 µm all elements are the specified tolerance.

6. Summary

As we could see, designing and manufacturing customized implants is a task relying on many factors and different scientific areas. The quality of SLM made metal human implants depends on many parameters which can be factors of the technology, storage, handling, structural design, the patient herself and so on. The material measurement results showed us key effects – e.g. patient-health-risking changes - occuring in the material composition during an SLM manufacturing cycle which could have

IOP Conf. Series: Materials Science and Engineering **448** (2018) 012063 doi:10.1088/1757-899X/448/1/012063

been unknown otherwise. We have to identify the parameters causing these effects, so the proper setting of these parameters can be the key to create implants that can ensure the better life quality of patients. That is why we are going to do further anlysis with greater attention on these specific parameters among the rest of them.

7. Acknowledgement

This project was supported and brought by the National Research and Innovations Office (Nemzeti Kutatási és Innovációs Hivatal) from the NKIH Fund, the project's name is: „Developing of the know-how of customized medical-biological human implants made by additive technologies" („Egyénre szabott orvos-biológiai implantátumok és segédeszközök új generációs gyártási folyamatának kidolgozása additív technológiákra"); application ID: NVKP_16-1- 2016-0022.

The researchers are grateful for the support.

8. References

[1] Herczeg Sz and Takács J 2004 Influence of scanning speed on the mechanical and structural properties of laser sintered prototypes. *Periodica Polytechnica Transport Engineering* **32/1-2**, pp.: 83-90, HU ISSN 0303-7800

[2] Keszte R 2000 Korszerű járműgyártás technológiák *OMFB TEP*

[3] Bankoff A D P 2012 Biomechanical characteristics of the bone. *INTECH Open Access Publisher.*

[4] Bartolo P, Kruth J P, Silva J, Levy G, Malshe A, Rajurkar K, ... and Leu M 2012 Biomedical production of implants by additive electro-chemical and physical processes. *CIRP Annals-Manufacturing Technology*, **61(2)**, 635-655.

[5] Bartolo P J, Chua C K, Almeida H A, Chou S M and Lim A S C 2009 Biomanufacturing for tissue engineering: present and future trends. *Virtual and Physical Prototyping*, **4(4)**, 203-216.

[6] Liu C Z and Czernuszka J T 2007 Development of biodegradable scaffolds for tissue engineering: a perspective on emerging technology. *Materials science and technology*, **23(4)**, 379-391.

[7] Sealy M P and Guo Y B 2011 Fabrication and characterization of surface texture for bone ingrowth by sequential laser peening biodegradable orthopedic magnesium-calcium implants. *Journal of Medical Devices*, **5(1),** 011003.

[8] Murr L E, Gaytan S M, Ramirez D A, Martinez E, Hernandez J, Amato K N and Wicker R B 2012 Metal fabrication by additive manufacturing using laser and electron beam melting technologies. *Journal of Materials Science & Technology*, **28(1)**, 1-14.

[9] Dai K and Shaw L 2005 Finite element analysis of the effect of volume shrinkage during laser densification. *Acta materialia*, **53(18)**, 4743-4754.

[10] Vandenbroucke B and Kruth J P 2007 Selective laser melting of biocompatible metals for rapid manufacturing of medical parts. *Rapid Prototyping Journal*, **13(4)**, 196-203.

[11] Kruth J P, Mercelis P, Van Vaerenbergh J, Froyen L and Rombouts M 2005 Binding mechanisms in selective laser sintering and selective laser melting. *Rapid prototyping journal*, **11(1)**, 26-36.

[12] Spektruma Analytik Gmbh. Fabrikzeile 21. 95028 Hof, Deutschland

[13] Ellis H A, McCarthy J H and Herrington J 1979 Bone aluminium in haemodialysed patients and in rats injected with aluminium chloride: relationship to impaired bone mineralisation. *Journal of clinical pathology*, **32(8)**, 832-844.

Research and Manufacture of Stretch Films using LLDPE blends

Dorottya Nagy[1], Károly Belina[1,2], Eszter Kókai[1,2]*

[1]Department of Materials Technology, GAMF Faculty of Engineering and Computer Science, John von Neumann University, H-6001 Kecskemét, P.O.B. 700, Hungary
[2]RotaPack Ltd., H-6728, Szeged, Dorozsmai út 35, Hungary

*kokai.eszter@gamf.uni-neumann.hu

Abstract. New linear low-density polyethylene (LLDPE) blends were investigated to use as stretch and breathable films. In our investigation different types of LLDPE blends were prepared by extrusion. The characteristic melting and crystallization temperatures of these LLDPE blends were determined. Specimens were injection moulded and their tensile properties were measured. The LLDPE blends with the best thermal and mechanical results were chosen to prepare film by film blowing in laboratory conditions. Tearing properties of the films were also investigated. Our aim was to determine the most appropriate LLDPE blend for stretch film usable under industrial conditions.

1. Introduction

Polyethylene (PE) is one of the most important and widely used polymer in packaging industry, film and tube forming technology [1]. They are used for a variety of applications and in different film processes: blown film, cast film and biaxially oriented films (double bubble process) [2]. Stretch film was made from many types of LLDPE. One publication mentioned that coating of calcium carbonate with stearic acid is an essential process for breathable film applications [3]. It was revealed in this paper that not only the $CaCO_3$ type and quantity but the stretching process has an enormous effect on breathability performance. The quality of porosity, the amount of interconnecting pores were determined to play an important role in breathability. A Croatian research group investigated many types of fillers, like zeolite, nanoclay and precipitated $CaCO_3$ in LDPE composites film for packaging to make films with better gas barrier and mechanical properties [4]. In our research new LLDPE blends were made and used to prepare film under laboratory conditions. Our aim was to determine the most appropriate LLDPE blend for stretch film usable also under industrial conditions.

2. Experimental

2.1. Materials

Three LLDPE materials were used for the preparations of the LLDPE blends: Braskem PE LL5400S produced by Nexeo Solutions, is an LLDPE developed for blown film extrusion with polyethylenes; Exceed™ 3518 Series produced by Exxon Mobil Chemical, is a metallocene ethylene-hexene copolymer and Dowlex™ 2107GC produced by Dow Chemical, is an LLDPE resin for high line speed processes. Nine LLDPE blends were made of the three LLDPE materials, the ratio of the materials is shown in Table 1.

IOP Conf. Series: Materials Science and Engineering **448** (2018) 012064 doi:10.1088/1757-899X/448/1/012064

Table 1. The ratio of the LLDPE blends.

	Exceed™ 3518 Series[a] A	Dowlex™ 2107GC[a] B	Braskem PE LL5400S[a] C
1.	6	4	—
2.	9	1	—
3.	1	9	—
4.	—	6	4
5.	—	9	1
6.	—	1	9
7.	6	—	4
8.	9	—	1
9.	1	—	9

[a]Abbreviations: A100 = Exceed™ 3518 Series, B100 = Dowlex™ 2107GC, C100 = Braskem PE LL5400S; LLDPE blends marked as follows: A60_B40 means the ratio is 6:4 = Exceed™ 3518 Series: Dowlex™ 2107GC etc.

The blends were made by a Collin Teach-line extrusion line with an E20T single-screw extruder: nozzle temperature was 250 °C, the zones of the extruder were 210 °C, 205 °C and 195 °C (from the capillary to the hopper), the speed of the extrusion was 120 1/min and the cooling system for the filament was a water bath. The granulates were made with an automatic grinder connected to the extruder after the water bath.

2.2. Methods

2.2.1. DSC analysis method. The characteristic melting and crystallization behavior of LLDPE blends were measured by Thermal Analysis TA Q200 (TA Instruments) heat-flux differential scanning calorimeter (DSC) according to ISO-11357-1 standard. Sample weights for the analysis were in the range of 3-5 mg. Sample holder was an aluminum pan and the reference was an empty pan. The instrument was calibrated by Indium. The measurements were performed in nitrogen atmosphere with 50 mL/min flow rate. The temperature program started from 30°C up to 220°C. A "heat-cool-reheat" method was used for the analysis [5]. The heating/cooling rate was 20°C/min.

2.2.2. Tensile properties analysis method. The test pieces for measurements of the tensile strength were made with a Demag IntElect 100-500 type injection molding machine according to ISO 294-1standard. The tensile properties were measured by Instron 3366 universal testing machine. The data was recorded at a crosshead speed of 1-500 mm/min, the speed limit was changed at 1% of tensile. All specimens were conditioned at room temperature (25°C) for at least 24 h. Five-five samples were tested, and the average values were reported for each composition.

3. Results and Discussion

During extrusion, materials are influenced by thermal and mechanical effect. That is why polymers were melted and cooled at controlled circumstances (from 220°C to ambient temperature with 20°C/min cooling rate) to measure the behavior characteristic to the raw materials. The melting DSC curves of Exceed™ 3518 Series (A) and Dowlex™ 2107GC (B) blends are shown in a heat flow (W/g) - temperature (°C) diagram (Figure 1.). The characteristic temperatures (starting (T_{eim}), peak(s) (T_{pm1} and T_{pm2}), end temperature (T_{efm})) and enthalpy change (ΔH_m) values were determined and summarized in Table 2. Most of the melting curves of the LLDPE blends have two peaks (T_{pm1} and T_{pm2}) which are typical for LLDPE copolymers. The melting of the materials is a complex process. There is a wide, shoulder like peak around 110 °C, and a sharp, well defined peak around 120 °C. For example A10_B90 has 109.4°C (T_{pm1}) and 123.6°C (T_{pm2}) for melting peak temperatures (Figure 1.), it is related to the different crystalline sizes and chain regularities.

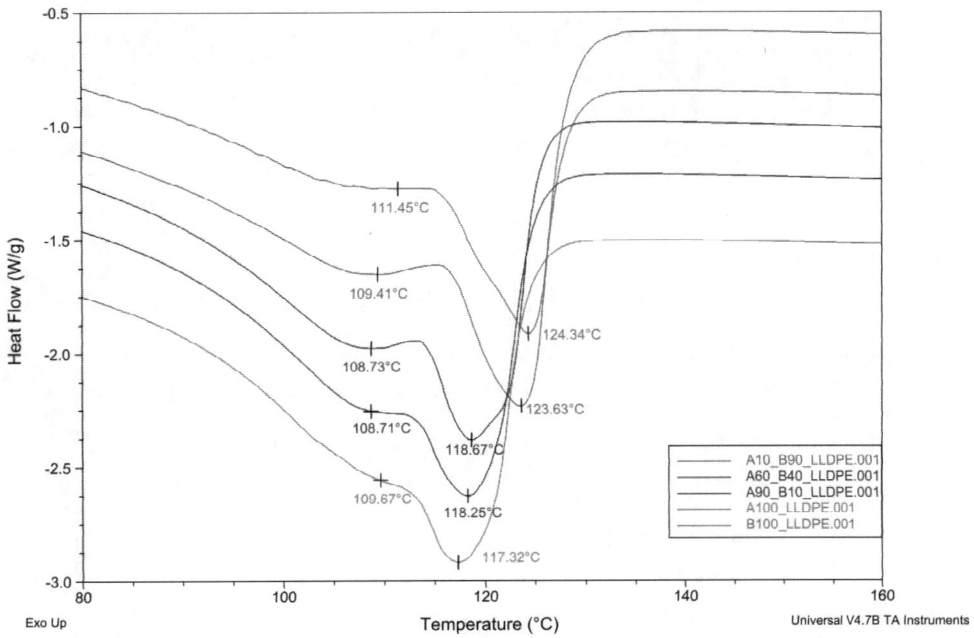

Figure 1. Melting DSC curves of Exceed™ 3518 Series (A) and Dowlex™ 2107GC (B) blends.

Table 2. Characteristic temperatures of the melting of the materials.

Sample	T_{eim} [°C]	T_{pm1} [°C]	T_{pm2} [°C]	T_{efm} [°C]	ΔH_m [kJ/kg]
A100	92.4	-	117.4	124.7	108
B100	81.6	106.7	124.3	129.1	99
A90_B10	92.8	109[a]	118.3	125.2	110
A10_B90	67.1	109.4	123.6	128.3	112
A60_B40	86.0	108.7	118.7	125.6	119
B90_C10	91.6	108.8	123.5	128.3	105
B10_C90	68.2	-	126.6	131.8	100
B60_C40	62.7	209[a]	124.0	128.3	110
A90_C10	90.7	109.0	118.7	125.1	118
A10_C90	74.3	-	125.3	129.6	100
A60_C40	86.3	110[a]	124.0	129.9	107

[a]manually marked temperature value

The starting (T_{eic}), peak (T_{pc}), end temperature (T_{efc}), and enthalpy change (ΔH_c) values were also determined during the crystallization sections and the datas are shown in Table 3. The characteristic crystallization peak temperatures of Exceed™ 3518 Series (A) and Dowlex™ 2107GC (B) blends are shown in a heat flow (W/g) - temperature (°C) diagram (Figure 2.).

The A10_B90 blend has the highest and the pure Exceed™ 3518 Series (A) has the lowest starting crystallization temperature. All of the characteristic melting and crystallization temperatures are shown in Table 2 and Table 3.

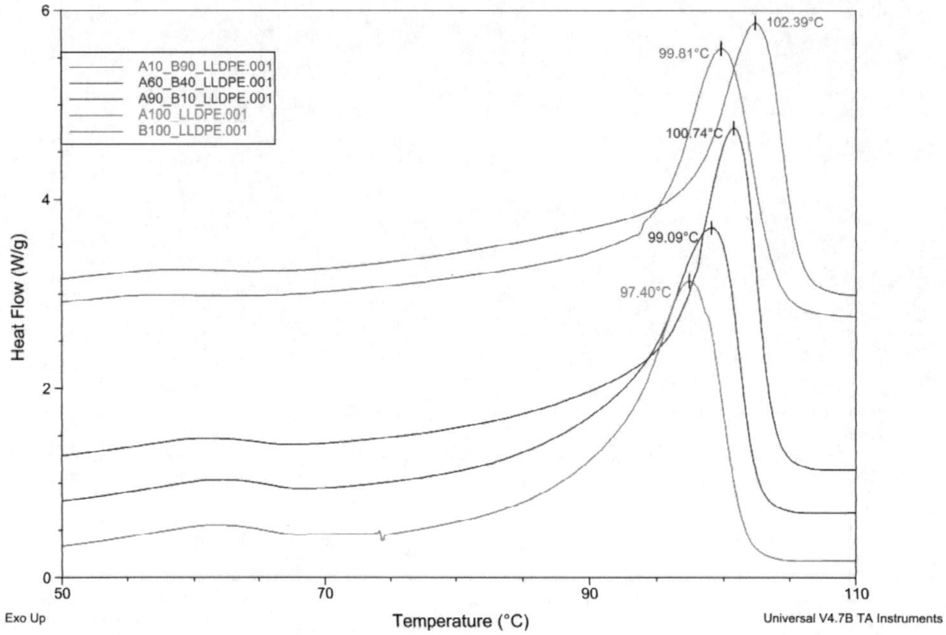

Figure 2. Crystallization DSC curves of Exceed™ 3518 Series (A) and Dowlex™ 2107GC (B) blends.

Table 3. Characteristic temperatures of the crystallization section.

Sample	T_{eic} [°C]	T_{pc} [°C]	ΔH_c [J/g]
A100	101.5	97.4	114
B100	104.0	99.8	107
C100	108.5	106.0	107
A90_B10	102.5	99.1	119
A10_B90	105.8	102.4	115
A60_B40	103.8	100.7	124
B90_C10	106.5	103.5	114
B10_C90	109.1	105.7	104
B60_C40	107.5	104.8	114
A90_C10	103.7	100.9	120
A10_C90	108.8	105.9	111
A60_C40	105.3	102.5	107

The elongation at contraction were 21%-65% (shown in Table 4, Figure 3). Two blends, A10_B90 and A10_C90 had ~20% and the others had around 60%. From the measurement of the tensile strength the most important value was the length of the orientation section because it is related to the flexibility of the film. The orientation length was almost the same of the different LLDPE blends, the best was 298% of A10-B90. The LLDPE blends had 404%-636% elongation at break, with 10.45-18.11 MPa tensile strength. The A90-B10 blend had the biggest value of elongation at break (636%).

Table 4. Mean values of the measured parameters with deviation.

Sample	Modulus E [MPa]	Elongation at contraction ε_Y [%]	Tensile stress at contraction σ_Y [MPa]	Orientation section length [%]	Elongation at break (cross-head) ε_B [%]	Tensile strength σ_B [MPa]
A10_B90	172±2	61.09±1.71	9.52±0.08	298±6	538.20±0.13	14.63±1.83
A10_C90	198±2	21.28±0.75	9.98±0.07	285±12	413.45±0.24	11.94±1.75
A60_B40	174±2	63.92±0.45	9.83±0.11	270±14	565.34±0.33	16.76±1.72
A60_C40	191±4	58.43±0.59	9.97±0.14	285±10	512.39±1.72	14.37±3.99
A90_B10	179±3	64.91±0.70	10.17±0.09	265±31	635.94±3.14	18.11±2.54
A90_C10	192±2	62.09±1.26	10.31±0.03	270±12	524.37±0.43	16.48±1.67
B10_C90	192±3	22.45±0.37	9.73±0.15	270±19	403.61±1.74	10.45±3.08
B60_C40	180±2	53.76±0.94	9.51±0.07	275±11	427.75±1.29	11.42±1.85
B90_C10	175±1.57	58.25±0.70	9.43±0.08	285±6	442.89±0.19	11.86±1.57

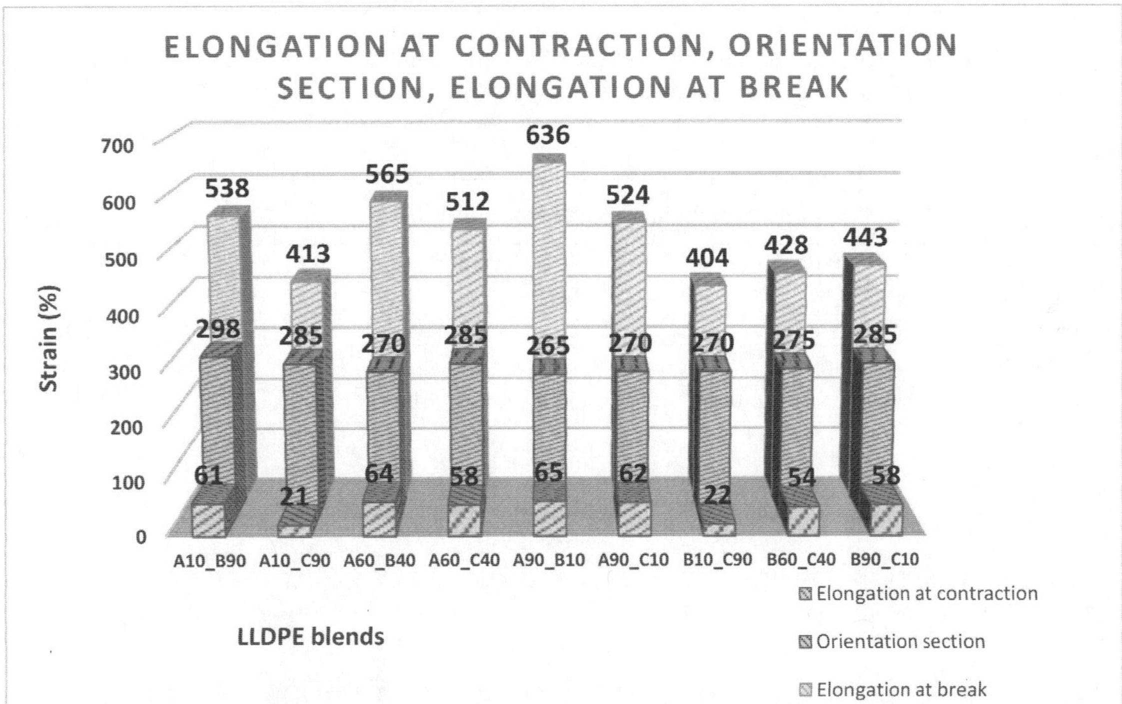

Figure 3. Elongation at contraction, orientation section, elongation at break.

According to the thermal and mechanical results A60_C40, A10_C90 and A10_B90 were chosen for manufacturing films. The parameters of blown film extrusion under laboratory conditions are shown in Figure 4. The equipment was Collin teach-line E 20 T extruder and Collin teach-line BL 50 T film blowing unit; T_{z1}, T_{z2}: temperature for zones of film blowing unit, T_{e1}, T_{e2}, T_{e3}: temperature for extruder zones, T_h: temperature of the extruder of the hopper, p_e: pressure of the extruder).

IOP Conf. Series: Materials Science and Engineering **448** (2018) 012064 doi:10.1088/1757-899X/448/1/012064

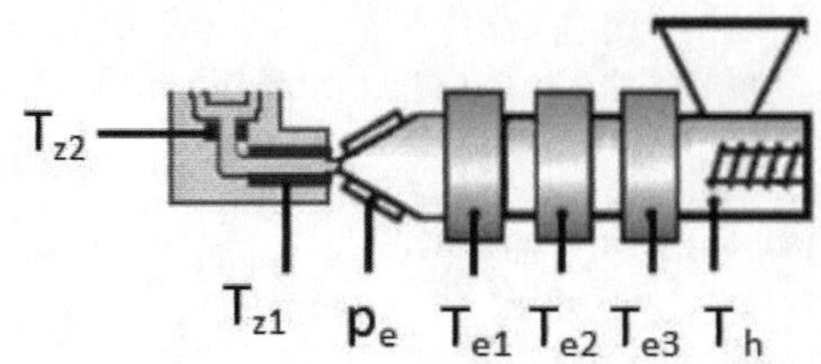

Figure 4. The parameters of blown film extrusion under laboratory conditions.

Table 5. The parameters of blown film extrusion under laboratory conditions.

	Temperature (°C)
T_{z1}	170
T_{z2}	170
T_{e1}	165
T_{e2}	160
T_{e3}	155
T_h	24

	p_e (bar)
A60_C40	101-112
A10_C90	152-184
A10_B90	156-162

Screw speed (1/min)	570

The tearing strength was measured by Instron 3366 equipment. Data were recorded at a crosshead speed of 500 mm/min at 24°C. The size of the specimens were 10 cm x 5 cm with a 2-5 mm long notch. Six-six notched specimens were tested and the average values were reported for each composition (Table 6.).

Table 6. Results of the tearing strength measurement of the notched specimens (MD=machine direction, TD=transverse direction).

Specimen	A10_B90		A10_C90		A60_C40	
Notch direction →	MD F [N]	TD F [N]	MD F [N]	TD F [N]	MD F [N]	TD F [N]
Mean value	5.99	6.05	6.07	7.12	7.73	8.94
Deviation	0.47	0.99	0.94	0.60	0.35	0.75

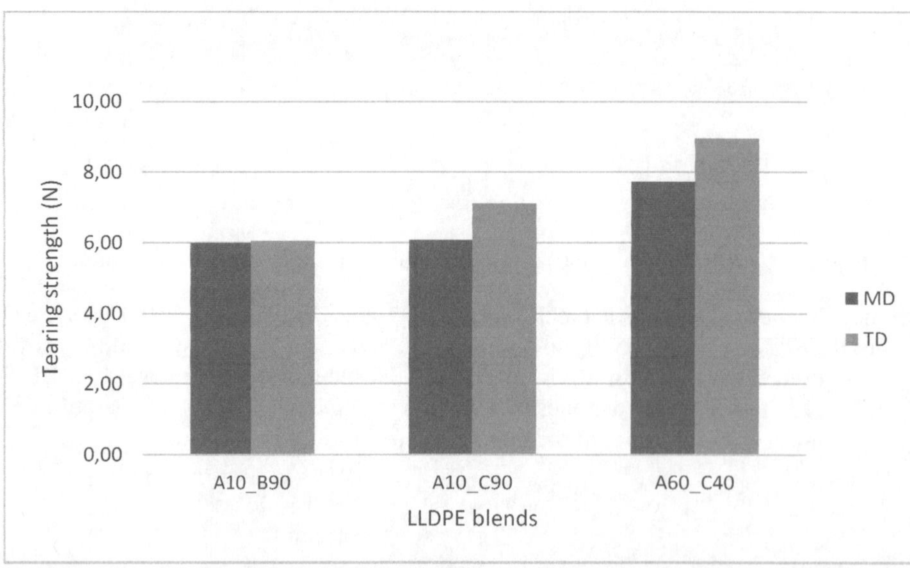

Figure 5. The tearing strength of the notched specimens (MD=machine direction, TD=transverse direction).

IOP Conf. Series: Materials Science and Engineering **448** (2018) 012064 doi:10.1088/1757-899X/448/1/012064

The toughest film was the A60_C40, because the force of the MD tear was 7.73 N and of the TD tear was 8.94 N, so this blend is the most suitable for stretch films (Figure 5.).

4. Conclusion

In our investigation different types of LLDPE blends were prepared by extrusion. The characteristic melting and crystallization temperatures of these LLDPE blends were determined. Most of the melting curves of the LLDPE blends have two peaks, which are typical for LLDPE copolymers. The first melting peak has a shoulder shape, while the second melting peak is a sharp, well defined peak. It is related to the different crystallite sizes and chain regularities. The A10_B90 blend has the highest and the pure Exceed^{TM} 3518 Series (A) has the lowest starting crystallization temperature. Specimens were injection moulded and their tensile properties were measured. The LLDPE blends had 404%-636% elongation at break, with 10.45-18.11 MPa tensile strength. The A90-B10 blend had the biggest value of elongation at break (636%). According to the thermal and mechanical results A60_C40, A10_C90 and A10_B90 were chosen for manufacturing films in laboratory conditions. Tearing properties of the films were investigated. The toughest film was the A60_C40, so this blend is the most suitable for making stretch films.

References

[1] Elvers B, Hawkins S and Ullman S 1992 Encyclopedia of industrial chemistry, 5th ed., *VCH Publishers: New York*, 487.

[2] Ajji A, Auger J, Huang J and Kale L February 2004 Biaxial stretching and structure of various LLDPE resins *Polymer Eng. and Sci.*, , Vol. **44**, No. 2

[3] Özen I and Sismek S 2015 Changing breathability of polyethylene composite films with different porous structures depending on stretching processes *Advances in Pol. Tech.*, **36** 2, DOI 10.1002/adv.21600

[4] Sirocic A P, Rescek A, Scetar M, Krehula L K, Hrnjak-Murgic Z 2014 Development of low density polyethylene nanocomposites films for packaging *Polym. Bull.* **71** 705–17.

[5] Ehrenstein G, Riedel G, Trawiel P 2004 Thermal analysis of plastics theory and practice *Carl Hanser Verlag GmbH & Co. KG*

Acknowledgments

This research is supported by GINOP-2.1.1-15-2015-00088 "Research for the development of permeable stretch films" project. The project is supported by the Hungarian Government and co-financed by the European Social Fund.

IOP Conf. Series: Materials Science and Engineering **448** (2018) 012065 doi:10.1088/1757-899X/448/1/012065

Examination of surface roughness of holes of plastic parts drilled under cryogenic cooling conditions

G Varga[1], S Ravai-Nagy[2], F Szigeti[3]

[1] Institute of Manufacturing Science, University of Miskolc, Miskolc – Egyetemváros, Hungary

[2] Department of Engineering and Management of Technology, Faculty of Engineering, Technical University of Cluj-Napoca, Baia Mare, Dr. V. Babes 62/A, Romania

[3] Institute of Engineering and Agricultural Sciences, University of Nyíregyháza, Nyíregyháza, Kótaji u. 9-11, Hungary

[1] gyula.varga@uni-miskolc.hu

Abstract. Machining performance can be improved by cooling with cryogenic assistance. This paper focuses on the influence of cryogenic cooling conditions using dry ice, solid carbon dioxide when drilling of PE-HD1000 plastic parts. The goal was to determine how the surface roughness of the drilled hole changes as a function of the cutting speed and the feed rate at room temperature and when cryogenic cooling is applied. The function relations were determined using Factorial Experiment Design. Improvement of the machinability of low temperature drilling results the improvement of roughness of the drilled hole.

1. Introduction

Industrial plastics are used in large quantities in different areas of industry for their various advantageous properties. Their dynamic development and the emergence of new materials is a continuous challenge for the industry, as the optimal cutting parameters vary considerably depending on the plastic material quality. Determining the optimal parameters for cutting industrial plastics is a continuous challenge for production engineering specialists. During machining, the realization of dimension accuracy and surface quality prescribed by the designers must be cost-effective.

During the processing of industrial plastics, it is known that the heating is the determining element of the machining due to the friction of the tool - chip, the work-piece - tool interface and the energy used to remove the chips. In the case of industrial plastics, due to warming, the machinability of material deteriorates, the plastics melts when machining, it stuck to the tool and the work-piece. One way of increasing the cutting parameters is the use of cooling and lubricating fluids, however, the coolants and lubricants have negative effects on the polymers, so the literature recommends dry machining and the use of cooling by compressed air. Industrial plastics cannot be machined over a certain temperature. To increase the cutting performance, heating of the work-piece must be limited. To keep the work-piece and the tool under a certain temperature, it is possible to use the cooling of the various parts of the machining system to reduce the heating of the work-piece, chip and tool.

IOP Conf. Series: Materials Science and Engineering **448** (2018) 012065 doi:10.1088/1757-899X/448/1/012065

Solution can be cryogenic cooling, which comprehensive examination has been carried out by Yildiz and Nalbant [1] in 2008 and Kaynak et al [2] in 2014. Benefits of application of cryogenic cooling: it can increase tool life and dimension accuracy during the cutting process; reduce the cutting temperature, surface roughness, and the energy used during the metal removal process, thereby improves the productivity. Jawahir et al summarized the most frequently applied liquids in cryogenic cooling [3]. During cutting processes, most of cryogenic cooling applications have been investigated mostly in turning operations [4-7], although other machining operations such as grinding [8-10], drilling [11-12] and milling [13-14] also occur.

The performance of cryogenic treated tools increased from 65% to 343%, depending on the cutting conditions [15]. In another experiment HSS twist drill for cryogenic treated twist drill was used for drilling stainless steel work-pieces. The results of the study showed a significant improvement in the lifetime of cryogenic treated drills [15]. Kim and Ramulu [16] used cryogenic treated carbide tools for drilling thermoplastic composites to inspect the machinability from the point of view the drilled hole quality and tool wear. Their results showed that while the cryogenic treated carbide tools produced better drill properties compared to conventional carbide tools, conventional carbide tools showed better wear resistance than cryogen-treated ones.

Based on an overview of the literature on processing in cryogenic environments, research can be divided into three main groups:

- Cooling tool (recommended for machining materials where overheating of tools causes premature wear, e.g. machining of carbides) [17];
- Cooling the work-piece (when aim is changing the properties of the material to be machined) [18];
- Cooling of work-piece and tool (for high-precision machining when working on the work-piece and preventing thermal expansion of the tool) [19].

Our paper presents the results of experiments done by cryogenic drilling of the PE-HD1000 polyethylene work-piece. During the experiment we have studied the effect of cooling on the surface roughness of the bore and the chips separation properties.

2. Experimental conditions

During the experiment, Ø12mm diameter holes were made according to DIN 388 standard with HSS twist drill where the length of the hole was different: L/D = 1; L/D = 2.5 and L/D = 5. At the previous two drilling, a blind hole was made, while in the third case it was a through hole.
Further data:

- Machined material: PE-HD1000
- Cutting speed:
  - v_1=6.78 m/min (revolution number of the spindle: n_1=180 rev/min);
  - v_2=18.85 m/min (revolution number of the spindle: n_2=500 rev/min);
- Feed rates:
  - f_1=0.32 mm/rev;
  - f_2=0.62 mm/rev;
- Temperature of the work-piece:
  - T_1=-35°C;
  - T_2=+20°C;
- Cooling medium: solid carbon dioxide (dry ice, SCO_2).

Elements of the technology system used for the drilling experiment:

- CNC controlled machine tool: Type NCT EmL510B;
- Thermometer, measuring range: -200C° ÷ 1200C°; K type probe, measuring accuracy: 0.1C°;

During the experiments, 8 experimental settings were performed using combinations of the above parameters. At each experimental setup 5 holes were made to meet the statistical requirements for the experiments. Variation of experimental data are listed in Table 1. Figure 1 shows the model of work-piece and the locations of holes machined during the experiments. Figure 2 shows the experimental setup at the different temperatures: T_1=-35°C and T_2=+20°C.

Table 1. Variation of experimental data

No.	Cutting speed v_c, $\left[\frac{m}{min}\right]$	Feed rate f $\left[\frac{mm}{rev}\right]$	Temperature T $[°C]$
1	6.78	0.32	-35
2	18.85	0.32	-35
3	6.78	0.62	-35
4	18.85	0.62	-35
5	6.78	0.32	20
6	18.85	0.32	20
7	6.78	0.62	20
8	18.85	0.62	20

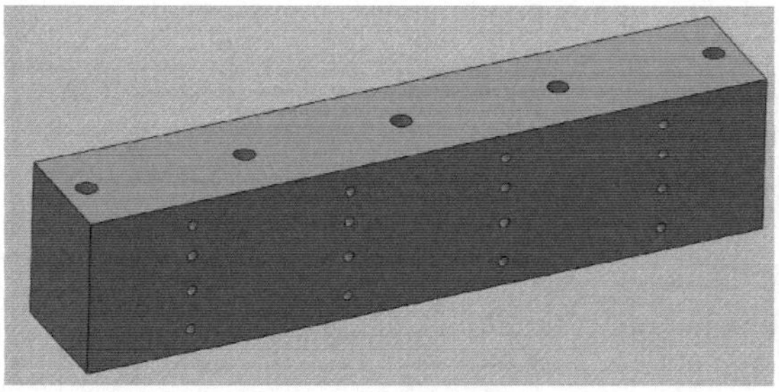

Positioning holes for the work-piece fixing at the work holding system

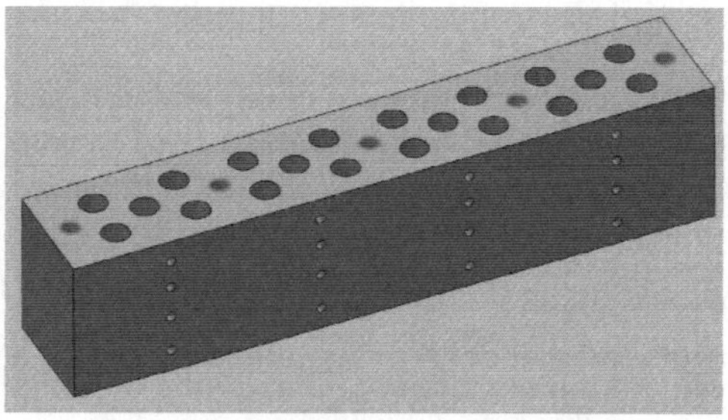

Machined holes after the experiment

Figure 1. 3D model of the work-piece before and after drilling

Machining in room temperature T_2=+20°C

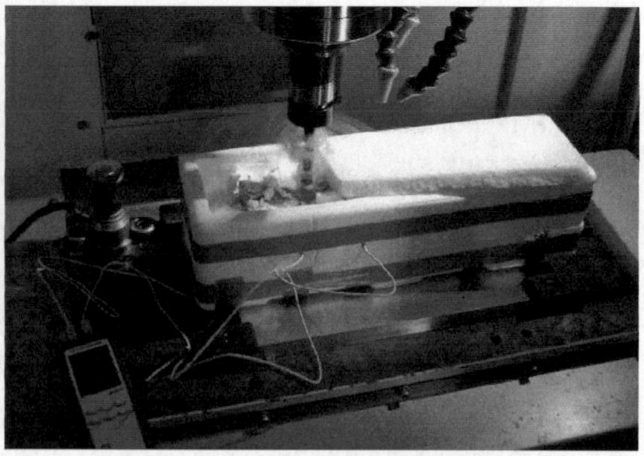

Machining in temperature T_1=-35°C

Figure 2. Using different fixtures at machining of different temperatures

According to Szabo [20] the effectiveness of a machining process can be expressed among the others by the following parameters: accuracy, surface roughness, complex surface quality (integrity), material removal rate, costs and productivity of the process. In this paper the surface roughness, the Arithmetical mean deviation of the profile (Ra) is examined. After the experiment we analysed the surface roughness of the drilled holes. The surface roughness of the holes was measured by the TalySurf520 3D surface roughness measuring equipment (Fig. 3). we examined the 2D roughness parameters (Ra, Rz, Rt) when the measuring length was set to 12.5 mm.

The roughness was measured with a CL2 white light confocal chromatic displacement sensor (CCS) and a probe with a magnifying glass MG140. When using the AltiSurf 520, the measurement parameters are set in the Phenix V2 measuring software running on the computer. Here we selected the type of measurement (profile or surface), then we set the measurement parameters: measurement position, measured profile or surface dimensions, resolution, sensor selection and settings, data file name and other data. Our measurement was profile measurement.

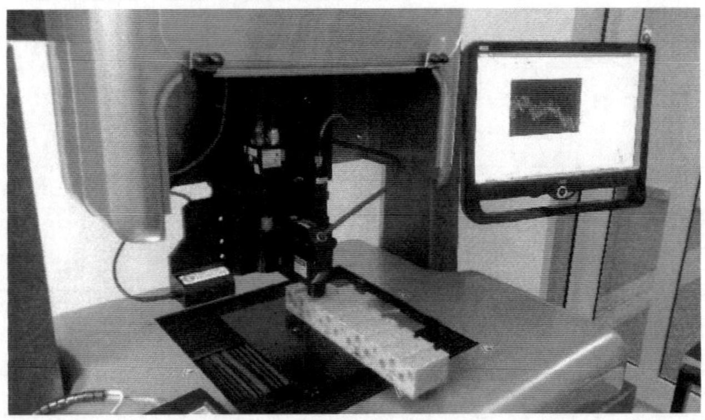

Figure 3. Measurement of surface roughness on AltiSurf 520 type measuring equipment

The holes having different lengths were measured times. Figure 4 shows the places of surface roughness measurements of the different holes.

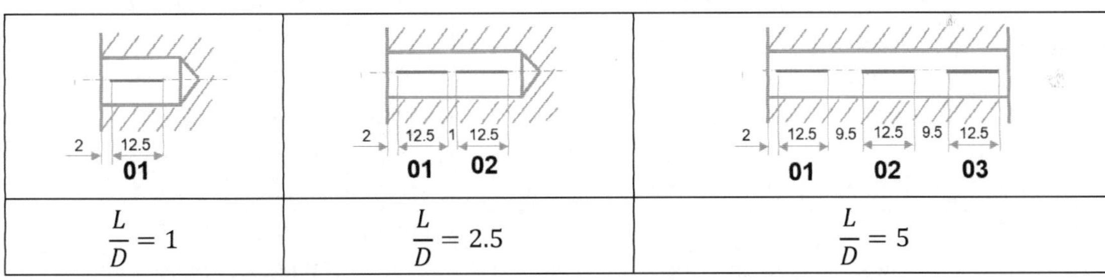

Figure 4. Measuring places of surface roughness of the different holes

After cutting the drilled work-pieces into two parts, the roughness was measured on both sides of the holes and the averaged was calculated. In addition, 2 holes were made with the same technological parameter variants and the average was calculated too.

3. Results of experiments of conventional and cryogenic drilling technology

3.1. Examination of the surface roughness

After creating measuring the surface roughness of the holes on different places according to Figure 4 the evaluation could be done. To make it properly the measured data are organised into Table 2. It contains the technological parameters (cutting speed, feed rate, temperature) applied when executing the experiments. Featuring characteristics of the Table 2 are the measuring places (can be 1, 2 or 3) and the type of the hole ($\frac{L}{D} = 1$, $\frac{L}{D} = 2.5$, and $\frac{L}{D} = 5$). When the hole was short ($\frac{L}{D} = 1$) than there are 2 zeros in the table, because there was surface roughness measurement only one place. This can be seen on the other places as well.

Table 2. Measured values of the Ra

No.	Cutting speed v c, $\left[\frac{m}{min}\right]$	Feed rate f $\left[\frac{mm}{rev}\right]$	Temp. T $[°C]$	Measuring place	Ra [μm] $\frac{L}{D}=1$	Ra [μm] $\frac{L}{D}=2.5$	Ra [μm] $\frac{L}{D}=5$
1	6.78	0.32	-35	01	5.1178	3.6075	3.3110
				02	0	3.1690	3.0285
				03	0	0	3.0272
2	18.85	0.32	-35	01	4.7049	3.2111	4.0046
				02	0	2.9395	3.6853
				03	0	0	3.7515
3	6.78	0.62	-35	01	3.0559	4.6259	4.9782
				02	0	5.2961	4.6710
				03	0	0	4.4323
4	18.85	0.62	-35	01	3.9561	3.7175	4.5925
				02	0	3.9801	4.6896
				03	0	0	4.6177
5	6.78	0.32	20	01	6.4884	7.1530	5.7206
				02	0	6.9983	4.4424
				03	0	0	4.2652
6	18.85	0.32	20	01	8.1934	7.8114	4.8942
				02	0	7.4200	3.9610
				03	0	0	3.9929
7	6.78	0.62	20	01	5.5749	4.4706	6.3571
				02	0	3.8248	5.8988
				03	0	0	5.9632
8	18.85	0.62	20	01	4.9095	5.3008	6.9399
				02	0	4.4423	6.5988
				03	0	0	5.5185

Featuring profiles can be seen in Figure 5 for the case of short holes, that is when $\frac{L}{D} = 1$. In Figure 5 it the differences can be compared in between drilling with T_1=-35°C and T_2=+20°C. The amplitudes are greater at drilling in room temperatures.

3.2. Evaluation of the measured values

The measured values of Table 2 can be demonstrated in stereoscopic square column diagrams. Figure 6 shows the differences of drilling in different temperatures. Figure 6 contains the measuring results belonging to the different L/D ratios, on the different measuring places (1, 2 or 3) on the base of the different technological parameter variation (Table 2). When the temperature is cooler (T_1=-35°C) the values of Ra roughness parameter is lower, so better. To give more precious values, it is suggested to calculate the average values of the Ra roughness parameters along the axis of the holes. These average values can be found in Table 3 where the transformed values of the technological parameters are shown as well. They are used when the application of Factorial Experimental Design [21-22].

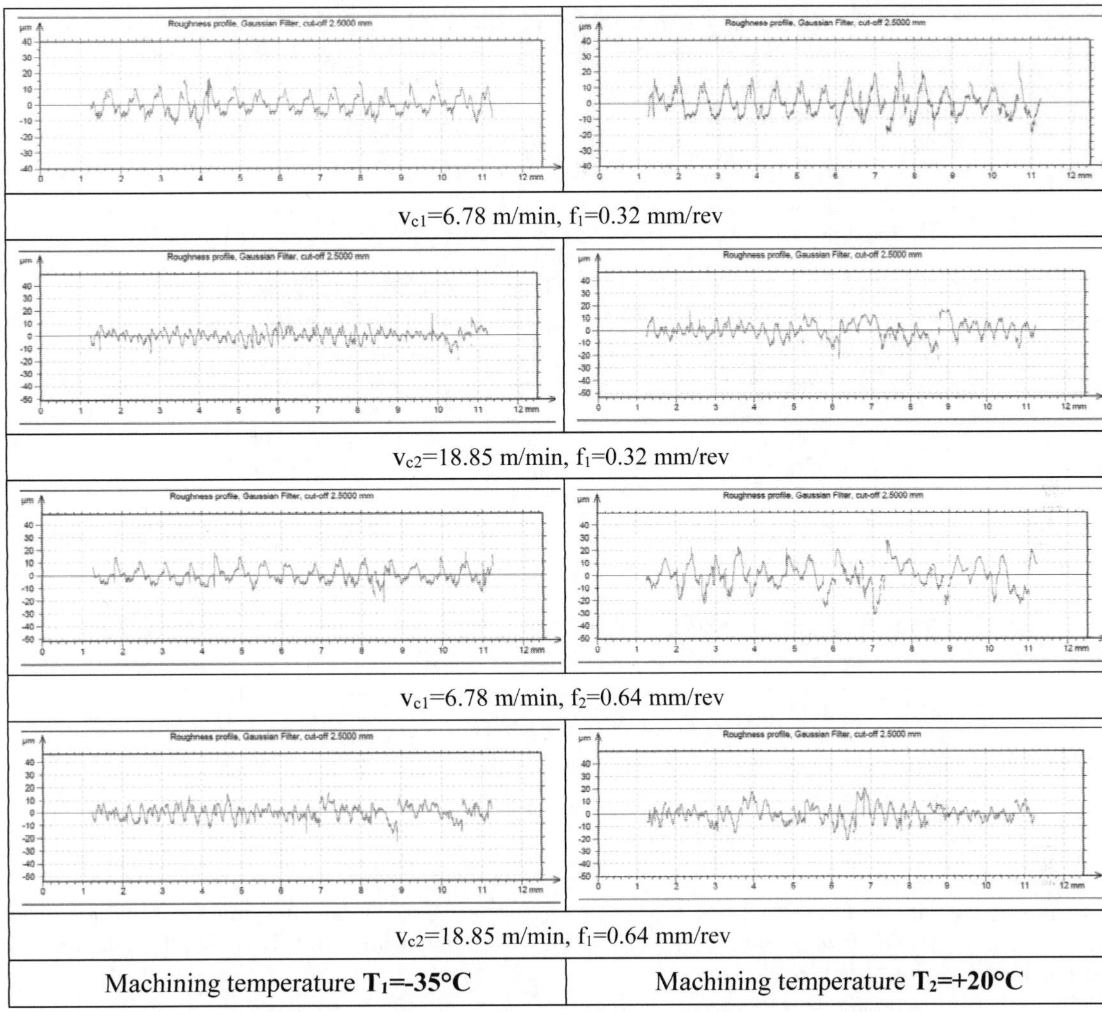

Figure 5. Comparing of 2D profile of the work-piece when (L/D=1)

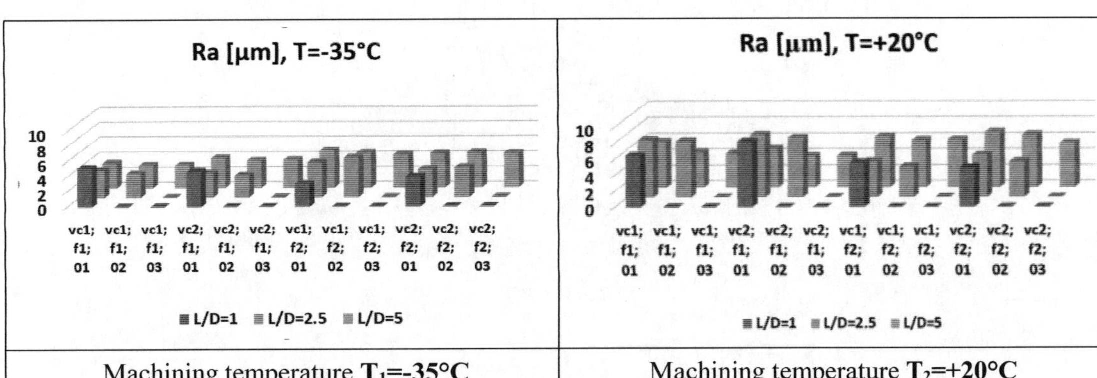

Figure 6. Comparing of Arithmetical mean deviation (Ra) of the profiles drilled in different temperatures

Table 3. Averages of the measured values of Ra

No.	Transformed parameters			Ra, $[\mu m]$		
	X_1	X_2	X_3	$\frac{L}{D}=1$	$\frac{L}{D}=2.5$	$\frac{L}{D}=5$
1	-1	-1	-1	5.1178	3.3759	3.1447
2	+1	-1	-1	3.0559	4.9610	4.6243
3	-1	+1	-1	4.7049	3.0753	3.9404
4	+1	+1	-1	3.9650	3.8488	4.4531
5	-1	-1	+1	6.4884	7.0750	4.8090
6	+1	-1	+1	5.5749	4.1477	5.0200
7	-1	+1	+1	8.1934	7.6158	4.2800
8	+1	+1	+1	4.9095	4.8715	6.3500

Using the method of Factorial Experiment Design and the software written in „Mathcad 15.0" empirical formulas (1)-(3) were determined from the average values can be seen in Table 3.

$$Ra_{\frac{L}{D}=1} = 5.1247 - 0.019v_c + 5.041f - 0.068T - 0.284v_c \cdot f + 7.663 \cdot 10^{-3} \cdot v_c \cdot T + 0.254f \cdot T - 0.019v_c \cdot f \cdot T \quad (1)$$

$$Ra_{\frac{L}{D}=2.5} = 6.0949 - 0.091v_c + 1.117f + 0.108T - 0.049v_c \cdot f - 8.396 \cdot 10^{-3} \cdot v_c \cdot T + 0.017f \cdot T + 4.994 \cdot 10^{-3} \cdot v_c \cdot f \cdot T \quad (2)$$

$$Ra_{\frac{L}{D}=5} = 4.3747 - 0.018v_c - 1.714f + 0.1T + 0.23v_c \cdot f - 6.452 \cdot 10^{-3} \cdot v_c \cdot T - 0.176f \cdot T + 0.014v_c \cdot f \cdot T \quad (3)$$

Substituting the values of the factors into formulas (1)-(3) surfaces can be determined and axonometric figures can be drawn. When using formula (1) the result regarding to L/D=1 can be get and its axonometric figure is shown in Figure 7.

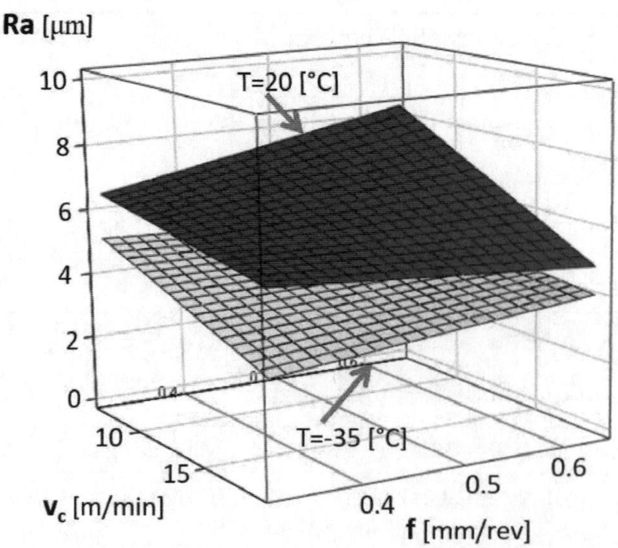

Figure 7. Changing the Arithmetical mean deviation (Ra) of the profile when L/D=1

After substitution into formula (2) Figure 8 relates to L/D=2.5 and Figure 9 belongs to formula (3) when L/D=5.

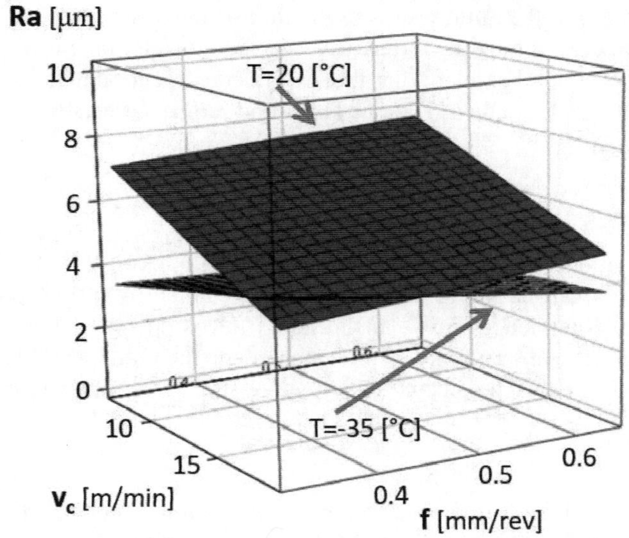

Figure 8. Changing the Arithmetical mean deviation (Ra) of the profile when L/D=2.5

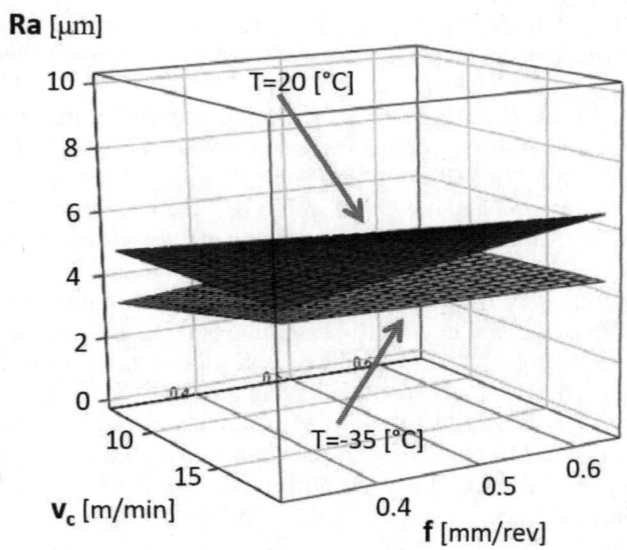

Figure 9. Changing the Arithmetical mean deviation (Ra) of the profile when L/D=5

4. Summary

Based on the analysis of the results of surface roughness tests, it can be stated:

- In case of cryogenic cooling, when L/D=1 and L/D=5, the values of the Arithmetical mean deviations of the profile (Ra) were lower for each cutting speed and feed rate variations.
- When the hole belonging to L/D=1 ratio was drilled, the best parameter combinations were: v_{c2}=18.85 m/min; f_1=0.32 mm/rev; and T_1=-35°C. In case of L/D=2.5 the best parameter combinations are: v_{c1}=6.78 m/min; f_2=0.62 mm/rev; and T_1=-35°C, and for L/D=5 these parameter combinations: v_{c1}=6.78 m/min; f_1=0.32 mm/rev; and T_1=-35°C.

- Although in the case of L/D=2.5 also the hole drilled on cryogenic way served the best surface roughness, but it is interesting to note that, when using high cutting speed (v_{c2}=18.85 m/min) and low feed rates (f_1=0.32 mm/rev) is used, almost identical surface roughness was obtained for both temperatures. It requires further investigation to determine its reason.
- In case of applying cryogenic cooling the Arithmetical mean deviation (Ra) of the profile is increasing when increasing the L/D ratio. This trend can be noticed drilling in room temperature as well.

Acknowledgments
"Project no. NKFI-125117 has been implemented with the support provided from the National Research, Development and Innovation Fund of Hungary, financed under the K_17 funding scheme."
"The described study was carried out as part of the EFOP-3.6.1-16-00011 "Younger and Renewing University – Innovative Knowledge City – institutional development of the University of Miskolc aiming at intelligent specialisation" project implemented in the framework of the Szechenyi 2020 program. The realization of this project is supported by the European Union, co-financed by the European Social Fund."

References

[1] Yildiz Y, Nalbant M 2008 *A review of cryogenic cooling in machining processes*, vol 48 (3) International Journal of Machine Tools & Manufacture pp 947-964
[2] Kaynak Y, Lu T, Jawahir I S 2014 *Cryogenic machining-induced surface integrity: a review and comparison with dry, MQL, and flood-cooled machining* Vol. 18 (2) Machining Science and Technology pp. 149–98.
[3] Jawahir I S, Attia H, Biermann D, Duflou J, Klocke F, Meyer D, et al. 2016 *Cryogenic manufacturing processes* vol 65 (2) CIRP Annals - Manufacturing Technology pp 713–736.
[4] Evans C 1991 *Cryogenic diamond turning of stainless steel* vol 40 (1) CIRP Annals pp 571-575
[5] Bagherzadeh A, Budak E 2018 *Investigation of machinability in turning of difficult-to-cut materials using a new cryogenic cooling approach* Vol 119 Tribology International pp 510–520
[6] Jerold B D, Kumar M P 2012 *Experimental comparison of carbon-dioxide and liquid nitrogen cryogenic coolants in turning of AISI 1045 steel* Vol 52 (10) Cryogenics pp 569–574
[7] Wang Z Y, Rajurkar K P 2000 *Cryogenic machining of hard-to-cut materials* Vol 239 (2) Wear pp 168–175.
[8] Paul S, Chattopadhyay A B 2006 *Environmentally conscious machining and grinding with cryogenic with cryogenic cooling* Vol 10 Machining Science and Technology pp 87-131
[9] Chattopadhyay A B, Bose A, Chattopadhyay A K 1985 *Improvements in grinding steels by cryogenic cooling* Vol 7 (2) Precision Engineering pp 93–98.
[10] Paul S, Chattopadhyay A B 1996 *The effect of cryogenic cooling on grinding forces* Vol 36 (1) International Journal of Machine Tools and Manufacture pp 63–72
[11] Ahmed L S, Kumar M P 2016 *Cryogenic Drilling of Ti–6Al–4V Alloy Under Liquid Nitrogen Cooling* Materials and Manufacturing Processes vol 31 pp 951–959
[12] Liska J, Kodacsy J 2012 *Drilling of glass fibre reinforced plastic* vol 472-475 Advanced Materials Research pp 958-961.
[13] Yu S, Ning H, Liang L, Xinlong L, Wei Z 2006 *Effects of cryogenic nitrogen gas jet on machinability of Ti-alloy in high speed milling* vol 17 (11) China Mechanical Engineering pp 1183-1187
[14] Fillipi A D, Ippolite R 1971 *Facing milling at -180°C* vol 19 (1) Annals of CIRP pp 399–406
[15] Chatterjee S 1992 *Performance characteristics of cryogenically treated high speed steel drills* vol 30 (4) International Journal of Production Research pp 773-786
[16] Kim D, Ramulu M 2004 *Cryogenically treated carbide tool performance in drilling thermoplastic composites* vol 32 Transactions of the North American Manufacturing Research Institute pp 79-85

[17] Pušavec F, Kopač J 2011 *Sustainability Assessment: Cryogenic Machining of Inconel 718* vol 57 (9) Journal of Mechanical Engineering pp 637-647
[18] Aldwella B, O'Mahonyb J, O'Donnella G E 2014 *The effect of workpiece cooling on the machining of biomedical grade polymers* vol 33 Procedia CIRP pp 305-310
[19] Akinuma Y, Kidani S, Aoyama T 2012 *Ultra-precision cryogenic machining of viscoelastic polymers* Vol 61 (1) CIRP Annals - Manufacturing Technology pp 79-82
[20] Szabó O 2014 *Goodness indicator and technological optimization of honing*, Key Engineering Materials, vol 581 pp 261-265
[21] Taguchi G 1987 *System of experiment design (New York)* UNIPUB Kraus International Publications
[22] Fridrik L 1987 *Chosen chapters from the topics of experimental design of production engineering (Budapest)* Műszaki Könyvkiadó (in Hungarian)

IOP Conf. Series: Materials Science and Engineering **448** (2018) 012066 doi:10.1088/1757-899X/448/1/012066

Optimize the joining technology of CFRP and aluminium sheets using hybrid and high strength adhesive

M Berczeli[1], R Kürtösi[2], A Tímári[2], J Liska[3]

[1]John von Neumann University, Department of Material Science
[2] John von Neumann University, GAMF Faculty of Engineering and Computer Science
[3]John von Neumann University, Department of Vehicle Technology

berczeli.miklos@gamf.uni-neumann.hu

Abstract. In order to reduce the weight of a vehicle the designers have to use different materials pairing. During the light building the vehicle needs to complete the mechanical and structural requirements, these quality properties are especially important in the competition racings. To fulfil the needs while minimize the weight hybrid material usage and bonding are used. One of the critical joining is the Carbon Fiber Reinforced Plastic (CFRP) and aluminium pairing. In this paper we investigated the adhesive joining technologies of these hybrid pairings using hybrid and high strength adhesive. Our goal is to optimize the hybrid bonding while different surface treatment methods are used on the aluminium substrate. The mechanical properties, the surface roughness and the breaking method were investigates using tensile test and macro geometry inspection. In the results the connection between the surface topography and the mechanical properties were analysed.

1. Introduction

Development of processes to join aluminum sheets with other plastic materials is necessary and important in order to allow aluminum to be applied broadly in the automotive industry, as the aluminum plates itself has a low bending and tensile strength. Other pairing materials as reinforcing the aluminum can be used dual phase steels [1], fiber reinforced composites [2], and any special usable materials [3]. There is a lot of method to join hybrid materials together, such as brazing [4], soldering [5-7], but this paper deals with the adhesive joining methods using two different type of glues.

Adhesive bonding is the process of binding two different components using a suitable binder. Applications of adhesives for joining elements made of dissimilar materials are commonly employed in aviation, automotive and building industries [8]. Joining of CFRPs with aluminum alloys via adhesive bonding is by far the most conventional method with both advantages and limitations. The investigated researches used, CFRP with aluminum, but we could use these experiments to joining aluminum foams with adhesives in the future. Since adhesive bonding is an irreversible process, attempts to dissemble the joints can be expensive, which results in the complete material damage involved in the joints [9,10].

Adhesive bonding not only seals the joints, but also prevents crevice and galvanic corrosion between two dissimilar materials. Almost any pair of dissimilar materials such as metals, polymers or ceramics can be joined with this method [11]. Adhesive bonding is the only viable method to achieve structures involving the joining of thin-walled elements, among which an element has substantial dissimilar

IOP Conf. Series: Materials Science and Engineering **448** (2018) 012066 doi:10.1088/1757-899X/448/1/012066

thickness. Adhesive bonding offers light-weighted structures with respect to other assembly technologies and developments, particularly in aviation industries. In addition, stress concentration becomes less significant without the requirement of bolt holes, thus avoiding structure weakening [12]. Adhesive joining technologies can be calculated using numerical methods such as in other manufacturing methods like deep drawing [13]. The adhesives as the main elements in adhesive bonding should have good wettability with respect to joining components, such as CFRPs and aluminum alloys.

2. Experimental procedure

The substrates for the research are non-coated AlSi1 sheet of 5 mm and Carbon Fiber Reinforced Plastic (CFRP) of 1.2 mm thickness. The CFRP was vacuum laminated during 24 hours using 2 layer of 1 x 1 mm weave carbon fiber and epoxy glue. After the lamination heat treatment was used to increase the adhesive and cohesion force in the CFRP, the temperature of the heating was 80°C and the time was 2 hours. The materials were mechanically sectioned to avoid heat input, into 30 x 50 mm, the CFRP into 15 x 50 mm. Before the joining method, the surface off the CFRP were grinded to remove the form separating layer. The CFRP has not modified with any treatment, only the grinding and the degreasing method was used. The aluminum and CFRP surfaces were all cleaned and degreased with cleaning spray. 4 different surface modification method were used on the AlSi1 during the research: the un-treated surface, grinded surface with a P10 grinding stone, sand blasted surface using 0.3-0.7 mm abrasive and chemical etched with hydrochloride acid mixed with distillated water in 1:5 ratios for 5 minutes.

The adhesive joining method of the overlapped joints was performed using two different type of adhesive, the firs was Loctite 4080 which is a 2 component cyanoacrylate / acrylic adhesive, it is recommended for adhesive joining of different type of materials. The second was Loctite 9466 2 component epoxy adhesive, with this adhesive high strength of bonding can be created. First step of the adhesive method was the cleaning of all the contact surfaces, add glue on the materials on a 15 x 10 mm area, the joints were overlapped (figure 1). The hybrid joints were squeezed over 24 hours in a bending machine, this time needs to reach the 100% strength of the adhesive.

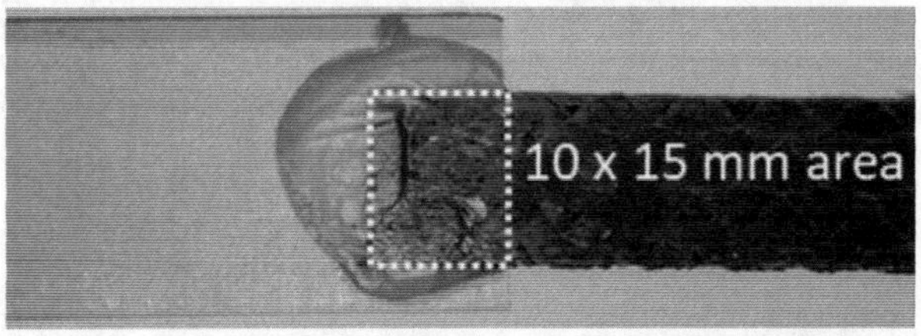

Figure 1. Adhesive procedure setup and connecting area.

In order to analyze the effect of the different surface preparations to the tensile strength of the hybrid CFRP and aluminum joints surface roughness measurement was performed. The macroscopic structure of the surfaces was investigated. Shear tensile test were carried out on Instron 5900R testing machine at a cross-head speed of 2 mm/min (figure 2). Five samples were tested for each surface modification method. The breaking method was investigated to find out which component has lower adhesion force to the adhesive.

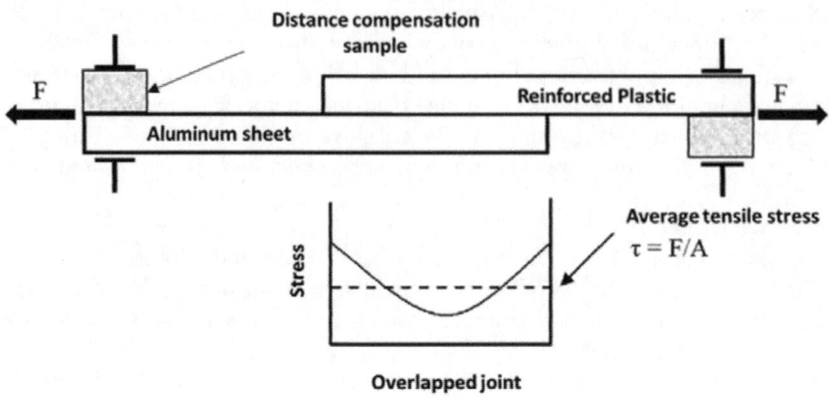

Figure 2. Test setup of tensile strength measurements.

3. Results

3.1. Effect of the different surface modification for the macroscopic topography and surface roughness

After the surface modification of the aluminum sheets the surface roughness measurement were performed. In figure 3 the macroscopic topography of the surfaces can be seen after the adhesive process.

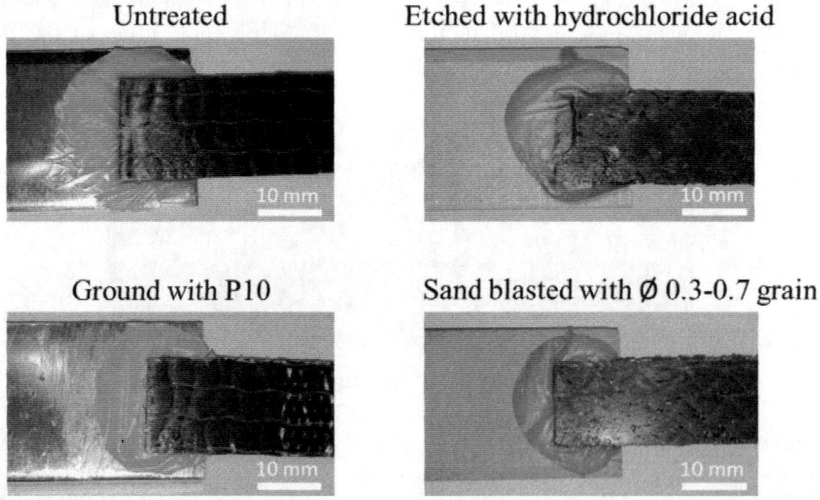

Figure 3. Surface macro topography and roughness using different preparations.

The manufactured and rolled surface of the AlSiMg1 is a very smooth, mirroring surface with a Ra= 0.5 μm and Rz= 5.8 μm, on this surface preparation the effect of the un-treated strength of adhesive joints were performed. The first mechanical prepared was the grinding with Ra= 7.5 μm Rz= 34.9 μm, unfortunately the grinding is not enough homogenous surface preparation and only can be used on flat surfaces properly. A homogenous preparation could be the diameter of 0.3 – 0.7 mm sand blasting on the aluminum substrate surface, it causes Ra= 5.6 μm Rz= 40.4 μm. It both take the surface homogenous and can be used on any type of surface. With these methods the oxide layer can be modified using commercial and standard mechanical equipments. The chemical etching in thin hydrochloride acid cause homogenous surface with Ra= 6.7 μm Rz= 57.7 μm, also the etching has effect on the oxide removal.

3.2. Effect of the value of the carbon fiber layers to the lamination method and the tensile force

In figure 4 and table 1 the tensile force of the different amounts of carbon fiber layers can be seen using vacuum lamination.

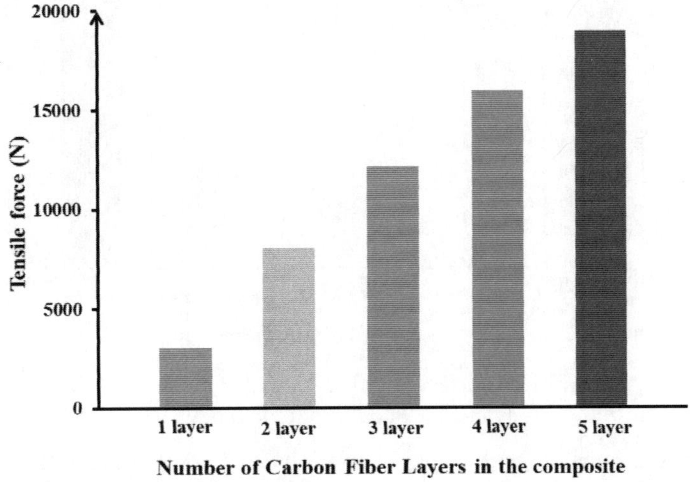

Figure 4. Effect of the carbon fiber layers for the strength of laminated plastics.

Table 1. The tensile force of the different amounts of carbon fiber layers.

Effect of Carbon Fiber layers to the Composite lamination	
Number of fiber layers	**Tensile force (N)**
1 layer	3030
2 layer	8017
3 layer	12149
4 layer	15938
5 layer	18935

The connection between the amount of carbon fiber layers and the tensile force of the plastics shows correlation coefficients. In the beginning adding one more layers increased the tensile force with about 5000 N, but adding additional carbon fiber layers decreases the increase of the force. From 2 to 3 layers it cause 4000 N, from 3 to 4 layers 3800 N and from 4 to 5 carbon fiber layers the force only increase with 3000 N. The laminated CFRP specimens broke in the vertical centre line of the plastics along the lamination glue, because the strength of the carbon fiber is about 1000-1200 MPa. The two component lamination glue is the weakest part between the carbon fibers in every case of the specimens. CFRP specimens with 2 layers of fiber can be used for the adhesive joining research, because the adhesion force between the substrates and the adhesive will be lower than in the base materials.

3.3. Effect of the different surface modification for the tensile force using cyanoacrylate / acrylic hybrid adhesive

The tensile force of the aluminium and CFRP hybrid joints can be seen in figure 5 and table 2 using cyanoacrylate / acrylic hybrid adhesive and different surface preparations.

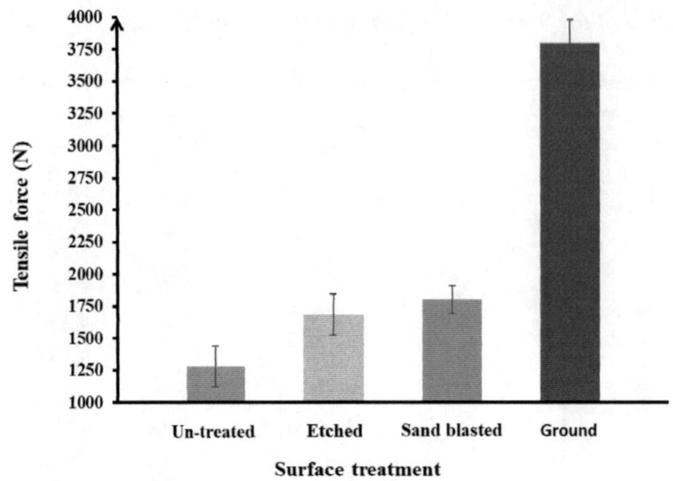

Figure 5. Effect of the different surface modifications for the Al – CFRP hybrid joints using cyanoacrylate / acrylic adhesive.

Table 2. The tensile force of the different hybrid joints using cyanoacrylate / acrylic adhesive.

Effect of surface treatments using Loctite 4080 hybrid adhesive				
Surface treatment	**Tensile force (N)**	**Tensile stress (MPa)**	**Ra (μm)**	**Rz (μm)**
Un-treated	1278	8.5	0.5	5.8
Etched	1685	11.2	6.7	34.9
Sand blasted	1800	12	5.6	40.4
Ground	3800	25.4	7.5	57.9

The chemical and mechanical surface preparations cause the increase of the tensile force in the adhesive joints. The control results are the un-treated Al – CFRP joints. Using chemical etching and sand blasting using 0.3 – 0.7 grain the tensile force only increasing with low level. In case of etching only a low oxide layer could remove during the 5 minutes hydrochloride acid treatment, it cause a homogenous surface texture and the adhesive force of the adhesive can increase. Using sand blasting on aluminium surface the oxide layer can be removed with low level of material removal and a very rough surface created. The wettability in these type of surfaces using mixed cyanoacrylate and acrylic adhesive were in wrong condition. The ground aluminium surface using P10 grinding stone shows the best adhesive force on the aluminium surface using these setups and adhesive. The grinding method removed 1 mm material from the surface, during the procedure the thickness of the oxide layer was decreased or totally removed. The relative surface area is increased compared to the untreated aluminium surface, because the grinding stone. The connection between the surface properties and tensile force shows that the relative surface area and the oxide removal has high effect to the adhesion force on the aluminium surface using cyanoacrylate / acrylic glue.

3.4. Effect of the different surface modification for the tensile force using epoxy high strength structural adhesive

The tensile force of the epoxy high strength adhesive depending on the AlMgSi1 sheets surface modification can be seen in figure 6.

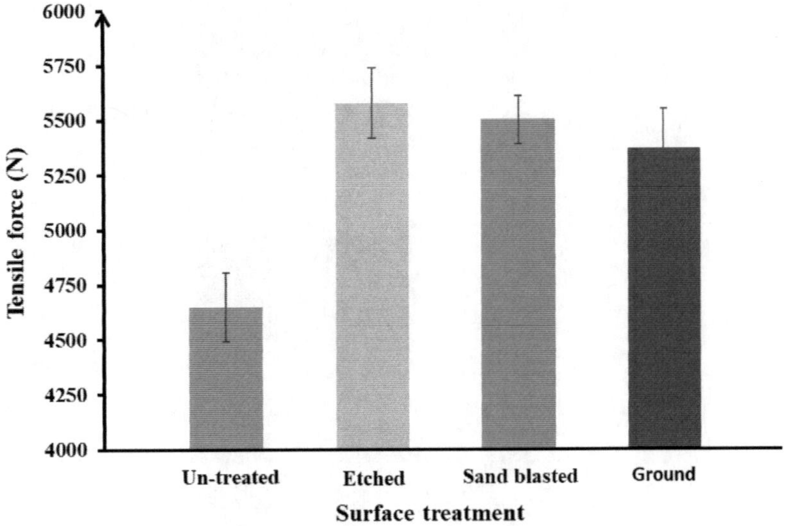

Figure 6. Effect of the different surface modifications for the Al – CFRP hybrid joints using epoxy high strength structural adhesive.

Table 3. The tensile force of the different hybrid joints using epoxy high strength structural adhesive.

Effect of surface treatments using Loctite 9466 high strength adhesive				
Surface treatment	**Tensile force (N)**	**Tensile stress (MPa)**	**Ra (µm)**	**Rz (µm)**
Un-treated	4650	31	0.5	5.8
Etched	5578	36.7	6.7	34.9
Sand blasted	5369	37.2	5.6	40.4
Ground	5502	35.8	7.5	57.9

The control results are the un-treated Al – CFRP joints with epoxy adhesive. The adhesive joint with the un-treated aluminium reached 4650 N, which is 360% higher than using the hybrid acrylic glue. The surface preparation methods changed the adhesive force between the aluminium and the glue, but the level of the increase is lower than using acrylic glue. There is no significant difference between the etched, sand blasted and ground aluminium surface, both of these surface preparation methods removed or changed the surface oxide layer on the AlMgSi1 surface and made homogenous surface topology.

3.5. Effect of the different surface modification for the breaking method

The breaking methods during the tensile test can be seen in figure 7.

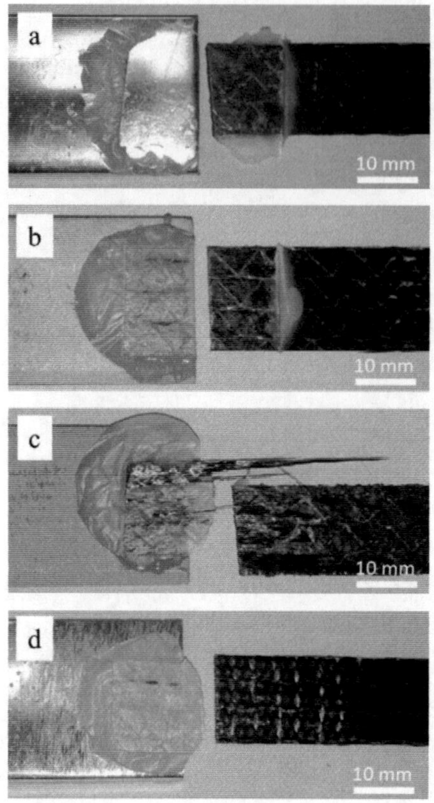

Figure 7. Breaking methods during the tensile test using epoxy adhesive and surface treatments: a: un-treated, b: etched with hydrochloride acid, c: sand blasted with 0.3 – 0.7 mm grain, d: ground with P10 grinding stone.

In case of untreated joints the aluminium surface has the original as received topology, roughness, oxide layer and these cause that the two component epoxy absolutely separated from the aluminium surface. The adhesive and wetting properties had wrong condition on the un-treated surface (figure 7, a). When the surfaces of the AlMgSi1 were modified the epoxy adhesive remained all over on the surface of aluminium (figure 7, b,c,d), during these test the adhesion force between the glue and the CFRP shows the weakest point of the hybrid joints.

4. Conclusion

In this research, the adhesive joining technology of Carbon Fiber Reinforced Plastic and AlMgSi1 sheets hybrid joints were investigated using cyanoacrylate mixed acrylic and high strength epoxy structural adhesive, to improve the strength of the joints 4 different surface preparation method was used only on the aluminum surface. The following conclusions can be drawn from this study:

- The highest tensile strength can be created with sand blasting surface preparation with this technical parameter. The increasing of the strength is more than 10%.

5. References

[1] G Béres, Z Weltsch, 2018 Estimation of Strength Properties from Microhardness Results in Dual Phase Steels with Different Martensite Volume Fraction, *PERIODICA POLYTECHNICA-TRANSPORTATION ENGINEERING*, 4 pp. 1-7.

[2] H Chung, K Rhee, B Lee, J H Lee, 2009 Effect of oxygen plasma treatment on the bonding

strength of CFRP/aluminum foam composite, *Journal of Alloys and Compounds*, Volume 481, Issues 1–2, Pages 214-219,

[3] Z Weltsch, K Klam, A Lovas, 2017. The Comparison of Hardness and Coercivity Evolution in Various Fe–B Based Glasses (Including FINEMET Precursor) during Relaxation and Crystallization, *ACTA PHYSICA POLONICA A* 131:(4) pp. 669-671.

[4] M Berczeli, Z Weltsch, 2018. Experimental Studies of Different Strength Steels MIG Brazed Joints, *PERIODICA POLYTECHNICA-TRANSPORTATION ENGINEERING* 46:(2) pp. 63-68.

[5] 2. G Gergely, D Koncz-Horvath, Z Weltsch, Z Gacsi, 2017 Development of crack detection method with 2 dimensionally generated 3 dimensionally reconstructed images in THT solder joints, *ARCHIVES OF METALLURGY AND MATERIALS* 62:(2) pp. 1033-1038.

[6] J Hlinka, Z Weltsch, 2013. Relation between the Wetting Property and Electrical Conduction of Silver-Gold (Ag-Au) Alloys, *PERIODICA POLYTECHNICA-TRANSPORTATION ENGINEERING* Vol 41:(No 2) pp. 95-98.

[7] I Salacz, Z Weltsch, 2013. Alloying Effects on Wetting Ability of Diluted Cu-Sn Melts on Graphite Substrates, *PERIODICA POLYTECHNICA-TRANSPORTATION ENGINEERING* Vol 41:(No 2) pp. 123-126.

[8] A Dezso, G Kaptay, 2016. On the configurational entropy of nanoscale solutions for more accurate surface and bulk nano-thermodynamic calculations", *Entr.*, vol. 19, no. 6, pp. 248-259,

[9] R D Hussein, D Ruan, G. Lu 2017. Cutting and crushing of square aluminium/CFRP tubes", *In Comp. Str.*, vol. 171, pp. 403-418,

[10] R Neugebauer, Ć C Lies, Ć Hohlfeld, Ć . Hipke, 2007. Adhesion in sandwiches with aluminum foam core *Prod. Eng. Res. Devel.*, vol. 1, pp. 271–278.

[11] G Kim, F Ajersch, 1997. Surface energy and chemical characteristics of interfaces of adhesively bonded aluminium joints, *J. Mater. Sci.*, vol. 29, no. 3, pp. 676–81.

[12] A Pramanik, A K Basak, Y Dong, P K Sarker, M S Uddin, G Littlefair, A R Dixit, S Chattopadhyaya, 2017. Joining of carbon fibre reinforced polymer (CFRP) composites and aluminium alloys – A review, *I. Comp. Part A: Apl. Sci. and Manuf.*, vol. 101, pp. 1-29.

[13] G Béres, Z Weltsch, Zs Lukács, M Tisza, 2018. Prediction of stress- and strain-based forming limits of automotive thin sheets by numerical, theoterical and experimental methods, *AIP CONFERENCE PROCEEDINGS* 1960: Paper 160002.

Acknowledgments
This research is supported by EFOP-3.6.1-16-2016-00014 project. The Project is supported by the Hungarian Government and co-financed by the European Social Fund.

Bending and tensile test of glued aluminium foams

A Timari[1], M Berczeli[2], J Liska[3], Z Weltsch[2]
[1]John von Neumann University, GAMF Faculty of Engineering and Computer Science
[2]John von Neumann University Department of Material Science
[3]John von Neumann University Department of Vehicle Technology

timari.attila@kenji.hu

Abstract. The hybrid material usage is an innovative way to improve the strength performance of the vehicles without causing useless weight increasing. One of the main goals of the competition sports are to lose more and more weight, meanwhile the vehicle needs to accepted on the certifications. That is the reason why new and special materials, like the aluminum foam what belongs to the metal matrix composites should be used. The purpose of this research is that the aluminum foam can be used safely as supporting element in the motorcycle body. In this paper a frame and swing arm of a racing motorcycle using aluminum foam and carbon fabric composite has been made. Adhesive bonding was used to create the hybrid material pairing. To investigate the strength properties of this hybrid aluminum foam and composites tensile and bending tests were performed. The results are used for calculations of the motorcycle body parameters, base information to the designing.

1. Introduction

1.1. Properties of aluminum foams

Composite and foam materials have begun to replace traditional materials because of the necessity for a more economical use of energy. Metal foams can be used in a wide range of fields from mechanical to thermal applications thanks to their superb strength/weight ratios.

Metal foams offer unique physical and mechanical combinations such as high resistance and lower specific gravities or high gas permeability and superior thermal conduction. Despite the many advantages features of metal foams for using in industry must have the appropriate assembly techniques. Research of foam materials are generally limited to production of foam, studies on secondary operations of foam are insufficient [1].

Metal foam is cellular material as wood, so they can be joined with techniques for developed woods joining such as wood screws, adhesive joints and embedded links. Also using joining techniques of metals is possible for metal foams [1-3].

Metal foams have properties which make them attractive in light-weight construction, for energy absorption devices and for acoustic or thermal control. All these fields are relevant for automotive industry, which has been extremely interested in them since they were first developed. Potential applications also exist in ship building, aerospace industry and civil engineering [1, 2].

Light-weight construction: foams can be used to optimize the weight-specific bending stiffness of engineering components. For example, the bending stiffness of flat foam panels of a given weight, width and length is approximately proportional to their thickness, and therefore inversely related to density.

1.2. Advantages of using aluminium foams in the competition sports

True optimization, however, calls for more elaborate solutions as will be discussed below. In any case, light-weight construction exploits the quasi-elastic and reversible part of the load-deformation curve [3].

Energy absorption: owing to their high porosity, foams can absorb a large quantity of mechanical energy when they are deformed, while stresses are limited to the compression strength of the material. Foams can therefore act as impact energy absorbers which limit accelerations in crash situations. This mode exploits the horizontal, irreversible part of the load-deformation diagram. As metal foams can have much higher collapse strengths than polymer-based foams – up to 20 MPa – they can find applications in areas not accessible to foams up to date [4].

Acoustic and thermal control: foams can damp vibrations and absorb sound under certain conditions. Moreover, their thermal conductivity is low. These properties are not outstanding – polymer foams are much better sound absorbers – but they could be useful in combination with other features of the foam. This application makes use of the internal configuration of a foam, namely the labyrinth of struts and the associated air-filled voids [5, 6].

In this paper the mechanical testing of these aluminum foams are usually the tensile and the bending test, the results could give us in this paper information about the energy absorption effect of the parts [7, 8]. Our goal was to using the mechanical test results, the competition race parts could be optimized and simulated.

2. Experimental procedure

2.1. Manufacturing of aluminum foams

Nowadays we meet with special hybrid material pairing in the race cars or motorcycles body. The aluminum foam is one of them, that belongs to the metal matrix composites groups. In this research the foam was made with molding process for building, because of this process we had to work with fix size aluminum foam tables. More over by the frame's production we could not work with just one aluminum foam table, because of the frame's size. The sizes of these frames have limitation because of the manufacturing processes. To increase the useful are of the aluminum foam frame's joining technologies should be used. After examining of advantages of the binding methods and considering the properties of this hybrid material agglutination was used [11]. It should be mentioned the aluminum foam have a way better damping effect so less sound damping material should be needed. The rider's safe is very important, but also the vehicle's safe too [12, 13]. To increase the mechanical properties of these foams carbon fiber reinforced composite (CFRC) should be used over the foam.

2.2. Preparation of tensile tested specimens of adhesive joined aluminum foams

In order to investigate the possibility of adhesive joining of aluminum foam frames Al 6061 aluminum with 0.6 gramm/ccm foam density was used for the tensile test specimen. Two types of glue and four types of connection profiles were used. The two type glue was: Loctite 9466 two component high strength adhesive and Terokal 5055 two component structural adhesive. The four different connection profile can be seen in figure 1.

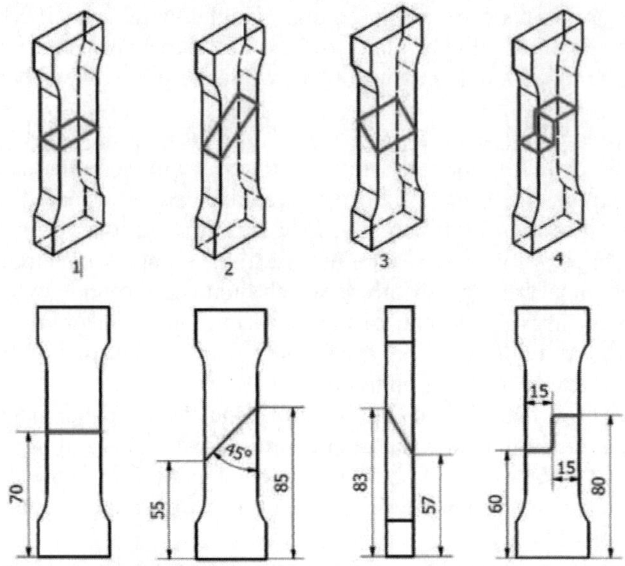

Figure 1. 4 different type of connection profile for adhesion possibility test, 1-4 type from left to right.

The motivation of the different cutting planes were to increase the useful surface for the adhesive bonding technology. The specimens were prepared for the tensile test using milling machine. The geometry of the standard tensile test specimen of the aluminum foam can be seen on figure 2. The final aluminum foam specimens with 4 different type can be seen on figure 3.

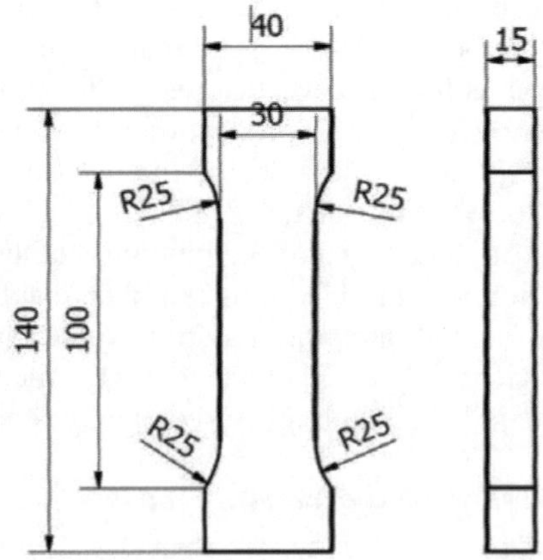

Figure 2. Geometry of tensile test specimen.

IOP Conf. Series: Materials Science and Engineering **448** (2018) 012067 doi:10.1088/1757-899X/448/1/012067

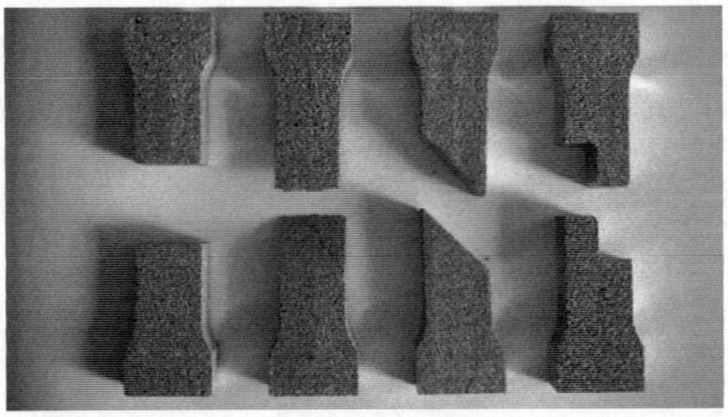

Figure 3. Tensile test specimens before the adhesive procedure, 1-4 type from left to right.

For the tensile test an Instron 4482 general tensile test machine was used with a cross head speed of 10 mm/min.

2.3. Preparation of bending tested specimens of CFRC coated aluminum foam wing arm

We used AlSi 10 and 0.5 gramm/ccm desity foam made by melt gas injection method.as core part of the wing arm (figure 4).

Figure 4. Motorcycle wing arm made from aluminum foam as core part.

The separated foam frames were connected using adhesive joining method. The used glue was a Loctite Terokol 5055 two component glue. After the joining of the foams the part was coated with a micro ballon layer. (UD) carbonfiber layers were used to increase the stiffness of the arm using longitudinal direction during the vacuum lamination process. Heat treatment was used to increase the adhesive and cohesion force between the connected parts.

The final profile of the cross sectioned arm can be seen in figure 5.

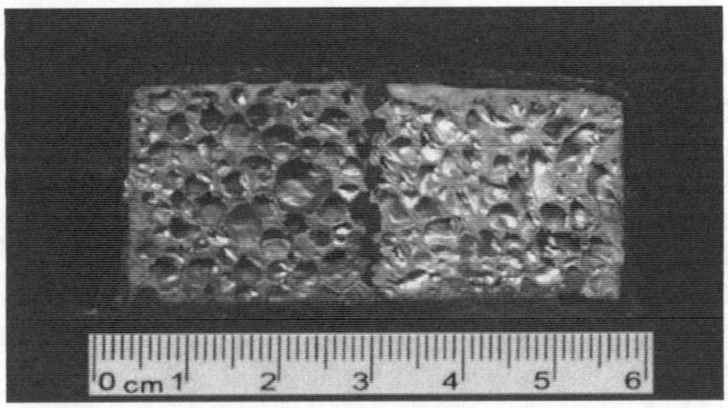

Figure 5. Motorcycle wing arm made from 2 aluminum foam frames as core part and CFRC as reinforce part.

IOP Conf. Series: Materials Science and Engineering **448** (2018) 012067 doi:10.1088/1757-899X/448/1/012067

For the bending test a smaller part was cutted out from the wing arm with a length of 400 mm. The bending test was made with Instrion 4482 general tensile machine using 3 point bending test. The figure 6 and 7 shows the setup of the bending procedure. The bending test was carried out until the specimen did not slip down on the support elements.

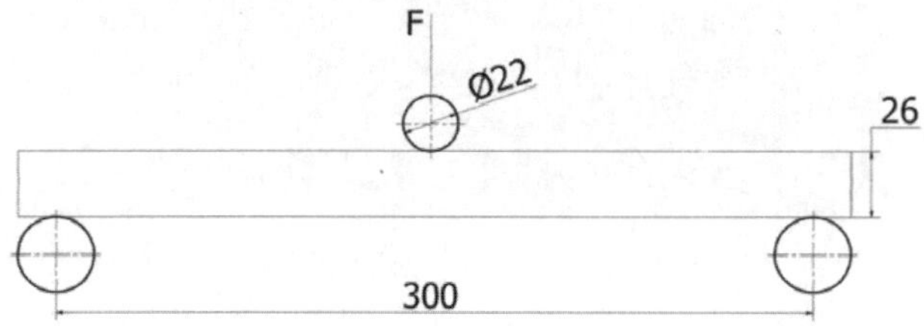

Figure 6. Setup of the 3 point bending test.

Figure 7. End point of the bending test.

3. Results

3.1. Effect of the different type of connection surfaces for the adhesive joining of aluminum foam specimens

The experiment shows that the chosen glues were stronger than the aluminium foam tensile strength even at the smallest cross-section of the specimens. Type 2 sectioned specimen break method can be seen in figure 8, the breaking plane is much above from the connected surfaces. As a result, both of the yield strengths and the tensile strengths were the Al 6061 0.6 gram/ccm aluminium foam mechanical properties, the differences could be due the preparation and production technology. It can be seen from the comparison of the samples' tensile strength with its weight that there is no significant connection between the weight and the tensile strength of the aluminium foams (table 1).

This is due to the fact that there might be some material defects inside the samples that weaken the structure of the samples at a given cross-section. Based on the measurements the tensile strength of the tested foam type is: Rm= 7.65 MPa, while its yield strength is Rha= 2.64 MPa.

Table 1. The summarizing results of the tensile test of the 4 different type of connection surfaces using 2 different glue.

Number of test specimens	Weight without glue (gramm)	Weight with glue (gramm)	Weight of glue (gramm)	Type of the glue	Geometry type of the connection surfaces	Yield point (MPa)	Tensile strenght (MPa)
1	45.27	45.53	0.26	Loctite 9466	1	2.01	6.91
2	40.69	40.93	0.24	Terokal 5055	1	2.21	7.32
3	33.51	34.49	0.98	Loctite 9466	2	2.82	5.94
4	34.72	35.39	0.67	Terokal 5055	2	3.05	6.38
5	46.02	46.37	0.35	Loctite 9466	3	2.51	8.63
6	44.79	45.03	0.24	Terokal 5055	3	2.47	8.87
7	37.98	38.64	0.66	Loctite 9466	4	2.83	7.58
8	40.01	40.32	0.31	Terokal 5055	4	2.96	7.74

Figure 8. The breaking line is above from the connected plane.

The aim of the test was to find out as to what extent it is necessary to extend the bonding surface so that they can be applied safely and securely. The test revealed that in the case of the 0.6 gram/cm foam – made with Al6061 aluminium alloy – the chosen glues did not require the extension of the bonding surface when the two parts were put together.

3.2. Bending properties of adhesive joined of aluminum foam frames with Carbon Fiber Reinforced Composite

The result of the bending test can be seen on figure 9.

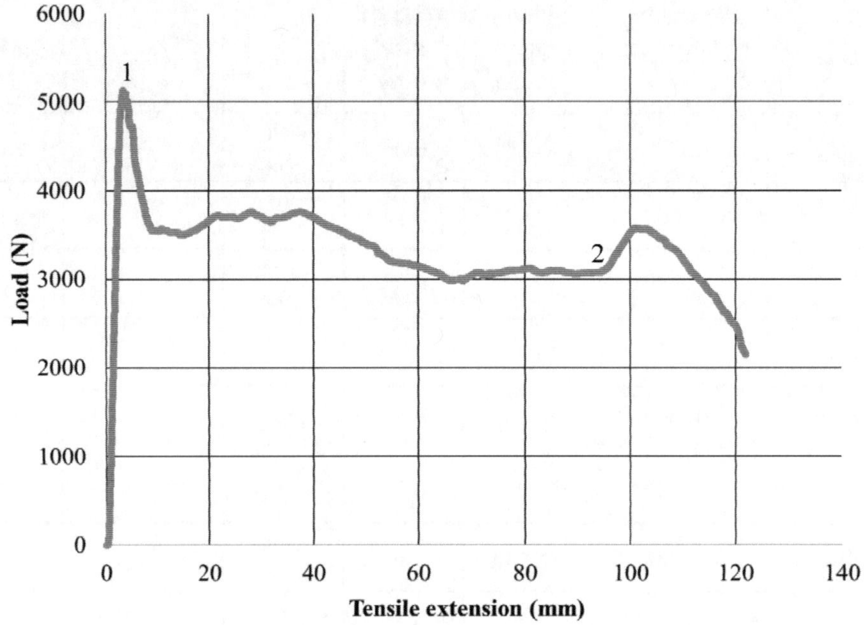

Figure 9. The results of the 3 point bending test.

The maximum load was 5140 N during the measurement, the R_{mh} bending strength can be calculated in the point 1. The calculation to measure the available maximum stress at fracture (1 point):

$$R_{mh} = \frac{M_{max}}{K} = \frac{384750\ \text{Nm}}{6480\ \text{mm}^3} = 59.4\ \frac{\text{N}}{\text{mm}^2} \approx 60\ \frac{\text{N}}{\text{mm}^2} \qquad (1)$$

,where R_{mh} the sought bending strength, M the maximum bending moment on the specimen, K is the factor of cross-section.

The calculation of $M_{max.}$(2), and K factor (3):

$$M_{max} = \frac{F_{max}*l}{4} = \frac{5130\ \text{N}\times 300\ \text{mm}}{4} = 384750\ \text{Nmm} \qquad (2)$$

$$K = \frac{a*b^2}{6} = \frac{57.5\text{mm}\times 26\text{mm}^2}{6} = 6478.23\ \text{mm}^3 \approx 6480\ \text{mm}^3 \qquad (3)$$

Using these results, the optimization of the wing arm for bending can be calculated, designed. At the bending marked points in figure 10 the effect of the load increasing can seen. In this optimization method and design of the hybrid pairing of wing arm, only the result datas before the highest load could be used. After the moment when the part cause deformation on the 3 point bending support elements, the results can cause mistakes.

The purpose of the test is that the determination of the flexural strength on the thinnest cross-section of the motorcycle swing arm. The flexural strength is Rmh = 59.4 N/mm^2, with this data the simulation and optimization of the swing arm can be reached.

Figure 10. Picture of the bended test specimen with the surface damage that made by the support element of the bending tool.

4. Future plan

In good engineering practices it is necessary to be aware of nature and attributions of the tested material, which we take into account in the engineering process. With these information in mind, we are able to develop products more effectively and economically. With the application of such innovative materials in the vehicle industry as aluminium foam, it is crucial that we use the engineered party safely. In the present test, we measured the possible weaknesses of an oversized construction. The tests justify that if we laminate the sample with appropriate carbon fibre layer and using these glues before tensing, the bonding would not have any weaknesses. These results were successful in the case of the prototype. The real challenge for the foams is to gain ground in a broader market. In order for it to be more affordable, further tests need to be implemented. In the case of composite materials, there is a great potential in the simplification of its production process, as well as in the more optimal choice of the fibre direction. In the case of bonding, it is more expedient to repeat the above test/experiment with cheaper glue.

5. Conclusion

In this research, the tensile properties of the different type of connection surfaces using two different type of glue and the bending properties of adhesive bonded aluminum frames coated with CFRC were investigated. The following conclusions can be drawn from this study:

- In case of adhesive joining of aluminum foam frames using the standard geometry of tensile specimens and different connection surfaces plains does not cause any joining problem.
- Both of the Loctite and Terokal glues were able to create an adhesive joining which has higher tensile strength than the aluminum foams.
- In every case of tensile test the aluminum foam broke far from the adhesive connected surfaces.
- The tensile strength of the tested Al 6061 0.6 gram/ccm foam is Rm= 7.65 MPa, the yield strength is Rha= 2.64 MPa
- The maximum bending moment on the cross section of wing arm was 384750 Nmm.

- The flexural strength of the arm is Rmh= 59.4 N/mm^2.
- Results from this paper can be used to optimize and simulate the adhesive bonded lightweight vehicle parts made from aluminum foam and carbon fibre reinforced composite.

6. References

[1] Banhart J. 2005 Aluminium foams for lighter vehicles, Int. J. Veh. Des., vol. 37, no. 2/3, pp. 114-125

[2] Kleiner M, Geiger M, Klaus A. 2003 Manufacturing of Lightweight Components by Metal Forming, CIRP Ann. - Manuf. Technol., vol. 52, no. 2, pp. 521–542

[3] Baumeister J, Banhart J, Weber M, 1997 Aluminium foams for transport industry, Mater. Des., vol. 18, no. 4–6, pp. 217–220.

[4] Sawei Q, Xinna Z, Qingxian H, Renjun D, Yan J, Yuebo H, 2015. Research Progress on Simulation Modeling of Metal Foams, In Rare Metal Mate. and Eng., vol 44, no 11, pp. 2670-2676.

[5] Wang Y, Zhai X, Wang W, "Numerical studies of aluminum foam filled energy absorption connectors under quasi-static compression loading", In Thin-Wall. Str., vol 116, pp. 225-233, 2017.

[6] Malte M, Nicolai M, William H, Sarah R, Kajta F, Reto K, David J F, Myles R A.2009. Greenhouse-gas emission targets for limiting global warming to 2°C. Nature. 459/909. pp. 1158-1162.

[7] Joeri R, David L M, Brian C O, Keywan R. 2013. 2020 emissions levels required to limit warmint to below 2°C. Nature Climate Change. 3. pp. 405-412.

[8] Budai D, Tiszai M. 2015. Development trends in aluminum car body production. Material Engineering, 148/5, pp. 29-36.

[9] http://automotive.arcelormittal.com/News/2791/AHSSforAudiA8 [Downloaded: 2018.05.26.]

[10] Boutar Y, Naimi S, Mezlini S, Carbas R J C, Silva L F M, Ben M. 2018. Fatigue resistance of an aluminium one-component polyurethane adhesive joint for the automotive industry: Effect of surface roughness and adhesive thickness. International Journal of Adhesion and Adhesives.

[11] Lekatou A, Gkikas N, Karantzalis AE, Kaptay G, Gacsi Z, Baumli P, Simon A. 2017. Effect of Wetting Agent and Carbide Volume Fraction on the Wear Response of Aluminum Matrix Composites Reinforced by WC Nanoparticles and Aluminide Particles. Archives of Metallurgy and Materials. 62/2. pp. 1235-1242.

[12] Sergio C, Vitor A, Luis R, 2018. Effect of surface treatment on adhesively bonded aluminium-aluminium joints regarding aeronautical structures. Engineering Failure Analysis. 84. pp. 34-45.

[13] Deepak M, Sebastian A, Yang Z, Jesper H, Jon S. 2018. A study of laser surface modification of polymers: A comparison in air and water, Journal of Manufacturing Processes.32.

Acknowledgments

This research is supported by EFOP-3.6.1-16-2016-00014 project. The Project is supported by the Hungarian Government and co-financed by the European Social Fund.

IOP Conf. Series: Materials Science and Engineering **448** (2018) 012068 doi:10.1088/1757-899X/448/1/012068

Measuring of drill bit vibration by laser Doppler methods

M Béres , B Paripás

Institute of Physics, University of Miskolc, H-3515, Miskolc, Hungary

mechbere@uni-miskolc.hu , fizpari@uni-miskolc.hu

Abstract. The beginning of the drilling process was measured on the drill bit by laser Doppler interferometric methods. The measurements of the velocities were based on the Doppler effect. An LDV (Laser Doppler Vibrometer) device was used to measure the translational vibrations, and then the torsional vibrations (angular velocity changes) of the drill bit were investigated by an LTV (Laser Torsional Vibrometer) device. The time series and frequency spectra were also analysed.

1. Introduction

One of the most common cutting operations is drilling. It occurs particularly often in the aerospace industry and vehicle industry sectors, where it can represent up to 30-40% of the working procedures. Because of its productivity and other beneficial properties, the twist drill is the most frequently applied tool [1]. In order to obtain adequate production quality, the parameters of the tool and its environment must be set to appropriate values. The drill has several technical parameters, such as point angle, chisel edge angle, cutting lip length, helix angle, etc.., which in different ways affect e.g. the cutting force. At the same time, the RPM and feed rate set on the machine, or even the design of the machine itself all have a direct effect on the quality of the boreholes. Figure 1 below shows some borehole images from the drilling experiments that we carried out earlier. It is a common characteristic for all of them that the discontinuous pattern observable on the machined surface was caused by self-excited vibrations. (It is worth to note the odd number of the recurrence pattern!)

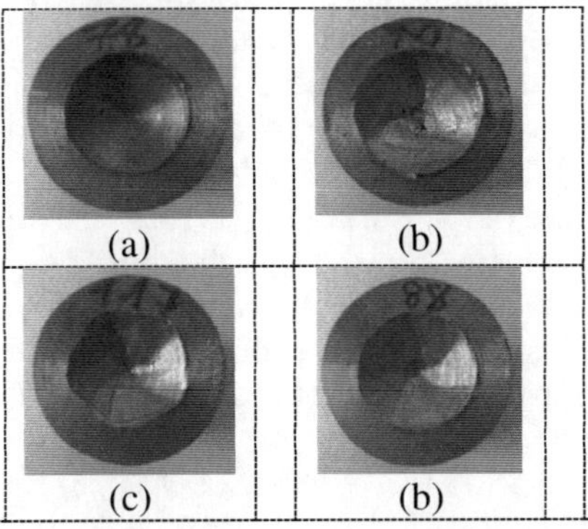

Figure 1. Drill pattern made in the experiments.

Consequences similar to the above can be the result of the whirling vibration of the tool, which contains the twisting or lateral oscillations, or both oscillations at the same time.

Understanding the phenomena behind machining is not only needed for the setting of the cutting parameters, but it is an essential safety interest as well. The simplest way to do this is to continuously monitor the process, i.e. the cutting forces, the cutting temperature, and the vibrations that develop on the tool. The acquired signal can have information about tool life, quality of cutting and defects in the workpiece.

Laser measuring instruments are often used to measure rotating bodies in industrial environments [2,3]. Nowadays we can also find them within the research projects in connection with metal cutting. Without the demand of completeness, we can mention an example for milling research [4], or turning studies [5].

In spite of the foregoing, the laser Doppler drilling studies are not too common in the literature [6,7,8]. Most of the instruments used during drilling are not laser equipment, but devices that measure some other physical parameters, such as torque or force. However, when these devices are used, most of the time special clamping, structural alterations are necessary. As opposed to this, the laser measurement technology - with its non-contact intervention - could provide a truer picture about the process.

2. Circumstances and methods

The measurements were performed on a DMG CNC general lathe in the workshop of the Department of Machine Tools, Institute of Machine Tools and Mechatronics. We chose a general twisted drill bit from Tivoly, contained 5% Co content (DIN338) with a diameter of 12.0 mm. Its point angle is 130°, and its cutting edge ratio is 1/3-2/3 (1/3 of the cutting lip remained).

Throughout these measurements the workpiece (or the tool) was rotated at 825 RPM. In case of the workpiece with Fe490-2 material quality, we did not prepare pilot bore before drilling. This was not necessary because of the point thinning of the cut chisel edge. The feed rate was f_z=0.1mm/rotation. We kept the machining speed (i.e. the RPM of the main spindle) low. In the course of the drilling process the tool arrives to the defined depth (3 mm), then it immediately comes back to the starting position. It only stays here just for a moment (few ms), and the operation sequence of the drilling is closed with the moving of the tool to the reference point. Most of the experiments lubricant fluid was used only to a limited extent (minimal quantity lubrication). It was justified by tool-material pairing also. During experimentation, several sets of measurements were made with various circumstances; however, in this article we describe only a few ones in detail.

The translational and the torsional vibrations were examined by laser interferometric methods. To monitor the process, we used two different non-contact measuring devices applied one after the other. We measured the lateral vibrations of the drill bit from the point outside of the workplace, using the LDV that was mounted on its own tripod (figure 2a). In this case the drill bit was standing, while the workpiece and the chuck were rotating. Although for such circumstances an axial alignment error may occur between the axis of the drill and the axis of the rotating workpiece, it has the advantage that the rotation of the tool does not have any effect on the measurement. By the investigations performed with the LTV (figure 2b) we observed the RPM changes of the drill bit. However, during these experiments - although the instrument is capable of measuring the angular speed changes of a stationary object as well [9] – the drill was rotating and the workpiece was standing. This setup also had the advantage that we could avoid the axis misalignment problem between the workpiece and the tool.

LDV and LTV devices are well known in the literature. Some publications demonstrate the LDV and LTV detection scheme [6,10,11], so here we do not deal with them.

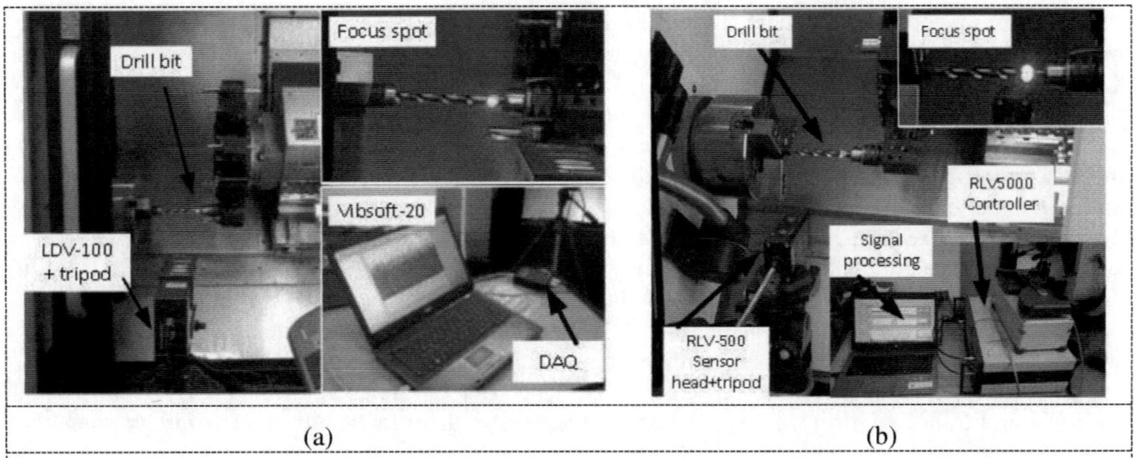

Figure 2. LDV and LTV measuring setup.

3. Results and discussion

Figure 3(a,b) shows the time domain representation of the LDV (a) and LTV (b) signal. We can notice, that has some difference between them. Two signals have same characteristics; the values grow along time. But the LTV plot has a relatively permanent phase (about 3.8s-4.38s), where the absolute mean of the signal does not change. This is because this period of the drilling was executed with full diameter of the drill bit. Furthermore, these plots do not really give us a lot of information. Thus we computed short time Fourier transformations, which returned frequencies in the signals as a function of time. Figure 3(c,d) visualize the output of the calculation, and displays the power spectral densities of the measured data. The colors in the spectrogram denote the power levels in the LDV (c) and LTV (d) signals. We have chosen appropriate settings to obtain a good time resolution without losing fitting frequency resolution. We can see here how the frequency of the signals changes as a function of time. The colors in the spectrogram denote the power levels in the measured signal. The blue color depicts low power level and the yellow one depicts high power level. If we closely look at the color pattern on the LDV PSD plot, a blue area is located in the high frequency range at the beginning of the cutting.

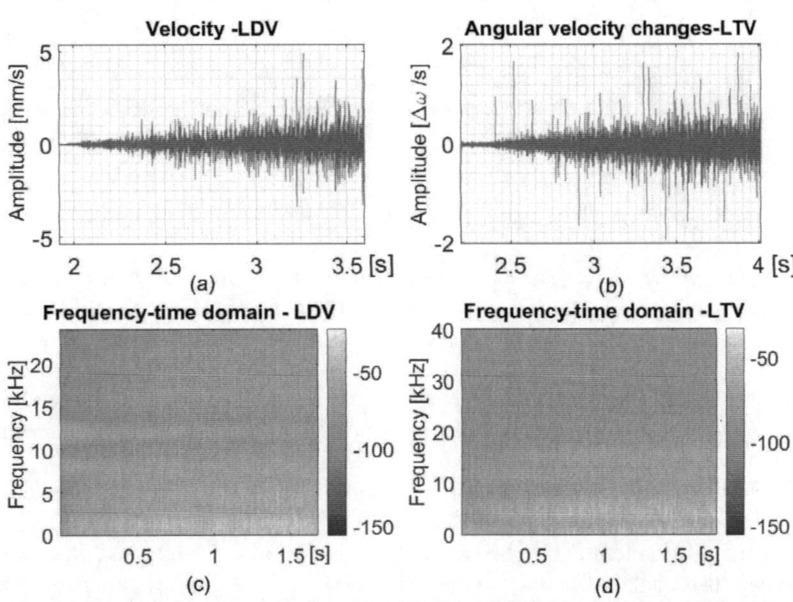

Figure 3. Time domain and Power spectral density of the signals from two laser device.

IOP Conf. Series: Materials Science and Engineering **448** (2018) 012068 doi:10.1088/1757-899X/448/1/012068

The intensity of these frequencies will be more dominant as time goes on. Some of the lower frequencies (for instance around 12 kHz) lose their significance, but some of them (for instance around 5 kHz) increase.

If we take a closer look at the spectrogram of the LTV, in the upper band of the spectrogram that is beyond 20kHz the color pattern is pretty much the same for the entire duration. The color intensity of these signals suggest that the power levels are not so high. So the presence of these components do not add any value to the analysis of the torsional vibration. In contrast to this, based on the color profile, in the lower area we see three or four distinct bands, which change in color. Some of them become wider, and others become thinner. So we need to get rid of the frequency components higher than 20kHz. We can conclude that there is a need to pre-process the LTV signal using a low-pass filter.

We can get a more accurate picture about the exact frequency components, if we make the FFT spectra. The vertical axes named as normalized amplitude, where 1 means the maxima of the FFT signal. On the lower plot, the name of the vertical axis is "synchronized", because of the differences between the LDV and LTV FFT-lines values, we had to synchronize them to each other, in order to show the FFTs together.

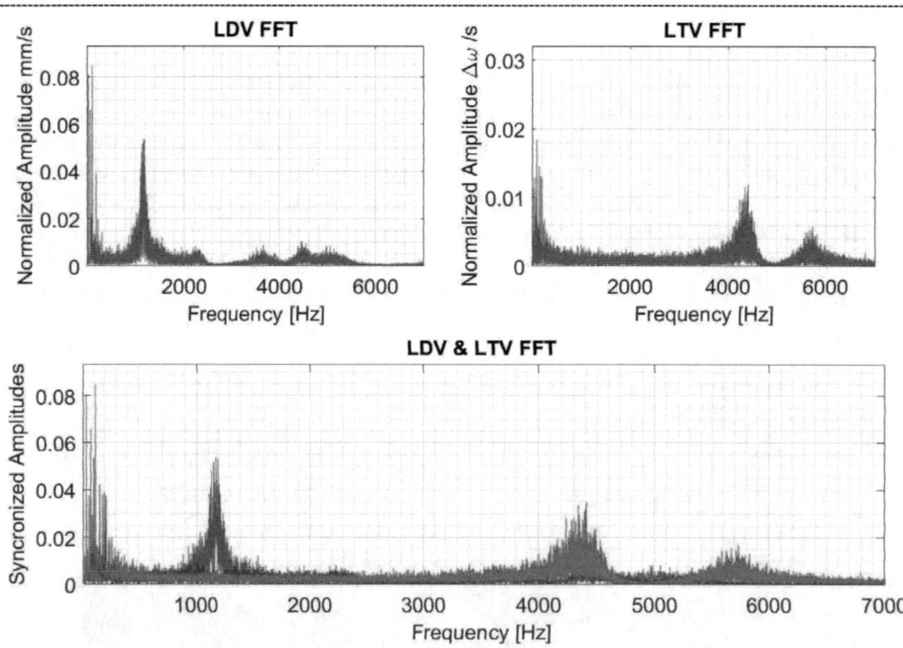

Figure 4. FFT spectra of the signals from LDV and LTV device (above), common spectra (below).

In the LDV spectrum the lower (at most 1500Hz) frequency components are more significant than the upper frequencies, not only in their numbers, but in their magnitudes as well. In the FFT-spectrum of the LTV signal, however, two higher frequency signals are markedly present: at around 4500Hz and 5700 Hz. From the spectral analysis, we can see that they are both permanently present at the time of metal cutting. At the same time two, even higher frequency oscillations can be recognized (at around 12kHz), out of which, however, one is becoming stronger and the other is becoming weaker in comparison with the others as time goes by.

Nevertheless, both FFT-spectrums agree on having definite peaks in the lowest frequency range. In figure 5, the low frequency part of the two FFT spectra can be seen together, and it is quite visible that some of the components are identical in the two vibrations. We also have to take into account, however, that the LTV signals also contain the noises caused by the pattern of the reflecting surface. The so called pseudo vibrations are the consequence of this, as it has been reported in the literature [12,13].

IOP Conf. Series: Materials Science and Engineering **448** (2018) 012068 doi:10.1088/1757-899X/448/1/012068

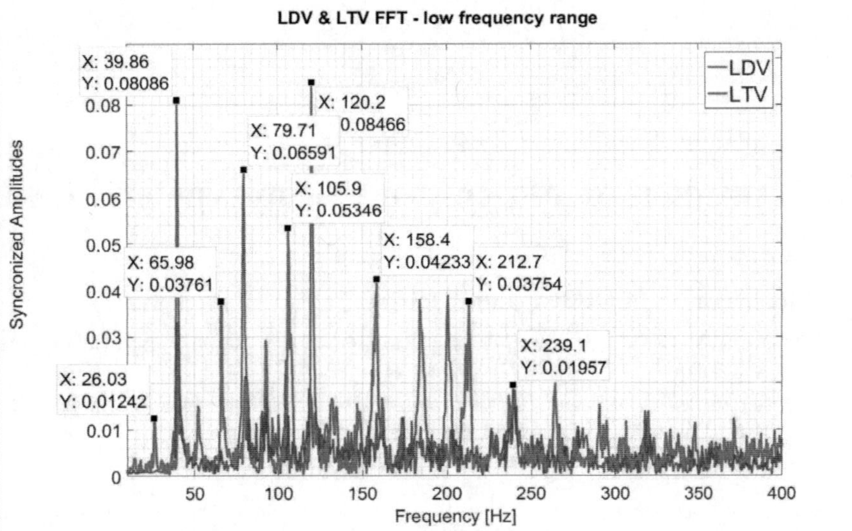

Figure 5. Low frequency peaks of LDV & LTV FFT

Acknowledgments
Project no. NKFI-125117 has been implemented with the support provided from the National Research, Development and Innovation Fund of Hungary, financed under the K_17 funding scheme.

References

[1] Dudás I, Lierath F and Varga Gy 2010 *Környezetbarát technológiák a gépgyártásban, Forgácsolás szárazon, minimális hűtéssel – kenéssel* (Budapest: Műszaki Kiadó)
[2] Rothberg S J et. al. 2017 An international review of laser Doppler vibrometry: making light work of vibration measurement *Optics and Lasers* **99**
[3] Castellini P, Revel G M and Tomasini E P 1998 *The Shock and Vibr. Dig.* **30** 6 pp 443–456
[4] Nakagawa H *et al* 2008 Experimental Analysis of Chatter Vibration in End-Milling Using Laser Doppler Vibrometers *Int. J. Aut. Techn.* **2** 6
[5] Prasad B S, Babu M P 2008 Correlation between vibration amplitude and tool wear in turning: Numerical and experimental analysis *Eng. Sci. Technol. Int. J.* **20** 1
[6] Béres M, Paripás B 2018 Measurements of Vibration by Laser Doppler Method in the Course of Drilling *Vehicle and Automotive Engineering 2* (Springer International Publishing AG, part of Springer Nature)
[7] Reddy Y R M, Prasad B S 2017 Analysis of vibration assisted drilling–A base for tool performance evaluation *J. Prod. Eng.* **20** 1
[8] Balajti M, Murthy B S N and Rao M 2016 Optimization of Cutting Parameters in Drilling of AISI 304 Stainless Steel Using Taguchi and ANOVA *Procedia Technol.* **25**
[9] Polytec GmbH RLV-5500 Rotational Laser Vibrometer Datasheet 2017
[10] Béres M, Paripás B 2016 Comparison of two laser interferometric methods for the study of vibrations *Vehicle and Automotive Engineering* (Springer International Publishing AG, part of Springer Nature)
[11] ed. Drain L E *The laser Doppler techique* 1980 (New York: Wiley-Chichester)
[12] Rothberg S J, Baker J R and Halliwell N A 1989. Laser vibrometry: pseudo-vibrations *J. Sound. Vibr.* **135** 3, pp 516–522
[13] Martin P and Rothberg S J Methods for the quantification of pseudo-vibration sensitivities in laser vibrometry 2011 *Meas. Sci.* Technol. **22** 035302

IOP Conf. Series: Materials Science and Engineering **448** (2018) 012069 doi:10.1088/1757-899X/448/1/012069

Overview of production of pre-preg, prototype and testing

S Rusnáková[1], M Kalová[2] and Z Jonšta[2]

[1]Department of Production Engineering, Faculty of Technology, Tomas Bata University in Zlín, Vavrečkova 275, 760 01 Zlín, Czech Republic
[2]Faculty of Metallurgy and Materials Engineering, VŠB – Technical University of Ostrava, 17. listopadu 15/2172, 708 33 Ostrava, Czech Republic
rusnakova@utb.cz

Abstract. The content of the scientific work is to evaluate the properties of matrix pre-preg systems of composite materials suitable for the production of high performance composite parts, for example for automotive or aerospace industry. The research work introduces composite materials, types, properties and fields of use of matrices and fibres. The main part deals with the description of the pre-preg production, the description of the state of the material conversion with a focus on the B-state of the resin, including the technologies for the production of composite materials from pre-preg systems. The practical part is focused on the testing of matrices intended for the manufacture of pre-preg systems of composite materials, it compares different types of epoxy matrix and glass fiber in several layers. The aim of the work is to evaluate the tested composite systems suitable for pre-preg applications, to compare the properties and parameters of the given application with the focus on the recommended production methods. The result of the work can then be used as the most suitable production application according to the customer's requirements for the product.

1.Introduction

Pre-pregs are often used to produce of composites. Pre-preg systems are called semi-finished products to produce fibre composites, the main component of which is pre-impregnated reinforcement with partially cured resins. Pre-preg is the abbreviated form of the English word preimpregnated fibres, which means "pre-impregnated fibres". Prepregs are semi-finished flats with specially developed resin matrices that are reinforced with artificial fibres such as carbon, glass and aramid [1].

These systems are stacked to the desired thickness and then molded into molds and subsequently pressurized with heat and pressure. Pre-pregs are most commonly used with unsaturated polyester resins, vinyl and epoxy systems. The resulting composite material is highly rigid, heat resistant, extremely stiff and extremely lightweight. One of the advantages of pre-preg technology is also a simpler manufacturing process where machine fabrics are pre-impregnated to produce a high-quality product with very low bulk density and excellent mechanical properties [2].

2.Pre-preg overview

Pre-preg technology is a technology based on the principle of using a semifinished product consisting of a saturated reinforcement by resin. During pre-preg production we distinguish three basic phases or states of resin - called. "A-state", which is an uncured resin, "B-state", in which the polymerization of resin (pre-preg) "C-state", which is the final polymerized matrix (composite) [3]. The B-state is a transition degree of polymerization between the uncured resin and the final polymerized matrix. The resin of the semifinished product in the so- B-condition means, that the material has been partially but

IOP Conf. Series: Materials Science and Engineering **448** (2018) 012069 doi:10.1088/1757-899X/448/1/012069

very sparingly crosslinked. The resulting pre-preg material, however, has a limited workability, especially at room temperature, because it is usually stored in a freezer box. The workability at room temperature is only a few days depending on the type of material. In Tab. 1 is a brief description of the advantages and disadvantages of individual manufacturing processes in the B-state of pre-preg production.

Table 1. Advantages and disadvantages of pre-preg manufacturing technique.

Process type	Advantages	Disadvantages
Solvent based	Easy impregnation Low viscosity Storage stability	Solvent vapour formation Residual solvent content Risk of sedimentation
Hot-melt based	Resin handling Stickiness Solvent-free	High viscosity Difficult impregnation
Chemical B-stage	Resin handling Adjustable reactivity Solvent-free	Exothermic risk Stickiness reproducibility

The production of pre-pregs in the so- B-state can be divided into three ways:

- Physical B-state: solvent based systems
- Physical B-state: hot-melt based systems
- Chemical B-state

3.Experimetal part

Two systems were choosing for pre-preg prototype production. The polymer matrix I/ System I is available from company Huntsman and are designed for high performance composite materials and pre-preg systems. The resin has an extremely long processability time and the resulting composite has a chemical resistance to acids at temperatures up to 176°C. Pre-preg processing is the chemical B-state method, and production methods of winding, pultrusion and die casting are recommended for the resulting composite material. The resin hardens only above 80°C to 160°C, which determines the heat resistance of the final system (it is not cured at room temperature).

Second polymer matrix II/System II, the Eporezit Epovil A / T-111 epoxy system has been chosen to for prototype pre-preg prototypes. An important factor of choice was its availability and features.

As composite sample reinforcement, we chose Aeroglass 200 (198g/m^2). It is a high-quality glass fabric used for most resins and is suitable as a universal fabric thanks to its construction, does not shrink and does not look like loose fibers, is resistant to defects in handling and molding.

3.1. Experimental testing according EN ISO 527-4

The tensile test was performed on the ZWICK test machine. Tensile test was done according to EN ISO 527-4: 1997, Plastics - Determination of tensile properties - Part 4: Test conditions for isotropic and orthotropic fiber reinforced plastic composites. Evaluation of test was done by using Test Expert II.

Samples were prepared by pressure molding technology (PCM). Pressure molding technology is popular manufacturing technique for composite parts. Pressure molding technology offer a high degree of automation, short cycle times, good reproducibility and excellent dimensional stability thermosets which is the reason for diverse applications in various industrial sectors including the automotive industry.

IOP Conf. Series: Materials Science and Engineering **448** (2018) 012069 doi:10.1088/1757-899X/448/1/012069

Table 1 and table 2 describes proposed pre-preg systems for prototype production and mechanical testing. Systems re with different processing times, different curing temperature and different number of layers.

Table 2. Proposal and processing conditions for pre-preg system I.

Pre-preg System I			
Components	System 1	System 2	System 3
	Weight ratio of components		
Araldite LY1556	100	100	100
Aradur 1571	23	23	23
Accelerator 1573	3	5	7
Hardener XB 3471	14	14	14

Processing by ratio of components		**System I/1**		
Curing temperature	Curing time	Sample - description		
		Number of layers of fabric		
		2 layers	4 layers	6 layers
110°C	18-22 min.	CH1-2-I.	CH1-4-I.	CH1-6-I.
120°C	8-12 min.	CH1-2-II.	CH1-4-II.	CH1-6-II.
130°C	4-6 min.	CH1-2-III.	CH1-4-III.	CH1-6-III.

Processing by ratio of components		**System I/2**		
Curing temperature	Curing time	Sample - description		
		Number of layers of fabric		
		2 layers	4 layers	6 layers
110°C	18-22 min.	CH2-2-I.	CH2-4-I.	CH2-6-I.
120°C	8-12 min.	CH2-2-II.	CH2-4-II.	CH2-6-II.
130°C	4-6 min.	CH2-2-III.	CH2-4-III.	CH2-6-III.

Processing by ratio of components		**System I/3**		
Curing temperature	Curing time	Sample - description		
		Number of layers of fabric		
		2 layers	4 layers	6 layers
110°C	18-22 min.	CH3-2-I.	CH3-4-I.	CH3-6-I.
120°C	8-12 min.	CH3-2-II.	CH3-4-II.	CH3-6-II.
130°C	4-6 min.	CH3-2-III.	CH3-4-III.	CH3-6-III.

IOP Conf. Series: Materials Science and Engineering **448** (2018) 012069 doi:10.1088/1757-899X/448/1/012069

Table 3. Proposal and processing conditions for pre-preg system II.

Pre-preg system II	
Components	Weight ratio of components
EPOREZIT EPOVILL-A	1000
EPOREZIT T-111	735

Processing by ratio of components		EPOREZIT		
Curing temperature	Curing time	Sample - description		
		Number of layers of fabric		
		2 layers	4 layers	6 layers
100°C	18-22 min.	CH1-2-I.	CH1-4-I.	CH1-6-I.
130°C	8-12 min.	CH1-2-II.	CH1-4-II.	CH1-6-II.
160°C	4-6 min.	CH1-2-III.	CH1-4-III.	CH1-6-III.

Comparison of modulus of elasticity E in composite with 2 layers of glass reinforcement - assessment of temperature influence and ratio of composite components is depict on figure 1 (on the left) and comparison of the tensile strength σ in the composite with 2 layers of the glass reinforcement - assessment of the influence of temperature and the ratio of composite components on figure 1 (on the right).

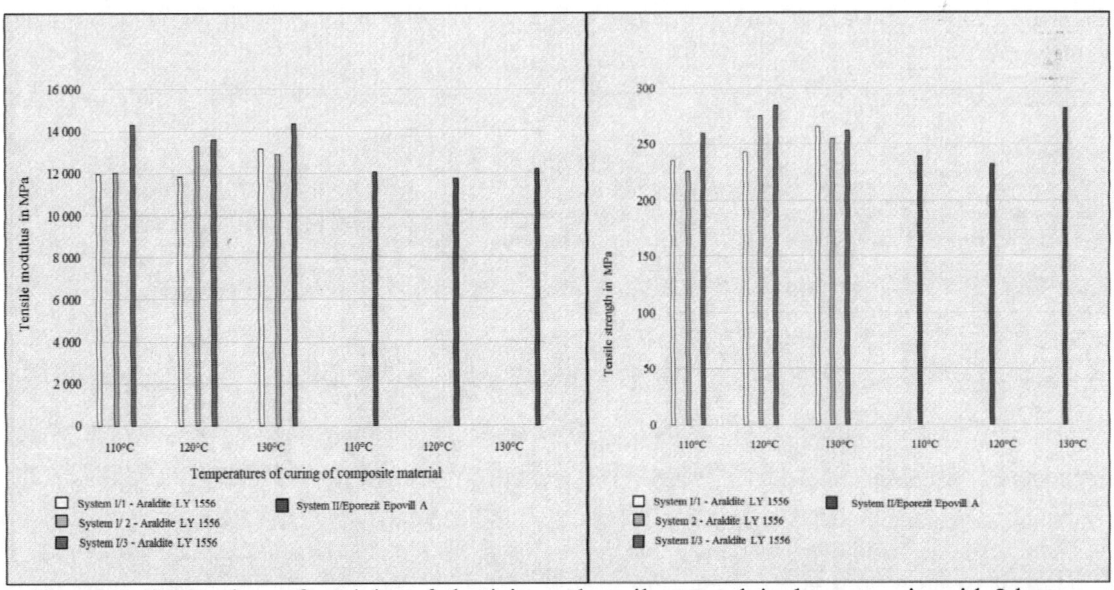

Figure 1. Comparison of modulus of elasticity and tensile strength in the composite with 2 layers.

Comparison of modulus of elasticity E in composite with 4 layers of glass reinforcement - assessment of temperature influence and ratio of composite components is depicting on figure 2 (on the left) and comparison of the tensile strength σ in the composite with 2 layers of the glass reinforcement -

assessment of the influence of temperature and the ratio of composite components on figure 2 (on the right).

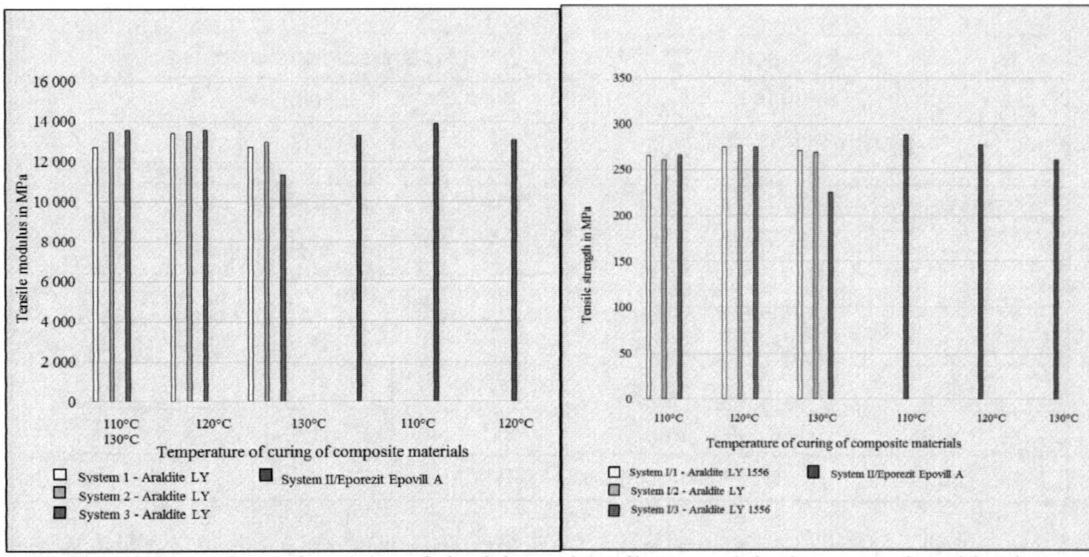

Figure 2. Comparison of modulus of elasticity and tensile strength in the composite with 4 layers.

Comparison of modulus of elasticity E in composite with 6 layers of glass reinforcement - assessment of temperature influence and ratio of composite components is depicting on figure 3 (on the left) and comparison of the tensile strength σ in the composite with 2 layers of the glass reinforcement - assessment of the influence of temperature and the ratio of composite components on figure 3 (on the right).

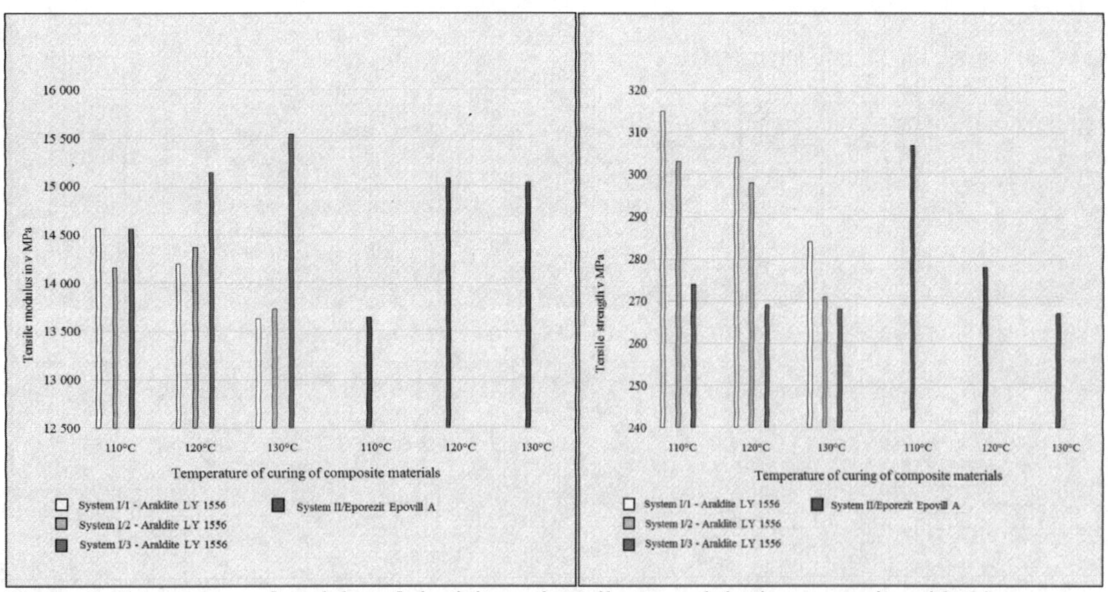

Figure 3. Comparison of modulus of elasticity and tensile strength in the composite with 6 layers.

IOP Conf. Series: Materials Science and Engineering **448** (2018) 012069 doi:10.1088/1757-899X/448/1/012069

4. Results

The main advantage of pre-preg production systems is the high fiber ratio with the optimum amount of matrix. During experimental prototype production of pre-preg systems with various number of laminas we used manual production. Commercial production line used machine production process, where is achieved perfect impregnation of reinforcement (unidirectional, woven) [4].

With multiple layers of glass reinforcement increase the modulus of elasticity, but the tensile strength decreases. This decrease of tensile strength is due to experimental production of pre-preg system [5].

The saturation of fabric is an important factor of mechanical properties of composites. If the impregnation is ideal, the appearance of delamination, formation of cavities is minimalized.

References

[1] MORGAN, P. (2005). Carbon fibers and their composites, 1153 p. Taylor & Francis Group, Boca Raton, Florida. ISBN-10 0-8247-0983-9.

[2] KRETSIS, G. (1981). A review of the tensile, compressive, flexural and shear properties of hybrid fibre-reinforced plastics. Composites, Vol. 18, No. 1, pp. 13–23.

[3] HULL, D., CLYNE, T. W. (1996). An Introduction to Composite Materials, 326 p. Cambridge University Press, Cambridge. ISBN 0-521-38855-4.

[4] Banks, R., Mouritz, A. P., John, S., Coman,F., Paton,R.(2004) Development of a new structural prepreg: characterisation of handling, drape and tack properties. Composite Structures, Volume 66, Issues 1–4, pp 169-174.

[5] Nobuyuki Odagiri, Hajime Kishi, Masaki Yamashita (2012) Development of TORAYCA prepreg P2302 carbon fiber reinforced plastic for aircraft primary structural materials, Advanced Composite Materials, 5:3, 249-254.

Acknowledgement

This paper was created with the financial support of the internal grant of TBU in Zlín No. IGA/FT/2018/004 funded from the resources of specific university research.

IOP Conf. Series: Materials Science and Engineering **448** (2018) 012070 doi:10.1088/1757-899X/448/1/012070

Investigation of aluminium composites by milling

J. Líska[1] and Zs. Kovács[2]

[1]Associate professor, John Von Neumann University, Kecskemét, Hungary
[2]Assistant, John Von Neumann University, Kecskemét, Hungary

E-mail: liska.janos@gamf.uni-neumann.hu

Abstract. The article deal with the attendants of milling aluminium composites. We can use aluminium composite with ceramic matrix. This type of material used in the industry at the produce of foam matierials. During cutting of basic material is not easy, because the high tool wear and some attendants harden the machining.

1. Introduction

Most materials are machined by cutting in the knowledge of the right technology. These knowledge applies not only to the cutting process but also to the correct tool selection and the application of the optimum technological parameters. This summary examines the cutting possibilities of aluminum composites which are similar to ALUFOAM (see in Fig. 1.). The aim of the authoress to get a comprehensive picture about the possibilities of machinability of aluminum composites in the world.

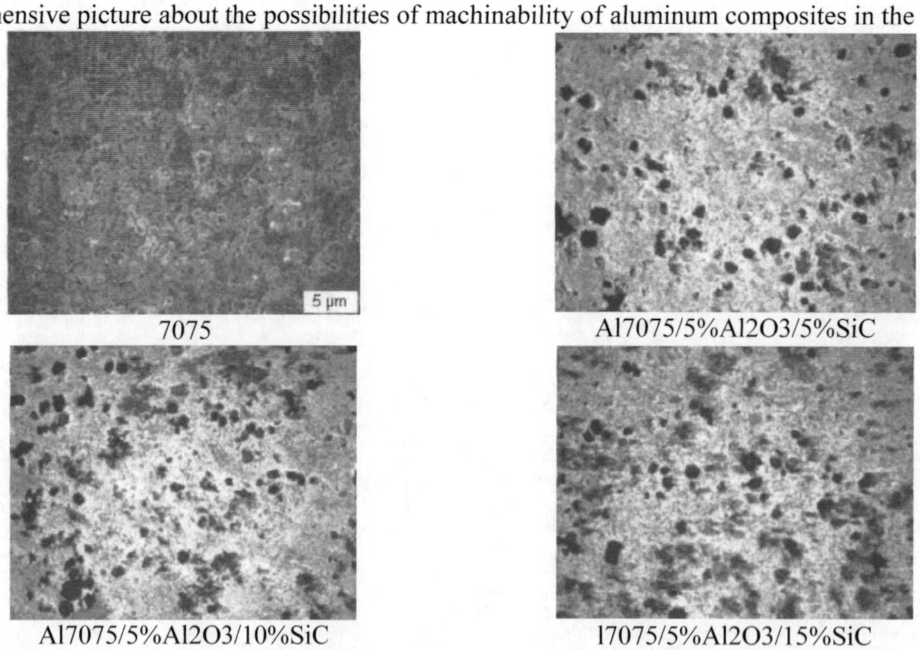

Fig. 1 The microstructure of Al and Al composites [1]

After studying foreign and domestic publications, it can be stated that not only the machinability of the composites were investigated, but also the machined surface' quality and the changings in the cutting zone. All this leads to the conclusion that the processing of aluminum composites requires a complex and comprehensive knowledge.

Among the cutting machining, milling is one of the most complex machining technology because there is a wide range of tool and machine selection.

Sener Karabulut at al. also investigated the surface roughness and cutting strength for aluminum alloy (AA7039/Al2O3). They also made several experiments which were evaluate by Taguchi's method. As a result, it has been found that the tooth per feed has the greatest effect on the cutting force (Figure 2), while the material composition has great influence on roughness (Figure 3) [2].

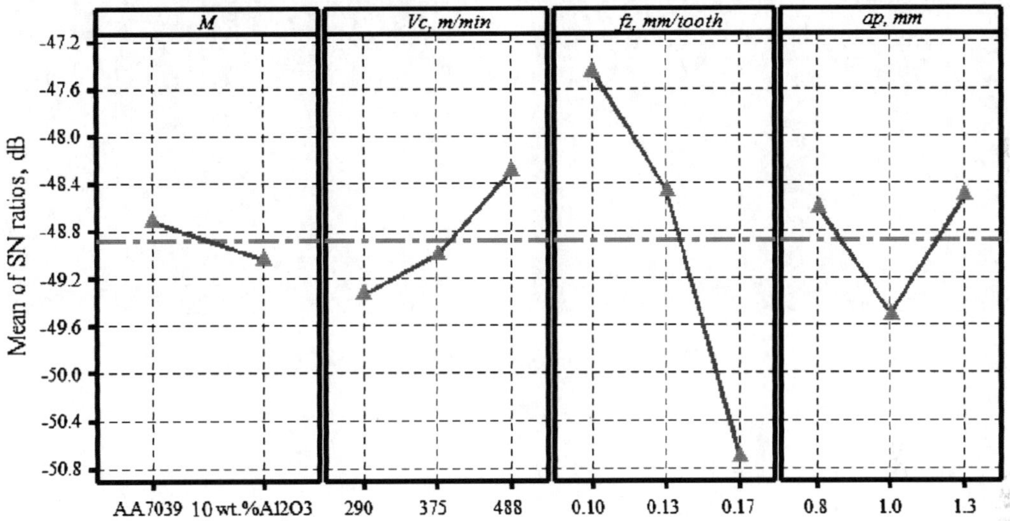

Fig. 2 Effect of cutting parameters on average S/N ratio for F [2]

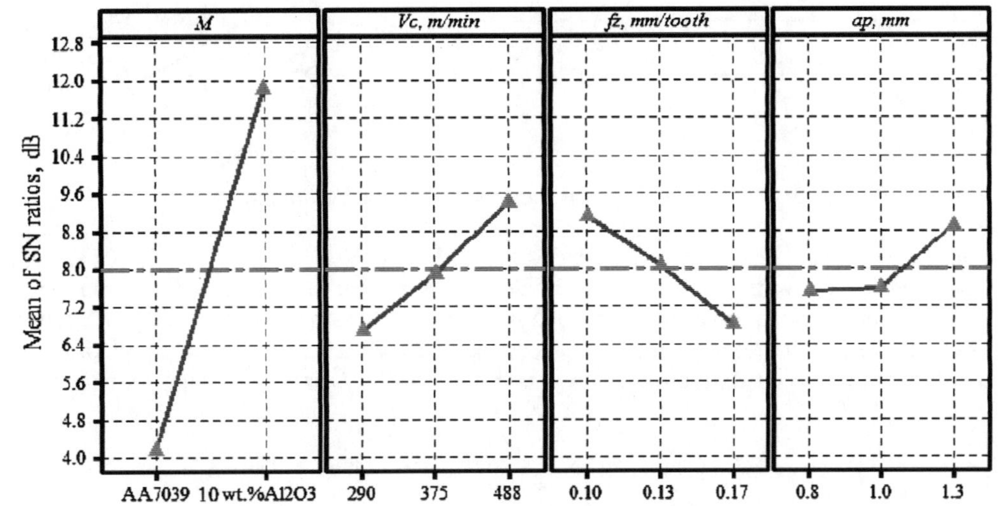

Fig. 3 Effect of cutting parameters on average S/N ratio for Ra [2]

IOP Conf. Series: Materials Science and Engineering **448** (2018) 012070 doi:10.1088/1757-899X/448/1/012070

2. Design of Experiment

During the experiments, two types of material were used - M6061 and C347 (hereinafter Material 1, Material 2). The Material 1 is a 6061 aluminum alloy and the Material 2 is a composite of the Material 1 (Fig. 4). This means that during the pre-fabrication processes of Al 6061 alloy was reinforced with 6 micron grain size Al_2O_3 at a density of 6%. The workpieces were 150x30x4 dimension (Fig. 5a). The free cutting technology was ensured, where the tip of the tool did not engage during cutting.

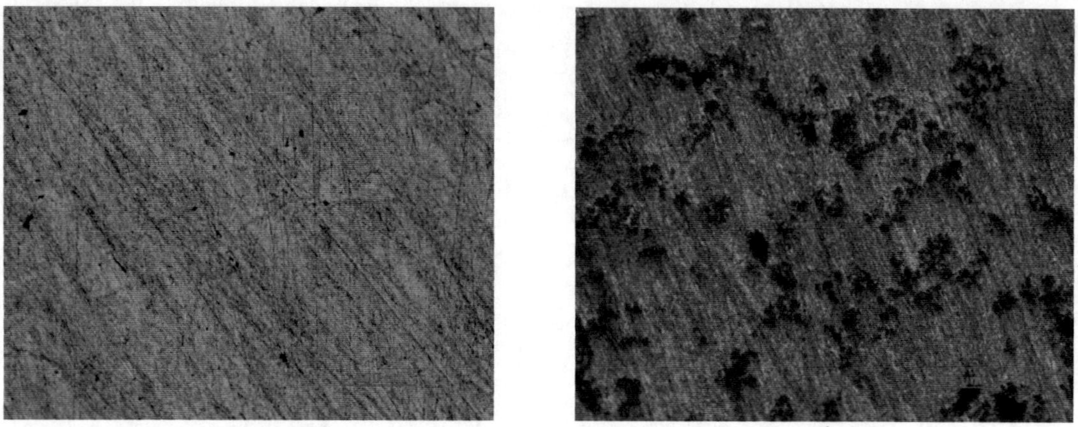

a b

Fig. 4 The used workpieces M6061 (a) and C347 (b)

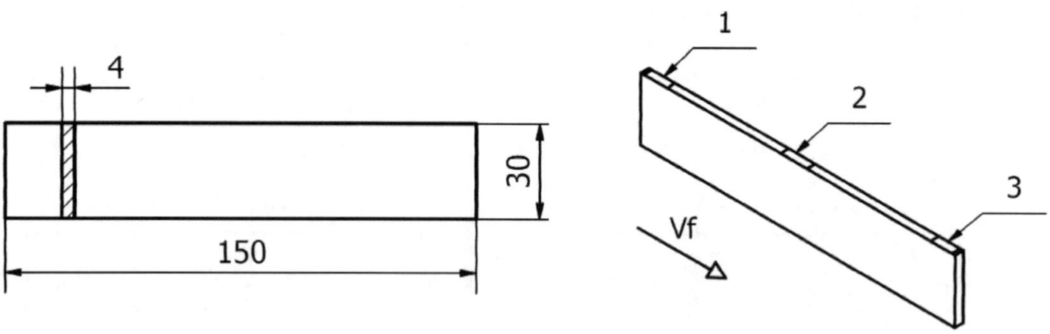

a b

Fig. 5 Dimension of workpieces (a) and machining strategy (b)

The applied tool was a milling cutter, type of SANDVIK COROMANT N331.1A-05 45 08H-NL 1130 with AlTiCrN PVD coated (Fig. 6). The diameter of the milling tool was Ø100 mm and at the same time, only one carbide inserts was attached in the tool. It was necessary, to can be safely record the chipping process. During the milling process straightforward technology were applied, in case the carbide insert start with the maximum chip cross parameter while at exit time is approximately 0.

Fig. 6 Fixed cutting insert into the tool holder

During the experiments, a total of 8 adjustments ($2^3 = 8$) were made based on the full factorial experiment, these values are summarized in Table 1. It is clearly seen that the effects of three variables have been investigated. To evaluate the effect of each variable, we have added a value of "1" for Material 1 and a value of "2" for Material 2. The Table 1. also contains the measured values (F_t and R_a), these will be later analyzed. A conventional milling machine was used for the experiments because to ensure more accurate measurement results and good accessibility. Because the applied the machine tool has fixed speed (n) and feed (v_f) rates, the cutting speed was constant. Its value was v_c=347 m/min.

It was examined two main parameters during the measurements. One of the tangential cutting force (F_t) and the other is the average surface roughness (R_a). For force measurements the torque was measured by a KISTLER rotary dynamometer. Because the F_t is a calculated value, it is specified from the torque. A large amount of data were recorded that can be detectable the moment of entry (Fig. 5 b "Zone 3") and exit (Fig. 5 b "1" zone) during the force measurements. The measuring frequency was 30000Hz.

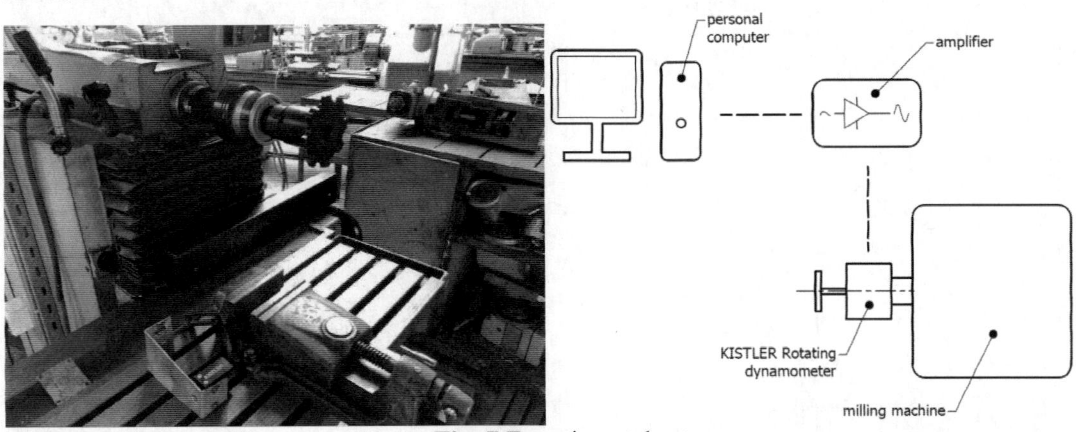

Fig. 7 Experimental set-up

For evaluation of surface roughness, measuring equipment (MITUTOYO Formtracer SV-C3000) was used. The average surface roughness parameters (R_a) were measured in zone "2" (see in Fig. 5b). Furthermore the "1" and "3" zones also were evaluated, but it does not represent relevant values. All measurements were carried out 5 times, the average of these shows the Table 1.

Table 1 Experiment set-up and measured values

No.	Material [ ]	f_z [mm/tooth]	a_e [mm]	F_t [N]	R_a [µm]
1	2	0,08	1	127,4	0,552
2	1	0,08	1	123,2	0,232
3	2	0,2	2	202,6	1,042
4	2	0,2	1	140,4	1,146
5	1	0,08	2	157,8	0,285
6	2	0,08	2	155,2	0,755
7	1	0,2	2	240,2	0,203
8	1	0,2	1	154,8	0,421

3. Evaluating of results

For the evaluations several software were used, such as Minitab Experiment Planner and Evaluation System for graphic evaluations and factorial experimental planning, Microsoft Excel for analytical evaluation and Origin software for fitting the measured values graphically.

3.1. Graphical evaluating

After the correct and thorough designing, a very important element the error-free evaluation. In the interest of it the Minitab statistical evaluation software was used. It is possible to display the results graphically in several ways by this system. The graphical evaluation of the tangential cutting forces is summarized in Fig. 8. Four types of appearance were used. As a result, it can be seen that there was no measurement error and deviation of the results were also appropriate during the measurements.

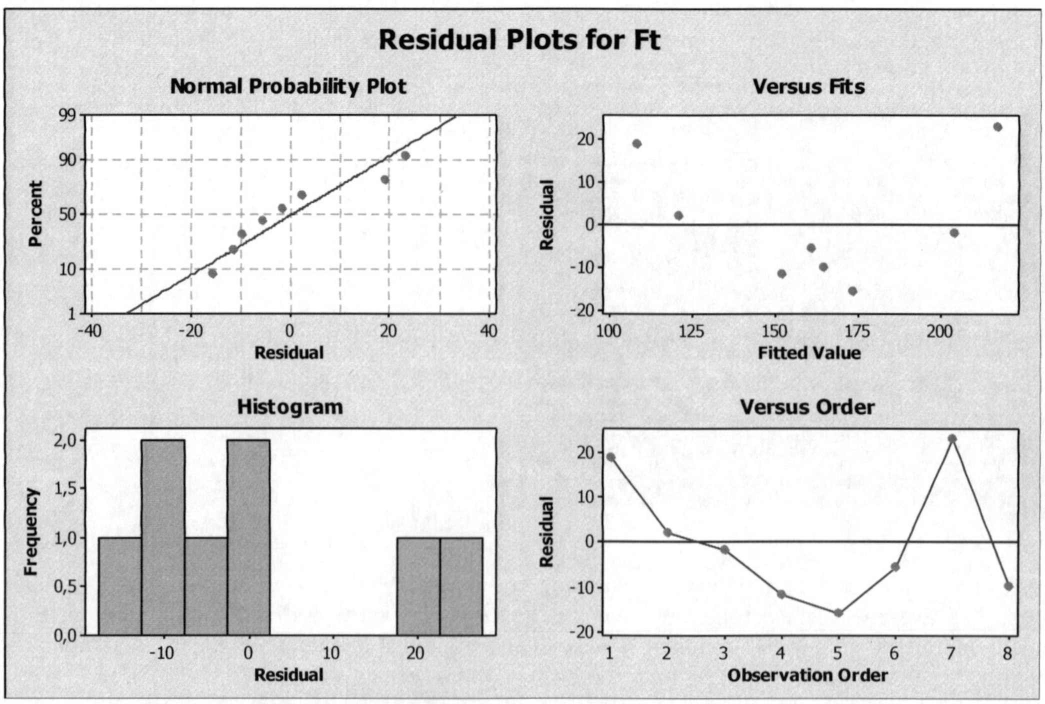

Fig. 8 Residual plots for tangential cutting forces

A similar evaluation also carried out with the surface roughness. Similarly to the normal distribution, the Fig. 9 shows that the measurements are well done. The results of the surface roughness, as described above, were measured in zone "2", practically where the tool works in a full chips cross section and the workpiece stability is the greatest.

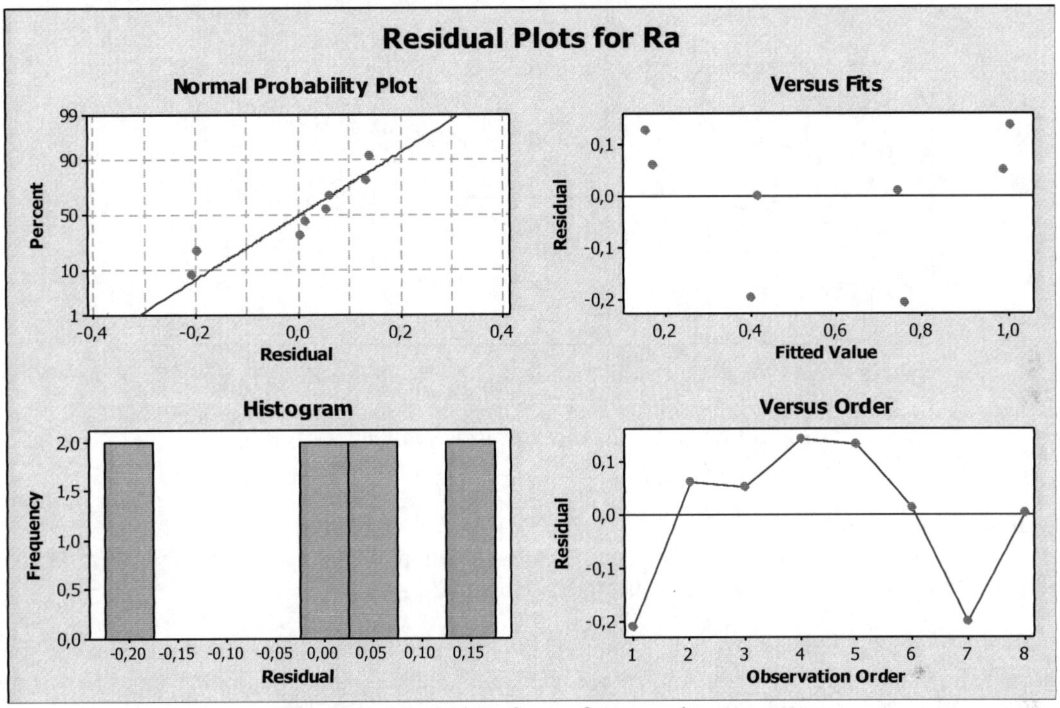

Fig. 9 Residual plots for surface roughness average

The main effects of the factors on the tangential cutting forces and surface roughness values after machining are shown in Fig. 10-11. Clearly visible in the Fig. 10-11, that the f_z and a_e has the greatest effects on tangential cutting forces (F_t), while the quality of the workpiece has the best effects on the surface quality and so the f_z has little effects on the surface quality too.

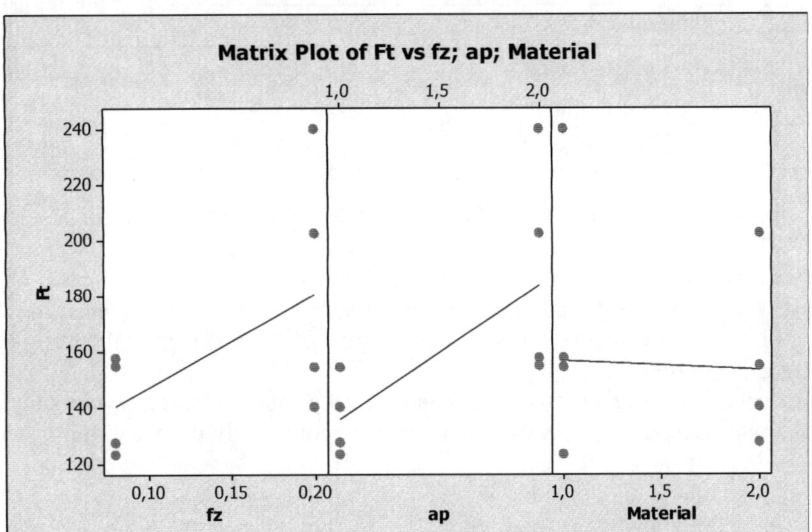

Fig. 10 Main effects plot of tangential cutting forces

IOP Conf. Series: Materials Science and Engineering **448** (2018) 012070 doi:10.1088/1757-899X/448/1/012070

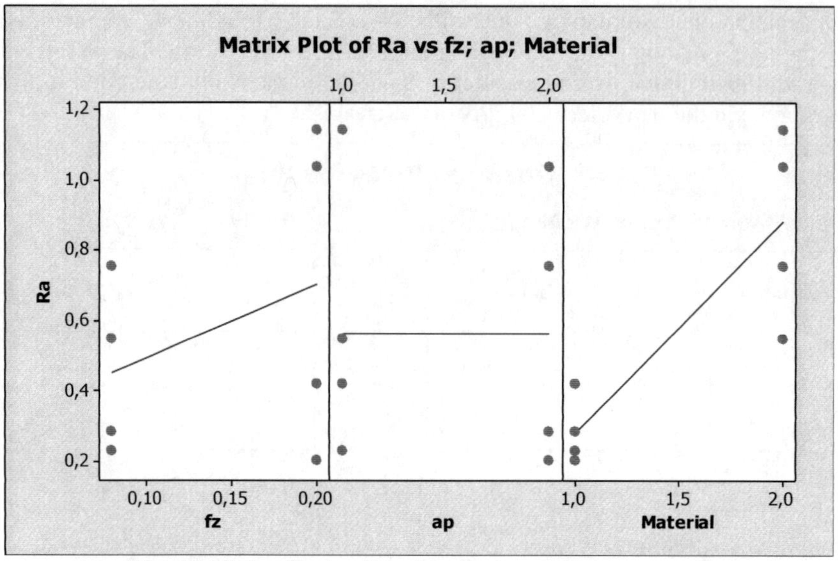

Fig. 11 Main effects plot of surface roughness average

3.2. Analytical evaluating

The next chapter was deal with the analytical evaluation methods. Its essence, to find those sensitive settings which has actually effects on desired value.

First, a regression analysis was carried out. This is summarized in Table 2, where in some cases only the influential values (R-square) were occupied. It is clearly visible that both values (Response) of all variables ("all variables") are sufficiently good values. For both of the requested values (Response) we got good values depending on all variables. This practically means that regression has been successful.

Table 2 Regression of values

Regression	Response		Variations
	F_t	R_a	
R-square	0,87	0,86	all variables
	0,84		only "f_z" "a_e"
		0,74	only "material"

However, if look at Table 3, it is clear that value of F-significant has not the best indicators in case "all variables". Table 4 shows the results of coefficients and p-values. Here we can more clearly see the fact that, why we did not get the best F-significant results in "all variables" case. The p-value in the case of surface roughness the f_z and a_e value are very high, while the Material has very promising result (see in Fig. 11). The values of f_z and a_e have a great effect on tangential cutting forces, while the Material is almost negligible.

From all of these, it is necessary to make a separate regression in cases where only the influencing factors are taken into account. In this case, you can see completely different, more tangible results in Table 3-4.

IOP Conf. Series: Materials Science and Engineering **448** (2018) 012070 doi:10.1088/1757-899X/448/1/012070

Table 3 Analysis of variance (ANOVA)

ANOVA	Response		Variations F-significant
	F_t	R_a	
F-significant	0,029	0,031	
	0,0097		
		0,0064	

Table 4 Coeficients and P-values in different values

	Coficents	p-value	
Constant	-0,566666667	0,141223	all variables of R_a
Material	0,5885	0,00932*	
f_z	2,058333333	0,119907	
a_p	-0,0165	0,90156	
Constant	-0,30325	0,230634	only material of R_a
Material	0,5885	0,006408	
Constant	51,98333333	0,188707	all variables of F_t
Material	-12,6	0,396717	
f_z	363,3333333	0,030469*	
a_p	52,5	0,016806*	
Constant	51,98333333	0,188707	only "f_z" and "a_p" of F_t
f_z	363,3333333	0,030469	
a_p	52,5	0,016806	

*significant values!

Taking into account the above facts, we can write the following equations (based on experimental values):

$$Ft = 51{,}98 + fz^{363{,}3} + ap^{52{,}5} \quad (1)$$

furthermore,

$$Ra = -0{,}303 + Material^{0{,}58} \quad (2)$$

4. Discussion

During the experiments it can be stated that composite materials (not only the polymer-based) are being used in more and more places so it is very important to do researches with this materials [3-5]. Also so important to be know the right technological methods and parameters.

The measurements have also shown that the well-chosen experimental method is important for machining tests. As a result, the results were not scattered and the corresponding values were shown.

We also highlighted, the evaluation should not only be graphic but also be analytical, because it has more accurate value. In addition, we have found that in the case of high-frequency signaling, we can observe those processes which are beyond the normal evaluations. Fig. 12 shows the moment of starting of chip removing. On the inserted curve, it is clearly visible that after the first touch, the forces begin to grow and then become almost constant and then again begin to grow. This will continue until you get the maximum value. This interesting phenomenon can be explained, that the carbide insert

IOP Conf. Series: Materials Science and Engineering **448** (2018) 012070 doi:10.1088/1757-899X/448/1/012070

does not cutting at the moment of starting of machining, only burnish the surface and when it reaches the limit of the shear strength of the material, it begins the chip removing. This phenomenon could only be achieved by forces alone. In the future we are planning to do vibrations and temperature tests too. We assume that different materials will behave differently, this will be show wider and deeper scene about the chip removing tests.

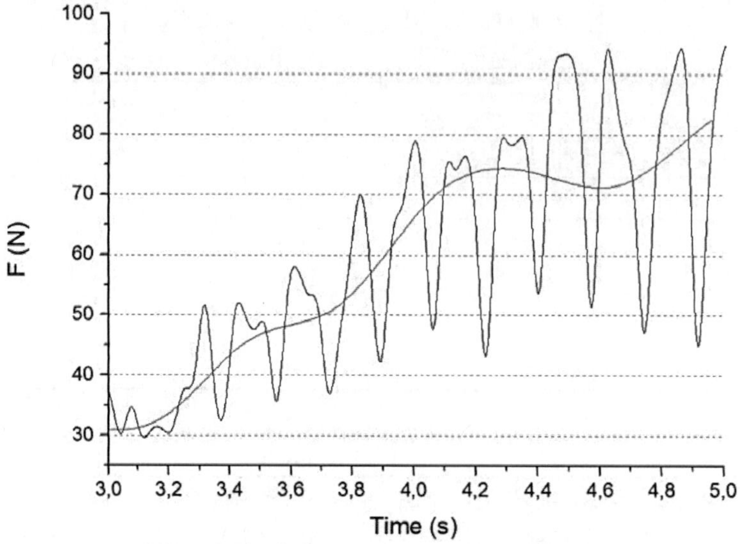

Fig. 12 The indent moment of the milling

References

[1] B. Rajeswari, K.S. Amirthagadeswaran: Experimental investigation of machinability characteristics and multiresponse optimization of end milling in aluminium composites using RSM based grey relational analysis. Measurement 105 (2017), pp.: 78–86

[2] Sener Karabulut: Optimization of surface roughness and cutting force during AA7039/Al2O3 metal matrix composites milling using neural networks and Taguchi method. Measurement 66 (2015), pp.: 139–149

[3] Z. Weltsch, A. Lovas: Wettability of Ceramic Substrates by Silver Based Alloys. ACTA PHYSICA POLONICA A Vol 124:(No. 1) (2013), pp.: 78-86

[4] Z. Weltsch, K Klam, A. Lovas: The Comparison of Hardness and Coercivity Evolution in Various Fe–B Based Glasses (Including FINEMET Precursor) during Relaxation and Crystallization. ACTA PHYSICA POLONICA A 131:(4) (2017), pp.: 669-671

[5] Z. Weltsch, A. Lovas: The Relation between the Surface Tension and the Bulk Properties of Diluted Silver Based Melts, MATERIALS SCIENCE FORUM Vol. 729 (2013), pp.: 19-24

AUTHOR INDEX

AUTHOR INDEX